Teaching Anatomy

Lap Ki Chan • Wojciech Pawlina
Editors

Teaching Anatomy

A Practical Guide

Second Edition

Springer

Editors
Lap Ki Chan
Department of Biomedical Sciences
Faculty of Medicine
Macau University of Science and Technology
Macao Special Administrative Region
People's Republic of China

Wojciech Pawlina
Department of Clinical Anatomy
Mayo Clinic College of Medicine and Science
Rochester, MN
USA

ISBN 978-3-030-43282-9 ISBN 978-3-030-43283-6 (eBook)
https://doi.org/10.1007/978-3-030-43283-6

This Springer imprint is published by the registered company Springer Nature Switzerland AG
The registered company address is: Gewerbestrasse 11, 6330 Cham, Switzerland

"To my parents, Laura, and my teachers and students, from whom I have learned much."

Lap Ki Chan

"To my anatomy teachers who guided me to the threshold of my optimism: Kazimierz Pawlina, MD my father and first anatomy teacher; Franciszek Jugowski, MD, PhD from the Jagiellonian University; Lynn H. Larkin, PhD from the University of Florida; Stephen W. Carmichael, PhD from Mayo Clinic; and to my past, present and future students."

Wojciech Pawlina

Preface

Seven years ago, when we began work on *Teaching Anatomy: A Practical Guide*, we wondered how the book would be received by the academic community. After all, there were already a large number of books offering tips on university teaching and also a few good books on teaching in medical programs. There was even a high-ranking journal dedicated to the field of anatomy education. Over time, the positive reception the book has received and its download figures have told us that there is a strong demand for concise, practical guidance on teaching anatomy.

Seven years is a long time in this fast-paced era. We have seen the introduction of so many novel ideas and practices in active learning pedagogies, faculty development, engagement of the public, ethical use of human bodies, assessment, development of core syllabi, and the application of technology in anatomy education, to name a few.

We therefore felt that there was a need to update the book. In the second edition of *Teaching Anatomy: A Practical Guide*, there are 17 completely new or rewritten chapters, and many chapters that were in the first edition have also been updated, for a broad survey of the field of anatomy education.

The aim of the book remains the same: to offer practical advice to teachers, both novice and experienced, to help them face the diverse and ever-changing educational situations that they commonly encounter, amid their diverse responsibilities. Theories are introduced to help teachers adapt to their local teaching contexts. The writing has deliberately been kept simple and concise, so that anatomy teachers without training in the field of education can easily understand the materials and transform them into actions to help their students learn.

The COVID-19 pandemic is one of those changes in educational environments faced by anatomy teachers. Suddenly, we all have to teach anatomy online. We hope this book can offer some ideas and practical advice to help the anatomy teaching community face the challenge together.

Macao SAR, China — Lap Ki Chan
Rochester, MN, USA — Wojciech Pawlina
May 2020

Contents

About the Editors

Lap Ki Chan is an anatomist and medical educator, with a background in orthopedics and physical anthropology. He received his medical degree from The University of Hong Kong and his PhD in physical anthropology from Duke University. He completed his training in orthopedics and moved into medical and anatomy education soon after that. He spent 12 years at the Li Ka Shing Faculty of Medicine, The University of Hong Kong, where he served as the lead of the Interprofessional Team-Based Learning Program, the coordinator of the Body Donation Program, deputy director of the Bau Institute of Medical and Health Sciences Education, and assistant dean, among other responsibilities. He is currently a professor in the Department of Biomedical Sciences, director of the MBBS program, and the head of Medical Education Unit, at the Faculty of Medicine, Macau University of Science and Technology, Macao Special Administrative Region of the People's Republic of China. His research interests include innovative pedagogies in anatomy education, problem-based learning (PBL), team-based learning (TBL), interprofessional education (IPE), and faculty development. His teaching excellence has been recognized by such awards as the Thomas Henry Huxley Instructorship from Duke University, Outstanding Teaching Award from The University of Hong Kong, and the Bronze Discipline Award (Life Sciences) in the Quacquarelli Symonds (QS) Stars-Wharton Reimagine Education Awards. He served as a visiting professor at Zhongshan School of Medicine and Hainan Medical University and as an associate editor for *Anatomical Sciences Education* from 2013 to 2019. He has served as the educator for the Asia Pacific region for the AO Foundation (Arbeitsgemeinschaft für Osteosynthesefragen) since 2011.

Wojciech Pawlina is an anatomist and medical educator with the background in obstetrics and gynecology. He earned his medical degree from the Copernicus Medical School in Krakow, Poland, where he was appointed as instructor in the Department of Descriptive and Topographical Anatomy. He completed his residency in Obstetrics and Gynecology. Since 1986 he worked as a Postdoctoral Associate at the University of Florida College of Medicine in Gainesville before joining the faculty of the Department of Anatomy and Cell Biology as an Assistant Professor. In 1999 he joined Mayo Clinic, where he is now appointed as a Professor of Anatomy and Medical Education at Mayo Clinic College of Medicine and Science in Rochester, Minnesota. He is past Chair of the Department of Anatomy (2005-2019) and past Director of the Procedural Skills Laboratory (2007-2019). He also served as Assistant Dean for Curriculum Development and Innovation at Mayo Medical School (2007-2013). He has extensive experience in teaching gross anatomy, histology, and embryology to undergraduate, medical, dental, and health care professions students, as well as residents and fellows. He received numerous teaching awards from both the University of Florida, Mayo Clinic, and professional anatomy associations. His research interest in medical education is directed towards strategies of implementing innovative teaching methodologies, teaching professionalism, leadership, and teamwork in early medical curriculum. He is an author of several histology and anatomy textbooks including the *Histology: A Text and Atlas with Correlated Cell and Molecular Biology*. He serves on several editorial boards of scientific journals and is also the Editor-in-Chief of the *Anatomical Sciences Education*, a journal of the American Association for Anatomy.

Contributors

Hamza Mohammad Abdulghani, MBBS, DPHC, ABFM, FRCGP, MMEd Department of Medical Education, Assessment and Evaluation Center, King Saud University, Riyadh, Saudi Arabia

Zubair Amin, MBBS, MHPE, FAAP Department of Pediatrics, Yong Loo Lin School of Medicine, National University of Singapore, National University Health System, Singapore, Singapore

Department of Neonatology, National University Hospital, National University Health System, Singapore, Singapore

Eng-Tat Ang, PhD, PT Department of Anatomy, Yong Loo Lin School of Medicine, National University of Singapore, National University Health System, Singapore, Singapore

Thierry R. H. Bacro, PhD, PT Department of Regenerative Medicine and Cell Biology, Center for Anatomical Studies and Education, Medical University of South Carolina, Charleston, SC, USA

Boon Huat Bay, MBBS, PhD Department of Anatomy, Yong Loo Lin School of Medicine, National University of Singapore, National University Health System, Singapore, Singapore

Esther M. Bergman, PhD Department of Educational Research and Development, Faculty of Health, Medicine and Life Sciences, Maastricht University, Maastricht, The Netherlands

Department of Anatomy, Radboud University Medical Center Nijmegen, Nijmegen, The Netherlands

John Biggs, PhD Tasmanian Institute of Learning and Teaching, University of Tasmanian, Hobart, TAS, Australia

Anja Böckers, Dr Med, MME Institute of Anatomy and Cell Biology, Ulm University, Ulm, Germany

Douglas C. Broadfield, PhD Department of Cell Biology, University of Miami, Miller School of Medicine, Miami, FL, USA

William S. Brooks, PhD Department of Cell, Developmental, and Integrative Biology, University of Alabama at Birmingham, School of Medicine, Birmingham, AL, USA

Cory A. Buenting Gritton, MS Cell & Developmental Biology, University of Colorado School of Medicine, Aurora, CO, USA

Thomas Campbell, BA, MB, BCh, BAO, MSc Discipline of Anatomy, Biomedical Section, School of Medicine, University College Dublin, Dublin, Ireland

Tomasz Cecot, PhD School of Biomedical Sciences, Li Ka Shing Faculty of Medicine, The University of Hong Kong, Hong Kong Special Administrative Region, People's Republic of China

Thomas H. Champney, PhD Department of Cell Biology, University of Miami, Miller School of Medicine, Miami, FL, USA

Lap Ki Chan, MBBS, PhD, FHKAM, FRCS(Ed) Department of Biomedical Sciences, Faculty of Medicine, Macau University of Science and Technology, Macao Special Administrative Region, People's Republic of China

Rocky Chun Chung Cheung, MSc School of Biomedical Sciences, Li Ka Shing Faculty of Medicine, The University of Hong Kong, Hong Kong Special Administrative Region, Hong Kong, SAR, People's Republic of China

Beth A. Cloud-Biebl, PT, PhD Program in Physical Therapy, Mayo Clinic, Rochester, MN, USA

Siobhan A. Connolly, RN, BSc. (Hons), MSc Department of Anatomy, Edinburgh Medical School: Biomedical Sciences, College of Medicine and Veterinary Medicine, University of Edinburgh, Edinburgh, UK

Jon Cornwall, PhD Centre for Early Learning in Medicine, Otago Medical School, University of Otago, Dunedin, New Zealand

John Dent, MMEd, MD, FHEA, FRCS(Ed) The Association for Medical Education in Europe (AMEE), Dundee, UK

Camille DiLullo, (Deceased), PhD Department of Bio-Medical Sciences, Philadelphia College of Osteopathic Medicine, Philadelphia, PA, USA

Richard L. Drake, PhD, FAAA Cleveland Clinic Lerner College of Medicine, Case Western Reserve University, Cleveland, OH, USA

Joyce El-Haddad, BMedSci, MRes Faculty of Science and Engineering, Macquarie University, Sydney, NSW, Australia

Darrell J. R. Evans, PhD, FSB, FRMS, FHEA, FAS, FAAA Faculty of Health and Medicine, University of Newcastle, Callaghan, NSW, Australia

Faculty of Medicine, Nursing and Health Sciences, Monash University, Clayton, VIC, Australia

Gabrielle M. Finn, BSc, PhD, PGDipELM, NTF, FAS Faculty of Biology, Medicine and Health, University of Manchester, Manchester, UK

Alexandria Garino, PA, MS Physician Associate Program, Department of Medicine, Yale University, New Haven, CT, USA

Laura Gorman, BSc, MSc Discipline of Anatomy, Biomedical Section, School of Medicine, University College Dublin, Dublin, Ireland

Matthew C. E. Gwee, PhD, MHPEd, BPharm(Hons) Centre for Medical Education (CenMED) Unit, Dean's Office, Yong Loo Lin School of Medicine, National University of Singapore, National University Health System, Singapore, Singapore

Frederic W. Hafferty, PhD Division of General Internal Medicine, Program in Professionalism and Values, College of Medicine and Science, Mayo Clinic, Rochester, MN, USA

Nathan J. Hellyer, PT, PhD Program in Physical Therapy, Mayo Clinic, Rochester, MN, USA

Catherine M. Hennessy, BSc (Hons), MSc, PCAP Brighton and Sussex Medical School, University of Sussex, Brighton, East Sussex, UK

Sabine Hildebrandt, MD Division of General Pediatrics, Boston Children's Hospital, Harvard Medical School, Boston, MA, USA

Jane Holland, PgDipEd, MD, PhD, MRCS, FAS Royal College Surgeon Ireland, Dublin, Ireland

David R. Hoodiman, LE, LFD Office of Medical Education, University of Miami, Miller School of Medicine, Miami, FL, USA

Susan Huff, BA Office of the Medical Education, Cooper Medical School of Rowan University, Camden, NJ, USA

Anneliese Hulme, BChiroSc, MChiro, MRes, SFHEA Faculty of Science and Engineering, Macquarie University, Sydney, NSW, Australia

Polly R. Husmann, PhD Anatomy and Cell Biology, Indiana University School of Medicine, Bloomington, IN, USA

Christian Jacobson, BSc, MSc, PhD Department of Biology, University of Waterloo, Waterloo, ON, Canada

James F. X. Jones, MB, BCh, BAO, BSc, MD, PhD Discipline of Anatomy, Biomedical Section, School of Medicine, University College Dublin, Dublin, Ireland

Toshihiro Kitama, PhD Center for Life Science Research, University of Yamanashi, Yamanashi, Japan

Brenda J. Klement, PhD Department of Medical Education, Morehouse School of Medicine, Atlanta, GA, USA

David A. Krause, PT, MBA, DSc, OCS Program in Physical Therapy, Mayo Clinic, Rochester, MN, USA

Nirusha Lachman, PhD Department of Clinical Anatomy, Mayo Clinic College of Medicine and Science, Mayo Clinic, Rochester, MN, USA

Michelle D. Lazarus, PhD Centre for Human Anatomy Education and Monash Centre for Scholarship in Health Education, Department of Anatomy and Developmental Biology, Faculty of Medicine Nursing and Health Sciences, Monash University, Melbourne, VIC, Australia

James Lister, PhD Division of Integrative Anatomy, Department of Pathology and Laboratory Medicine, David Geffen School of Medicine at UCLA, University of California, Los Angeles, CA, USA

Xiaoyang Mao, PhD Interdisciplinary Graduate School, University of Yamanashi, Yamanashi, Japan

Joanna Matthan, MB BS, MA, PGDipClinEd, SFHEA Faculty of Medical Sciences, School of Dental Sciences, Newcastle University, Newcastle upon Tyne, UK

Jennifer M. McBride, PhD Cleveland Clinic Lerner College of Medicine, Case Western Reserve University, Cleveland, OH, USA

Conan McCaul, MB, BCh, BAO, FFARCSI, HDip, MD Department of Anaesthesia, Rotunda Hospital and Mater Misericordiae University Hospital, Dublin, Ireland

Stephen McHanwell, BSc, PhD, FRSB, FAS, FRSA School of Medical Education and School of Dental Sciences, Faculty of Medical Sciences, Newcastle University, Newcastle upon Tyne, UK

Cheryl Melovitz-Vasan, PT, DPT, PhD Department of Biomedical Sciences, Cooper Medical School of Rowan University, Camden, NJ, USA

Christian Moro, PhD, BEd, BSc, MBus, SFHEA Faculty of Health Sciences and Medicine, Bond University, Gold Coast, QLD, Australia

David A. Morton, PhD Department of Neurobiology and Anatomy, University of Utah School of Medicine, Salt Lake City, UT, USA

Valerie Dean O'Loughlin, PhD Department of Anatomy and Cell Biology, Indiana University School of Medicine - Bloomington, Bloomington, IN, USA

Nalini Pather, PhD School of Medical Sciences, Medicine, University of New South Wales, Sydney, NSW, Australia

Douglas F. Paulsen, PhD, FAAA Department of Pathology and Anatomy, Morehouse School of Medicine, Atlanta, GA, USA

Wojciech Pawlina, MD, FAAA Department of Clinical Anatomy, Mayo Clinic College of Medicine and Science, Mayo Clinic, Rochester, MN, USA

Linda H. Pellico, PhD, MSN, CNS-BC, RN School of Nursing, Yale University, New Haven, CT, USA

Andrew Phillips, MD, MEd University of North Carolina Chapel Hill, Department of Emergency Medicine, Chapel Hill, NC, USA

Mark Pickering, BA, PhD Discipline of Anatomy, Biomedical Section, School of Medicine, University College Dublin, Dublin, Ireland

Danielle Rhodes, BSc (Hon), PhD candidate Centre for Human Anatomy Education and Biomedicine Discovery Institute, Department of Anatomy and Developmental Biology, Faculty of Medicine Nursing and Health Sciences, Monash University, Melbourne, VIC, Australia

Lawrence J. Rizzolo, PhD Department of Surgery, Yale University, New Haven, CT, USA

Danielle F. Royer, PhD Department of Cell & Developmental Biology, University of Colorado School of Medicine, Aurora, CO, USA

Dujeepa D. Samarasekera, FAcadMEd(UK), FAMS, MHPE, MBBS Centre for Medical Education (CenMED) Unit, Yong Loo Lin School of Medicine, National University of Singapore, National University Health System, Singapore, Singapore

Stephen Schettler, PhD Division of Integrative Anatomy, Department of Pathology and Laboratory Medicine, David Geffen School of Medicine at UCLA, University of California, Los Angeles, CA, USA

Patrick Schiller, MD University of Chicago, Department of Medicine, Ellis Ave, IL, USA

Christopher See, MB BChir, MA, PGCE, PhD, FHEA School of Biomedical Sciences, Li Ka Shing Faculty of Medicine, The University of Hong Kong, The Hong Kong Special Administrative Region of People's Republic of China

Ronnie Homi Shroff, PhD The Hong Kong Polytechnic University, Hong Kong Special Administrative Region, People's Republic of China

Claire F. Smith, BSc (Hons), PGCEA, PhD, FAS, NTF Brighton and Sussex Medical School, University of Sussex, Brighton, East Sussex, UK

Dinesh Kumar Srinivasan, MBBS, PhD Department of Anatomy, Yong Loo Lin School of Medicine, National University of Singapore, Singapore, Singapore

Jane Stewart, PhD, MSc, Cert Ed School of Medical Education, Faculty of Medical Sciences, Newcastle University, Newcastle upon Tyne, UK

William B. Stewart, PhD Department of Surgery, Yale University, New Haven, CT, USA

Allan C. Stirling, MBChB, MRCS, MClinEd Faculty of Health Sciences and Medicine, Bond University, Gold Coast, QLD, Australia

Christopher Straus, MD University of Chicago, Department of Radiology, Ellis Ave, IL, USA

Goran Štrkalj, PhD School of Medical Sciences, Faculty of Medicine, UNSW Sydney, Sydney, NSW, Australia

Atsushi Sugiura, PhD Center for Life Science Research, University of Yamanashi, Yamanashi, Japan

Catherine Tang, PhD Tasmanian Institute of Learning and Teaching, University of Tasmanian, Hobart, TAS, Australia

Samuel Sam Wah Tay, BSc(Hons), MSc, PhD Department of Anatomy, Yong Loo Lin School of Medicine, National University of Singapore, National University Health System, Singapore, Singapore

Adam Michael Taylor, BSc (Hons), PhD, SFHEA, NTFS Lancaster Medical School, Faculty of Health and Medicine, Lancaster University, Lancaster, UK

Jill E. Thistlethwaite, MBBS, PhD, MMEd, FRCGP, FRAVGP Faculty of Health, University of Technology Sydney, Ultimo, Sydney, NSW, Australia

Andrew R. Thompson, PhD Department of Medical Education, University of Cincinnati College of Medicine, Cincinnati, OH, USA

Masahiro Toyoura, PhD Interdisciplinary Graduate School, University of Yamanashi, Yamanashi, Japan

Robert B. Trelease, PhD Division of Integrative Anatomy, Department of Pathology and Laboratory Medicine, David Geffen School of Medicine at UCLA, University of California, Los Angeles, CA, USA

Miriam Uhlmann, MSc, PhD Faculty Development, AO Foundation, AO Education Institute, Davos, Switzerland

Evelyn H. Vargas, LE, LFD Office of Medical Education, University of Miami, Miller School of Medicine, Miami, FL, USA

Nagaswami S. Vasan, DVM, MVSc, PhD Department of Biomedical Sciences, Cooper Medical School of Rowan University, Camden, NJ, USA

Quenton Wessels, PhD, MSc (Clin Ed), FHEA Department of Anatomy, School of Medicine, University of Namibia, Windhoek, Namibia

Adam B. Wilson, PhD Department of Cell and Molecular Medicine, Rush University, Chicago, IL, USA

Timothy D. Wilson, PhD Department of Anatomy and Cell Biology, Schulich School of Medicine and Dentistry, CRIPT – Corps for Research of Instructional and Perceptual Technologies, Western University, London, ON, Canada

Lawrence E. Wineski, PhD Department of Pathology and Anatomy, Morehouse School of Medicine, Atlanta, GA, USA

Jian Yang, PhD, MB School of Biomedical Sciences, Li Ka Shing Faculty of Medicine, The University of Hong Kong, Hong Kong Special Administrative Region, People's Republic of China

1 Elements of Successful Adult Learning

Lap Ki Chan and Miriam Uhlmann

This book is mostly about how teachers can design teaching and learning activities that engage their learners, so as to give them the best anatomy learning experiences. The activities may take many forms, in various settings, and use different methods and tools. Despite these variables, there are some common elements that may lead to better learning experiences. This chapter discusses these elements in general.

To understand the elements, it is important to know how adults learn. Adult learning theory (also called "andragogy"), in the tradition of Malcolm Knowles [1], is defined as "the art and science of helping adults learn." It is based on four assumptions about adult learners:

1. Adults need to know why they need to learn something.
2. Adults need to learn experientially.
3. Adults approach learning as problem-solving.
4. Adults learn best when the topic is of immediate value to their training or work.

L. K. Chan (✉)
Department of Biomedical Sciences, Faculty of Medicine, Macau University of Science and Technology, Macao Special Administrative Region, People's Republic of China
e-mail: lapkichan@gmail.com

M. Uhlmann
Faculty Development, AO Foundation, AO Education Institute, Davos, Switzerland
e-mail: miriam.uhlmann@aofoundation.org

Knowles [1] contrasted andragogy with pedagogy, which he defined as "the art and science of teaching children" (note that it is different from the general usage of the term nowadays) wherein the learners are assumed to be more dependent on the teachers in determining what and how they learn, have little personal experience to bring to the learning process, have learning needs largely determined by someone other than themselves, and are more subject centered. However, andragogy and pedagogy should be considered as two separate sets of assumptions that can sometimes be applied to learners of any age under different situations [1].

For adults to learn successfully, educational activities usually need to match their learning needs, motivate by triggering their internal drivers, provide clear goals or outcomes that they are expected to achieve, engage through active learning, stimulate reflection, and create connections with existing experiences. One element that is missing in adult learning theory is feedback, which will also be discussed in this chapter.

Based on Needs

A theory of motivation based on human needs was described by Abraham Maslow and is known by many as Maslow's pyramid/hierarchy of human needs [2]. This defined the term "need" in a broad general sense of human biological and

L. K. Chan, W. Pawlina (eds.), *Teaching Anatomy*, https://doi.org/10.1007/978-3-030-43283-6_1

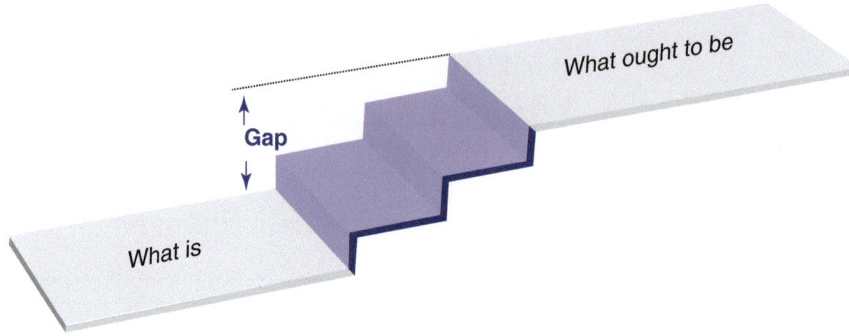

Fig. 1.1 The elements of needs. (Adapted from Fox and Miner [3])

Table 1.1 Advantages/disadvantages of several need assessment methods

Method	Advantages	Disadvantages
Informal discussion	Convenient, inexpensive, rich in details	Lack of methodology, interviewer bias
Formal interviews	Standardized, quantitative, and qualitative information	Needs trained interviewers, costly
Focus group discussions	Efficient, learn about group behavior, qualitative data	Needs skilled facilitator, time, and financial costs
Questionnaires	Standardized questions, quantitative and/or qualitative data, easy to use (especially online), large samples possible	Question-writing skills needed, response rate issue, time intense for data collection and analysis
Direct observation	Best method for assessing skills and performance	Time-consuming, development of guidelines
Tests	Objective measure of knowledge or skills	Requires time, effort, and skill to construct valid test questions

Adapted from Kern et al. [4]

psychological requirements. The relation to education was described by Knowles in 1980 [1]: "These basic needs have relevance to education in that they provide the deep motivating springs for learning, and in that they prescribe certain conditions that the educators must take into account if they are to help people learn… An educational need, therefore, is the discrepancy between what individuals (or organizations or society) want themselves to be and what they are; the distance between an aspiration and a reality."

In other words, the educational need can be described as a gap between the present level of an ability (what is) and the desired level of the same ability (what ought to be) required for effective performance as defined by the learners, their organization, or society (Fig. 1.1).

Clarification of who the target learners are, their needs, and their environment is crucial to assure that learning takes place. Kern et al. [4] outlined several methods to assess learners' needs (Table 1.1).

Motivation

Motivation to learn arises from an educational need. Perception plays a major role in motivation and "…is the cornerstone of understanding why health professionals may have different levels of motivation related to similar topics and programs…" [3]. It is important to understand that the gap that affects each learner's motivation is always the gap between the perceived present level of ability of the learner and the desired level of ability and that the extent of discrepancy one perceives between these two affects the extent to which one is motivated to learn. The interactions between perceived and actual needs are summarized in Fig. 1.2.

The best situation is where we find a high perceived need and a high actual need, which results in the learner being highly motivated to learn. On the other hand, very large discrepancies are associated with a high anxiety level, which may lead to feelings of aversion rather than attraction and

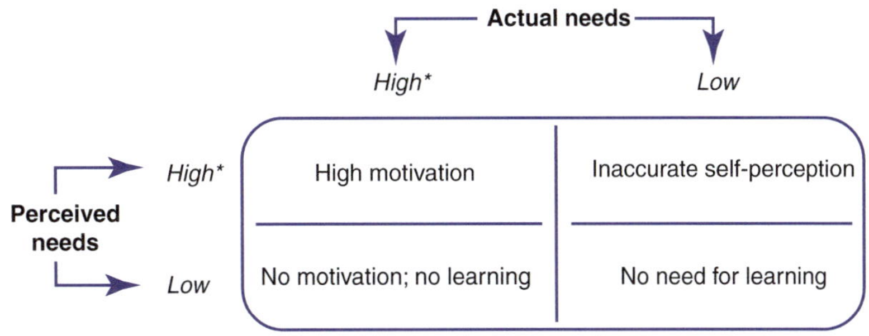

Fig. 1.2 Perceived vs. actual needs. (After Fox and Miner [3]). *Not extremely high

therefore a lack of motivation to learn. The most difficult situation is when learners believe that their performance is close to the standard but it is not, so the perceived need is low, but the actual need is high. In such a situation, the learners will not see any need to learn. The question is now how we can motivate these learners.

For adults, it is also important to be self-directed and to decide how they want to close an identified gap. Fox and Miner stated that "Motivation to participate in a specific learning activity will be greatest when the physician perceives strong or many goals, that those goals are important, that participating in the specific learning activity is personally satisfying, and that participating will result in achieving goals" [3].

For practical application, it is crucial to consider the following two points to motivate your learners:

1. Help your learners to realize their gaps in knowledge and performance by, for example:
   - Online self-assessments: Based on your defined outcomes, ask your learners about their perceived present level and their desired level. This can easily be done with an online survey tool. It is important that learners can see their results to recognize their gaps.
   - On-site small group discussions: During small group discussions, you can find out about the present level of knowledge, and you can help learners to understand where they are and where they should be. This is important in situations where learners think they already know a lot (although they in fact do not) and would therefore not be highly motivated to learn.
   - Reflection (see also section on "Reflection"): Reflective practice helps learners to identify their gaps.
2. Help your learners to stay motivated by using a variety of teaching methods:
   - Use interactive methods for teaching such as interactive lectures and small group discussions.
   - Use new technologies to allow for self-directed learning, e.g., provide online resources such as readings or recorded lectures/webinars.
   - Provide learners with clear goals and outcomes.
   - Provide time and opportunities for reflection.
   - Blend traditional strategies with technology, e.g., self-assessment tests can be completed online and linked to discussion forums.

Outcome Driven

Many teachers declare the objectives before starting a teaching and learning activity. However, these are often the objectives of the teacher: "in the next hour, I am going to tell you ABC, then do DEF…." Such objectives may give learners an idea of what the teaching/learning process will be like. A clear articulation of the learning product, i.e., what the learners are expected to be able to do after the activity, would be more helpful to learners. These expectations, written from the

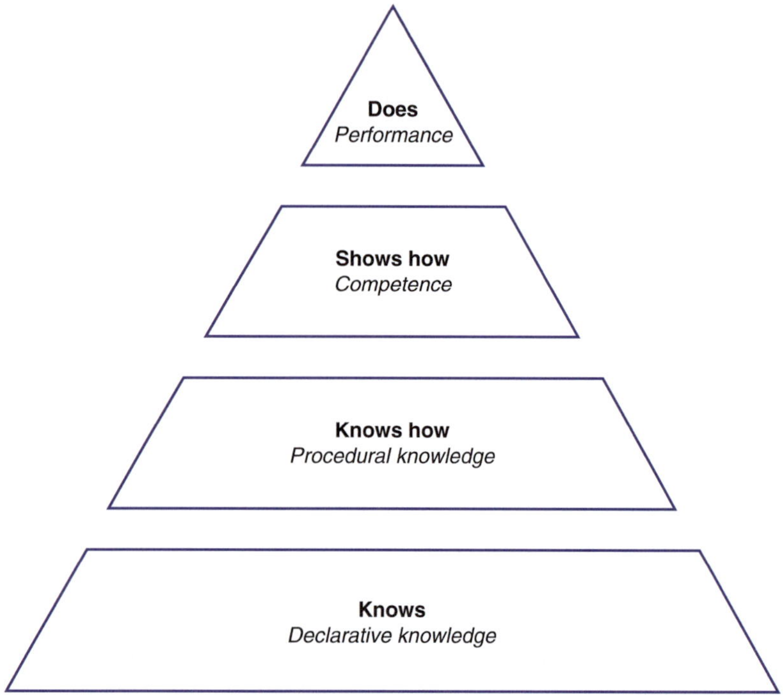

Fig. 1.3 Miller's pyramid. Framework for clinical assessment. (After Moore [7])

perspective of the learners, are called the intended learning outcomes.

There are several models to help teachers to develop learning outcomes. One is the SOLO (Structure of Observed Learning Outcomes) taxonomy [5], which describes several levels of complexity in the learner's understanding of a subject: prestructural (learners have unconnected information), unistructural (learners are able to make simple and obvious connections between facts), multistructural (learners see more connections but miss the significance to the whole), relational (learners appreciate the significance of the parts to the whole), and extended abstract (learners make connections beyond the subject and are able to generalize). Another model is the revised Bloom's taxonomy [6], which identifies six subcategories in the cognitive domain of learning activities: knowledge, comprehension, application, analysis, synthesis, and evaluation. Whether these six subcategories are hierarchical is debatable, but the subcategory "knowledge" here is defined as remembering and recalling (not knowledge in the general sense) and is usually considered the simplest level of intellectual activity. Miller's pyramid can also help teachers to formulate learning outcomes for their learners (Fig. 1.3) [7]. The pyramid shows the ideal stages of the development of clinical competence but can also be applied to learning in other areas. The first stage is that the learner knows what to do, and then he/she knows how to do it (i.e., he/she can describe the process but might not be able to do it). The next level is that the learner shows how it can be done in a safe environment, and the highest level is to apply it in actual practice. Similar to Bloom's taxonomy, Miller's pyramid also distinguishes learning that consists of memorizing facts (declarative knowledge) from learning that enables one to apply procedural knowledge in real-life situations.

After teachers have decided on the intended learning outcomes, they plan backward. They need to decide on the teaching/learning activities that will best help learners achieve the outcomes. They also need to decide on the assessment methods and standards with the intended outcomes in mind. Such an alignment of teaching/learning activities and assessment with the outcomes is called constructive alignment [8] and will be

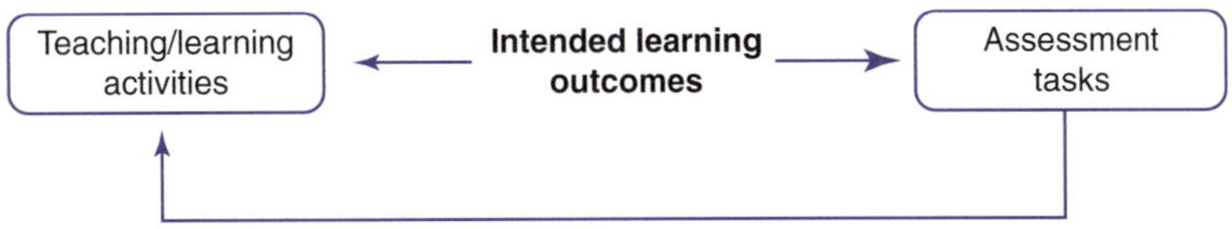

Fig. 1.4 Aligning teaching/learning activities and assessment tasks with the intended learning outcomes

discussed in much greater detail in Chap. 3. If the outcomes are not met by the learners, the teachers may need to reconsider and adapt the teaching/learning activities and the assessment the next time they engage in the same activity, until the intended learning outcomes are achieved to a satisfactory level (Fig. 1.4).

In an outcome-based approach to learning, the intended learning outcomes take a central role. If the outcomes are not properly articulated, it will not be clear what the teaching/learning activities are trying to help the learners to achieve and what the assessment will be measuring. Thus, the articulation of the intended learning outcomes must be carefully done. Each outcome should begin with a verb that describes an observable and assessable action. The action indicates not only whether the learners are able to do certain things but also at what level they are expected to do it after the teaching/learning activity and under what conditions. For example, "describe" and "hypothesize" are appropriate verbs in outcomes since they both indicate not only whether the learners are able to understand certain content but also the levels the learners are expected to understand the content (lower level for "describe" than "hypothesize"). Assessment can also be focused on these actions. On the other hand, "understand" will not be an appropriate verb, because it is not observable and can only be indirectly assessed. If it was used in an intended learning outcome, the teacher may have difficulty designing the appropriate teaching/learning activity that will help the learners to achieve it, because it is not clear how much understanding is expected of the learners. The learners will also not know how, and to what level, they will be assessed on that outcome.

An outcome-based approach sets clear goals for learners. If the learners perceive a gap between their current level of ability and the goals, they will be more motivated to take part in the learning activities if they believe the activities have been designed to help them achieve the goals. An outcome-based approach also helps teachers and administrators to cooperate to achieve the same goals, especially if the activity involves more than one teacher. It ensures that the right products are delivered. It is particularly important in the healthcare field because the amount of knowledge learners are required to know is rapidly expanding, while the length of healthcare training programs remains more or less the same [9]. An outcome-based approach ensures that the teaching/learning activities in a program will produce graduates with a set of intended competencies. This approach also encourages debate over the set of intended competencies because the teachers now need to explicitly articulate it [10, 11]. Doing so also enhances the transparency and quality assurance of healthcare training programs.

Active Learning

There is evidence that adults learn better with active learning, which can generally be defined as a learning process in which the learners are engaged in meaningful activities in the classroom and are mindful of what they are doing [12, 13].

The design of a teaching and learning activity determines the tasks that the learners need to engage in and how much active learning is possible. A traditional lecture, which is delivered in a unidirectional manner without interactions between the teacher and the learners, is often used to illustrate what learning is like when active learning is absent. In this kind of learning, the learners passively receive knowledge in a form already determined by the teacher. There are no tasks that the learners need to be involved in. In fact, the learners do not even need to be there for the lecture to take place. In active learning, the learners do not just sit and passively receive information. They are engaged in such activities as discussion in small groups, think–pair–share activities, short writing exercises, debate, or gaming. Some of these activities can be integrated into lectures, thereby introducing some element of active learning into this relatively passive form of teaching and learning. But the lecture format makes these active learning opportunities difficult. Other activities that have been specifically designed to stimulate active learning include collaborative learning [14], cooperative learning [15], and problem-based learning [16], and active learning can be integrated into most other activities, including anatomical dissection (see Chap. 26).

Tasks that stimulate active learning are those that encourage learners to take ownership of the learning, which stimulates them to think critically and creatively in order to accomplish certain tasks. By engaging in these tasks, learners reflect on their prior or newly acquired knowledge, identify gaps in their knowledge, seek out relevant information, assess current problems, analyze facts and opinions, etc.

Although the main responsibility of learning rests with the learners in active learning, the teacher also has a very important role to play. The teacher should cease to be the "sage on the stage" as in a lecture and should become a "guide on the side" by providing a safe and inquisitive environment for the learners to explore and construct knowledge. A friendly and supportive environment encourages the learners to articulate their thoughts and to ask and answer questions in front of others without the fear of feeling embarrassed when they make any mistakes.

Reflection

People learn from experience by reflecting on it. Reflection is thus an integral part of learning. Before a meaningful discussion can proceed, "reflection" must first be defined because this term is used in everyday life and has different meanings in specific circumstances. Moon [17] defined it as "a form of mental processing with a purpose and/or anticipated outcome that is applied to relatively complex or unstructured ideas for which there is no obvious solution," while Boud et al. [18] defined it as "a generic term for those intellectual and affective activities in which individuals engage to explore their experiences in order to lead to a new understanding and appreciation." A more inclusive definition is given by Sanders [19]: "Reflection is a metacognitive process that occurs before, during and after situations with the purpose of developing greater understanding of both the self and the situation so that future encounters with the situation are informed from previous encounters." It is thus considered a process of thinking about thinking (metacognition) that involves not only the acquisition of new knowledge or skills but also an understanding of both the self and the situation, so that the learner will respond differently in future encounters.

The significance of reflection can be described using Kolb's cycle or the learning cycle [20]. The cycle consists of four stages (Fig. 1.5). Experience is just one of the four stages, and it alone is not sufficient for learning to occur. One needs to reflectively observe the experience ("reflective observation") and then formulate and integrate the new "skills, knowledge, attitudes and values with the learners' cognitive framework" [21] ("abstract conceptualization"). Based on the new cognitive framework after reflection on a previous experience, the learner will respond differently when he or she encounters similar situations in the future ("action"). The new response is itself an experience that the learner can reflect on, leading to further modification of the cognitive framework.

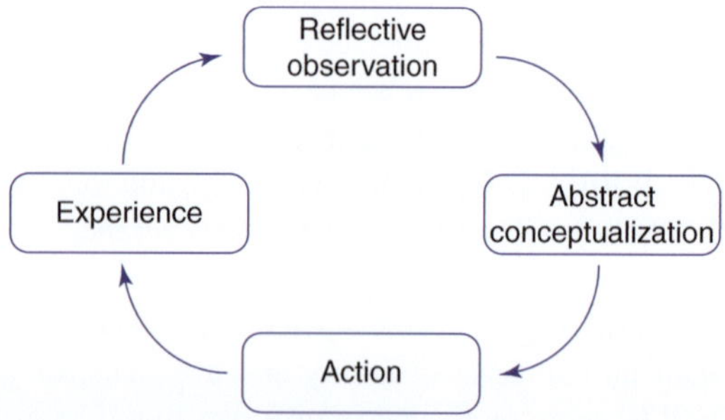

Fig. 1.5 Kolb's cycle. (After Kolb [20])

Learners going through the stages in Kolb's cycle are more likely to achieve deep learning because numerous links are formed between the new facts and ideas and their existing cognitive framework, in contrast to surface learning, in which new facts and ideas are isolated and unconnected [22]. The linking process also gives more meaning to the new knowledge, skills, attitudes, and values by relating them to the larger context.

Given the important role of reflection in learning, it is surprising that it does not spontaneously occur as often as desired and needs to be actively promoted. Moon [23] pointed out that learner reflection can be promoted when the tasks are challenging and ill structured (e.g., real-life examples), demand ordering of thoughts (e.g., following exposure to disorganized data), involve evaluation, and require integration of the new into previous learning. Chapter 26 illustrates how teaching and learning around anatomical dissection can be structured to promote reflection.

Feedback

Feedback is an essential part of medical education. It is "...specific information about the comparison between a trainee's observed performance and a standard, given with the intent to improve the trainee's performance" [24, 25]. It helps learners to maximize their potential at different stages of their lifelong learning path, raise their awareness of strengths and areas for improvement, and identify actions to be taken to improve performance. Therefore, the purpose of feedback is to improve performance [26] and reflection [25], not to criticize or judge.

A common model for giving feedback in clinical education settings was developed by Pendleton et al. [27]. Pendleton's rules consist of the following steps:

1. Check if the learner wants and is ready for feedback.
2. Allow the learner to give comments/background to the material that is being assessed.
3. The learner states what was done well.
4. The observer states what was done well.
5. The observer states what could be improved.
6. The teacher states how it could be improved.
7. An action plan for improvement is made together.

Pendleton's rules are structured in such a way that the positives are highlighted first (steps 3 and 4) in order to create a safe environment. In addition, step 3 forces the learner to reflect on his/her action. This is followed by the observer reinforcing these positives and adding some more if needed. "What could be done differently?" is then suggested, first by the learner and then by the observer. Again, it gives the learner the opportunity to reflect and to decide what to do next time. The advantage of this method is that the learner's strengths are discussed first. Avoiding a discussion of weaknesses right at the beginning prevents defensiveness and allows reflective behavior in the learner. The most crucial step is step 7, the action and follow-up plan, where the learner agrees with the observer on changes he/she will make for the next time.

Although this model provides a useful framework, there have been some criticisms of its rigid and formulaic nature, and a number of other models have been developed for giving feedback in a structured and positive way. One of these is the "sandwich" model, which starts with identifying the learner's strength, is followed by identifying the learner's areas in need of development, and concludes by reinforcing the strengths again.

Elements of Effective Feedback

- Provide a culture of giving feedback between learners and teachers. Feedback should be given *frequently*.
- Give feedback only when asked to do so or when your offer is accepted.
- Schedule formal feedback sessions that are convenient for the learner and the teacher, and adequate time should be given for both parties to prepare. Give feedback in a *timely* manner (not too soon or too late after the event).
- Select a location that is as private as possible.
- Measure the learner's performance against *standards* and well-defined goals and objectives.
- Provide *specific* and *accurate* information, including examples, not generalizations.
- Focus on the *positive*.
- Focus on *behaviors* that can be changed, not personality traits.
- Be *sensitive* to the impact of your message. Feedback is for the recipient, not the giver.
- Consider the content of the message, the process of giving feedback, and the congruence between your *verbal and nonverbal messages*.
- Encourage *reflection*.
- Be clear (have a goal) about what you are giving feedback on, and link this to the learner's *overall development* or intended program outcomes.
- Do *not overload*—identify two or three key messages to outline at the end. Be sure that the learners themselves identify the changes they want to make.

Summary

To promote effective learning, teachers need to understand the needs of learners and to motivate them by enabling them to perceive any gaps that exist between their present level of ability (what is) and the desired level (what ought to be). Learners can be motivated to learn better if they are informed regarding the specific learning outcomes of the activities and how achieving these outcomes will help them to bridge their gaps. Teaching and learning activities and assessment methods and standards should all be designed to help learners achieve these outcomes. Learners should be actively engaged in the learning process, instead of passively receiving information. They should be given frequent, accurate, and specific feedback at the appropriate time and be given time and opportunities for reflection.

References

1. Knowles M. The modern practice of adult education: from pedagogy to andragogy. 2nd ed. Chicago: Association Press/Follett; 1980.
2. Maslow AH. Motivation and personality. 2nd ed. New York: Harper and Brothers; 1970.
3. Fox RD, Miner C. Motivation and the facilitation of changing, learning, and participation in educational programs for health professionals. J Contin Educ Heal Prof. 1999;19:132–41.
4. Kern DE, Thomas PA, Hughes MT. Curriculum development for medical education: a six-step approach. 2nd ed. Baltimore: The Johns Hopkins University Press; 2009.
5. Biggs J, Tang C. Teaching for quality learning at university. 3rd ed. Berkshire: McGraw-Hill; 2007.
6. Anderson LW, Krathwohl DR, Airasian PW, et al. A taxonomy for learning, teaching and assessing: a revision of bloom's taxonomy of education objectives. 2nd ed. New York: Longman; 2001. p. 336.
7. Moore DE Jr, Green JS, Gallis HA. Achieving desired results and improved outcomes: integrating planning and assessment throughout learning activities. J Contin Educ Heal Prof. 2009;29:1–15.

8. Biggs JB. Enhancing teaching through constructive alignment. High Educ. 1996;32:1–18.
9. Harden RM, Laidlaw JM. Essential skills for a medical teacher. Edinburgh: Elsevier Churchill Livingstone; 2012.
10. Swing SR. The ACGME outcome project: retrospective and prospective. Med Teach. 2007;29:648–54.
11. Simpson JG, Furnace J, Crosby J, Cumming AD, Evans PA, Friedman Ben David M, et al. The Scottish doctor—learning outcomes for the medical undergraduate in Scotland: a foundation for competent and reflective practitioners. Med Teach. 2002;24:136–43.
12. Bonwell CC, Eison JA. Active learning: creating excitement in the classroom. ASHE-ERIC higher education reports (ERIC document reproduction service no. ED 336 049). 1991.
13. Prince M. Does active learning work? A review of the research. J Eng Educ. 2004;93:223–31.
14. Barron B. Achieving coordination in collaborative problem-solving groups. J Learn Sci. 2000;9:403–36.
15. Millis B, Cottell P Jr. Cooperative learning for higher education faculty. American council on education. Phoenix: ORYX Press; 1998.
16. Hmelo-Silver CE. Problem-based learning: what and how do students learn? Educ Psychol Rev. 2004;16:235–66.
17. Moon JA. A handbook of reflective and experiential learning: theory and practice. Routledge Falmer: Abingdon; 2004.
18. Boud D, Keogh R, Walker D. Reflection: turning experience into learning. London: Kogan Page; 1985.
19. Sanders J. The use of reflection in medical education: AMEE guide no. 44. Med Teach. 2009;31:685–95.
20. Kolb DA. Experiential learning: experience as the source of learning and development. Englewood Cliffs: Prentice Hall; 1984.
21. Branch WT, Paranjape A. Feedback and reflection: teaching methods for clinical settings. Acad Med. 2002;77:1185–8.
22. Houghton W. Engineering subject centre guide: learning and teaching theory for engineering academics. Loughborough: Loguhborough University, Engineering Subject Centre; 2004.
23. Moon J. Reflection in learning and professional development. London: Kogan Page; 1999.
24. van de Ridder JM, Stokking KM, McGaghie WC, ten Cate OT. What is feedback in clinical education? Med Educ. 2008;42:189–97.
25. Gordon J. BMJ ABC, of learning and teaching in medicine: one to one teaching and feedback. Br Med J. 2003;326:543–5.
26. Oxman AD, Thomson MA, Davis DA, Haynes RB. No magic bullets: a systematic review of 102 trials of interventions to improve professional practice. Can Med Assoc J. 1995;153:1423–31.
27. Pendleton D, Scofield T, Tate P, Havelock P. The consultation: an approach to learning and teaching. Oxford: Oxford University Press; 1984.

2 Learners of a New Generation

Camille DiLullo

It has been proposed that learning is expedited when clearly defined "learning outcomes" are stipulated in advance. This concept and an approach for implementation are more fully addressed toward the end of this chapter. To exemplify the process, I am providing two outcomes you should expect to attain after completion of this reading. The anticipated outcomes to be acquired are the ability to (1) identify issues that can impact learning of the next generation and (2) coordinate content delivery that can facilitate next-generation learning.

The Learner Persona

Differentiate the Constituents of Multigenerational Learning Cohorts

Learner cohorts in higher education are more likely to be comprised of individuals from multiple generations as compared to the learner cohorts in K-12 education which are predominantly composed of individuals from a single generation. Whether a particular cohort is comprised of learners from one or more than one generation, it can be predicted that individual learning cohorts will include people with varied learning characteristics. Strauss and Howe [1] defined distinguishing characteristics for specific generations of the twentieth century in the USA. The authors proposed that identifiable generational traits are in part shaped by global, national, and societal events that occur during the time period of the previous generation. It is likely that generalized traits of individual generational cohorts from other countries can also be defined but will vary based on their respective national and societal events. When examining charts that delineate generational descriptions, individuals commonly feel they do not personally possess every characteristic assigned to their particular generation or alternatively that personality traits assigned to previous or later generations are more apropos. Appreciation of how your personality reflects the attributes associated with your own generation can help elucidate the premise that learners within any generational cohort are not a clonal population. The exercise provided in Table 2.1 is designed to help you ascertain how in synch you believe your personality traits are with those categorized for a single generation. Peruse the characteristics listed in Table 2.1 for four American generations. If you are an American, select the generational cohort to which you belong. If you are not part of an American gen-

C. DiLullo (Deceased) (✉)
Department of Bio-Medical Sciences, Philadelphia College of Osteopathic Medicine, Philadelphia, PA, USA

L. K. Chan, W. Pawlina (eds.), *Teaching Anatomy*, https://doi.org/10.1007/978-3-030-43283-6_2

Table 2.1 Attributed generational traits

1925–1942 silent generation	√	*1943–1960 baby-boom generation*	√
Adaptive		Antiestablishment	
Cautious		Confident	
Conformist		Entrepreneurial	
Due process-oriented		Free-spirited	
Fair		Idealistic	
Hard-working		Individualistic	
Mediator		Independent	
Nonviolent		Personal growth	
Reserved		Revolutionary	
Risk-averse		Self-directed	
Socially conscious		Self-indulgent	
Solitary		Self-motivated	
1961–1981 generation X	√	*1982–2003 millennial generation*	√
Cynical		Accepts authority	
Detached		Achiever	
Determined		Civic	
Enigmatic		Considered special	
Informal		Family-oriented	
Pragmatic		Happy	
Quick		Pressured	
Reactive		Optimistic	
Realistic		Protected	
Self-protective		Self-assured	
Self-reliant		Service-minded	
Street-smart		Team player	

eration, select the group which has a majority of traits that you consider most appropriately represent your particular generation. In the column to the right of the group you selected, check how many of the listed traits you feel characterize your persona. Now examine the traits in columns of other generational cohorts with which you did not identify. Again, in the right column, check those traits which you believe can fairly be ascribed to your persona. You may not feel strongly bound to distinctive traits attributed to the group with which you primarily identified. Alternatively, you may feel closely aligned with traits linked to other generational cohorts. Members of any generation may possess many of their generations' ascribed traits, but they are just as likely to possess traits typical of other generational cohorts.

Appreciate Individual Learner Characteristics

It is essential that we view generational traits as trends in a given learner cohort rather than as inflexible characteristics. It is incumbent upon educators to tease out what traits charged to the newest generation of learners might influence their learning keeping in mind that many of these things might impact learners of previous generations as well. Global evolution has created the culturally diverse and technologically rich environment in which the millennial generation has been raised. Millennial learners come from a multitude of cultural, religious, and socioeconomic backgrounds [2]. They have grown up in a world dominated by digital communication both in their personal lives and in their educational experience. Diversity and engagement with technology will be factors that affect the success of millennial as well as next-generation learning into the foreseeable future. Societal circumstances have also been suggested to affect millennial generation learning including the protected nature of their upbringing, the way they have all been made to feel special, the pressure they have been put under to become high achievers, and the tendency for them to engage in team activities. In our global environment of competition with goals for ever-increasing productivity and efficiency, many traits that have been ascribed to the millennial generation are likely to remain the same for next-generation learners. Despite overarching trends in generational traits, learning cohorts will encompass individuals with unique personalities. In the process of learning, that which we are inherently interested in or excited about we learn best. Guiding learners to explore new knowledge through the lens of their distinctive professional passions within real-world contexts, which I refer to as interrelational learning (IRL), will more deeply engage students in the learning process.

Much has been written regarding the transformation in learning style of the millennial generation [2–5]. Suggested learning style changes include multitasking, a preference for learning using technology and working in groups as well as the elimination of reading. Alternative views assert that the learning style of the millennial generation may not be as radically different from previous generations as is often proposed [6, 7]. Studies have shown that many millennial learners continue to be engaged with traditional teaching methods and do read although it may be with e-books rather than textbooks [8]. While individual learners can absorb information in multiple ways, they generally demonstrate a learning style preference. In other words, they learn more easily in one particular modality as compared to others. Learners process information in one of several ways and can be defined as visual, aural, read/write, or kinesthetic [9]. Individuals will vary in their predilection for specific modalities, so within any learner cohort there will be a mix of preferred learning styles.

Support Varied Learning Styles with a Balanced Educational Approach

The recognition of different learning styles [10, 11] has advanced in tandem with the evolution of available educational resources. For centuries, learners had to adapt their learning style to the resources that were available. In the last century, resources included primarily lecture, handouts, books, films, personal notes, and a physical library. The technological explosion has exponentially increased the overwhelming array of educational resources beyond the traditional to include digital pedagogy, lecture capture, computer animation, YouTube, e-books, web searches, virtual programs, a digital library, blogs, e-communities, and so on. This vast assortment can substantially support a greater variety of learning styles. It may now be more appropriate to view the need for evolution in teaching modalities to be an outcome of the increased selection of educational resources rather than the inherent learning differences of next-generation learners. The pedagogical pendulum has for centuries been shifted toward using traditional teaching methodologies that favor aural and read/write learners. Collectively, our expanded learning resources offer more opportunity to directly engage visual and kinesthetic learners. However, the introduction of innovative delivery methods with the concurrent elimination of traditional pedagogy could swing the pendulum to the other extreme and limit the learning environment for aural and read/write learners. Facilitation of learning must take a balanced approach in content delivery that incorporates multiple and varied learning paradigms to accommodate learners of all types. Some educators have embraced the idea that each learner should have a special curriculum designed exclusively for them to meet their particular learning needs [12, 13]. Embracing a comprehensive approach to content delivery would eliminate the need for custom learner curricula. Next-generation learners would have the opportunity to self-select instructional modalities that provide them optimum success in developing expertise.

Ensure Learner Competence with Innovative Technology

Millennial generation learners sometimes referred to as digital natives [14, 15]—individuals whose development has been infused with technology—have had extensive experience with digital exploration, gaming, and communication and are purported to be adept with user-friendly digital devices. Despite the pervasive use of technology by this generational cohort, data indicates that their proficiency with commonplace digital devices does not necessarily translate into an aptitude for educational technology [16, 17]. Educators should not presume that all next-generation learners, let alone learners from previous generations, will be skilled in the use of educational technology such as integrated educational content platforms, anatomical simulators, and virtual anatomy programs. Sufficient instruction must be available to appropriately prepare learners in the use of technology that is employed to deliver content. Time management is crucial to millennial generation learners who feel under great pressure to achieve and are always attempting to expedite successful completion of their requirements. Clear directives on the use of educational technology will maximize learner engagement with and benefit from these resources. Upfront instruction on technological applications can reduce invested effort and help learners facilitate time management.

Keep Learners Focused with Active Learning and Frequent Feedback

The incessant simultaneous use of multiple digital devices by millennial generation learners has them convinced they are capable of unrestrained multitasking. They are consistently in the habit of rapidly switching tasks. The nature of multitasking frequently manifests in learner dispositions that are easily distracted and have difficulty maintaining focus. Millennial learners have little difficulty in overlapping texting, web surfing, and gaming tasks. However, data indicates that while multitasking may not appreciably impact activities such as gaming that can be classified as habit learning [18, 19], it is likely to significantly disrupt the acquisition of knowledge aimed toward subsequent integration and reasoning [20–24]. Learners must be sufficiently engaged to maintain their focus on learning. The use of active learning and a broad variety of content delivery paradigms is likely to promote more consistent engagement for learners of all styles. Sustained focus on learning can also be encouraged with the provision of frequent feedback. Feedback that triggers discovery and reflection of personal strengths and weaknesses allows learners to target areas where improvement would benefit overall success. In addition to feedback from formative and summative testing, learners should be directed to gain feedback through interactive social discourse. As millennial learners are already adapted to collaboration in teams, feedback via group interaction whether physical or virtual will stimulate enhanced learner motivation.

The Way They Learn

Using "CER" Cycles to Develop Expertise

The concept of "teaching around the cycle" is founded on the premise that in processing information learners proceed sequentially through the following specific phases: (1) concrete experience, (2) reflective observation, (3) abstract conceptualization, and (4) active experimentation [25]. However, it can also be theorized that learners initiate the intellectualization of new knowledge with the "conceptualization" of information and follow this by related "experience" and "reflection." A cyclical progression of *c*onceptualization, *e*xperience, and *r*eflection designated "CER cycles" can prepare learners for interactive experimentation in the development of expertise. Learners advance through multiple CER cycles before participating in relevant interactive experimentation. The concept of "CER cycles" differs from that of "teaching around the cycle" in the learners' approach to the assimilation of new content. For "teaching around the cycle," learners would be absorbed with continuous sequential participation in the aforementioned phases of learning. For "CER cycles," learners would be engaged in prescribed cyclical phases of conceptualization, experience, and reflection prior to a phase of interactive experimentation. Programs can deliver CER cycles through (1) provision of content (conceptualization), (2) experiential challenges to apply conceptualized knowledge (experience), (3) opportunity to reflect upon acquired knowledge (reflection), and periods of simulated or actual real-world scenarios (interactive experimentation). CER cycles can be created for defined blocks of information that are sequentially arranged to build programmatic knowledge. Progression through a CER cycle is exemplified with the learning outcome in the following text box.

Learning Outcome: Obtain a Blood Pressure Measurement with a Blood Pressure Cuff

- *Conceptualization*
  Individual learner appreciation of critical arm anatomy as well as the relationship of individual arm structures one to another
- *Experience*
  Independent learner examination of (1) discreet arm structure and surface anatomy relationships, (2) brachial artery pulse localization, and (3) blood pressure cuff operation
- *Reflection*
  Group review of pertinent information in a laboratory setting, a relevant case-based session, or a personal study group
- *Interactive Experimentation*
  Engagement in a team exercise in which students practice taking their partner's blood pressure with a blood pressure cuff

Conceptualization and Unistructural Understanding

During the initial phase of a CER cycle, "conceptualization," learners absorb new information. They acquire isolated facts that they may not necessarily link together. It is a one-dimensional understanding of information. In terms of Biggs' "Structure of Observed Learning Outcomes" or SOLO taxonomy, this would be considered unistructural understanding [26]. The disparate nature of newly conceptualized information is often more obvious at the beginning of a course or program. The fragmented understanding of new information learned in the conceptualization phase of a CER cycle may be masked by accumulating knowledge. However, information conceptualized at the beginning of any CER cycle will be integrated into prior knowledge during the remaining "experience" and "reflection" phases of that cycle. Generally, focused conceptualization of information occurs independently rather than in a group format.

In an unfamiliar learning environment, each learner commonly approaches conceptualization of knowledge in ways, whether visual, aural, read/write, or kinesthetic, that they have determined from previous experience best facilitates their learning. For example, the initial steps in learning how to take a blood pressure measurement would involve conceptualization of basic facts. Learners on their own, depending upon their options, may choose to attend a lecture, watch a podcast or animation, read a text, study an atlas, etc. They will continue to review this information until they believe it is adequately understood. Frequent opportunity for independent self-quizzing will provide feedback that can illuminate for learners areas in which they may be weak. Independent self-quizzing immediately after initial conceptualization has been shown to improve long-term retention of information [27–30]. Actively directing learners toward alternative styles of learning with which they might not be as familiar can provide a form of early intervention which may guide students toward enhanced conceptualization of information they find confusing and could improve success in assimilating program content.

Experience Through Real-World Contexts

In the second phase of a CER cycle, "experience," learners independently relate their newly acquired knowledge to contextual scenarios. This experiential application of conceptualized knowledge supports the development of expertise as learners are stimulated to connect acquired knowledge to practical situations. Learners begin to recognize the significance of how initially disparate pieces of information are interrelated to each other as well as to real-world applications. In this phase, new knowledge can be linked to knowledge acquired through previous CER cycles. Learners begin to develop a two-dimensional or multistructural understanding of information. The experiential application of new knowledge, like initial conceptualization, generally occurs independently. In general, the types of individual experiential learning in which students can participate are multimodal. Visual and kinesthetic learners might gravitate toward using physical or virtual cadavers and laboratory specimens to help them connect various aspects of their conceptualized knowledge. Aural and read/write learners may be more engaged using podcast demonstrations and online or paper-based case problem solving with self-quizzing.

Reflection Through Interactive Discourse

Reflection is the final phase in a CER cycle. Once learners have independently acquired knowledge through conceptualization and experience, they need time to reflect on what they have learned. The reflection that completes a CER cycle is reflection that determines how successful the phases of conceptualization and experiential learning have been rather than reflection on personal feelings on individual experiences. In this phase of the cycle, learners should participate with others in interactive discourse. Interactive social discourse provides direct feedback from others that affords opportunity for learners to validate that their comprehension of assimilated knowledge is both accurate and adequate. Learners can identify information that they may have overlooked or not have fully understood. With reflection, learners will advance to more fully integrating facts, concepts, and principles and evolve a more three-dimensional or relational perspective of knowledge. The knowledge will become more interrelated. Ample opportunity within each CER cycle for reflection and personal authentication of learning can effectively prepare students for successful engagement in the next CER cycle. Both group study and team activities offer face-to-face social interaction that can promote reflection on learning. Subsequent to a program-defined number of CER cycles, learners will be prepared to engage in interactive experimentation through which they can utilize their accumulated learning in real-world contexts.

"CER" Cycle Success Will Vary By Learner

Figure 2.1 provides a visualization of the progressive advancement of learner knowledge through sequential CER cycles. The graph illustrates a projected series of CER cycles in a prescribed learning period designed to lead to attainment of all defined program learning outcomes. The 45° stepped line represents programmatic CER cycles that are sequentially presented to learners throughout the prescribed program learning period. Four individual learners are represented by unique shapes that symbolize their individuality. The color changes for each learner indicate at what point along the program timeline they have successfully integrated a specific percentage of program learning outcomes. It is expected that on average learners will progress through programmatic CER cycles within the prescribed program learning period. However, learners begin the prescribed learning period with disparate prerequisite knowledge in addition to having varied learning style preferences and academic proficiency. These among other factors can influence variable learner progression. On the graph, the individual slope of each learner indicates the rate at which they successfully pass through the series of CER cycles. Most learners will advance through the defined CER cycles in a relatively predictable pattern and successfully attain all expected program learning outcomes using the entire prescribed learning period. However, even among these cohorts, learners will vary to some degree in how quickly they acquire learning outcomes (learners 2 and 3). The slope of their lines will be clustered but may not be identical. A learner slope that is significantly different than the 45° stepped line indicates learner divergence from average. Some will quickly grasp information and make connections among program content and concepts allowing them to rapidly advance through CER cycles to succeed in mastering the prescribed learning outcomes (learner 1). Others may struggle with one or more CER cycles, delaying their integrated understanding of program content and achievement of program success (learner 4).

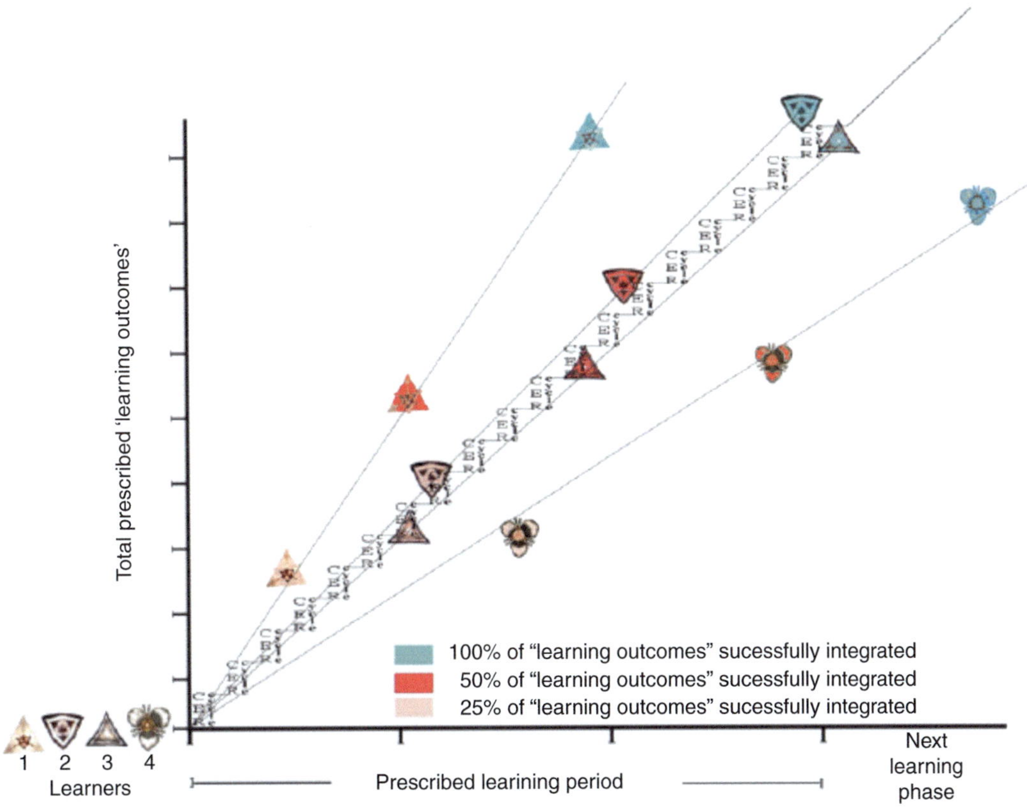

Fig. 2.1 This figure illustrates the progression of four individual learners through a prescribed program period. The four learners (1–4) are represented by unique shapes that symbolize the variability of learners that can exist within a particular learner cohort. The program timeline is delineated along the *x*-axis. Contiguous with the "prescribed learning period" is inserted the "next learning phase." The total "prescribed learning outcomes" are represented along the *y*-axis. Sequential programmatic CER cycles that are projected to facilitate learner acquisition of "prescribed learning outcomes" are represented in relation to time with a 45° stepped line. The color changes for each learner indicate what percentage of expected program knowledge the learner has attained (peach = 25%, red = 50%, turquoise = 100%). Factors specific to individual learners will influence the pace of successful progression through the program. Average progression through the program will produce a learner slope line that is similar to the 45° stepped CER cycle line. The progression slope of the majority of learners should be clustered around this CER cycle line. A learner slope that is significantly different than the 45° stepped line indicates learner divergence from average

Strategies to Facilitate Success in Next-Generation Learning

Communicate Clearly Defined Learning Outcomes

Subsequent to the determination of appropriate content, clear definition and communication of expected learning outcomes are the next major step in facilitating learning. Unequivocal documentation regarding the knowledge that is to be acquired will keep learners focused and promote effective time management. The terminologies "learning outcomes" and "learning objectives" while often considered to be synonymous can be defined as distinctly different entities. "Learning objectives" generally list what knowledge will be covered within a specific block of learning. "Learning outcomes" details demonstrable results learners are expected to manifest at the completion of the course of study [31, 32]. In the context of interrelational learning, learning outcomes should reflect intended real-world applications of successfully assimilated knowledge. Learning outcomes should be written using active and measureable verbs such as *align*, *correlate*, *trace*, *connect*, and *predict* which can suggest appropriate assessment methodologies. Ideally, assessments should be directly correlated to the stated learning outcomes [32, 33].

The difference between a "learning outcome" and a "learning objective" can be illustrated in anatomy using a typical heart learning module. "Understand the structure and function of components in the four chambers of the heart" would be a "learning objective" which informs learners about what information should be studied. Learner efforts would be focused on memorizing the names of all the heart structures that are stipulated. Using the "learning outcome"—"correlate normal and abnormal heart sounds with typical and atypical heart structure and function"—for the same module would guide learners toward a deeper understanding of how heart structure can impact heart function. Learner efforts would be guided toward better integrating heart structure as it relates to patient auscultation and cardiac function in an authentic context.

Learning Outcome Versus Learning Objective

- *Learning Outcome*
  Correlate normal and abnormal heart sounds with typical and atypical heart structure and function.
- *Learning Objective*
  Understand the structure and function of components in the four chambers of the heart.

A multimodal exercise to fulfill the stated learning outcome would be one in which learners actively correlate normal and abnormal heart sounds with appropriate cardiac structure/function through animations, audio files, radiographic imaging, and anatomical specimens. The learning outcome directly suggests an assessment strategy through the active verb *correlate*. A typical assessment could be based on evaluation of learners correlating normal and abnormal heart structure with normal and abnormal heart sounds. Unequivocal direction with learning outcomes will assist learners in efficient navigation throughout their educational journey.

Challenge Learners Through Multimodal Learning Paradigms

Recognizing that multiple learning style preferences exist within our student cohorts, we should ensure that learners have a variety of content delivery modalities from which to choose. It is important to advance learning through a multimodal approach which will provide opportunity for individual learners, whether visual, aural, read/write, and kinesthetic, to self-select the educational paradigm which most readily facilitates their learning. Various resources will support different types of learning. Cadaver dissection, videos, animations, models, plastinated specimens, and virtual dissection will favor visual learners. Lecture (in-class, podcasts, or lecture capture) as well as small group discussion may be more suited to aural learners. Reading assignments, lecture (note-taking), and writing assignments will benefit read/write learners. Simulation workshops, virtual reality programs, simulated patients, cadaver dissection, models, and anatomical or plastinated specimens are effective for kinesthetic learners. Some content delivery paradigms will be equally effective for different learning styles. For example, physical or virtual dissection can engage both visual and kinesthetic learners.

However, to enhance lifelong learning skills, it is most advantageous for the learner to be able to assimilate information in any modality that might be encountered. Therefore, in addition to learning in their preferred learning style, it is important for students to learn with modalities in which they feel less adept. Learners should be encouraged if not required to acquire information in ways that are a "match" as well as a "mismatch" to their learning style preferences. Advancing skills in absorbing information from a wide range of communication modalities will advance learner proficiency for information integration and enhance the prospect for success with future independent learning.

Provide Guidance for Identification of Authenticated Information

In the process of conceptualizing information, learners are challenged by the overwhelming number of educational resources available to them. In addition to traditional resources, there has been a proliferation of creative digital resources and various technologies that can be utilized to facilitate learning including e-books, podcasts, lecture capture, animations, videos, virtual programs, audience response systems, simulators, course management systems, e-portfolios, and social media. In addition to required resources, learners usually search for supplemental resources that can facilitate their learning. Learners invest significant effort in attempting to ascertain which learning resources will best serve their needs. Educators can provide guidance to learners in sorting through the countless available choices by designating a targeted collection of resources that will best explicate the program-defined learning outcomes. Expeditious selection of auxiliary educational material that will most appropriately support an individual's knowledge acquisition can positively impact effective time management and optimally facilitate learning.

Next-generation learners are accustomed to exploring digital resources in order to identify information they believe will advance knowledge comprehension. They are, however, often naïve about the quality of the resources they uncover [34–37]. Beyond the educational environment, individuals will frequently be expected to self-select reliable information germane to any number of topics. Within the context of the educational program, learners should receive appropriate instruction to cultivate their ability to independently search for new authoritative information. They should be directed away from resources that may be inaccurate or unvetted. Learners must be prepared to recognize the subtle messages conveyed in digital communication known as second-order information [38]. Understanding how to evaluate indirect cues like the purpose of a web page or the source of information can guide learners in evaluating the objectivity of content. Proficiency in information literacy will provide another facet in the progression of learning that can facilitate knowledge acquisition, further successful time management, and enhance lifelong learning skills.

Offer Opportunity for Interactive Social Discourse

Next-generation learners, who have routinely received frequent feedback both in and out of the educational environment, continue to look for feedback in learning. Generally, feedback is crucial in helping identify areas of strengths and weaknesses which once recognized can be addressed. Using formative and summative testing, as well as course instructor feedback, learners attain fundamental information regarding their success with tested or queried knowledge. However, learners can obtain more dynamic feedback through interactive social discourse. The penchant that next-generation learners have for working in groups makes interactive feedback very effective. Learners can construct a personal learning network (PLN) [39] sometimes referred to as a professional learning network. The PLN is a place for individual learners to consolidate social interactions that offer the most productive feedback to advance their learning. The interactive social discourse engaged in through the PLN provides committed periods for learners to pass through the final reflective phase of a CER cycle. Educators should guide learners in the selection of participants for their particular PLN.

A variety of educational delivery paradigms including problem-based learning and team-based learning can support social discourse and be included in the students' PLN. More recent methodologies like the flipped classroom also require students to independently conceptualize information delivered digitally and then subsequently discuss it in a more interactive forum. Study groups in which learners participate after independent conceptualization and experiential learning become part of the learners' PLN. The learners' PLN can be expanded with online venues such as Facebook, particularly a Facebook site linked to the educational institution.

Communication avenues in which learner content can be critiqued by others like small group discussions, discussion boards, blogs, e-learning communities, e-portfolios, Twitter, and Pinterest can act to support interactive educational social discourse. As part of linking knowledge acquisition to real-world scenarios and individual learner professional interests in the context of interrelational learning, ongoing professional extracurricular activities can also generate interactive discussion which can corroborate learning and contribute to a learner's PLN.

Conclusions

Next-generation learners are engaged by a variety of learning styles in the process of absorbing information. Programmatic content delivery should be designed utilizing the broadest possible array of learning modalities to facilitate knowledge acquisition for individual learners. Innovative methodology and digital technology blended with traditional pedagogy can be used to create advancing sequential cycles of conceptualization, experience, and reflection which will advance the learner in the development of expertise. To maximize efficient learner time management, enhance their learning experience, and foster academic success, the following educational practices should also be integrated into the instructional paradigm: (1) identification of clearly defined learning outcomes correlated to real-world applications, (2) adequate instruction for technology platforms and programs, (3) guidance in the use of learning resources, (4) provision of frequent feedback including appropriate assessments and evaluations linked to learning outcomes, and (5) opportunity for interactive social discourse.

References

1. Strauss W, Howe N. Generations: the history of America's future, 1584 to 2069. New York: William Morrow and Company; 1991. p. 279–335.
2. Keeter S, Taylor P. Millennials: a portrait of generation next. Washington, DC: Pew Research Center; 2009. http://www.pewsocialtrends.org/files/2010/10/millennials-confident-connected-open-to-change.pdf.
3. Howe N, Strauss W. Millennials go to college: strategies for a new generation on campus. Great Falls: American Association of Collegiate Registrars and Admissions Officers; 2003. p. 51–63.
4. Oblinger D. Boomers, Gen-Xers, and Millennials: understanding the new students. Educ Rev. 2003;38:37–47. http://net.educause.edu/ir/library/pdf/ERM0342.pdf.
5. Twenge JM. Generational changes and their impact in the classroom: teaching generation me. Med Educ. 2009;43:398–405.
6. Hoover E. The millennial muddle: how stereotyping students became a thriving industry and a bundle of contradictions. Chron Higher Educ. 2009, 56: Oct 11. http://chronicle.com/article/The-Millennial-Muddle-How/48772.
7. DiLullo C, Morris HJ, Kriebel RM. Clinical competencies and the basic sciences: an online case tutorial paradigm for delivery of integrated clinical and basic science content. Anat Sci Educ. 2009;2:238–43.
8. National Endowment for the Arts. To read or not to read: a question of national consequence, research division report #47. 2007. 99. http://www.nea.gov/research/ToRead.pdf.
9. Fleming ND, Mills C. Not another inventory, rather a catalyst for reflection. Improv Acad. 1992;11:137–47.
10. Coffield F, Moseley D, Hall E, Ecclestone K. Learning styles and pedagogy in post-16 learning: a systematic and critical review, vol. 174. London: Learning and Skills Research Centre; 2004. http://www.academia.edu/1597391/Learning_Styles_a_systematic_and_critical_review.
11. Young T. How valid and useful is the notion of learning style? A multicultural investigation. Procedia Soc Behav Sci. 2010;2:427–33.
12. Ozuah PO. First, there was pedagogy and then came andragogy. Einstein J Biol Med. 2005;21:83–7.
13. Gillani BB. Using the web to create student-centered curriculum. In: Cole RA, editor. Issues in web based pedagogy. Westport: Greenwood Press; 2000. p. 161–82.
14. Prensky MH. Sapiens digital: from digital immigrants and digital natives to digital wisdom. Innovate. 2009;5(3):1–9. http://www.wisdompage.com/Prensky01.html.
15. Margaryana A, Littlejohn A, Vojt G. Are digital natives a myth or reality? University students' use of digital. Comput Educ. 2011;56:429–40.
16. Kvavik RB, Caruso JB. Convenience, connection, control, and learning. Students and information technology. Boulder: EDUCAUSE Center for Applied Research (ECAR); 2005. http://net.educause.edu/ir/library/pdf/ERS0506/ecm0506.pdf.
17. Kennedy GE, Judd TS, Churchward A, Gray K, Krause K-L. First year students' experiences with technology: are they really digital natives? Australas J Educ Technol. 2006;24:108–22.

18. Dye MW, Green CS, Bavelier D. Increasing speed of processing with action video games. Curr Dir Psychol Sci. 2009;18:321–6.
19. Dye MW, Green CS, Bavelier D. The development of attention skills in action video game players. Neuropsychologia. 2009;47:1780–9.
20. Cheyne JA, Carriere JS, Smilek D. Absent-mindedness: lapses of conscious awareness and everyday cognitive failures. Conscious Cogn. 2006;15:578–92.
21. Foerde K, Knowlton BJ, Poldrack RA. Modulation of competing memory systems by distraction. Proc Natl Acad Sci U S A. 2006;103:11778–83.
22. Erickson KI, Colcombe SJ, Wadhwa R, Bherer L, Peterson MS, Scalf PE, Kim JS, Alvarado M, Kramer AF. Training-induced functional activation changes in dual-task processing: an fMRI study. Cereb Cortex. 2007;17:192–204.
23. Ophir E, Nass C, Wagner AD. Cognitive control in media multitaskers. Proc Natl Acad Sci U S A. 2009;106:15583–7.
24. Watson JM, Strayer DL. Supertaskers: profiles in extraordinary multitasking ability. Psychon Bull Rev. 2010;17:479–85.
25. Kolb DA. Experiential learning: experience as the source of learning and development. Upper Saddle River: Prentice Hall; 1984. p. 256.
26. Biggs JB, Collis KF. Evaluating the quality of learning: the SOLO taxonomy. New York: Academic; 1982. p. 245.
27. Karpicke JD, Zaromb FM. Retrieval mode distinguishes the testing effect from the generational effect. J Mem Lang. 2010;62:227–39.
28. Zaromb FM, Roediger HL, Karpicke JD. Comprehension as a basis for metacognitive judgements: effects of effort after meaning on recall and metacognition. J Exp Psychol Learn Mem Cogn. 2010;36(2):552–7.
29. Roediger HL, Karpicke JD. The power of testing memory: basic research and implications for educational practice. Perspect Psychol Sci. 2006;1(3):181–210.
30. Roediger HL, Karpicke JD. Test enhanced learning: taking memory tests improves long term retention. Psychol Sci. 2006;17(3):249–55.
31. Driscoll A, Wood S. Developing outcomes-based assessment for learner-centered education. Sterling: Stylus Publishing, LLC; 2007. p. 1–18.
32. Biggs J, Tang C. Teaching for quality learning at university. Buckingham: Society for Research into Higher Education and Open University Press; 2007. pp. 64–91 and 163–95.
33. Harden RM, Crosby JR, Davis MH, Friedman M. AMEE Guide No. 14: outcome-based education: part 5—from competency to meta-competency: a model for the specification of learning outcomes. Med Teach. 1999;21(6):546–52.
34. Weiler A. Information-seeking behavior in generation Y students: motivation, critical thinking, and learning theory. J Acad Libr. 2004;31:46–53.
35. CIBER Group, University College London. CIBER briefing paper: information behavior of the researcher of the future. London: University College London; 2008. http://www.jisc.ac.uk/media/documents/programmes/reppres/gg_final_keynote_11012008.pdf.
36. Considine D, Horton J, Moorman G. Teaching and reading the millennial generation through media literacy. J Adolesc Adult Lit. 2009;52:471–81.
37. Hargittai E, Fullerton L, Menchen-Trevino E, Yates TK. Trust online: Young adults' evaluation of web content. Int J Commun. 2010;4:468–94.
38. Gershon I. The Breakup 2.0: disconnecting over new media. Ithaca: Cornell University Press; 2010. p. 123–65.
39. Richardson W, Mancabelli R. Personal learning networks. Bloomington: Solution Tree Press; 2001.

Constructive Alignment: An Outcomes-Based Approach to Teaching Anatomy

3

John Biggs and Catherine Tang

Outcomes-Based Education and Outcomes-Based Teaching and Learning

In order to improve the quality of student learning, we advocate a form of outcomes-based education. But first, we need to make an important distinction between two kinds of outcomes-based education.

The first refers to broad institutional outcomes, such as averaged student performances, that are used for such purposes as quality assurance, benchmarking, accreditation requirements, and the requests of external stakeholders such as employers and policymakers [1]. The question here is for administrators: "How well is the institution performing in relation to its own mission statement and performance indicators and in relation to other institutions?" There is no *necessary* connection here between these externally driven managerial concerns and the quality of the teaching that supposedly produced those outcomes.

The second kind of outcomes-based education is intended directly to improve student learning, so we refer to it as "outcomes-based teaching and learning," which addresses the learning outcomes that students are expected to achieve at both program and especially course levels. The question here is for teachers: "What do I want my students to be able to do as a result of their having learned specified topics?" The intended outcomes not only nominate the topic content but additionally what the student is intended to do with that content. The intended outcomes for any given course may be determined by the individual teacher or the teaching team, together with any relevant external input.

Once the outcomes have been defined, teaching should be designed to engage students in learning activities that are likely to achieve those outcomes. Assessment then addresses how well those outcomes have been achieved by students and is therefore criterion referenced. Such assessment is best achieved by rubrics or statements that specify the standards for different levels of student performance. Assessment tasks should also allow for any unexpected or unintended but desirable outcomes. While unintended but desirable outcomes cannot be specified in advance, they can be allowed for by asking students to place any evidence they think appropriate for their achieving unintended outcomes in a portfolio that is submitted for assessment. In outcomes-based teaching and learning, students should not be assessed according to how their performances compare with each other and then graded according to a predetermined distribution such as the bell curve.

These issues of linking teaching and assessment to achieving the intended learning outcomes

J. Biggs (✉) · C. Tang
Tasmanian Institute of Learning and Teaching, University of Tasmanian, Hobart, TAS, Australia
e-mail: jbiggs@bigpond.com; ckctang@bigpond.com

L. K. Chan, W. Pawlina (eds.), *Teaching Anatomy*, https://doi.org/10.1007/978-3-030-43283-6_3

(ILOs) are specifically addressed in the version of outcomes-based teaching and learning known as constructive alignment, as discussed below.

Constructive Alignment

Some years ago, Thomas Shuell summarized the implications for teaching and learning of cognitive psychology thus [2]:

> If students are to learn desired outcomes in a reasonably effective manner, then the teacher's fundamental task is to get students to engage in learning activities that are likely to result in their achieving those outcomes.... It is helpful to remember that what the student does is actually more important in determining what is learned than what the teacher does.

Embedded in this seemingly obvious statement is a powerful design for teaching that draws on two important principles:

- Knowledge is not transmitted by a teacher but is constructed by students through their own learning activities.
- The intended outcomes of teaching need to be stated upfront, and teaching methods and assessments need to be aligned to what those outcomes require if they are to be met.

These principles are the basis for constructive alignment [3, 4]. Constructive alignment is a form of outcomes-based teaching and learning in which both teaching and assessment are aligned to the ILOs, which specify what the student is expected to do with the content taught. The traditional topic-based curriculum only specifies what content the teacher is to address.

Accordingly, the ILOs need to state not only what the student is intended to learn but how, to what level, and in what context. Thus, if the topic in question is the anatomy of the upper and lower limbs, the learning outcome focuses on what the student is supposed to be able to do with that anatomical knowledge, for example, to *list* the muscles, bones, joints, nerves, and blood supply to the limbs or at a higher level to *explain* how the different anatomical components work and coordinate to produce movement or higher still to *predict* what functional disability will result from a specified injury to a certain part of the anatomy and to *design* a management program or a treatment protocol for such an injury. Thus, the student will not only *know about* the anatomy of the upper and lower limbs but be able to act properly on the basis of that knowledge—which is after all the purpose of professional education.

Good teachers have always taught with a view to what students should be able to do after being taught and how well they should do it—that is one reason why they are good teachers. In constructively aligned teaching, we simply make this explicit.

These verbs in the ILOs are used to prescribe the teaching/learning activity (TLA) the students need to engage in order to optimally achieve the outcome. If the ILO refers to *explain* the functioning of the musculoskeletal system, the TLA would require the students themselves to explain how the system functions, say to another student or in a presentation, not just to listen and take notes from a teacher who is doing the explaining.

The assessment task in turn addresses that verb in the ILO; if the verb is *explain*, the assessment is in terms of how well the explanation is carried out. Often, the most appropriate assessment task is the TLA itself, making alignment perfect. This is the case in problem-based learning (PBL), in which the ILO is solving a particular professional problem, the teaching/learning activity is solving the problem, and the assessment is how well the problem is solved.

High-level verbs in the ILO, such as *hypothesize*, *reflect*, *solve unseen problem*, or *create*, leave the outcome quite open. In that case, the assessment task needs to allow for the unexpected, as does an assessment portfolio.

Several writers have mentioned the utility of constructive alignment in teaching different subjects: computing science [5], earth sciences [6], designing e-learning [7], engineering education [8], health sciences [9], plant physiology [10], physiology [11], social work [12], statistics [13], teacher education [14], and veterinary sciences [15]. Typical advantages of using constructive alignment referred to in these studies include the following:

- Students being able to focus more effectively on the key learning goals
- Fairer and more reliable assessment
- Improved learning outcomes, including critical thinking and depth of student work
- Greater transparency leading to easier and more accurate interuniversity and international comparisons
- Greater coherence in programs and more effective evaluation of modules and courses

The major disadvantages refer to staff and student workloads. The major staff workload is in the initial transition from traditional to constructively aligned teaching, while that for students is often self-inflicted—they work harder in constructively aligned courses.

The principles of constructive alignment are used as frameworks for quality assurance agencies in the UK and Hong Kong [16]. Edström [17] writes: "course evaluation should be regarded as a component of constructive alignment, together with the ILOs, learning activities and assessment." Constructive alignment is widely regarded as a key idea on postgraduate certificates in higher education [18].

There are three main stages involved in applying constructive alignment to the teaching of anatomy: designing the ILOs, designing the teaching/learning activities (TLAs), designing and aligning assessment tasks (ATs), and obtaining a final grade.

Designing the Intended Learning Outcomes (ILOs)

ILOs contain a content topic or topics, action verbs that help specify the level of performance the students are expected to achieve, and a context for the learning to take place. Verbs such as *identify*, *name*, *describe*, *explain*, *predict*, *plan*, *design*, *hypothesize*, *generate*, and *create* are examples of increasing levels of understanding. Thus, to use the verb *understand* itself in an ILO is inadequate because it does not identify the kind or level of understanding required. The ILO verbs give students a clear signpost as to what and how they are to learn and on which they will be assessed. Ideally, there should be no more than five or six ILOs for a one semester course.

The SOLO taxonomy [4] provides a useful guide in deciding the levels of understanding or performance. The first two levels, *unistructural* and *multistructural*, see understanding as a quantitative matter of *knowing more*. The next two levels focus on qualitative changes as understanding deepens. At the *relational* level, students are able to *put things together* by interrelating the multiple aspects to create a structure, while at the *extended abstract* level, students are able to go *beyond the given* and take the argument or application of a principle to a new dimension. When writing ILOs, teachers need to decide what levels of understanding or performance students are expected to achieve and clearly indicate those levels by the ILO verbs. Other taxonomies such as the Bloom taxonomy may also used as a source of verbs, but we have found SOLO more useful as it provides a hierarchy of levels of understanding.

Let us take the example when students are learning the anatomy of the upper and lower limbs. Some of the relevant ILOs could be for the students to be able to do the following:

1. *Identify* and *name* all the muscles, bones, ligaments, joints, nerves, and blood supply to the upper and lower limbs.
2. *Describe* the origin, course, and insertion of the muscles in the upper and lower limbs.
3. *Describe* the courses of the nerves and blood vessels in the upper and lower limbs.
4. *Explain* how movements are produced in the upper and lower limbs.
5. *Compare* and *contrast* the mobility and stability of a ball and socket joint and a hinge joint in relation to their anatomy, using the shoulder joint and the knee joint as examples, respectively.
6. *Predict* what functional disability will result from injuries to a particular anatomical structure of the upper and lower limb.

These ILOs are generic and only meant here to serve as examples: readers will need to design their own (see To Do below).

When designing ILOs, teachers need to consider the context within which the course is being taught, the aims of the course, the level of the course, the target students, the reasons for which they are taking the course, and the needs of the profession which the students will be engaged in after graduation. Where clinical applications are desirable, ILOs might contain verbs such as *assess*, *diagnose*, *plan (management/treatment program)*, *implement*, *make a prognosis*, and *evaluate*.

Apart from the ILOs that are specified in the curriculum, there may well be outcomes that are a positive outcome of teaching but that were not intended by the teacher. Teachers need to allow for such unintended but desirable outcomes with open-ended assessment tasks so that the students themselves may provide evidence of such outcomes.

To Do

- Consider a topic in anatomy that you are going to teach. Design up to 5 ILOs that you intend your students to achieve.
- We will return to these ILOs in later exercises.

Designing the Teaching/Learning Activities (TLAs)

Teachers need to design activities that will engage students in activating the ILO verbs. The traditional lecture usually provides a one-way transmission and teacher-centered mode of teaching, with little opportunity for student engagement. If students are expected to *explain* how movements are produced in the upper and lower limbs, it is insufficient for the teacher to explain this in a lecture for here the students are not engaged in explaining but in listening, taking notes, and perhaps asking a question. For the students to achieve the ILO *explain*, they need to do the explaining themselves if that is what it is intended that they should be able to do.

Take the example in the previous section when students are learning the anatomy of the upper and lower limbs; to be able to explain, students can acquire the background information from books or other sources, doing dissection in anatomy lab, and then engage in *explaining* to fellow students—in pairs or in a group in class—on how movements are produced, say by using a model skeleton, by pulling on the muscles on a dissected cadaver in the anatomy lab, or by drawing a concept map to explain how the various anatomical components work in producing movements. An assignment would allow the students to explain in written form. These are some aligned TLAs for the ILO *explain*, and teachers can decide on the appropriate ones that would suit the particular teaching sessions.

A higher-level ILO might require students to *predict* what functional disability would result from injuries to a particular anatomical structure of the upper and lower limb. Again, students need to do the predicting themselves. Students doing background reading on relevant information, discussing with fellow students, role-playing the resulting disability, or working on a case study are aligned teaching and learning activities for the achievement of the ILO to *predict*.

Aligned TLAs can be conducted in large or small class situations and can also be either managed by the teacher or students or as self-learning activities, consistent with available resources. There are many suitable TLAs other than lectures and tutorials.

To Do

- Select one ILO from the list you have written in the previous section. Design some TLAs that would engage your student in achieving the outcome.
- When designing these TLAs, you may consider what you need to do as a teacher, but it is more important to consider what your students will be required to do themselves.
- You will also need to consider any required and available resources to implement those TLAs.

Designing and Aligning Assessment Tasks (ATs)

Assessment tasks (ATs) for a given ILO or set of ILOs are aligned to the target ILO(s) by presenting the student with tasks that require them to enact the verbs in the ILOs—or closely related ones—and provide evidence of the level of performance that they have achieved. The best form of alignment is where the TLA is itself the assessment, as in problem-based learning. There are two steps in designing a suitable assessment task:

1. Selecting a practicable task that embodies the target verb
2. Judging how well that task has been performed

In designing ATs, we need to select practicable tasks that embody the target ILO verb(s), taking into consideration the workload for the teacher and students and the available resources in implementing the tasks. The invigilated examination provides a very restrictive assessment environment that will very likely not be able to address many ILOs, especially the higher-level ones.

Take the two ILOs that we have been using: "*Explain* how movements are produced in the upper and lower limbs," and "*predict* what functional disability will result from injuries to a particular anatomical structure of the upper and lower limb." A written assignment, an oral presentation, a concept map, and a case study are all possible and aligned ATs. Where the TLAs are the assessment tasks, alignment is perfect.

To Do

- Using the same ILO that you have designed TLAs for in the previous exercise, design ATs that would be appropriate to assess the student's achievement of that ILO.
- When designing these ATs, you may consider first whether the TLAs used would be appropriate for perfect alignment.
- Also consider any required and available resources for implementing these ATs.

In outcomes-based teaching and learning, assessment is criterion referenced. Student performances can be assessed by judging them against established grading criteria or rubrics, instead of the more usual practice of marking quantitatively by accruing marks bottom-up. In constructive alignment, the logic of assessment is holistic, not analytic. The sample rubrics in Tables 3.1 and 3.2 enable the whole to be assessed and awarded a qualitative assessment category,

Table 3.1 Example of rubrics for *explain* (this is only a generic example. Rubrics for "explain" in a particular content need to be developed in context)

	Subgrades (scale score)	Evidence
A	A- A A+ 3.7 4.0 4.3	As in "B" but also provides views on possible alternative causes and/or results depending on changes of conditions Able to link current reasoning to situations in real-life and/or professional contexts
B	B- B B+ 2.7 3.0 3.3	Able to identify a full range of relevant points with details. Support by relevant literature Points are organized to provide a comprehensive and cohesive reasoning or causality
C	C- C C+ 1.7 2.0 2.3	Able to identify a number of relevant points with some details Use these points to provide a fair reasoning of causality, but little or no evidence of a comprehensive overview of reasoning or causality
D	1.0	Able to identify and briefly write about limited points Very little evidence of using these points to provide reasoning to why they are interrelated
F	0.0	The information is sparse and is mainly inaccurate Explanation lacks relevant reasoning or is based on irrelevant information

Table 3.2 Example of rubrics for *predict* (This is only a generic example. Rubrics for "predict" in a particular content need to be developed in context)

Grade	Subgrades (scale score)	Evidence
A	A- A A+ 3.7 4.0 4.3	As in "B" but also provides a comprehensive and holistic view on how and why the prediction has been made Able to provide possible implications of prediction to situations in real-life contexts Evidence of attempting to provide suggestions for intervention to amend or counteract any possible adverse effects as a result of the prediction. In the case of medical and allied health students, this may be suggestion of any possible management or treatment programs (the merits of which need to be considered in relation to the level of study of the students)
B	B- B B+ 2.7 3.0 3.3	Prediction is accurate and realistic Able to identify a full range of relevant points to support the prediction Evidence of integrating the details in a cohesive manner with some evidence of a comprehensive overview of various factors or reasons leading to the prediction
C	C- C C+ 1.7 2.0 2.3	Prediction is largely accurate Able to provide some details of relevant information to support the prediction Evidence of providing a fair reasoning to support the prediction but little or no evidence of a comprehensive overview of various factors or reasons leading to the prediction
D	1.0	Prediction is incomplete Able to provide limited information Very little evidence of using the information to support the prediction
F	0.0	Information is sparse and mainly inaccurate Prediction is inaccurate and irrelevant to the context given

such as A, B, C, or D. However, for logistic purposes such as obtaining a final grade, a quantitative scale can be allotted to each category or subcategory (see section "Obtaining a Final Grade").

These criteria are made clear prior to the assessment process to all parties involved in assessment, students, teacher, and any other parties such as the assessment panel or external assessor. Examples of rubrics for assessing the two ILOs for *explain* and *predict* are provided in Tables 3.1 and 3.2, noting these are generic examples and such criteria need to be developed in the context of the content area and the discipline concerned.

Traditionally, the teacher has been the sole party responsible for assessing and grading the student performance. Research has shown that students could be involved in the assessment process [19, 20]. Students assessing their own work (self-assessment) or assessing their peers' work (peer assessment) can be used as a TLA or an AT. To be able to assess either their own or their peers' work, students need to be very clear beforehand about the ILO(s) that the ATs address and the criteria with which the performance is to be assessed, thus reinforcing their own understanding and enabling them to better achieve the ILOs. Making a judgment about whether a performance meets the given criteria is vital for reflective professional practice, when professionals need to judge their own and their peers' performance and to identify how their performance can be improved. These are important lifelong learning skills that many professionals say are most lacking in their undergraduate education [21].

To Do

- Now try to develop rubrics for the ILO that you have selected and used in the previous exercises.
- If you have selected the same ILO that we have used in this chapter (explain/predict), you will need to consider how the sample rubrics need to be modified to suit the content and context of your teaching.
- You will also need to consider any required and available resources to implement them.

Obtaining a Final Grade

Individual assessments of a student's different performances need to be combined to form a final grade for the course. Qualitative assessments can be converted into a number scale that can then be dealt with arithmetically to yield a final grade. Thus, in Tables 3.1 and 3.2, a quantitative number may be allocated to each category and subcategory; thus, A+ is given 4.3, A 4.0, and A− 3.7. There is a larger gap across categories than within categories: B+ is 3.3, that is, 0.4 difference from A− rather than 0.3 as within categories (these figures are taken from the policy of one university). The final grade can thus be determined by averaging these figures, and when the average has been computed, it can be converted back to the nearest letter grade.

Alternatively, it is possible to derive an a priori system by which the final grade is awarded according to a predetermined pattern of letter grades. For example, the final grade of A is awarded if most other performances are A grade with none lower than B; a similar procedure is sometimes used to award the class of honors in an undergraduate program.

If a pass/fail system is used, then the teacher needs to decide what minimal levels of outcome are needed to meet the particular professional requirements.

Conclusion

We have outlined here a design for teaching that is eminently suited to teaching anatomy. Clearly, the statements of ILOs will be specific to a particular course and the purposes for which it is being taught. We have given examples of TLAs and ATs for two different ILOs, *explain* and *predict*, in the hope that readers will be able to construct their own ILOs and align suitable TLAs and ATs to them. Further details about the theory, and examples of implementation of constructive alignment from several disciplines, are available [4].

References

1. Miller MA, Ewell P. Measuring up on college-level learning. San Jose: The National Center for Public Policy and Higher Education; 2005.
2. Shuell TJ. Cognitive conceptions of learning. Rev Educ Res. 1986;56:411–36.
3. Biggs JB. Enhancing teaching through constructive alignment. High Educ. 1996;32:1–18.
4. Biggs J, Tang C. Teaching for quality learning at university. Maidenhead: McGraw-Hill and Open University Press; 2011.
5. Colvin J, Phelan A. Evaluating student opinion of constructivist learning activities on computing undergraduate degrees. In: 1st annual workshop on constructive alignment. Nottingham Trent University; 2006.
6. Boyle A. Using alignment and reflection to improve student learning. Elements. 2007;3(2):113–7.
7. Lebrun M. Quality towards an expected harmony: pedagogy and technology speaking together about innovation. Assoc Adv Comput Educ J. 2007;15(2):115–30.
8. Nightingale S, Carew A, Fung J. Application of constructive alignment principles to engineering education: have we really changed? AAEE - Annual Conference of Australasian Association for Engineering Education. Melbourne, Australia: Department of Computer Science and Software Engineering, University of Melbourne. Melbourne; 2007. pp.1–9.
9. Raeburn P, Muldoon N, Bookallil C. Blended spaces, work-based learning and constructive alignment: impacts on student engagement. In: Same places, different spaces. 2009. p. 820–831. http://www.ascilite.org.au/conferences/auckland09/procs/.
10. Hoddinott J. Biggs' constructive alignment: evaluation of a pedagogical model applied to a web course. In: Bourdeau J, Heller R, editors. Proceedings of *ED-MEDIA 2000*–world conference on educational multimedia, hypermedia and telecommunications. Montreal, Canada: Association for the Advancement of Computing in Education (AACE). Chesapeake; 2000. pp. 1666–1667. https://www.learntechlib.org/primary/p/16428/.
11. Ladyshewsky R. Aligning assessment, rewards, behaviours and outcomes in group learning tasks. In: Evaluation and assessment conference: enhancing student learning. Bentley: Curtin University; 2006.
12. Moulding NT. Intelligent design: student perceptions of teaching and learning in large social work classes. High Educ Res Dev. 2010;29(2):151–65.
13. Morris MM. Evaluating university teaching and learning in an outcome-based model: replanting bloom. PhD Thesis. Wollongong: University of Wollongong; 2008.
14. Brook V. Learning-focused curriculum development: the redesign of elements of a PGCE science (subject year) Programme. Investig Univ Teach Learn. 2006;3(2):27–35.

15. Taylor R, Canfield P. Learning to be a scholarly teaching faculty: cultural change through shared leadership. In: Brew A, Sachs J, editors. The transformed university: scholarship of teaching and learning in action. Sydney: Sydney University Press; 2007.
16. Rust C. The impact of assessment on student learning: how can the research literature practically help to inform the development of departmental assessment strategies and learner-centred assessment practices. Act Learn High Educ. 2002;3:145–58.
17. Edström K. Doing course evaluation as if learning matters most. High Educ Res Dev. 2008;27(2):95–106.
18. Kandlbinder P, Peseta T. Key concepts in postgraduate certificates in higher education teaching and learning in Australasia and the United Kingdom. Int J Acad Dev. 2009;14(1):19–31.
19. Harris D, Bell C. Evaluating and assessing for learning. London: Kogan Page; 1986.
20. Boud D. Enhancing learning through self-assessment. London: Kogan Page; 1995.
21. Boud D. Implementing student self-assessment. Green guide no. 5. Sydney: Higher Education Research and Development Society of Australasia; 1986.

Using Teaching Assistants in Anatomy

4

Darrell J. R. Evans

Introduction

The recent resurgence in the need for and perceived relevance of anatomy in many medical and allied programs has left anatomy departments with a dilemma as there is a relative lack of suitable educators available to inspire and develop appropriate student knowledge in anatomy [1–3]. Such an issue has been exacerbated by a reduction in the number of clinicians being recruited to anatomy, changes in clinical trainee availability and the demands of basic scientific research [3, 4]. In response, many anatomy departments have therefore looked closely at the methods they use to teach anatomy to ensure they are efficient as well as effective. Combined with this has been a need to identify those best placed to provide the learning guidance to students. For many, this has resulted in the renaissance of the teaching assistant.

Teaching assistants have been used for centuries within the field of anatomy, although the assistant has taken on a number of guises. For most, though, the assistant has been exemplified by the junior qualified doctor or resident who intends to follow a career in surgery or radiology and who takes on a part-time or fixed-term role as an anatomy demonstrator [5–9]. It has been suggested that the success of this type of teaching assistant in anatomical education is due to the incorporation of attributes of peer teaching as well as providing opportunities for mentoring students [7–8]. The revitalized utilization of demonstrators is widely seen as one of the primary and favored methods of providing teaching assistants; however, other departments of anatomy have investigated using alternative approaches. This chapter will outline a number of the ways teaching assistant programs can be integrated within anatomy courses and how such programs can contribute positively not only to student learning but also in the development of knowledge and skills in many of the assistants, providing excellent training opportunities.

Examples of the Teaching Assistant

The Traditional Anatomy Demonstrator

The anatomy demonstrator has traditionally been a recently medically qualified individual, with some clinical experience and who is in residency training to be a surgeon, anesthesiologist or radiologist [6, 9]. Sometimes retired clinicians (and academics) are recruited to act as demonstrators and bring with them a wealth of knowledge and

D. J. R. Evans (✉)
Faculty of Health and Medicine,
University of Newcastle, Callaghan, NSW, Australia

Faculty of Medicine, Nursing and Health Sciences,
Monash University, Clayton, VIC, Australia
e-mail: darrell.evans@newcastle.edu.au

L. K. Chan, W. Pawlina (eds.), *Teaching Anatomy*, https://doi.org/10.1007/978-3-030-43283-6_4

front-line experience to the teaching and learning environment. Demonstrators can also be those that are undertaking postgraduate training in biomedically related fields at PhD and post-doctoral levels [10]. The demonstrator's participation in anatomy teaching can range from an informal approach using a 'drop-in when available' style to a formalized program whereby demonstrators are appointed for a specified period to contribute to an anatomy course and may take on official titles [11]. Positions can have a salaried or unsalaried basis and can be an integral part of the teaching assistant's own formal medical or postgraduate training (through residency rotations, elective periods or defined teaching modules) [10, 12]. Such versatility in the available approaches to including demonstrators as teaching assistants may provide the academic team with enhanced scope when designing anatomy teaching sessions but also requires careful and deliberate planning.

The clinically trained anatomy demonstrator is ideally placed to highlight to students the importance of integrating clinically relevant anatomy into their learning, and this is recognized by students they teach [12]. Using such an approach allows students to realize the importance and relevance of anatomy to their later learning and practice. It is, however, important that the contribution of the clinician to the teaching environment be appropriately guided by the course leaders to ensure it is appropriately aligned in terms of the level, content, depth and applicability. Demonstrators are also in an appropriate position to act as role models for students and should be able to exhibit a range of desirable attributes such as leadership, professionalism and support and can provide opportunities for challenging the ethical awareness of students [11]. If programs are designed so that demonstrators have regular contact with particular students, an enhanced mentor role can develop with demonstrators dealing with pastoral as well as academic issues [6]. Results suggest that students find demonstrators approachable, effective and reliable and therefore a positive addition to their learning [11–13].

During tenure as a demonstrator, individuals can be exposed to a range of developmental opportunities in addition to those of deeper understanding of anatomy and teaching experience. Some programs also offer demonstrators to engage with anatomical or educational research projects while others have demonstrators learn the skills of dissection through preparation of prosected specimens for teaching or during practical sessions [6, 11]. Together these elements provide competitive career-enhancing opportunities for demonstrators, with specific potential to continue to engage with academic activity into the future.

The Senior Anatomy Student: Near-Peer Assistants

Doctors have been recognized as experts in the discipline they teach but not for how they teach it; therefore there has been a call for the provision of appropriate skills development training for students [14]. Unfortunately until recently students received little or no formal training opportunities to develop their own teaching skills [15]. This is disappointing as evidence suggests that the inclusion of opportunities for students to have experience as teachers during undergraduate years is beneficial for the student, the student tutees and the program as a whole [13, 16, 17]. Reports suggest that students who have acted as teachers may have increased retention and understanding of the subject material they have taught [18].

The concept of the near-peer teacher is an educational model which utilizes students who are more advanced in their studies as teachers in learning activities aimed at students in earlier phases of their learning. Near-peer teaching is an accepted approach that provides opportunities for students to develop and practice teaching skills and has been implemented in anatomy courses in a variety of ways [7, 8, 19–22].

In most cases, near-peer teaching assistants are best trained using the attributes of facilitators of learning rather than didactic teachers [15]. This is critical to ensure an active approach to the teaching-learning process and one that is focused on appropriate pedagogy and not tradition. Because near-peer teaching assistants are not separated from the learner in the same way as a member of academic staff will usually be, they are more easily able to explain complex topics and concepts but with a new-found deeper knowledge themselves [10]. The recent experience of the material and the more effective peer-based communication by the near-peer teaching assistant appears to be at the heart of the success of this approach [23]. In addition, if learners see that a near-peer assistant has been able to begin to 'master' the discipline when only a few years ahead, they are likely to feel they are able to achieve the same success [20].

With appropriate training and supervision, it has been reported that near-peer teaching assistants have increased their communication and teaching abilities. This along with increased knowledge and understanding in the discipline has an impact on their own future career development and progression and can act as a motivator for students to become involved [4, 7, 8].

Desirable Features of Teaching Assistant Programs

Development of any teaching assistant program requires the inclusion of features designed to ensure the most effective outcome for student learning and the continual professional development of the assistant.

Appropriate Selection and Training

Recruitment of appropriate individuals is a key factor for many of those responsible for developing programs that use teaching assistants. Some have used a selection procedure based on the practical demonstration of teaching skills or knowledge of anatomy, whereas some base recruitment decisions on motivation statements, personal interviews, past teaching experience enthusiasm for and interest in anatomy and previous performance in anatomy [7, 8, 20, 24]. Once selection has been made, teaching assistants should be provided with appropriate training opportunities. Training programs for future anatomy educators are not new [25] and have been developed in a number of ways to provide the desired skills for effective teaching in anatomy. The approach to training should be tailored to the level of the assistant, the period of tenure and the resources available to provide training. Some training can be offered in short bursts [26], while other training can take place over a prolonged period. Training can be a formalized activity which includes timetabled sessions, practice opportunities and self-directed tasks or less formal with assistants learning from experienced academic staff through observation or by asking questions [6–8, 10, 19]. In most cases, the development of an 'in-house' localized training program is most appropriate; however, more recently, national and international programs have also been developed although these appear more suited to longer-term anatomical appointments rather than more 'transient' assistants [3]. Appointing practicing clinicians or retired clinicians and academics will require appropriate tailoring of or different approaches to training. Such training might be focused on the particular approach of curriculum delivery within the course and the pedagogies that have been included, rather than teaching practice per se.

An additional component of training which should be considered is the inclusion of opportunities for defined reflection. Reflection is an essential part of developing teaching practice and enables an educator to assess areas of their teaching that have gone well and those that may require further training and development. Reflection should become an everyday part of an individual's teaching practice and will help the assistant to grow in confidence and become established as an effective educator and hopefully a thinking performer [27].

Opportunities for Curriculum Design and Development

Where teaching assistants have a more substantive period of tenure, it is possible for them to be included in the design of teaching activities and where appropriate be given guided opportunities to develop and plan particular elements of the course or session. Creativity and innovation are important elements in establishing oneself as an effective educator and providing a good learning experience for students. Assistants can be given chances to create supplementary teaching materials and in-class activities or devising student-selected elective components, but in each case this should be accompanied by integrated methodologies for evaluating the effectiveness of the resulting output [8, 28]. Such opportunities provide assistants with greater insight into the pedagogy underpinning various approaches to teaching, can improve the interest and abilities of assistants and can lead to conference presentations and publications [29]. In all cases, these types of opportunities must be integrated into the overall curriculum design led by the course leaders to ensure appropriate alignment with the desired learning outcomes and chosen pedagogies avoids potential overloading of the curriculum. Involvement of active clinicians and also retired clinicians and academics in this process can require careful management as there may be a tendency for some to want to focus on approaches that are not as applicable to a modern anatomy curriculum or teaching pedagogy.

Peer Observation

The inclusion of defined peer observation or review activity within a teaching assistant program can help assistants enhance the quality of their teaching by providing them with the opportunity to reflect on what they do, bringing the same sense of enquiry and curiosity that drives much of the best research into the teaching situation and learn from colleagues [7, 8]. An effective peer observation activity should be aimed at helping the assistant to develop and enhance their teaching by exchanging ideas, with an observer identifying areas of good practice and highlighting any development needs. Observation should not only be directed at the effectiveness of the presentation of material but also effectiveness within the teaching team and contributions made to the course [8]. For such an approach to have positive effects on the learning journey of the students being taught, observations should be made early and at other periodic stages during the tenure of the assistant so that a purely summative approach can be avoided. Opportunities for the more experienced assistant to act as an observer will provide an additional and beneficial perspective.

Assessment

Teaching assistant programs that utilize clinical trainees or senior students provide opportunities for assistants to develop their own learning of anatomy. Programs can include tutorials from academic staff that are designed for the level of the assistant and can be directed towards their own professional examinations and portfolio development. Where possible, it is important to help assistants to assess the level of their learning and their improvement during the course of their tenure. This can be done using formal or informal viva voce opportunities or other explicit forms of assessment, although the provision of subsequent feedback to the assistant is key. Making sure assistants have an appropriate level of knowledge and understanding is essential for an effective student learning experience and avoids the worry of the assistant just being ‘one page ahead’ of the student. Their teaching and allied skills can also be assessed using a variety of approaches, although it is most effective when this has a formative basis and one that is focused on further development.

Support and Feedback

The provision of a supportive environment for teaching assistants is a key attribute to consider when devising an assistantship program. One way is to appoint experienced mentors (outside of the typical anatomy teaching team) for teaching assistants who are able to advise, guide and provide appropriate feedback [6, 30]. Alternatively academic staff can act as the main providers with input from others within the team and even the student tutees. Feedback should be provided often and using a variety of approaches including debriefing sessions, informal conversations, staff evaluation systems and student evaluation questionnaires [7, 8]. Without appropriate positive support and feedback to develop effective teaching and mentoring skills, the learning experience of the students may be compromised and the future ability of the assistant to supervise and train others within their own clinical teams may be undermined. The success of assistant facilitated programs relies on the motivation and enthusiasm of the teaching assistant, and therefore, it is important that support is directed in a way that enhances these attributes where possible, which will be of benefit to all [13].

Reward

Teaching assistants can make significant contributions to anatomy teaching programs and in return receive the rewards of enhanced teaching skills, deeper understanding of anatomy and other developmental opportunities such as confidence building, involvement in curriculum design and research projects [7, 11, 16, 26]. Depending on the nature of the program, however, additional reward mechanisms may need to be considered as necessary elements of a teaching assistant program. These may include monetary rewards such as salaries, stipends and bursaries or simply having the prestige of an academic appointment [6, 11, 12]. Certificates for the completion of training programs and events or official documents detailing any assessment of teaching including observation sessions, appraisal forms and recommendation letters should be encouraged so that assistants can make additions to their portfolios. Previous terms of appointment as teaching assistants have been viewed as a favorable attribute when applications are made for clinical posts such as those in surgery [31]. For those engaging in biomedical research or those with academic track aspirations, a deepening of anatomical knowledge and understanding will serve them well and act as a reward in itself if applied effectively.

Desirable Features of a Teaching Assistant Program

- Appropriate selection and training
- Opportunities for curriculum design and development
- Peer observation
- Assessment
- Support and feedback
- Reward

Advantages of Using a Near-Peer Approach

The inclusion of teaching assistants who are students or trainees that are at more advanced stages of their training (i.e. near-peer teachers) provides certain advantages for the tutor-tutee relationship:

Increased Numbers of Available Tutors

The inclusion of near-peer teaching assistants within anatomy courses can provide an additional pool of tutors and facilitators for teaching sessions and provide a means for reducing teaching pressures [13]. Such a group of available assistants is particularly useful when there are large numbers of students to teach and when the staff-to-student ration is usually low [20]. The availability of these assistants may require particular management in order to accommodate the assistant's own timetabling arrangements and workload, but it is clear that students feel they benefit from increased numbers of assistants [7]. When increasing numbers of assistants, the course team must ensure there is capacity for training and oversight of the assistants.

Conducive Environment for Questions and Queries

The principle of near-peer teaching has largely been based on the premise that it provides an environment that encourages enhanced interaction between tutors and tutees. This is because near-peer teaching assistants are not identified as members of academic staff, and therefore, interaction is less intimidating and there is more freedom to ask questions. This 'mediator' role may allow the students to focus on issues that they may be less willing to share with members of the academic team in case they are perceived as ignorant. The style and approach of the near-peer teaching assistants are therefore critical to ensure they provide an open and non-threatening learning environment. Some near-peer teaching assistants may find this challenging early on.

Opportunities for Mentoring and Role Modeling

Professionalism is an integral feature of medical training, including in anatomy [30]. Using students as near-peer teaching assistants can provide an opportunity to enhance the mentoring that students receive and allow them to easily identify desirable attributes of those in later stages of their course. Both students and near-peer teaching assistants appear to believe that role modeling is an appropriate feature of a near-peer program [15].

Increased Knowledge and Understanding in Anatomy of Teaching Assistants

The nineteenth-century philosopher, Joseph Joubert, said that 'to teach is to learn twice'. Near-peer teaching therefore provides teaching assistants with an excellent opportunity to develop their own knowledge of anatomy to a level that enables them to communicate understanding to students. In addition, revisiting anatomical material and working closely with other members of the teaching team allow teaching assistants to have a deeper knowledge of the subject area, which can be directed to their own future needs. This is particularly important for those entering specialties such as surgery or radiology and is supported by feedback from former teaching assistants [4].

Skill Development of Teaching Assistants

It is expected that students in many professional courses are provided with opportunities for developing and assessing their teaching and mentoring skills so that they contribute to their ability to be effective health-care practitioners and leaders. The implementation of a program that allows at least some students to have face-to-face contact as educators and facilitators of anatomical learning is therefore one such way. Primary skills covered include those of gauging the level of the learners, demonstrating appropriate information, practicing effective communication and providing encouragement and motivation for learning [4, 6, 9]. In addition, such opportunities also allow the assistant to become familiar with other aspects of the teaching agenda such as providing useful and appropriate feedback and developing ways of eliciting student understanding and focusing on the principles of pedagogy. Near-peer teaching assistants have reported that in addition to their anatomical knowledge, through their teaching experiences, they improve their professional behavior and an array of skills, including those that importantly contribute to resilience [32, 33].

Generally Less Expensive

The near-peer teaching assistants' main motivation for wanting to be part of such a program is usually to develop their teaching and communication skills and their own understanding of anatomy [9]. Consideration may be given as to whether payments can be offered to these teaching assistants for the teaching they provide; however, this will be at a level that enables a core teaching team to be developed at relatively little cost. The adoption of near peer-teaching assistants into an anatomy program should, primarily, be for pedagogical reasons and not as a cost-saving measure, although using near-peer teaching assistants can also be an advantageous supplementary outcome for those departments with little financial or staffing resource [10]. The hidden costs of using near-peer assistants should not be forgotten though such as time for training, supervision, observation, etc.

Effective Outcomes

Students that are taught by near-peer teaching assistants do not have significantly different outcomes compared to those taught by experienced educators, suggesting the inclusion of such assistants does not diminish the student learning experience [22, 34–37]. Evidence suggests that student learners believe near-peer teaching assistants to be effective, particularly in small group settings, although their overall knowledge base and experience both in teaching and providing clinical relevance may be perceived as less when compared to more experienced educators [7, 15, 20]. To ensure the outcomes for both the student learner and teaching assistant are maximized and appropriate, it is key that suitable expectations are set with student learners and assistants at the start of the course.

Advantages of Using Near-Peer Teaching Assistants

- Increased numbers of available tutors
- Conducive environment for questions and queries
- Opportunities for mentoring and role modeling
- Increased anatomical knowledge and understanding in teaching assistants
- Skill development of teaching assistants
- Generally less expensive
- Effective outcomes

Designing a Teaching Assistant Program

The design and implementation of an effective teaching assistant program involve a number of considerations:

Identifying Outcomes

Like any course development, the inclusion of teaching assistants within anatomy programs should be based on outcomes needing to be achieved. As outlined in this chapter, there may be a number of particular drivers for including assistants such as lack of educators in the anatomy program, the desire to develop the skills of trainees or the need to provide role models for students. When recruiting assistants, it is important that the academic team's expectations are fully evident and the outcomes that the assistant might achieve are clearly presented and agreed. The outcomes should be incorporated appropriately into assessment and reward mechanisms.

Ensuring the Program Is Workable

While ad hoc approaches for incorporating teaching assistants into anatomy teaching are used and can be beneficial, approaches that involve well-designed programs appear to have more extensive and measured outcomes and the reliability of increased assistant numbers. The design of the program should incorporate as many of the desirable features that have been outlined as is practical but should be as manageable as possible to enhance the student learning experience. Therefore, a careful balance is required between enacting a quality program that improves the anatomy course and one that becomes a burden to the academic team. Using different levels of assistants within the teaching activities can help provide a self-management system freeing up academic staff time to spend on student interaction rather than assistant supervision.

What Will You Do?

- You have constructed a teaching assistant program which involves near-peer assistant teaching within the dissecting room. For one of the assistants, you observe effective teaching skills; however, you notice that they have given several inaccurate anatomical definitions and functions during demonstrations. What mechanism will you use to feed this back to the assistant, and how will you inform the students of the mistakes without undermining their confidence in the assistant?
- You develop a near-peer teaching program, and you are inundated with applications to become a near-peer teaching assistant. What process will you develop to select the most appropriate assistants?

Cross-Linking

- See also Chapter 5 "Engaging Residents and Clinical Faculty in Anatomy Education".

When and Where to Use Assistants

Depending on the format of the course, the course team must decide on what type of sessions in which to use the assistants. This may be activities such as practical sessions, tutorials, seminars or assessments and sessions which include large or small groups. It is important to consider the level of experience and skills of the chosen assistants to ensure they are able to deliver in the desired environment and do not feel out of their depth, which could be damaging to the student learning experience and development of the assistant. It is usual for assistants to be given increased responsibility and further opportunities for teaching different formats as their experience develops, although additional training might be necessary.

What to Include in the Training

The content of the training program can include aspects that are largely generic to good teaching practice as well as those specific to the discipline of anatomy including particular approaches. Training that uses a pedagogical underpinning provides a framework for assistants to understand the principles which lead to effective teaching and learning. Therefore, when developing the training for teaching assistants, course teams should consider theoretically based as well as practical activities if possible. Sets of skills and attributes that might be considered are importance of a sound knowledge base, ability to be supportive and encouraging, when and how to highlight clinical or functional relevance, demonstrating professional behavior, developing a facilitating role and knowing one's limitations. Other skills that should have a particular focus include the educator's ability to gauge the knowledge level of their students and be able to adapt their teaching accordingly during a session, as well as deal with students having difficulties and maintain the motivation and interest of students and aspects such as time management [38]. An apprenticeship model for teaching assistant training, which includes the provision of basic teaching instructions, but is primarily based on the teaching assistant actively developing their skills through research of appropriate literature and extensive practice, is favored by some [7, 8, 29, 39]. The key to the success of this type of model is active and ongoing mentoring by experienced staff aimed at nurturing assistants to become effective independent educators [29]. Such an approach also requires a helpful evaluation system whereby trainees receive appropriate feedback on their performance, which can include debriefing sessions, peer assessment and student evaluation [7, 8].

Evaluating, Reflecting and Improving

As with the implementation of any new approach to a teaching program, it is essential that the inclusion of teaching assistants into teaching activities be appropriately evaluated to ensure that the approach is effective for both learners and educators. Such analysis may be confined to informal observation or feedback or may include seeking formalized qualitative and quantitative feedback from student learners, defined peer observation activities by staff and other forms of assessment. As well as the academic staff reflecting on the success of the initiative and students providing their opinions, the teaching assistants must be given the opportunity to provide input into the reflection process to ensure their views are actively considered. For instance, teaching assistants indicate that near-peer teaching experiences motivate them to do more teaching suggesting a self-perpetuating model can be introduced [32, 40]. Taken together, such rich evaluation will provide the foundation for developing a program further and maximizing outcomes and impact.

Conclusions

Inclusion of teaching assistants in the delivery of anatomy teaching activities has proved to be a popular and effective approach for many years. When devising a program, it is important that the specific drivers for including teaching assistants be clearly identified so that a chosen approach is appropriate and delivers the most effective outcomes for that course. The design of a teaching assistant program should consider a range of features such as a selection and training approach, a system for observing and assessing progress, methods of providing feedback and support and identification of appropriate reward mechanisms used. The success of such programs depends on aspects such as the motivation and commitment of the assistant; appropriate training and development opportunity, using effective pedagogy including clinical and relevant insight; giving useful feedback; and demonstrating appropriate professionalism. When assessing the impact of using assistants in anatomy teaching activities, the program leads need to gauge the opinions of the teaching assistants as well as student learners and other members of the teaching team. Results suggest that the utilization of teaching assistants can contribute positively not only to student learning but also in the development of knowledge and skills in the assistants.

> Using teaching assistants in anatomy courses and activities provides a positive opportunity not only to support student learning but also for developing the skills and knowledge of the next generation of educators and mentors.

References

1. McKeown PP, Heylings DJA, Stevenson M, McKelvey KJ, Nixon JR, McCluskey DR. The impact of curricular change on medical students' knowledge of anatomy. Med Educ. 2003;37:954–61.
2. McCuskey RS, Carmichael SW, Kirch DG. The importance of anatomy in health professions education and the shortage of qualified educators. Acad Med. 2015;80:349–51.
3. Fraher J, Evans DJR. The training of the next generation of anatomy teachers. Anat Sci Educ. 2009;2:119–25.
4. Ocel JJ, Palmer BA, Wittich CM, Carmichale SW, Pawlina W. Outcomes of the gross and developmental anatomy teaching assistant experience. Clin Anat. 2003;16:526–30.
5. Lee ES, Shew RL, Harris IB. Improving gross anatomy education by using residents as co-teachers. Acad Med. 1999;74:586.
6. Lockwood AM, Roberts AM. The anatomy demonstrator of the future: an examination of the role of the medically qualified anatomy demonstrator in the context of tomorrow's doctors and modernizing medical careers. Clin Anat. 2007;20:455–9.
7. Evans DJR, Cuffe T. Near-peer teaching in anatomy: an approach for deeper learning. Anat Sci Educ. 2009;2:227–33.
8. Lachman N, Christensen KN, Pawlina W. Anatomy teaching assistants: facilitating teaching skills for medical students through apprenticeships and mentoring. Med Teach. 2013;35:e919–25.
9. Koshi R, Holla SJ. Experience of anatomy demonstrators in Tamil Nadu. India Clin Anat. 2001;14:278–81.
10. Cheng X, Wang L, Guo K, Lui S, Chu G, Zhou L-H. Postgraduate fellows as teaching assistants in human anatomy: an experimental teaching model at a Chinese research university. Anat Sci Educ. 2011;4:29–32.

11. McBride JM, Drake RL. Rewarding the resident teacher. Anat Sci Educ. 2011;4:227–30.
12. Gossage JA, Horner B, Ellis H. The anatomy demonstrator has an important role in medical student education. Med Educ. 2003;37:1055–6.
13. Rashid MS, Sobowale O, Gore D. A near-peer teaching program designed, developed and delivered exclusively by recent medical grauates for final year medical students sitting the final objective structured clinical examination (OSCE). BMC Med Educ. 2011;11:11.
14. MacDougall J, Drummond MJ. The development of medical teachers: an enquiry into the learning histories of 10 experienced medical teachers. Med Educ. 2005;39:1213–20.
15. Bulte C, Betts A, Garner K, Durning S. Student teaching: views of student near-peer teachers and learners. Med Teach. 2007;29:583–90.
16. Ten Cate O, Durning S. Dimensions and psychology of peer teaching in medical education. Med Teach. 2007;29:546–52.
17. Jackson TA, Evans DJR. Can medical students teach? A near-peer-led teaching program for year 1 students. Adv Physiol Educ. 2012;36:192–6.
18. Gaudet AD, Ramer LM, Nakonechny J, Cragg JJ, Ramer MS. Small-group learning in an upper-level university biology class enhances academic performance and student attitudes toward group work. PLoS One. 2010;5:e15821.
19. Shiozawa T, Hirt B, Celebi N, Bur F, Weyrich P, Lammerdong-Koppel M. Development and implementation of a technical and didactical training program for student tutors in the dissection course. Ann Anat. 2010;192:355–60.
20. Duran CEP, Bahena EN, Rodriguez MAG, Baca GJ, Uresti AS, Elizondo-Omana RE, et al. Near-peer teaching in an anatomy course with a low faculty-to-student ratio. Anat Sci Educ. 2012;5:171–6.
21. Hall S, Lewis M, Border S, Powell M. Near-peer teaching in clinical neuroanatomy. Clin Teach. 2013;10:230–5.
22. Tayler N, Hall S, Carr NJ, Stephens JR, Border S. Near peer teaching in medical curricula: integrating student teachers in pathology tutorials. Med Educ Online. 2015;20:27921.
23. Allen AR, Boraks N. Peer tutoring: putting it to the test. Read Teach. 1978;32:274–8.
24. Youdas JW, Hoffarth BL, Kohlwey SR, Krammer CM, Petro JL. Peer teaching among physical therapy students during gross anatomy: perceptions of peer teachers and students. Anat Sci Educ. 2008;1:199–206.
25. Prentice ED, Metcalf WK, Sharp JG, Hard WL. Training teachers in the anatomical sciences. J Med Educ. 1976;51:1006–9.
26. Berkowitz LR, McDermott CH, White K, Saavedra A, Hafler JP. Residents as teachers of the pelvic floor and perineal anatomy. Female Pelvic Med Reconstr Surg. 2012;16:268–71.
27. Evans DJR, Canavan A. Designing a clinical skills programme: a partnership between students, patients and faculty. Int J Clin Skills. 2008;2:130–4.
28. Pearce M, Evans DJR. Developing multimedia resources: for students, by students. Clin Teach. 2012;9:178–82.
29. Erie AJ, Starkman SJ, Pawlina W, Lachman N. Developing medical students as teachers: an anatomy-based student-as-teacher program with emphasis on core teaching competencies. Anat Sci Educ. 2013;6(6):385–92.
30. Camp CL, Gregory JK, Lachman N, Chen LP, Juskewitch JE, Pawlina W. Comparative efficacy of group and individual feedback in gross anatomy for promoting medical student professionalism. Anat Sci Educ. 2010;3:64–72.
31. Jones OM, Parkin IG. Consultant's opinions of anatomy demonstrators. Ann R Coll Surg Engl. 1990;81(Suppl):10–2.
32. Hall S, Harrison CH, Stephens J, Andrade MG, Seaby EG, Parton W, et al. The benfits of being a near-peer teacher. Clin Teach. 2018;15:403–7.
33. Alvarez S, Schultz JH. Professional and personal competency development in near-peer tutors of gross anatomy: a longitudinal mixed-methods study. Anat Sci Educ. 2019;12(2):129–37.
34. Rees EL, Quinn PJ, Davies B, Fotheringham V. How does peer teaching compare to faculty teaching? A systematic review and meta-analysis. Med Teach. 2016;38:829–37.
35. Abay ES, Turan S, Odabasi O, Elcin M. Who is the preferred tutor in clinical skills training: physicians, nurses, or peers? Teach Learn Med. 2017;29:247–54.
36. Harrison CH, Elmansouri A, Parton W, Myers MA, Hall S, Stephens JR, et al. The efficacy of frontline near-peer teaching in a modern medical curriculum. Anat Sci Educ. 2019;12:236–44.
37. Viana RB, Campos MH, Santos DAT, Xavier ICM, Vancini RL, Andrade MS, et al. Improving academic performance of sport and exercise undergraduate students in gross anatomy using a near-peer teachign program. Anat Sci Educ. 2019;12(1):74–81.
38. Alvarez S, Nikendei C, Schultz JH. Development of a didactical training concept for peer-tutors in gross anatomy. Anat Sci Educ. 2017;10:495–502.
39. Johansson E, Holmin TE, Johansson BR, Braide M. Improving near-peer teaching quality in anatomy by educating teaching assistants: an example from Sweden. Anat Sci Educ. 2018;11(4):403–9.
40. Curlewis K, Sharp E. Near-peer teaching as medical undergraduates. Clin Teach. 2019;16:169.

5 Engaging Residents and Clinical Faculty in Anatomy Education

Jennifer M. McBride and Richard L. Drake

Traditionally, the role of residents as teachers in medical school curricula has been limited to the core clinical areas including internal medicine, family medicine obstetrics–gynecology, pediatrics, surgery, and psychiatry. The crucial role residents play in educating medical students is emphasized by the Accreditation Council for Graduate Medical Education (ACGME) Common Program Requirements for residency education which among other activities requires residents to engage in clinical teaching [1]. This requirement has precipitated the emergence of several residents-as-teachers curricula in multiple disciplines, with positive results [2]. Given the number of curricula established, a recent literature review identified several key articles pertaining to the subject of residents-as-teachers programs, which is an insightful resource for faculty tasked with creating or revising residents-as-teachers programs [3]. Inherent in their responsibility as clinical educators, residents also serve as role models, guiding medical student career choices, albeit not as strongly as their staff counterparts [4]. In addition to their clinical teaching role and perceived role model status, residents can also play a part in the basic science years of the medical school curriculum.

Residents can also play a part in the basic science years of the medical school curriculum.

In the first 2 years of their education, medical students are immersed in a series of basic science courses to help build a foundation of knowledge which they can later draw upon in the clinical years. Often presented separately, with little to no integration between subjects, many of these courses create anxiety among the students as they scramble to memorize endless pages of details and essentially a new language of terms and definitions. In the last few years, the Liaison Committee on Medical Education (LCME) included in their accreditation guidelines that "The faculty of a medical school ensure that the medical curriculum includes self-directed learning experiences and unscheduled time to allow medical students to develop the skills of lifelong learning" [5]. This meant that long hours of didactic lecture were no longer acceptable and that other methods of instruction and learning were needed. While this change elicited a visceral reaction from anxious faculty, it also opened the door to creative and innovative thinking. Since the announcement of this guideline, many schools and course directors have implemented a modified curriculum and written reports of their experiences and results. One area in which several

J. M. McBride (✉) · R. L. Drake
Cleveland Clinic Lerner College of Medicine,
Case Western Reserve University, Cleveland, OH, USA
e-mail: mcbridj@ccf.org; DRAKER@ccf.org

L. K. Chan, W. Pawlina (eds.), *Teaching Anatomy*, https://doi.org/10.1007/978-3-030-43283-6_5

groups have written about is their experiences in the areas of the anatomical sciences [6–9].

Often perceived by students as a formidable course, gross anatomy is an area of the basic science curriculum which harbors a rich opportunity for adding clinically relevant correlations, the introduction of which promotes students' acquisition of knowledge. In addition to the more obvious ways in which to introduce these types of correlations such as the use of imaging, case studies, disease process, or structures damaged by penetrating wounds, a higher-yield option and more memorable one for the students is with the use of residents in the gross anatomy laboratory. Several factors define residents as suitable teachers in this environment including their proximity of age to the average medical student and their prior experience in clinical medicine. The inclusion of residents in this learning environment can also benefit the residents as it offers an opportunity to review anatomical structures and provides them with an opportunity to teach and improve their teaching skills.

Advantages of Using Residents as Teachers

- Their proximity of age to the average medical student
- Prior experience in clinical medicine
- Gives residents a chance to review anatomical structures
- Provides an opportunity for residents to improve teaching skills

Regardless of course hours, residents from various disciplines can be utilized for instruction of anatomy in several different ways and with students at various levels of education. For example, in a first-year medical student course, residents can participate in the completion of cadaveric dissections and later present the dissection during the laboratory session. In this scenario, it is best to draw from a pool of residents working in the same system. For instance, contact orthopedic residents during the appendicular skeleton or musculoskeletal section of the course and ear, nose, and throat residents during discussion of the head and neck section. Beyond a basic introduction to anatomy of the region, they can also describe the clinical importance of the anatomy. This same collection of residents can be called upon in second-year or more advanced anatomy courses. In addition to the review of material students were exposed to in the first year, the residents can increase the amount of clinical material they relay in their presentation. In an advanced-year anatomy course, such as the third year or beyond, residents can again review the anatomical structures as well as clinical correlations and add in details on current procedures utilized to correct common pathologies.

Key to any of these options is a time to meet with the residents for discussion of the session goals and develop learner appropriate clinical correlations that emphasize the learning objectives. For example, more detailed discussion of surgical techniques and the like may be a topic best suited for advanced anatomy electives or surgical anatomy courses. On the other hand, viewing a laparoscopic cholecystectomy can be a powerful visual way to solidify a novice student's understanding of the relationships between biliary system structures and surrounding vasculature. Perhaps the most important group of residents to meet with before inclusion in an anatomy course are those scheduled to participate in first-year courses in which the students have little and most often no anatomical knowledge. It is vital to spend additional time with these residents to clarify the educational goals of the session, specify what level of detail the students should be getting from the prosection station, review the structures dissected by the residents, and/or review PowerPoint slides if the residents are preparing an oral presentation. All of these checks will help ensure that the students have a meaningful educational experience and avoid dissemination of misinformation or interpretation errors. When preparing for these meetings, it is important to remember that the residents are also students in the early stages of specialized training, and for many, anatomy is not a subject they have thoroughly reviewed since their first year of medical school.

To Do

- Schedule meeting with residents to discuss session goals.
- Identify learning objectives appropriate to student knowledge base.
- Discuss clinical correlations which will complement learning objectives.

Rewarding the Resident Teacher

A major consideration to keep in mind when planning the inclusion of residents in an academic program, particularly if they are going to be asked to participate in multiple sessions, is an appropriate reward system. Residents have many obligations to their programs with very little time for other activities, so monetary compensation, access to resources to improve their teaching skills, opportunity to work toward an academic title of clinical instructor, and the prospect of receiving a teaching award are a few examples of ways to acknowledge and reward their participation. The reward system which requires a little more effort but certainly carries more prestige is the appointment of clinical instructor. For this reward system, as outlined by McBride and Drake, the first step is to establish appropriate guidelines for this type of appointment in cooperation with your Committee on Appointments, Promotions, and/or Tenure [10]. Several things are important to consider such as limiting the program to trainees within a particular hospital system and establishing that the residents have a significant role which would directly impact student knowledge. From an evaluative standpoint, the residents must demonstrate proficiency in their academic or clinical field, exhibit effective teaching, and participate in a minimum number of pre-established hours.

Tips on Engaging Residents in Your Academic Program

- Post a synopsis of teaching opportunities on affiliate GME websites.
- Contact residency directors and discuss benefits of resident participation.
- Identify residency programs which include a research year.
- Develop or participate in a resident-as-teacher program which includes instruction on teaching methods, opportunities for teaching practice, and faculty feedback on strengths and areas for improvement.
- Establish some form of resident reward system such as monetary compensation, opportunity to work toward an academic title, or prospect of receiving a teaching award.

Recruiting Residents

As mentioned previously, residents have very little time so another thing to consider is how to contact them or make them aware of teaching opportunities available. One of the simplest ways is through a short synopsis of how residents are included in your course posted on the Graduate Medical Education (GME) site of your affiliated hospitals. Another means of reaching them is by contacting the residency directors. Residency directors can be a powerful advocate if they support your cause; unfortunately, the reverse is also true. Lastly, if you are aware of residency programs at your affiliated hospitals which still include a research year, these residents are the easiest to work into your schedule as their time is somewhat less constrained. Again, contacting the GME office is a great resource for obtaining program-specific details like this.

Clinical Faculty

Another opportunity to include instructors with a clinical background is with the participation of clinical faculty. With more experience in their specialty, this group of instructors can offer real-world examples in the form of patient cases, surgical perspectives on anatomy, clinical methods to address pathologies, history of previous and current thinking regarding treatment, and existing areas of research.

From a learning perspective, it is most beneficial to schedule a clinical person after the students have had a basic introduction to the region of interest, accomplished with introductory lectures and/or seminars or with assigned reading. This foundation allows the clinician to emphasize salient points and provide perspective on the application of this knowledge. In addition, the students will be better prepared and feel more confident to participate in discussion and ask questions of the speaker. There are several ways to structure sessions which include clinical faculty members such as with discussion of a clinical case or through demonstration of common anatomically driven procedures in the laboratory. In the example of discussing a clinical case, it is best to include a common scenario with accompanying questions, all of which the students can review before the speaker comes. Again, the questions will help start the discussion and give the students some direction of what the case is about. For the procedures, they should be anatomically driven, which will allow them to review the relationships of the structures they have studied.

Conclusions

In summary, there are multiple areas in the instruction of gross anatomy where both resident and faculty clinicians can contribute to medical student learning. Throughout their educational career, medical students continue to build on their knowledge base of anatomy and its role in function, diagnosis, and treatment. Inclusion of these more senior health-care professionals plays a crucial role in their acquisition of knowledge with the use of level-appropriate activities, facilitated discussion, and application of knowledge.

References

1. ACGME. Common Program Requirements 2017. https://www.acgme.org/Portals/0/PFAssets/ProgramRequirements/CPRs_2017-07-01.pdf. Accessed 14 Feb 2019.
2. Post RE, Quattlebaum RG, Benich JJ III. Residents-as-teachers curricula: a critical review. Acad Med. 2009;84:374–80.
3. Krzyzaniak SM, Cherney A, Messman A, Natesan S, Overbeck M, Schnapp B, et al. Curated collections for educators: five key papers about residents as teachers curriculum development. Cureus. 2018; 10:2154.
4. Sternszus R, Cruess S, Cruess R, Young M, Steinert Y. Residents as role models: impact on undergraduate trainees. Acad Med. 2012;87:1282–7.
5. Liaison Committee on Medical Education. Functions and structure of a medical school: standards for accreditation of medical education programs leading to the M.D. Degree. 2019. http://lcme.org/publications/#Standards. Accessed 9 May 2019.
6. Drake RL. A unique, innovative, and clinically oriented approach to anatomy education. Acad Med. 2007;82:475–8.
7. McBride JM, Prayson RA. Development of a synergistic case-based microanatomy curriculum. Anat Sci Educ. 2008;1:102–5.
8. Vasan NS, Defouw DO, Compton S. Team-based learning in anatomy: an efficient, effective, and economical strategy. Anat Sci Educ. 2011;4: 333–9.
9. Klement BJ, Paulsen DF, Wineski LE. Anatomy as the backbone of an integrated first year medical curriculum: design and implementation. Anat Sci Educ. 2011;4:157–69.
10. McBride JM, Drake RL. Rewarding the resident teacher. Anat Sci Educ. 2011;4:227–30.

A Significant Role for Sessional Teachers in the Anatomy Education Landscape

6

Michelle D. Lazarus and Danielle Rhodes

Introduction

There are a wide variety of reasons medical school programs may engage sessional teachers; these reasons may be as varied as the sessional teacher terminology (Fig. 6.1). Sessional anatomy teachers (SATs) are cost-effective [1], help increase staff-to-student ratios [2–6] and may serve to offset the global anatomy instructor deficit [7–10]. SATs can also provide unique benefits to anatomy education programs. SATs who are proximal to anatomy students (in terms of age, life stage and/or experience) may share social and/or cognitive congruence with students. Social congruence refers to an instructor's ability to relate to, or empathise with, their students' learning experiences. SATs who are socially congruent with their students can reduce student anxiety, improve student motivation and create learning environments wherein students feel comfortable asking questions and revealing knowledge deficits [11–15]. SATs may also share cognitive congruence with students if they are at similar, or slightly advanced, educational levels because they likely share a similar knowledge base. Cognitive congruence theory posits that instructors with a similar knowledge base to students (i.e. SATs) can be more effective teachers than expert instructors who possess an advanced or encapsulated knowledge base (i.e., academic anatomists) [11, 16]. SATs' educational proximity to students means they are more aware of prior learning, less likely to make assumptions about student knowledge, and thus able to anticipate the cognitive difficulties students may encounter [11–13, 17, 18]. SATs can also draw on their recent anatomy student experiences to explain concepts at an appropriate level by using learner-accessible language [12–14, 17, 18]. Both social and cognitive congruence theories suggest a *purposeful and positive role* for SATs in anatomy programs, a role which established anatomy academics may struggle to fill due to their educational distance from their learners. With these social and infrastructural benefits in mind, this article will serve as a guide on effective SAT qualifications, SAT curricular roles, the support and management needed for SATs to thrive and potential impacts of effective SATs on relevant stakeholders.

M. D. Lazarus (✉)
Centre for Human Anatomy Education and Monash Centre for Scholarship in Health Education, Department of Anatomy and Developmental Biology, Faculty of Medicine Nursing and Health Sciences, Monash University, Melbourne, VIC, Australia
e-mail: michelle.lazarus@monash.edu

D. Rhodes
Centre for Human Anatomy Education and Biomedicine Discovery Institute, Department of Anatomy and Developmental Biology, Faculty of Medicine Nursing and Health Sciences, Monash University, Melbourne, VIC, Australia
e-mail: danielle.rhodes@monash.edu

L. K. Chan, W. Pawlina (eds.), *Teaching Anatomy*, https://doi.org/10.1007/978-3-030-43283-6_6

Fig. 6.1 The terminology used to describe sessional anatomy teachers varies widely across the globe. Some more common terms are bolded. (Figure designed in collaboration with John Allen, Kat Orgallo and Kristeen Barker, Teaching Resource Support Unit, Faculty of Medicine Nursing and Health Sciences, Monash University)

Qualifications

By definition all sessional teachers are engaged on a short-term, temporary or casual basis [19]; however, the qualifications and educational distance between SATs and their students can vary widely (Table 6.1). Rhodes et al. [20] identified three variants of SATs: reciprocal-peer teachers, near-peer teachers and cross-level teachers (Fig. 6.2). These designations primarily refer to the SATs' level of expertise and proximity to the student learning experience. Reciprocal-peer SATs are at equivalent educational levels and enrolled in the *same year* as their student peers; near-peer SATs are at a similar educational level but *one or more years advanced* from their students; and cross-level SATs are at a different (typically more advanced) educational level (Table 6.1). While reciprocal-peer and near-peer SATs are often recruited from within the anatomy program and/or healthcare degree, cross-level SATs tend to come from two main pools: i) graduate students enrolled in a PhD or master's degree and ii) newly practicing healthcare workers (junior doctors or residents, physiotherapists or nurses, etc.) [20]. If SATs are recruited from the latter pool, their experience in authentic healthcare practice may compliment the skills and knowledge of the permanent academic teaching team. It is important, however, that healthcare workers be aware that their knowledge may be encapsulated [21]; this means that while they may be able to discuss the clinical applications of the basic science knowledge, they may struggle to communicate the underlying related anatomy to a novice learner. The qualifications and experience of SATs are thus important considerations when hiring or integrating SATs into a teaching program (Fig. 6.2).

Table 6.1 Defining the three variants of sessional anatomy teacher—reciprocal-peer, near-peer and cross-level

SAT designation	Definition	Example
Reciprocal-peer SATs	Are at an *equivalent educational level* and enrolled *in the same year* as the students they teach	Healthcare students teaching their peers, e.g. *second-year medical students teaching second-year medical students*
Near-peer SATs	Are at an *equivalent educational level* but are *one or more years advanced* from the students they teach	Senior healthcare students teaching junior healthcare students, e.g. *fourth-year medical students teaching second-year medical students*
Cross-level SATs	Are at a *more advanced educational level* than the students they teach	Healthcare professionals or graduate students teaching healthcare students, e.g. *junior doctors or residents teaching medical students*

Recruitment and Selection

The recruitment and selection of SATs is a critical step that influences the quality of sessional teaching in healthcare programs. After identifying a suitable local source of SATs (reciprocal-peer, near-peer or cross-level), the anatomy program must recruit and select SATs with the desired skills, knowledge and personal attributes. A recent systematic review identified anatomical

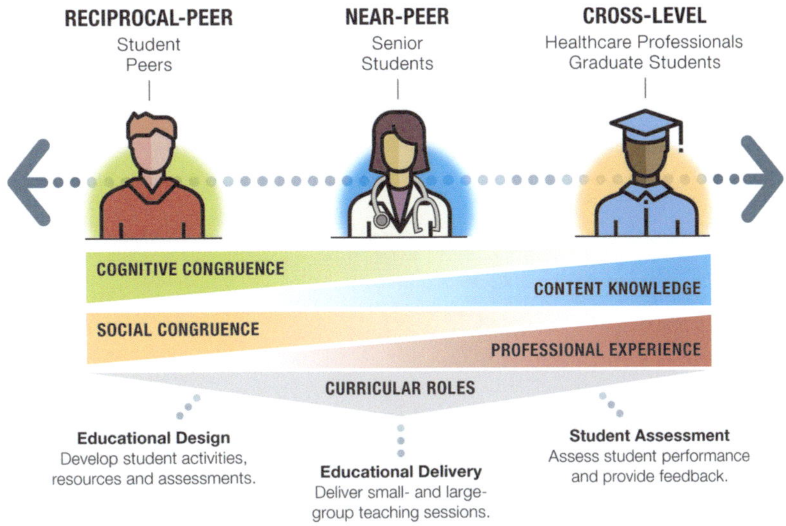

Fig. 6.2 The educational spectrum of sessional anatomy teachers impacts SAT qualifications, experiences, and thus influences potential curricular roles. (Figure designed in collaboration with John Allen, Kat Orgallo and Kristeen Barker, Teaching Resource Support Unit, Faculty of Medicine Nursing and Health Sciences, Monash University)

knowledge, enthusiasm to teach, communication skills and teaching ability as key selection criteria for SATs within medical programs [20]. The authors propose that interpersonal skills, self-management and the ability to engage in reflective practice are also key selection criteria. The ability to self-evaluate and reflect on one's own knowledge/performance is a particularly important, but often overlooked quality in SATs. SATs who can recognise their own limitations and admit when they 'don't know' are highly valued members of sessional teaching teams. This behaviour helps build trust between the SAT and supervising academic(s), and sets a positive example by communicating acceptance of knowledge limitations to learners.

Assessing the skills and attributes of potential SAT recruits can occur through multiple modalities (Table 6.2), and typically relies on proxy measurements. Anatomical knowledge can be assessed indirectly by reviewing examination scores and prior academic performance of potential reciprocal-peer and near-peer SATs [4, 22–24] or academic qualifications of cross-level SATs. More direct assessments of anatomical knowledge may occur during the interview process by interviewers posing content-based questions or asking candidates to engage in tasks demonstrating their anatomical knowledge (e.g. presentation of an anatomical specimen) [6]. In both cases, it is important to note that anatomical knowledge can often be adequately refreshed prior to teaching; therefore, a greater focus on prospective SAT learning aptitude and enthusiasm may be warranted. SAT enthusiasm may be evaluated via motivational statements or cover letters [23, 24] but is likely best assessed throughout the candidate's interview. For reciprocal-peer and near-peer SATs, classroom performance (i.e. leadership during peer learning activities) and prior interactions with academic staff may also be indicative of their enthusiasm for teaching. Communication skills, teaching ability and interpersonal skills are best suited to direct assessments during the interview process. Some interview techniques may include providing recruits with an anatomy topic and allowing them the opportunity to demonstrate their teaching ability to their fellow recruits, interview panel or undergraduate students [6]. This approach allows the interviewers to assess the accuracy and depth of a candidate's anatomical knowledge, the clarity of their explanations, their ability to build a rapport with others and the quality of their teaching. Posing a question for which there is no known answer, or one that has an ambiguous answer, is another valuable technique which may help assess how a candidate responds to uncertainty and being asked questions beyond their knowledge base. Ultimately, the skills and attributes of prospective SATs can be ascertained from a variety of sources including their curricu-

Table 6.2 Recommended assessment approaches for key sessional anatomy teacher selection criteria during the selection phase. (CV = curriculum vitae)

Key selection criteria	Assessment	Candidate selection/interview recommendations and tips
Anatomical knowledge	Academic performance and transcript, CV and qualifications, interview	Ask relevant curricular content-based interview questions Require candidates to prepare a presentation/demonstration on relevant discipline material.
Enthusiasm	Motivational statement, cover letter, prior interactions, interview	Request motivational essays from candidates with application. Be wary of candidates who focus only on what they will *gain* as opposed to what they will *bring* to the position.
Teaching ability	Teaching experience, CV, prior interactions, letters of recommendation, interview	Ensure requested presentation is given to relevant stakeholders (interviewers, fellow candidates and/or students). Ask candidates about their *learner* experiences (good and bad). Inquire what was learned from these experiences.
Communication skills	CV, prior interactions, letters of recommendation, interview	Note the verbal and non-verbal communication styles of the candidate during selection phase(s). Request stakeholder feedback following the candidates' presentation.
Interpersonal skills	Prior interactions, letters of recommendation, interview	Note the interpersonal skills and personal attributes of the candidate to stakeholders during selection phase(s). Conduct group interviews and note candidates' ability to interact and establish rapport with others.
Self-management	Prior interactions, letters of recommendation, interview	Provide the candidate with a task that must be completed prior to the interview and assess their preparedness and ability to meet the deadline. Ask the candidate to describe how they effectively manage time and plan to juggle an SAT position alongside their other personal and professional commitments.
Reflective practice	Prior interactions, letters of recommendation, interview	Ask the candidate to describe a time they engaged in reflective practice and the outcomes. Their ability to communicate what they have learned from their student experiences will also give insight.
Recognition of limitations	Prior interactions, letters of recommendation, interview	Push the limitations of expected knowledge of the demonstrator, asking them a question they are UNLIKELY to have knowledge of. Observe whether they ad-lib or whether they recognise their limitations, admitting they 'don't (may not) know'.

lum vitae, cover letter, motivational statement, letters of recommendation and interview (Table 6.2). Importantly, the skills, qualities and attributes of SATs need to match their curricular roles within an academic anatomy program.

Curricular Roles

The curricular roles occupied by sessional anatomy teachers are many and varied. SATs primarily act as laboratory instructors and guide student learning during human donor dissection alongside other laboratory-based activities [20]. The engagement of SATs can improve the teacher-to-student ratio in laboratory classes [4–6] and relieve teaching pressures faced by academics [3–5, 25]. SAT's curricular responsibilities have diversified as anatomy education has evolved and can be broadly divided into three categories: delivery, design and assessment.

Sessional teachers in anatomy programs primarily provide small-group instruction to healthcare students. The heavy reliance on SATs in small-group settings is underpinned by several factors including a low academic-to-student ratio in many anatomy programs [4] and a desire to engage small-group instructors who share social and cognitive congruence with the students [12]. Numerous studies outline the social and cognitive benefits associated with utilising senior medical students (near-peer SATs) and junior doctors (cross-level SATs) as small-group instructors [6, 12, 13, 17, 26, 27]. SATs who are close in proximity to anatomy students (in terms of age or career stage) can create effective learning environments where students feel comfortable asking questions, engaging in discussion and disclosing knowledge-based misconceptions [12, 13, 17]. Evidence of SAT effectiveness in small-group settings is substantiated by the diversity of small-group teaching roles that SATs occupy both within and beyond the anatomy laboratory. These roles include instructing human donor dissection/prosection [2, 4, 22–25, 28–31], teaching living/surface anatomy [3, 32, 33], reviewing medical imaging [4, 22–24, 34], facilitating clinical case studies [4, 22] and moderating team-based student projects [35].

As modern healthcare curricula move away from traditional didactic lectures towards more effective student-centred active-learning approaches [36], new roles for SATs are emerging within large-group teaching settings. SATs may act as additional facilitators during 'flipped' or 'active' lectures providing academics with the additional support required to integrate active-learning techniques into these traditionally didactic classes. Reports indicate that this SAT support empowers academics to modernise medical curricula and implement new teaching strategies [25].

Several anatomy programs describe a role for SATs in healthcare curriculum design and development [13, 23–25, 37–42]. SATs contribute to the development of small-group activities [13, 23–25, 37], student assessments [23, 24], review sessions [12, 17, 25] and development of other educational resources. Senior medical students (near-peer SATs) recall the pitfalls associated with learning anatomy and can use this knowlege to help develop activities that are tailored to the needs of their junior peers [13], a practical example of their cognitive congruence. Conversely, healthcare professionals (cross-level SATs) are familiar with the anatomical knowledge required for learners' future clinical practice and can bring a vocational perspective to curricular design. Involving sessional teachers in curricular development is a mutually beneficial arrangement potentially enhancing SATs' sense of empowerment and job satisfaction [43].

Ultimately, the curricular roles occupied by SATs vary depending on the local needs of the anatomy program and the sessional teacher profile. While SATs at all levels of the educational spectrum facilitate laboratory-based activities and provide small-group instruction to medical students [20], higher-order tasks such as curriculum development are typically reserved for SATs with advanced anatomical knowledge and/or prior teaching experience [20]. While there is no clear consensus regarding how curricular roles are to be assigned to sessional teachers, the qualifications, anatomical knowledge, healthcare experience and teaching expertise of SATs are key factors which are taken into consideration when assigning SAT tasks.

Training and Support

The training and support of sessional anatomy teachers are essential to their success. SATs may be offered and accept teaching roles with little or no prior teaching experience. This is particularly true for reciprocal- and near-peer sessional teachers who are themselves enrolled as students within the educational program. Healthcare professionals may have some teaching experience as participants in educating patients, other healthcare professionals and the general community; however, reports suggest that they often lack the skills required to facilitate effective learning [44, 45] and may be unprepared to teach in undergraduate courses. Providing adequate training and support to SATs at all levels is thus critical for effective integration of this group of educators.

Induction

All SAT programs benefit from commencing with SAT role induction. This is particularly important for sessional teachers who are recruited externally and may be unfamiliar with the university systems, policies and procedures. There are two key components to a successful induction: i) university induction and ii) course induction. An induction to the university includes an introduction to relevant policies and procedures, induction to laboratories, buildings and other key work areas and training in administrative procedures (i.e. use of the learning management system and timesheet submission). An induction to the course provides an overview of the anatomy program including discipline content, course format, teaching philosophy, student assessments and the role of SATs. The key learning objectives for students should be highlighted and the SATs should be provided with an introduction to the student cohort (background, prior learning, etc.); the extent to which each of these domains is covered will depend on the autonomy given to the SAT within the curriculum, their familiarity with the university and their knowledge of the anatomy program.

Training

Training for SATs should cover a wide variety of topics while being tailored to the curricular role, qualifications and experience level of the SAT team. Ideally, SATs will receive training in instructional strategies, communication skills, classroom organisation and management, behaviour management and how to provide effective student feedback. Training in discipline content is not necessary to include in an SAT training program if the expectation is that sessional teachers know, or are able to independently revise, the discipline content covered in the course. Depending on the curricular role of the SAT, training in technical skills such as basic and advanced dissection techniques may also be required [4, 22, 30].

The timing and implementation of SAT training varies depending on curricular goals and available infrastructural support. While many institutes report single SAT training interventions [29, 30, 46, 47], the broader literature supports a move towards regular, ongoing training with built-in opportunities for reflection, personalised guidance and feedback [48, 49]. This ongoing, regular SAT training may include structured training sessions [4, 6, 25, 27, 30, 46] and informal mentorship or apprenticeship-style programs [6, 23, 24]. While structured training sessions allow academics to efficiently train large teams of SATs, mentorship and apprenticeship-style arrangements permit a more personalised approach to SAT training and professional development; some institutions may provide a hybrid approach with an initial induction for the larger team and ongoing, individual, longitudinal support for each SAT. While reports indicate that mentoring SATs is a task primarily undertaken by academics [6, 23, 24], allowing experienced SATs to mentor their more inexperienced or junior colleagues is also a valuable model. This latter arrangement has an additional benefit as it provides senior SATs leadership opportunities within the anatomy course. Given the highly transient nature of sessional teaching positions and the sessional workforce, the sharing of expertise between more and less experienced SATs can help to ensure that there are sufficient numbers of appropriately trained SATs in the anatomy course. Thus, SAT-led mentorship and apprenticeship is a mutually beneficial arrangement for academics, the mentors and mentees.

In addition to pre-semester induction and training, we strongly advocate for regular in-semester meetings. In-semester meetings and training sessions can be tailored to upcoming teaching sessions and provide SATs an opportunity to view teaching materials, ask questions and consolidate their knowledge (in a manner aligning with the local pedagogical paradigm) prior to teaching the students. Additionally, these regularly scheduled meetings allow for dedicated reflection time, discussion and plans for management of any difficulties encountered (with the learning environment or content), as well as the opportunity to highlight aspects of the previous weeks' educational successes. Critical to the success of the reflexive component is the regular two-way conversation between the lead curricular academic(s) and the SATs; thus, effective ongoing support includes all educator parties being present at these meetings.

Management

Managing sessional teachers can encompass a wide range of different activities. All SATs benefit from a dedicated supervisor who is responsible for managing the SATs and acts as a go-to person if they have any questions, concerns or queries. Supervising sessional teachers is a task that is often overlooked by the university infrastructure, but can be managed through explicit listing as a key academic responsibility. Managing working conditions is a key role of the SAT supervisor. SATs may have heavy-duty restriction hours that limit their teaching contact time capacity (e.g. current students with educational commitments and healthcare professionals with clinical responsibilities). The individual availabilities and external commitments of SATs must therefore be considered when managing sessional teams. Supervisors must also determine the extent to which compensation will be provided to SATs and what form this will take. SATs in anatomy programs may receive financial compensation, course credit, and fee waivers or be engaged on a voluntary basis [2, 4, 30]. The incentives and/or compensation offered to SATs depends on the nature of the sessional position and the SAT profile. Current students (e.g. reciprocal-peer and near-peer SATs) may benefit from, and find motivation through, performance evaluation which is linked to their grades or coursework. In this case, outcome measurements may come from the SATs' requirement to submit reflective writing on their teaching experience, peer review of teaching, evidence of curricular development and action research [24, 50] as well as quantifiable outputs such as attendance, timeliness and number of teaching activities. Cross-level SATs (e.g. graduate students and healthcare professionals) may be more motivated by financial reimbursement. Paid employment for SATs may be hourly or salaried. Having observed cross-level SAT integration into anatomy programs using both options, the salaried option (providing a 0.5-1.0 fractional equivalent) appears to provide greater commitment and investment by the cross-level SATs. This financial incentive approach covers face-to-face class time as well as preparation, set-up, training sessions and meetings. Often those that are clinically qualified are also maintaining some level of commitment in their chosen healthcare field; the fractional financial investment fosters greater commitment to the SAT position by 'professionalising' the role. Since moving towards a fractional position, away from hourly pay, we have observed a positive change in timeliness, commitment in the classroom and SAT-driven curricular developments; thus the institutional payback far outweighs the financial payout. Ultimately, the working conditions and compensation provided to SATs are likely regulated at both a local and institutional level.

Quality control is also central to sessional teacher management, and is paramount to successful SAT curricular integration. Working with large numbers of sessional teachers who have varying degrees of knowledge, experiences and teaching skills has the potential to generate inconsistencies within a course [51, 52]. Providing sessional teachers with clear learning objectives and detailed instructional materials and assessment criteria can help to reduce this variability, particularly when in tandem with regular SAT meetings. Evaluating the performance of sessional teachers and providing them with resultant feedback are also integral in maintaining consistent high sessional teaching standards. While some anatomy programs evaluate SAT performance [12, 13, 24, 26, 30], few close the loop on this feedback by providing SATs with individual performance feedback and mentorship for improving identified deficits. We strongly encourage SAT programs to provide opportunities for ongoing feedback from all relevant stakeholders and to ensure that this feedback is part of an evolutionary process of SAT improvement similar to the 360-degree approach used in healthcare professional development [53].

Conclusion

When sessional teaching is effective, there are broad impacts on relevant stakeholders including the learners, SATs, permanent academics and, in turn, the hiring institution. Learners benefit from increased educator exposure with increasing SAT numbers; an improved learning environment, particularly through the social and cognitive congruence SATs can provide; and a potential for understanding how the discipline knowledge may be authentically applied in the workplace, chiefly when SATs are practitioners in the field (i.e. cross-level teachers). The academics benefit from many of the same outcomes as students, with an additional benefit of a decreased workload allowing for increased time to focus more on higher-order and complex curricular tasks such as education research, curriculum development and teaching innovations [3, 5, 25]. Importantly, SATs themselves also benefit from sessional teaching positions. SAT roles provide opportunities for interpersonal and professional skill improvement in teaching, leadership and critical thinking, while also affording knowledge refreshers within the anatomy discipline [2, 6, 22, 23, 37, 46]. While it may appear costly to engage SATs (both monetarily and temporally) within healthcare curricula, when properly executed, SATs provide widespread, evidence-based, positive impacts on the pedagogy and curriculum within which they are employed and typically the return on investment is high.

References

1. Manyama M, Stafford R, Mazyala E, Lukanima A, Magele N, Kidenya BR, et al. Improving gross anatomy learning using reciprocal peer teaching. BMC Med Educ. 2016;16:95.
2. McBride JM, Drake RL. Rewarding the resident teacher. Anat Sci Educ. 2011;4(4):227–30.
3. Johnson EO, Charchanti AV, Troupis TG. Modernization of an anatomy class: from conceptualization to implementation. A case for integrated multimodal-multidisciplinary teaching. Anat Sci Educ. 2012;5(6):354–66.
4. Duran CEP, Bahena EN, Rodriguez MDG, Baca GJ, Uresti AS, Elizondo-Omana RE, et al. Near-peer teaching in an anatomy course with a low faculty-to-student ratio. Anat Sci Educ. 2012;5(3):171–6.
5. Findlater GS, Kristmundsdottir F, Parson SH, Gillingwater TH. Development of a supported self-directed learning approach for anatomy education. Anat Sci Educ. 2012;5(2):114–21.
6. Evans DJR, Cuffe T. Near-peer teaching in anatomy: an approach for deeper learning. Anat Sci Educ. 2009;2(5):227–33.
7. Santana S. Wanted: more anatomy instructors - institutions explore ways to cope with shortage. AAMC Rep. 2003;13:6–7.
8. McCuskey RS, Carmichael SW, Kirch DG. The importance of anatomy in health professions education and the shortage of qualified educators. Acad Med. 2005;80(4):349–51.
9. Holden C. Anatomy classes face gross shortage. Science. 2003;299:1309.
10. Wilson AB, Barger JB, Perez P, Brooks WS. Is the supply of continuing education in the anatomical sciences keeping up with the demand? Results of a national survey. Anat Sci Educ. 2018;11(3):225–35.
11. Ten Cate O, Durning S. Dimensions and psychology of peer teaching in medical education. Med Teach. 2007;29(6):546–52.
12. Lockspeiser TM, O'Sullivan P, Teherani A, Muller J. Understanding the experience of being taught by peers: the value of social and cognitive congruence. Adv Health Sci Educ Theory Pract. 2008;13(3):361–72.
13. Hall S, Stephens J, Andrade T, Davids J, Powell M, Border S. Perceptions of junior doctors and undergraduate medical students as anatomy teachers: investigating distance along the near– peer teaching spectrum. Anat Sci Educ. 2014;7(3):242–7.
14. Singh S. Near-peer role modeling: the fledgling scholars education paradigm. Anat Sci Educ. 2010;3(1):50–1.
15. Heidenreich MJ, Musonza T, Pawlina W, Lachman N. Can a teaching assistant experience in a surgical anatomy course influence the learning curve for non-technical skill development for surgical residents? Anat Sci Educ. 2016;9(1):97–100.
16. Cornwall M. Students as teachers: peer teaching in higher education. PhD Thesis, University of Amsterdam, Amsterdam, 1979.
17. Rengier F, Rauch PJ, Partovi S, Kirsch J, Nawrotzki R. A three-day anatomy revision course taught by senior peers effectively prepares junior students for their national anatomy exam. Ann Anat. 2010;192(6):396–9.
18. Han ER, Chung EK, Nam KI. Peer-assisted learning in a gross anatomy dissection course. PLoS One. 2015;10(11):e0142988.
19. Percy A, Scoufis M, Parry S, Goody A, Hicks M, Macdonald I, et al. The RED report, recognition - enhancement - development: the contribution of sessional teachers to higher education. Sydney: Australian Learning and Teaching Council; 2008.
20. Rhodes D, Fogg QA, Lazarus MD. Dissecting the role of sessional anatomy teachers: a systematic literature review. Anat Sci Educ. 2018;11(4):410–26.
21. Rikers R, Schmidt HG, Moulaert V. Biomedical knowledge: encapsulated or two worlds apart? Cogn Psychol. 2005;19(2):223–31.
22. Reyes-Hernandez CG, Pulido JMC, Chapa RID, Vazquez RPS, Briones RDA, Banda PMP, et al. Near-peer teaching strategy in a large human anatomy course: perceptions of near-peer instructors. Anat Sci Educ. 2015;8(2):189–93.
23. Erie AJ, Starkman SJ, Pawlina W, Lachman N. Developing medical students as teachers: an anatomy-based student-as-teacher program with emphasis on core teaching competencies. Anat Sci Educ. 2013;6(6):385–92.
24. Lachman N, Christensen KN, Pawlina W. Anatomy teaching assistants: facilitating teaching skills for

medical students through apprenticeship and mentoring. Med Teach. 2013;35(1):E896–902.
25. Cheng XA, Wang L, Guo KH, Liu S, Li F, Chu GL, et al. Postgraduate fellows as teaching assistants in human anatomy: an experimental teaching model at a chinese research university. Anat Sci Educ. 2011;4(1):29–32.
26. Stephens JR, Hall S, Andrade MG, Border S. Investigating the effect of distance between the teacher and learner on the student perception of a neuroanatomical near-peer teaching programme. Surg Radiol Anat. 2016;38(10):1217–23.
27. Walser J, Horneffer A, Oechsner W, Huber-Lang M, Gerhardt-Szep S, Boeckers A. Quantitative and qualitative analysis of student tutors as near-peer teachers in the gross anatomy course. Ann Anat. 2017;210:147–54.
28. Miller SN, Neal DE. Surgical trainees as anatomy demonstrators. Ann R Coll Surg Engl. 1994;76:187–90.
29. Whelan A, Leddy JJ, Mindra S, Hughes JDM, El-Bialy S, Ramnanan CJ. Student perceptions of independent versus facilitated small group learning approaches to compressed medical anatomy education. Anat Sci Educ. 2016;9(1):40–51.
30. Shiozawa T, Hirt B, Celebi N, Baur F, Weyrich P, Lammerding-Koeppel M. Development and implementation of a technical and didactical training program for student tutors in the dissection course. Ann Anat. 2010;192(6):355–60.
31. O'Neill LD, Morcke AM, Eika B. The validity of student tutors' judgements in early detection of struggling in medical school. A prospective cohort study. Adv Health Sci Educ Theory Pract. 2016;21:1061–79.
32. Aggarwal R, Brough H, Ellis H. Medical student participation in surface anatomy classes. Clin Anat. 2006;19(7):627–31.
33. Davis CR, Bates AS, Ellis H, Roberts AM. Human anatomy: let the students tell us how to teach. Anat Sci Educ. 2014;7(4):262–72.
34. Celebi N, Zwirner K, Lischner U, Bauder M, Ditthard K, Schurger S, et al. Student tutors are able to teach basic sonographic anatomy effectively - A prospective randomized controlled trial. Ultraschall Med. 2012;33(2):141–5.
35. Anstey LM, Michels A, Szymus J, Law W, Ho MHE, Qu F, et al. Reflections as near-peer facilitators of an inquiry project for undergraduate anatomy: successes and challenges from a term of trial-and-error. Anat Sci Educ. 2014;7(1):64–70.
36. Wilson AB, Brown KM, Misch J, Miller CH, Klein BA, Taylor MA, et al. Breaking with tradition: a scoping meta-analysis analyzing the effects of student-centered learning and computer-aided instruction on student performance in anatomy. Anat Sci Educ. 2019;12(1):61–73.
37. Ocel JJ, Palmer BA, Wittich CM, Carmichael SW, Pawlina W. Outcomes of the gross and developmental anatomy teaching assistant experience. Clin Anat. 2003;16(6):526–30.
38. Reidenberg JS, Laitman JT. The new face of gross anatomy. Anat Rec. 2002;269(2):81–8.
39. Chen LP, Gregory JK, Camp CL, Juskewitch JE, Pawlina W, Lachman N. Learning to lead: self- and peer evaluation of team leaders in the human structure didactic block. Anat Sci Educ. 2009;2(5):210–7.
40. Gregory JK, Lachman N, Camp CL, Chen LP, Pawlina W. Restructuring a basic science course for core competencies: an example from anatomy teaching. Med Teach. 2009;31(9):855–61.
41. Cantwell S, Bonadurer GF, Pawlina W, Lachman N. Near-peer driven dissection selective: a primer to the medical school anatomy course. Clin Anat. 2015;28(8):985–93.
42. Shields RK, Pizzimenti MA, Dudley-Javoroski S, Schwinn DA. Fostering interprofessional teamwork in an academic medical center: near-peer education for students during gross medical anatomy. Anat Sci Educ. 2015;8(4):331–7.
43. Flaherty A, O'Dwyer A, Mannix-McNamara P, Leahy JJ. Evaluating the impact of the "teaching as a chemistry laboratory graduate teaching assistant" program on cognitive and psychomotor verbal interactions in the laboratory. J Chem Educ. 2017;94(12):1831–43.
44. Crowe P, Harris P, Ham J. Teaching on the run: teaching skills for surgical trainees. ANZ J Surg. 2000;70(3):227–30.
45. Rotenberg BW, Woodhouse RA, Gilbart M, Hutchison CR. A needs assessment of surgical residents as teachers. Can J Surg. 2000;43(4):295–300.
46. Shiozawa T, Hirt B, Celebi N, Werner A, Weyrich P, Lammerding-Koeppel M. Does a combined technical and didactical training program improve the acceptance of student tutors in the dissection course? A prospective controlled randomized study. Ann Anat. 2010;192(6):361–5.
47. Shiozawa T, Hirt B, Lammerding-Koeppel M. The influence of tutor training for peer tutors in the dissection course on the learning behavior of students. Ann Anat. 2016;208:212–6.
48. Steinert Y, Mann K, Centeno A, Dolmans D, Spencer J, Gelula M, et al. A systematic review of faculty development initiatives designed to improve teaching effectiveness in medical education: BEME guide no. 8. Med Teach. 2006;28(6):497–526.
49. Hendry GD. Problem-based learning tutors' conceptions of their development as tutors. Med Teach. 2009;31(2):145–50.
50. Mills GE. Action research: a guide for the teacher researcher. 6th ed. London: Pearson Education; 2017.
51. Johansson E, Holmin TE, Johansson BR, Braide M. Improving near-peer teaching quality in anatomy by educating teaching assistants: an example from Sweden. Anat Sci Educ. 2018;11(4):403–9.
52. Muzaka V. The niche of graduate teaching assistants (GTAs): perceptions and reflections. Teach High Educ. 2009;14(1):1–12.
53. Hammerly ME, Harmon L, Schwaitzberg SD. Good to great: using 360-degree feedback to improve physician emotional intelligence. J Healthc Manag. 2014;59(5):354–65.

7 Preparing the Next Generation of Anatomists Through Training Programs and Professional Development

William S. Brooks and Adam B. Wilson

Defining the Need for Training and Professional Development

Training programs, continuing education (CE), and professional development (PD) activities in the anatomical sciences afford budding and seasoned educators the opportunity to develop and maintain their professional competence and to explore how new topics and approaches may benefit their learners. Due to a decline in the number of graduate programs in the anatomical sciences over the past few decades [1–3], the need to develop a cadre of qualified anatomy educators is potentially the greatest it has ever been [4–6]. The American Association for Anatomy's website presently lists only 53 institutions in the United States (US) offering graduate degrees in the anatomical sciences, though the total number of annual graduates is unknown [7]. Although many anatomy educators feel there is a need for postdoctoral training [3] and continuing education in anatomy [8], relatively few programs exist to provide educators with formal opportunities for professional growth. For example, Indiana University is one of only a few institutions in the US known to offer formal postdoctoral training in anatomical sciences education.

In 2019, a research project assessed the current state of the US anatomy educator job market and reaffirmed that anatomy educators remain in short supply [6]. Of concern is the anticipation that the number of anatomy educator job vacancies may deepen given the continued expansion of medical schools and health science programs [6]. Wilson and colleagues also concluded that departmental leaders seeking anatomy educators prefer job candidates with a Ph.D. in anatomical sciences education over other degree types (e.g., traditional basic science Ph.D., M.D. degree, or Ph.D. in anthropology) [6]. Highly sought-after qualities/experiences in anatomy educator applicants include experience teaching the anatomical sciences, teaching versatility across multiple anatomical disciplines, and versatility in teaching methods/pedagogies [6]. Unfortunately, these desired skillsets are often underdeveloped during typical biomedical graduate and postdoctoral training experiences due to a general lack of teaching exposure and a scarcity of education-focused instruction.

W. S. Brooks (✉)
Department of Cell, Developmental, and Integrative Biology, University of Alabama at Birmingham, School of Medicine, Birmingham, AL, USA
e-mail: wbrooks@uab.edu

A. B. Wilson
Department of Cell and Molecular Medicine, Rush University, Chicago, IL, USA
e-mail: adam_wilson@rush.edu

L. K. Chan, W. Pawlina (eds.), *Teaching Anatomy*, https://doi.org/10.1007/978-3-030-43283-6_7

Top Six Qualities or Experiences for Job Applicants

1. Anatomy teaching experience
2. Versatility teaching multiple anatomy disciplines
3. Versatility in teaching methods
4. Peer-reviewed publication record
5. Experience in assessment and evaluation
6. Experience in curricular design

Given these findings and job market projections, there is a clear need to further develop individuals at all levels of the training continuum. More specifically, there is a need to better align the goals of educational programs with current recruitment preferences and the evolving needs of the healthcare training profession. With these considerations in mind, the remainder of this chapter provides guidance on how to plan, design, and maintain educational programs with a common mission of training up future educators to replenish and preserve the anatomy educator profession.

Program Planning

Tips for Conducting a Needs Assessment

The development of any new educational program ought to begin with a focused needs assessment, the results of which will drive the refinement of the program's mission and the planning process. A *needs assessment* is the systematic process of examining the current training landscape and the needs of targeted populations to determine whether ample interest and sufficient resources exist to fill a gap in training services. Needs assessment findings are often used to (1) justify whether a real need exists, (2) further direct the focus of the training mission, (3) guide informed decision making, and (4) determine resource and training priorities. Surveys, interviews, and focus groups, including town hall-style meetings, are the most common methods for gathering needs assessment information.

When beginning a needs assessment, foundational and relevant questions to consider include the following: (1) How does the new program's mission fill a lingering or unique gap? (2) Who is the target audience? (3) How will information from the target audience be gathered? (4) What are the anticipated learning needs of the targeted population? (5) What teaching format or approach is likely to be used? (6) What resources are needed to effectively implement the designed program? and (7) What are the broader societal needs, if any? To better contextualize the role of a needs assessment and the importance of these questions, consider the following example.

The Gross Anatomy for Teacher Education (GATE) program was initiated in 2014 at the University of Alabama at Birmingham as a 3-day hands-on workshop in gross anatomy for high school and undergraduate anatomy and physiology teachers [9]. By coupling anatomy lectures with cadaveric dissection and educational topics, this annual workshop offers a unique experience that this population of educators would not have had access to otherwise (Fig. 7.1). Each summer a group of 35–40 attendees from across the continental US convene to cover topics related to one of four anatomical regions: (a) trunk, (b) back and upper limb, (c) lower limb, and (d) head, neck, and brain. Each summer attendees focus on a different region such that a complete full-body dissection can be achieved by attending four consecutive workshops.

Due to the success and popularity of this growing program, a needs assessment was conducted to evaluate the feasibility of commercializing GATE for large-scale implementation and to discern whether there was general interest among potential attendees in other regions. The needs assessment collected national data using a survey to determine whether the available supply of continuing education in the anatomical sciences was keeping up with the demand [8]. The data revealed there is a demand for CE

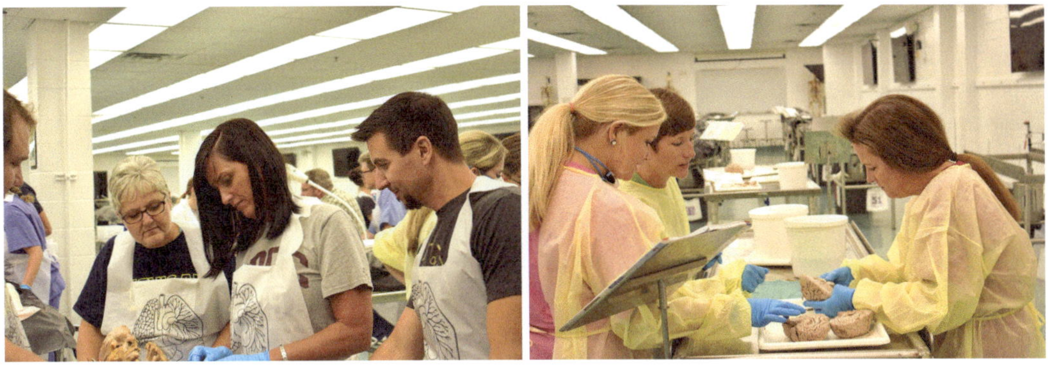

Fig. 7.1 Attendees perform gross dissection of the head, neck, and brain in the Gross Anatomy for Teacher Education (GATE) course

among anatomy educators at a national level. More specifically, the population with the greatest need for CE consisted of early career educators who teach anatomy to high school and college-level learners [8]. The information gleaned from this assessment allowed the program designers to tailor the GATE experience to include more education-focused content (e.g., an introduction to teaching pedagogies) and more clinically relevant topics such as the use of newer anatomy imaging modalities. Given that time constraints are typical of working professionals, survey results reinforced the need for GATE to be delivered as a succinct, hands-on workshop with supplemental asynchronous online modules [8]. While this particular needs assessment was conducted at a national level, most are conducted to gauge local and/or regional needs.

Preferred Format of Anatomy-Related Continuing Education in Rank Order

1. Asynchronous online videos or learning modules
2. Intensive hands-on workshops (e.g., 1–2 days)
3. Extensive hands-on workshops (e.g., 3 days to 1 week)
4. Graduate coursework leading to a certificate

Seeking External Funding to Support an Educational Program

In the current US economic climate, external funding can be an invaluable resource by which graduate, postdoctoral, and CE/PD programs are made financially feasible. Some professional societies will offer educational outreach and innovation grants that can be used to offset the costs of professional development activities; the American Association for Anatomy's Outreach Grant Program is one example. The National Science Foundation (NSF) also offers a number of grant mechanisms largely targeting the development of STEM educators of the K-12 sector. Federal funding through the National Institutes of Health (NIH) may also be a viable option for funding postdoctoral training programs.

The Scientist-Educator Program at Vanderbilt University was established in 2005 through an NIH Institutional Research and Academic Career Development Award (IRACDA), which provided funding for postdoctoral programs that combined mentored basic science research at a research-intensive institution with academic teaching at an institution with demonstrated commitment to underrepresented groups [10]. Postdoctoral fellows of the Scientist-Educator Program entered with doctoral-level training in a basic science discipline but often lacked formal training in the anatomical sciences. During this 3-year fellowship, postdoctoral fellows devoted 60% of their time to traditional bench research under the guidance of a faculty mentor and 40% of their time to anatomy education under the supervision of anatomy teaching faculty. In the first year, fellows took coursework in gross anatomy and embryology alongside medical students. In the second and third years of the fellowship, trainees taught in the gross anatomy laboratory and gave lectures in the medical school. Through a combination of mentored bench research, training in anatomy content, and lecture and laboratory teaching experiences, the Scientist-Educator Program provided training for Ph.D. bench scientists to become anatomy educator-scholars and produced a number of high-caliber academic anatomists.

It is important to point out that programs primarily funded through external resources run the risk of never achieving independence or longevity unless other funding mechanisms are established in parallel. The Scientist-Educator Program fell victim to this dependency; when the flow of funds ceased, so too did the program.

Additional Planning Considerations

Once fundamental questions regarding a programs's purpose have been answered, securing buy-in from relevant stakeholders is often the next crucial step in the planning process. In most instances, successful deployment of a new program will require the collective effort of a team of dedicated educators and buy-in from administrators to support faculty time and staff resources. Preparing a business plan which outlines expenses and the anticipated return-on-investment, via tuition dollars or registration fees, can help to provide administrators with important context and projected fiscal outcomes. For face-to-face CE/PD workshops, anatomy educators indicate that approximately US $100 per day is an appropriate registration cost [8]. GATE registration fees, for example, total US $450 for a 3-day workshop. This registration cost includes course materials, cadaveric dissection, and one meal each day. Using this as a baseline cost estimate, projections can be made regarding the number of attendees needed to recoup program costs and potentially generate a profit.

Finally, when planning a new program, the importance of learning from others who have gone before you cannot be overstated. Networking with fellow colleagues who have successfully developed similar programs can be invaluable for gaining critical insights and for avoiding common pitfalls. In particular, navigating legal compliance issues, approval processes, and program credentialing can be daunting without direct guidance from experienced individuals and legal experts. By capitalizing on the knowledge of fellow colleagues and by taking advantage of society supported communities (e.g., AAA's Anatomy Connected), you will surely enhance the development and design of your educational program.

Program Design

Elements of Effective Professional Development

The Learning Policy Institute, which conducts research to improve education policy and practice, outlines several features that are common to effective professional development offerings [11]. Similar research by others also affirms these key elements [12–14]. To summarize, effective professional development:

- Is *content focused*, meaning there is intentionality in aligning teaching strategies with content attendees are most likely to engage with. PD sessions covering broad, nonspecific, and/or less relevant topics, are far less effective.
- Incorporates *active learning* in a way that allows attendees to directly engage with and practice the application of the presented content, which instills deeper learning and improves contextualization.
- Supports *collaboration* by creating a community of practice for the exchange of ideas, innovations, and struggles.
- *Models effective instructional practices* so that attendees have "a clear vision of what best practices look like" [11].
- Provides time for *reflection and feedback* so that attendees can contemplate their current

practices and receive input on individual needs through *coaching and expert support*.

- Is of *sustained duration* to allow participants "adequate time to learn, practice, implement, and reflect upon new strategies that facilitate changes in their practice" [11].

One take-away from these principles is that traditional professional society meetings with numerous and disparate one-off lecture style presentations do not constitute high-quality professional development. While national conferences offer learning opportunities, they simply are not as rich as offerings that follow these elements of effective professional development, such as the Anatomy Education Research Institute.

Funded by an Innovations Grant from the American Association for Anatomy, the inaugural Anatomy Education Research Institute (AERI) was held in 2017 at Indiana University, Bloomington [15, 16]. This 5-day, face-to-face workshop provided instruction on foundational concepts in educational research. The attendees were interested anatomy educators who lacked relevant experience and expertise to initiate educational research projects on their own. In total, 62 participants from 8 countries took part in this PD opportunity. A combination of plenary sessions and hands-on workshops covered a range of topics including developing research ideas, study design, quantitative and qualitative research methodologies, and much more. Given this context, it is evident that all AERI content was held together by a core theme – learning how to conduct quality educational research. A hallmark feature of AERI, which incorporates the elements of reflection, feedback, and coaching, was the pairing of novices with expert anatomy/medical education research mentors who provided valuable coaching during and after the institute. Sprinkled throughout the AERI conference were a number of dedicated mentor/mentee sessions where mentees directly applied their newfound knowledge, an example of active learning. The pairing of two to three mentees per mentor created micro-communities of practice where research ideas and obstacles could be freely discussed. Throughout the institute, blocks of time were deliberately dedicated to learning new content, applying newfound knowledge, and strategic planning in order to promote research productivity long after the completion of AERI.

Establishing a Niche/Unique Mission

It is important that any new program set itself apart from others by offering something unique and marketable to its targeted audience. This could include a unique educational experience such as travel to Germany to learn plastination techniques as part of the Contemporary Human Anatomy master's degree program at Eastern Virginia Medical School [17]. It might also involve a unique programmatic structure such as asynchronous distance education offered by the Anatomy Training Program [18]. Regardless of the one-of-a-kind learning activity or unique program structure, having distinctive, marketable programmatic features can position a program for successful learner recruitment.

Co-sponsored by the American Association for Anatomy and the Anatomical Society (United Kingdom), the Anatomy Training Program (ATP) provides an avenue for trainees, junior faculty, or those undertaking a career change to obtain the necessary knowledge and attitudes to teach anatomy at the undergraduate and postgraduate levels [18]. This program was initially established in 2008 in response to the perceived shortage of anatomy educators. The ATP utilizes a hybrid structure involving distance learning and face-to-face instruction. Through distance learning, participants complete independent online modules throughout the academic year under the guidance of a mentor at one's home institution. Four discrete modules are available: limbs, head and neck, trunk, and neuroanatomy. Each module incorporates medical imaging, embryology, and case studies in addition to core clinical anatomy content. Participants then travel to the University of Oxford for an intensive, 1-week, face-to-face workshop designed to consolidate knowledge and impart expertise in aspects of teaching and pedagogy. Two modules are offered each year and trainees receive an ATP Certificate upon completion of all four modules. This model is uniquely attractive to western participants because it incorporates European travel as a requirement. It is also a flexible hybrid model that allows professionals to schedule program learning activities around their own academic and personal obligations. Together these unique features make for an attractive and flexible program that never has a shortage of interested and hopeful applicants.

Anatomy Training Program (ATP) Modules

1. Limbs
2. Head and Neck
3. Trunk
4. Neuroanatomy

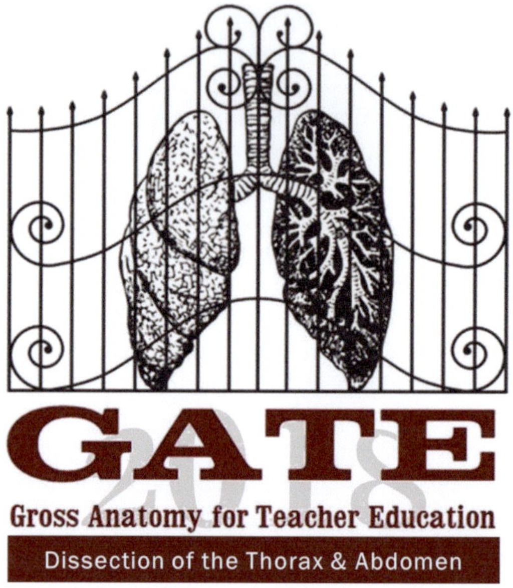

Fig. 7.2 GATE logo and advertisement

Initiating and Maintaining Programs

Sound Advertising Strategies

Advertising is an essential preliminary step preceding program implementation. Good advertising broadens a program's reach to new learners and reminds former participants of what your program uniquely offers. Effective advertisements share a few commonalities. Knowing that visually and audibly engaging advertisements catch the interest of onlookers is common sense. What you may not realize is that advertisements with some unexpected feature are actually superior at grabbing one's attention and are easier to recall [19]. Take, for example, the GATE logo with a pair of lungs situated behind an ornate rod iron fence. This organ-gate pairing is certainly unexpected and entices a viewer to take a closer look (Fig. 7.2). Successful advertisements also appeal to emotion and reason simultaneously. These advertisements often play first to emotion, coaxing an individual to *want* a product. The emotional appeal is then followed with reason, which convinces the individual that he or she *needs* the product. Suppose a national society is advertising a silent auction fundraiser where a set of original, hand-painted Netter Anatomy plates are available for bid. The ad shows the plates hanging in a classic professorial office, not too dissimilar from your own. Now that an emotional tie exists due to your innate interest in the subject matter and the appeal of hanging the art in your own office, the ad follows with verbiage highlighting the good cause for which the event is raising funds, giving the onlooker even more reason to place a bid. This tactic is surely to draw interest from the targeted population followed by action, such as enrolling in a professional development program. Lastly, successful advertisements are often frequent and coordinated. Marketing research suggests that consumers need to be exposed to ads or messaging about three times before they become fully aware of what is being offered. Coordinating the distribution of ads to occur repeatedly across multiple platforms, while still maintaining a central theme, may be necessary for maximum exposure and effect [20].

Maximize Your Reach and Keep Registration Simple

Soliciting the help of departmental or institutional communications staff is pivotal. These individuals can assist with logo design, website creation, and the production and distribution of marketing materials. When seeking avenues for advertising, consider alumni email lists (usually held by each institutional degree program) and national or state professional societies, which usually sell advertising space on their websites and within conference booklets. Working with an institutional business office to establish payment methods is also important for streamlining the registration process. Consider how attendees will be able to pay any associated registration fees (e.g., credit card payments or checks) and how those revenues will filter back to the program.

Capitalize on Available Resources

Perhaps the oldest trick in the proverbial book for developing a successful program is capitalizing on available resources. From a program development perspective, necessary resources often range in variety from human resources (e.g., intellectual capital, expertise, and manpower) to financial and physical resources (e.g., facilities, equipment, and incidentals). Capitalizing on unique resources can strengthen a program and can help to further establish a program's niche. Sharing valuable and/or expensive resources can help to broaden a resource's impact and can reduce costly overhead. Here is one success story of how a locally available resource put a new program on the map.

In 2008, the Department of Anatomy and Cell Biology at Indiana University (IU) School of Medicine collaborated with the IU School of Education to develop an education-focused PhD program [1]. Graduates of this program receive training in the anatomical sciences from anatomical sciences faculty. Simultaneously, they are trained by PhDs and EdDs in higher education in subjects including instructional pedagogies, educational psychology, and statistics in educational measurement and evaluation, to name a few. Graduates are also required to complete an anatomy-/medical education-focused research dissertation which is often chaired by a School of Education faculty member. The IU School of Education and its scholarly faculty are well known for their state, national, and global impact in training quality teachers. In developing the anatomy education track PhD program, Brokaw et al. [1] keenly recognized how the strength of a local resource (i.e., easy access to renowned education faculty) would nicely complement the limitations of a highly specialized biomedical department. As a result of capitalizing on this instrumental resource, the IU anatomy education program has now graduated more than a dozen gainfully employed academicians who are often highly sought-after for their unique educational expertise.

Longevity Depends on Champions and Financial Sustainability

The sustainability of a training or CE/PD program for anatomy educators can be achieved but is often reliant upon two factors. First, the importance of a passionate individual serving as the program champion cannot be understated. Because these educational programs are typically small scale and offered annually, they can easily get lost in the milieu of high-enrollment programs at an institution. The champion ensures that programming occurs regularly and serves to promote the importance of education for the continued development of professionals locally, regionally, and nationally. Secondly, the survival of any educational program is a function of its financial stability. A business model which strives for a program to become financially self-sustaining must be established early on and periodically revisited. In an era of economic difficulty and less state funding for higher education, activities that drain financial resources from a department are often those that meet the chopping block first. Programs that fund themselves are best positioned to maintain longevity.

Conclusions

Training opportunities at the graduate, postdoctoral, and professional levels are essential for continuously developing professionals to teach the anatomical sciences to future healthcare providers. Careful planning, based on a needs assessment and business plan, allows relevant stakeholders to make informed decisions regarding a program's mission, structure, and financial stability. Advertising a new program by highlighting one or more unique program features and working with institutional marketing and communications staff can aid in the recruitment of students and/or attendees. Finally, successful programs are often led by an enthusiastic program director who capitalizes on available resources and works to achieve financial self-sustainability to help ensure a program's longevity.

References

1. Brokaw JJ, O'Loughlin VD. Implementation of an education-focused PhD program in anatomy and cell biology at Indiana University: lessons learned and future challenges. Anat Sci Educ. 2015;8(3):258–65.
2. McCuskey RS, Carmichael SW, Kirch DG. The importance of anatomy in health professions education and the shortage of qualified educators. Acad Med. 2005;80(4):349–51.
3. Schaefer AF, Wilson AB, Barger JB, Azim HM, Brokaw JJ, Brooks WS. What does a modern anatomist look like? Current trends in the training of anatomy educators. Anat Sci Educ. 2019;12:225–35.
4. Anatomy classes face gross shortage. Science. 2003;299:1309.
5. Santana S. Wanted: more anatomy instructors-institutions explore ways to cope with shortages. AAMC Rep. 2003;13:6–7.
6. Wilson AB, Notebaert AJ, Schaefer AF, Moxham BJ, Stephens S, Mueller C, et al. A look at the anatomy educator job market: anatomists remain in short supply. Anat Sci Educ. 2020;13(1):91–101.
7. American Association of Anatomists: graduate programs in anatomy. 2019. http://www.anatomy.org/graduate-programs-in-anatomy.html . Accessed 26 Feb 2019.
8. Wilson AB, Barger JB, Perez P, Brooks WS. Is the supply of continuing education in the anatomical sciences keeping up with the demand? Results of a national survey. Anat Sci Educ. 2018;11(3):225–35.
9. Doss AA, Brooks WS. Gross anatomy for teacher education (GATE): educating the anatomy educator. HAPS Educ. 2016;20(2):46–50.
10. Bader D, Dalley AF, Norden JJ, Nanney LB, Pettepher CC, Richmond AW, et al. Development of a scientist-educator program. FASEB J. 2010;24:S827.7.
11. Darling-Hammond L, Hyler ME, Gardner M. Effective teacher professional development. Palo Alto: Learning Policy Institute; 2017.
12. Ronfeldt M, Farmer SO, McQueen K, Grissom JA. Teacher collaboration in instructional teams and student achievement. Am Educ Res J. 2015;52(3):475–514.
13. Jacques C, Behrstock-Sherratt E, Parker A, Bassett K. Investing in what it takes to move from good to great. Washington: American Institutes for Research; 2017.
14. Gulamhussein A. Teaching the teachers: effective professional development in an era of high stakes accountability. In: Center for Public Education. VA, Alexandria: National School Boards Association; 2019. http://www.centerforpubliceducation.org/research/teaching-teachers-effective-professional-development. Accessed 18 Feb 2019.
15. O'Loughlin VD, Husmann PR, Brokaw JJ. Development and implementation of the inaugural anatomy education research institute (AERI 2017). Anat Sci Educ. 2019;12(2):181–90.
16. Husmann PR, O'Loughlin VD, Brokaw JJ. Knowledge gains and changing attitudes from the anatomy education research institute (AERI 2017): a mixed methods analysis. Anat Sci Educ. 2020;13:192–205.
17. Eastern Virginia Medical School: contemporary Human Anatomy master's program. https://www.evms.edu/education/masters_programs/contemporary_human_anatomy/curriculum. Accessed 18 Feb 2019.
18. Fraher JP, Evans DJ. Training tomorrow's anatomists today: a partnership approach. Anat Sci Educ. 2009;2(3):119–25.
19. Heath C, Heath D. Made to stick: why some ideas survive and others die. New York: Random House; 2007.
20. Red Crow Marketing. Five key principles to creating ads that work. In: Red Crow Marketing Inc; 2013. https://www.redcrowmarketing.com/2013/01/11/five-key-principles-to-creating-ads-that-work/. Accessed 18 Feb 2019.

8 Gamification in Anatomy Education

Christopher See

It is likely that both anatomy teachers and their students will have some experience in the playing of games in the general sense. This could be as simple as rolling a die and moving pieces on a game board or as sophisticated as two world-renowned grandmasters of chess pitting their wits against one another. The scale of such endeavours ranges from one-on-one games to massive multiplayer online (MMO) video games with thousands of participants contributing simultaneously. Gamification of education attempts to incorporate game-playing into education, or indeed education into game-playing, in order to generate learning opportunities of a more immersive, complex or interesting manner.

This chapter describes the learning theory relating gamification to anatomy education and proposes ways in which anatomy teachers can harness the power of play in their teaching in order of reap pedagogical and motivational rewards for their students.

C. See (✉)
School of Biomedical Sciences, Li Ka Shing Faculty of Medicine, The University of Hong Kong, The Hong Kong Special Administrative Region of People's Republic of China
e-mail: christophersee@cuhk.edu.hk

What Is Gamification?

In the broad context of health professions education, gamification may be considered the application of game elements to non-game contexts [1]. De Freitas and Liarokapis [2] propose a list of key game elements which are typical of gamification: the presence of a reward structure, an overarching theme or storyline, competition between participants, interactivity between the participants and the game and progression within the game itself. Other authors have developed a list of game attributes which may be adapted individually or as a group to gamify a learning experience [3] (Table 8.1).

The term gamification in education is increasingly associated with digital games played via a student's computer, tablet or mobile device [4]. As time progresses, the digital emphasis may become even more important to the digital natives whom we teach [5]; nevertheless, it is important to recognise low-to-no technology games which have potential for use in the context of anatomy education [6, 7].

Gamified learning can extend beyond incorporating game elements to being almost entirely taught via a fully formed game [8]. Such experiences may still be characterised as gamified, but indeed can be also be described as game-based learning (GBL), serious games or educational games. There is considerable variation between the terms used for education with differing degrees of gamification [9], but anatomy teachers may wish to consider the broad question of

L. K. Chan, W. Pawlina (eds.), *Teaching Anatomy*, https://doi.org/10.1007/978-3-030-43283-6_8

Table 8.1 Types of game elements

Attribute	Definition
Action language	The method and interface by which communication occurs between a player and the game itself
Assessment	The method by which accomplishment and game progress are tracked
Conflict/challenge	The problems faced by players, including both the nature and difficulty of those problems
Control	The degree to which players are able to alter the game and the degree to which the game alters itself in response
Environment	The representation of the physical surroundings of the player
Game fiction	The fictional game world and story
Human interaction	The degree to which players interact with other players in both space and time
Immersion	The affective and perceptual experience of a game
Rules/goals	Clearly defined rules, goals and information on progress toward those goals, provided to the player

whether their aim is to add game elements to an existing teaching method or to create an entire game for learning. This chapter will address considerations for both approaches.

Pedagogy and Learning Theory in Gamification

Pedagogy

Systematic reviews of gamification in higher education and medical education characterise the potential advantages of improvement of student motivation, engagement and enjoyment of the learning experience [10, 11]. Specific game elements may have a particular utility in enhancing student learning. The role of competition between students in learning anatomy has been recognised as helpful in improving student academic performance [12] and is a common feature of gamification. Educational games can encourage students to apply their knowledge to new scenarios within games, be they authentic or fantastical. This allows the learning processes to move beyond memorisation to higher functions such as application and reasoning [13]. A characteristic feature of many educational games is interaction between participants to generate collaborative learning, defined as "a situation in which two or more people learn or attempt to learn something together" [14]. This type of learning has seen growing traction in both clinical contexts [15] and basic sciences instruction [16] in the form of peer teaching and near-peer teaching, as well as in problem-based learning (PBL) [17] and team-based learning (TBL) [18] settings. Facilitation of this process is another pedagogical feature of gamification of education.

Learning Theory

From the perspective of learning theory, Lave and Wenger [19] emphasise the role of communities and culture in learning. They characterise the journey of learning as joining a community of practitioner or thinkers, and as such learners develop by observing the actions of the community, with interpretation and reflection playing key roles in their education. This description mirrors the role students may take in serious games where they might see the actions of other players and reflect on the results of their actions by their outcomes in the game. In the context of gamification, Kapp [8] describes the concept of cognitive apprenticeship, whereby players can observe and interpret the actions of fellow players and take on an increasing role in the game-playing community as their learning progresses. The process of knowledge construction, with the activation of prior knowledge activation, is thought to be a feature of game-based learning, paralleled by the more familiar sphere of problem-based learning [17].

Potential benefits of gamifying anatomy education

- Allowing students to fail in a safe environment
- Enhancing student motivation
- Providing instantaneous feedback
- Promoting collaborative or peer learning
- Applying knowledge to novel situations

Examples of Gamified Learning in Anatomy

In anatomy education, the examples of gamified learning may be considered in digital and non-digital subgroups.

Digital Gamified Learning

Ngan et al. [20] developed an electronic professional study (EPS) courseware containing gamified quizzes and narrative animations for anatomy education. Their intervention featured a student scoring system and progressive advancement through different levels of quizzes on aspects of the cardiovascular system such as the coronary circulation. Stirling et al. [21] incorporated multimedia into a digital e-book through an interactive video-game engine (Unity 3D) for students to explore a 3D heart model which could be manipulated with gestures. This feature had options for formative learning and self-assessment. Web-based simulated patients for applied anatomy learning [22], virtual reality education for forearm structures [23] and augmented reality mobile application for the learning of neuroanatomy [24] are further examples of gamified, active learning in the digital sphere.

Non-digital Gamified Learning

Physical board games have proven to be a source of gamification ideas for anatomists, exemplified by Burleson et al. [25], who generated a game entitled "Clue Connect", based on an array of descriptive tiles and diagrams for the purposes of learning anatomical directions, actions and their relationship to structures. Another study used a range of popular board game formats of Bingo, Archipelago and Kahoot deliver quizzes on anatomy on topics such as head and neck and thoracic anatomy [26].

Gameshows such as "Who Wants To Be A Millionaire?" have been adapted for the training of anatomical knowledge [27] in a classroom setting. In this case, there are several interactive elements as participants are allowed to "phone a friend" or "ask the audience" before giving their answer. In the utilisation of gameshow formats, it may be prudent to adopt one which is culturally appropriate and familiar with the cohort, rather than importing an idea from abroad or the past which students may be unfamiliar with. When introducing students to an educational game for the first time, it may be easier for them to understand the rule system if they have some idea of the format from their own experience. Asking students what types of games they play or watch for leisure can also be a helpful starting point for educators in thinking about gamification of their teaching.

Coming Up with Ideas for Gamifying Your Teaching

Inspiration for gamification may come from a range of different sources such as board games, video games, sports or other physical games or real-world scenarios, which may be evaluated for game characteristics you wish to consider for your classroom (Table 8.2).

This may then help you to match the academic content, skills and behaviours you wish to teach to the format of the educational game.

Activities which are not directly considered games could be another source of inspiration for

Table 8.2 Evaluation characteristics for educational games

Game characteristics	Example considerations
Interaction between players	Is it competitive or collaborative? For individuals or teams?
Interaction between student and teachers	Is there an expert, referee or mediator involved in the game?
Repeat value	Is it suited for single play-through or is repeated gameplay possible?
Player metrics	Points scoring? Leader board? Single winner or team winners?
Game environment	Is it fully or partially immersive?

the inquisitive anatomy educator. McCarroll et al. [28] devised a role-playing game involving muscle identification based on the model of speed dating. Students sat across from one another and introduced themselves for 2 minutes, after which their counterpart would have to identify the muscle based on the information given. The only equipment needed for implementation was classroom desks, with no boards, cards or technology required.

To Do

Consider your current teaching activities and identify any areas you would like to make more interactive or engaging. Envisage how you might transform that section by converting it into an interactive quiz, board game, video game or game-like activity.

Incorporating Game Elements into Your Teaching

Not every educational activity can or should be a full-blown game. Teachers can employ elements of gamification to enhance an existing teaching format. In essence, a pedagogical mix-and-match approach may be adopted by deliberately choosing game elements such as instantaneous feedback, competition between peers or any other game attribute described in Table 8.1. This is particularly true when creating a novel teaching experience if teachers are unsure of how students may react to a full-blown game.

Dickson et al. [29] employed the physical play in the form of miming to a lecture, allowing students to mimic actions and make "funny faces" in the learning of the cranial nerves. This game-like element was included in an otherwise traditional format of a lecture and created a simple, effective enhancement with improvements in student perception and examination performance. Herling et al. [30] incorporated some gamified sessions of anatomy teaching such as competitive quizzes ("Anatomy Jeopardy") into an anatomy boot camp which also featured traditional teaching formats. They demonstrated benefits to both the academic and social transition to medical school for participants.

Game elements may be most appropriate when a teacher wishes to improve an existing teaching format, by bringing in some aspect such as physical correlation to the learning material, or competition between participants, as per the above examples.

Beyond Game Elements: Creating an Educational Game

Educational experiences which centre around a game can become entities unto themselves, rather than being small part of a traditional learning format. This type of activity is an educational game, a self-contained learning experience which contains the academic content, and itself provides feedback to the learners. Examples of full educational games include a digital game designed by Janssen et al. [31] entitled "They Know: Anatomy", which was a competitive quiz-based game for desktop computers. This involved individual or groups of students progressing through a virtual board consisting of nodes by virtue of answering quiz questions, in a race against others.

"Vascular Invaders" is another example of a digital game which is played as its own, complete learning experience in which students would explore the human cardiovascular system, exploring and correctly identifying a pathway of vessels [32]. Anyanwu et al. [7] created a complex board game superimposed over a human body with 72 questions and its own set of in-game economics, entitled "Anatomy Adventure". These examples are sophisticated and comprehensive learning experiences. The development of such games requires a considerable amount of time and effort on the part of the educators.

To aid in the planning of the creation of such experiences, Olszewski et al. [33] propose a framework for developing serious games in medical education. It breaks the process down into three phases: Preparation and Design;

Development; and Formative Evaluation. A summary of the steps condensed and adapted for anatomy teaching is presented below.

Phase 1: Preparation and Design

Team assembly is the gathering of parties with the relevant interest and expertise, including those with game design expertise. *Anatomy concepts transfer* is the process by which the anatomical knowledge and concepts are shared with all parties in development. *Content production* involves the creation of the anatomical substance which forms the learning material, including any text, diagrams or video, and is approved by the anatomists. *Learner experience mapping* is a process in which the team plans the journey of the students through the game, demonstrating the flow through tools such as storyboards.

Phase 2: Development

The first step in development is production of a visual representation of the actual appearance of the game as you wish it to be. It is helpful to create *wireframes*, which are simplified diagrams representing game elements. These are important visualising content in both digital games and physical games and can be used in the same way that a still photograph might help to plan the filming of a video. *Prototype* is an early product which may contain a small phase of the game which can be tested by anatomists to identify errors and indicate improvements. This feedback is incorporated into the *iterative prototype* for further rounds of refinement and testing.

Phase 3: Formative Evaluation

Usability testing involves small groups of students testing the game and the collection of their feedback through survey, focus group interview or other means for further improvement. *Final product delivery* to the target audience involves the implementation and evaluation of the game.

How to Implement Gamification

There are several practical considerations for the implementation of gamification in anatomy teaching.

Timing of the Game

Consider the relationship of your educational game to class time with students.

- Pre-class: as a flipped classroom experience where students discuss their game experience and learning issues encountered during a class following their gameplay
- Intra-class: as a resource to enhance student interaction in the learning environment
- Post-class: as a self-evaluation game, to extend learning beyond class time or as a tool to stimulate reflection

Pilot Testing

Consider doing a pilot test for the following reasons:

- Aligning gameplay with students' interests and backgrounds by asking what they currently play or enjoy
- Identifying rule ambiguity, areas of potential exploitation or cheating and students playing the game in a manner not anticipated by designers
- Titrating distraction so as to allow balancing of the "fun" and learning components

Creating a Narrative Game

One factor that differentiates a game from a series of tasks is an underlying story which weaves these activities into a coherent narrative which can attract the interest of students. If your intention is to generate a narrative for an anatomy game, consider employing one of the following frameworks to build your plot:

- Solving a mystery/deciphering of puzzles
  Placing students in the role of an academically orientated detective may be a way of engaging them in anatomy learning. If your educational game employs codes, anagrams and crossword puzzles, this type of narrative could be particularly suitable. You may wish to introduce a sense of conflict by creating villain for the game against which the students pit their anatomical wits.
- Exploration of new environment
  One way to achieve this is through the sense of scale. For example, a game might position the students as miniaturised humans within an organ system, whose landmarks can then be explored from a new perspective, e.g. from the inside. A board game might be constructed around the alimentary system, and players would need to avoid the various functional dangers (acid, pepsin, bile) as they attempt to identify key anatomical features before leaving the body.
- Following a character or community storyline
  The personification of anatomical structures might be one method of introducing a character to an educational game. A game might follow the story of an intrepid red blood cell, from its maturation on the bone marrow and its journey through the arterial and venous system. This character could be guided by an elderly and wise erythrocyte who sadly is phagocytosed in the spleen, leaving our hero or heroine to find their own (vascular) path through the rest of the game unaided.

Pedagogy and Implementation

Do not be so distracting by making the creative and fun elements that you neglect the core pedagogy of your game. For example, when constructing a game, consider how you can facilitate knowledge scaffolding for participants: by providing clues or hints, teacher feedback or help from peers.

Learning Outcomes

As with any learning activity, learning outcomes should be well defined and carefully constructed so that students can understand what is expected of them beyond the "playing" element. With serious games, the outcomes could be demonstrated by the students being able to complete the game. It may also be appropriate to incorporate briefing and debriefing elements to your game.

An Example of Gamification in Anatomy Education

The following example demonstrates the use of a gamified learning environment for first year medical students at the University of Hong Kong. It was styled as an escape room, where students worked together in groups of 6–8 and had to explore puzzles which included those relating to thoracic anatomy, surface anatomy and understanding of the coronary arteries in order to escape from a locked laboratory.

The narrative was that the they had been captured by a crazed professor and held hostage in his laboratory to be experimented on; they had to escape within an hour before he returned from a lecture. Figure 8.1 demonstrates the wireframes (simplified drawings) of the game elements and character design. Figure 8.2 gives a sense of the gamified environment.

In Fig. 8.2a, students had to answer questions on the surface markings on the thorax model, which was labelled with small fluorescent letters. Identifying the correct sequence of letters gave rise to a password which allowed students to move on to the next learning activity. Since the model could not be moved and the questions were in another room, students had to communicate with other group members in another room via a hand-held radio, testing their ability to describe anatomical positions with words only.

In Fig. 8.2b, questions relating to the coronary arteries could be found in the small chest of drawers on either side of the heart. Answering them by exploring the model generated a four-digit code, which was used to unlock a padlock on the wall, revealing the hidden path to escape.

For further details, please see this TEDx Talk entitled "Gamification in Higher Education" at https://youtu.be/d8s3kZz1yQ4?t=8.

Fig. 8.1 Wireframes (simplified drawings) of the game demonstrating the design of the game elements and character

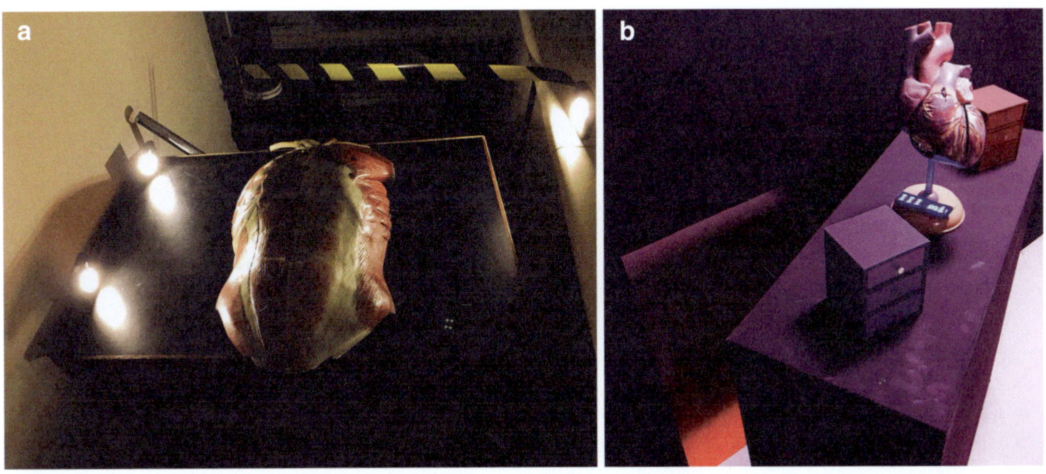

Fig. 8.2 Photographs of the gamified learning environment showing the puzzles on thoracic anatomy and coronary arteries

The Limitations of Gamification

Gamification is not a panacea for poor learning design. An ill-conceived or executed educational game can be a terrible chore for participants. There may be students who do not enjoy particular types of games or those who do not enjoy games at all. McDougall aptly draws the analogy of chocolate-covered broccoli for poor educational games which appear sweet but are destined to be disappointing [1]. Furthermore, whilst the benefits of gamification have been demonstrated in anatomy education, there is still a lack of conclusive evidence for improvement of long-term benefits for students [4] or unequivocal outcomes such as changes in clinical practice or patient outcomes [9].

Conclusions

Gamification is a potentially powerful tool for anatomy teachers to engage students in learning. It can be applied at a number of levels, from simple game elements to full-blown immersive educational games. The spectrum of complexity in gamification is broad, and digital and non-digital solutions for gamification may be considered. When used appropriately with careful design and iterative evaluation and refinement, gamification can produce highly engaging learning environments.

References

1. McDougall A. When I say… gamification. Med Educ. 2018;52(5):469–70.
2. de Freitas S, Liarokapis F. Serious games: a new paradigm for education? In: Ma M, Oikonomou A, Jain LC, editors. Serious games and edutainment applications. London: Springer; 2011. p. 9–23.
3. Landers RN, Landers AK. An empirical test of the theory of gamified learning: the effect of leaderboards on time-on-task and academic performance. Simul Gaming. 2014;45(6):769–85.
4. Dichev C, Dicheva D. Gamifying education: what is known, what is believed and what remains uncertain: a critical review. Int J Educ Tech Higher Ed. 2017;14(1):9.
5. Roberts DH, Newman LR, Schwartzstein RM. Twelve tips for facilitating Millennials' learning. Med Teach. 2012;34(4):274–8.
6. McIntire CL. Corpus Morphus: the human anatomy board game. Am Biol Teach. 1995;57(8):538–43.
7. Anyanwu EG. Anatomy adventure: a board game for enhancing understanding of anatomy. Anat Sci Educ. 2014;7(2):153–60.
8. Kapp KM. The gamification of learning and instruction: game-based methods and strategies for training and education. San Francisco: Wiley; 2012.
9. Akl EA, Kairouz VF, Sackett KM, Erdley WS, Mustafa RA, Fiander M, et al. Educational games for health professionals. Cochrane Database Syst Rev. 2013;3:Cd006411.
10. Subhash S, Cudney EA. Gamified learning in higher education: a systematic review of the literature. Comput Human Behav. 2018;87:192–206.
11. Graafland M, Schraagen JM, Schijven MP. Systematic review of serious games for medical education and surgical skills training. Br J Surg. 2012;99(10):1322–30.
12. Van Nuland SE, Roach VA, Wilson TD, Belliveau DJ. Head to head: the role of academic competition in undergraduate anatomical education. Anat Sci Educ. 2015;8(5):404–12.
13. Miller SA, Perrotti W, Silverthorn DU, Dalley AF, Rarey KE. From college to clinic: reasoning over memorization is key for understanding anatomy. Anat Rec. 2002;269(2):69–80.
14. Dillenbourg P. What do you mean by collaborative learning? In: Dillenbourg P, editor. Collaborative-learning: cognitive and computational approaches. Bingley: Emerald Group Publishing; 1999. p. 1–19.
15. Aba AB. Near-peer-assisted learning (NPAL) in undergraduate medical students and their perception of having medical interns as their near peer teacher. Med Teach. 2015;37(1):S33–9.
16. Tayler N, Hall S, Carr NJ, Stephens JR, Border S. Near peer teaching in medical curricula: integrating student teachers in pathology tutorials. Med Educ Online. 2015;20(1):27921.
17. Wood DF. Problem based learning. BMJ. 2003;326(7384):328–30.
18. Parmelee D, Michaelsen LK, Cook S, Hudes PD. Team-based learning: a practical guide: AMEE guide no. 65. Med Teach. 2012;34(5):e275–87.
19. Lave J, Wenger E. Situated learning: legitimate peripheral participation. Cambridge: Cambridge University Press; 1991.
20. Ngan OMY, Tang TLH, Chan AKY, Chen DM, Tang MK. Blended learning in anatomy teaching for non-medical students: an innovative approach to the health professions education. Health Prof Educ. 2018;4(2):149–58.
21. Stirling A, Birt J. An enriched multimedia eBook application to facilitate learning of anatomy. Anat Sci Educ. 2014;7(1):19–27.
22. Allen EB, Walls RT, Reilly FD. Effects of interactive instructional techniques in a web-based peripheral nervous system component for human anatomy. Med Teach. 2008;30(1):40–7.
23. Codd AM, Choudhury B. Virtual reality anatomy: is it comparable with traditional methods in the teaching of human forearm musculoskeletal anatomy? Anat Sci Educ. 2011;4(3):119–25.
24. Küçük S, Kapakin S, Göktaş Y. Learning anatomy via mobile augmented reality: effects on achievement and cognitive load. Anat Sci Educ. 2016;9(5):411–21.
25. Burleson KM, Olimpo JT. ClueConnect: a word array game to promote student comprehension of key terminology in an introductory anatomy and physiology course. Adv Physiol Educ. 2016;40(2):223–8.
26. Ang ET, Chan JM, Gopal V, Li Shia NJCA. Gamifying anatomy education. Clin Anat. 2018;31(7):997–1005.
27. Choudhury B, Gouldsborough I, Gabriel S. Use of interactive sessions and e-learning in teaching anatomy to first-year optometry students. Anat Sci Educ. 2010;3(1):39–45.
28. McCarroll ML, Pohle-Krauza RJ, Martin JL. Active learning in the classroom: a muscle identification game in a kinesiology course. Adv Physiol Educ. 2009;33(4):319–22.
29. Dickson KA, Stephens BW. It's all in the mime: actions speak louder than words when teaching the cranial nerves. Anat Sci Educ. 2015;8(6):584–92.

30. Herling PJ, Mohseni BT, Hill DC, Chelf S, Rickert JA, Leo JT, et al. Impact of anatomy boot camp on students in a medical gross anatomy course. Anat Sci Educ. 2017;10(3):215–23.
31. Janssen A, Shaw T, Goodyear P, Kerfoot BP, Bryce D. A little healthy competition: using mixed methods to pilot a team-based digital game for boosting medical student engagement with anatomy and histology content. BMC Med Educ. 2015;15(1):173.
32. Gauthier A, Corrin M, Jenkinson J. Exploring the influence of game design on learning and voluntary use in an online vascular anatomy study aid. Comp Educ. 2015;87:24–34.
33. Olszewski AE, Wolbrink TA. Serious gaming in medical education: a proposed structured framework for game development. Simul Healthc. 2017;12(4):240–53.

Anatomy Education to the Public

9

Adam M. Taylor

Background

Anatomy has a history of being in the public forum. The origins of public interaction with anatomy align with the history of medicine, dating back to Ancient Greece and dissection of human corpses [1]. It saw a resurgence in the Middle Ages with the public anatomical dissection of criminals by de Luzzi in the 1300s [2]. This public dissection of criminals continued in the centuries that followed [3], until the legislation saw changes in the legal system(s) and the way bodies were received by institutions [4].

In the twenty-first century, there is a growing demand for public engagement, knowledge exchange and better integration of academic institutions in their local environments. This shift has seen an increase in the amount of engagement activity, with academics expected to utilize their expertise in the field to deliver this activity. Any form of civic engagement should have purpose and some form of benefit to the community [5].

Anatomy and anatomists are often at the forefront of engagement for three reasons:

1. Anatomy is a visual subject, regardless of resources used to deliver the teaching [6, 7]. It is easy to see anatomy and many of the structures are large enough for the public to observe.
2. Every person on Earth has anatomy. Each person has (within reason) the same anatomical structures and these form part of their day-to-day lives.
3. Anatomy is intrinsically linked to health. Normal anatomy is taught to medical and allied health professionals and then abnormalities present with anatomical deviation from the considered norm [8, 9]. The public, often without realizing it, see normal anatomy in their day-to-day life: skin colour, eye colour, facial structure and many other externally visible things. It is the change in these external (and internal) structures that present as pathology, and when this occurs, these anatomical changes become of significance and can be used as a proxy for detrimental change in health.

Anatomy education has formed the cornerstone of medical curricula for centuries and continues to do so. In more recent times, it has also become central to dental and allied health professional curricula. Whilst the integrity of anatomy to these courses is key, the continual shift in curricular content has reduced the time for anatomy teaching [10, 11]. Many anatomists are now finding their knowledge base and expertise requested outside the traditional anatomy teaching spaces that are associated with their institutional courses.

A. M. Taylor (✉)
Lancaster Medical School, Lancaster University, Lancaster, Lancashire, UK
e-mail: a.m.taylor@lancaster.ac.uk

L. K. Chan, W. Pawlina (eds.), *Teaching Anatomy*, https://doi.org/10.1007/978-3-030-43283-6_9

Today, there is a growing demand for greater understanding of the body from the public, which is in part due to curiosity but also the continual need for better public health and understanding of science [8, 12–15]. The availability of plastic three-dimensional models, innovative imaging technologies, three-dimensional computer models and three-dimensional printing, as well as those exhibitions which bring cadaveric material into the public light, has resulted in a vast increase in the availability of resources that the public can engage with [16]. Historically, anatomical education and learning was undertaken by many of the pioneers of anatomy and medicine. In Victorian England, the exposure of the public to anatomical dissection was punishment for the criminals, rather than an educational opportunity for the public [4].

The need for public engagement and communication of science is a growing area and is partly driven by the availability of information on media platforms, much of which is not peer-reviewed or presented by experts. This causes problems with how science is interpreted and understood by the public. Specific high-profile examples include climate change and vaccination programs. These examples show the power of poor communication and the damage it can do to science and public health that lasts across generations.

Reasons for Engaging the Public

Engaging the public in anatomy education fills many desirable functions: it promotes the civic duties of institutions, it offers educational opportunities to attendees, and it can be used to inform curricular design and teaching necessities for students and present access to novel and harder-to-reach research participants.

There are multiple reasons for engaging:

1. Accountability of institutions to their local communities and beyond.
2. Values of the institution as a civic leader and altruistic approach to others.
3. Opportunities for individuals and groups who would find it hard to access expertise or equipment to be able to interact and learn and also opportunity for development and education opportunities for staff involved.
4. Relevance of institutional priorities to local and national areas of focus.
5. Career progression and strengthening networks for those involved. Many institutions now link engagement to probation, promotion and tenure.
6. Research opportunities to utilize public engagement as a tool for answering research questions and informing teaching.

Designing Objectives for Engaging the Public

Objectives of specific events should be broad enough to encompass all potential participants and be linked to the strategic objectives of the institution. The objectives should follow the EARS outline: explicit, appropriate, realistic and straightforward.

Designing the objectives may be as simple as asking yourself, "what one piece of information do you want participants to take away from the event and how to convey this information to them with the resources provided".

The designing and utilization of objectives ensures that there is a uniform dissemination of knowledge to participants. This works well in themed events, such as those around the heart or brain. Where the content of the engagement event is broader, the number of objectives may increase, and it may not be possible to cover all objectives with all participants, particularly in interactive sessions.

Topics which lend themselves to easy objectives for engagement events:

- The skeleton
- The heart
- The lungs
- The muscles of the upper limb
- The muscles of the lower limb
- The brain
- The upper gastrointestinal system

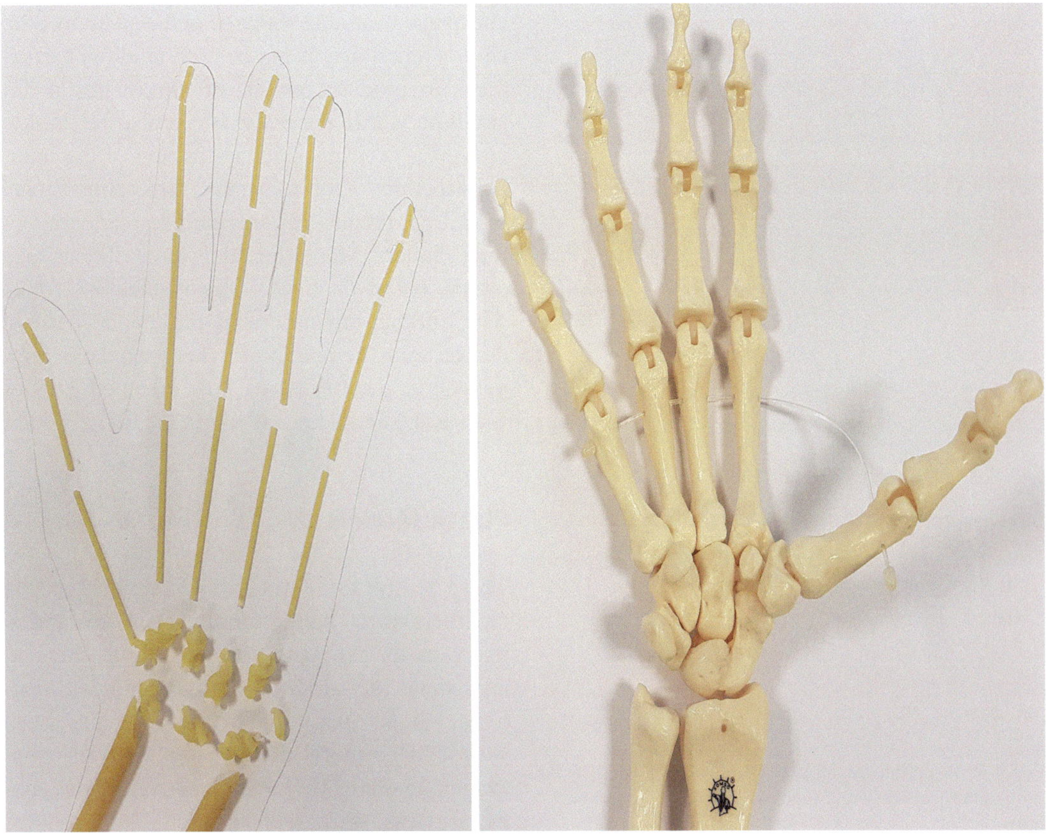

Fig. 9.1 An image showing the construction of a pasta hand, where a 5-year-old participant has drawn round her own hand and then used a plastic model as a template to place the pasta (the "bones")

Resources for Engaging the Public

There are a variety of potential resources that lend themselves to engaging the public. Although anatomy educators have a vast knowledge of the body, many members of the public may have a limited interest or knowledge and benefit from having visual and physical resources to interact with.

Craft-Based Stations

There are endless resources on the Internet demonstrating some of the craft-based stations that can be used to engage the public. Craft models are excellent for engaging children and give them something to take away. Simple paper cutouts of hands, with the opportunity to stick the metacarpal bones and phalanges onto them, in the form of pasta or straws, give an interesting and rigid hand (Fig. 9.1). If you utilize straws, then the threading of string through the straws, knotting it at the fingertips and taping down towards the wrist, demonstrates how the tendons assist in moving the fingers.

Questions you may pose to get participants to consider anatomical structure when making hand models:

- How many bones do you think there are in each hand?
- How many bones do you think there are in each finger? Count your own if unsure.
- How many bones are there in each thumb? Count your own if unsure.

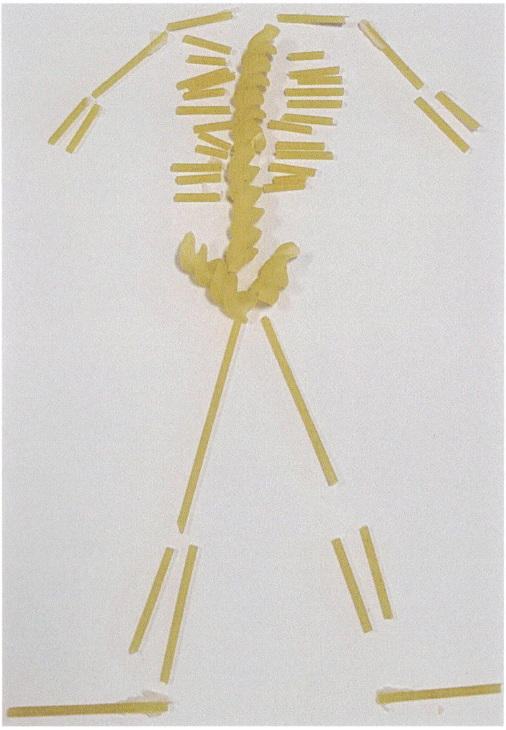

Fig. 9.2 A partial skeleton constructed by a 5-year-old participant from different types of pasta at an engagement event

Another inexpensive and engaging activity is the pasta skeleton. Images of skeletons can be left openly available for participants to view, and participants are given the opportunity to "build their own skeleton" using different types of pasta to be adhered to paper or cardboard. This activity gives the additional bonus of being able to demonstrate the properties, strengths and weaknesses of different bones as well as the basic structure of the skeleton. Fusilli pasta used for the vertebral column and pelvis replicates the strength of the vertebral column and pelvis in supporting other parts of the skeleton. Similarly, pieces of spaghetti are useful for long bones, demonstrating a small degree of flexibility but break on over-exertion (Fig. 9.2).

Play-Doh or Plasticine

Play-Doh or plasticine represents a time-tested toy and teaching tool; it lends itself to being a cheap and re-usable resource in the anatomy educator's repertoire. It can be used to allow participants to produce models of almost any major structure within the body. It can also be used to convert a two-dimensional drawing of the internal structures of the body into a three-dimensional model. Examples would include vertebrae or long bones. Modelling the three-dimensional organs based on a two-dimensional body map with various organs drawn on can be used to demonstrate the relationship between the organs, as well as demonstrating how cramped cavities such as the abdomen are within the body.

Plastic Models

These plastic three-dimensional models enable members of the public of all ages to see, and in many cases get hands on, to enable them to appreciate the structure of the displayed anatomy. Whilst this type of resource has been shown to be more effective for students' anatomical learning [17], the objectives of the public are less well defined. Many arise out of curiosity, or the opportunity to access resources that are unavailable to them in their day-to-day life or during their own education. The durability and size of plastic three-dimensional models make them ideal for engaging all ages of the public (Fig. 9.3).

Microscopy

Microscopes are useful resources for engagement; they are portable and give a clear and distinct look into the ultrastructure of many tissues. Many people will have had some exposure to these instruments during their formal education. The ability to demonstrate an individual cell, its basic organelles such as those that store genetic material and the extracellular matrix that exist around the cells are useful points for discussion and generating interest. Simple tissues such as connective tissues, bone, cartilage and tendon, as well as muscle, all provide easily recognizable structures and

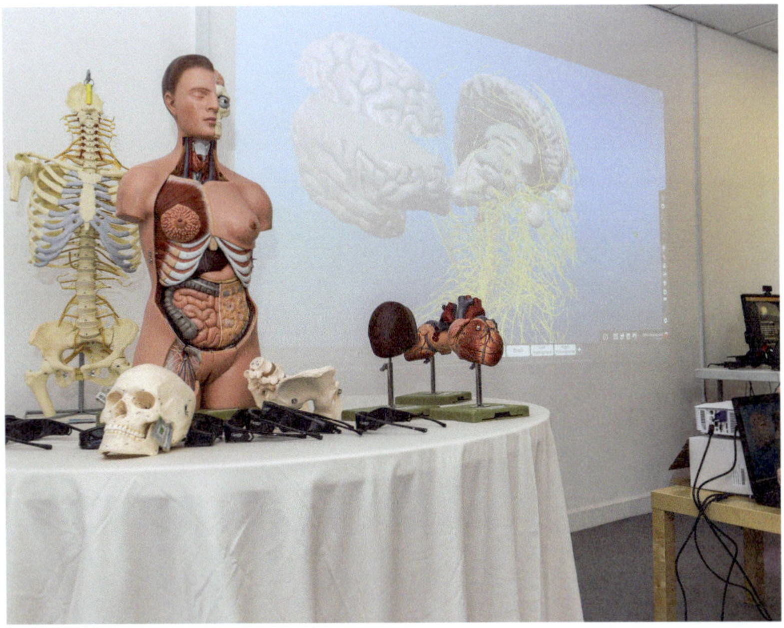

Fig. 9.3 A typical set-up of an engagement event, as part of "Campus in the City" (a key part of Lancaster University's outreach and public engagement program), using a blend of plastic models and an interactive three-dimensional projection system

functional considerations in day-to-day activities. One of the difficulties in using microscopy for engagement is the ability for the staff and participants to see the same thing at the same time. Having printed, laminated images of the tissue to talk through and show participants what to look for is beneficial (Fig. 9.4). A whiteboard marker enables you or the participants to label the laminated images and point to important structures. Having a ruler allows you to convert recognizable measurements of distance, such as millimetres, which are visible to the naked eye, to micrometres for cell size. This conversion enables participants to get an appreciation of the scale. For example, the educator can ask, "Did you know that the osteocyte that you are looking at in the bone is so small that you could fit up to 100 of them in a line, within the millimetre gradation on a ruler?" Alternatively, "The slice of tissue you are looking at is so thin that we could stack between 100 and 200 of them on top of each other and it would just fill the millimetre gradation on the ruler". Additionally, having pencils and paper available for them to draw their observations helps consolidate their learning and enables them to take something away from the experience.

Recent innovation has developed a smartphone microscopy platform, bringing microscopy to the fingertips of anyone with a mobile phone [18].

Interactive Media

The availability of software on computers, tablets and smartphones now makes anatomical knowledge widely available to non-specialist users. However, many of these are unable to give context to the information displayed to them. Three-dimensional projection systems give a "wow-factor" to engagement, with users of all ages keen to engage and immerse themselves in the human body (Fig. 9.5). Specialist software comes at an expense that can only be met by the department or institution and requires portability for off-campus engagement events [8]. Examples of this include Cyber-Anatomy 3D software, laptop and projector or smartphone-projected apps, such as the Enlightenment app which converts your smartphone into a microscope platform for education and engagement [18]. On-campus engagement events allow larger teaching resources, such as Anatomage and Sectra tables, to be available for attendees since these systems

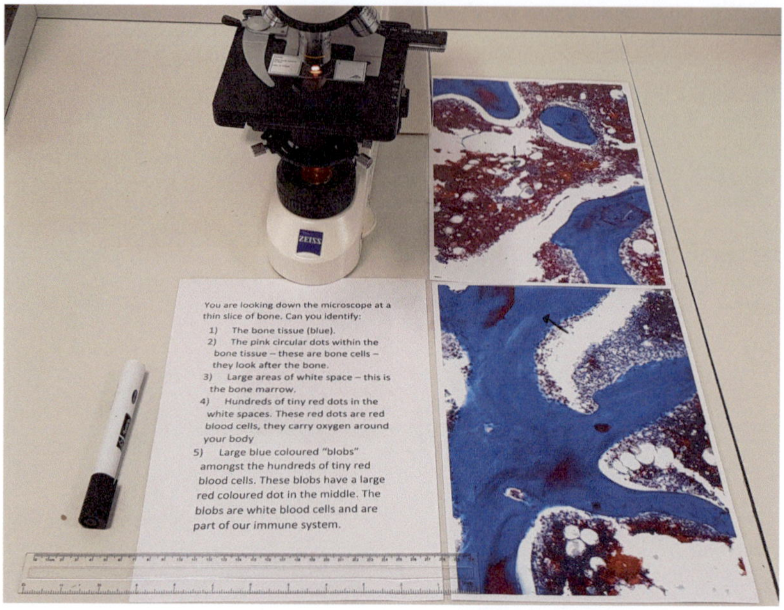

Fig. 9.4 A microscopy station at a public engagement event showing a microscope with accompanying laminated printouts for explaining observations to participants. Note the presence of the ruler which helps with explaining the scale of cells and tissues being observed and the whiteboard marker to allow annotations on the laminated images

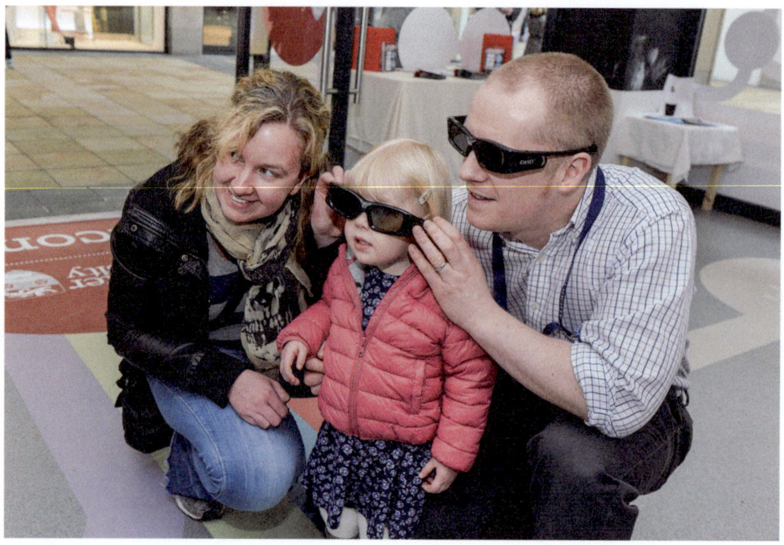

Fig. 9.5 Participants utilizing Cyber-Anatomy 3D Projector and software as part of "Campus in the City" (a key part of Lancaster University's outreach and public engagement program)

are typically impractical to move from designated teaching spaces.

When undertaking engagement activities around a specific area of the body or health-related area, it may be useful to have a video of the anatomy or clinical problem to introduce the public to the issue. These introductory vignettes make potential participants aware of the day-to-day context, or the clinical circumstances that may necessitate an understanding of this piece of anatomy as well as introduce them to something that may be of interest to them. The educator can then present the engagement activity and invite participants to be involved in them.

Rapid prototyping techniques, such as three-dimensional printing, have allowed the reproduction of normal and abnormal anatomical models that can be displayed in public, at relatively low cost [19–21]. The production of these models has given skeletal anatomy and more complex regional anatomy new resources for study [22, 23]. Small-scale printed models of vertebrae or long bones may also be useful for giving away to people as educational tools.

Some ultrasound machines and phantoms for teaching are portable and can be a useful and eye-catching public engagement tool. The clinical application of the machinery and the ability to "look inside the body" in real time makes ultrasound a resource with a wow-factor. Machines which have the Doppler function also enable participants to view blood flow through vessels or organs, bringing the functioning internal anatomy to life as part of any event. We have found that utilizing pre-screened volunteers, such as students, or ultrasound phantoms are the best way to utilize ultrasound. It is not recommended to screen willing volunteers on the day due to potential for injury or detection of abnormal anatomy or pathology.

Body Painting

Body painting has been demonstrated as an engaging and effective means to teach anatomy to students [24, 25]. The depiction of organs on the surface makes for easy recognition of location to any observer. The utilization of body paint enables additional understanding of the location of a structure in a moving person [24, 25]. Appropriate expertise and resources may allow for painting of attendees at the event (see Chap. 20).

Most recently, additional impact has been seen by utilizing ultraviolet body paints. This technique has recently been added to the medical education field, but caution has been advised that overuse of this technique lessens its impact in teaching settings [26]. This avenue of medical education and interactive anatomical drawing makes it ideal for high-impact engagement activity; the effects of being glow in the dark and the vibrant colours add to the eye-catching appeal for prospective participants.

Tips on designing activities and resources

- Think big. Most participants will be comfortable and familiar with organ-level size of structures.
- Find a way to catch your participants' attention.
- Pitch activities at the right level.
- Consider whether there can be something participants can physically take away with them at the end of the activity.

Ethical Issues

When undertaking public engagement activities, it is important to consider the ethical issues that may arise, separate to the requirement to have appropriate ethical approval in place for any research affiliated with the event(s).

There are likely to be smartphones present and so appropriate consent should be obtained for any pictures that are being taken, particularly by staff. Similarly, the public should be notified that photographs of exhibits or staff either are or are not permitted. This should be explicitly stated on signage around the venue. If activities involve tasks, it is important that any images taken by individuals protect the anonymity of anyone involved, particularly if any of the images are for research purposes.

There are ethical and legal governance frameworks in many countries regarding the use of tissues of human origin in public. It is important to ensure that all guidelines are followed and to ensure that there is no confusion. All publicity materials and materials at the events should contain information clearly stating the origin species of any tissues in use.

Venues

The opportunity for public engagement in anatomy can take place in many environments: on campus, in city centre venues such as bars and restaurants, libraries and many other places. Many institutions have dedicated departments, teams and budgets to help support engagement activities, including the identification of appropriate venues.

The benefit of undertaking engagement as part of a university- or institution-led initiative is the resources and support it brings. Venues, logistical arrangements and publicity in the run-up to and on the day of an event are often taken care of by other university staff, with anatomists left to focus on the content planning and delivery.

In deciding whether a venue is suitable for a public engagement event, it may be useful to consider the following aspects of the venue and the event:

- Size and layout: In a lecture-based event, does it need raked seating or a projection system for all to observe? If there are activities, can participants undertake them whilst others can still clearly see the educational resources?
- Seated versus standing: Is there adequate space for all those who need to sit to be able to do so, whilst allowing others to stand and observe, or take part in activities and move around resources freely and unimpeded?
- Structured activities versus stations for free interaction: Are there activities such as craft-based stations for participants to undertake or educational models or resources for them to observe and interact with?
- Availability of power supply: When any type of digital media is available, do you have sufficient power sockets and can access them safely with wiring not causing a hazard?
- Facilities: Where there are craft-based activities using things such as glue or Play-Doh, are there sinks for handwashing afterwards?
- Natural lighting, particularly if utilizing slides or interactive media: Are there large windows which may allow natural light to reduce the desired effect of any digital media used? Can you test this equipment using appropriate "blocking" of windows using paper or other materials before the event?

An example of public engagement event in the city centre:

Taylor and colleagues utilized an intuitional engagement program to undertake anatomical engagement in the city centre [8]: Campus in the City (CITC).

> It is a key part of Lancaster University's outreach and public engagement program, bringing Lancaster's world-class research and projects to life in an informal and accessible setting – a shop in the city center.
> CITC brings local people, academics and students together through a program of free and interactive activities, which explore a wide range of topics, including the human body and child development. Aiming to challenge, inspire and inform, it also provides opportunities for the local community to get involved with research projects, helping to shape the world around us and inform teaching and research.
> Academic contributors to the project are supported to develop activities which are engaging and interesting to the general public, whilst all the planning, logistics and publicity are managed by the University's public events team. 40% of academic contributors to CITC 2018 plan to use outputs from their activities for future funding bids.
> (Head of Stakeholder Relations and Events, Communications and Marketing, Lancaster University)

Funding

Funding support for engagement can be provided via institutions, as with the example above. Alternatively, many professional societies and charities recognize the value of public engagement to their members and subject areas and so provide funding for activities in these areas, usually for resources such as craft materials or consumables.

In funding a public engagement event, it may be useful to consider the following aspects of the event:

- Eligibility: member versus non-member. Some societies only provide funding to its members to run events; other societies will provide funding through a competitive process to anyone.

- Costs covered: resources, venue costs and refreshments. There may be some restrictions on what you can spend this funding on. These restrictions vary by funder.
- Time constraints: money to be used by a certain date. Funders may stipulate time limits on funding. Ensure that your event and claim for reimbursement of costs is submitted in time.
- Report: some funders require a report. If yours does, ensure that you are able to complete all post-activity analysis in time. Funders are keen to learn more about how the funding is spent and the impact of the event. This is usually done by submission of a report within a set number of days after the award or event. It is important to ensure any data is collected and collated into the report to demonstrate the impact of how the money has been used.

Assistance with Engagement

Undertaking engagement requires significant human resources in terms of number of people and person-power, which presents opportunities for students to become involved in the process and aid their own learning as well as that of others. The level of knowledge of the public on anatomy and health-related subjects is varied [8]. The level required to demonstrate anatomical structures to the public and explain basic function should be within the competencies of all medical and allied health professional students. This is, in part, an opportunity for peer-to-peer teaching, which has been shown to benefit students' knowledge of anatomy and other medically related subject material [27].

Where these students are from a healthcare program, such as medicine or an allied health professional course, where they will be interacting with patients, the benefits of being involved in anatomical engagement are multiple. The activities help students in learning how to and how not to use their medical language in communicating medical knowledge with the public. The opportunity also improves their knowledge, and all of these skills will increase their competence in their future profession(s) [28]. The insight into the thinking of students in these circumstances may also prove useful in understanding student perspectives on subject material that will later help educators in adopting explanations and examples of difficult concepts to take back into the classroom to assist with anatomical teaching to students.

Students can also be utilized to help participants move through interactive stations and to generate interest, but the opportunity for learning and teaching for them would not be as beneficial as having them actively involved. Students can also be helpful in "dummy running activities" to look for potential errors or difficult concepts that may occur in relation to the activities.

Evaluation of Public Participation

The evaluation is a key component of your engagement: it helps assess your impact and enables you to reflect and learn from every engagement experience undertaken.

Depending on the exact aims of the engagement activity, evaluation of effectiveness can be difficult to ascertain [29]. It is important to have clear aims for the engagement activity, and developing the evaluation criteria at the same time may be beneficial. Be aware that undertaking engagement activities as part of an institutional or public body engagement event may present different challenges in measuring effectiveness. Many of these larger events have their own feedback and evaluation measures relating to more strategic aims that may have reduced relevance to the exact content of your activity. It is important to approach discussions of collecting this information as early as possible to ensure any specific measures or criteria you require may be incorporated. This also helps in reducing the amount of feedback the public may be asked to give and can also reduce the workload on the day of the activity as dedicated individuals from public engagement teams will often take responsibility for collecting this information.

Incorporating Research into Your Engagement Activities

The involvement of the public in engagement activities also presents research opportunities. Fonseca and colleagues have shown that exhibits such as the "Body Worlds" have the effect of increasing the anatomical knowledge of the public attending these opportunities [30]. Similarly, Taylor and colleagues have shown that there is a demand from the public for anatomical knowledge and that their knowledge level is varied [8]. Incorporating research opportunities into engagement events, particularly those with the general public, gives a more representative reflection of the level of knowledge across the population.

Research can be undertaken in many forms, but one of the most common would be to evaluate how well an activity improves participant knowledge following involvement by using a pre- and post-test. Another common form is to have participants demonstrate their level of knowledge to inform research questions or build cases for student learning or curricular design.

Appropriate ethical review for incorporating research is a mandatory step in the process; planning early to meet committee deadlines and opportunities to address comments is particularly important. Ensuring you have appropriate research questions, methodology and all participant material prepared is key. It is also necessary to ensure appropriate consenting mechanisms are in place and to overcome any issues around anonymity. Finally, all legal, health and safety regulations must be met, which will vary depending on the type and size of event you are planning.

Tips in Seeking Ethical Approval

- Allow plenty of time for the committee to consider your application and for you to respond to comments or changes needed.
- Plan a clear research question.
- Have a robust methodology.
- Ensure all instructions are clearly articulated.
- Limit time spent on participant activities, to ensure a full data completion, rather than participants losing interest and giving up part way through, reducing the strength and validity of data collected.
- Adhere to all laws and legal governance around utilization of human tissues. When using tissues in the form of slides, teaching resources or specimens, it must be made clear that these are non-human in origin. The display and use of human tissues are strictly regulated in some countries; it is important to ensure that all legal and ethical frameworks are adhered to so that the donors' wishes are respected and the governing laws are followed.

During the Engagement Event

Invite people to interact with the resources. The overall purpose is meant to enable anyone to participate, and welcoming them to look or interact with resources may break the ice.

Ask participants if there is anything in particular that they were looking to see or learn about. This may be as simple as explaining what they are looking at. Others may come with a set interest in a particular region of anatomy or structure due to personal experience.

Avoid offering "medical" advice unless qualified to do so. There is a fine line between educating about anatomy and the body and offering medical advice. It is important to distinguish anatomical information from that which may be interpreted as medical advice by any participant. Be aware that participants may share their own medical issues with you, and this should be treated in confidence and discussion limited to relevant normal, rather than abnormal, anatomy.

Allow participants to take photographs of themselves interacting with the resources to allow them to reflect on and revisit their experience away from the setting. Caution and commonsense should be applied to appropriateness of photographs and the protection of others in and around the exhibits who have not consented to being in the image, particularly those participants who are children. Signs warning of the use of photography or prohibiting photography clearly marked around the event help to set the tone and manage expectations.

Conclusion

Anatomical education through public engagement is a key opportunity for staff, students and participants. Those staff and students have the opportunity to showcase educational content, research and activities that take their usual subject matter out into the wider forum. These events generate impact, knowledge exchange and accountability of institutions to their communities on a local, national and international scale.

To make the most of these opportunities, anatomists must be interactive, motivate people to participate, pitch information at the right level and welcome questions and feedback.

Participants do not necessarily come with the intention to learn, but many learn facts, figures and information that they potentially did not know existed. Individuals come to see content that they might be curious about or have never had exposure to. Engagement provides unique opportunities to disseminate academic work and initiatives more widely. There is a fundamental link between anatomy and engagement. Anatomists are in a unique position to undertake engagement. Their broad and in-depth knowledge of the body enables them to connect with any person on the planet because everyone has a body and at some point in their life, something changes that causes them to want to understand more about it.

References

1. Elizondo-Omaña RE, Guzmán-López S, García-Rodríguez ML. Dissection as a teaching tool: past, present, and future. Anat Rec B New Anat. 2005;285(1):11–5.
2. Mavrodi A, Paraskevas G. Mondino de Luzzi: a luminous figure in the darkness of the middle ages. Croat Med J. 2014;55(1):50–3.
3. Ghosh GK. Human cadaveric dissection: a historical account from ancient Greece to the modern era. Anat Cell Biol. 2015;48(3):153–69.
4. Mitchell PD, Boston C, Chamberlain AT, Chaplin S, Chauhan V, Evans J, et al. The study of anatomy in England from 1700 to the early 20th century. J Anat. 2011;219(2):91–9.
5. Stanton TK. New times demand new scholarship: opportunities and challenges for civic engagement at research universities. Educ Citizsh Soc Justice. 2008;3(1):19–42.
6. Wilkinson T. Primal pictures anatomy teaching resources: 3D anatomy software and 3D real-time viewer. J Anat. 2012;220(1):118–9.
7. Collett T, Kirvell D, Nakorn A, McLachlan JC. The role of living models in the teaching of surface anatomy: some experiences from a UK medical school. Med Teach. 2009;31(3):e90–6.
8. Taylor AM, Diggle P, Wessels Q. What do the public know about anatomy? Anatomy education to the public and the implications. Anat Sci Educ. 2018;11(2):117–23.
9. Drake RL. Anatomy education in a changing medical curriculum. Anat Rec. 1998;253(1):28–31.
10. Gogalniceanu P, O'Connor EF, Raftery A. Undergraduate anatomy teaching in the UK. Ann R Coll Surg Engl (Suppl). 2009;91:102–6.
11. Turney BW. Anatomy in a modern medical curriculum. Ann R Coll Surg Engl. 2007;89(2):104–7.
12. Lanie AD, Jayaratne TE, Sheldon JP, Kardia SL, Anderson ES, Feldbaum M, et al. Exploring the public understanding of basic genetic concepts. J Genet Couns. 2004;13(4):305–20.
13. Heijmans M, Uiters E, Rose T, Hofstede J, Devillé W, van der Heide I, et al. Study on sound evidence for a better understanding of health literacy in the European Union. In: Policies, information and services. European Commission. 2015. https://ec.europa.eu/health/sites/health/files/health_policies/docs/2015_health_literacy_en.pdf. Accessed 14 Feb 2019.
14. Chen GD, Huang CN, Yang YS, Lew-Ting CY. Patient perception of understanding health education and instructions has moderating effect on glycemic control. BMC Public Health. 2014;14:683.
15. Gillingwater TH, Findlater GS. Anatomy: back in the public spotlight. Lancet. 2015;385(9980):1825.
16. Singh D. Scientist or showman? BMJ. 2003;326(7387):468.
17. Preece D, Williams SB, Lam R, Weller R. "Let's get physical": advantages of a physical model over 3D computer models and textbooks in learning imaging anatomy. Anat Sci Educ. 2013;6(4):216–24.
18. Wicks LC, Cairns GS, Melnyk J, Bryce S, Duncan RR, Dalgarno PA. EnLightenment: high resolution smartphone microscopy as an educational and public engagement platform. Wellcome Open Res. 2017;2:107.
19. AbouHashem Y, Dayal M, Savanah S, Štrkalj G. The application of 3D printing in anatomy education. Med Educ Online. 2015;20:29847.
20. Wu AM, Wang K, Wang JS, Chen CH, Yang XD, Ni WF, et al. The addition of 3D printed models to enhance the teaching and learning of bone spatial anatomy and fractures for undergraduate students:

a randomized controlled study. Ann Transl Med. 2018;6(20):403.
21. Su W, Xiao Y, He S, Huang P, Deng X. Three-dimensional printing models in congenital heart disease education for medical students: a controlled comparative study. BMC Med Educ. 2018;18:178.
22. McMenamin PG, Quayle MR, McHenry CR, Adams JW. The production of anatomical teaching resources using three-dimensional (3D) printing technology. Anat Sci Educ. 2014;7(6):479–86.
23. Adams JW, Paxton L, Dawes K, Burlak K, Quayle M, McMenamin PG. 3D printed reproductions of orbital dissections: a novel mode of visualising anatomy for trainees in ophthalmology or optometry. Br J Ophthalmol. 2015;99(9):1162–7.
24. McMenamin PG. Body painting as a tool in clinical anatomy teaching. Anat Sci Educ. 2008;1(4):139–44.
25. Finn GM. Current perspectives on the role of body painting in medical education. Adv Med Educ Pract. 2018;9:701–6.
26. Finn GM, Bazira PJ, Bateman J, Sanders KA. Ultra-violet body painting: a new tool in the spectrum of anatomy education. Eur J Anat. 2018;22(6):521–7.
27. Bretelle F, Shojai R, Brunet J, Tardieu S, Manca MC, Durant J, et al. Medical students as sexual health peer educators: who benefits more? BMC Med Educ. 2014;14:162.
28. Sbayeh A, Qaedi Choo MA, Quane KA, Finucane P, McGrath D, O'Flynn S, et al. Relevance of anatomy to medical education and clinical practice: perspectives of medical students, clinicians, and educators. Perspect Med Educ. 2016;5(6):338–46.
29. Rowe G, Horlick-Jones T, Walls J, Pidgeon N. Difficulties in evaluating public engagement initiatives: reflections on an evaluation of the UK GM Nation? Public debate about transgenic crops. Public Underst Sci. 2005;14:331–52.
30. Fonseca GR, Finn GM. The impact of body worlds on adult visitors' knowledge on human anatomy: a preliminary study. Clin Anat. 2016;29:439–45.

10 Giving a Lecture

Lap Ki Chan

> So, evidence is mounting that readjusting the focus of education from information transfer to helping students assimilate material is paying off. My only regret is that I love to lecture. –Mazur [1]

A lecture has been defined as "a process by which information is transferred from the notes of the lecturer to the notes of the student without going through the minds of either" (by Sir Joseph Barcroft, cited by Book [2]). This humorous definition reflects the fact that some lectures are like monologues, possible even in the absence of students. The students do not actively take part in learning, and there is little time for them to think and reflect. There is little feedback to either the students or the teacher. But lectures do not need to be monologues. It is possible to deliver lectures that promote effective student learning.

What to Do Before a Lecture

Formulate the Intended Learning Outcomes

Adopt the Outcome-Based Approach

"It's not what we do, but what students do that's the important thing" [3, p., 19]. Many teachers will focus on what they need to do when they plan a lecture ("what we do"). Instead, they should focus on what the students need to learn and acquire ("what students do"). In the outcome-based approach, the aims of teaching and learning activities (in this case, a lecture) are explicitly defined as what students are expected to be able to do after the activities (i.e., the outcomes), and the teacher plans activities and assessments accordingly to help students achieve these outcomes. The aim of the assessments is to find out whether students are able to perform the actions described in the outcomes, at the desired level and context. This alignment of activities and assessments with the intended learning outcomes is called *constructive alignment*.

Write Intended Learning Outcomes

For a lecture, three to six outcomes are appropriate. Each of these outcomes should be expressed as a phrase beginning with a verb that describes an action that is observable and assessable.

L. K. Chan (✉)
Department of Biomedical Sciences, Faculty of Medicine, Macau University of Science and Technology, Macao Special Administrative Region, People's Republic of China
e-mail: lapkichan@gmail.com

L. K. Chan, W. Pawlina (eds.), *Teaching Anatomy*, https://doi.org/10.1007/978-3-030-43283-6_10

Bloom's taxonomy [4] provides a useful list of such verbs for cognitive learning, which are classified into the following categories:

1. Knowledge, which is no more than memorizing facts (such as *describe*, *define*, *state*).
2. Comprehension (such as *explain*, *interpret*, *demonstrate*).
3. Application (such as *apply*, *solve*, *generalize*).
4. Analysis (such as *analyze*, *compare*, *infer*).
5. Synthesis (such as *combine*, *compose*, *hypothesize*).
6. Evaluation (such as *compare*, *defend*, *criticize*). Lectures in general are more useful for achieving the lower levels of Bloom's taxonomy (i.e., knowledge, comprehension, and application). If a higher level of learning is desired, other forms of teaching and learning activity are usually more appropriate. The outcomes should be declared to the students at the beginning of the lecture, so that students have a clear idea of what they are expected to be able to do afterward.

Cross-Linking
See Chapter 3 "Constructive alignment: An outcome-based approach to teaching anatomy".

Determine the Content of a Lecture

Help Students to Achieve the Outcomes

The content of a lecture should help students to achieve its intended learning outcomes. All other goals are secondary. If previous assessment results show that students are not achieving the intended learning outcomes, the teacher should modify the teaching and learning activities until they do.

Put the Lecture in Its Context

The teacher needs to know how his/her lecture fits into the course or the whole program: what the students learned before that lecture, how it relates to students' later study, and how the lecture helps students to achieve the outcomes of the whole course or program. The students can thus build on what they have already learned, in the context of their future study. It also helps the teachers taking part in that course or program to collaborate to achieve the overall outcomes.

Avoid Overpacking

Good teachers arouse students' interest, organize complex information, and provide the framework on which students can construct future knowledge. Teachers are not simply providers of information. Moreover, overpacking a lecture with information may actually decrease learning. Russell et al. [5] found that students in lectures with low information density actually learned more than those in lectures that conveyed more information. They suggested that only half of the time in a lecture should be used for introducing new information and the other half should be devoted to explanations, reinforcement, and applications.

Rehearse

Rehearsals enable the teacher to time the prepared lecture, which should be shorter than and not equal to the allocated time, to allow for time for questions and other interaction.

If presentation slides are used, rehearsals help the teacher to become familiar with them, so that there will be a smooth flow from one slide to another. If the teacher is not completely familiar with each slide, he or she may not know how to make the transition from one slide to another. Such a lack of preparation reflects negatively on the teacher, who is now being led by the slides, instead of leading the slides and the lecture. Teachers who are familiar with the slides move from one slide to the next with a continuity of ideas. They can even jump from one slide to another in response to the audience's needs or questions (by pressing the slide number and then "enter"). But the aim of rehearsal is not to enable the teacher to recite the entire lecture from the memory. Instead, rehearsing allows the teacher to become confident, spontaneous, and responsive and closer to the learners. It also helps to reduce any stage fright.

Rehearsal should include testing the equipment for the lecture and preparations for possible unexpected events in a lecture, which can include those arising from problems with the projector (e.g., burnt-out light bulb, low graphic resolution), the computer (e.g., not displaying certain graphics and movies, no network connection), the lighting (e.g., too bright for showing fine photos, too dark for keeping students alert), the audio system (e.g., faulty microphone, noises from outside), and the students (e.g., student behavioral problems). Checking the projector, computer, lighting, and audio system before the lecture can prevent many of these problems. The teacher should prepare for "unexpected" situations that cannot be prevented (e.g., student behavioral problems) and think of appropriate ways to react to these situations to minimize disruptions to the lecture and avoid wasting precious time. This helps the teacher to remain composed and in control. It also keeps the teacher focused on teaching and the students on learning.

What to Do During a Lecture

How to Present Yourself

Show Enthusiasm

An enthusiastic teacher can arouse the students' interest and motivate them to learn more about the subject. Enthusiasm will also encourage a social response from the students—it would be rude not to pay attention when someone is talking to them enthusiastically. Your enthusiasm is reflected in your voice, eye contact, gestures, movements, and the way you interact with the students.

Speak Enthusiastically and Clearly

A lecturer can speak more effectively by:

1. Conveying enthusiasm. Variations in the tone and speed, among other things, convey enthusiasm.
2. Emphasizing points. You can emphasize important points by saying them more loudly, at a higher pitch; by lengthening certain syllables; or by pausing before and after the important words.
3. Speaking clearly. A good teacher should avoid halting speech, false starts, redundancy, and fillers (e.g., *um*, *er*, *ah*, *uh*, *eh*, *right*, *like*, *you know*). The best way to self-diagnose filler is to listen to a recording of one's own lecture.
4. Pausing appropriately. Pauses are important, not only for emphasis but also to leave room for students to think, to digest, and to raise questions. Pausing is also important after you ask a question. Most teachers do not pause long enough before they give the answers themselves. Such self-answering behavior indicates to the students that they do not need to answer your questions since they know you will give them the answers.
5. Being humorous. Appropriate use of humor can relax both the teacher and the student and shorten the perceived psychological distance between them.
6. Using a wireless microphone. When the audience is large, you need a microphone. However, speaking into the microphone fixed to the podium immobilizes and limits your performance on stage. A wireless microphone is much preferred.

How to Present Yourself

- Show enthusiasm.
- Speak enthusiastically and clearly.
- Make eye contact.
- Gesture.
- Walk around.
- Dress appropriately.

Cross-Cultural Considerations

The practices discussed in this and other chapters must be applied with great cultural sensitivity, especially in interactions between teachers and students coming from different cultures. For example, eye contact is not appropriate between men and women in some cultures. Different cultures also have different interpretations of what constitute enthusiastic behaviors, appropriate humor, gestures, movements, and attire.

Make Eye Contact

Appropriate eye contact instills enthusiasm and confidence and invokes social responses. The teacher should not stare at the screen, the computer, the podium, or the air above the students. However, the teacher should not always look at the same student, since that may make that student feel uncomfortable.

Gesture

Hand gestures can be used to emphasize important points. The left and right hands can be used in sequence to visually display "on the one hand" and "on the other hand." Head nodding conveys approval and can help students feel more confident in interacting with the lecturer. Opening your arms after a question invites answers. Leaning against the side of the podium can relax your students (and yourself). The size of the audience should be considered in using gestures. For smaller lecture venues, gesture with hands and forearms. For larger audiences, make slightly larger gestures than you would in ordinary conversation so that everyone can see them.

There are, however, some gestures that you should avoid, e.g., pointing at the students (people in general do not like being pointed at), holding onto the side of the podium, putting your hands in your pockets (both give the impression that you feel insecure), or any gestures and mannerisms that might distract students' attention, such as jiggling coins or keys in your pocket.

Walk Around

A podium can be compared to a turret during battle: it defends the speaker from the audience. Standing behind a podium transforms the relationship between the lecturer and the students into one of opposition. The lecturer should walk around on stage or among the students, so that students in different parts of the room get an equal share of the teacher's voice, eye contact, and enthusiasm. Walking around the stage or room will decrease the physical and psychological distance between the teacher and students. Students are more likely to listen to the teacher, respond to questions, and feel that the teacher is more approachable.

Dress Appropriately

Professional attire can lend credibility but can also establish a distance between the teacher and the students. If you are giving a series of lectures, you can wear more formal clothing for the first few lectures to establish your credibility and then switch to more relaxed attire when you want to decrease the perceived power distance between you and the students. But in any case, a teacher's appearance should not distract the students' attention from their learning.

Use Tools

There are several tools you can use to help make a lecture run more effectively:

1. A pointer allows the teacher to indicate the point of interest on the slide. The pointer can be a laser pointer, the cursor on the computer (if one is used for the lecture), or even just a finger. However, whatever type of pointer you use, it should be clearly visible on the screen, or else students will be too busy looking for the pointer to pay attention to what you are saying. Therefore, the laser pointer needs to be bright (but remember not to shine it at the audience!), and the computer cursor needs to be large and visible at all times.
2. A presenter (or remote presentation clicker) remotely controls the progression of computer slides. Some have built-in laser pointers or even timers. It is a very useful tool if the teacher wants to walk around during the lecture.
3. A visual presenter (or document camera) consists of a video camera connected to the projector or a computer. The teacher puts an object of interest, such as a bone, under the camera and projects it on the screen to illustrate points in the lecture. One difficulty of using this method is that the teacher must pay attention to the orientation of the specimen on the projector screen since what the students see on screen is quite different from what the

teacher sees when looking directly at the specimen.
4. An on-screen drawing device allows the teacher to digitally draw on the screen of the computer at the podium or even on the projected image. If PowerPoint is used on a PC, pressing "control-p" changes the cursor arrow to a pen for drawing on the screen. Pressing "control-a" changes it back to a cursor arrow. On a Mac, press "command" instead of the "control" key.
5. Audience response systems (ARS) or clickers allow all students in a lecture to indicate their responses to, say, a question. Students have access to a device that is either handheld and wireless or mounted on their seats. The students' anonymous responses will then be analyzed almost instantaneously. Such formative feedback allows the students to compare their performance to their peers and the faculty to identify at-risk students [6]. The anonymous nature of the responses motivates the reluctant students to interact during a lecture [7]. The responses of the students also allow the teacher to get an overview of the audience's understanding as the lecture progresses, thus enabling him or her to adjust the lecturing strategies, e.g., by spending more time on any parts of the lecture that students have difficulties with.

What to Do in Different Parts of a Lecture

A lecture can generally be divided into three parts, introduction, body, and conclusion [8, 9], each with different aims.

The Introduction of a Lecture

The first few minutes of the lecture are very important since you establish your relationship with the students and prepare them for learning. A self-introduction will assure the students of the teacher's knowledge of the lecture content if it is the first meeting between the teacher and the students. An informal conversation with the students will relax both the teachers and students and is a good way to establish a closer relationship. In the introduction, the teacher should state the lecture topic. A short review of how the current lecture relates to previous ones can trigger relevant prior knowledge, thus facilitating knowledge building. The intended learning outcomes should be stated. People learn better if they know what they are expected to achieve [2]. For some complex lectures, a framework at the beginning can help students to organize the materials.

The Body of a Lecture

The body of the lecture needs to be well organized and presented. Some strategies for successful presentation of a lecture are:

1. *Divide it into key points.* Brown and Tomlinson [10] describe three methods of organizing the materials in a lecture in the medical and health sciences. In the *classical method*, a lecture is divided into sections and each section into subsections, each with its own key points. In anatomy lectures delivered using this method, a region is divided into subregions or a structure into its parts, each with its own elaboration and summary. The *problem-centered method* starts with a problem that forms the focus of the lecture, with the solutions forming the parts of the lecture. This method can be intellectually stimulating and motivating, can stimulate students to reflect on their prior knowledge, and can be used for illustrating the clinical aspects of anatomy. The *sequential method* consists of a series of linked statements, eventually leading to the conclusions. This method is commonly used in lectures on management of clinical problems. These three methods of organizing a lecture are not mutually exclusive. One can have a classical lecture, with one of the sections being organized by the problem-centered method or the sequential method. Dividing a lecture into parts may make it more manageable, but the linkage between the parts must be made very clear to the students.
2. *Elaborate the key points.* A well-delivered lecture should not be completely packed with key points. Each key point needs elaboration,

which can be explanations, examples, stories, qualifications, photos, applications, etc. Such elaborations help the students to learn by repeating and reinforcing the new information, by presenting the same information in different forms, and by linking with students' prior knowledge. Furthermore, if the elaboration involves applications in students' future study, students are more likely to be motivated to learn, and they are more likely to remember the information when they come across the same context in the future.

3. *Link the key points.* The linking of the key points needs to be made explicitly to the students, by using such words as "thus," "therefore," and "consequently." For an audience of nonnative English speakers in a lecture delivered in English, these linkages need to be clearly signposted by using such phrases as "the above is true because…" rather than more subtle markers like "because," "then," "since," etc. [2]. The linked chain of key points can help the students to understand the big picture.
4. *Manage the pace.* The pace of a lecture is proportional to the amount of material to be covered and inversely proportional to the time available. The more materials to be covered and the less time that is left, the faster the teacher will deliver the lecture. As the pace increases, the teacher tends to sacrifice elaboration in order to deliver all the key points. The best way to avoid getting into this situation of being forced to go fast at the expense of student learning is to avoid overpacking the lecture.

Parts of a Lecture

Introduction (not sequential steps)
- Establish a relationship with students.
- Introduce yourself.
- Announce the topic.
- Relate the lecture to previous ones.
- State expected outcomes.
- State the framework of the lecture.

Body (not sequential steps)
- Divide into key points.
- Elaborate the key points.
- Link the key points.
- Manage the pace.

Conclusion (not sequential steps)
- Summarize.
- Review intended outcomes.
- Facilitate self-directed learning.

The Conclusion of a Lecture

A lecture should not end abruptly. The attention of the students is usually the highest in the last few minutes of the lecture. The teacher should summarize the key points and explain how the lecture should have helped the students to achieve the outcomes. The students may be expected to use the lecture materials as a foundation for further self-directed learning. The conclusion of a lecture is the appropriate time to state this idea explicitly. The teacher can provide problems or resources to help the students.

Interaction Between the Teacher and Students

Interactions are possible and should be encouraged in lectures. But students are more likely to interact with teachers who are perceived as having less physical and psychological distance from them (called *teacher immediacy*), i.e., those who speak enthusiastically, walk around in the lecture hall, make eye contact, adopt open gestures, etc.

Interactions with Students
- Questions
- Think–pair–share
- Write–pair–share
- Demonstration
- Role-playing
- One-minute paper

Interactions can renew listeners' attention spans and can let students actively participate in the learning process. Interactions also provide feedback to the teacher on how well the students are learning, so that the lecturing strategies can be adjusted in time. More importantly, interactions can help the students to achieve some of the intended learning outcomes (ILOs). For example, if one of the ILOs is for the students to be able to explain the function of certain muscles, the lecture should include some activities in which at least some of the students explain the functions of at least some of these muscles. This will demonstrate to the students what they are expected to learn and be able to do. Some types of interaction are described below:

1. *Questions.* The simplest interaction between the teacher and students is questioning. Questions can be used by the teacher to check student understanding, to trigger active thinking, and to stimulate students to reflect on prior or newly acquired knowledge or be used as the lead-in for a new section or even a whole lecture. But after asking a question, the teacher should pause for at least 10 s, to allow students to think and prepare the answer. If there is no response, the teacher should restate the question differently or more specifically. The teacher can move closer to the students, lean forward, and open his/her arms to invite answers. When an answer comes, the teacher should be supportive and nonjudgmental (verbally and in your facial expression and gestures), even if the answer is wrong, in which case the teacher can guide the students to the correct answers by asking the appropriate questions. The anonymity of responders in audience response systems can motivate more students to respond to questions.

   Questions raised by the students should be encouraged. When a student asks a question, repeat the question through the microphone to the whole class, so that everyone can think about it. If you do not know the answer to the question, you can say that you need more time to think or even put it back to the students.
2. *Think–pair–share* and *write–pair–share*. The teacher first asks a question which the students need to think about. Then the students form groups of two to exchange their responses. After 2 or 3 min, the teacher asks for the responses from some or all of the groups. Write–pair–share is similar except that the individual students write down their thoughts before discussing them with their neighbors.
3. *Demonstration.* Demonstration can take many forms. In anatomy teaching, the teacher can demonstrate surface anatomy on his/her own body or on another volunteer student or use tools to illustrate the anatomical relationships being discussed (see Chap. 29). In the teaching of some diseases, a demonstration can be an interview or physical examination of a patient.
4. *Role-playing*. The teacher and the students can both take part. Role-playing can be used for teaching communication skills, ethical issues, or clinical manifestations of illnesses.
5. *One-minute paper*. In the last few minutes of a lecture, the teacher asks the students to write down on a piece of paper their answers to such questions as follows: (a) What are the most important points that you learned today? (b) What are the most confusing points? (c) Are there points that are not clear to you? There are many variations. This is a method for the teacher to find out how well the students have been following the lecture and which parts of the lecture they have difficulties with, so that some of these difficulties can be addressed in future lectures, if necessary.

How to Prepare Presentation Slides

Presentation programs (such as PowerPoint and Keynote) are commonly used for lecturing. But it is important to stress that these programs are just aids to the teacher, whose messages should be the focus of the lecture. Therefore, the preparation of these presentation slides must be carefully done so as not to distract the audience from the teacher's messages [11, 12]. In fact, some anatomy teachers do not use presentation slides at all. Tufte argued that programs like PowerPoint should be abandoned because their linear and hierarchical structure and preoccupation with format and decoration make them inherently unsuitable for rich and complex content [13]. But PowerPoint or other similar programs are just tools. If the user uses these tools correctly, they can be helpful [12]. The most important thing is to keep the slides simple. However, making slides simple does not mean making them devoid of content. It means that all the elements of the slides must help the students to learn and there are no unnecessary or poorly constructed elements to distract the students. It is important to maintain the focus on student learning and not be tempted to add extras such as unusual typefaces and animated elements, which are more distracting than helpful to learners.

Background

The main function of the background is to keep the attention of the students on the slide content. Therefore, a solid color background is the best. A picture should not be used as a background. When a picture must be used, it should be dimmed or blurred. One should use a consistent background for a set of slides, since changing the background could be distracting. Using a light-colored background in a dark lecture hall will attract students' attention.

Text

Short Bullet Points

The messages of the lecture should be delivered by the teacher, not by the presentation slides. Therefore, the text in a presentation slide should serve only as textual anchors of what the teacher is delivering and should be very short (no complete sentences), in bullet points arranged to reflect conceptual structures. But the "six-by-six" rule, which states that a slide should contain at most six lines of text, each with at most six words, should not be taken too literally in presenting complex content to students. On the other hand, one should not put everything one wants to say on the slides and use them as script (a presentation style jokingly referred to as "PowerPoint karaoke"). Not only do slides presented this way become densely packed, but in reading material off the slides, one's voice becomes monotonous and lacking in enthusiasm. However, teachers may want to include more text in the slides if the presentation is in a language that is not the usual language of the teachers or the students.

Simple and Consistent Text Format

The text in presentation slides should not be distracting. Therefore, a teacher should use only one or at most two fonts. Sans serif fonts are preferred, since serif fonts may not display well, especially when using projectors with lower image resolutions. The point size should be of adequate size (at least 32 points, but it needs to be even larger if the lecture hall is large), and the text should contrast with the background and should be in only one or two colors, or else students will be distracted from the lecture content by focusing too much on trying to read the text or to decode the meaning of the colors. Incorrectly spelled words can also be distracting and indicate that the teacher is careless. Text that is written all in upper case should also be avoided since it is more difficult to read and gives the impression of the speaker shouting at the audience.

Platform Compatibility

If the slides are created on a Mac but will be presented on a PC (or vice versa), it is best to use a font which is available on both platforms, such as Arial, Times New Roman, or Courier. But even these fonts may be rendered slightly differently. Therefore, the placeholder (the box into which you enter text) should not be too tight, or else the alignment of the text may be affected. If other fonts, especially nonstandard ones, are used in the slides, one must make sure that the computer used for the presentation also has those fonts installed. Otherwise, the nonstandard font will be replaced by another font, and text may be misaligned. Misaligned text will distract students from the lecture, since they will need to mentally realign the text.

Multimedia

The use of multimedia, especially graphics, is important in anatomy lectures. The *multimedia principle* of learning says that people learn better when the instructional messages are delivered in both words and graphics, rather than in words alone.

Use Good and Relevant Graphics

Use high-quality graphics that can help students to achieve the intended learning outcomes. The graphics should be of high resolution, have good colors and contrast, be properly aligned and cropped, and be appropriately labeled to show the structure of interest (with the other non-relevant labels removed or covered). Their sources should also be acknowledged in a non-distracting manner. Decorative graphics are distracting and should be kept to the minimum.

Elaborate and Progressively Reveal Complex Graphics

The interpretation of complex anatomy graphics takes time. The teacher should help the students to comprehend anatomy graphics by first stating the directions of the three-dimensional axes and what structures have been sectioned, displaced, or removed to expose the structures in the pictures. When there are many labels, the teacher should reveal only those that are pertinent at the moment. The teacher can then reveal the others as the lecture moves on. This method of progressively revealing a complex anatomy graphic helps students to stay focused.

Avoid Animation

Excessive use of animation should be avoided, since it takes up precious lecture time, distracts the students without helping them to learn, and can be irritating.

Use Video

Video can be more powerful than static graphics in reinforcing student learning, e.g., in showing the anatomy of clinical procedures or the anatomical basis of some diseases. Many relevant videos are on the Internet and can simply be embedded into the presentation file as links without downloading (but make sure that the presentation computer has Internet access). Nevertheless, teachers should be aware of the massive amount of information presented in videos that may distract students from key learning points. A video that is suitable for more advanced students may overwhelm novices and not facilitate learning.

Avoid Chart Junk and Junk Charts

Chart junk is elements of a chart that do not contribute to its message and therefore serve only to distract. Junk charts are charts that are poorly designed to convey their intended message. For example, they may be unnecessarily complicated, difficult to read, impossible to understand, or just odd. They should be avoided.

Slides

Integrate Text and Multimedia

It is common for teachers to put text on one side of a slide and graphics on the other. But in the seconds spent on the slide, students will need to mentally integrate the text and graphic by reading the text and then finding the relevant parts in the graphic, all while they are listening to the teacher. Integrating text and graphics means doing this mental integration for the students by putting the parts of the text next to the corresponding parts of the graphic. It helps student learning by saving the students' cognitive resources, so that they can focus on the lecture content. This recommendation is called the *contiguity principle* [14].

Explain Graphics Orally

The *modality principle* tells us that students learn better when the graphics are explained in spoken words than with written text accompanying the graphics [14]. The working memory has a limited capacity and has one storage area for visual information and another for phonetic information. When graphics are explained orally, both the visual and phonetic areas are being utilized, thus gaining maximum access to the working memory. The *redundancy principle* says that when a graphic is explained orally and is accompanied by textual explanation, student learning may be impaired [14]. The reason is that the graphic and the textual explanation are both jammed into the visual component of the working memory, which may thus be overloaded.

Leave Space

Overpacking the slides does not help student learning. If the teacher wants to provide additional materials, he/she can put them in a separate handout or in the "Notes" area of the slide in PowerPoint.

Use a Consistent Layout

The layout refers to the position of the text, graphics, titles, etc., on the slide. Using a consistent layout will help students to adapt to your presentation quickly. Every time the layout is changed, the students need to adapt, which distracts them from learning.

Use B and W Keyboard Functions

To focus students' attention totally on what the teacher has to say, temporarily black out the projector or press the "b" or the "w" key in presentations using PowerPoint and Keynote. Pressing "b" will turn the whole screen black while pressing "w" will turn it white. When the slide suddenly disappears, the students' attention will immediately be directed to the teacher.

Preparation of Simple Slides

Background

- Use solid color background.
- Avoid using pictures.
- Use one background throughout a presentation.

Text

- Use short bullet points.
- Use one or at most two sans serif fonts.
- Use only one to two colors; avoid bizarre colors.
- Use a large font (at least 32 point size).
- Use fonts that are available on both PC and Mac platforms.
- Run spell check.
- Avoid putting everything you want to say onto the slides.
- Avoid all caps.

Multimedia

- Use clear, simple, and relevant graphics.
- Elaborate and progressively reveal complex graphics.
- Avoid animation.
- Use video appropriately.
- Avoid chart junk and junk charts.

Slides

- Integrate text and multimedia.
- Explain graphics orally.
- Leave space.
- Use a consistent layout.
- Use B and W.
- Avoid animated slide transitions.

Avoid Fancy Slide Transitions

Presentation programs offer many different ways to go from one slide to another. Such fancy transitions serve only to waste precious lecture time and distract the audience.

Conclusions

Lectures, when appropriately delivered, can promote effective student learning. During preparation of a lecture, an outcome-based approach helps the teacher to plan the content and activities to help students achieve the intended learning outcomes. The teacher also needs to consider the relationship of the lecture to the rest of the course or program so that the lecture builds on what students have learned and prepares them for further study. In the introduction of a lecture, the teacher needs to establish a closer relationship with the students and prepare them for learning. In the body of a lecture, the teacher needs to organize the content into key points and link them to give students the big picture. But the teacher must not pack too many key points into a lecture, sacrificing elaborations. There should also be an effective conclusion in which the teacher can summarize the key points and outcomes and stimulate further self-directed learning. Apart from the careful organization of the lecture content, the teacher's enthusiasm will also significantly affect student learning and attitude toward the subject, and it is reflected in the way that the teacher speaks, moves, makes eye contact, and interacts with the students. The teacher also needs to prepare the presentation slides carefully, if he/she chooses to use them, so that they do not distract the students from the content and the interactions with the teacher.

References

1. Mazur E. Farewell, lecture? Science. 2009;323:50–1.
2. Book CL. Lecturing. In: Vangelisti AL, Daly JA, Friedrich GW, editors. Teaching communication: theory, research, and methods. Mahwah: Lawrence Erlbaum Associates; 1999. p. 333–46.
3. Biggs J, Tang C. Teaching for quality learning at university. 3rd ed. Berkshire: McGraw-Hill; 2007.
4. Bloom BS, Krathwohl DR, Masia BB. Taxonomy of education objectives: handbook I. cognition domain. New York: Mackay; 1956.
5. Russell IJ, Hendricson WD, Herbert RJ. Effects of lecture density on medical student achievement. J Med Educ. 1984;59:881–9.
6. Alexander C, Crescini W, Juskewitch J, Lachman N, Pawlina W. Assessing the integration of audience response system technology in teaching of anatomical sciences. Anat Sci Educ. 2009;2:160–6.
7. Collins J. Audience response systems: technology to engage learners. J Am Coll Radiol. 2008;5:993–1000.
8. McLaughlin K, Mandin H. A schematic approach to diagnosing and resolving lecturalgia. Med Educ. 2001;35:1135–42.
9. Dent J. Lectures. In: Dent JA, Harden RM, editors. A practical guide for medical teachers. Edinburgh: Elsevier Churchill Livingstone; 2005.
10. Brown G, Tomlinson D. Improving lecturing. Med Teach. 1979;1:128–35.
11. Holzl J. Twelve tips for effective PowerPoint presentations for the technologically challenged. Med Teach. 1997;19:175–9.
12. Harden R. Death by PowerPoint—the need for a 'fidget index'. Med Teach. 2008;30:833–5.
13. Tufte E. The cognitive style of powerpoint: pitching out corrupts within. 2nd ed. Cheshire: Graphics Press LLC; 2006.
14. Clark R, Mayer R. E-learning and the science of instruction: proven guidelines for consumers and designers of multimedia learning. San Francisco: Wiley; 2011.

11 The Flipped Classroom: Starting with the End in Mind

David A. Morton

Introduction

Imagine walking into your music performance class with your trumpet under one arm and the new piece of music just handed to you under the other. You start talking with your neighbors as you place the trumpet across your lap and start skimming through the sheet music. Your teacher greets the class and spends the 50-minute period teaching you and your fellow performance students about the piece of music, the key changes, the dynamics, and how the rhythm of alternating quarter and sixteenth notes should sound. The bell rings, the class period is over, and you never once put the trumpet to your lips.

This seems like a missed opportunity, doesn't it? You and your classmates are music performance students after all, and it seems rather silly that you never had a chance to perform for your instructor and that he did not have a chance to provide feedback on your tone, dynamics, and ability at "hitting the right notes." Consequently, it is not hard to imagine the frustration that occurs as you practice the new music for the first time at home and realize that "*talking* about the right notes in the right rhythm" is a world away from actually "*playing* the right notes in the right rhythm." And how do you know if you are playing it right or wrong? And if you are playing it wrong, how would you know? Does this not seem odd that during the most crucial time of performance preparation (the practicing) the instructor is not around to assist? Additionally, the first time you will have the opportunity to play this new piece for your instructor is on Monday … when your music class performs a concert for the school. Isn't receiving feedback from your teacher on your concert performance after the performance a little too late to know what could have been done to better prepare and improve for that session?

Now imagine you are a medical student walking into anatomy class with your laptop under one arm and a PowerPoint (PP) lecture that was just provided by the professor under the other. You start talking with your neighbors as you place the laptop on your desk and start skimming through the PP slides. Your anatomy professor then starts class and spends the 50-minute period teaching you and your fellow medical students about heart chambers, valves, great vessels, blood oxygen content, and directional flow. The bell rings, the class period is over, and you never once had a chance to practice putting this new anatomy knowledge to your brain.

This seems like a missed opportunity, doesn't it? You and your classmates are medical students after all, and it seems rather silly that you never had a chance to apply this new anatomy

D. A. Morton (✉)
Department of Neurobiology and Anatomy,
University of Utah School of Medicine,
Salt Lake City, UT, USA
e-mail: david.morton@hsc.utah.edu

L. K. Chan, W. Pawlina (eds.), *Teaching Anatomy*, https://doi.org/10.1007/978-3-030-43283-6_11

knowledge in the context of medicine and receive feedback from your professor. Consequently, it is not hard to imagine the frustration that occurs as you try studying heart anatomy for the first time at home and realize that "*talking* about the heart" is a world away from actually "*solving* clinical problems relating to the heart." And how do you know if your answers to questions about the heart are right or wrong? And if your answers are wrong … how would you know it? Does this not seem odd that during such crucial moments of learning (the application) the professor is not around to assist? Additionally, the first time you will have the opportunity to show your newly learned anatomy knowledge is on Monday … when you and your fellow medical students take a test on the heart. Isn't receiving feedback from your professor on your heart anatomy knowledge after the test a little too late to know what could have been done to better prepare and improve for that test?

> Anatomy in medical school should be like a music class, with the majority of the time spent "practicing" the material and not just "talking" about it.

How can the time professors and medical students spend together become more about "practicing" the material and receiving feedback from the professor? An answer to this question is the flipped classroom.

What Is the Flipped Classroom?

The flipped classroom (FC) model is a teaching method that takes advantage of technology and promotes students to play a more active role in their learning. Student involvement in the learning process helps not only with memorization of anatomy information but also in the application of anatomy in the form of solving problems [1]. In essence the FC does two things:

I. Moves the traditional in-class lecture (delivery of content) to outside of the classroom
II. Moves activities that have traditionally been considered homework (practice problems) into the classroom

The FC method also caters to the principle that students become better at what they practice doing [2]. As such, if we want medical students to be better problem-solvers, then does it not make sense that we should help them practice solving problems with our content? Therefore, in the FC some of the time that would have been spent by the student studying after class is flipped to occur before the in-class session (hence, the name the "flipped-classroom") in order to prepare them to apply their learning through problem-solving activities. Research demonstrates the following benefits of the FC:

- Improvement of higher-order thinking and problem-solving skills [3]
- Identification by the professor of student understanding of course content [4]
- Promotion of peer-to-peer and professor-to-student interactions [5]
- Development of analytical skills and application of course content rather than memorization of general facts [6]
- Engagement of students in the class, course content, and interaction with peers [7]

The FC puts the focus of educational responsibility on the students. In other words, the spotlight is on the student, not the professor.

> The flipped classroom moves some of the time students were already using to study the material after class and flipping it to occur before class.

How Does the Flipped Classroom Work?

The FC consists of the following three parts: pre-class work, in-class work, and post-class work.

Pre-class Work

The pre-class work is where the initial delivery of content occurs and is to be completed by each student before the appointed classroom session. In order to effectively prepare pre-class work for the students, it is necessary to have the learning objectives detailed. Learning objectives:

- Describe what students will learn after completing the session
- Link tightly with the pre-class work
- Provide the scaffolding upon which all content for that session hangs
- Prepare the students for assessments

The pre-class work provided to the students teaches the material that would have been originally covered through a classroom lecture. Most pre-class work consists of one of the following:

- *Video Tutorials:* The ease with which video is accessed and viewed through the Internet has made it so the FC model has come to be identified with it. Students prefer video tutorials because they control the speed of their learning. If Max does not know a word, he can pause the video and find a definition; he can stop the video to take notes and rewind if he missed concept or fast-forward if already knows a concept.
- *Assigned Reading:* Pre-assigned reading most often comes in textbook pages or class notes. The content within textbooks is vetted, especially if it has undergone a number of editions. The images are numerous, professional, and tightly linked with the text. However, students must have access to the textbook, and sometimes the content does not dovetail as nicely with the courses learning objectives. For example, when teaching a topic, you emphasize "A" but not "B," but the textbook does not cover "A" and emphasizes "B" as well as "C," "D," and "E." Ultimately, if assigned reading is used for pre-work, it must knit tightly to the learning objectives and ultimately the active learning exercises used during the in-class session. Assigning dozens and dozens of textbook pages to students is less effective for a couple of reasons. Firstly, students are novices at the material and do not yet have the understanding to know what information is important and what is unimportant. Secondly, it is not feasible for a student to learn the information covered over 25 textbook pages for a 50-minute class [8]. Remember, this is only one class session of the day. The students may have 5 or 6 additional hours of large classroom content that day. If each class assigned 25 pages from the textbook to prepare for class, that could be over 125–175 pages of textbook reading per day. Therefore, keep readings concise (under 11 pages), targeted (i.e., read pages 126–129 and 150–152), and closely linked to the learning objectives.

Completing the pre-work in the FC does not require that students become experts on the topic. However, students are expected to have experienced focused and concentrated study of the content at least once prior to time with the professor. As such, the up and coming class time becomes a time for students to retrieve knowledge they have tried to learn—and that is where the power of the FC comes.

> The pre-class work provided to the students (usually in the form of videos) teaches the material that would have been originally covered through a classroom lecture.

In-Class Work

The in-class session is where the application of content occurs and is done with the class and professor together solving problems. The value of a FC is in the repurposing of class time into a space where students can ask questions, test their skills in applying knowledge, and interact with one another. During in-class sessions (sometimes called face-to-face time), the professor functions as a facilitator encouraging students to collaborate together in activities that promote the application and synthesis of the material covered in the pre-class work.

Types of In-Class Activities

Multiple-Choice Questions The students are not expected to be experts at this point in their learning as they have just completed the pre-class work and still quite novice in their understanding. MCQs provide a chance to challenge students' understanding but provide a scaffolding of choices to choose their answer from. It is important to remember that the ultimate goal is not that the student gives the correct answer. The educational power comes when the student is asked to defend her answer—to explain why she chose "B" instead of "E." It is important to note that MCQs that test a higher level of understanding generate more discussion and thus provide a more rigorous learning experience for the students. For example, an MCQ that requires straight recall will not provide students with an activity that demonstrates their understanding of the material and will likely not generate much discussion ("What nerve innervates the biceps brachii muscle?"). However, placing the concept into a clinical context provides the students with a question that requires them to apply their knowledge, will more likely stimulate discussion, and demonstrates to the professor the level of student understanding ("A patient presents with weakness in supination and elbow flexion. Assuming a nerve injury, where would the patient most likely have a lack of sensation?"). The use of polling systems is another way to help gauge the level of students' understanding in MCQs, especially with larger classes.

Open-Ended Questions This is a more challenging activity for students because they do not have answers to choose from. Instead, they must work toward an answer relying upon their understanding of the pre-class work and information provided to them in the question. It is helpful to use clinical application problems which include clinical features (signs and symptoms, laboratory values, etc.) and radiographic imaging. In this way students must interpret information in the stem (which is less familiar to them) with information that they learned in the pre-class work (which is more familiar to them) and put them together to solve a problem. This type of question often provokes more discussion among the students.

Case Studies Providing the students with a patient case study that includes the anatomy covered in the pre-class work helps to put the learning objectives into context and provides an opportunity to integrate material from other disciplines (i.e., clinical medicine and physiology). Case studies also provide motivation to learn the material because they can see its applicability and enables them to link their newly learned knowledge to other branches of medicine.

Think-Pair-Share When conducting discussion in the classroom, the think-pair-share model is recommended. The "think-pair-share" approach is a method whereby once a problem is posed, students first "think" through a solution on their own , then "pair" with a couple of fellow students to discuss their answers, and then after a designated period of time, one or two groups are selected to "share" their discussion with the class. Hence the term "think-pair-share." This method provides students with the opportunity to study a question individually, in small groups, and then with the entire class.

Class Size

Can the in-class portion of the FC be managed for a class of 10, 50, 100, or 400 students? The answer to this question is "Yes." For larger classrooms where it is difficult to hear a student give an answer, a roaming microphone can be utilized or perhaps a throwable microphone. Additionally, audience-response systems are popular for polling students for their answers. The think-pair-share method works regardless of the number of students. The pairing up of students results in groups of 2–4. The only difference between a big class and small class is the number of these groups of 2–4.

Classroom Environment

Imagine an anatomy professor starts class off by posing a question and the students raise their hands to answer. Jose's answer is partially correct but has some errors in it. The professor responds, "Wrong! The answer is …." At this point the professor does not know why Jose gave the answer he did, and Jose did not have an opportunity to explain why he provided his answer. This less-than-ideal interaction between professor and student creates two problems: students will be more hesitant to answer a question in class and the response resulted in no feedback to the professor on the student's understanding of the material. In video games, the gamers get better when they fail, realize what they did wrong, and start again. In musical instruments, performance improves when students play in front of their teacher and receive feedback. In anatomy, students' knowledge, retention, and ability to apply information improves when they practice applying their anatomy knowledge in front of their professor and receive correction and encouragement. When the educational environment caters to a student feeling supported and encouraged to share his or her thought process, the FC model will succeed.

Responding to Students Providing Answers

Our goal is to engage our students in a dialogue or discussion of their knowledge. So, how do we do this when they provide an answer to a question? I propose the following straightforward response [9–11]: "Thank you … why did you say that?". Here is why I suggest this phrase:

- *Correct Response:* If the student provides the correct answer to a question, do you know why she gave her response? Did she get the right answer because she knows the topic inside and out? Or did she guess? The learning for the student, and the rest of the class, occurs in the *defense* of an answer, not in the *giving* of it. For example, if she says, "I think the answer is 'D'" and you respond, "Correct, next question," it is as unsatisfying as a sneeze that did not happen. However, if you respond, "Thank you … why did you choose 'D'?", you will find out if she really understands the content. It will also generate discussion about why "C" was less correct than "D."
- *Partially Correct Response:* If the student provides an answer that is partially correct and partially wrong, do you know why he gave his response? What was his reasoning? If you do not know the answer to these questions, there is a straightforward response, "Thank you … why did you say that?". In this way you can hear the reasoning behind the response and now have evidence to highlight his correct reasoning and know what he needs correction with. It is always good to identify and commend students for when they are right. This helps to create an environment where they have the courage to share their answers with their classmates. It is also important to ensure wrong ideas and poor thinking are corrected before moving on.
- *Incorrect Response:* What if the student gives a response that is wrong? Again, the question you should ask yourself is if you know why the student responded with this answer. If you do not know why, then ask the question, "Why did you say that?". It is amazing what teaching gems are revealed when a student explains why and how she came to a wrong conclusion. Sometimes it is because of something she thought she heard from an earlier comment. Sometimes it is because of a wrong concept she understood during the pre-class work that is easily corrected. And sometimes it is because she is ignorant of only a part of a concept but understands a good chunk of the rest. It provides you, the professor, the opportunity to

praise the student for the part of her understanding that she got right. It is always important to reinforce correct logic. It then provides you with ample evidence of where she is void of understanding and can make a correction or ask the class to fill in the blank. Another benefit of always responding to a student's response of "Thank you … why did you say that?" is that class soon learns that you, as the professor, are more interested in their logic and reasoning behind an answer than the answer itself. Plus, you do not give away if their answer is correct or not until they provide evidence. The class understands that it is the defense of an answer that is most important, not just a one-word response. This form of interchange between "professor and student" and "student with student" encourages an environment of "higher learning" and not just "higher memorization." Is not this where we ultimately want our students to be at? For students to use information to solve problems and then defend their answers? Active learning methodologies like the FC or team-based learning (TBL) provide a platform to accomplish this end. In contrast, a PowerPoint lecture as the primary teaching tool cannot ultimately provide this higher-order learning. Students get better at what they practice day after day. If they practice solving problems every day, then they become better at solving problems.

> The learning for the student, and the rest of the class, occurs in the defense of an answer, not in the giving of a one-word response.

Post-class Work

Following a class session together, it is beneficial to provide additional problems to further facilitate study, application, and understanding. Once the students have had some practice solving problems, they feel more comfortable to solve more problems on their own. This further reinforces their knowledge and skill in application.

Pros and Cons of the Flipped Classroom

Pros

The FC has a number of advantages.

Provides Students with More Control of Their Learning

In the traditional lecture model, Jasmine may be bored because the professor spends too much time on concepts she already knows. In contrast, Noah cannot keep up with the professor because he moves through the slides too quickly to understand and capture notes. A benefit of the FC is that when initially learning the content through a video, it does not matter if Jasmine needs 30 minutes to complete the pre-class work and Noah needs 90 minutes. Both students have control over their learning and control the pace as they prepare for the in-class activities [12, 13].

Increases Accessibility of Course Content

Students miss in-class activities for a variety of reasons (i.e., illness, conferences, global pandemics). The FC not only provides the freedom for students to learn the content at their own pace but also in their chosen location. As long as the technology is available (i.e., Internet, computer), the student can catch up on material because of its convenient digital access. Additionally, if a student cannot remember a concept, he can always go back to the videos and re-watch them to clarify misconceptions [14, 15].

Promotes Student-Centered Learning

When questions are used in class, the focus of learning falls on the shoulders of the students. As a result, the FC promotes more student-centered learning when compared to a traditional lecture. Additionally, during lecture students sit side-by-side but have little interaction as they are paying attention to the professor. However, during the FC students continually engage with their classmates, collaborate on answers, and discuss their reasoning. The FC takes the spotlight away from the professor and places it on the students [5, 16].

Increases Efficiency of Study Time

Some criticize the FC by claiming it creates more study time than compared to lecture. In other words, it may take a student 90–120 minutes to prepare for a large classroom session. If a professor gave that same content through lecture, then it would have only taken 50 minutes. However, time in the lecture hall does not necessarily translate to learning. Learning is the process of not only consuming information but also making connections with other topics, organizing the content, and drawing connections. Additionally, the more students are prepared for the in-class session, the more they take out of it. It has been the author's experience that students take less time studying content in the FC because the process of pre-class work and in-class activities solidifies the material, makes the material more memorable, and helps the students know what to focus their studying attention on. Some videos are created by the professor, which has its advantages as the videos are more tightly knit to the learning objectives and in-class activities. Additionally, 50-minute lectures are often 60–80% shorter when they are turned into video tutorials.

Supports Professional Behavior

The teaching and delivery of information is a behavior that is expected at an institution of higher learning. Before the advent of technology, the primary and basically only means of providing the teaching of course content was through the lecture. Now there is a seemingly never-ending supply of content delivery through the Internet. As such, students can get the class content even without the professor. However, many professors continue to deliver content, and they miss the golden opportunity to promote the other behaviors characteristic of an institution of higher learning—to challenge students through problems, engage them in discussions, encourage correct understanding, and identify errors in their logic.

Develops and Improves Problem-Solving Skills

In traditional lecture so much time is spent teaching the concepts that little time is left over to apply that information. In contrast, with the FC method, the majority of class is spent diving deeper into each learning objective because the concepts were covered before class. As such, students develop and improve their problem-solving skills and thus not only are able to retain and recall information better when compared to students in lecture but also can solve higher-order problems.

Cons

The literature contains some negative outcomes regarding the FC [17]. Some negative publications may be the result of poor implementation of one or more elements of the FC. These FC elements (pre-work, active learning, peer-to-peer teadching, etc.) have a long history and robust backing of strong educational outcomes. However, professors need training, time, and experience to deliver each of these elements effectively. When one or more of these elements is lacking or poorly implemented, a negative experience may result. Some of the negative publications are a result of the following.

Relies on Preparation and Trust

For the FC to succeed, the professor must provide effective pre-class work and the students must complete it. In other words, the students trust that the pre-class material provided to them is effective and the professor trusts that the student will complete it. If the pre-class work is not effective and/or the students do not do the pre-work, this will result in a less-than-ideal in-class portion of the FC. This is not so much a con as it is more of a contract that if one party does not keep up their end of the bargain, it results in a less than satisfactory experience.

Requires Time and Practice to Produce Good Pre-class Work

The pre-class work provided to students, which is often in the form of videos, must knit tightly with the learning objectives, in-class activities, and assessments. To create these videos, professors often provide a voice-over to their PowerPoint presentations, which requires a good-quality microphone and thoughtful approach in organizing the text and images. Locating suitable, non-copyrighted images to include in the presentation is often difficult. A regular PowerPoint lecture used in class does not necessarily translate over to a high-quality educational video. As such, time, effort, and practice are required to produce a good-quality video for pre-class work.

Requires Time and Practice to Locate Good Pre-class Work

If the professor decided against creating videos for pre-class work, then it is necessary to find alternatives. There are numerous no-cost resources on line (i.e., YouTube and Khan Academy) that may be referenced, but it takes time to find and view content. Additionally, videos produced by someone else may not form a perfect fit with the learning objectives, in-class work, and assessments.

Requires an Investment of Time and Effort by the Students

For most traditional lectures in medical school, there is little time and effort needed by the student to prepare for class. In contrast, the FC method requires students to spend considerably more time before class watching and re-watching videos and referencing textbooks to prepare for the in-class activities. Pre-class work is especially problematic when students are required to be in class morning and afternoon of each day of the week and thus hard to find time for any out-of-class work (Table 11.1).

Table 11.1 Pros and cons of the flipped classroom

Pros	Cons
Provides students with more control of their learning	Relies on preparation and trust
Increases accessibility of course content	Requires time and practice to produce good pre-class work
Promotes student-centered learning	Requires time and practice to locate good pre-class work
Increases efficiency of study time	Requires an investment of time and effort by the students
Supports professional behavior	
Develops and improves problem-solving skills	

Conclusion

The FC is one of many new teaching methods to be "in vogue" at this time and has been adopted into many health sciences programs (including anatomy). However, the FC is not new; it is just new to many in the medical education community. The sound educational principles espoused by the FC have been around for years. Students have always been expected and, in fact, required to gain some foreknowledge of material prior to coming to class. Before the term FC was coined, the "pre-class" preparation was simply called "required reading." The manner in which a professor engages students in dialogue and assists them in working through problems has always been the expectation of higher education. Before the FC this professor-to-student interaction was called "the Socratic method." Furthermore, the FC enables students to re-study and review material such as the video tutorials and problem sets used in the classroom, thus augmenting and distilling the material the student must ultimately master. Prior to the FC, this was simply referred to as "homework." What the FC method ensures is that the elements of sound, time-tested, educational principles such as preparing before class, open discussion with the professor in solving problems, and post-class work are implemented into the classroom. The FC is a formal method of accomplishing these sound principles.

References

1. Moravec M, Williams A, Aguilar-Roca N, O'Dowd DK. Learn before lecture: a strategy that improves learning outcomes in a large introductory biology class. CBE Life Sci Educ. 2010;9:473–81.
2. Michael J. Where's the evidence that active learning works? Adv Physiol Educ. 2006;30:159–67.
3. Crouch CH, Mazur E. Peer instruction: ten years of experience and results. Am J Phys. 2001;69:970–7.
4. Prober CG, Khan S. Medical education reimagined: a call to action. Acad Med. 2013;88:1407–10.
5. Ramnanan CJ, Pound L. Advances in medical education and practice: student perceptions of the flipped classroom. Adv Med Educ Pract. 2017;8:63–73.
6. Morton DA, Colbert-Getz J. Measuring the impact of the flipped anatomy classroom: the importance of categorizing an assessment by Bloom's taxonomy. Anat Sci Educ. 2017;10(2):170–5.

7. O'Connor EE, Fried J, McNulty N, Shah P, Hogg JP, Lewis P, et al. Flipping radiology education right side up. Acad Radiol. 2016;23(7):810–22.
8. Klatt EC, Klatt CA. The reading rates of medical students. Acad Med. 2011;86(9):1079–83.
9. Chan LK, Yang J, Irby D. Application of the one-minute preceptor technique by novice teachers in the gross anatomy laboratory. Anat Sci Educ. 2015;8(6):539–46.
10. Chan LK, Wiseman J. The use of the one-minute preceptor as a teaching tool in the gross-anatomy laboratory. Anat Sci Educ. 2011;4(4):235–8.
11. Chan LK, Sharma N. Effects of training experienced teachers in the use of the one-minute preceptor technique in the gross anatomy laboratory. Anat Sci Educ. 2014;7(2):124–9.
12. Missildine K, Fountain R, Summers L, Gosselin K. Flipping the classroom to improve student performance and satisfaction. J Nurs Educ. 2013;52:597–9.
13. McLaughlin J, Roth M, Glatt D, Gharkholonarehe N, Davidson C, LaToya G. The flipped classroom: a course redesign to foster learning and engagement in a health professions school. Acad Med. 2014;89(2):236–43.
14. Mason G, Shuman T, Cook K. Comparing the effectiveness of an inverted classroom to a traditional classroom in an upper-division engineering course. IEEE Trans Educ. 2013;56(4):430–5.
15. Prober CG, Heath C. Lecture halls without lectures - a proposal for medical education. N Engl J Med. 2012;366:1657–9.
16. O'Flaherty J, Phillips C. The use of flipped classrooms in higher education: a scoping review. Internet High Educ. 2015;25:85–95.
17. Wong TH, Ip EJ, Lopes I, Rajagopalan V. Pharmacy students' performance and perceptions in a flipped teaching pilot on cardiac arrhythmias. Am J Pharm Educ. 2014;78(10):1–6.

12 Preparing and Recording Lectures for Online Delivery

Thierry R. H. Bacro

Historical Background

Despite technological advances in the last few decades, traditional lecture delivery remains one of the effective teaching approaches used to deliver a significant amount of information in anatomy courses in the USA and throughout the world. Although criticized in the last few years by educational specialists because it is not perceived as the "ideal" method to help students acquire and retain large amounts of information required to successfully pass required licensing exams. This mode of lecturing, however, is a necessary evil in curricula that are facing a shortage of trained anatomy teachers, significantly increasing their class size, coupled with diminishing hours dedicated to anatomy in an ever-changing academic environment weakened by decreased state and federal funding. However, while accepting that lecturing is still a reality of teaching anatomy in a modern fast-paced curriculum, the way one uses lecture in modern education has changed in order to remain as an effective teaching and learning modality and to appeal to what is now described as the millennium student.

T. R. H. Bacro (✉)
Department of Regenerative Medicine and Cell Biology, Center for Anatomical Studies and Education, Medical University of South Carolina, Charleston, SC, USA
e-mail: bacrotr@musc.edu

Historically, the classic anatomy lecture was taught by a lecturer speaking in the front of students seating in a lecture hall, listening attentively (or not), asking maybe a few questions, and trying to take notes while the lecturer was drawing anatomical structures on a blackboard with colored chalks. To complete these visual resources, lecturers would sometimes provide rudimentary notes created using a stencil duplicator also called a mimeograph machine. Interestingly, to successfully learn the material, the students had to team up to capture all the drawings presented during the lecture and had to meet after the lecture to share their notes. This was an early form of what would now be labeled team approach to learning, collaborative learning, and peer learning. Although lecturing is now often considered passive teaching, these students had to remain very engaged and very attentive to the lecturer to benefit from this experience. Occasionally, the lecturer would project some slides or even an 8 mm movie to illustrate the lesson. At some point, plastic overheads showing drawings of anatomical structures were used by the lecturer to avoid having to draw during every lecture, allowing a faster delivery of the lecture content. The appearance of instant development of slide films in the 1970s fostered the adoption of slides for lecturing. These slides were created using a variety of software at the time, but PowerPoint (PPT) progressively started to dominate the educational market, and with the birth of

L. K. Chan, W. Pawlina (eds.), *Teaching Anatomy*, https://doi.org/10.1007/978-3-030-43283-6_12

the Course Management Systems (CMS), PPT became the standard to deliver large amount of anatomy materials through a live lecture and/or for online delivery. At that stage, very few faculty knew how and/or had the time to record their voice over the visual aids presented in PPT when using a CMS. Even with the use of complementary notes provided to support this format, these anatomy lectures were definitely delivered in a very dry and passive format, with very little interaction between teachers and learners. In parallel to the changes described above in the last decade or so, anatomy educators started experimenting with other forms of teaching online using audio–video recordings and started to realize that the use of online technology did not imply passive learning. As technology evolved and as educators became more creative with the existing educational tools, one started to see the reemergence of interactive learning/teaching, integrated in delivery models of synchronous teaching and/or for online delivery in asynchronous teaching models. These models were tailored to learning styles, providing for small and large groups of students, allowing for questions and discussion of materials as well as to ensure the acquisition of the knowledge to pass standardized licensing exams. The most recent development in terms of didactic lectures is the popularization of automatic lecturing recording systems (LRS).

How Does It Work?

The purpose of a lecture recording system is to capture the visual information displayed on the screen as well as the sound associated with it during the lecture. The software synchronizes the visual information with the sound and saves these data into a file format that can be displayed later in the chosen online model. The visual information captured can range from very simple to very sophisticated: simple text and images created with basic software, PowerPoint presentations, simple or very complex animations, or even audio–video files. The sound can be the voice of the speaker recorded through a microphone set on the computer station or any sounds displayed in a file played during the lecture, including from an audio–video file. So, a lecture recording can range from a simple short presentation of some text or images with a voice discussing these resources to a very complex mix of highly sophisticated learning objects. An additional important feature is that the visual information recorded will capture the mouse used by the speaker to point out elements of the presentation. The quality of the recordings, both in terms of the visual and the sound, will depend on the quality of the screen and sound capture hardware installed, the processor speed of the computer station used, as well as the type of LRS employed.

Lecture Recording Systems

Quite a few LRS software can be purchased commercially, but some are also available through open-initiative communities (see Table 12.1).

Now, the systems listed in Table 12.1 ranged from a very simple application for a single station to very large and complex systems to be distributed throughout large academic centers. In some cases, the recording can be done on a

Table 12.1 Lecture recording systems (in alphabetical order)

Accordent Technologies	http://www.polycom.com
Arrive ViewPoint	http://www.arrivesys.com/arrive/arrive-viewpoint.php
Echo360	http://echo360.com
Camtasia	http://www.techsmith.com/camtasia.html
Galicaster[a]	http://wiki.teltek.es/display/Galicaster/Galicaster+project+Home
Media transformation	http://www.cisco.com
Opencast Matterhorn[a]	http://opencast.org
Panopto	http://www.panopto.com
Presentations	http://www.presentations2go.eu
Presentations 2Go	http://www.presentations2go.eu
ProfCast	http://www.profcast.com
Tegrity	http://www.tegrity.com

[a]Available through open-initiative communities

single station equipped with the proper hardware (see next section) on which the software has been installed. The lecture can be recorded on the station itself, and at the end of the session, the packaged file can be moved using a USB drive (or any other mode of transfer) to be distributed to the students directly or through a CMS. In that case, an Internet connection is not required when recording the lecture. With some more advanced software, the lecture can be recorded on the fly after an initial log-in on the software. Other software allows the faculty to stream the lecture to students viewing it online at other locations while also recording it to save and distribute later through a chosen mode of distribution. In that case, the viewing of the streamed lecture is delayed a few seconds (5–10 s usually) compared to the real lecture. Regardless of streaming or not, at the end of the lecture, the faculty might be prompted to upload it automatically (Internet connection required) or to open the file to view it, edit it (or delete it) before, and then upload it in a system where students can access it. Some of this software will allow you to do some editing of the file created, while others include very advanced features that will allow the faculty or the educational specialist to perform some sophisticated editing, splicing, removing, or adding contents (visual and auditory) before final rendering and distribution of the file. This last approach is not recommended for the novice faculty as it is extremely time consuming and requires a high level of technical expertise of the chosen software. Many of these software programs have advanced settings which, when set properly, can automate and simplify the tasks required from the faculty to produce quality recording. Several of these programs can produce several different file formats (i.e., MP3 podcast for MP3 player or Linux; enhanced podcast and enhanced video podcast for iPod, iTunes, or Linux; video streaming to iOS and Android apps), and the faculty will need to spend some time familiarizing themselves with the advantages of each file type before deciding on the settings used: limitation in terms of the distribution through a given CMS, final size, necessity to have or not a specific application on the student side to read the file, and most importantly quality.

Getting Started

Besides securing the proper LRS software, the teacher will need to secure the proper hardware: a microphone, a camera, an up-to-date computer, and a fast Internet connection. The specifics of this hardware may vary greatly from station to station, but to ensure good audio–video recording, the teacher should choose good- to high-quality capture sound and video cards that are compatible with his/her own computer station as well as the specifications of the LRS chosen. The vendor for the LRS chosen should be able to provide the faculty with the details of the required minimum specs to ensure proper recording output. Even when these specs are met, creating quality recordings might be tricky sometimes, and the cooperation of the local technical support at the home institution might be necessary to optimize the hardware for success. Although it is sometimes absolutely necessary for some demonstrations, the camera can be optional as many teachers will use a still picture of themselves embedded in the resources and not use a live feed of the teaching itself. This greatly reduces the overall size of the final file (see later). Once the software and the hardware have been chosen and the computer recording station is properly set, the faculty will be able to start recording the lectures. To be effective when using a recorded lecture, the teacher will need to review and critically assess a number of teaching parameters before starting the recording process.

Creating a Good Recording

Although it may seem obvious, a teacher must first create a good lecture to generate a good recording. The principles of good teaching described in earlier chapters in this book must be respected when creating the lecture, and the lecturer will need to ensure that the lecture adheres to the following:

- Covers the objectives listed in the course syllabus.
- Follows a well-structured lecture plan that flows logically, with an introduction; with a main body with breakpoints in the form of questions, illustrations, or examples; and ending with a summary and a take-away message.
- Uses the proper visual objects (background type, size of the font, color patterns) to induce learning in the population taught, i.e., younger undergraduate degree students (millennium students) versus graduate degree residents or even more advanced and older clinicians. Because the recorded screen is smaller than the image usually projected in a classroom, the faculty need to have good contrast in the images used for the lecture, avoid having cluttered and too detailed visual aids, and must use visual information in a somewhat simpler format than a traditional lecture, meaning try to present one or at the most two key concepts by visual aid.

Another component to consider for new faculty recording a lecture is the voice component. Some of us may have a tendency to speak faster when we enthusiastically teach a subject we love. Creating a good recording however requires the teacher to, in fact, slow down the pace, enunciate clearly, and fully appreciate the fact that the audio component of recording is as equally important as the visual information. Anecdotal evidence indicates that many students will speed up the recording and use it in fast learning mode when the technology allows it. The true extent of that practice, and whether it is beneficial, has not been studied, but recording the lecture with quality sound is usually appreciated by the students. In addition to the voice issue, an additional level of difficulty in recording a lecture is created by the fact that the faculty member not only needs to think about what he/she is going to say during the lecture and monitor the pace and the quality of his/her speech but must also point at the visual aids using the cursor on the computer throughout the lecture, rendering the live recording a very difficult practice, especially for the less seasoned faculty. In many cases, when starting new with this process, rehearsing the lecture several times, in its full length, on the equipment used for the recording in question, and examining the recording in its entirety for problems, errors, and quality issues are very effective ways to improve the quality of the final recording.

> Creating a good recording however requires the teacher to slow down the pace, enunciate clearly, and fully appreciate the fact that the audio component of recording is as important as the visual information.

Use of a Lecture in a Curriculum

Besides the obvious preparation of the lecture itself already described, the speaker needs to understand the issues associated with the distribution of online recording of even a well-designed lecture. The faculty will need to consider the environment in which the recording will be distributed which in fact can be one of two environments:

1. The lecture recording is distributed in a curriculum in which the students are given the choice to attend the live lecture and/or access the recording.
2. The lecture recording is made available only through an online environment without the possibility of attending the live lecture.

In the first environment, the students can decide if they will attend the live lecture or access the recording only or use both. Much has been written in the literature about how they choose, but after a certain of amount of time, some students, anywhere from about 10 to 30% of the class, will continue to attend the live lecture, while others will stop attending the lectures and will only use the recorded lecture to study the information. Of course, some students will switch from the first or second approach and vice versa depending on the time of the day, the lecturer, or the topic. This situation creates a special set of problems for the faculty who choose (or are coerced) to record their lecture. It is very difficult to create quality recordings (slow speech, proper pointing with the mouse to visual elements of the lecture, etc.) during a live lecture. The students who decided to attend the lecture on that occasion expect the faculty to attend to their needs in the live classroom. They will resent the faculty who stay behind the podium in order to move the mouse (that one cannot see well on the main projected screen anyway) to point at elements of the presentation for the recording. So, this setting often makes it nearly impossible to create a high-quality recording in the given time allotted for the live lecture (often 45–50 min at the most). In order to alleviate this problem, faculty who have the time and the propensity to care about providing quality live teaching as well as creating quality lecture recordings have adopted a hybrid approach. They will teach the live lecture focusing on the students attending the live lecture, using a laser pointer to point directly at the screen where the visual aids are being projected, and answer live questions on the spot. At a later time (or, in some cases, in advance of the lecture), they will, just like a faculty member who teaches only in the second environment described above (recording made available only through an online environment), create a high-quality recording to be distributed. Although working very well, this approach is time consuming. It can, however, present significant advantages for both the students and the faculty.

Advantages of Recording Lectures

The recording of a lecture presents a number of advantages for the student unrelated to whether the recording session is live or not. The student can do the following:

- Play it back as many times as needed to acquire the information.
- Skip or speed up parts of the lecture that are easy or already understood.
- Pause the lecture for convenience or, by necessity, to make notes, search for additional information using a textbook or online search engines, write an e-mail asking a question to the faculty, and doing so without missing information.
- Study it alone or in a small or large group.
- Catch up on the lecture material after valid absences due to personal or family illness and death in the family or absence due to weather conditions.

The recording of a lecture in non-live settings presents additional advantages for the faculty:

- The lecture recording, if needed, can be longer than the time allotted in a standard live lecture.
- The lecture, although possibly longer in total length, can be recorded by pausing every 10–15 min or at logical points in the lecture, allowing the faculty to pace themselves, achieving higher quality, and rendering the task of recording easier.
- The sound quality, an important component, of the recording is usually higher, with less interferences.

> LRS has been an invaluable tool for students who experience valid absences and is now part of our institutional response plan in case of significant emergency that could disrupt the teaching schedule.

Disadvantages of Recording Lectures

The use of recorded lectures in a curriculum may also present significant challenges. In most cases, these issues can be addressed as follows:

- The use of recorded lectures may impact attendance to lectures. To address this issue, faculty could propose more interactive and participatory classes to engage learners of all learning preferences and/or also use an audience response system (ARS or clickers) which has been shown to be effective in engaging students during lectures [1–3]. In that case, a single question per lecture presented to the audience through the ARS allows the faculty to determine the attendance of the class and then take action, if needed or required, to address the attendance issue. So, the attendance issue alone should not be used as a deterrent to the implementation of an LRS.

- The recorded lecture does not allow opportunity for questions. In this case, faculty may want to set additional time for questions and answer sessions or use e-mail, chat, forum, or blogs to answer the students' questions synchronously or asynchronously.

Fear of decreased attendance should not be a deterrent in terms of using LRS in a curriculum. If attendance is an issue, mechanisms such as integrating clickers in the lectures can be used to monitor the presence of students in lecture hall.

Creative Use of Recordings

Some faculty use LRS for other purposes than lecturing. For example, LRS has now been used to perform the following:

- Prepare recordings for students to view to introduce them to the dissection laboratory sessions.
- Record mini educational modules on a given topic, lymphatic system of the head and neck, for example.
- Summarize information taught across several lectures.
- Answer FAQs.
- Use in what is now called the "flipped lecture" where students are provided with information to prepare for a session where a given topic is discussed.

In all the abovementioned cases, the faculty can take advantage of the recording capabilities of the software chosen, by being more creative than in a live lecture, by creating resources in a format that appeal to the millennium student, by integrating information in manner not possible in classical teaching, and even by distributing the created recording though a medium that will entice the student, such as YouTube or Facebook.

Proper Use of Recordings

The published research in the usefulness of lecture recording clearly indicates that the use of LRS is usually well received by students [4–11]. However, recent published evidence indicates that not all students in a course will spontaneously use LRS. Bacro and colleagues reported in 2010 that 30% of the medical students in the first-year class at the Medical University of South Carolina did not use the LRS at all and that an additional 41% used it very little (less than 10 times for the semester) even though the survey of the students' perception showed that 74% of the students agreed/strongly agreed that the recordings were useful with 6% disagreeing/strongly disagreeing and 11% undecided [12]. Also, the pedagogical value of LRS has not been clearly documented and is somewhat unclear. In one instance, a research study showed that a group of psychology students watching a single lecture recording scored significantly higher on examination compared to a group watching the matching live lecture [13]. In another case, a report showed students who chose to use an LRS called Lectopia instead of attending live lectures scored lower on continuous summative assessment, examinations, and final marks in a medical pharmacology course in Australia [14]. McNulty and his group showed that students found the recordings of basic science lectures useful in a medical curriculum but also found that the students who accessed the lecture recordings more frequently scored significantly lower [15]. Other researchers, however, have found that the use of LRS does not impact grades, negatively or positively. Bridge conducted a 5-year retrospective review to assess the impact of lecture recordings on scores on the United States Medical Licensing Examination (USMLE) for Step 1 for 1736 stu-

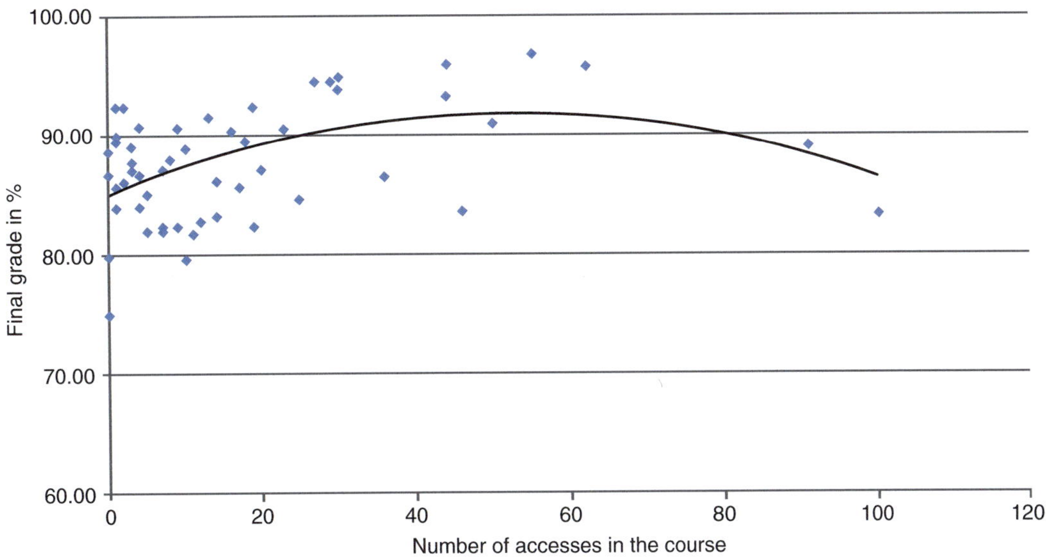

Fig. 12.1 Plot of the number of accesses per student versus the final grades in a dental gross and neuroanatomy course with 55 lectures. A significant but poor correlation was detected between final grades and the number of times the student accessed the lecture recordings ($r = 0.33$ with $P = 0.01$). The *black line* represents a best curve fit of the data using a quadratic polynomial

dents and concluded that, after correcting for the national trend over the same period, the effect of LRS on the measured outcomes was neutral [16]. Similarly, Bacro and colleagues did not find a significant correlation between final grades in three medical basic science courses (cell biology/histology, physiology, and neurosciences) and usage of LRS when examining the number of time students accessed the LRS as well as time spent using the recordings [12]. More recently, the same group showed that students who self-rate themselves as having an auditory preference component when studying and using the LRS on average for less than 10 min per access had an average final grade of 16.43% higher than the students ranking themselves as non-auditory and using the LRS for the same amount of time per access [17]. In the same study, the data also indicated that the students accessing the recordings more than once per lecture were at risk of scoring lower in the course (Fig. 12.1).

Conclusion

In our experience, the LRS has been an invaluable tool for students who experience valid absences due to personal or family illness or a death in the family or absence due to weather conditions. It is also now part of our institutional emergency response plan in the case of a pandemic flu, major weather disaster such as a hurricane, or any other significant emergency that could disrupt the teaching schedule. However, the consensus among faculty who have significant experience in this area is that students should be educated about the proper use of LRS to benefit from it, i.e., use it one time per lecture on average with the most common time of use being about 10 min per lecture recording. The decision to use it and how to use it should also remain with them and not be decided centrally by administrators, as some successful students clearly refuse to use LRS when given the choice.

References

1. Alexander CJ, Crescini WM, Juskewitch JE, Lachman N, Pawlina W. Assessing the integration of audience response system technology in teaching of anatomical sciences. Anat Sci Educ. 2009;2:160–6.
2. Wait KR, Cloud BA, Forster LA, Jones TM, Nokleby JJ, Wolfe CR, Youdas JW. Use of an audience response system during peer teaching among physical therapy students in human gross anatomy: perceptions of peer teachers and students. Anat Sci Educ. 2009;2:286–93.
3. Hoyt A, McNulty JA, Gruener G, Chandrasekhar A, Espiritu B, Ensminger D, Price R Jr, Naheedy R. An audience response system may influence student performance on anatomy examination questions. Anat Sci Educ. 2010;3:295–9.
4. Dev P, Rindfleisch TC, Kush SJ, Stringer JR. An analysis of technology usage for streaming digital video in support of a preclinical curriculum. Arch Proc AMIA Symp. 2000:180–4.
5. Nieder GL, Nagy F. Analysis of medical students' use of web-based resources for a gross anatomy and embryology course. Clin Anat. 2002;15:409–18.
6. Copley J. Audio and video podcasts of lectures for campus-based students: production and evaluation of student use. Innov Educ Teach Int. 2007;44:387–99.
7. Brecht HD, Ogilby SM. Enabling a comprehensive teaching strategy: video lectures. J Inf Technol Educ. 2008;7:IIP71–86.
8. Gosper M, Green D, McNeill M, Phillips R, Preston G, Woo K. The impact of web-based lecture technologies on current and future practices in learning and teaching. Sydney: Australian Learning and Teaching Council; 2008. https://www.mq.edu.au/ltc/altc/wblt/dissemination.htm.
9. Craig P, Wozniak H, Hyde S, Burn D. Student use of web based lecture technologies in blended learning: do these reflect study patterns? In: Proceedings of the 26th annual conference of Australasian Society for Computers in Learning in Tertiary Education—Same Places, Different Spaces (ASCILITE 2009), Auckland. 2009, p. 159–167. Australasian Society for Computers in Learning in Tertiary Education, Figtree, NSW, Australia. http://www.ascilite.org.au/conferences/auckland09/procs/craig.pdf.
10. von Konsky BR, Ivins J, Gribble SJ. Lecture attendance and web based lecture technologies: a comparison of student perceptions and usage patterns. Aust J Educ Technol. 2009.;Tech;25:581–95.
11. Jalali A, Leddy J, Gauthier M, Sun R, Hincke M, Carnegie J. Use of podcasting as an innovative asynchronous e-learning tool for students. US China Educ Rev. 2011;A6:741–8.
12. Bacro TR, Gebregziabher M, Fitzharris TP. Evaluation of a lecture recording system in a medical curriculum. Anat Sci Educ. 2010;3:300–8.
13. McKinney D, Dyck JL, Luber ES. iTunes university and the classroom: can podcasts replace professors? Comput Educ. 2009;52:617–23.
14. Fernandes L, Maley M, Cruickshank C. The impact of online lecture recordings on learning outcomes in pharmacology. J Int Assoc Med Sci Educ. 2008;18:62–70.
15. McNulty JA, Hoyt A, Gruener G, Chandrasekhar A, Espiritu B, Price R Jr, Naheedy R. An analysis of lecture video utilization in undergraduate medical education: associations with performance in the courses. BMC Med Educ. 2009;9:6.
16. Bridge PD, Jackson M, Robinson L. The effectiveness of streaming video on medical student learning: a case study. Med Educ Online. 2009;14:11.
17. Bacro TR, Gebregziabher M, Ariail J. Lecture recording system in anatomy: possible benefit to auditory learners. Anat Sci Educ. 2013;6(6):376–84.

13 Evaluating Your Own Performance in a Lecture

John Dent

> Although everybody thinks that they can lecture adequately it is probably true that meaningful, constructive feedback will be valuable to everyone. –Dent [1]
>
> A good lecture, like a good meal, should contain a light starter to whet the appetite, a main course containing a variety of tastes and textures that should be properly chewed and digested, and a light dessert to complement what has gone before. –McLaughlin and Mandin [2]

Introduction

Probably the most difficult thing for a lecturer to obtain is a truthful, objective critique of their lecture. Whoever it comes from, there is always the risk of it being deferential or biased, subjective, or prejudiced. This is why personal reflection is often put forward as the main way to evaluate our performance. While this may well be true, it can nevertheless be improved on if we add to that reflection a glance at the opinion of students and colleagues. So we have three methods of evaluating our performance as lecturers that we can take into account:

- Student feedback
- Peer review
- Personal reflection

Of course, although the aim may be to evaluate the *lecturer* and his or her performance, it is not really possible to do this without also evaluating the *lecture* and its content and construction.

Student Feedback

The use of student opinion to rate teaching, introduced in the 1960s/1970s, has become the most common method used to provide feedback to lecturers and may also be used to inform faculty decisions for staff retention, tenure, and promotion. Although not universally popular with faculty members, it is nevertheless more widely used than peer feedback as a method of evaluating teaching. Studies have shown the following:

- It can provide reliable and valid evidence of teaching effectiveness across groups of raters, time periods, and courses.
- It is only slightly affected by class size and severity of grading.
- It correlates with comparable ratings from colleagues.
- It is positively related to more objective measures of teaching [3].

J. Dent (✉)
The Association for Medical Education in Europe (AMEE), Dundee, UK
e-mail: j.a.dent@dundee.ac.uk

L. K. Chan, W. Pawlina (eds.), *Teaching Anatomy*, https://doi.org/10.1007/978-3-030-43283-6_13

Table 13.1 To evaluate the lecturer

	Strongly agree	Slightly agree	Slightly disagree	Strongly disagree
Was enthusiastic				
Was clearly audible				
Seemed confident				
Gave clear explanations				
Encouraged participation				

Adapted from Brown and Manogue [5]

Student feedback may be verbal or written. Cantillon [4] lists four methods of acquiring student feedback:

- Ask a sample of students if you can read their lecture notes.
- Ask for verbal feedback from individual students.
- Ask students to complete a 1-min paper.
- Ask students to complete an evaluation questionnaire.

Verbal feedback may be acquired casually from private conversations or discussion with groups of students, or it may be more formally sought from focus group events. While personal conversations run the risk of omitting negative comments, group sessions are at risk of being dominated by the opinion of the most vocal individuals.

Written feedback from students, if anonymous, may encourage more truthful comments and so highlight both good and poor aspects of a lecture. However, this approach is unlikely to give any thoughts about how the lecture could be improved unless this is specifically prompted, for instance, by an open-ended question. Students can be asked to complete an evaluation form to rate the general characteristics of a lecturer (see Table 13.1) [5]. Student opinion can also be sought on the lecture itself, and again a table by Brown and Manogue [5] may be used to focus comments on the lecture (see Table 13.2). Examples of student feedback questionnaires can be found online. One from University College London [6] asks students to answer 12 specific questions on a three- or four-point scale and then to give an overall rating for the lecture.

Table 13.2 To evaluate the lecture

	Strongly agree	Slightly agree	Slightly disagree	Strongly disagree
Clear				
Interesting				
Easy to take notes from				
Thought provoking				
Relevant to the course				

Adapted from Brown and Manogue [5]

However, student opinion must be treated with caution as it is more likely to be swayed by the performance value of the lecturer than by the content [7] and may be influenced by a variety of other factors [8, 9]. We must not fall into the mistake of thinking that we must respond to every student comment.

Other ways of collecting student feedback may include the use of a personal response system (PRS) in the lecture theatre to ask students to vote on various aspects of the lecture as it progresses. Alternatively, a discussion board can be set up in the medical school's virtual learning environment (VLE) for feedback, comments, and discussion of ongoing problems or to provide answers to frequently asked questions.

Peer Review

So, is there a way of obtaining a more objective critique? A valuable adjunct to student feedback is peer review [3]. This may include rating scales, checklists, and open comments. Peer review can be required for part of the faculty's quality assurance (QA) process in evaluating its program, as a component of your own continuing professional development (CPD) or to contribute to an argument for academic advancement in your institution.

Peer checklists may include scorings for:

- Presentation
- Relationship with the class
- Material
- Illustration
- Overall impression

Implementation

A system whereby two lecturers give mutual feedback to each other after lecturing is a relatively non-threatening example. The risks of this approach are a potential loss of honesty or objectivity, but it can lead to open, profitable discussion and be a stimulus for personal reflection later. A team approach to reviewing a lecture, on the other hand, has a risk of appearing threatening and judgmental. It is good practice, before the event, to have a discussion between the observer and the observed to define the content of the assessment and the objectives which are to be judged. Staff development sessions are probably necessary to prepare faculty to participate in these activities successfully; to warn of pitfalls, especially when giving feedback; and to maximize the positive benefits that can be achieved. Not everyone may be able to participate in the exercise however. Time constraints, a heavy workload, as well as fear of scrutiny and criticism may all deter participation [10].

Giving Feedback

When considering giving feedback, it is good to remember Pendleton's advice [11]:

1. Clarify any points of information/fact.
2. Ask the lecturer what she/he did well—ensure that she/he identify the strengths of the performance at this stage and do not stray into weaknesses.
3. Discuss what went well, adding your own observations (if there is a group observing the performance, ask the group what went well; again, keep them to the strengths).
4. Ask the lecturer to say what went less well (weaknesses) and what she/he would do differently next time.
5. Discuss what went less well, adding your own observations and recommendations (if there is a group observing the performance, ask the group to add their observations and recommendations).

The strengths of Pendleton's rules (http://www.gp-training.net/training/educational_theory/feedback/pendleton.htm) include the following:

- Offers the lecturer the opportunity to evaluate his/her own practice and allows even critical points to be matters of agreement
- Allows initial lecturer observations to be built upon by the observer(s)
- Ensures strengths are given parity with weaknesses
- Deals with specifics

To help us provide good peer review, Siddique and colleagues [12] suggest 12 tips:

1. Choose the observer carefully. It is important that there is good rapport between the observer and the observed who should have comparable rank within the faculty and a shared empathy.
2. Set aside time for peer observation, which should include pre-observation discussion, the observation itself, and post-observation reflection.
3. Clarify expectations of the format of the event and the roles and expectations of the participants.
4. Familiarize yourself with the course, the type of lecture, its objectives, and the resources available.
5. Select the instrument wisely, e.g., a checklist that matches the session format.

6. Include students and let them know that an observer is going to be present, especially if their numbers are small.
7. Be objective: Take the opportunity to observe the students' attitude to the lecture and their reaction to the lecturer.
8. Resist the urge to compare with your own teaching style: No one style provides a gold standard.
9. Do not intervene: Concentrate on making your observations.
10. Follow the general principle of feedback. Encourage initial self-reflection which creates a positive learning climate and can lead to discussion.
11. Maintain confidentiality. Avoid making any judgmental conclusions.
12. Make it a learning experience. Give supportive feedback and constructive advice. Learn how to do this well!

Causes of Lecturalgia in the Teachers' Opinion

Misjudged learner/context	45%
Flawed implementation of teaching strategy	30%
Lack of preparation	25%
Difficulty with audiovisuals	25%
Too much content	20%
Use of wrong strategy	15%
Lack of purpose	10%
Inflexibility	10%

From Pinsky and Irby [13]

Personal Reflection

Reflection is the cornerstone of continuing professional development. At the end of your lecture, ask yourself: What went well and what did not go so well? Especially did you keep to time, did you lose your audience at any stage, and did you cover everything you intended to? A video recording of your lecture may provide a shocking insight into any annoying mannerisms you may have been unaware of and provides a good way of making objective decisions about which aspects of the lecture can be improved. However, finding time to do this after the event may be difficult for many. It would be good practice to review your lecture against the students' learning objectives for the course and to ask yourself whether you could improve the content or delivery of any part and possibly reconsider any activities your students could do in it.

A schematic approach may help us to organize our thoughts for this reflective exercise by giving us some questions to ask ourselves. According to McLaughlin and Mandin [2], the problem with a lecture ("lecturalgia"—a painful lecture) may be due to three primary causes: poor judgment, poor organization, and poor delivery (see Fig. 13.1).

1. *Was the problem with the lecture due to poor judgment?*

   The format: Was a lecture the best method to choose to present this information, or would a small group discussion or student activity have been better?

   The lecturer: Were you the right person for the job, or may someone else have been better? Did you have the right knowledge and the right attitude to motivate the audience?

   The audience: Did you make the subject relevant to the audience? Did you facilitate learning by activating prior knowledge or by using encoding specificity providing information in a sequence which mimics the way it will be used in practice? Did you use examples and elaboration to facilitate the recall of knowledge?
2. *Was the problem with the lecture due to poor organization?*

   The introduction: Did you attract attention, establish rapport, identify knowledge base, provide objectives/advance organizers, and indicate the use of the content?

   The body of the lecture: Did you select the right content, did you sequence the key points in a logical progression, and did you link these together in an engaging way?

   The conclusion: Did you review the key points and objectives, provide take-home tasks or revision handouts, and ask for any feedback?

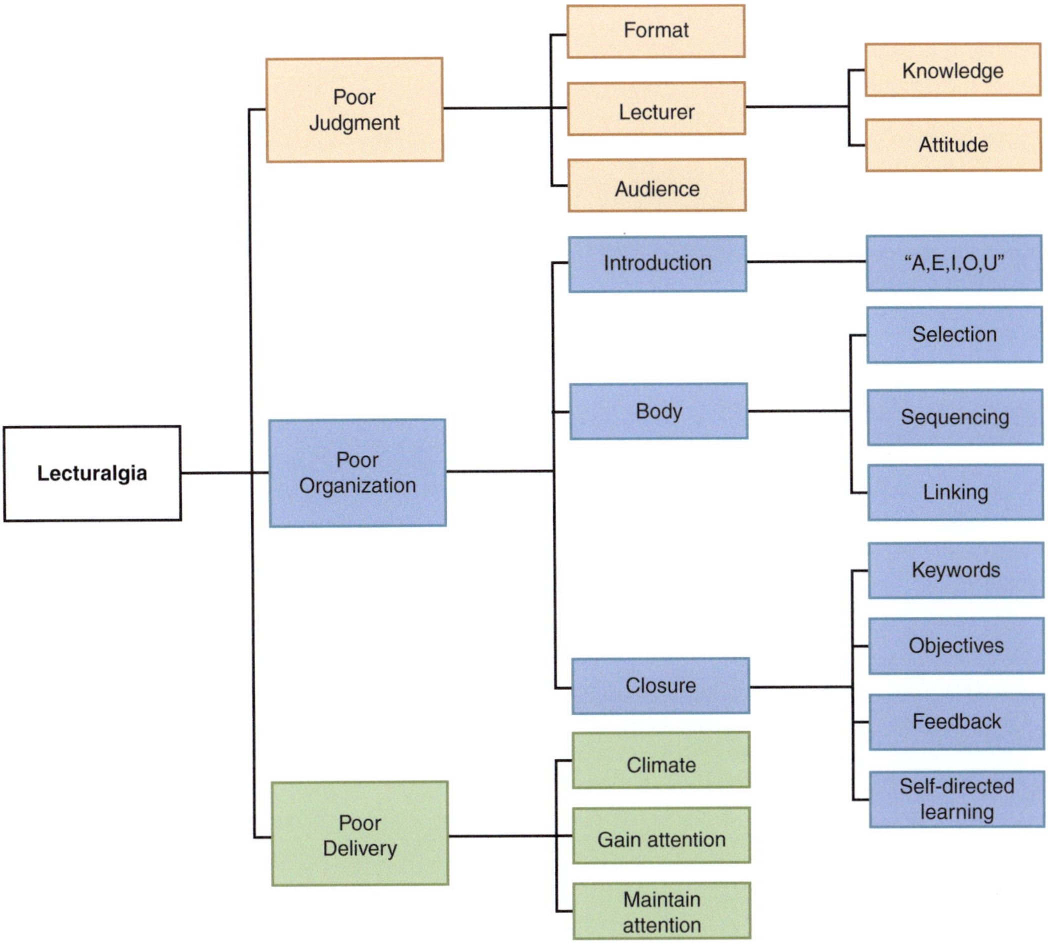

Fig. 13.1 A schematic way to identify the possible problems causing a poor lecture. Key: "A,E,I,O,U": "Attracts attention, Establishes rapport, Identifies knowledge base, provides advance Organizer, and indicates Usefulness". (Adapted from McLaughlin and Mandin [2])

When Deciding on the Content of a Lecture

- Provide objectives and key points.
- Choose contents congruent with the objectives.
- Limit the content rather than seeking to "cover the ground."
- Elaborate on the key points using illustrations and examples.
- Plan to facilitate learning after the lecture is finished.

From McLaughlin and Mandin [2]

3. *Was the problem with the lecture due to poor delivery?*
   The climate: Was there a problem with the physical or emotional climate for the students? Was there a problem gaining or maintaining student attention?

   At this point you might choose to reflect on whether your lecture achieved the objectives that you planned for it before you started.

   You might also like to compare your reflections on your performance in the lecture with feedback from students and colleagues. Analyze all the evidence available and ask

Table 13.3 Lecture evaluation checklist

Introduction/set	Score yes/no	Comments
Attracts attention Establishes rapport Identifies knowledge base Provides advance organizer Indicated usefulness		
Body	Score yes/no	Comments
Structured the content Controlled the amount of content Emphasized major points Summarized after each sections Sequencing clear and logical Relevant/interesting examples provided Kept "with" the audience Interacted with/engaged the audience Appropriate amount of material covered encouraged and responded to questions		
Conclusion/close	Score yes/no	Comments
Summarized main points Related back to advance organizer Indicated further reading etc. Conclusions clearly stated Did not introduce new material		
General presentation	Score yes/no	Comments
Appropriate mannerisms/body language Maintained eye contact Spoke audibly and fluently Ensure clarity and modulation of speech Varied the speed of presentation		
Used silence Used appropriate AV aids Asked questions Used humor Kept to time Showed enthusiasm		

yourself if the weak points you identified were the same ones your students or colleagues observed. Reviewing a comprehensive checklist completed by students or peers might be good to review at this stage (see Table 13.3). This example asks questions about the introduction, the body, and the closure of the lecture as well as generic questions about the presentation.

Finally, do you need to make a substantial change? The flipped classroom [14, 15] remains a popular topic in the literature and at medical education conferences. Would this approach be a faculty's appropriate method for students to learn the topic normally presented by your lecture?

Two Bigger Questions Remain

Has Student Learning Been Improved by This Lecture?

A useful list of tips to improve student learning from lectures is given by Brown and Manogue [5]:

- Outline the structure of the lecture. Describe signposts used and how students can recognize key points and be sensitive to verbal cues such as "However…," "Nevertheless…," and "So…."
- Provide listening, observing, and note-taking exercises and get the students to compare their notes after these events.

- Discuss what constitutes "good" lecture notes and how to use them.
- Provide a list of lecture topics at the beginning of the course and indicate the links between them.
- Encourage them to review and compare their notes by setting mini-tasks at the beginning and end of some lectures.

A specific way of evaluating a lecture in terms of whether student learning has been helped is described in the "1-min paper" by Sinclair and colleagues [16] which asks students to review their notes and then to answer three questions:

- What was the most useful, meaningful thing you learned during the lecture?
- What questions remain uppermost in your mind as we end this session?
- What was the "muddiest" point in this lecture?

Lecturers can respond to these comments at the beginning of the next lecture and correct any misunderstandings of the content.

An unannounced open notebook quiz can be used at the end of a lecture to test factual recall of key points of content as well as to challenge critical thinking on the topics covered. The PRS can be used at the end of a lecture to ask students to answer a few multiple-choice questions (MCQs) on the topic. It can also be used to ask them which parts of the lecture, or which of the styles used (passive/active learning), they found most helpful [17].

Does Evaluation of a Lecture Lead to Improvement of Lecturing?

Can student feedback actually lead to an improvement in a lecture? Murray [3] suggested that under certain conditions, this can be the case. Evidence from faculty surveys, field experiments, and longitudinal comparisons suggests that student evaluation of teaching has contributed to improvement of teaching and the contribution of student evaluation is enhanced when combined with contributions from the institution's experts in staff development. Evaluation of a lecture, especially if carried out during the session, may guide a lecturer about when to introduce a more active learning component.

So there is some evidence that student evaluation can lead to improvement in teaching and that this information is probably best used when combined with other sources of data, but perhaps most significantly, as McKeachie [18] commented, improvement in lecturing is more likely to occur when the lecturers themselves want to improve their teaching.

Conclusions

Information from a variety of sources is available to help us evaluate our performance in a lecture. Reviewing data from the various sources available may take a little time but generates a culture of personal questioning, reflection, adaptation, and improvement [19]. So do you think you can improve your next lecture? You might like to consider attending a staff development session in your institution on presentation skills or on "How to give a good lecture" [20]. You may look at some of the literature quoted in this chapter, observe a senior colleague lecturing, and, of course, discuss with your colleagues what they consider makes a lecture good.

References

1. Dent JA. Lectures. In: Dent JA, Harden RM, editors. A practical guide for medical teachers. 2nd ed. London: Elsevier; 2005.
2. McLaughlin K, Mandin H. A schematic approach to diagnosing and resolving lecturalgia. Med Educ. 2001;35:1135–42.
3. Murray HG. Does evaluation of teaching lead to improvement in teaching? Int J Acad Dev. 1997;2:8–23.
4. Cantillon P. ABC of learning and teaching in medicine. Teaching large groups. Br Med J. 2003;326:437–40.
5. Brown G, Manogue M. Refreshing learning a guide for lecturers. AMEE Medical Education Guide No. 22: refreshing lecturing: a guide for lecturers. Med Teach. 2001;23:231–44.

6. UCL Department of philosophy, lecture evaluation form. http://www.ucl.ac.uk/philosophy/docs/lecture-evaluation-form.pdf.
7. Economics network. Evaluating your lecture. https://www.economicsnetwork.ac.uk/handbook/lectures/4.
8. Husbands CT. Variations in students' evaluation of teachers' lecturing in different course on which they lecture: a study at the London School of Economics and Political Science. Higher Educ. 1997;35:51–70.
9. Ramsden P. Learning to teach in higher education. London: Routledge; 1992.
10. Adshead I, White P, Stephenson A. Introducing peer observation of teaching to GP teachers: a questionnaire study. Med Teach. 2006;28:67.
11. Pendleton D, Scofield T, Tate P, Havelock P. The consultation: an approach to learning and teaching. Oxford: Oxford University Press; 1984.
12. Siddiqui ZS, Jonas-Dwyer D, Carr SA. Twelve tips for peer observation of teaching. Med Teach. 2007;29:297–300.
13. Pinsky LE, Irby DM. "If at first you don't succeed": using failure to improve teaching. Acad Med. 1997;72:973–6.
14. Flipped learning. In: Knowledge Hub, Higher Education Academy. 2017. www.heacademy.ac.uk/knowledge-hub/flipped-learning-0. Accessed 7 Feb 2019.
15. Acedo M. Ten pros and cons of a flipped classroom. In: TeachThought. 2019. www.teachthought.com/learning/10-pros-cons-flipped-classroom. Accessed 7 Feb 2019.
16. Sinclair M, Rowe K, Brown G. The minute paper: a quick guide to assessing student learning. Nurs Times. 1998;2:4–5.
17. University of Minnesota, Evaluating lectures. http://www.umn.edu/ohr/.
18. McKeachie WJ. Teaching tips: a guidebook for the beginning college teacher. 9th ed. Lexington: Heath; 1994.
19. Peel D. Peer observation as a transformatory tool? Teach High Educ. 2005;10:487–504.
20. Nicoll T, Schofield S, Ibrahim M. "Getting started…" with large group teaching. In: Dent JA, Davis MH, editors. "Getting started…" a practical guide for clinical teachers. 3rd ed. Dundee: Centre for Medical Education, University of Dundee; 2008.

14 Choosing Between Lecture and Briefing Sessions

Nirusha Lachman and Wojciech Pawlina

Introduction

Teaching practices that once impressed undergraduate and graduate medical students can no longer sustain a generation of technologically skilled students for whom access to information is no longer confined to the classroom. The World Wide Web, online course material, electronic databases, and e-books present a unique challenge to traditionally approached curriculum delivery. For the anatomy professor, the time for lamenting is over. The focus is no longer on how knowledge should be *transferred* but rather on how knowledge should be *managed*.

One may argue that in the era of technological domination, massive open online courses, virtual and augmented reality platforms, webinars, flipped classrooms, video streaming, and, among others, social media, students are able to independently direct their own learning, eliminating the need for formal class time. While this is true in the general sense of the underlying principle of technology-based learning, the need for student-to-teacher and student-to-student interaction remains an important driver for intellectual and social development that meets competency expectations of future health-care professionals. The challenge therefore is embedded in the question of how to incorporate desired global outcomes and maximize the skill of teaching for a group of individuals who have free access to multiple sources of information but lack direction on how to interpret and develop clinical reasoning.

> The challenge should not be to determine supremacy of one methodology over another but to maximize the learning benefit available from the different methods. –Turney [1]

In this chapter, we explore core elements of what could be included in a teaching and learning session that is designed to target both expected outcomes and the skill of clinical reasoning.

Objective of Teaching Anatomy

Anatomy still remains one of the most valued basic science subjects in medical education among practicing physicians and residents [1, 2]. There are few, if any, anatomy programs that still provide a content-heavy course [3, 4]. The decrease in course hours, the reduction in laboratory time, and the positive move away from passive, didactic, detailed courses toward more clinically and functionally relevant anatomy [1, 5] have transformed traditional delivery of ana-

N. Lachman (✉) · W. Pawlina
Department of Clinical Anatomy, Mayo Clinic College of Medicine and Science, Mayo Clinic, Rochester, MN, USA
e-mail: Lachman.nirusha@mayo.edu; pawlina.wojciech@mayo.edu

L. K. Chan, W. Pawlina (eds.), *Teaching Anatomy*, https://doi.org/10.1007/978-3-030-43283-6_14

tomical knowledge into a more contemporary student-facilitated experience.

In a contemporary curriculum, a clinically integrated anatomy course for medical and allied health-care students should be designed to provide opportunities to develop clinical reasoning skills based on the (1) understanding of core anatomical concepts as it relates to their practice (through various modalities, e.g., radiology, ultrasonography, embryology) and (2) reflection, critical thinking, and clinical reasoning as it relates to application of basic science to patient care. In addition, non-discipline-related competencies that foster social awareness, successful team interaction, communication skills, leadership, and professionalism [6] should be incorporated through assignments and student-centered activities that contribute toward the overall course grade.

Target Audience

Effective strategies directed at millennial learners have highlighted the importance of a (1) learning environment that is relaxed and welcoming and (2) professional rapport with learners while reinforcing core concepts with clear presentation of learning objectives. Attention to current influences and trends using evidence-based educational methods has proved successful in connecting students with information that is relevant and resonant [7]. With clearly outlined objectives, the millennial student has little trouble providing deliverables for assigned tasks. They have little trouble accessing resources and will present you with completed projects having exhausted the information available on the World Wide Web. As independent learners, however, students are often challenged by their ability to critically evaluate or validate their knowledge. Furthermore, their capacity to determine what key concepts are important to facilitate their understanding of anatomy and apply its basic science principles for proficient patient care in the future remains sufficient. These academic themes provide the fundamental guiding principle that determines how students can be enabled to overcome steep learning curves and successfully meet the competencies expected of them [8].

Now several books and analyses on millennial generation have been published [9, 10]. Raines and Arnsparger [11] provide a comprehensive feedback-based report as to what promotes success for millennials in the workplace. When applied to the educational setting, a few key thoughts may be considered:

- *Interaction with peers and near-peer teachers is as important inside the classroom as it is outside.*
  In the current learning climate, medical schools are increasingly challenged by the lack of student interest in classroom attendance [12]. In many preclinical courses, replacement of traditional classroom approaches by learning technologies may indeed serve as justification to drive policies in favor of non-compulsory attendance. However, relinquishing opportunities to problem-solve with classmates and removing connectedness and the ability to be part of a team working toward a common goal is tangential to authentic health-care practice [7, 13]. In a team-based setting, students' opportunity for social interaction within the learning environment is met. As evidenced in many studies, millennial students are able to work effectively in teams as well as independently. The team arrangement characteristic of anatomy laboratory dissection tables may also be applied to the classroom setting. Organizing the classroom so that students are able to physically sit together in small groups encourages discussion and enables facilitators to lead question-based teaching sessions. In addition, physical proximity allows students within the group to gain better appreciation for team member's strengths and areas for improvement when providing peer feedback. Participation of teaching assistants [14] adds to the learning process as students are able to connect with their near-peer teachers at a level different than the way they would with faculty.

- *Active learning appeals to learning preferences.*
  To bring the best out of students, it is important to build into a curriculum learning activities that are relevant, current, and appealing to their learning preferences. Students who have been exposed to technology at an early age have an expectation that the curriculum they learn from will implement technologies. They appear technologically savvy and able to multitask with exceptional skill—keep them actively engaged. For anatomy curricula that include a radiology component, scanning whole cadavers and implementing CT-based activities in the laboratory keep students engaged as it combines the use of technology with opportunity to practice spatial reasoning skills. In addition, the use of ultrasound and other scanning modalities may benefit students' appreciation for understanding anatomy as it relates to clinical practice [15]. As health-care industry shifts focus to use of platforms that include artificial intelligence and virtual, augmented, and mixed reality [16, 17], classroom activities that integrate such modalities will soon become an expectation. While not yet pervasive, the growing trend as well as administrative pressure to incorporate 3D printing in addition to AR/VR platforms will remain both a challenge and an expectation within anatomy curricula [18–20].
- *Students value good feedback and are goal and achievement oriented.*
  As young adult learners, millennial students are driven by achievement of tangible goals. They set high standards for themselves and are determined to succeed. Providing them with regular quizzes and test scores and feedback encourages their learning drive. Provide them with meaningful feedback [21, 22] and keep them motivated within an academically challenging environment. Feedback may be elicited during the course from peers through formal evaluations (leadership and peer evaluations) and from near-peer teachers and faculty through informal reflective sessions (see Chap. 19). In addition, soliciting feedback from students reinforces the teaching and learning partnership and enhances meaningful delivery of anatomical concepts—one that also meets the needs of the student.
- *Students need direction and reasonable structure, relevance, and reassurance.*
  While flexibility is important to them, millennials have a distinct reliance on guidance and defined objectives. Providing clear objectives and achievable outcomes creates a sense of reassurance in their ability to succeed in meeting learning expectations. This becomes increasingly important for courses that provide 120 hours of teaching over a short time period (e.g., 6-week courses). With high volume of information, teachers have a greater responsibility to define specific objectives that are relevant to clinical interpretation. While students may be able to memorize with a reasonable level of understanding, actual depth of understanding depends largely on how they are guided through the material in order to internalize core concepts. In addition, information accessed through iPad and iPhone applications, the Internet, and anatomy websites may contain inaccuracies and points of focus different from courses in which they may be enrolled. Encouraging students to use a prescribed textbook and other faculty-endorsed resources helps provide students with the most reliable information. The expectation that the course will provide students with the most relevant and accurate core knowledge is a minimum expectation. Course material provided to students should be reviewed, vetted, and updated regularly to ensure that basic science concepts remain aligned with current practice [23].

> Given that these students are much more likely to have been exposed to new technologies compared to previous generations of students it is more likely that these students will expect academic staff to be comfortable with and utilize a wide range of technologies in their teaching. –Jonas-Dwyer and Pospisil [24]

Enabling Students to Achieve Outcomes

The past decade has seen extensive debate with the threat of de-emphasizing (and the fear of eliminating) the traditional approach to teaching anatomy [25]. In response, anatomy teachers turned to innovative ways in maintaining traditional approaches to teaching medical students about the human body [15, 26–29]. Regan de Bere and Mattick [27] further explored the complexity of anatomy as a subject and what is involved in achieving competency in understanding structure and its clinical application. Based on participant feedback, they defined anatomical competence as a multilevel expertise that combines factual knowledge; application of the knowledge; hands-on skills, insight, and understanding; personal awareness; and aesthetic appreciation. The study also affirms rote memorization as an unsuitable way of learning about a complex subject, the understanding of which depends on inquiry and creative thinking.

Traditional Lectures vs. Short Lectures

Traditional lecturing (at least 1-hour duration) remains one of the most commonly used methods of delivering information. It is a teacher-centered approach that involves continuous periods of exposition with the primary focus on information relay. While content of the lecture may vary, the style of delivery has little flexibility and is often disengaging and ineffective in promoting clinical reasoning. Therefore, a lecture would be most effective in the event that resources are limited, and the goal is to provide students with information from reliable sources. On the positive side, time designated for presentations may be used more effectively to explain concepts that are not easily grasped through diagrams, animation, or case studies. Either way, the approach remains one-dimensional and teacher centered.

With reduced emphasis on information transfer, anatomy teachers may choose to present shorter, more objective-driven presentations that help students determine important concepts while studying the material from assigned readings. The "short lecture" is designed to highlight the major anatomical points with greater focus on core objectives than factual information. When compared with traditional lecture approach, the short lecture reduces student listening time and provides more direction for both conceptual understanding and hands-on learning. However, it too is not effective for engaging student participation or promoting student thought.

How Do We Promote Clinical Reasoning and Critical Thinking?

Eric Mazur's enlightening viewpoint in *Science* (2009) emphasizes the notion that "*education is so much more than just information transfer*" [30]. His philosophy of "*teaching by questioning*" instead of "*teaching by telling*" forms the basis for the design of a teaching and learning model that de-emphasizes content, promotes critical thinking, and reinforces core concepts that can be applied in clinical reasoning. The term "briefing session" will be used to describe an interactive teaching and learning session designed to reinforce anatomical knowledge through a process of clinical reasoning and the promotion of thought.

> In hindsight, the reason for my students' poor performance is simple. The traditional approach to teaching reduces education to a transfer of information. –Mazur [30]

The Briefing Session

Elements of a Briefing Session

The framework for delivery is based on a team setting. The expectation is that students would have read and become familiar with basic anatomy of the region under study, aligned with the principles of team-based learning [25, 31, 32]. Access to resources in the form of prescribed texts, suggested reading, and online sources may be used at the discretion of the student, thereby cultivating a sense of independence in information gathering. This may appeal to and benefit students with different approaches to learning, thereby promoting knowledge retention more effectively than a time-limited traditional lecture presentation. It aims to prepare students for deep-level learning through questioning and critical evaluation as opposed to superficial memorization of material. In addition, it provides students with a form of immediate feedback such as formative feedback sessions using the audience response system (see Chap. 19) enabling them to evaluate their knowledge interpretation and retention. By reducing the time spent on information transfer, briefing sessions can allocate time for orientating students for the laboratory exercises allowing students with untrained dissection skill progress with more ease through laboratory assignments:

- *Team-based setting.*
  The briefing session is most effective when used in a team-based setting but is also useful for larger classroom settings. By promoting a partnership in teaching and learning, students are responsible for acquiring information under guidance of teachers and near-peer teachers, through endorsed online material and prescribed texts. As mentioned earlier, arranging the classroom so that students are able to sit in groups further facilitates team interaction both academically and in terms of promoting non-discipline-related skills.
- *Use a combination of resources.*
  The briefing session is designed to promote a culture of personal responsibility for acquisition of knowledge. Choose texts that are peer reviewed from well-established authors. The choice of text will depend on the target audience. Consider the academic level and academic expectation for the group. Anatomy courses for medical students will require a text that is clinically oriented with simple diagrams and high-yield highlighted concepts, whereas an anatomy course for surgical residents may benefit from a more detail-oriented text focused on anatomical landmarks and relations as well as surgical approaches.
- *Search the Internet for reliable online course material.*
  The most reliable online course material is generally accessed from established university sites or a source affiliated with a learning center or educational institution. Despite being skilled at searching the Internet, students value teacher-endorsed material [33, 34].
- *Provide a database of electronic material and leverage learning management systems.*
  Lecture notes, slides, short video clips, animations, and clinical images if available may be uploaded onto websites or accessed through course management systems such as Blackboard.
- *Encourage practice of peer sharing of resources and learning material.*
  Peer study notes, online material, assignments, projects, dissection findings, student-generated video clips, reliable anatomical images, and radiologic images are valuable shared resources.

> Instead of teaching by telling, I am teaching by questioning. –Mazur [29]

Delivering a Briefing Session

1. *Objective-driven orientation.*
   The briefing session begins with a review of core concepts that are re-emphasized through an objective-driven orientation. Students are reminded of their reading assignments, and objectives are clearly projected (verbally and visually).
2. *Clinically oriented team question.*
   The session immediately continues with a clinically oriented multiple-choice ques-

tion developed around a key concept. Students engage in a 2- to 3-minute group discussion to reach a consensus, and answers are recorded using the audience response system [35, 36]. Answer statistics are then revealed, and through a process of clinical reasoning, each answer option is then discussed through student-teacher dialogue to assess answer possibilities. Through a process of reasoning, the correct answer is derived. In doing so, students maintain a partnership in determining the correct answer and are given the opportunity to question and understand the concept more clearly. It is also important to look at every option and not just present the correct answer. This way, students are encouraged to think and assess their own knowledge and understanding of the concept.

3. *Reinforce concept with two to three good teaching slides.*
   A few good teaching slides may be used to reinforce understanding of the concept. Choose clear images with simple points. Avoid scans of text images with multiple labels in annotations. Concept maps or flow diagrams may also be useful. Include where possible 3D printed material and additional opportunity for use of virtual classroom spaces to reinforce learning and accommodate a diverse population of our students.
4. *Relate core concepts to objectives for teaching within the context of the laboratory session.*
   At the end of the *briefing session*, students receive a pre-lab orientation highlighting important relations based on a list of anatomical structures relevant to assigned dissection.
5. *Formative feedback session and debriefing after laboratory session.*
   End the day's teaching by a short 15-minute debriefing and quick test questions (e.g., audience response system) for individual student feedback [35, 37].

Core Elements of a Briefing Session

Design the presentation around the following guiding principles:

- Concepts should have an underlying clinical principle.
- Formulate multiple-choice questions around the concept.
- Identify key anatomical structures easily located through dissection.
- Focus on anatomical relations.
- Avoid the tendency to present everything.

Choosing Content and Material for Briefing Session Slides

Figure 14.1 is an example of what may be included in a briefing session. The clinical concept (slide 1) may be determined from a prescribed clinically oriented text and should relate to commonly encountered clinical scenarios. The question (slide 2) should be written around the concept with answer possibilities that all relate to the concept. Two to three clear, simple diagrams (slides 3 and 4) accompanied by actual images (slide 5) should be used to explain the concept. Steer away from heavy textual information, but refer students to resources for further details. Provide tips for dissection and highlight the most important dissection outcome (slide 6). A 1-hour time frame should be sufficient to discuss at least five important concepts.

Saroyan and Snell [38] characterize three principle-driven lecture types: *content driven (teacher centered)*, *context driven (objective centered)*, and *pedagogy driven (student-centered active learning)*. While content-driven lectures received the lowest rating among participants (students and teachers), both context- and pedagogy-driven lectures were seen as useful methods of helping students get the most

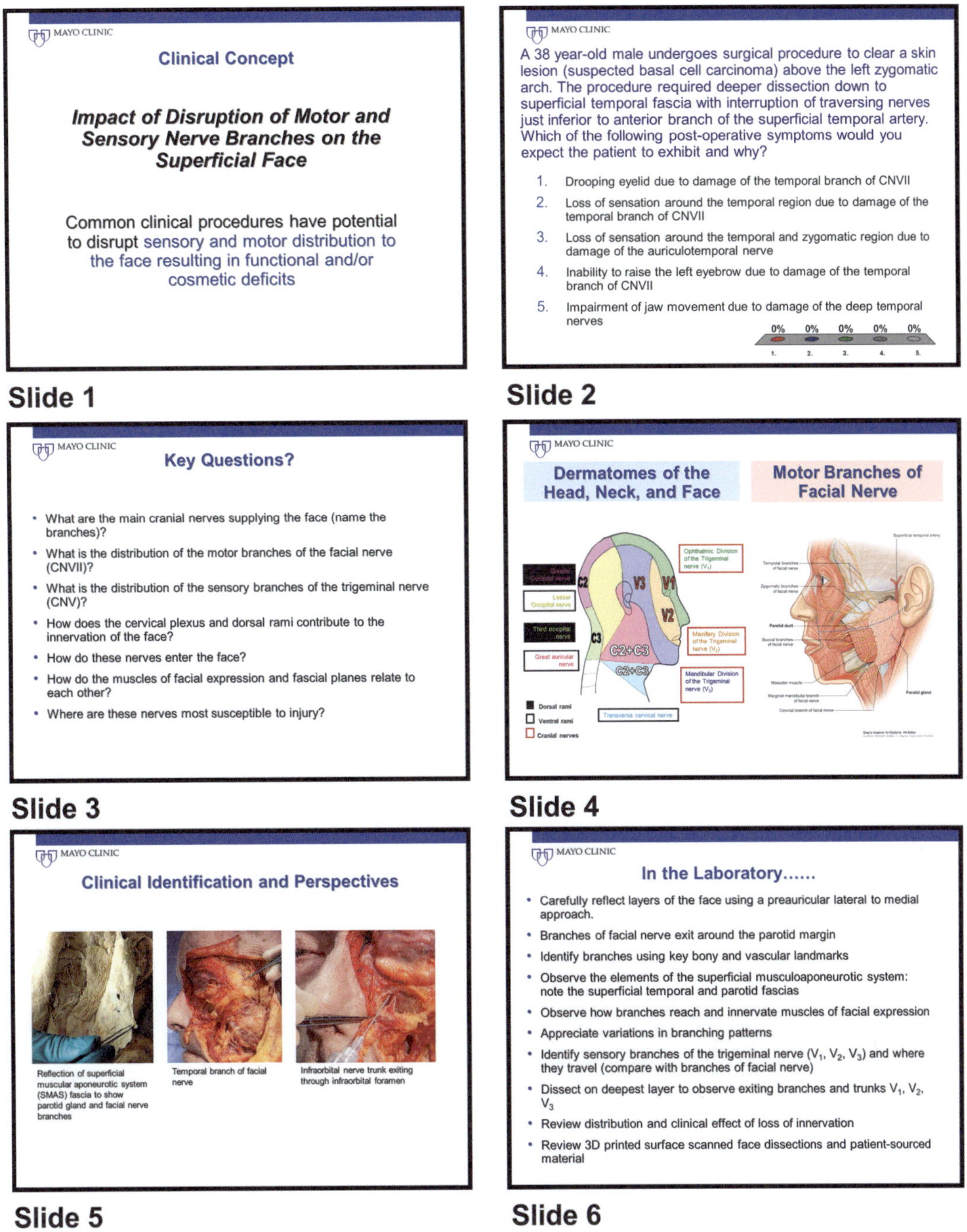

Fig. 14.1 Briefing session on innervation of the head, neck, and face: slide content

out of learning in the classroom. Pedagogy-driven lectures however seemed to be most valued as it targeted learning of clinically relevant material and provided opportunity to apply understanding by offering active learning opportunities during lectures.

Reviving Anatomy Teaching
Philosophical thoughts:

- Old style anatomy teaching has not survived a contemporary curriculum.
- Environment and medium of teaching anatomy must evolve to meet the requirements of the twenty-first-century learners [13, 39].
- Welcome teaching innovations that appeal to millennial students [7, 26].
- Minimize telling and promote discussion [30].

In this chapter, we have presented a method for conveying core concepts that are driven by inquiry, critical thinking, reflective learning, and student centeredness. While the design of the briefing session is embedded within a team-based learning framework, its principles may be as effectively applied to larger classroom settings. The implication of adopting a purely pedagogy-driven lecture is that not only is there greater student involvement but also greater demand on teacher creativity and planning and a change in the nature of the teaching mission—teaching by "questioning instead of teaching by telling" [30]. Millennial students will continue to inspire transformation [7, 40, 41] of educational tradition. For the anatomy professor, this challenge will continue to inform how anatomy courses and, more importantly, the delivery of anatomical knowledge will adjust to cater for the dynamic authentic learning environment, patient-driven health-care practice, and artificial intelligence platforms [42].

References

1. Turney BW. Anatomy in a modern medical curriculum. Ann R Coll Surg Engl. 2007;89(2):104–7.
2. Heylings DJ. Anatomy 1999-2000: the curriculum, who teaches it and how? Med Educ. 2002;36(8):702–10.
3. Drake RL, McBride JM, Lachman N, Pawlina W. Medical education in the anatomical sciences: the winds of change continue to blow. Anat Sci Educ. 2009;2(6):253–9.
4. McBride JM, Drake RL. National survey on anatomical sciences in medical education. Anat Sci Educ. 2018;11(1):7–14.
5. Sugand K, Abrahams P, Khurana A. The anatomy of anatomy: a review for its modernization. Anat Sci Educ. 2010;3(2):83–93.
6. Gregory JK, et al. Restructuring a basic science course for core competencies: an example from anatomy teaching. Med Teach. 2009;31(9):855–61.
7. Ruzycki SM, Desy JR, Lachman N, Wolanskyj-Spinner AP. Medical education for millennials: how anatomists are doing it right. Clin Anat. 2019;32(1):20–5.
8. Borges NJ, Manuel RS, Elam CL, Jones BJ. Comparing millennial and generation X medical students at one medical school. Acad Med. 2006;81(6):571–6.
9. Tulgan B. Not everyone gets a trophy: how to manage the millennials. 2nd ed. Hoboken: Wiley; 2016.
10. Jenkins R. The millennial manual: the complete how-to guide to manage, develop, and engage millennials at work. 1st ed. Atlanta: Ryan Jenkins LLC; 2017.
11. Raines C, Arnsparger A. Millennials at work. Wichita Falls: Claire Raines Associates; 2010. http://www.generationsatwork.com/millennials-at-work/. Accessed 20 June 2019.
12. Marzuk PM. More about medical students' attendance at lectures. Acad Med. 2013;88(2):149.
13. Pawlina W, Drake RL. Authentic learning in anatomy: a primer on pragmatism. Anat Sci Educ. 2016;9(1):5–7.
14. Lachman N, Christensen KN, Pawlina W. Anatomy teaching assistants: facilitating teaching skills for medical students through apprenticeship and mentoring. Med Teach. 2013;35(1):e919–25.
15. Kotze SH, Driescher ND, Mole CG. The translucent cadaver: a follow-up study to gauge the efficacy of implementing changes suggested by students. Anat Sci Educ. 2013;6(6):433–9.
16. de Faria JW, Teixeira MJ, de Moura Sousa Júnior L, Otoch JP, Figueiredo EG. Virtual and stereoscopic anatomy: when virtual reality meets medical education. J Neurosurg. 2016;125(5):1105–11.
17. Moro C, Štromberga Z, Raikos A, Stirling A. The effectiveness of virtual and augmented reality in health sciences and medical anatomy. Anat Sci Educ. 2017;10(6):549–59.
18. Izard SG, Juanes Méndez JA, Palomera PR. Virtual reality educational tool for human anatomy. J Med Syst. 2017;41(5):76.
19. McMenamin PG, Quayle MR, McHenry CR, Adams JW. The production of anatomical teaching resources using three-dimensional (3D) printing technology. Anat Sci Educ. 2014;7(6):479–86.
20. Young JC, Quayle MR, Adams JW, Bertram JF, McMenamin PG. Three-dimensional printing of archived human fetal material for teaching purposes. Anat Sci Educ. 2019;12(1):90–6.

21. Camp CL, Gregory JK, Lachman N, Chen LP, Juskewitch JE, Pawlina W. Comparative efficacy of group and individual feedback in gross anatomy for promoting medical student professionalism. Anat Sci Educ. 2010;3(2):64–72.
22. Herrmann-Werner A, Loda T, Erschens R, Schneider P, Junne F, Gilligan C, Teufe M, Zipfel S, Keifenheim KE. Face yourself! - learning progress and shame in different approaches of video feedback: a comparative study. BMC Med Educ. 2019;19(1):88.
23. Cotofana S, Lachman N. The superficial face dissection as an example for integrating clinical approaches, authentic learning, and changing perspectives in anatomy dissection. Anat Sci Educ. 2020;13(1):117–21.
24. Jonas-Dwyer D, Pospisil R. The millennial effect: implications for academic development. Proceedings of the 2004 annual international conference of the Higher Education Research and Development Society of Australasia (HERDSA). 2004. pp. 356–366.
25. Vasan NS, DeFouw DO, Compton S. Team-based learning in anatomy: an efficient, effective, and economical strategy. Anat Sci Educ. 2011;4(6):333–9.
26. Nanjundaiah K, Chowdapurkar S. Body-painting: a tool which can be used to teach surface anatomy. J Clin Diagn Res. 2012;6(8):1405–8.
27. Regan de Bere S, Mattick K. From anatomical 'competence' to complex capability. The views and experiences of UK tutors on how we should teach anatomy to medical students. Adv Health Sci Educ. 2010;15(4):573–85.
28. Trelease RB, Nieder GL. Transforming clinical imaging and 3D data for virtual reality learning objects: HTML5 and mobile devices implementation. Anat Sci Educ. 2013;6(4):263–70.
29. Trelease RB. From chalkboard, slides, and paper to e-learning: how computing technologies have transformed anatomical sciences education. Anat Sci Educ. 2016;9(6):583–602.
30. Mazur E. Education. Farewell, lecture? Science. 2009;323(5910):50–1.
31. Vasan NS, DeFouw DO, Holland BK. Modified use of team-based learning for effective delivery of medical gross anatomy and embryology. Anat Sci Educ. 2008;1(1):3–9.
32. Vasan NS, DeFouw DO, Compton S. A survey of student perceptions of team-based learning in anatomy curriculum: favorable views unrelated to grades. Anat Sci Educ. 2009;2(4):150–5.
33. Miller GW, Lewis TL. Anatomy education for the YouTube generation: technical, ethical, and educational considerations. Anat Sci Educ. 2016;9(5): 496–7.
34. Barry DS, Marzouk F, Chulak-Oglu K, Bennett D, Tierney P, O'Keeffe GW. Anatomy education for the YouTube generation. Anat Sci Educ. 2016;9(1):90–6.
35. Alexander CJ, Crescini WM, Juskewitch JE, Lachman N, Pawlina W. Assessing the integration of audience response system technology in teaching of anatomical sciences. Anat Sci Educ. 2009;2(4):160–6.
36. Moorleghen DM, Oli N, Crowe AJ, Liepkalns JS, Self CJ, Doherty JH. Impact of automated response systems on in-class cell phone use. Biochem Mol Biol Educ. 2019;47(5):538–46.
37. Youdas JW, Krause DA, Hellyer NJ, Rindflesch AB, Hollman JH. Use of individual feedback during human gross anatomy course for enhancing professional behaviors in doctor of physical therapy students. Anat Sci Educ. 2013;6(5):324–31.
38. Saroyan A, Snell LS. Variations in lecturing styles. High Educ. 1997;33(1):85–104.
39. Johnson EO, Charchanti AV, Troupis TG. Modernization of an anatomy class: from conceptualization to implementation. A case for integrated multimodal-multidisciplinary teaching. Anat Sci Educ. 2012;5(6):354–66.
40. DiLullo C, McGee P, Kriebel RM. Demystifying the millennial student: a reassessment in measures of character and engagement in professional education. Anat Sci Educ. 2011;4(4):214–26.
41. Twenge JM. Generational changes and their impact in the classroom: teaching generation me. Med Educ. 2009;43(5):398–405.
42. Hamet P, Tremblay J. Artificial intelligence in medicine. Metabolism. 2017;69S:S36–40.

Facilitating Small Group Learning 15

Boon Huat Bay, Samuel Sam Wah Tay,
and Dinesh Kumar Srinivasan

A proficient knowledge in human anatomy is not only essential to prepare medical students for their future training in the clinical disciplines but also is necessary for good clinical practice [1–3]. In a survey of medical schools in the United States [4], gross anatomy classroom hours in 2016–2017 have increased by approximately 24%, while laboratory hours in gross anatomy have reduced by 16% compared to a similar survey published in 2014 [5]. Therefore, there is a need to develop the best instructional methods for teaching gross anatomy.

Patel and Moxham [6] observed that cadaveric dissection by students was the most preferred method of teaching by anatomists, followed by prosection classes, living (surface) anatomy and radiological anatomy, computer-aided learning, didactic lectures alone, and the use of anatomical models [1–6]. Although the didactic lecture may not always produce learning [7] or promote long-term retention [8], this approach is still commonly used by many teachers.

B. H. Bay (✉) · S. S. W. Tay
Department of Anatomy, Yong Loo Lin School of Medicine, National University of Singapore, National University Health System, Singapore, Singapore
e-mail: antbaybh@nus.edu.sg; anttaysw@nus.edu.sg

D. K. Srinivasan
Department of Anatomy, Yong Loo Lin School of Medicine, National University of Singapore, Singapore, Singapore
e-mail: dineshkumar@nus.edu.sg

Small group teaching has been extensively applied in the teaching of human anatomy [9, 10]. This teaching methodology was reported to be an effective method for laboratory teaching of human gross anatomy, especially when combined with the appropriate use of other modalities such as plastinated specimens and radiology-based imaging [11]. In the same study, 82.4% of the students found that anatomy learning objectives were better understood in the laboratory than those derived from didactic lectures.

Small group teaching is not only restricted to the study of human structures but has also been utilized to enhance students' understanding and application of physiological concepts [12] and for the teaching of biochemistry [13]. However, among the preclinical subjects, human anatomy classes tend to have considerable faculty–student interaction, as teachers have the opportunity to work with students directly in small laboratory groups and also during tutorial classes [14].

Teaching Pedagogies Using Small Group Discussions

Dissection/Prosection Laboratories

The dissection/prosection classes are highly valuable as they provide hands-on experience, enable the appreciation of three-dimensional relationships and anatomical variations in the human body, and help to

L. K. Chan, W. Pawlina (eds.), *Teaching Anatomy*, https://doi.org/10.1007/978-3-030-43283-6_15

develop fine motor control skills as well as promote teamwork and professionalism [1, 15–18].

Tutorials

Tutorials are usually given after the students have attended didactic lectures on a region of the human body and completed their practical classes [19]. Here, individual tutors review and discuss the tutorial objectives (which have been made available earlier) with small groups of students.

Team-Based Learning (TBL)

Michaelsen and colleagues originally developed team-based learning (TBL) for business courses [20]. TBL, an instructional strategy that involves discussions in small groups, has been credited for nurturing an active learning environment by promoting student engagement and peer teaching [21]. TBL emphasizes the importance of teamwork, mastery of content, and solving clinically relevant problems. Hence, TBL has emerged as an attractive strategy to adopt for imparting human gross anatomical knowledge, because it requires students to be conversant in gross human structures, from which they construct anatomical concepts and principles that could help them to solve relevant clinical problems [10, 21, 22].

Problem-Based Learning (PBL)

The integration of basic science knowledge with medical data pertaining to patients plays an important role in clinical assessment and patient diagnosis [23]. Problem-based learning (PBL) helps to consolidate integrative thinking among medical students [24]. First implemented by the McMaster University Medical School in 1969, the PBL approach focuses on collaborative learning in small groups of students based on a prescribed clinical scenario [25]. The students drive the learning process and are responsible for the acquisition of knowledge (which includes recalling prior knowledge), thereby not only promoting active learning but also preparing them for self-directed lifelong learning, which is in light of the burgeoning medical information that is made available, absolutely essential in the medical profession [26, 27].

Teaching with Simulators

This is a powerful method for teaching clinical skills and appropriate application of knowledge. The advantage of this type of teaching method largely centers around the opportunity for the learner to practice and learn in an environment using simulators which is as close to reality as possible. An important consideration for facilitators involved in teaching with simulators is that it requires the facilitator to be completely familiar with the simulation scenario and procedures [28].

Advantages of Small Group Discussions

Encourages Participation of Students

In a small group setting, students are more willing to share their knowledge and speak up as they would be in a less intimidating environment. Some students are reserved and shy and would therefore be more reluctant to speak out in a large class situation.

Improves Communication Skills

Students will learn to organize their thoughts before presenting to their peers, thereby developing their interpersonal and communication skills [11]. Effective communication between a doctor and his/her patients is regarded as an integral part of medical practice [29].

Reaches Out to Students with Different Learning Styles

Students come from diverse backgrounds and culture and hence may have dissimilar learning styles. The words "learning style" have been defined as "a particular set of behaviors and attitudes related to learning context" [30]. The tutorial caters to both active students and reflective learners and also enable the teacher to tailor specific strategies for each student [31].

Inculcates Responsibility

Small group discussions help to develop a professional attitude of responsibility and accountability towards each other. The students are expected to come prepared for the tutorials and must actively contribute to the learning process. They are aware that if they frequently come unprepared for tutorials, they would leave an unfavorable impression of themselves not only to the tutor but to their peers as well.

Advantages of small group discussions:
- Encourages student participation
- Improves communication skills
- Reaches out to students with different learning styles
- Inculcates responsibility

Features of a Tutorial Class

In this chapter, more emphasis will be placed on tutorial classes. The tutorial reinforces what has been taught during the lectures and what students have observed during their laboratory classes. Tutorial classes are particularly useful for discussing the clinical relevance of the anatomical knowledge learned and also to achieve analysis type of outcomes, with special reference to clinical scenarios.

Typically, at the start of a tutorial, some teachers may set quizzes for the students so as to assess their understanding of the topic and others may give an overview of what is to be achieved during the class. The bulk of the time in a tutorial class should be spent on discussing the tutorial objectives.

Tutorial Objectives

The compiled tutorial objectives for a region (for instance, the upper limb) would have been made available to the students at the start of their study for that region.

Types of Objectives

The objectives can be divided into general objectives which enable the students to have an overview of the conceptual framework and anatomical structures under consideration in the region of interest. The specific objectives review the details of structures in the region under study. For instance, for a topic such as "Nerves of the Upper Limb," an example of a "general objective" would be to "understand the formation of the brachial plexus and distribution of its major branches," and an example of a "specific objective" would be to "trace the course of the ulnar nerve and enumerate the functional loss if there is an injury to this nerve at the elbow and at the wrist."

Student Preparation

Some tutors may assign specific tutorial objectives to individual students while others may not. The pitfall of the former approach is that students may not prepare for the other objectives of the topic under study other than those specifically allocated to them.

Mode of Presentation

Tutors may also request the students to prepare PowerPoint presentations or may themselves use PowerPoint or the more sophisticated HTML 5 (current version of the Hypertext Markup Language [HTML] standard) for creating web applications. Currently there are also many PowerPoint alternatives, such as Panopto, Prezi, SlideShark, Haiku Deck, Keynote, Visme, Emaze, Projeqt, Slides, Slidedog, Slidebean and Zoho Show, and animation presentation software such as Powtoon and GoAnimate. With advances in educational technology, multimedia platforms, augmented reality and virtual reality (VR) can be incorporated into the tutorial sessions and may have benefits such as increased student engagement and interactivity [32]. As VR can be categorized into two types depending on the level of interaction and immersive environment [33], there could a dedicated virtual room or simply computer-generated 3D anatomical structures represented on a personal computer in a non-immersive VR environment [32, 33].

Active Participation of Students

One of the important ingredients of a well-conducted tutorial is active student participation. Students should be encouraged to present facts and information on the topic under study, answer questions, and seek clarification if they have any doubts. The tutor should only act as a facilitator during the tutorial session and should not conduct a mini-lecture, thereby providing answers to all the stated objectives.

At the closure of the tutorial class, some tutors give quizzes to find out how much knowledge the students have imbibed and whether they have acquired a better understanding of difficult concepts, while others may highlight take-home messages to the students.

Features of a tutorial class:

- Tutorial objectives are distributed before the class.
- Students are required to make adequate preparation for the classes.
- Students are expected to participate actively in the discussion.

Tips for Facilitating Group Dynamics in a Tutorial Class

As pointed out by Lee et al. [34], tutorial skills can generally be divided into group dynamics and discussion content. Group dynamics has been defined as the flow of discussion and interpersonal interaction [35]. Lee and colleagues [34] have reiterated four norms, namely, (a) iteration of the principles for the class, (b) delegating the responsibility of learning to students, (c) creating a good discussion forum, and (d) developing a conducive learning environment. These principles have been incorporated into the steps below for facilitating group dynamics.

Setting Up the Environment

Depending on the size and shape of the tutorial room, the tutor may want to arrange the chairs either in a circle or in two or three rows so that student–student and student–teacher interactions can flow better.

Establishing Ground Rules

The tutor may want to set ground rules during the first tutorial class such as punctuality, respect, and showing courtesy to the tutor and tutorial mates by not interrupting unnecessarily or talking among themselves during class.

Iterating the Principles of the Tutorial Class

Students must understand that a tutorial class is not a mini-lecture delivered by the tutor. They should realize that they would benefit most if they have read up and come prepared for classes and be willing to participate actively. But they should not feel stressed when they make mistakes, since the tutor should in correcting their mistakes, create a safe, explorative learning environment.

Delegating the Responsibility of Learning to Students

Tutorial sessions should be student-centered and the students should be the driving force. Students should be motivated to share their knowledge with their peers and be given the opportunity to clarify any difficult concepts and principles with the tutor. As student-centered learning is rooted in Western culture, one must be cognizant that cultural factors may have implications on student behaviors, such as perception of group relations, competition, and achievement [36]. The dimensions of culture have been categorized by Joy and Kolb into "in-group collectivism, institutional collectivism, power distance, uncertainty avoidance, future orientation, performance orientation, humane orientation, assertiveness and gender egalitarianism" [37].

Creating a Good Discussion Forum

Things *not* to do:
- Conduct a mini-lecture
- Speak all the time
- Be too authoritative

Time management is very important and each student should be given a chance to answer questions. Students should be encouraged to speak up and given the opportunity to clarify any difficult concepts and principles during the course of their study. It is also essential that tutors prepare ahead of time to ask varied and purposeful questions, answer potential learner questions, and respond appropriately to comments. At the end of discussing each objective, the tutor should encourage students to reflect on what they have learned. Tutors can also gauge how each student is performing and identify the weak students for remedial action early.

Engaging Students

Some students may be shy and may not talk unless the tutor specifically directs questions to them. For students who are assertive and always want to be in the limelight, the tutor should speak to them discretely and explain to them that they have to give others a chance to participate in the class. There may also be students who show lack of interest during the class and the teacher should speak to them after the class or at an appropriate time, to find out what is bothering them and counsel them if necessary. What happens if none of the students are able to give an answer and all have become quiet? Rather than giving the answer straightway, the tutor can ask related questions that would lead them to the correct answer.

Developing a Conducive Learning Environment

It is important for the group to be in an environment where the atmosphere is relaxed and comfortable. A tutor may want to introduce some humor or tell a story (viz., a patient with an interesting case history) at an appropriate time and ensure that there is a balanced interaction among the group members.

Things to do that will improve group dynamics:
- Know your students by their names.
- Be student centered.
- Try to involve every student in the discussion.
- Ensure balanced interaction among students.
- Allow students to clarify difficult concepts.
- Identify the weaker students.
- Make sure the atmosphere is relaxed.

What will you do?

- You have just traced the course of the ulnar nerve with the students and discussed its distribution.
- Going to the next objective, you asked the student what functional loss can result from a lesion to the nerve at the medial condyle and he replied "wrist drop." The student next to him looked perplexed. What would you do?

Specific Teaching Methods Used for Facilitating a Tutorial Class

There are also specific methods that have been used within traditional small group teaching sessions (seminars, tutorials and workshops), such as role-play, fishbowl training, lecturing, brainstorming, step-by-step discussion, tutorless groups, and free discussion [38]. Other methods include gamification, which is utilizing game mechanics in a non-gaming environment for learning specific content and varying the intensity of the learning experience [28]. In fishbowl training, the basic structure is that students are divided into two groups: a working (discussion) group and an observing group [39]. Students in the working group discuss a given assignment (a relevant issue, topic, or case) which is observed by the other group, who will then assess the validity and merits of arguments proposed [38]. Once the working group's allotted time is over, the observing group has the opportunity to ask questions, make comments, and offer feedback [39]. The facilitator who is monitoring the session will provide a summative feedback at the end [38]. Fishbowl training is believed to encourage team building and improve intergroup communications and relations [40–44].

An emerging platform used to facilitate small group teaching is the numerous audience response software (ARS) that has been made available which allows for tutor–student interaction, such as Poll Everywhere, Slido, Glisser, Crowdpurr, OMBEA, Pigeonhole Live, TurningPoint, Bravura, Crowd Connected, MeetingPulse, and many more. For instance, Poll Everywhere (https://www.polleverywhere.com/), a bring-your-own-device (smartphone, laptop, or tablet) classroom response system, has a user-friendly application that can be downloaded from the app store. An example of responses to a multiple-choice question (MCQ) answered by a group of students is shown in Fig. 15.1. Other than MCQs, instructors also have the choice of asking open-ended questions.

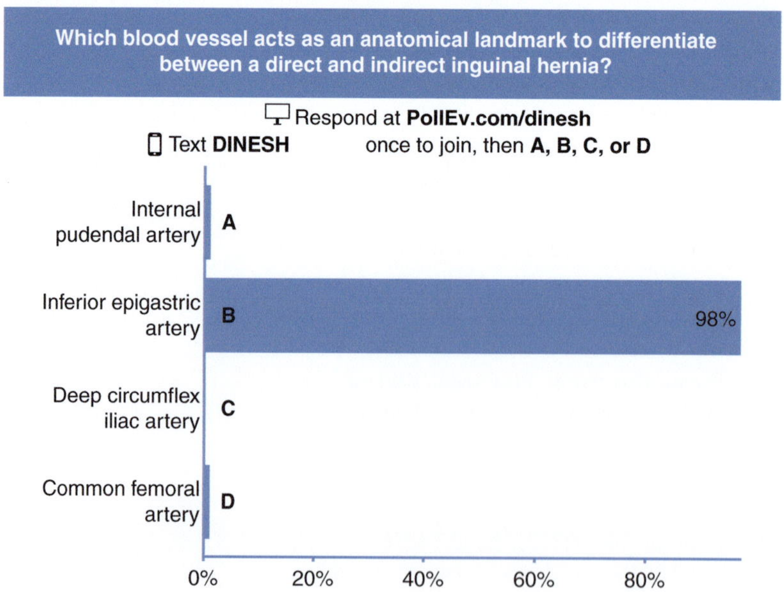

Fig. 15.1 A screenshot of an MCQ answered by a group of students using the Poll Everywhere software

In a study on the utility of Poll Everywhere in students taking the Investigative Biology Laboratory course at Cornell University, the investigators observed that the advantages of using web-based response systems prevailed over the challenges faced and concluded that this form of digital pedagogy can help establish a productive channel of communication with students in the classroom, albeit large group teaching [45]. What about the advantages of using ARS in the small group setting? Because of anonymity, students would be more inclined to be more open with their responses without fear of being judged [46]. Moreover, tutors are able to ascertain if the students are keeping up with the tutorial in real time [45]. In a study of 76 dental students who were broken up into two groups, for two different topics within the Orthodontics curriculum to test the use of ARS in a crossover trial, the authors concluded that ARS improved student concentration and enhanced knowledge retention [47].

Conclusions

There is no doubt that small group discussions are an effective instructional method in learning the anatomy of the human body. An important facet of small group teaching is the tutor's ability to facilitate the discussions [48]. Small group teaching would imply that there would be a high teacher–student ratio so as to facilitate optimal teacher–student interactions. Faced with a manpower crunch, some medical schools have tried to alleviate the shortage of teachers by using experienced students (at more advanced stages of their training) as tutors [49] with what is termed as "near-peer" teaching [50]. For medical students and residents, small group teaching would enable them to learn more effectively the "information and skills that are necessary for competent clinical practice of medicine" [51]. It is clear that small group discussion, as a teaching modality in human anatomy, is here to stay. A well-organized small group discussion session is dependent not only on the content for discussion but also on the group dynamics. In this respect, the tutor plays a vital role in facilitating a fruitful small group discussion.

> Small group discussions, if facilitated by a good tutor, enhances the students' understanding of the subject, improves interpersonal and communication skills, reaches out to students with different learning styles, and helps to develop a sense of accountability to others.

References

1. Bay BH, Ling EA. Teaching of anatomy in the new millennium. Singap Med J. 2007;48(3):182–3.
2. Ahmed K, Rowland S, Patel V, Khan RS, Ashrafian H, Davies DC, et al. Is the structure of anatomy curriculum adequate for safe medical practice? Surgeon. 2010;8(6):318–24.
3. Ang ET, Sugand K, Hartman M, Seow CS, Bay BH, Abrahams P. Singapore's anatomical future: quo vadis? Anat Sci Educ. 2012;5(4):234–40.
4. McBride JM, Drake RL. National survey on anatomical sciences in medical education. Anat Sci Educ. 2018;11(1):7–14.
5. Drake RL, McBride JM, Lachman N, Pawlina W. An update on the status of anatomical sciences education in United States Medical Schools. Anat Sci Educ. 2014;7(4):321–5.
6. Patel KM, Moxham BJ. Attitudes of professional anatomists to curricular change. Clin Anat. 2006;19(2):132–41.
7. Rizzolo LJ, Rando WC, O'Brien MK, Haims AH, Abrahams JJ, Stewart WB. Design, implementation, and evaluation of an innovative anatomy course. Anat Sci Educ. 2010;3(3):109–20.
8. Michael J. What makes physiology hard for students to learn? Results of a faculty survey. Adv Physiol Educ. 2007;31(1):34–40.
9. Arroyo-Jimenez Mdel M, Marcos P, Martinez-Marcos A, Artacho-Perula E, Blaizot X, Munoz M, et al. Gross anatomy dissections and self-directed learning in medicine. Clin Anat. 2005;18(5):385–91.
10. Nieder GL, Parmelee DX, Stolfi A, Hudes PD. Team-based learning in a medical gross anatomy and embryology course. Clin Anat. 2005;18(1):56–63.
11. Chan LK, Ganguly PK. Evaluation of small-group teaching in human gross anatomy in a Caribbean medical school. Anat Sci Educ. 2008;1(1):19–22.
12. Carvalho H, West CA. Voluntary participation in an active learning exercise leads to a better understanding of physiology. Adv Physiol Educ. 2011;35(1):53–8.
13. Bobby Z, Koner BC, Sridhar MG, Nandeesha H, Renuka P, Setia S, et al. Formulation of questions followed by small group discussion as a revision exercise at the end of a teaching module in biochemistry. Biochem Mol Biol Educ. 2007;35(1):45–8.

14. Ward PJ. First year medical students' approaches to study and their outcomes in a gross anatomy course. Clin Anat. 2011;24(1):120–7.
15. Bockers A, Jerg-Bretzke L, Lamp C, Brinkmann A, Traue HC, Bockers TM. The gross anatomy course: an analysis of its importance. Anat Sci Educ. 2010;3(1):3–11.
16. McLachlan JC, Patten D. Anatomy teaching: ghosts of the past, present and future. Med Educ. 2006;40(3):243–53.
17. Older J. Anatomy: a must for teaching the next generation. Surgeon. 2004;2(2):79–90.
18. Pawlina W. Professionalism and anatomy: how do these two terms define our role? Clin Anat. 2006;19(5):391–2.
19. Wong WC, Tay SS. The teaching of anatomy: the first hundred years (1905–2005). Ann Acad Med Singap. 2005;34(6):72C–8C.
20. Michaelsen LH, Fink RH, Knight A. Designing effective group activities: lessons for classroom teaching and faculty development. In: Dezure D, editor. To improve the academy, vol. 16. Stillwater: New Forums Press and the Professional and Organizational Development Network in Higher Education; 1997. p. 373–97.
21. Vasan NS, DeFouw DO, Holland BK. Modified use of team-based learning for effective delivery of medical gross anatomy and embryology. Anat Sci Educ. 2008;1(1):3–9.
22. Vasan NS, DeFouw DO, Compton S. A survey of student perceptions of team-based learning in anatomy curriculum: favorable views unrelated to grades. Anat Sci Educ. 2009;2(4):150–5.
23. Woods NN, Neville AJ, Levinson AJ, Howey EH, Oczkowski WJ, Norman GR. The value of basic science in clinical diagnosis. Acad Med. 2006;81(10 Suppl):S124–7.
24. DiLullo C, Morris HJ, Kriebel RM. Clinical competencies and the basic sciences: an online case tutorial paradigm for delivery of integrated clinical and basic science content. Anat Sci Educ. 2009;2(5):238–43.
25. Spaulding WB. The undergraduate medical curriculum (1969 model): McMaster university. Can Med Assoc J. 1969;100(14):659–64.
26. Bate E, Taylor DC. Twelve tips on how to survive PBL as a medical student. Med Teach. 2013;35(2):95–100.
27. Gurpinar E, Kulac E, Tetik C, Akdogan I, Mamakli S. Do learning approaches of medical students affect their satisfaction with problem-based learning? Adv Physiol Educ. 2013;37(1):85–8.
28. Jones RW. Learning and teaching in small groups: characteristics, benefits, problems and approaches. Anaesth Intensive Care. 2007;35(4):587–92.
29. De Haes JC, Oort FJ, Hulsman RL. Summative assessment of medical students' communication skills and professional attitudes through observation in clinical practice. Med Teach. 2005;27(7):583–9.
30. Brown B. Myths and realities No. 26. Teaching style vs learning style. Educational Resources Information Center: Columbus; 2003.
31. Alghasham AA. Effect of students' learning styles on classroom performance in problem-based learning. Med Teach. 2012;34(Suppl 1):S14–9.
32. Moro C, Štromberga Z, Raikos A, Stirling A. The effectiveness of virtual and augmented reality in health sciences and medical anatomy. Anat Sci Educ. 2017;10(6):549–59.
33. Lee EA-L, Wong KW. A review of using virtual reality for learning. In: Pan Z, Cheok AD, Müller W, editors. Transactions on edutainment I. Berlin: Springer-Verlag; 2008. p. 231–41.
34. Lee GH, Lin CS, Lin YH. How experienced tutors facilitate tutorial dynamics in PBL groups. Med Teach. 2013;35(2):e935–42.
35. Azer SA. Facilitation of students' discussion in problem-based learning tutorials to create mechanisms: the use of five key questions. Ann Acad Med Singap. 2005;34(8):492–8.
36. Frambach JM, Driessen EW, Chan LC, van der Vleuten CP. Rethinking the globalisation of problem-based learning: how culture challenges self-directed learning. Med Educ. 2012;46(8):738–47.
37. Joy S, Kolb DA. Are there cultural differences in learning style? Int J Intercult Relat. 2009;33(1): 69–85.
38. Edmunds S, Brown G. Effective small group learning: AMEE guide No. 48. Med Teach. 2010;32(9):715–26.
39. Kane CM. Fishbowl training in group process. J Spec Group Work. 1995;20(3):183–8.
40. Jaques D. Teaching small groups. BMJ. 2003;326(7387):492–4.
41. Eitington JE. The winning trainer. 4th ed. New York: Routledge; 2011.
42. Fry H, Ketteridge S, Marshall S. A handbook for teaching and learning in higher education. 3rd ed. New York: Routledge; 2009.
43. Race P. The lecturer's toolkit. 3rd ed. Abingdon: Routledge; 2007.
44. Tricio J, Montt J, Orsini C, Gracia B, Pampin F, Quinteros C, et al. Student experiences of two small group learning-teaching formats: Seminar and fishbowl. Eur J Dent Educ. 2019;23:151–58.
45. Sarvary MA, Gifford KM. The benefits of a real-time web-based response system for enhancing engaged learning in classrooms and public science events. J Undergrad Neurosci Educ. 2017;15(2):E13–6.
46. Kerns SC. Technological tools for library user education: one library's experience. Med Ref Serv Q. 2007;26(3):105–14.

47. Dhaliwal HK, Allen M, Kang J, Bates C, Hodge T. The effect of using an audience response system on learning, motivation and information retention in the orthodontic teaching of undergraduate dental students: a cross-over trial. J Orthod. 2015;42(2):123–35.
48. Barrows HS. The tutorial process. Springfield: Southern Illinois University School of Medicine; 1988.
49. Evans DJ, Cuffe T. Near-peer teaching in anatomy: an approach for deeper learning. Anat Sci Educ. 2009;2(5):227–33.
50. Duran CE, Bahena EN, Rodriguez Mde L, Baca GJ, Uresti AS, Elizondo-Omana RE, et al. Near-peer teaching in an anatomy course with a low faculty-to-student ratio. Anat Sci Educ. 2012;5(3):171–6.
51. Joslin S. Perceptions of anatomy education—a student's view. Anat Sci Educ. 2008;1(3):133–4.

Teaching and Learning Anatomy in a PBL Curriculum

16

Esther M. Bergman

This chapter describes good practice and pitfalls when teaching anatomy in a problem-based learning (PBL) curriculum. This should be distinguished from including PBL in anatomy education, which means applying the learning principles underlying PBL—constructive, collaborative, contextual, and self-directed learning [1]—in one or more anatomy courses, even when PBL is not included elsewhere in the program. However, some paragraphs may be useful for teachers who want to teach anatomy in a PBL format, and there are other valuable publications for them to draw from [2, 3]. Additionally, this chapter draws attention to some outdated notions and misconceptions concerning anatomy education in PBL.

Definitions

Constructive learning principle. Learners should be involved actively and should be stimulated towards activation of prior knowledge, elaboration, and deep learning, because this leads to deeper and richer understanding and better use of knowledge.

Collaborative learning principle. Learners should be stimulated to interact with each other because these interactions may positively influence learning.

Contextual learning principle. Learners are preferably exposed to a professionally relevant context and confronted with cases or problems from multiple perspectives, because this stimulates transfer of knowledge.

Self-directed learning principle. Learners should be stimulated to be aware of their prior knowledge and should be stimulated to regulate or direct their learning process both from a motivational and a cognitive perspective.

Anatomy Education in a PBL Curriculum

To clarify the teaching of anatomy in a PBL curriculum, it is helpful to compare it (somewhat simplistically) with teaching anatomy in a

E. M. Bergman (✉)
Department of Educational Research and Development, Faculty of Health, Medicine and Life Sciences, Maastricht University, Maastricht, The Netherlands

Department of Anatomy, Radboud University Medical Center Nijmegen, Nijmegen, The Netherlands
e-mail: e.bergman@maastrichtuniversity.nl

L. K. Chan, W. Pawlina (eds.), *Teaching Anatomy*, https://doi.org/10.1007/978-3-030-43283-6_16

"traditional" curriculum. Within a traditional curriculum, there is a "preclinical phase" in which basic sciences are presented in an isolated, department-based course. There is little or no integration across disciplines, little deliberate instruction in the application of basic science to clinical problems or the use of basic sciences in clinical skills, and limited student–patient interaction [4–6]. Courses consist mainly of lectures and teacher-centered instruction in laboratories.

In PBL and other "innovative" curricula, basic sciences are taught simultaneously and in an interrelated manner. This is called *horizontal integration*. Furthermore, clinical sciences are frequently introduced in the early years of the curriculum and integrated with the basic sciences. And basic sciences are revisited in later years in the context of clinical sciences. This is vertical integration [4]. Consequently, traditional curricula can be represented by an *H shape* and innovative curricula by a *Z shape* (see Fig. 16.1).

Curriculum integration can be achieved by providing interdisciplinary courses aligned by organ system or multidisciplinary courses aligned by theme (e.g., by disease) [7]. A PBL curriculum aligned by organ system should not be confused with a systems-based curriculum. Within a PBL curriculum, learning is centered around problems (see section "General Aspects of PBL"), and education is based on the four aforementioned learning principles. A systems-based curriculum may still be lecture based and teacher centered.

Influence of PBL on Anatomical Knowledge

Individual studies and meta-analyses examining the effectiveness of PBL have shown students to have more favorable attitudes towards and opinions about their education but no clear benefits or drawbacks of PBL curricula on students' basic science or clinical knowledge [8–15]. Therefore, it has been hypothesized that students' perceived and actual basic science (anatomical) knowledge is less determined by the general educational approach of the curriculum than by educational strategies such as time on task, repetition, and teaching in context [13, 16]. This shows the importance of a vertically integrated and spiral curriculum design.

Vertical integration within inter- or multidisciplinary courses is attained by integrating the teaching of basic sciences with clinical sciences. While there is widespread support among clinicians for more vertically integrated anatomy teaching in undergraduate curricula [17], vertical integration is often unidirectional. In other words, clinical topics are integrated into the early years (traditionally the time for

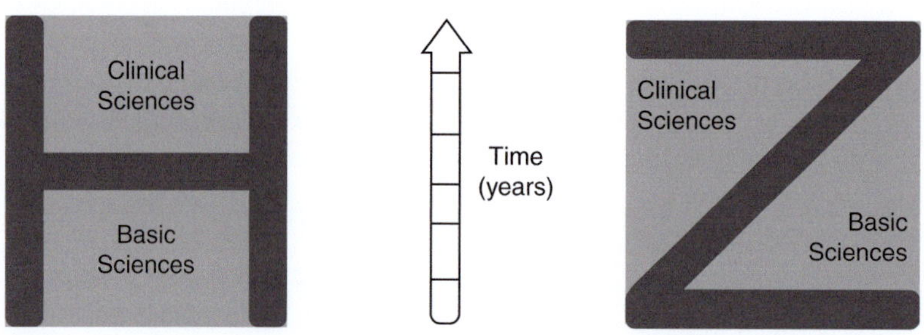

Fig. 16.1 The traditional curriculum can be represented by an H shape, with all clinical sciences later in the curriculum. The innovative curricula can be represented by a Z shape, where students are introduced to clinical sciences at the beginning of the curriculum

teaching basic sciences such as anatomy), but basic sciences are far less commonly taught in the later years [18–21]. A vertically integrated curriculum provides an opportunity to teach basic sciences in the context of the clinical sciences.

"A spiral curriculum is one in which there is an iterative revisiting of topics, subjects or themes throughout the course. A spiral curriculum is not simply the repetition of a topic taught. It requires also the deepening of it, with each successive encounter building on the previous one" [22, p. 141]. Knowledge is presented in a logical sequence from simple to complex in a spiral curriculum. Studies have shown this to be motivating because it activates and reinforces prior knowledge and stimulates a more advanced level of application and integration of knowledge, which increases expertise and feelings of competence [22, 23]. A spiral curriculum provides the opportunity for repetition and scaffolding of basic science knowledge.

To ensure anatomical knowledge is repeated and taught in context, anatomy teachers need to take on educational coordination tasks and play an active role in curriculum development committees. Furthermore, they need to be proactive in establishing collaborations with clinicians to teach basic sciences in the later years of the curriculum.

To Do

- Ascertain whether you know the options within your medical school through which you could integrate anatomy vertically or spirally.
- Furthermore, think of one clinician you know who is teaching in the later years of the curriculum and how you could collaborate with him or her to teach anatomy during that later clinical phase of the curriculum.

Teaching Anatomy in a PBL Curriculum

General Aspects of PBL

In order to strengthen connections between and within students' knowledge networks, PBL curricula are student centered and focus on active learning [1, 24]. Furthermore, clinical exposure is scheduled very early in the curriculum: even first-year students may, for example, attend non-invasive procedures such as angioplasty.

Basic to PBL is that student learning is organized around problems (patient cases, patient problem scenarios, professional problems). Tutorial groups of, on average, ten students and their tutor or facilitator meet about twice a week to discuss problems. These problems consist of a description of some phenomena, which students try to explain in terms of underlying biomedical or psychosocial processes, using their prior knowledge, opinions, and preexisting ideas. During this discussion, students discover what they already know and what they do not know. They then develop learning objectives to which they seek answers during the self-study phase by making use of faculty resources (lectures, laboratory sessions, skills training, etc.) and/or studying relevant literature. The initial discussion helps them with both the direction and extent of study they need to undertake [1, 25]. During the first part of the next tutorial group meeting, students discuss their answers to the learning objectives and thus acquire a deep understanding of the problem. They are then presented with a new problem and the cycle starts again.

Most anatomists have (at least) two roles in a PBL curriculum: PBL tutor and anatomy teacher. The challenges and opportunities of these two roles are described in the following paragraphs.

Learning Anatomy in Tutorial Groups

The notion that tutors should do nothing more than guarding the group process is outdated. What often happens is that students have information (the information on their learning goals or "the ingredients" in a cooking analogy) but are unable to make a complete story of it (use the information to explain the phenomena in the patient case, or use the ingredients to "bake the cake"). A tutor really needs to have some knowledge of the subject to help students create links between concepts (tutor as "recipe"). Furthermore, when problems are constructed to serve as a starting point for learning, the phenomena described are chosen with certain learning objectives in mind. With a well-written problem and a good discussion, students tend to formulate the intended learning objectives by themselves. Sometimes, however, students overlook or overemphasize a piece of information, which results in the learning objectives becoming too superficial or deep. The writers of a problem under discussion should create a "tutor manual," which would help address both types of issues. This tutor manual often includes the intended learning objectives so that tutors can steer the discussion and formulation of the learning objectives in the correct direction. Furthermore, it contains some concise information about the discussion topics. It is possible, nevertheless, that an anatomist has to do some studying—for example, on immunology if this is involved in the tutorial they are attending. The reverse may also be true: an immunologist attends a tutorial group in which he/she has to guide students discussing anatomical learning goals. It is therefore preferred that anatomists play an active role in writing the tutor manual to help non-anatomist tutors.

Within tutorial groups, learning can be hindered by the quality of the written problem, failure by tutors to act as facilitators, or dysfunction in the groups themselves. There is an abundance of literature on these issues; see, for example, Dolmans et al. [1]. Another very practical article for tutors is "Goals and strategies of a problem-based learning facilitator" by Hmelo-Silver and Barrows [26]. This section therefore focuses on issues that can be encountered when addressing anatomical knowledge in tutorial groups.

Practical Tips for Tutorial Groups

In your role as a tutor in tutorial groups, pay attention to the following:

- Using problems that fit with students' prior knowledge but are realistic, open ended, and *not* too simple or well structured
- Orientating your guidance towards the students' learning process
- Giving more guidance in the beginning and less towards the end
- Knowing when and how to intervene
- Monitoring students who show lack of elaboration, lack of cohesion, and lack of motivation

With regard to addressing anatomical knowledge in tutorial groups, pay attention to the following:

- Creation of *not* too general anatomical learning objectives
- Discussion of anatomical learning objectives to repeat and apply knowledge

Formulation of Anatomical Learning Objectives

Students have a tendency to skip the initial exploration of probable causes and underlying mechanisms of patient problems and not to bother formulating appropriate learning objectives related to basic sciences. Instead, they confine themselves to looking up a diagnosis and its associated symptoms and treatments [27].

When formulating anatomical learning objectives, students often jot down "anatomy of the heart" when the problem involves a heart attack or "anatomy of the ankle" when the problem involves a sprained ankle. The challenge is to get them to be more specific by, for example, asking questions about what they already know about the heart or ankle, what structures they need to study or they think they will encounter while studying, why these structures are there, what their function is, and how they relate to the phenomena described in the problem.

A word of caution: students often need to learn more of the anatomy of the region under discussion than is strictly necessary to "solve" the problem. For example, the anatomy department may choose to include information about the large arteries in lectures and dissection room sessions that accompany the "heart attack tutorial group." As often occurs, this may be the only opportunity to widen students' knowledge of the cardiovascular system within a packed curriculum. Some students do not understand why they need to learn about the large arteries when the tutorial group has discussed the anatomy of the heart, so it is up to anatomy teachers to show the relevance of this knowledge. That can be done by, for example, putting the large arteries in the context of the physical examination of a patient with a heart attack (measuring blood pressure: brachial artery; feeling pulse: radial artery) or treatment of a heart attack (angioplasty: entering the vascular system through the femoral artery).

Discussion of Anatomical Learning Objectives

"[The] potential disadvantage [of PBL] is that students may become more interested in the clinical aspects of a problem and neglect the underlying basic science knowledge" [19, p. 608]. When an anatomical learning objective is up for discussion in a tutorial group, students often adopt a "been there, done that" attitude. Everybody has attended the lectures and dissection room sessions so they tend to move to the next learning objective if nobody has a specific question. A tutorial group discussion, however, is an excellent chance to repeat information ("repeat to remember"; opportunities to repeat information are scarce in today's packed curricula) and practice the application and transfer of the knowledge. Possible ways to do this are to have students draw on the whiteboard, interpret X-rays or MRI scans, and bring skeletal material/plastinated specimens or low-tech models (described in a separate chapter in this book) into tutorial groups.

Role of the Anatomy Teacher

As an anatomy teacher within a PBL curriculum, you are still responsible for "teaching anatomy." On-demand education (which is often associated with PBL) is not really possible in larger medical schools (100+ students). Anatomy teachers should know when certain problems are to be discussed in tutorial groups so they can estimate when students will need to study a certain topic and lectures and dissection room sessions need to be planned. In a true PBL curriculum, sessions are almost never compulsory, adhering to the self-directed learning principle. If students feel they can get the information they need to pass their assessment in a different way, they are free to do so. Therefore, it is *not usual* to register attendance. An exception is when, for example, students have signed up to attend sessions or when failure to attend has consequences.

It is a misconception that "teaching" is not part of PBL. Especially in dissection room sessions, anatomy is still "taught" in the sense that anatomists help students in every possible way (often similar to more traditional curricula) to attain ("learn") a complete and accurate understanding of the structure of the human body.

Role of Anatomy Lectures in a PBL Curriculum

It is a misconception that lectures are not included in PBL. They are certainly part of the curriculum; however, they have different goals from lectures in more traditional curricula. A lecture in a traditional curriculum often summarizes what the lecturer believes to be the core course content [28], which can result in a series of lectures, each of which lasts up to 3 h. Lectures within a PBL curriculum serve three very different goals:

1. To introduce students to a new topic [28] and organize their study effort [29]
2. To explain particularly difficult concepts or problems or to address frequently occurring misconceptions [29]
3. To place the study material in context [28] and put the content into a broader perspective (place in the curriculum and how it prepares the students for professional practice) [29]

A lecture that aims to reach goal 1 (and 3) is usually scheduled after the initial tutorial group discussion but before other faculty resources or self-study. A lecture that aims to reach goal 2 (and 3) is usually scheduled after other faculty resources and self-study and either before or after the second tutorial group meeting. As the lectures are aimed at "understanding the whole," rather than memorizing content by rote learning [28], they last a maximum of 1.5 h.

Practical Tips for Anatomy Lectures in a PBL Curriculum

When creating a lecture in a PBL curriculum, do *not* summarize the course content, but focus on the following:

- Introducing students to a new topic and organizing their study effort. Lecture content can be a short overview of the anatomy of the to-be-studied region and/or their aim to familiarize students with the Terminologia Anatomica of the region before a dissection room session.
- Explaining particularly difficult concepts or problems or addressing frequently occurring misconceptions. Lecture content could recapitulate complex anatomical structures after a dissection room session.
- Placing the study material in context. Lecture content can illustrate the clinical relevance of the anatomy under study either before or after the dissection room session.

Dissection Room: Studying Prosected Specimens and Cadaveric Dissection

In Winkelmann's [30] extensive review, although confounded by the included studies differing in more than one variable, cadaveric dissection appeared to offer a slight added benefit compared to studying prosected specimens. A review study comparing cadaveric dissection to other teaching tools also slightly favored dissection. More important, however, a combination of teaching tools appeared to give the best results [31]. Furthermore, recent research has shown that a dissection course is not a uniformly positive learning experience [32]. Different students may have different approaches to dealing with a dissection course (or other teaching methods for that matter) and therefore undergo divergent learning experiences. This may result in a difference in the amount and form of knowledge between individual students. Further research needs to be conducted on *what* and *how* students learn from dissection and other teaching tools before drawing conclusions in favor of either method. For more information, see Chap. 27.

The choice to use prosected specimens in PBL curricula is often a practical one. When the problem under discussion in the tutorial groups is a heart attack, there is not enough time to dissect a cadaver from the skin down to the heart. Therefore, several cadavers are prosected and made available for students to study from, often guided by specific clinically relevant assignments to guide them (contextual learning principle). Furthermore, it is common to schedule multiple dissection room sessions. Students are not assigned to a specific one but can select a convenient session, so they can choose to attend either early or later during the self-study phase (self-directed learning principle).

To offer students the experience of cadaveric dissection, many medical schools with PBL curricula schedule extracurricular or elective courses. In these courses, different approaches may be taken. For example, an anatomy department can schedule a musculoskeletal course in the morning (with four students working on a limb each) and a head/thorax/abdomen course in the afternoon (with up to five students working on a specific region). In that case, students often dissect from skin to skeleton, after which the remains of the cadaver are of no further use and are cremated. Another option is for an anatomy department to have students create prosected specimens for use in future teaching. This is of course a better use of the cadaveric materials, especially when their supply is scarce. Even though students learn a lot, it still does not give them a full dissection experience. A problem with extracurricular or elective courses is that there is often a restriction on the number of students who can attend them, so it still does not allow all students to have the experience of dissection.

Teaching Dilemma
How can extracurricular (elective) dissection courses be made available for a large student population (many medical schools accommodate over 300 students per year)?

Surface Anatomy, Body Painting, and E-Learning

Other laboratory-based teaching methods such as surface anatomy, body painting, and e-learning may also be part of a PBL curriculum. Surface anatomy and/or body painting sessions are scheduled after dissection room sessions and, if possible, before the second tutorial group meeting, whereas e-learning sessions can be scheduled at any time depending on their aim. As body painting and e-learning are discussed in other chapters in this book, the focus here will be on surface anatomy sessions.

Studying surface anatomy stimulates elaboration on the static anatomy of the cadaver by enabling students to see phenotypical structures (particularly those of the musculoskeletal system) move and function in a living human being. Whereas surface anatomy sessions in more traditional curricula could involve demonstration of structures by a teacher on a model, surface anatomy sessions in a PBL curriculum include constructive, collaborative, contextual, and self-directed learning [33], by, for example, making use of peer examination (collaborative learning), referring to information studied in the dissection room (constructive learning), or elaborating on anatomical structures by relating them to physical examination, medical imaging, or symptoms and signs in common diseases (contextual learning).

Teaching Materials for Anatomy Education in PBL

Most of the differences between a traditional curriculum and a PBL curriculum is in the availability of materials during the self-study phase. The self-directed learning principle implies that students should be able to decide from which resource they want to study. Therefore, students in a PBL curriculum do not have one core textbook to study from (although most of them do buy one to study at home). Instead, they are encouraged to use many different resources, which is facilitated by the different anatomical textbooks and atlases in the medical school's library. In the case of anatomy, books are often supplemented with digital atlases, e-learning material, and a "study landscape" with models, plastinations, medical imaging, etc., which students can use on request. This practice is also reflected in the teaching materials created by anatomical staff. For example, images from many different resources are used to illustrate the subjects of lectures.

Learning Anatomy in a PBL Curriculum

Two main issues arise when students learn anatomy in a PBL curriculum: not taking a deep approach to learning and losing a coherent mental representation of the anatomy of the whole body.

Reproduction of Names and Structures Versus Understanding Signs and Symptoms

In a study by Mattick and Knight [23], students said that anatomy is unique in that it represents a huge set of facts, codified in a specialized language, which calls for different learning strategies than other basic sciences. Because of this, learning anatomy relies on both surface and deep learning approaches. Memorizing information is generally referred to as a surface approach to learning. A deep approach, in contrast, is characterized by efforts to understand the structure of material to be learned and manipulating the information to make sense of it in relation to what a student already knows about the subject matter [34]. Because anatomy has a complex vocabulary, it has been hypothesized that a deep approach to anatomy learning needs to build on a preliminary stage of rote learning, which is difficult to distinguish from a surface approach [35].

PBL curricula should facilitate many aspects of learning. With respect to the basic sciences, PBL should, among other things, stimulate intrinsic interest in the subject matter. Research suggests that if students recognize the relevance of basic sciences, their learning is enhanced. In traditional curricula, students may have the impression that they are studying basic sciences because they must, without being aware of their clinical relevance. On the other hand, PBL stimulates students to learn basic sciences in order to understand clinical problems, thus fostering a deep approach to learning [13, 18, 19].

However, what is often seen is that students perceive and describe memorization of details as an endpoint of learning anatomy rather than a stage in the process leading to understanding signs and symptoms [23, 34, 36]. The patient problem presented in the tutorial group does not seem to be enough of a stimulus to adopt a deep learning approach during an anatomy session later in the week. One solution is to apply the constructive, collaborative, contextual, and self-directed learning principles of PBL in anatomy education. Do not send students into the dissection room with a list of structures they need to know, but create challenging assignments in which they need to apply anatomical knowledge to explain signs and symptoms, interpret medical imaging, etc.

Loss of Coherent Mental Representation of the Anatomy of the Whole Body

Monkhouse and Farrell [37] pointed to the danger that systems-based curricula prevent students from gaining a coherent view of the anatomy of the whole body, as anatomy is likely to be taught piecemeal with no regard to the whole organism. Inevitably, students are left with only a fragmented knowledge of anatomy. Experience suggests that this is true of all integrated curricula, including PBL. Some students struggle with the random order in which body regions are studied, while others refer to having to learn nerves and blood vessels in the lower leg without having any idea where they originated from or not recognizing a stomach herniated through the diaphragm on a CT scan because the thorax and abdomen were treated as two totally unrelated regions [38].

It is up to anatomy teachers to come up with creative solutions to the problems described above. For example, to get a better overview of the peripheral nervous system, always go back to the same pictures in your lectures. Start at the spinal cord and the nerve plexus of the region to be studied. If applicable, show nerves that have already been studied and then zoom in on the nerves under discussion, their origins from the plexus, and their spatial relationships to previously studied nerves.

To Do

- Think of another way of stimulating students to adopt a deep approach to learning anatomy.
- Ascertain whether your medical school facilitates students to gain a coherent view of the anatomy of the whole body and how you as an anatomy teacher could facilitate students to accomplish an improved mental picture.

Conclusion

It is basic to PBL that student learning is organized around problems. The biggest difference between anatomy education in PBL and any other curriculum is that anatomy is taught in tutorial groups. The correct formulation and discussion of anatomical learning objectives may rest on the shoulders of non-anatomists. It is therefore preferred that anatomists play an active role in writing the tutor manual to help non-anatomist tutors. Furthermore, anatomy education in PBL curriculum is shaped by the learning principles underlying PBL: constructive, collaborative, contextual, and self-directed learning. As repetition and teaching in context seem to influence students' anatomical knowledge strongly, PBL seems to have an advantage when a curriculum is truly spiral and vertically integrated.

Lectures in a PBL curriculum serve a different purpose, but dissection room sessions do not really differ from those in other curricula with a student-centered, active learning approach. PBL curricula struggle with the same issues as do other curricula when it comes to stimulating students to take a deep approach to learning. More unique to PBL curricula is having to make several learning resources available to students and having to facilitate students to create a coherent mental picture of the anatomy of the whole human body.

References

1. Dolmans DH, De Grave W, Wolfhagen IH, van der Vleuten CP. Problem-based learning: future challenges for educational practice and research. Med Educ. 2005;39(7):732–41.
2. Wang J, Zhang W, Qin L, Zhao J, Zhang S, Gu J, et al. Problem-based learning in regional anatomy education at Peking University. Anat Sci Educ. 2010;3(3):121–6.
3. Yiou R, Goodenough D. Applying problem-based learning to the teaching of anatomy: the example of Harvard Medical School. Surg Radiol Anat. 2006;28(2):189–94.
4. Dahle LO, Brynhildsen J, Behrbohm Fallsberg M, Rundquist I, Hammar M. Pros and cons of vertical integration between clinical medicine and basic science within a problem-based undergraduate medical curriculum: examples and experiences from Linkoping, Sweden. Med Teach. 2002;24(3):280–5.
5. Drake RL. Anatomy education in a changing medical curriculum. Anat Rec. 1998;253(1):28–31.
6. Swanson DB, Case SM. Assessment in basic science instruction: directions for practice and research. Adv Health Sci Educ. 1997;2:71–84.
7. Bolender DL, Ettarh R, Jerrett DP, Laherty RF. Curriculum integration = course disintegration: what does this mean for anatomy? Anat Sci Educ. 2013;6(3):205–8.
8. Adibi I, Hasani N, Ashoorioun V, Sadrearhami S, Monajemi AR. Integrating physical examination and trunk anatomy; a new course for second year medical students. Med Teach. 2007;29(9):975–7.
9. Alleyne T, Shirley A, Bennett C, Addae J, Walrond E, West S, et al. Problem-based compared with traditional methods at the Faculty of Medical Sciences, University of the West Indies: a model study. Med Teach. 2002;24(3):273–9.
10. Colliver JA. Effectiveness of problem-based learning curricula: research and theory. Acad Med. 2000;75(3):259–66.
11. Hinduja K, Samuel R, Mitchell S. Problem-based learning: is anatomy a casualty? Surgeon. 2005;3(2):84–7.
12. Newman M. A pilot systematic review and meta-analysis on the effectiveness of problem based learning. Newcastle: Learning and Teaching Subject Network for Medicine, Dentistry and Veterinary Medicine; 2003.
13. Prince KJ, van Mameren H, Hylkema N, Drukker J, Scherpbier AJ, van der Vleuten CP. Does problem-based learning lead to deficiencies in basic science knowledge? An empirical case on anatomy. Med Educ. 2003;37(1):15–21.
14. Vernon DT, Blake RL. Does problem-based learning work? A meta-analysis of evaluative research. Acad Med. 1993;68(7):550–63.
15. Woloschuk W, Mandin H, Harasym P, Lorscheider F, Brant R. Retention of basic science knowledge: a comparison between body system-based and clinical presentation curricula. Teach Learn Med. 2004;16(2):116–22.
16. Bergman EM, Prince KJ, Drukker J, van der Vleuten CP, Scherpbier AJ. How much anatomy is enough? Anat Sci Educ. 2008;1(4):184–8.
17. Waterston SW, Stewart IJ. Survey of clinicians' attitudes to the anatomical teaching and knowledge of medical students. Clin Anat. 2005;18(5):380–4.
18. McCrorie P. The place of the basic sciences in medical curricula. Med Educ. 2000;34(8):594–5.
19. O'Neill PA. The role of basic sciences in a problem-based learning clinical curriculum. Med Educ. 2000;34(8):608–13.

20. Prince KJAH, Scherpbier AJAA, van Mameren H, Drukker J, van der Vleuten CPM. Do students have sufficient knowledge of clinical anatomy? Med Educ. 2005;39(3):326–32.
21. Whitcomb ME. The teaching of basic sciences in medical schools. Acad Med. 2006;81(5):413–4.
22. Harden RM, Stamper N. What is a spiral curriculum? Med Teach. 1999;21(2):141–3.
23. Mattick K, Knight L. High-quality learning: harder to achieve than we think? Med Educ. 2007;41(7):638–44.
24. Regehr G, Norman GR. Issues in cognitive psychology: implications for professional education. Acad Med. 1996;71(9):988–1001.
25. Dolmans DH, Schmidt HG. What drives the student in problem-based learning? Med Educ. 1994;28(5):372–80.
26. Hmelo-Silver C. Goals and strategies of a problem-based learning facilitator. Interdiscip J Probl Based Learn. 2006;1(1):21–39.
27. Prince KJ, Van De Wiel M, Scherpbier AJ, Can Der Vleuten CP, Boshuizen HP. A qualitative analysis of the transition from theory to practice in undergraduate training in a PBL-Medical School. Adv Health Sci Educ Theory Pract. 2000;5(2): 105–16.
28. Fyrenius A, Bergdahl B, Silen C. Lectures in problem-based learning–why, when and how? An example of interactive lecturing that stimulates meaningful learning. Med Teach. 2005;27(1):61–5.
29. van Berkel H, Schmidt H. On the additional value of lectures in a problem-based curriculum. Educ Health (Abingdon). 2005;18(1):45–61.
30. Winkelmann A. Anatomical dissection as a teaching method in medical school: a review of the evidence. Med Educ. 2007;41(1):15–22.
31. Bergman EM, van Der Vleuten CP, Scherpbier AJ. Why don't they know enough about anatomy? A narrative review. Med Teach. 2011;33(5):403–9.
32. Winkelmann A, Hendrix S, Kiessling C. What do students actually do during a dissection course? First steps towards understanding a complex learning experience. Acad Med. 2007;82(10):989–95.
33. Bergman EM, Sieben JM, Smailbegovic I, de Bruin AB, Scherpbier AJ, van der Vleuten CP. Constructive, collaborative, contextual, and self-directed learning in surface anatomy education. Anat Sci Educ. 2013;6(2):114–24.
34. Pandey P, Zimitat C. Medical students' learning of anatomy: memorisation, understanding and visualisation. Med Educ. 2007;41(1):7–14.
35. Entwistle NJ, Ramsden P. Understanding student learning. London: Croom-Helm; 1983.
36. Miller SA, Perrotti W, Silverthorn DU, Dalley AF, Rarey KE. From college to clinic: reasoning over memorization is key for understanding anatomy. Anat Rec. 2002;269(2):69–80.
37. Monkhouse WS, Farrell TB. Tomorrow's doctors: today's mistakes? Clin Anat. 1999;12(2):131–4.
38. Bergman EM, de Bruin AB, Herrler A, Verheijen IWH, Scherpbier AJ, van der Vleuten CPM. Students' perceptions of anatomy across the undergraduate problem-based learning medical curriculum: a phenomenographical study. BMC Med Educ. 2013;13:152.

Learning and Teaching Anatomy Through Case-Based Learning (CBL)

17

Jill E. Thistlethwaite

Defining Cased-Based Learning

Case-based learning (CBL) is a well-established method of learning and teaching across many disciplines such as law, medicine, and education. Law and medicine are indeed case-based professions, and logically trainees and qualified professionals will learn from and about cases throughout their careers. However, as CBL is delivered in many formats and at different stages of programs, it lacks a universally agreed definition. Indeed many educators and the literature also refer to the approach as case study teaching and case method learning. The underlying commonality however is the "case," which in medical and health professional education refers to a patient or client case.

CBL is not a new approach and pre-dates other forms of educational interventions by many decades. In 1912, James Lorrain Smith, the first full-time pathology professor at the University of Edinburgh in Scotland, introduced the "case method of teaching pathology" [1]. Medical students attending postmortems had to look up the deceased patient's history, clinical symptoms, and physical signs from the medical record and relate these to the anatomical and pathological findings of the autopsy. Smith's underlying aim was for students to link their scientific knowledge with clinical practice. A best evidence medical education (BEME) systematic review of CBL showed that this linkage between theory and practice continues to be a goal and a hallmark of effective CBL [2].

The BEME Definition of CBL

- CBL is a learning and teaching approach that aims to prepare students for clinical practice, through the use of authentic clinical cases.
- These cases link theory to practice, through the application of knowledge to the cases, and encourage the use of inquiry-based learning methods.

The Evidence for CBL

The BEME review [2] of 104 papers describing CBL across a diverse range of health professional education settings and programs concluded that typically CBL takes place in small face-to-face groups (2–15 students) but may also take place online and, less commonly, is undertaken by individuals or by large groups of more

J. E. Thistlethwaite (✉)
Faculty of Health, University of Technology Sydney, Ultimo, Sydney, NSW, Australia
e-mail: jill.thistlethwaite@uts.edu.au

L. K. Chan, W. Pawlina (eds.), *Teaching Anatomy*, https://doi.org/10.1007/978-3-030-43283-6_17

than 30 participants. Cases obviously often feature in lectures attended by many more learners than this, but for CBL to be inquiry based, learners need to be involved actively in looking up information, discussing with their peers, and monitoring their learning outcomes. CBL fits between the structured and guided levels of inquiry-based learning as defined by Banchi and Bell [3].

Both learners and teachers have very positive views about CBL, but overall, while it helps promote learning, the evidence is patchy and inconclusive that it is more effective than other learning and teaching methods. Its efficacy appears to be due to the cases enhancing the relevance of the basic and biomedical sciences and integrating these with clinical management [4] in addition to learners' understanding of concepts. In relation to anatomy teaching, student evaluation has indicated that CBL does help make anatomy more relevant [5–7]. CBL is therefore a complementary process to lectures, dissection, and online learning. The learning process helps foster a deep approach to learning [8]; students gain greater understanding through the application of the knowledge they are learning to the cases presented rather than merely reproducing facts learned at a more superficial level. Student performance in the Step 1 USMLE (United States Medical Licensing Examination) has been enhanced through the use of CBL for cell biology, histology, and pathology when compared to the results of students taught anatomy in a more traditional way [9].

Developing a Case-Based Approach

As with any educational activity, it is important to define the learning outcomes that educators wish the students to achieve. Established anatomy programs will have a set of learning outcomes that students and faculty are aware of: which of these should be the focus of the CBL? Of course in most instances, the CBL will be complementing other methods of learning and teaching, as described in this book, and will be in addition to, not instead of, other teaching modalities. In some institutions, however, the medical program is delivered primarily as a case-based approach, in a similar way to those universities that have developed problem-based learning (PBL) curricula. Note that PBL is "learning that results from the process of working toward the understanding of a resolution of a problem. The problem is encountered *first* in the learning process" [10]. Students define their own learning outcomes in relation to the problem. In contrast, in the guided inquiry method of CBL, learning outcomes are defined for the students and the approach is more structured than PBL.

If CBL is not to be the principal curricular approach, then cases may be used to complement or substitute in some instances for other educational methods such as lectures, laboratory work, and dissection. If you are not considering changing the delivery of your entire anatomy curriculum, then you will need to decide how best the cases may be incorporated into existing teaching. Cases may follow didactic instruction to help students apply knowledge to practice; students may work on their own or in groups after lectures. In the flipped classroom model, learners independently watch short videos or other material online, which could include cases, and then come together with a facilitator to review what they have learned, to work in more depth through cases, and to engage collaboratively with the aid of a facilitator [11]. In clinical settings, the cases are the patients with whom the students interact, but CBL may be used in tutorial settings to integrate anatomy with clinical presentations and help remind students of their prior learning.

The learning outcomes for a case will dictate the focus of the case itself. Cases are best created from the stories of (de-identified) real-life patients, for whom authentic investigation results such as blood work, X-rays, and scans are available. Patient consent is required depending on how much the case is altered to mask identity.

According to the National Center for Case Study Teaching in Science, Cases Should

- Be aligned with defined learning outcomes for the course
- Be authentic (based on real patient stories)
- Involve common conditions
- Tell a story
- Have educational value
- Be interesting and enhance students' participation
- Create empathy with the patients
- Include quotations in the patient voice to add drama and realism
- Have general applicability
- Serve as memory aids
- Reduce cognitive overload
- Be easy and inexpensive to produce (Adapted from Herreid [12])

Cases may be developed for a paper-based delivery, but more usually these days, cases are in an electronic format involving virtual patients (though these are usually based on individual real patients or a combination of different patients). Virtual patients are interactive computer simulations of real-life situations for the purpose of health professional education [13], whereas real patients are "flesh and blood people" with whom students interact in clinical or community settings. Real patients may also be invited into the classroom and lecture theater to tell their stories or to demonstrate physical signs. Simulated patients are also "real" people, but they work to a script for the purpose of specific learning or assessment purposes and are able to adapt their roles to enhance learning. Many are also able to give feedback to learners. For clinical examinations such as the OSCE (objective structured clinical examination), standardized patients do not deviate from prepared histories.

To Do—Developing CBL

- Do you already use cases in your teaching? Do they match the criteria in the previous box? What should you change about your existing cases?
- How might you incorporate a CBL approach into your teaching?
- Consider the necessary steps to introduce CBL: where, when, how, and who?
- Consider how CBL could complement your existing teaching; maybe you could substitute some of your existing delivery with cases, in particular to achieve outcomes related to application of knowledge and clinical reasoning.

Writing Cases

Basic scientists are advised to work with clinicians to write cases in order to ensure their authenticity and to obtain material that can be used to enhance the delivery of the case such as MRI scans and even videos of surgery. Writing cases is not easy and can be time-consuming. It is more fun developing and discussing newly written cases in groups within your department. If you and your colleagues work alone, you may find that some cases are too similar. You may be able to use one case across several learning sessions as the patient's story unfolds. Reading a case to your colleagues allows for feedback. You can discuss where longer cases can be split so they can be given to students in digestible parts rather than overloading them with too much information at one time. Your colleagues can check whether the case is congruent with the stated learning outcomes. The clinicians can vouch for the reality of the patient experience. You may also want to consider how you will assess that the students have met the learning outcomes and how this assessment fits with the overall assessment of anatomy in your program.

To Do—After Writing Your Case

- Check that the case will meet the defined learning outcomes.
- Consider if the case is at the right level for your learners.
- Ensure the case is realistic and interesting.
- Determine if the material presented will stimulate the necessary questions and/or answers from the students.
- Decide what other resources are required—for example, X-rays, models, and websites.

Teaching with Cases

As mentioned above, CBL is usually a small group activity. Approached in this way, CBL also promotes collaborative learning. Groups may be self-directed but often have a facilitator. The facilitator should encourage learners to think critically about the material presented. The case scenario is given to the students and includes the learning outcomes and any questions that the students are required to answer. Cases can be given to students prior to the sessions so they can come prepared through background reading or handed out on the day, which is more challenging. Students should work in their groups with minimal input from the facilitator, brainstorming answers, looking up further required information, and reaching a consensus about the answers to the questions. The session draws to an end by students presenting to the facilitator and receiving feedback on their work. In some models, all the small groups meet together, present to each other, and then give feedback on both the content and the presentation itself.

The teacher/facilitator acts as a guide, asking students to explain or clarify their thinking, encouraging discussion and debate. When necessary, the facilitator asks for evidence of any statements: what is the evidence for that? Finally, the learning outcomes for the session are revisited to check that they have been covered and what further reading or work is required.

Senior students may not need questions in addition to the learning outcomes but should be thinking of their own questions and developing hypotheses to explain the unfolding of the case. If there are questions, these should be formulated to build on knowledge students have already assimilated but that also require some additional research, pushing the students in new directions and encouraging them to apply their knowledge in new situations. For example, to follow up a session on the anatomy of the arm, a case may involve a patient requiring to have veniepuncture with the students considering where the blood should be taken from and why and how. The session could be combined with students practicing the clinical skill of venepuncture on an appropriate model.

The same cases may be revisited several times in a year or program, but the questions are different and more challenging to reflect the stage of the students' prior learning. Complexity can be increased as the case develops. When learners are near the beginning of their anatomy journey, the questions should be focused on normal anatomy and function, with pathology and disease becoming a feature as students mature and the "patient" ages and/or develops further problems.

Faculty Development

Traditional teaching of anatomy has been didactic with students gaining practical experience through dissection. CBL is a much more facilitated approach, and therefore, its development, introduction, and delivery will require many educators to teach in a different way. Some may be anxious; some without a clinical background may find it difficult to write and engage with clinical cases. Therefore, it is important to carry out a needs assessment of staff in terms of their experience and skills in order to provide appropriate professional development activities.

It is vital when initiating a curricular change that all staff have a chance to discuss the implementation, the educational rationale, and their concerns. Faculty expressing personal negative or ambivalent opinions about CBL when facilitating undermine student confidence. Skills required for CBL, depending on how cases are used, include but are not limited to: engaging all students, giving just the right amount of guidance but allowing students to collaborate and set their own pace, giving formative feedback, and time management.

Suggested Steps for Students in Working with Cases

1. Read the learning outcomes for the case. Are there new learning outcomes? Are some related to previous work that you can now build on?
2. Read the case and clarify any terms or concepts with which you are not familiar.
3. Consider the members of the group—does there appear to be a difference in the level of understanding of the group members and why might this be so?
4. Read the questions, which are to help you meet the learning outcomes.
5. What systems of the body are involved in this case?
6. What anatomical concepts and other related topics need to be explored and/or explained?
7. Work through the questions, dividing up tasks between group members. Where might you find information to help?
8. Regroup to summarize your answers and prepare your case analysis.
9. Have you met the learning outcomes?
10. What other work do you need to do now?

To Do—Facilitation

- Consider what skills you and your colleagues require for facilitation.
- How will you develop these skills if you need to do so?
- What size groups will you work with?
- How will you ensure that all students are involved with each case?

The Learning Environment and Context

When preparing for CBL, you need to consider the learning environment and the context of CBL within the curriculum. Is the learning space conducive to small group work? How many students are in the class and how should they be divided into groups? Have they worked in groups before? How you deliver the cases will be dependent on your course structure. Does the CBL session follow a lecture? How does it fit with other teaching such as surface anatomy or dissection? One case may be completed in one session or may carry over to a second session with students continuing to work on it in the intervening days.

Students also need to be orientated to CBL especially if they are not used to guided learning methods compared to being "spoon-fed" information. If working in groups for the first time, they need to start by discussing ground rules such as respect, use of mobile phones, letting others speak, everyone actively contributing, etc.

CBL and Learning Outcomes in Anatomy

As anatomy courses involve students learning the vocabulary of the human body and recognizing and describing the structures of the body initially from the perspective of "normality," teachers need to be cautious when writing cases that they do not focus solely on pathology and thus disease. This can be a challenge particularly as students may be undertaking a spiral curriculum, which focuses on the normal in year 1 before tackling the abnormal in year 2. However, as CBL helps students understand the rationale for learning anatomy in relation to their potential profession, introducing pathology at this stage is acceptable as long as the underlying learning outcomes for normality are still met.

As the majority of health professional courses are now integrated through system-based learning, CBL may in fact involve the integration of physiology, biochemistry, and anatomy through interdisciplinary cooperation. This allows more interesting cases to be developed, which have learning outcomes across the sciences. Faculty need to understand the whole curriculum and what is learned where in order to develop cases that do build on prior knowledge and integrated learning.

Examples of Cases

Case One: Max and His Muscles

Max Lucas is a 23-year-old engineering student who has decided to "get fit" this year. His current fitness regime consists of walking to and from university five times a week, which takes about 45 min a day. He is slightly overweight and has started to reduce the size of his meals and the amount of beer he drinks. Max has no health problems. He joins the university gym and asks a personal trainer for some advice. He wants to enhance his muscle definition, bulk up his upper body, and gain a six pack.

Learning Outcomes

- Name the muscles of the arms, shoulders, upper back, chest, and abdomen.
- Identify the origins and insertions of these muscles on the articulated skeleton.
- Name the associated tendons.
- Describe the movements of muscles using the correct anatomical terms.
- Outline the surface anatomy of the muscles.
- Outline the processes by which muscle size and power is increased.

Questions

- Which muscles does Max need to work on to build his upper body strength?
- What do these muscles do?
- What exercises would you recommend to Max?
- What advice would you give Max about his exercise regime to avoid muscle injuries?
- What is the evidence base for your prescribed exercise regime and advice?

To Do—Max and His Muscles

- In relation to your own teaching, what other learning outcomes and questions might you add?
- What resources may the students require to complete this case?
- How would you assess student learning?

Comments

This is a fairly simple case, which should resonate with the student population. This case could complement a lecture on the musculoskeletal system of the upper body and torso, or it could be solely a small group learning activity with the addition of prosections and anatomical models. The learning outcomes are the same but are reinforced through the case, which puts the knowledge into a clinical context. Resources are also the students themselves in terms of surface anatomy and perhaps even a set of hand weights to demonstrate muscle contraction and relaxation with different forces.

Extension of This Case

Six months later, Max is fitter and stronger. He begins to play soccer in the local park on weekends though he is not particularly skillful at the game; he enjoys the team element of the sport. One Saturday afternoon, he twists heavily on his right knee, which gives way with an audible crack. He limps off in some pain and watches the rest of the match from a bench. By the time the whistle blows, his knee is swollen.

Learning Outcomes

- Describe the anatomy of the knee joint.
- Describe the mechanics of movement of the knee.
- Explain any symptoms, signs, or loss of function in the knee due to ligament damage.
- Explain any symptoms, signs, or loss of function in the knee due to cartilage damage.

Questions
- Why was there an audible crack?
- Why is Max's knee swollen?
- What investigations may you require to help diagnose the anatomical disruption to Max's knee?

Comments
The case has moved onto a clinical presentation, but students should be able to find out the answers and meet the learning outcomes. This is a common problem. It would be interesting to provide MRI scans of a normal knee joint and one with collateral ligament damage or even show footage of an arthroscopy.

Case Two: Lisa's Abdominal Pain

Lisa Balotelli is a 25-year-old teacher who presents to the emergency department one Saturday evening complaining of lower abdominal pain for 6 h. She has vomited once and looks distressed. The resident elicits a full history and examines Lisa, who is tender in the right lower abdomen and centrally above her pubic bone. Based on the history and external examination, the doctor also carries out a digital rectal examination (DRE) and bimanual pelvic examination (PE).

Learning Outcomes
- Describe the normal anatomy of the lower abdomen and female pelvis.
- Outline the structures that the resident is palpating externally and suggest possible causes of Lisa's pain based on the identified structures.
- Describe the anatomy examined with the DRE and PE.
- Describe the differences in the anatomy examined with a DRE in a male patient.

Questions
- What are the possible causes for Lisa's pain given your knowledge of the anatomy of the lower abdomen and pelvis?
- What investigations may be helpful in this case in relation to examining any abnormalities of the anatomy?
- If Lisa gives a classical history of appendicitis, with central abdominal pain later localizing to the right iliac fossa, explain the change in the location of the pain.

Comments
The learning outcomes relating to knowledge can be attained through lectures and other didactic methods, but again the CBL, either as facilitated group work or independent study, helps link the anatomy to clinical presentations and enhances the relevance of the material.

> **To Do—Writing Cases**
> - Choose a learning outcome for your student cohorts and develop a case to stimulate their learning around this outcome.
> - Write relevant questions for them to answer through research and discussion.
> - Decide on the format—are you able to develop your case online?

E-Learning and CBL

Online CBL is becoming more common within medical education as e-learning becomes more prevalent. Electronic cases range from simply uploading paper cases onto an educational platform to fully developed avatars living in virtual worlds. Very sophisticated cases are expensive to develop and maintain. Virtual patient cases do not need to be high fidelity though students (and tutors) used to costly video games and sophisticated avatars may be disappointed to begin with. For example, CT images from an embalmed cadaver have been readily constructed into a VP [14]. It is important to remember CBL is about learning and not technical prowess. There are open-access VPs and cases online, which may be used to complement an institution's own resources. Enthusiastic students can be directed to these sites for further learning.

Conclusion

CBL is a learning and teaching method well suited to anatomy as it helps bridge the gap between knowledge and practice. There are a variety of approaches that may be used to deliver CBL. CBL is effective for groups but requires faculty development of authentic cases. However, there are now many open-access cases that students may use to complement the resources provided within their own departments.

References

1. Sturdy S. Scientific method for medical practitioners: the case method of teaching pathology in early twentieth-century Edinburgh. Bull Hist Med. 2007;81(4):760–92.
2. Thistlethwaite JE, Davies D, Ekeocha S, Kidd JM, MacDougall C, Matthews P, Purkis J, Clay D. The effectiveness of case-based learning in health professional education: a BEME systematic review. Med Teach. 2012;34:e421–44.
3. Banchi H, Bell R. The many levels of inquiry-based learning. Sci Child. 2008;46:26–9.
4. Bowe CM, Voss J, Aretz HT. Case method teaching: an effective approach to integrate the basic and clinical sciences in the preclinical medical curriculum. Med Teach. 2009;31(9):834–41.
5. Hansen JT, Krackov SK. The use of small group case-based exercises in human gross anatomy: a method for introducing active learning in a traditional course format. Clin Anat. 1994;7(6):357–66.
6. Peplow P. Videoconferencing for small-group discussion sessions of a case-based learning programme in anatomy. Med Teach. 1996;18:309.
7. Scott TM. A case-based anatomy course. Med Educ. 1994;28:68–73.
8. Entwistle N. Teaching for understanding at university. Deep approaches and distinctive ways of thinking. Basingstoke: Palgrave Macmillan; 2009.
9. McBride JM, Prayson RA. Development of a synergistic case-based microanatomy curriculum. Anat Sci Educ. 2008;1(3):102–5.
10. Barrows H, Tamblyn R. Problem-based learning: an approach to medical education. New York: Springer; 1980.
11. Herreid CF, Schiller NA. Case studies and the flipped classroom. J Coll Sci Teach. 2013;42:62–6.
12. Herreid CF. What makes a good case? J Coll Sci Teach. 1997/1998;27:163–5.
13. Ellaway R, Candler C, Greene P, Smothers V. An architectural model for MedBiquitous virtual patients. Baltimore: MedBiquitous; 2006.
14. Jacobsen J, Epstein SK, Albright S, Ochieng J, Griffiths J, Coppersmith V, Polak JF. Creation of virtual patients from CT images of cadavers to enhance integration of clinical and basic science student learning in anatomy. Med Teach. 2009;31:749–51.

Teaching Anatomy: Effective Use of Modified Team-Based Learning Strategy

18

Cheryl Melovitz-Vasan, Susan Huff, and Nagaswami S. Vasan

Introduction

"To teach is to engage students in learning" [1]. This quote defines the premise that engaging students is vital to learning. Millennial students prefer efficient and effective teaching. With the introduction of adult learning theories [2], we are now witnessing the development of team-based learning (TBL) as a sustainable pedagogical strategy. TBL is a successful instructional strategy that was first developed and tested in 1970s by Professor Larry Michaelsen in business school [3]. The "classic" TBL is a three-phase cycle of (1) pre-class preparation, (2) in-class readiness assurance testing, and (3) application-focused problem-solving exercise.

Presently, TBL strategy is commonly used in all fields of health-care education, both in undergraduate and postgraduate education (medicine, dental, veterinary, pharmacy, nursing, physical therapy, and other allied health programs), engineering and business schools, continuing education, undergraduate college teaching, and secondary education [4]. As more and more programs adopted TBL, it became necessary to modify the three-step classical TBL to meet the local culture and curricular needs; thus, the concept of "modified TBL" evolved in the literature, which enabled more programs to adopt this pedagogy [5–8] (see Table 18.1).

C. Melovitz-Vasan · N. S. Vasan (✉)
Department of Biomedical Sciences, Cooper Medical School of Rowan University, Camden, NJ, USA
e-mail: melovitz-vasan@rowan.edu; vasan@rowan.edu

S. Huff
Office of the Medical Education, Cooper Medical School of Rowan University, Camden, NJ, USA
e-mail: huffsm@rowan.edu

The Four Essential Principles of TBL

The four essential elements or principles for successfully implementing TBL are: team, assignment design, accountability, and feedback [3].

Properly Formed and Managed Teams

To form an effective functional team, the faculty must ensure that each group has comparable resources and prospects to develop and grow into a functional learning team. When forming teams, we use basic parameters, such as science versus nonscience major, postgraduate education, prior TBL experience, previous experience of anatomy courses, gender and diversity (including underrepresented minorities), and whether a student is in the 3-year undergraduate plus 3-year medical school accelerated 6-year program. As much as possible, team members should not be from the same undergraduate college to avoid coalition among members. Lastly, attention should be paid when incorporating nontraditional student in

L. K. Chan, W. Pawlina (eds.), *Teaching Anatomy*, https://doi.org/10.1007/978-3-030-43283-6_18

Table 18.1 Modified five phases of team-based learning (TBL) and the sequence of steps involved

Phase	Step
Phase 1: Out of class preparation	*Step 1*: Self-directed pre-class individual preparation using focused study guide.
Phase 2: Readiness assurance (in-class)	*Step 2*: Discussion of study guide content, learning objectives among the team members—interactive peer teaching (need monitoring of "freeloading" team members).
	Step 3: iRAT (individual readiness assurance test). Students individually take a graded MCQ test based on the contents of learning objectives.
	Step 4: tRAT (team readiness assurance test). Each team as a whole retakes the same MCQs. This involves intra-team discussion. This needs close monitoring to assure all members of the team participate.
	Step 5: Faculty feedback and clarifications. The MCQs are discussed among the teams, allowing feedback from the entire participating class. Faculty may need to clarify some questions in the discussion. If student successfully challenges a question, the class receives credit.
Phase 3: Intra-team discussion followed by inter-team discussion	*Step 6*: Intra-team discussion. Each team discusses all the subsets of question and gets ready for inter-team discussion.
	Step 7: Inter-team discussion. Following step 6, the cases are discussed among the teams and the discussion is moderated by the faculty. These cases allow comprehension of the module content and the learning objectives. While this is not graded, it allows the class to get ready for the summative clinical vignette-based MCQ examination.
Phase 4: Module exam: application of factual knowledge to clinical vignettes	*Step 8*: iRAT. The summative clinical vignette-based MCQ examination covers the contents of the entire organ system. Students first take it individually (iRAT).
	Step 9: tRAT. Following the iRAT, each team as a whole retakes the same MCQs. This involves only individual team discussion closely monitored by the faculty.
Phase 5: Clarification and explanation	*Step 10*: This phase follows immediately after step 9. A brief time period is set aside for students to request clarification and explanation of specific questions. In general, since all these questions are vetted, the discussions does not result in additional points being awarded.

Abbreviations: *iRAT* individual readiness assurance test, *tRAT* team readiness assurance test
As described in the text, the steps are different from the classical three-phase TBL
Note: In any phases that are graded, students are not allowed to access outside resources

teams. Nontraditional students are older in age, returning to school (second career) after working for few years, have other personal commitments, and have more life experience. It is important to distribute them among various teams.

A functional team should have balanced knowledge, in addition to other demographic aspects mentioned earlier. The size of the team is very important. Larger teams may lead to formation of mini teams within a team, which can possibly cause decreased team cohesiveness. Larger teams may offer more diversity, but they may also encourage more social loafing. Contrarily, too small a team results in poor total resource and knowledge structure. A productive team consists of 5–6 members; it allows the team to quickly go through the phases of team formation ("forming, storming, norming, and performing") and more easily transform into a very supportive functional team [9]. Such a team is more likely to have a clear strategy and shared vision and to stay motivated, conversant, and resolves issues positively. Team members look after each other. Students remain in the same team for the duration of the course, which allows members to develop trust in each other and interdependence. As noted earlier, the faculty should pay attention to whether team members form a coalition based on prior association.

Assignments That Promote Team Development and Learning

In TBL, the adage is that assignment drives group process and learning. We find that, when an assignment is simple and can be done easily by individual effort, there is no team interaction. When the assignment is complex and truly requires the group to work together, there is more group interaction and greater team satisfaction.

Student Accountability for the Quality of Individual and Group Work

A critical component of TBL is accountability of individual team members for the quality of pre-class preparation. In lecture-based teaching programs, individual effort rarely affects peer performance or the class as a whole and the individual student is only accountable to the teacher and him/herself. With TBL, lack of individual preparation limits individual learning and contributions to the team effort. The TBL strategy ensures individual pre-class preparation through the use of individual readiness assurance test (iRAT).

Frequent and Timely Feedback

Immediate feedback is crucial to TBL since it is essential to the learning process, content retention, and its influence on team development. Feedback sessions with the teams also help the faculty to teach strategy for acquisition of conceptual skills to embark upon solving complex problems. Such feedback allows teams to mature and helps to foster respect among team members.

Essential Principles of TBL

TBL shifts the focus of instruction away from the teacher and places it on students.

The essential principles of TBL are as follows:

- Properly formed and managed team
- Assignment that promotes team development and learning
- Student accountability for the quality of individual and group work
- Frequent and immediate feedback

These principles provide a pedagogical framework so that even novice teachers can comfortably lead an active learning session with clinical application.

The Modified TBL

Students entering medical school are heterogeneous with respect to knowledge, skill, and attitude. For many, TBL will be a new pedagogical classroom experience and concerns such as competition for better grades, sharing knowledge with others, or uneasiness about revealing knowledge inadequacies are anticipated and natural. The progression of the phases of TBL depends on students' full understanding of the TBL process. Hence, a mini workshop on the first day of the course to move the students through the required steps of TBL and its mechanics can be very useful. Included in this training session are also clerical staff and faculty. In addition, during this TBL orientation, we share how TBL has improved overall class performance in the past and invite experienced upperclassmen to share their experiences. These hands-on processes have profound effects on student's preparedness and acceptance of TBL pedagogy.

Phase 1: Individual Pre-Class Work

Characteristically, medical students are highly competitive, with strong personal desires for high achievement. Thus, establishing rules and guidance for the team process that emphasize the value of being on time and providing meaningful contributions to each TBL session is key. In addition to emphasis on knowledge and skill, importance is also placed on being respectful during the sessions. We also affirm that high-performing teams consist of members who are mutually accountable for each other's performance and symbolize a collective action that arises from task interdependency [10]. Task interdependence is a set of rules and requirements that determine how information, materials, and expertise will be shared between team members assigned to interdependent tasks. It is a functional way to plan structured work by defining interdependencies between tasks and elaborating roles for people involved in the work.

For each TBL session, students are expected to read a set of preparatory materials provided. These may be in the form of assigned readings, didactic slides/audio lectures, and/or video lectures. An essential component of TBL is the accountability of individual team members to self-directed pre-class preparation. Self-directed learning (SDL) is "a process in which individuals take the initiative without the help of others in identifying their learning needs, formulating goals, identifying resources, and evaluating learning outcomes" [11]. For the remaining TBL phases to be successful, students must have the ability to focus on and complete the preparatory assignments. In TBL pedagogy, an individual's poor quality of pre-class preparation affects their own learning and their ability to contribute to team discussion and team members' learning.

It is important to provide session goals, learning objectives, and reading assignments that connect with the previous and following sessions while, at the same time, linking with the rest of the curricular framework. Providing concrete guides, which can include specific pages in textbooks, didactic slides/audio/video lectures, concepts to develop, and clinical relevance, is particularly important in broad subjects like human anatomy and embryology. Additionally, the teacher writes application exercises, constructs readiness assurance tests, selects resources, and chooses appropriate pre-class assignments. The success of a TBL session depends on the guide, so a well-organized study guide is imperative to help students build baseline facts into a framework of conceptual interpretation and understanding [5, 6, 12]. This offers students a clear idea of what knowledge is expected and what effort is needed to be successful in the course. Furthermore, if a session assignment is overly simple and not robust, individual effort and team interaction will become meaningless. When the assignment is complex and truly requires the group to work together, there will be more group interactions, a sense of accomplishment, and greater team satisfaction.

Phase 2: Readiness Assurance

Prior to iRAT, the team gets in-class time to discuss the study guide contents, clarify doubts—an interactive peer teaching—and get ready for the readiness assurance test (step 2, Table 18.1). This needs monitoring of "freeloading" team members (step 2, Table 18.1).

Individual Readiness Assurance Test (iRAT)

The number and quality of multiple-choice questions (MCQ) in the readiness assurance tests for each TBL session should correspond to the content of the preparatory assignment (Phase 1). The questions should be challenging enough to stimulate critical thinking, analysis, synthesis, and evaluation. Team members gain an opportunity to observe and learn from each other on how to analyze and critically approach a question to generate a correct answer. Equally important is that the difficulty level and quality of iRAT questions should be similar to those in high-stake summative examinations (see Appendix). If the iRAT questions are simple and easy when compared to the summative questions, two things are likely to happen: first, students will perform poorly in the summative assessment and, second, students will think that the iRAT is of no value.

TBL strategy ensures individual self-directed pre-class preparation through the use of iRAT. The iRAT generally consists of 6–8 MCQs, based on the amount of pre-class preparatory materials. It is important that the nature of these iRAT questions is comparable to what the students will be tested in their final examination. To a large extent, the MCQs test a higher order of understanding, as described in the levels of Bloom's taxonomy [13]. In our institution, the iRAT carries a small percentage (2–3%) of the course grade. IRAT performance helps to identify students with academic difficulties, enabling faculty to offer appropriate guidance and other assistance.

Team Readiness Assurance Test (tRAT)

After submitting the iRAT, the team takes the same test, which is therefore called the team readiness assurance test (tRAT) or the group readiness assurance test (GRAT). The team submits answers on a scratch card, called the Immediate Feedback Assessment Technique form (IF-AT, Epstein Educational Enterprises, Cincinnati, OH) or using TBL-enabled software, some of which is available commercially (e.g., InteDashboard). This approach helps students identify incorrect answers immediately and discuss questions with more depth, thus facilitating deeper understanding of the materials. Individual programs choose the point percentages that iRAT and tRAT count toward the student's final grade. At the authors' institution, the tRAT carries a lower percentage (1%) of the course grade than the iRAT because the same tRAT score is given to all the students in the same team irrespective of their individual comprehensive knowledge. Following the iRAT, detailed discussion of the questions and the learning objectives during tRAT enables students to act both as teachers and learners, clarifying and explaining the issues. After submitting the tRAT grade sheet, the faculty participate in the ensuing discussions to provide additional comments.

Clarification Session

After completing both the iRAT and tRAT, students have the opportunity to ask the faculty to clarify any misconceptions and give explanations. Students are also allowed to challenge a question's clarity or argue that it is beyond the assigned readings. If a question is successfully challenged, the whole class receives a point. In addition to challenging questions, the clarification session allows students to enhance their own understanding of the content. This is essential, since each TBL session is built on the knowledge of previous content and carries onto the next so that knowledge is accumulated continuously and recurrently over a period of time. Such layering of knowledge is important when students are engaged in the application phase, which assesses comprehension.

Phase 3: Inter-Team Case Application Exercises

Phase 3 involves application exercises that consist of 4–6 case scenarios with a subset of questions related to the case. The cases presented and discussed must be limited to the content of that particular organ system module. Phase 3 starts with intra-team discussion of the cases (step 6, Table 18.1) followed by inter-team discussion (step 7, Table 18.1). All the teams present answers to each question to facilitate inter-team discussion. Open-ended question allows the students to apply and expand on the knowledge that they have learned and been tested on (readiness assurance tests). It triggers students to discuss and solve clinical problems under the facilitation of the instructor. It is important to emphasize that unlike Phase 3.4, the activities in this phase is not graded. The case content involves the entire organ system and acts as a proxy to the summation of overall knowledge. Since this phase is followed by a graded summative final MCQ examination of the module, it allows the students to gain comprehensive knowledge.

Phase 4: End of Module Summative Examination

To further strengthen information and concept retention, the traditional three-phase TBL has been modified at the authors' institution to include an additional application-oriented summative examination. Phase 4 encompasses clinical vignette-based MCQ examinations that occur at the end of each organ system module (e.g., cardiopulmonary, gastrointestinal, musculoskeletal, and others). As in Phase 2, this summative examination includes iRAT and tRAT. Unlike iRAT that contains 10 MCQs, Phase 4 examination has 40 MCQs since it covers the entire modular content. The depth and breadth of the examination content (questions) acts as a proxy to the assessment of overall comprehensive knowledge of module content. After completing the iRAT, students come together as a team and take the same MCQ examination to generate a single answer. As in Phase 2, after submitting the group test, the students request clarification on certain questions and are allowed to challenge an answer. All such challenges are evaluated and, if a challenge is accepted, the class receives points.

Oftentimes, peer evaluation is optional, since many students consider it contemptuous. However, peer evaluation as part of professional identity development is now being implemented in all schools including the author's institution using TBL pedagogy. It includes both cognitive and noncognitive assessments, which students consider a valuable part of their professional training. The cognitive assessment takes into consideration content-specific knowledge, intelligence, and how one connects the current knowledge with past and future learnings. The noncognitive assessment includes one's motivation, self-efficacy, willingness to persevere, interpersonal skills, ability to think and convey ideas cogently, to persist through difficulties, and show continuous progress and productivity.

Modification to the Classical TBL

TBL is often modified to suite the course structure, faculty acceptance, and local curricular needs. The flexible modification of the classical TBL allows programs ranging from those in high schools to those in professional schools to successfully adopt this pedagogy. Oftentimes the number of phases varies initially. Once the faculty are comfortable with the pedagogy, more phases can be added as appropriate for the course.

Implementing TBL

Important Questions to Consider Before Planning TBL

Prior to establishing a formal TBL strategy, consider the following. Is it the best approach for teaching the subject? Is there enough time available in the course for students to work both in class and out of class to evolve into a cohesive team? Is the course content complex enough to justify adopting this strategy? Will it help development of skills such as efficiency, critical thinking, higher-level reasoning, and teamwork? Are there resources and support from the senior management at the institution? Additionally, there are other questions that also require careful consideration. Are there teachers who will not do well with TBL strategy, but will be successful lecturing? Will the team-based method negatively affect development of students' individual skills, such as motivation, self-efficacy, and perseverance? Does TBL lead to commitment for creating and maintaining an ongoing, self-motivated, positive attitude to lifelong learning for both personal and professional development?

Constructing a TBL Session

In developing a TBL session with an anatomy component, other interconnected elements to be considered are (1) the type and amount of session materials in relation to cadaveric dissection that may be going on in parallel with the TBL sessions, (2) the amount of time required to prepare for each session, (3) the time requirements of each session, and (4) importantly, demand from other concurrent courses in the curriculum. Faculty should be mindful of the curricular structure as a whole and not focus on their own courses.

Preferably, the length and complexity of reading assignments should be such that a student need not spend more than 2–3 hours of preparation time for each session (Phase 1). Exceeding that time limit may lead to the student splitting the work and perhaps not having the required knowledge to perform satisfactorily on the iRAT. Additionally, team discussion will be suboptimal. Ideally, TBL sessions should not exceed 2 hours in order to keep students focused on the task and avoid straying of the discussion. Conversely, if the session is too short and the students are not able to complete the content with meaningful learning, students either assume that the topic is less important or that the faculty are not passionate about the TBL. Thus, a visible faculty passion for the pedagogy is critical for ongoing program success. This highlights the importance of preparatory assignments and session planning.

Implementing TBL Strategy for the First Time: Starting Small

Adopting a TBL pedagogy needs careful evaluation and planning. Piloting a small segment of a course or single organ system module provides experience and, at the same time, allows students, staff, and faculty to provide feedback. Importantly, in a pilot model, it is quick and easy to catch and correct mishaps. After piloting TBL this way and making modifications expeditiously, scaling up to the entire course will be smooth, but will also need continuous monitoring and revisions.

Faculty and Staff Development

At least one member of the faculty team must have hands-on experience from a TBL workshop like the ones conducted at the annual meeting of Team-Based Learning Collaborative. The Team-Based Learning Collaborative is an organization of educators from around the world that encourages and supports the use of TBL at all levels of education. The trained person can provide education and guidance for the other faculty and staff, so they gain experience in all aspects of the TBL pedagogy. It is essential to have faculty who are interested in this pedagogy, since an uninterested faculty can derail the process. Faculty may also express apprehension for new pedagogy and concern about covering the course content. Conducted properly, TBL saves time, since much of what would be covered in lectures is converted into reading or other preparatory assignments (syllabus) for students' self-directed independent study.

A committed and enthusiastic staff is also critical to the TBL program. The course director should not assume clerical responsibilities. Before the course begins, the staff ensures that all course materials are sent to students. Additionally, staff types the team lists, creates team folders with members' names on the cover, and makes copies of the iRAT and tRAT for timely placement in the team folders. Importantly, all testing materials are color-coded to make the sequence of activities clear and, since the quizzes and exams are protected, each question booklet, Scantron answer sheet, and IF-AT form is numbered to ensure accountability and security. Following the iRAT, the staff collects the grading sheets and immediately scores them for faculty to identify unprepared poor performers. Staff also scores the tRAT and continuously maintains the iRAT and tRAT grade book. These activities, which are detail-oriented and time-consuming, are typically done several times during the course. Recently, some schools have started using web-based platforms that automatically do much of the above work.

Advantages of TBL Pedagogy

In the best interest of the group, every team member does his/her best to engage and contribute continuously to developing the collective knowledge. TBL also results in better class attendance and participation, since it includes graded iRAT and tRAT. There is improved learner engagement with the course content that allows development of higher levels of learning: application, analysis, and evaluation. Repetitive process and layering of content-specific knowledge has been shown to result in long-term retention and better performance on end of course examinations [6, 14–17]. Repetitive process involves repeating all the phases (1–4) on a given content and layering it over prior knowledge. When this process continues over a period of given lessons, it also improves knowledge gain.

TBL is increasingly used in many health-related programs such as nursing [18] and medical school programs [5, 6, 16, 19, 20] because it employs active learning to promote in-depth understanding that enhances student adaptability in problem-solving situations [21]. In-depth learning through TBL exercises results in greater retention of the material, likely because students understand and make personal sense of the material, rather than simply adopting superficial learning and memorizing to later reproduce information [19]. Through the use of TBL, faculty members can shift lectures to self-directed pre-class preparation, leaving more class time for active learning and integration of new learning with the knowledge gained before class [22]. TBL improves performance in both academically weak and strong students [12, 23]. When TBL is employed, students perform better on examination questions, indicating their increased mastery of course content [6, 12, 16]. Faculty members find that TBL may impact student behaviors, such as being better prepared for class, being more engaged during class, and taking more responsibility for their own learning [24]. Team-based learning also provides greater student-to-instructor engagement than traditional lecture during the learning process [5, 6, 12, 25–27].

Conclusion

Successful implementation of the TBL strategy depends on the sincerity of the faculty and staff and support of the administration to foster positive student perception of the process. TBL can advance academic success through teamwork, trust, mutual respect, interdependence, and attitudes toward learning. Postsecondary education should be an active learning process that engages students to work in teams. Use of team-based learning not only allows faculty to cover course content, despite reduction in course time, but also improves student understanding and retention. TBL facilitates active learning and peer teaching, leading to higher student engagement. TBL also provides opportunities for self-directed learning, which fosters the skills necessary for lifelong learning and professional development, skills that include self-assessment of learning needs; the identification, analysis, and synthesis of related information; and the evaluation of the reliability of information sources. These are all skills that are increasingly vital and emphasized by the twenty-first-century employers.

Appendix

Completion of this TBL will require the use of designated figures and text in Clinically Oriented Anatomy (COA), Moore et al., 8th edition [28], and Medical Embryology, Sadler, 14th edition [29].

Phase 1: Reading Assignment, Pagination from Textbooks

Heart: Anatomy to Comprehend

After completing the section on pericardium, you should be able to:

- Describe the fibrous pericardium.
- Distinguish parietal and visceral layers of the serous pericardium.
- Define the transverse pericardial sinus.

Content-Specific Learning Topics: Fibrous Pericardium and Serous Pericardium

Use Fig. 1.43, p. 129 (COA), to observe that the tough external fibrous pericardium covers the cellular serous pericardium. Recognize that mesothelium (simple squamous epithelium) forming the parietal layer of serous pericardium lines the inner surface of the fibrous pericardium. Observe at the origin of the great vessels that it reflects onto the heart as the visceral layer of serous pericardium.

Use Fig. 1.48 (B), p. 131 (COA), to recall that fusion of the pleuropericardial membranes divides the thoracic cavity into the pericardial cavity and bilateral pleural cavities, and use Fig. 1.48 (C) to observe after the division that the membranes form the fibrous pericardium.

Use Fig. 1.32, p. 110 (COA), to recognize that the pericardium (aka pericardial sac) fuses to the central tendon of the diaphragm, and use Fig. 1.33, p. 111 (COA), to envision continuity of the parietal and visceral layers of serous pericardium at the roots of the great vessels.

Use Fig. 1.46, p. 131 (COA), to recognize that the transverse pericardial sinus (arrow) is a pathway posterior to the intrapericardial parts of the aorta and pulmonary trunk and anterior to the intrapericardial parts of the superior vena cava (SVC) and pulmonary veins.

Embryology to Comprehend

After completing the section on the heart tube, you should be able to the following:

- Summarize the formation of the interatrial septum
- Summarize the formation of the interventricular septum

Content-Specific Learning Topics: Formation of the Atrial and Ventricular Septa

Use Fig. 13.15, p. 172 (ME), to determine that formation of interatrial and interventricular septa completes the differentiation of the primitive atria and ventricles. Recognize that formation of the septa involves mesenchymal cell-derived endocardial cushions (ridges) and narrow tissue strips that completely or incompletely separate the chambers.

Phase 2: In-Class Discussion, iRAT and tRAT

Questions for Application of Knowledge

1. From Cardiac Tamponade and Pericardiocentesis, pp. 133–134 (COA): Why can cardiac tamponade be fatal and how is pericardiocentesis normally performed?
2. From Surgical Significance of Transverse Pericardial Sinus, p. 133 (COA): How do cardiac surgeons utilize the transverse pericardial sinus?

Examples of questions (based on reading assignments, topics for discussion, and application of knowledge) in the module exam:

Case A 21-year-old college student on spring break fell from the balcony of his hotel and sustained blunt chest trauma. He was rushed to the ER, and very weak heart sounds, reduced cardiac output with declining blood and pulse pressures, bilateral jugular distension, and respiratory distress were detected.

The patient's symptoms most likely resulted from:

A. Hemothorax
B. Cardiac tamponade*
C. Cor pulmonale
D. Deep vein thrombosis
E. Pancoast's syndrome

Case A 61-year-old female complains to her physician about tiring easily and shortness of breath on exertion. Auscultation of the chest detected a diastolic murmur and a collapsing pulse was detected.

The patient is most likely suffering from:

A. Aortic valve insufficiency*
B. Mitral valve stenosis
C. Diseased left ventricular papillary muscle
D. Pulmonary valve insufficiency
E. Heart bundle block

Case An underdeveloped 3-year-old child was brought to this country to correct a congenital heart defect. The child had dyspnea and often suffered from pneumonia. Surgery corrected the congenital defect; however, several days later, the child was diagnosed with a heart block.

The surgery was most likely performed to correct:

A. An atrial septal defect
B. A ventricular septal defect*
C. A patent ductus arteriosus
D. A double aortic arch
E. Coarctation of the aorta

Phase 3: Inter-Team Case Discussion (Example of a Case)

Case A 67-year-old man, who has been a smoker since the age of 15, is seen at the family clinic.

History of present illness Numbness and tingling on the medial side of his right forearm and hand and swelling in the right supraclavicular region. He also started to notice that his voice was becoming hoarse.

Physical exam Edematous face and neck, puffiness around the right eye, and right jugular venous distension. Exhibits ptosis of the right eye and pupillary constriction.

Imaging Chest X-ray shows large tumor of the right apical lobe.

Discuss and explain the reason(s) for the following:

1. Numbness and tingling on the medial side of his right forearm and hand
2. Voice becoming hoarse
3. Swelling in the right supraclavicular region
4. Edematous face and neck, puffiness around the right eye, and right jugular venous distension

Phase 4: Module (Thoracic Structures) Exams with MCQ (Examples)

Case A third-year medical student was assisting in "C" section at the university hospital. The attending physician asked the student to describe the blood pressure changes that occur in a neonate.

Right atrium	Right ventricle	Pulmonary trunk	Left atrium	Left ventricle	Aorta
A. Increases	Increases	Increases	Increases	Increases	Increases
B. Increases	Increases	Increases	Decreases	Decreases	Decreases
C. Increases	Increases	Decreases	Increases	Increases	Decreases
D. Decreases	Decreases	Decreases	Decreases	Decreases	Decreases
E.* Decreases	Decreases	Decreases	Increases	Increases	Increases

Case A 65-year-old male, living alone, was found dead in his apartment. Postmortem examination of the heart showed necrotic changes limited to the anterior two-thirds of the interventricular septum.

If chronic coronary artery disease had resulted in arterial occlusion, the artery most likely occluded was the:

A. Left coronary artery
B. Left circumflex artery
C. Right coronary artery
D. Marginal artery
E. Left anterior descending artery*

Case A 52-year-old female with dyspnea came to the ER. During examination, crackling sounds (rales) were heard when the stethoscope was positioned on the right midclavicular line at the level of the fourth intercostal space, and acute bronchitis was suspected.

The location of the acute inflammation and resulting rales was most likely the:

A. Lateral bronchopulmonary segment*
B. Anterior bronchopulmonary segment
C. Apical bronchopulmonary segment
D. Anterior basal bronchopulmonary segment
E. Lateral basal bronchopulmonary segment

References

1. Christensen CR, Garvin DA, Sweet A. Education for judgment: the artistry of discussion leadership. Cambridge: Harvard Business School; 1991.
2. Merriam SB, Caffarella RS, Baumgartner LM. Learning in adulthood. San Francisco: Jossey-Bass; 2007.
3. Michaelsen LK, Knight AB, Fink LD, editors. Team-based learning: a transformative use of small groups in college teaching. Sterling: Stylus; 2004.
4. Haidet P, Kubitz K, McCormack W. Analysis of the team-based learning literature: TBL comes of age. J Excell Coll Teach. 2014;25:303–33.
5. Vasan NS, DeFouw D. A successful initial experience with modified team learning strategy in medical gross anatomy course. Med Educ. 2005;39:524.
6. Vasan NS, DeFouw DO, Holland BK. Modified use of team-based learning for effective delivery of medical gross anatomy and embryology. Anat Sci Educ. 2008;1:3–9.
7. Addo-Atuah J. Performance and perceptions of pharmacy students using team-based learning (TBL) within a global health course. Inov Pharm. 2011;2:37.
8. Inuwa IM, Al-Rawahy M, Roychoudhry S, Taranikanti V. Implementing a modified team-based learning strategy in the first phase of an outcome-based curriculum--challenges and prospects. Med Teach. 2012;34:e492–9.
9. Tuckman B. Developmental sequence in small groups. Psychol Bull. 1965;63:384–99.
10. Blue A. Creating Collaborative Care (C3), Medical University of South Carolina. J Allied Health. 2010;39:e127–8.
11. Knowles MS. Self-directed learning: a guide for learners and teachers. New York: Cambridge Books; 1975.
12. Vasan NS, DeFouw DO, Compton S. A survey of student perceptions of team-based learning in anatomy curriculum: favorable views unrelated to grades. Anat Sci Educ. 2009;2:150–5.
13. Bloom BS, Krathwohl DR, Masia BB. Taxonomy of educational objectives: the classification of educational goals. New York: D. McKay; 1956.
14. McInerney MJ, Fink LD. Team-based learning enhances long-term retention and critical thinking in an undergraduate microbial physiology course. Microbiol Educ. 2003;4:3–12.
15. Levine R, O'Boyle M, Haidet P, Lynn DJ, Stone M, Wolf D, et al. Transforming a clinical clerkship with team learning. Teach Learn Med. 2004;16:270–5.
16. Koles P, Stolfi A, Borges N, Nelson S, Parmelee D. The impact of team-based learning on medical students' academic performance. Acad Med. 2010;85:1739–45.
17. Rania N, Rebora S, Migliorini L. Team-based learning: enhancing academic performance of psychology students. Procedia Soc Behav Sci. 2015;74:946–51.
18. Dearnley C, Rhodes C, Roberts P, Williams P, Prenton S. Team based learning in nursing and midwifery higher education: a systematic review of the evidence for change. Nurse Educ Today. 2018;60:75–83.
19. Entwistle NJ, Thompson S, Tait H. Guidelines for promoting effective learning in higher education. Edinburgh: Centre for Research on Learning and Instruction, University of Edinburgh; 1992.
20. Fatmi M, Hartling L, Hillier T, Campbell S, Oswald AE. The effectiveness of team-based learning on learning outcomes in health professions education: BEME Guide No. 30. Med Teach. 2013;35(12):1608–24.
21. Whitley HP, Bell E, Eng M, Fuentes DG, Helms KL, Maki ED, et al. Practical team-based learning from

planning to implementation. Am J Pharm Educ. 2015;79(10):149.
22. Ofstad W, Brunner L. Team-based learning in pharmacy education. Am J Pharm Educ. 2013;77(4):70.
23. Nieder GL, Parmelle DX, Stolfi A, Hudes PD. Team-based learning in a medical gross anatomy and embryology course. Clin Anat. 2005;18:56–63.
24. Nelson M, Allison SD, McCollum M, Luckey SW, Clark DR, Paulsen SM, et al. The Regis model for pharmacy education: a highly integrated curriculum delivered by Team-Based Learning (TBL). Curr Pharm Teach Learn. 2013;5:555–63.
25. Kelly PA, Haidet P, Schneider V, Searle N, Seidel C, Richards B. A comparison of in-class learner engagement across lecture, problem-based learning, and team learning using the STROBE classroom observation tool. Teach Learn Med. 2005;17(2):112–8.
26. Vasan NS, DeFouw D, Compton S. Team based learning in anatomy: an efficient, effective and economical strategy. Anat Sci Educ. 2011;4:333–9.
27. Frame TR, Cailor SM, Gryka RJ, Chen AM, Kiersma ME, Sheppard L. Student perceptions of team-based learning vs traditional lecture-based learning. Am J Pharm Educ. 2015;79(4):51.
28. Moore KL, Dalley AF, Agur AMR. Clinically oriented anatomy. 8th ed. Baltimore: Lippincott Williams & Wilkins; 2017.
29. Sadler TW. Langman's medical embryology. 14th ed. Baltimore: Lippincott Williams & Wilkins; 2018.

Giving Feedback to Students 19

Nirusha Lachman

Introduction

In a truly integrated, student-centered teaching and learning environment, successful achievement of competencies is strongly dependent on a dynamic exchange of performance-related communication. While the student-centered approach to teaching places greater emphasis on self-directed learning, it places an even greater emphasis on *teacher-dependent feedback*. Students cannot be expected to know what they *do* or *do not* know if they have nothing to measure it against.

For anatomy teachers, understanding that whatever the design, whatever the grading structure of the course, providing students with meaningful communication regarding performance is imperative if one is to expect them to progress through their learning experience and achieve the desired competencies [1]. In a competency-based educational environment, assessing student performance is crucial to the achievement of core objectives and successful implementation of the curriculum.

Within the framework of a teaching and learning partnership, the use of meaningful *feedback*, as it relates to student performance, is no longer an optional exercise but rather an essential element in helping students achieve successful learning outcomes. With transition into future practice, especially true for health care, performance measures are crucial drivers of professional development and relate directly to clinical competencies [2–4].

Two central terms relate to feedback: (1) summative and (2) formative. Discussion surrounding the interpretation and application of these terms can be somewhat confusing [5]. The simplest way to understand the difference is to recognize that the goal of summative feedback is to provide an evaluation of student learning at the end of an instructional section using a benchmark (e.g., midcourse assessment of knowledge of anatomy of the back, upper limb, and thorax; final assessment of anatomy covered over entire course), whereas the goal of formative feedback is to supervise and guide student learning on an on-going basis *during* the course (e.g., daily quizzes, practice practical tests, feedback on professional development) to improve student learning. While the information gathered from summative feedback is important, it is limited in its ability to enhance student competency. *Summative* feedback is an effective way of evaluating effectiveness of course delivery and achievement of course objectives through measuring overall student performance. *Formative* feedback on the other hand is part of the instruc-

N. Lachman (✉)
Department of Clinical Anatomy, Mayo Clinic College of Medicine and Science, Mayo Clinic, Rochester, MN, USA
e-mail: Lachman.nirusha@mayo.edu

L. K. Chan, W. Pawlina (eds.), *Teaching Anatomy*, https://doi.org/10.1007/978-3-030-43283-6_19

tional process, informing students and teachers about student comprehension so that pedagogical adjustments can be made to enhance overall student performance [6].

This chapter will focus on how to effectively apply formative feedback strategies within a contemporary anatomy curriculum. We will discuss how directive and facilitative feedback can enable students to achieve competencies embedded within the core objectives of the anatomy course. The terms "students" and "learners" will be used interchangeably when relating directly and generally to the learning participants. While many of the principles embedded throughout the discussion are based on classroom design for team-based learning, the key skills to providing effective feedback may be applied to any classroom setting.

Formative Feedback

Formative feedback is a representation of assessment and observation that provides students with meaningful commentary *relating to performance* for the *purpose of improving performance*. When correctly applied as an educational strategy, feedback can be a powerful tool, not only in enhancing student performance but also in augmenting student motivation. For it to be effective, formative feedback should be nonjudgmental, student specific, timely, supportive, and multidimensional [7, 8].

The literature pertaining to formative feedback is abundant in the field of education and business [9–13]. However, its use in medical education and, in particular, in the teaching and learning of basic sciences is a more recently introduced topic [10, 14–17]. While anatomy courses are by nature designed to promote the understanding and retention of factual knowledge, they are no longer confined to providing students with such [18]. In fact, very few anatomy courses still maintain such a one-dimensional approach to the teaching and learning of the anatomy. Courses that have evolved to include more nontraditional discipline-independent skills and competencies have placed greater expectation on the critical evaluation of these skills. Black and William [19] discuss provision of feedback by distinguishing two main categories which may be applied in the anatomy course assessment structure: (i) *directive feedback* used to provide students with information on knowledge deficits that may place them at risk for achieving required grades and (ii) *facilitative feedback* used to provide students with verbal or written commentary on their performance relating to nontraditional discipline-independent skills with the intention of guiding them through their professional development. The "Educational Testing Services Research Report" on formative feedback [20] unfolds three important directives for effective and useful feedback: (1) the understanding that the student needs the feedback, (2) the detection of opportunities to provide timely feedback, and (3) the mindset that the student has the ability and is open to using the feedback to strengthen learning.

Directive Feedback

With the shift away from content-heavy courses, the contemporary anatomy curriculum [18] focuses on facilitating student learning of anatomical material (knowledge-based competency) within the framework of two main objectives:

- *The ability to demonstrate knowledge of normal function* (i.e., relating structures of the human body to its function)
- *The ability to apply knowledge of basic sciences to patient care* (i.e., identify and apply anatomical knowledge to clinical problems, apply principles of anatomical sciences to clinical scenarios, translate basic science knowledge into clinical decision-making processes, and possess effective laboratory observation skills)

Use of Audience Response System

The use of the audience response system (ARS) has been well described with significant data to support its effective use among students [21, 22] and its ability to reinforce learning by promoting knowledge retention and clinical reasoning [23, 24]. In addition, studies [21] show that the use of classroom clickers promotes both active learning [25] and social interaction within the team-based classroom environment [26].

One of the biggest challenges students face in anatomy relates to information overload. Apart from being overwhelmed, students are often unsure of where they stand with regard to application of newly acquired information. As a result, such uncertainty tends to decrease student motivation which further impacts student performance [20]. Summative evaluation is a valuable tool often used in providing a measure of what a student has retained and understood by the end of a given period. While this information may be used to determine what has been successfully *learned*, it does very little for the *process of learning*. The responsibility for facilitating student learning is coupled with the responsibility for providing well-timed, regular feedback. Providing timely (through daily debriefing or weekly consolidation sessions) and quality feedback (reviewing answers through a process of critical evaluation) reinforces concept understanding and integration of the material. The use of ARS as a method of providing directive feedback enables students to individually comprehend the question, determine an answer, share their thoughts with teammates, and, through a process of shared feedback and critical evaluation, agree on the most likely answer. The dynamic approach to this exercise encourages clinical reasoning and helps students determine for themselves where they stand in terms of their performance. With regular scheduled use, ARS scores have shown to be good predictors of final grades and good identifiers of students who may be experiencing academic challenges [21]. However, the key to making this source of feedback meaningful for student improvement lies in the skill of determining the most likely answer. Studies show directive feedback to be far more effective when it promotes thinking and reasoning in deriving at a correct answer as opposed to students simply receiving a correct answer [27, 28].

Use of Electronic Feedback in Midcourse or Quarterly Assessments

In the absence of ARS technology, the use of midcourse or quarterly assessments through curriculum management systems such as Moodle and Blackboard can be used as an electronically accessible tool to provide directive feedback and to clarify student questions on learning concepts. Supporting discussion with thoughtful, scientifically sound explanations with good references reinforces the learning partnership by placing value on the student's responsibility for their own learning. Also commercially available survey software platforms, e.g., Qualtrics, can be used to provide feedback [4].

Despite being less socially interactive and not immediate, pretest questioning is an indispensable assessment instrument in determining student assimilation of anatomical material and progress in applying their knowledge in clinical-based scenarios. The challenge, however, and one most teaching faculty would face with using this system, is that it is time-consuming to provide electronic feedback to each student—a challenge that is accentuated with the prospect of large class sizes.

One option for dealing with this challenge is to implement a near-peer teaching [29] approach through delegation of tasks to teaching assistants (TAs) [30]. Based on the peer feedback model described by Camp et al. [31], expanding the TA role as student tutors [30] is an effective way of integrating knowledge-based feedback on a more structured level. The type of feedback provided by the TAs need not be elaborate. Simple presentation of a score/grade/results from practice practical tests combined with a few short comments or a brief description highlighting areas for improvement with helpful resources determined by the TA would serve as an acceptable form of feedback. By receiving this feedback, students would feel equally noticed and value the teaching team's investment in the progress of their learning.

Use of Dissection Grading and Formative Practical Tests

One of the more underemphasized and often overlooked areas for providing student feedback revolves around activities in the anatomy laboratory. For courses that include a dissection component, providing feedback to teams is crucial in ensuring that students maximize the invaluable hands-on experience of dissection. It is not surprising, given the lack of expertise, that students in general are unable to make efficient use of laboratory time. More often than not, with the focus on "getting through" the laboratory objectives, cadavers end up being poorly dissected, structures remain unexposed, and teams trudge on leaving most if not all of their practical learning for the end of the course.

To encourage students to dissect in a thoughtful and efficient manner, a diminutive percentage of their final grade should be allocated to the quality of their dissection. Despite its low impact on overall performance, providing students with directive feedback on task quality provides them with the motivation needed to ensure that the resource offered to them is valued. Students should receive clear objectives and a list of key structures for identification [32], be informed in a timely manner of when the evaluation will occur, and supported throughout the laboratory session by faculty and TAs. Again, the responsibility for providing the feedback may be easily assigned to TAs [30]. Subjective directive feedback using a grading plan may be given via electronic communication (email correspondence to each team or curriculum management system) with a dissection grade and explanation of where the team's dissection falls short and what the team dissected well. In addition, helpful techniques and tips for uncovering structures as well as structures that have been clearly demonstrated should be provided to the class as a whole.

> Feedback is the cornerstone of effective clinical teaching. Without feedback, good practice is not reinforced, poor performance is not corrected and the path to improvement not identified. –Cantillon and Sargeant [17]

Grading Quality of Dissection

Provide feedback on how well the cadaver was dissected:

- Very well dissected (structures easily identified, attention to detail, all required portions of dissection completed in an excellent manner).
- Well dissected (most structures easily identified, all required portions of dissection completed).
- Adequately dissected (most if not all required portions of dissection completed).
- Poorly dissected (multiple structures destroyed or unidentifiable, visibly incomplete dissection).
- Very poorly dissected (no time or thought spent on isolating structures, incomplete).
- An example of feedback on how dissection can be improved: *Uncover the recurrent branch of the median nerve; the superficial branch of the radial nerve is often obscured by fascia as it passes beneath the brachioradialis….*

In addition, the use of practice practical tests with an opportunity for students to review the correct answers for tagged structures keeps students on task and track, eliminating the uncertainty associated with simply receiving a grade at the end of the test. During the review, faculty and TAs should be available for immediate feedback.

The fundamental nature of *directive feedback* lies in its capacity to (1) *verify* student knowledge of anatomy and (2) *elaborate* on the process of deriving the most likely answers through clinical reasoning. Studies agree that this type of feedback not only encourages student understanding of subject-specific concepts but also promotes student thinking, decreases student uncertainty, and enhances student motivation [33].

Characteristics of Directive Feedback

- Focuses specifically on knowledge-based competency (anatomical concepts)
- Objectively measures knowledge and understanding (application of anatomical concepts)
- Identifies knowledge paucity
- Reduces uncertainty for students regarding their own progress during the course
- Encourages self-evaluation and clinical reasoning
- Motivates students to improve performance
- Helps teachers identify class challenges in understanding material and reinforce anatomical concepts
- Verifies student knowledge and demonstrates use of critical thinking in deriving answers

Facilitative Feedback

Use of Facilitative Feedback to Promote Holistic Student Development

Facilitative feedback may be used to promote holistic student development. Within the integrated medical curriculum, anatomy courses provide an ideal platform for creating awareness of the importance of nontraditional discipline-independent skills in the current health-care system. These skills—social, interprofessional, creative, communication, leadership, and professionalism—are vital to successful integration within the team-based environment. While anatomists are no longer hesitant to incorporate these ideals as measurable outcomes within the curriculum [18, 34], they are expectedly apprehensive about providing feedback relating to nontraditional discipline-independent skills [17]. With facilitative feedback involving provision of verbal or written commentary to students with regard to these skills, the task poses a daunting challenge for teachers who have had no training in assessing students with the intent of guiding them through their own professional development [1].

For students, four main objectives require feedback during the course as well as upon completion: (1) *consistent demonstration of professional and ethical behavior*, (2) *effective communication*, (3) *display of effective teamsmanship*, and (4) *display of leadership*.

Communicating Facilitative Feedback

For experienced anatomy teachers, a new and very real challenge lies in learning how to work with the "millennial student." Awareness of generational characteristics is important if providing feedback is meant to be meaningful. Millennial students, reputed to be technologically savvy, confident, and optimistic with high expectations for themselves, are goal oriented and self-driven [35, 36]. Studies have shown that for students to remain engaged and motivated, teachers must be able to create cohesion between what is expected of them and what they are able to achieve for the level of training they are at [37].

Communicating facilitative feedback is no easy task. In fact, the majority of teachers avoid the use of verbal or written communication to effect student improvement simply because they feel uncomfortable doing it [16].

Without the proper tools, facilitative feedback can come across as being judgmental and subjective, oftentimes causing the recipient to become defensive or sensitive. If feedback is too detailed or complex [38], students may not pay attention to it or may diminish the message. If comments are too complimentary and focused on student embellishment, the opportunity for improvement is lost. The most meaningful and effective feedback must include a balance of (1) *direct observations* (positive behaviors and limitations for learning), (2) *points for reflection*, and (3) *guidelines for improvement*.

It is important to keep in mind that the students to whom you are providing feedback are at a very early stage in their career. They cannot be expected to have mastered nontraditional discipline-independent skills. For most, this may well be the first time they are expected to demonstrate such skills. Research has shown that for a learner to remain motivated and engaged, a lot depends upon *reaching a balance between a learner's goals and the expectations that these goals can be met*. If goals are set so high that they are unattainable, the learner will likely experience a sense of failure and become discouraged. If goals are set low so that students are certain that they will achieve them, the need for further effort is diminished [20].

Providing Verbal Feedback to Students

The teaching and learning environment of an anatomy course presents a multidimensional platform, providing a variety of opportunities for student-teacher interaction. However, the issue of how to provide verbal feedback is as important as when to provide it. Set expectations at the right level—remember that these are students learning to develop skills by being introduced to them during an already demanding anatomy course.

The classroom setting is a good arena to provide students with global feedback (general feedback relayed to the class as a whole) relating to good practice and demonstration of professionalism or opportunities for improvement, e.g., faculty observation of overall student disposition and attitude in the anatomy laboratory, team cohesiveness, and peer interaction. Feedback may be addressed at the end of each week in the form of a 10-minute debriefing session at the end of class.

Scheduled time for feedback allows for reinforcement of course expectations and invites students to openly reflect and/or share their thoughts that strengthen nontraditional discipline-independent skills.

In addition to providing group feedback, it is important to set time for students to meet with faculty for individual feedback. Leadership evaluations provide ideal opportunities for student feedback sessions. Leadership evaluations represent both assessments by peers and self (leader being assessed) for the same set of criteria: respect, integrity, responsibility, compassion, problem-solving, commitment to excellence, and overall professionalism. According to Garrison and Ehringhaus [6], self and peer assessment supports the awareness of a learning community within a classroom. It illustrates that students who are able to reflect (on their own and on their peers' skills) while being engaged in higher-level thinking maintain responsibility for their learning. Comparing self- and peer-evaluation scores elicits deeper reflection and strongly influences the outcome for self-improvement.

Students should be invited to meet with faculty to discuss their experience and reflect on their peer evaluation as team leader. Meetings should be scheduled in advance and need not take more than 15 minutes [39].

Preparing for the feedback session is as important as time spent engaging with the student in discussion. While the notion that using feedback as an academic intervention is always linked to positive relationship with student performance, the potential for the feedback to negatively affect student performance also exists. In the current social and learning climate, even the most experienced teacher can never be fully prepared for how the student may relate to the feedback. Cultural influences, expectations, norms and mindset are unpredictable variables that impact motivational mechanisms [40]. A helpful approach in preparing for the feedback session would be to lean on the "Feedback Intervention Theory" that recommends keeping feedback relevant to the performance measure and not directed toward the student. The student is then more likely to internalize the feedback as a meaningful impetus for behavioral change or skill improvement [41].

In conducting a verbal feedback session:

1. Always begin by stating the purpose of the meeting: "Thank you for meeting with me/us. The purpose of our meeting is to provide you with feedback regarding your role as team/dissection group leader."
2. Present the student with a paper copy of a graph plot of peer evaluation vs. self-evaluation with a summary of the results and invite the student to reflect on it: "Here is a

copy of your peer evaluation mapped against your self-evaluation. (i) Overall you have received good/very good/excellent scores that match your self-evaluation scores; congratulations on the good work. (ii) Judging from the graph, it appears that you have rated yourself much higher/lower than peers have. Would you like to reflect on this?"

3. Review peer comments, if any, and summarize so that only the most valuable feedback is presented to the student. Avoid handing over handwritten peer comments to the student. Keep in mind that the majority of students are not skilled at providing feedback and unedited comments from peers have the potential to create discomfort for the student working within the team.
4. Begin with comments that strengthen the student's self-esteem and then point out skills that the student could benefit from improving upon: "Student X is a valued team member, he is always prepared for class and always makes sure we understand the material without pushing us all to be right," "Good listener, open to others' points of view." Be aware and knowledgeable about characteristics of the skills being assessed and translate student comments so that they relate to the positive markers for the skill.
5. Approach weaker attributes with caution: "Student X has difficulty identifying most important or high-yield content, tends to be less focused." "Student X is insensitive and lacks respect when speaking to some members of the team. Student X is always unprepared and late for class." More often than not, peers do not provide examples of incidents that describe an action or behavior. Steer direct statements into more guiding comments so that student attention can be channeled toward working on behaviors that will benefit them: "It may benefit you to be more focused so that you can support your team members in identifying more high-yield knowledge." "It is important to remember that when working in a team, the way we present ourselves through what we say or how we say it may be sometimes be misperceived by others to come across as lacking respect." In the absence of information pertaining to a specific incident, ask the student to reflect on an occasion where his/her behavior could have led to formation of the perception. By doing this, the teacher indicates to the student that he/she is equally responsible for the learning experience. It creates a less discouraging environment, enabling the student to accept the constructive feedback and make the effort to improve on the skills concerned.
6. For more challenging reviews, include a TA or another faculty member in the feedback session. While not always logistically possible, it is still important as the nature of verbal feedback is such that what is discussed during the session may be susceptible to misconceptions from the student. When it comes to providing verbal feedback, teachers must ensure that comments are always thoughtful, directed toward positive learning opportunities, and justifiable.

Elements of Effective Feedback

Key Points

- The skill of giving feedback is central to good teaching.
- The process of giving feedback is linked to promotion of learning.
- Creating a partnership with students helps feedback to be received more positively.

Good feedback is

- Formative and occurs regularly
- Well timed and encourages and promotes the learning process
- Supportive and delivered with sensitivity and objectivity
- Directed toward behavior and not focused on the student
- Able to identify strengths, areas of deficiency, and areas for improvement
- Focused on behaviors that can be changed

Providing Immediate Feedback

With the day-to-day course of activities and the close interaction students and faculty share within the laboratory setting, faculty may need to provide immediate feedback to ensure student learning and progress, e.g., to prevent the poor dissection of a team or individual students from damaging important anatomical structures. In this situation, in order to minimize student stress, it is important to invite the student or team to reflect on the error or potential error before providing feedback. Under these circumstances, feedback should be nonevaluative, tactful, and purposeful so that the student or team may be directed back on task.

Providing Student-Elicited Feedback

Frequently, students seeking advice may initiate feedback sessions with faculty. However, some students may have difficulties in seeking feedback due to their perceptions of faculty approachability and concerns about being perceived as "weak" [42]. Therefore, emphasizing importance of the feedback during didactic activities in the classroom and laboratory for students' personal growth and development should be encouraged.

Providing feedback relating to conflict within teams, far-field behavior, or student-initiated concerns can be more challenging. Under these circumstances, verbal feedback (*delayed feedback*—feedback given after a certain amount of time has passed) should be given in a more neutral setting such as in an office space or common study area with an additional teaching faculty or TA in order to create a culture of safety. Feedback should start with a diplomatic articulation of the conflict/observation followed by an invitation for the student(s) to express their thoughts followed by guidelines for improvement or resolution. Avoiding judgmental language and opinions is an effective way of creating a feeling of acceptance and enabling students to improve [43]. Stay with the observed behavior and focus on the incident, not on the student. Know what institutional resources (e.g., Office of Dean for Student Affairs) are available to the student (such as tutoring, social development, student support, wellness programs, etc.) while understanding your own limitations (limit your involvement and deal with emotional situations in an objective manner). Create a culture of safety for the student and remind the student that whatever is discussed is confidential and that you are there to encourage and work with the student to overcome challenges.

Providing Written Feedback to Students

As is the case with many medical school courses, anatomy faculty is required to submit a written evaluation for each student at the end of the course. Included in these evaluations are observations on students' knowledge-based as well as nontraditional discipline-independent competencies described earlier in this chapter. Submission of the written feedback should be the responsibility of the anatomy course director, complied through a consensus of opinions of the teaching faculty and TAs. Designating time to meet is essential to ensure that comments are well documented. This should be done at scheduled debriefing sessions during the course [30] and at a formal meeting that includes all members of the teaching team, at the end of the course once collective data for the entire course are available.

It is important to remember that written feedback [38] has a strong impact on student psyche. Not only is it a permanent unit of constructive commentary, it also serves as official documentation of student strength and weaknesses within the integrated medical curriculum. Students should be made aware at the start of the course that they will receive formal written evaluations in addition to a course grade upon completion of the course. Faculty and TAs should be aware of their responsibility in identifying areas for providing students with feedback. Weekly debriefing sessions with TAs and faculty [30] afford good opportunity for highlighting student-related issues and performance. In this way, feedback can be cumulative rather than a cluster of comments being generated at the end of the course. It also allows faculty to assess change in student behavior and comment positively on lessons learned, considering that the student has just completed the anatomy course and that post-course feedback is meant to promote student learning as they continue on to courses that will assess the same competencies.

When compiling written feedback, always begin with the most positive comments and end with a positive or encouraging statement, including more constructive commentary in between. The choice of words matters; the tone with which they are written matters—clarity is crucial. There are many factors that influence the effect that feedback may have on students. Self-esteem, integrity, maturity, and willingness to self-improve, to name a few, are aspects of learner characteristics that impact the effectiveness of faculty-generated feedback [1, 44]. Providing students with an open invitation to discuss written feedback with faculty helps in clarifying student interpretation of the feedback and dissipates any misperception that could be a potential hindrance to student self-improvement [9].

Promotion of cultural competency and diversity remains a fundamental value in providing patient care. Understanding that the typical anatomy class of the twenty-first century contains a unique cultural blend of students [45] necessitates at least some degree of cultural competency [46] among anatomy teachers. With this in mind, anatomy teachers should recognize that when assessing student attitudes and behavior, what a teacher may perceive as being unacceptable may not occur to the student (who may have grown up under different cultural influences) as being unacceptable. Aligning attitudes and behaviors with what is necessary for functioning within the health profession should remain the only guiding principle for skill transformation.

The following helpful techniques may be considered when compiling written feedback:

1. Do not present feedback that discourages the learner or threatens the learner's self-esteem: "Student X is overconfident, disrespectful, and difficult to work with in the anatomy laboratory. Student X is culturally insensitive when working within a team."
2. Focus feedback on the skill, not the learner: "Student X may benefit from strengthening leadership skills by being more inclusive and aware of team members' strengths and weaknesses. Being sensitive to others' styles of learning and needs and open to sharing responsibilities is an important aspect in developing good teamworking skills."
3. Present elaborated feedback in manageable units. Use paragraphs to get specific observations across with guidance on how to strengthen learning [38]: "Student X would benefit from learning to reflect on her actions and work on her ability to self-improve. We suggest that the student contact the Dean for Student Affairs to enlist resources that will help improve this skill."
4. Be specific and clear with feedback messages, avoiding the use of emotive language and personal judgment: "There was room for improvement in student X's performance during dissection and attitude toward cadavers. As with any task, producing the best result possible is in keeping with putting the needs of the patient first" instead of "It was disappointing to see student X's reaction toward his assignments, his lack of enthusiasm toward dissection and poor treatment of cadavers, as well as the quality of his presentations, all of which were extremely poor."
5. Keep feedback as simple as possible, giving unbiased objective feedback: "Student X would benefit from having more confidence in her work and herself, being more assertive with team members and interacting more closely with teaching assistants and faculty," or "Student X would benefit from being aware that her attitude and manner of communicating can come off as being short and abrupt. She would benefit from improving her communication skills and being more aware of her verbal interaction with teammates and faculty."
6. Keep positive comments simple and tone down on exaggerated statements [1]:

"Student X was a great leader. He was an exceptional student with a wonderful attitude and great personality! Faculty and TA's loved having him in class" may be stated more effectively as "Student X showed strong leadership skills. His grasp of the material was impressive and his participation in the course was positive for both peers and faculty."

7. Avoid drawing comparisons regarding course performance: "After failing all practice practical examinations and receiving ARS grades below 50%, student X was unconcerned about his knowledge level and even when asked, his answer was that he did well…! Student X had the lowest score on the final examination in the class (348 points) 10th percentile rank nationwide." For providing information relating to scores, simply state overall grade: "Student X received a grade of 348 points (10th percentile rank)."
8. Be sure to include feedback acknowledging improvements made during the course: "Student X was receptive to feedback from TAs and faculty and made the effort to improve on the way in which she communicated with peers and teachers."

Students are eager to receive feedback. They are eager to learn and to succeed in their future careers and determined to achieve goals set out for them [36]. But they are also eager for support and mentorship. What students gain through the social interaction that comes with active participation in an anatomy class is unparalleled. Anatomy teachers should not underestimate their ability to influence students early in the medical curriculum. Incorporating the practice of providing regular formative feedback while teaching anatomy speaks to the needs of the student. As with any skill, the art of giving effective feedback to students takes effort, time, and practice [12, 14–16, 34, 46–49]. It takes a belief that the learning of anatomy demands active partnership between the teacher and the student—a partnership built on trust, honesty, and the genuine endeavor to promote and improve learning.

> The important things to remember about feedback in clinical medical education are that (1) it is necessary, (2) it is valuable, and (3) after a bit of practice and planning, it is not as difficult as one might think. Ende [12]

References

1. Clynes MP, Raftery SEC. Feedback: an essential element of student learning in clinical practice. Nurse Educ Pract. 2008;8(6):405–11.
2. Spiotta AM, Kalhorn S, Patel S. Millenials in neurosurgery: is there hope? Neurosurgery. 2018;83(2):E71–3.
3. Garner MS, Gusberg RJ, Kim AW. The positive effect of immediate feedback on medical student education during the surgical clerkship. J Surg Educ. 2014;71(3):391–7.
4. Hughes DT, et al. A novel minute feedback system for medical students. Am J Surg. 2017;213(2):330–5.
5. Harlena W, James M. Assessment and learning: differences and relationships between formative and summative assessment. Assess Educ Princ Policy Pract. 1997;4(3):365–79.
6. Garrison C, Ehringhaus M. Formative and summative assessments in the classroom. https://www.amle.org/portals/0/pdf/articles/Formative_Assessment_Article_Aug2013.pdf. Accessed 19 June 2019.
7. Brophy JE. Teacher praise: a functional analysis. Rev Educ Res. 1981;51(1):5–32.
8. Schwartz F, White K. Making sense of it all: giving and getting online course feedback. In: White KW, Weight BH, editors. The online teaching guide: a handbook of attitudes, strategies, and techniques for the virtual classroom. Boston: Allyn and Bacon; 2000. p. 57–72.
9. Cleary ML, Walter G. Giving feedback to learners in clinical and academic settings: practical considerations. J Contin Educ Nurs. 2010;41(4):153–4.
10. Kruidering-Hall M, O'Sullivan PS, Chou CL. Teaching feedback to first-year medical students: long-term skill retention and accuracy of student self-assessment. J Gen Intern Med. 2009;24(6):721–6.
11. Bienstock JL, Katz NT, Cox SM, Hueppchen N, Erickson S, Puscheck EE. To the point: medical education reviews—providing feedback. Am J Obstet Gynecol. 2007;196(6):508–13.
12. Ende J. Feedback in clinical medical education. JAMA. 1983;250(6):777–81.
13. London M. Job feedback: giving, seeking and using feedback for performance improvement. 2nd ed. Hillsdale: Lawrence Erlbaum Associates, Incorporated; 1997.
14. Hewson MG, Little ML. Giving feedback in medical education: verification of recommended techniques. J Gen Intern Med. 1998;13(2):111–6.

15. McKimm J. Giving effective feedback. Br J Hosp Med (Lond). 2009;70(3):158–61.
16. Thomas JD, Arnold RM. Giving feedback. J Palliat Med. 2011;14(2):233–9.
17. Cantillon P, Sargeant J. Teaching rounds: giving feedback in clinical settings. BMJ. 2008;337:a1961.
18. Lachman N, Pawlina W. Basic sciences and curriculum outcomes. In: Dent J, Harden R, editors. A practical guide for medical teachers. Edinburg: Churchill Livingstone: Paperback; 2009. p. 260–4.
19. Black P, William D. Assessment and classroom learning. Assess Educ Princ Policy Pract. 1998;5(1):7–74.
20. Shute VJ. Focus on formative feedback. Educational Testing Service. 2007. https://www.ets.org/research/policy_research_reports/publications/report/2007/hslv. Accessed 19 June 2019.
21. Alexander CJ, Crescini WM, Juskewitch JE, Lachman N, Pawlina W. Assessing the integration of audience response system technology in teaching of anatomical sciences. Anat Sci Educ. 2009;2(4):160–6.
22. Mazur E. Farewell, lecture? Science. 2009;323(5910):50–1.
23. Premkumar K, Coupal C. Rules of engagement—12 tips for successful use of "clickers" in the classroom. Med Teach. 2008;30(2):146–9.
24. Rubio EI, Bassignani MJ, White MA, Brant WE. Effect of an audience response system on resident learning and retention of lecture material. Am J Roentgenol. 2008;190(6):319–22.
25. Meyers C, Jones TB. Promoting active learning. Strategies for the college classroom. San Francisco: Jossey-Bass; 1993.
26. Fies C, Marshall J. Classroom response systems: a review of the literature. J Sci Educ Technol. 2006;15(1):101–9.
27. Bangert-Drowns RL, Kulik CC, Kulik JA, Morgan MT. The instructional effect of feedback in test-like events. Rev Educ Res. 1991;61(2):213–38.
28. Pridemore DR, Klein JD. Control of practice and level of feedback in computer-based instruction. Contemp Educ Psychol. 1995;20(4):444–50.
29. Nelson AJ, Nelson SV, Linn AM, Raw LE, Kildea HB, Tonkin AL. Tomorrow's educators ... today? Implementing near-peer teaching for medical students. Med Teach. 2013;35(2):156–9.
30. Lachman N, Christensen KN, Pawlina W. Anatomy teaching assistants: facilitating teaching skills for medical students through apprenticeship and mentoring. Med Teach. 2013;35(1):e919–25.
31. Camp CL, Gregory JK, Lachman N, Chen LP, Juskewitch JE, Pawlina W. Comparative efficacy of group and individual feedback in gross anatomy for promoting medical student professionalism. Anat Sci Educ. 2010;3(2):64–72.
32. Hofer RE, Nikolaus OB, Pawlina W. Using checklists in a gross anatomy laboratory improves learning outcomes and dissection quality. Anat Sci Educ. 2011;4(5):249–55.
33. Corbett AT, Anderson JR. Locus of feedback control in computer-based tutoring: impact on learning rate, achievement and attitudes. In: Proceedings of ACM CHI 2001 conference on human factors in computing systems. New York: ACM Press; 2001. p. 245–52.
34. Harden RM, Laidlaw JM. Be FAIR to students: four principles that lead to more effective learning. Med Teach. 2013;35(1):27–31.
35. DiLullo C, McGee P, Kriebel RM. Demystifying the millennial student: a reassessment in measures of character and engagement in professional education. Anat Sci Educ. 2011;4(4):214–26.
36. Jonas-Dwyer D, Pospisil R. The millennial effect: implications for academic development. In: Transforming knowledge into wisdom: holistic approaches to teaching and learning HERDSA Conference Proceedings; 2004. p. 195–207.
37. Fisher SL, Ford JK. Differential effects of learner effort and goal orientation on two learning outcomes. Pers Psychol. 1998;51(2):397–420.
38. Glover C, Brown E. Written feedback for students: too much, too detailed or too incomprehensible to be effective? Biosci Educ. 2006;7(1):1–16.
39. Branch WT Jr, Paranjape A. Feedback and reflection: teaching methods for clinical settings. Acad Med. 2002;77:1185–8.
40. Coultas CW, Bedwell WL, Burke CS, Salas E. Values sensitive coaching: the DELTA approach to coaching culturally diverse executives. Consult Psychol J Pract Res. 2011;63(3):149–61.
41. Anderson E, Buchko KJ. Giving negative feedback to millennials. Manag Res Rev. 2016;39(6):692–705.
42. Milan FB, Dyche L, Fletcher J. "How am I doing?" Teaching medical students to elicit feedback during their clerkships. Med Teach. 2011;33(11):904–10.
43. Varlander S. The role of student emotions in formal feedback situation. Teach High Educ. 2008;13(2):145–56.
44. Perrella A. Room for improvement: palliating the ego in feedback-resistant medical students. Med Teach. 2017;39(5):555–7.
45. DeLisa JA, Lindenthal JJ. Commentary: reflections on diversity and inclusion in medical education. Acad Med. 2012;87(11):1461–3.
46. Hawala-Druy S, Hill MH. Interdisciplinary: cultural competency and culturally congruent education for millennials in health professions. Nurse Educ Today. 2012;32(7):772–8.
47. Brinko KT. The practice of giving feedback to improve teaching: what is effective? J High Educ. 1993;64(5):574–93.
48. Vickery AW, Lake FR. Teaching on the run tips 10: giving feedback. Med J Aust. 2005;183(5):267–8.
49. Bing-You RG, Trowbridge RL. Why medical educators may be failing at feedback. JAMA. 2009;302(12):1330–1.

20 Using Body Painting and Other Art-Based Approaches for the Teaching of Anatomy and for Public Engagement

Gabrielle M. Finn

Body painting is considered by some to be the most ancient form of art. Origins of body painting stem from tribal cultures and are ceremonial. Body painting is now most frequently seen on the face, usually at sporting events or children's parties. However, body painting is now being used in the medical setting as a way of projecting the inside on to the outside.

Body painting is an innovative method of anatomy teaching. Op Den Akker et al. [1] first used body painting, described as painting internal structures on the surface of the body with high verisimilitude, during teaching in 1999. Body painting within medical education is described as painting internal structures on the surface of the body with a high degree of detail [1]. This contrasts with simple line drawings as used in conventional practice, which can be dated back for at least a century [2]. Body painting proves particularly useful as a method for introducing surface anatomy in medical teaching and complements the teaching of clinical skills and peer examination, including techniques such as palpation and auscultation [1, 3, 4]. The value of body painting as a teaching tool is frequently attributed to its kinaesthetic nature [3–5]. Furthermore, it is thought that the active and kinaesthetic nature of body painting, coupled with the strong and highly memorable visual images of underlying anatomy, contribute to its success as a learning tool [1, 3, 4, 6]. Body painting has also been reported to be a valuable tool for diminishing the apprehension often exhibited by students when conducting peer physical examinations [3]. Since it has been suggested that a fear of death may be oppressive for students studying in the dissecting room and may be correlated to poor academic performance, the use of alternative teaching methods, such as body painting, may therefore be beneficial to students who struggle with cadaveric work [4, 7, 8]. Body painting has also been attributed to a mechanism by which students can be taught by stealth, with deliberate utilization of the hidden curriculum described [9, 10].

Within this chapter, the practicalities and advantages of using body paint and other art-based approaches to teaching and learning anatomy are described. In addition, the place of anatomy pedagogy within public engagement is discussed. Examples are provided, along with recommendations for anyone considering developing their own body painting session. This process of painting internal structures onto the body promotes knowledge retention, creates an exciting teaching environment and improves student engagement in their anatomical studies.

G. M. Finn (✉)
Faculty of Biology, Medicine and Health, University of Manchester, Manchester, UK
e-mail: gabrielle.finn@manchester.ac.uk

L. K. Chan, W. Pawlina (eds.), *Teaching Anatomy*, https://doi.org/10.1007/978-3-030-43283-6_20

Examples of Anatomical Body Painting

Muscles of Facial Expression and Associated Neurovasculature

Painting the muscles of facial expression is a fun activity for students. It is of particular relevance for medical, dental, physiotherapy and speech therapy students. By painting the muscles onto the face, looking in a mirror or at a painted peer and pulling various expressions such as winking, puffing out the cheeks or raising the eyebrows, students are able to immediately see the muscles in action. Other structures such as the parotid gland and neurovasculature can be painted simultaneously (Figs. 20.1 and 20.2).

Dermatomes

Mapping dermatomes onto skin is a great way to learn this invisible map of the body. Not only do students learn the dermatomes, but they also revise their bony landmarks, as they must palpate the relevant bones in order to correctly map the dermatomes. Dermatomes can be painted from head to toe simultaneously (Fig. 20.1), or region by region as encountered through the curriculum. Dermatome painting lends itself to the use of bright, bold blocks of colour, which is highly memorable for students. Similarly, innervation from peripheral nerves also can be mapped.

Abdominal Regions, Quadrants and Areas of Referred Pain

A quick and easy activity for students is to paint regions of the abdomen alone, or better still, the associated abdominal viscera can then be painted in situ. This activity has much impact when students observe the relative size and positions of organs. Viscera can be painted using an appropriate anatomical colour palette, or using vibrant colours to promote subsequent

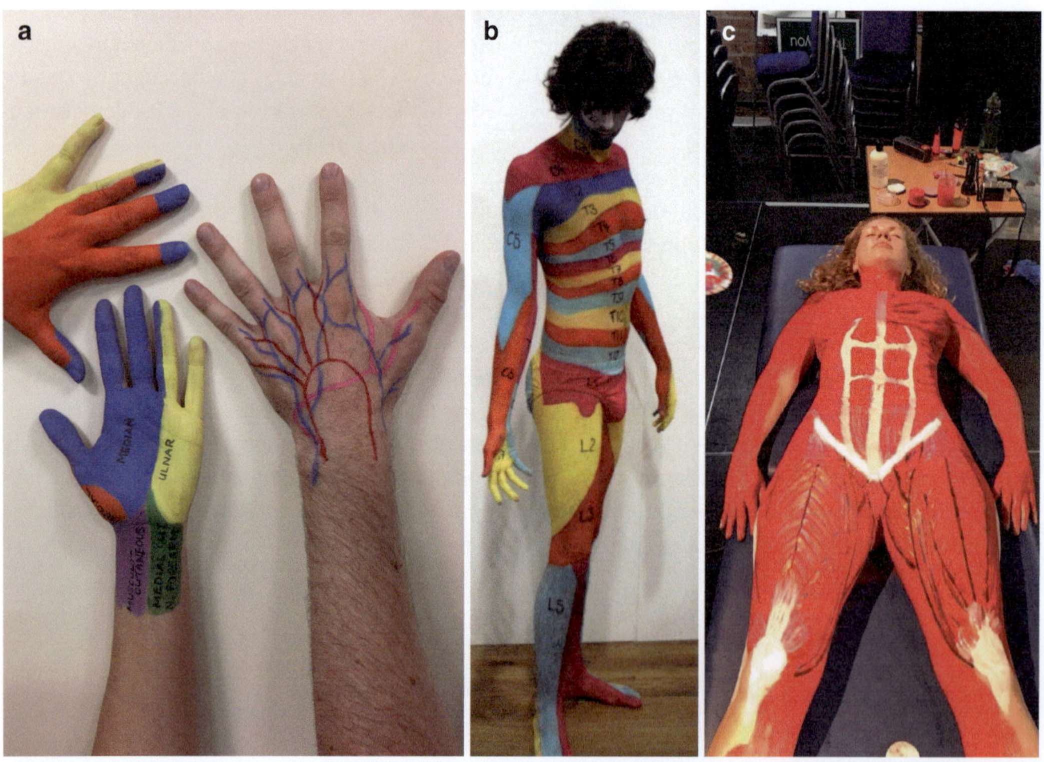

Fig. 20.1 Normal body paint examples: (**a**) hand anatomies, (**b**) dermatomes, (**c**) musculoskeletal anatomy

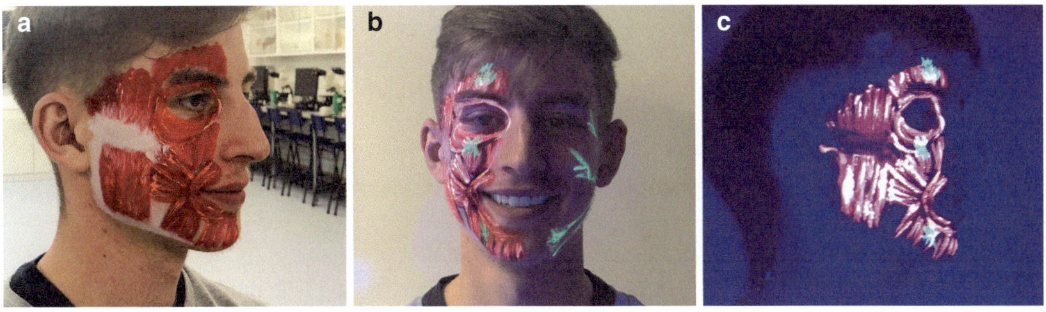

Fig. 20.2 UV facial anatomy: (**a**) daylight, (**b**) daylight + UV light, (**c**) blackout + UV light

recall. Students can paint blocks of colour to represent the referred pain from corresponding organs.

> **Topics Which Lend Themselves To Body Painting**
>
> - Abdominal viscera
> - Facial muscles and neurovasculature
> - Areas of referred pain
> - Musculature
> - Dermatomes
> - Bones
> - Borders of the heart and the position of valves
> - Position of the lungs and thoracic osteology

Skeletal Anatomy

Osteological knowledge is imperative for students. Painting the skeleton is a memorable task. Regions can be painted in isolation, such as the bones of the hand or an entire limb. Musculature and neurovasculature can be added if required. Bones of the hand are something that individuals can paint on to themselves, rather than having to work in pairs. An easily accessed area, such as the hand, lends itself to a painting activity with school groups, recruitment fairs or for other such public engagement events.

> Body painting has helped me to interact with individuals in a manner that is outside the normal socially acceptable level of conduct. This has helped me develop a more professional attitude towards patients and I see this as very useful. –A medical student at Durham University, UK

Advantages of Using Body Painting

Although body painting has origins in ancient tribal customs, it has a number of educational advantages within contemporary curriculum, and body painting has a number of contemporary advantages within an anatomy curriculum [4, 5]:

Creation of Learning Landmarks

Students have reported body painting as a highly motivating exercise [4]. Its main advantage appears to be the creation of what could be called "learning landmarks": vivid experiences which are memorable in themselves and which then provide access to the educational content associated within that context [4]. In particular, students acquire a good understanding of dimensions and positions of anatomical structures using this method while studying [1]. Literature is in accordance with the idea that body painting is a highly memorable experience, which gives students an appreciation of the links between the visual, tactile and auditory aspects of human anatomy.

Emphasis on Living Anatomy

Historically, anatomy has been taught by dissection. Cadavers are useful for studying the anatomy of large organs [11] and give an overview of spatial orientation [12, 13]. As anatomy is most commonly encountered by medical practitioners in the form of living anatomy and medical imaging [6], students should therefore also be encouraged to engage in teaching activities with an emphasis upon living anatomy and body painting is one such activity. "Clinicians often blame anatomists for teaching students too many details and not enough clinically relevant structures" [14]. Body painting allows the clinical relevance of the gross anatomy to be emphasized to students. Surface anatomy is a way of bringing cadaveric anatomy to life [15] and body painting falls into this category.

Cost-Effectiveness

Body painting is deemed to be a cost-effective tool [9, 16–18]. Body paints and paint brushes are readily available for purchase and are relatively inexpensive [5]. Old containers can be used for storage of water; the paint brushes do not need to be specialist. Large numbers of students can engage with the activity simultaneously and require little direction once instruction sheets have been produced, making the activity cost-effective in terms of both staff time and physical resources.

Positive Learning Environment

Even for the most reserved student, it is rare for students not to positively engage with body painting sessions. This fun activity becomes a break from the tedium of the dissecting room. Student enjoyment results in a positive learning environment and peer-peer teaching often results. As body painting is not a didactic modality for delivering teaching, its use creates positive and more relaxed relationships within the classroom between faculty and students [4, 5]. Body painting diffuses the formal learning environment and the sometimes challenging dissecting room context [9, 10, 17].

Retention of Knowledge

By actively engaging in the painting process, rather than passively learning in a didactic teaching session, students' learning becomes deep rather than surface [4, 5]. The use of bright colours aids students' memory of the structures that they paint. The painting process is multisensory and utilizes all learning approaches simultaneously; students visualize, students paint (kinaesthetic), students read instructions aloud (auditory) and students feel the paint on their own skin when acting as the canvas (sensory). Thus, retention of knowledge is promoted. Furthermore, students often photograph their painting for revision and sentiment – both positives in terms of subsequent recall. Recent research supports the notion that body painting helps students learn by reducing cognitive load; in other words, it enables information to be efficiently processed by working memory and stored in long-term memory, thus increasing information retention and recall [9].

Emphasis on Future Clinical Practice

During a body painting session, students must palpate bony landmarks in order to demarcate the associated anatomical structures. This has direct learning benefits for future clinical practice. Moreover, regions such as boundaries of the lungs and positions of heart valves can be painted on and lend themselves to subsequent clinical examination with the stethoscope. When dealing with peers in a state of relative undress and completing palpation and examination – students must communicate appropriately, empathize with their peer canvas and approach their peer in a professional manner; thus, skills required in the clinical environment are developed. Aka et al. explained this as the opportunity for students to form and practice their professional scripts [10]. These body painting sessions allow integration between anatomy and clinical skills and enable the future patient to be the focal point [4, 5, 9, 18].

An Alternative to Cadavers

One of the major advantages of using body paint is that it can be used as an alternative approach to cadaveric study on occasions where living and surface anatomy are being studied, or for students who struggle with cadavers [4]. The dissecting room environment can be one in which students who have emotional difficulties with cadavers find troublesome. Learning out of the dissecting room context can be both supportive and educationally beneficial for students who struggle emotionally. Similarly, surface anatomy is difficult to demonstrate on cadavers; for this reason, body painting is useful – it emphasizes living anatomy throughout. When time is short for dissection, or cadavers are in limited supply, faculty may wish to consider implementing body painting in their curricula. Body painting cannot entirely replace cadaveric study, but it is a powerful tool and adjunct for emphasizing the living nature of anatomy [4, 5].

Use Outside of the Classroom

In addition to anatomy teaching, body painting can be used for a number of other uses and institutional activities. Quick and easy activities are often required for departmental open days, tours and public engagement events. Painting the hand lends itself to such an occasion as no undressing is required and participants can paint their own limbs. Bones of the hand, the dermatomes of the hand or neurovasculature can be painted within a few minutes. Prior anatomical knowledge is not necessary; participants can work from images or simple instruction sheets. Similarly, body-painted models can be used for other public engagement events, such as museum exhibitions. Entire body musculoskeletal paintings, in a non-anatomical colour palette, are popular.

Using Ultraviolet Body Paints

UV light, invisible to the human eye, causes certain substances to glow, a capability capitalized upon to create readily available UV paints. Finn and colleagues have recently introduced the use of UV body painting into the medical curriculum [17, 19] with examples shown in Figs. 20.2, 20.3 and 20.4. The applications and advantages are as follows:

- UV paint can be layered on top of or beneath the normal body paint. This means that structures that are superficial or deep can be painted in the same view. Examples could be painting musculature in normal paint with UV innervation that will only be visible with UV light. Hiding structure names in UV also works well (Figs. 20.2, 20.3 and 20.4).
- UV paint has a wet appearance, especially with the reflection of the UV light, which gives painted structures a more visceral appearance.
- The reflection of light on painted musculature makes the structure appear to be under tension. An example of this is movement of the arm when the muscles of the chest have been painted showing tension at the points of origin and insertion. Actions of the muscles appear more obvious with this effect.
- UV is excellent for showing "hidden" maps of the body such as dermatomes, Langer lines and Blaschko's lines. These features can be mapped on to the body and only revealed once the UV light is on.
- Student users have remarked that UV paint is more inclusive as the bolder pigmentation enables it to be better seen on darker skin tones.
- UV body painting is the most striking medium of painting and therefore is of great use when significant impact needs to be made.

Caution must of course be taken when using ultraviolet light due to the health implications. For this reason, it is recommended that UV body painting is used sparingly [17, 19]. UV paint is visible in natural light so the use of the UV lamps can be avoided but the pigmentation is more remarkable with UV lighting turned on [17, 19].

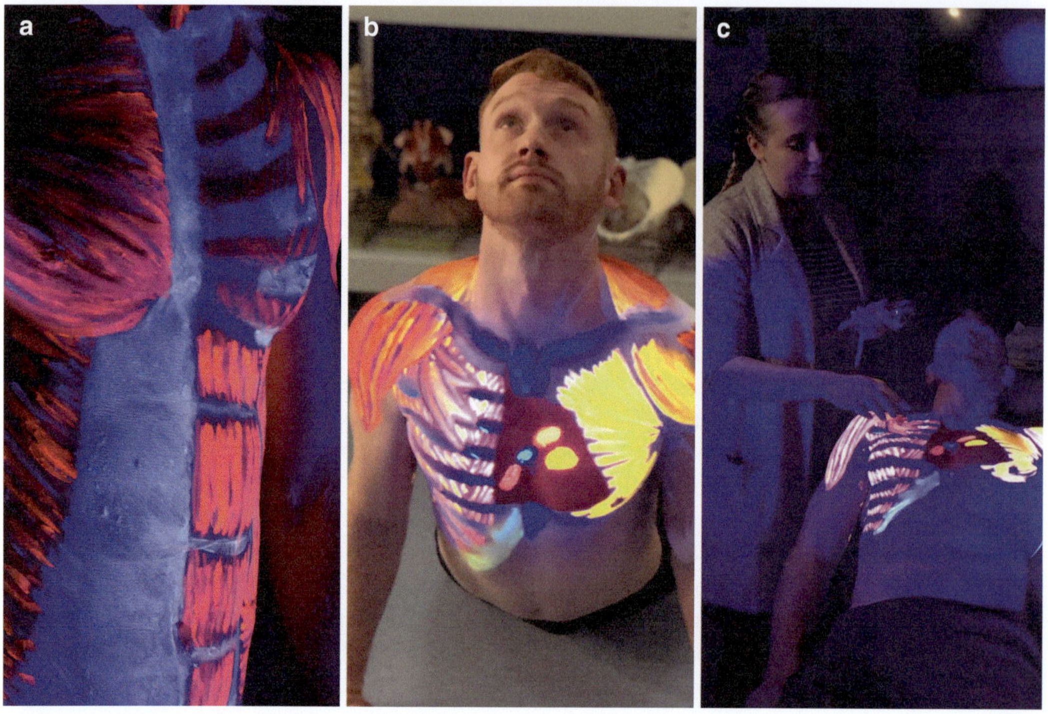

Fig. 20.3 UV thoracic anatomy: (**a**) thorax muscles, (**b**) thoracic anatomy, (**c**) cardiovascular examination

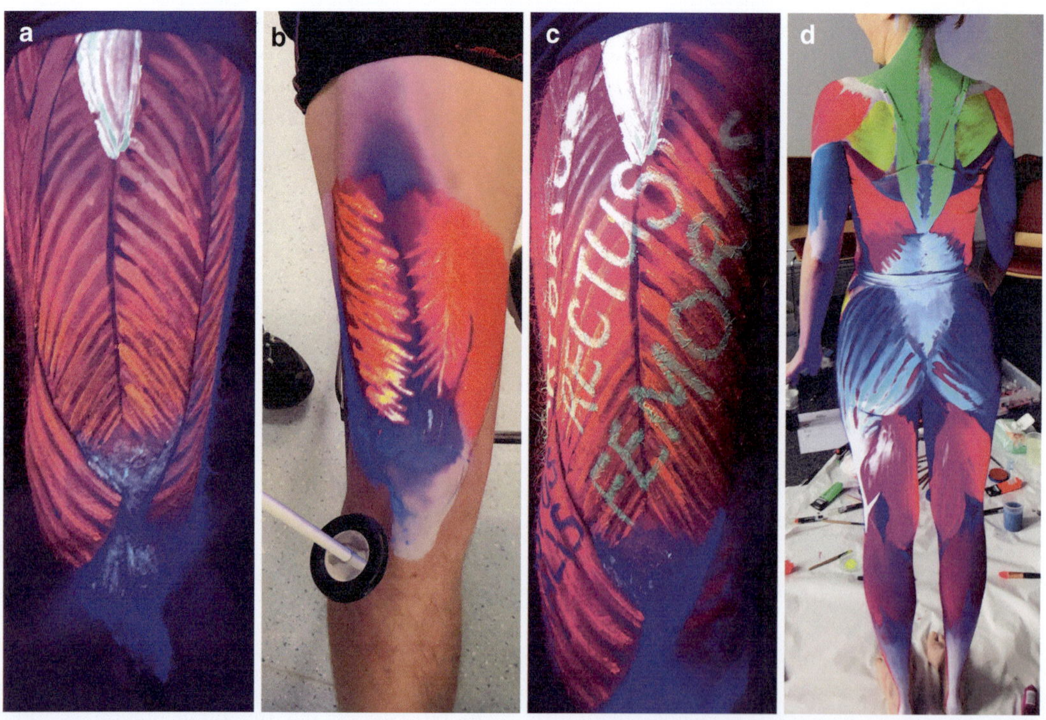

Fig. 20.4 (**a**) Quadriceps, (**b**) patellar tendon reflex, (**c**) UV labels, (**d**) UV musculoskeletal anatomy

> UV body painting appeared so realistic to me. It looked wet and as if I was literally viewing the muscles move beneath transparent skin. –Audience member at an anatomy demonstration who worked as a healthcare professional.

Designing Body Painting Teaching Sessions

Planning body painting teaching sessions can be daunting. By following these simple steps [4, 5, 17, 19], body painting sessions can be easily designed and implemented.

Identify Your Chosen Teaching Session and Learning Outcomes

All sessions need to be constructively aligned with specific learning outcomes [20, 21], teaching activities and, where appropriate, assessment. Give careful consideration to which structures you wish the students to learn. What is the take-home message of your session and can that be achieved using body painting as opposed to dissection or a lecture? Develop one session as a starting point, and then following evaluation, move on to develop further sessions. Body painting sessions can form part of timetabled anatomy and clinical skills teaching. Body painting fits particularly well into a multi-station practicum whereby each station addresses one aspect of the body region being taught. For example, a teaching session on the anterior thigh may have the following stations: musculature, nerve supply, blood supply and osteology. When body painting is introduced into a practicum, it may form one of the stations, or become a double station if a clinical skill is being taught simultaneously. The stations for the anterior thigh session may then become a musculature, neurovasculature, osteology and a body painting station which teaches the distribution of dermatomes. Alternatively, a clinical skills session on lower limb examination or neurological assessment would work as a place to paint. In this example, students might paint nerves or muscles with tendons and then elicit the patellar reflex, or just visualize such structures while performing a basic examination.

> **To Do**
>
> - Think about a teaching session with either gross anatomy or surface anatomy learning outcomes. How could you incorporate body painting into that class?
> - Consider the environment, the resources and which body region and structures will be painted.
> - Can you devise easy-to-follow instructions? Will you include images in your instruction sheets?

Design Your Instruction Sheet

Students need specific instructions to follow to enable them to achieve the desired learning outcomes. Instructions need to be formatted as a step-by-step guide which is easy to follow. Photographic images of a pre-painted model or diagrams are helpful. Copies of the instruction sheets, presented as a checklist which students can mark off, are a good aid for regions which are more complex and have combined clinical examination. Following their first use, ask students for feedback as to whether the instructions were easy to follow. From a practical perspective, laminating instruction sheets is useful due to surrounding water and paint. With a detailed set of instructions, body painting sessions require little facilitation [16].

Choosing Your Teaching Environment

Body painting is easily adapted to a number of teaching and learning environments. Consideration needs to be given to safety: are you using a laboratory where students can come into contact with irritants or other hazardous substances? Consider student dignity: is the interior of the room easily viewed from outside? Are the surfaces wipeable after painting? Will you be using UV paint? If so, are there blackout blinds and sockets available for lights to be plugged in. As the paints are water-based, classrooms, other than laboratories, can be utilized for painting. Students need a surface for their equipment and instructions, as well as enough space to move around their chosen student canvas.

Sourcing Your Equipment

Body paints are often sold as children's face paints. The most important considerations are that the paints you buy are water-based and hypoallergenic. The colours you select depend on whether you wish to paint in an anatomically correct palette, such as beige, red and brown, or whether you wish to use bolder colours. Paints are available from art and craft suppliers, toy shops and the Internet. Paint brushes do not need to be specialist. Brushes can be natural fibres or synthetic – depending on your budget. It is advisable to purchase a selection of sizes of brush as some areas, such as dermatomes, require large block painting, whereas others, such as nerves, require finer lines to be painted. Makeup brushes, bought in bundles, are an economical way to get a range of sizes easily; they wash well and produce a less streaky effect than normal paint brushes. Any plastic pots will suffice for washing brushes and wetting paints during the session. Disposable wipes are ideal for removing paint from skin or work surfaces after teaching, although often students wish to keep their paint on! Wipes do not need to be for makeup but can be bought from childcare sections of grocery stores – these are much more cost-effective.

UV body paints are readily available in either a tube or crayon – they do not require any water for use. Crayons are quick and ideal for mapping neurovasculature. UV paints are premixed and applied directly to the skin; they can also be applied over normal paint for layering effects or colour saturation.

Assigning Student Roles and Groups

Where possible, allow students to self-select their partner or group members. This eases any discomfort associated with undressing and physical examination. Some students may have a preference for the role they undertake, whether that is the painter or the canvas. Students can be encouraged to alternate roles as there are advantages to each [8]. To ensure that the canvas engages, they can read the instructions to the painter and observe the painting process in a mirror.

Equipment Needed to Use Body Paints

- A range of coloured paints, makeup or paint brushes and/or sponge applicators.
- Eyeliner pencils for outlines.
- Body paints should be water-based and hypoallergenic.
- UV paints and crayons for an additional resource.
- Water or wipes need to be available for paint removal/mixing.
- Mirrors should be available so that the student canvas can observe the painting.
- Screens or cubicles may be needed for dignity.
- UV lamps or black light if using UV paint.

Cross-Linking

See also the following chapters:

- Chapter 9 "Anatomy Education to the Public"
- Chapter 19 "Giving Feedback to Students"
- Chapter 26 "Promoting Active Learning in the Gross Anatomy Laboratory"
- Chapter 33 "The Use of Low-Fidelity Models to Enhance the Learning of Anatomy"
- Chapter 43 "Core Syllabi in Anatomy"
- Chapter 47 "Exploring the Hidden Curriculum and Anatomy Education"

Tips for Using Body Painting Within Anatomy Teaching

Use Within Both Large and Small Group Settings

Body painting use is not limited to small group settings. Sessions can easily be adapted for large numbers of students. The number of paints available and space requirements are the only limiting factors. Resource wise, students can easily share paints and brushes. Students can work in pairs, or larger groups, depending on the amount of anatomical structures to be painted and the time frame in which they must work.

Consider Students' Sensitivities and Cultural Perspectives

The majority of students will have no hesitations in participating in a body painting session. However, as with peer examination, students may have concerns with being in a state of undress. Faculty members are advised to be sensitive to these issues, particularly to students who are body image conscious. Providing screens or cubicles helps students feel more comfortable. Some students may choose to paint on top of clothing, which works well. Allowing students to self-select the groups they work in also encourages their participation.

Allow Photography

Permitting students to take photographs of their paintings promotes reflection and revision. Of course, caution must be used, and students should be encouraged to seek consent before photographing others. Advise students that photography is permitted in advance of the session.

Introduce and Demonstrate

Students may be reluctant to start painting without an introduction to the topic or a quick demonstration of surface anatomy palpation. A brief overview gives students greater confidence. This can include the instructor painting a model or volunteer if time allows. Instructors should circulate as much as possible in order to reassure students. In doing so, instructor should be mindful of student dignity and comfort, especially if regions such as the thorax or abdomen are being painted.

Give Students Feedback

During painting sessions, faculty can observe the painting and palpation being conducted by students. Timely and appropriate feedback should be given but with sensitivity and encouragement. Use this as an opportunity to engage the students in a dialogue about the learning objectives and, if appropriate, the relevance to their future clinical practice.

Reflect on the Success of the Session

Being a reflective practitioner is an important part of the teaching process. Following painting sessions, teachers should take the time to consider the successful elements of the session and the aspects that could be improved for future iterations. Was the time devoted to the activity appropriate? Did the session fulfil the learning outcomes? Did the students engage with the painting process? Was student comfort and dignity maintained? Did students receive adequate instruction and feedback? Was the equipment provided sufficient?

Important Considerations When Using Body Paints in Anatomy Teaching Sessions

- Provide easy to follow instructions.
- Allow students to self-select the peers they wish to work with.
- Involve the student canvas in the process by encouraging them to read the instructions aloud and by providing mirrors for them to observe through.
- Encourage students to use bold colours – this promotes retention of knowledge.
- Ensure sufficient time is provided.
- Encourage students to take photographs of their painting – this promotes subsequent recall.

Other Art-Based Approaches to Teaching Anatomy

Body painting is just one of many art-based activities which can be readily adapted for use in anatomical teaching.

Life Drawing

Engaging students in life drawing is particularly useful for teaching surface anatomy. Drawing provides a medium by which students can consider differing body morphologies and surface anatomy. Drawing trains students to observe the asymmetry in a body – a skill particularly useful for those going on to clinical practice. When taking the time to draw something, the artist must study the object in great detail, whether that is drawing anatomical prosections or plastic models. Due to this study, life drawing is therefore also a beneficial process for learning, in particular when considering spatial relationships between anatomical structures.

Plasticine (Clay) Modelling

Children's reusable modelling clay (plasticine) lends itself to anatomical teaching. As with body paints, the clay is cheap, readily available for purchase, reusable and can be adapted for use in large group sessions. Examples of its use include modelling the muscles of facial expression on to a plastic skull (Fig. 20.5), vertebrae or embryological structures. The advantages are similar to body paint in terms of appeal to a variety of learning styles, diffusion of emotional responses to cadavers and memorability.

Anatomical Cake Baking

Studying anatomy does not need to be a task confined to the classroom; students can enjoy learning anatomy at home too. Baking and decorating foodstuffs such as cakes, bread or cookies to resemble anatomies is a fun way to study (Fig. 20.6) without reading textbooks. By depicting the anatomy, students must orientate themselves with the anatomy and relevant spatial relationships. As this task takes time, the students are reinforcing the anatomy over a prolonged period, but often without realizing they are doing so. Holding an event where students can bring in their baked items makes a great end of semester revision session.

Pipe Cleaners

Readily available in art stores, pipe cleaners serve a number of craft purposes. Lightweight, easy to manipulate, inexpensive and available in an array of colours and effects, including patterned and glitter, make pipe cleaners ideal for quick modelling. Examples include mapping out arterial networks such as the circle of Willis or use with a plastic skull to either depict cranial nerves as they traverse foramina, or the venous sinuses (Fig. 20.7). Figure 20.7 also illustrates their use for a quick conceptual map of the brachial plexus. This activity is easy to transport and adaptable to non-laboratory environments such as a lecture theatre.

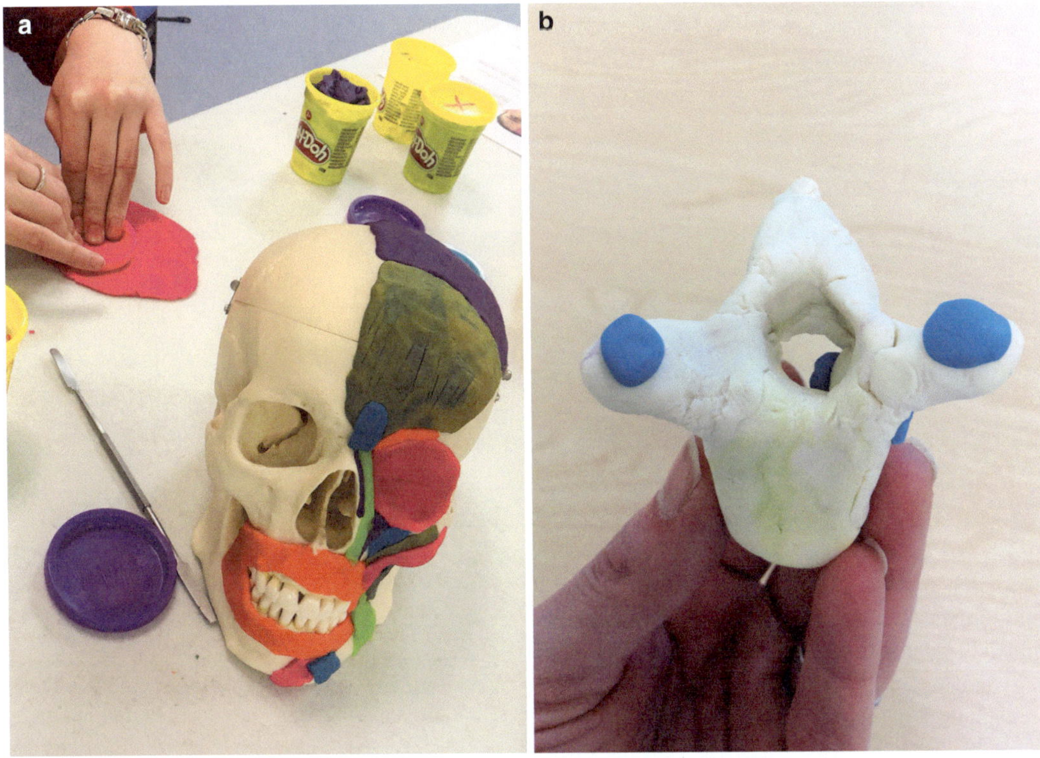

Fig. 20.5 Modelling clay: (**a**) muscles of facial expression, (**b**) vertebrae

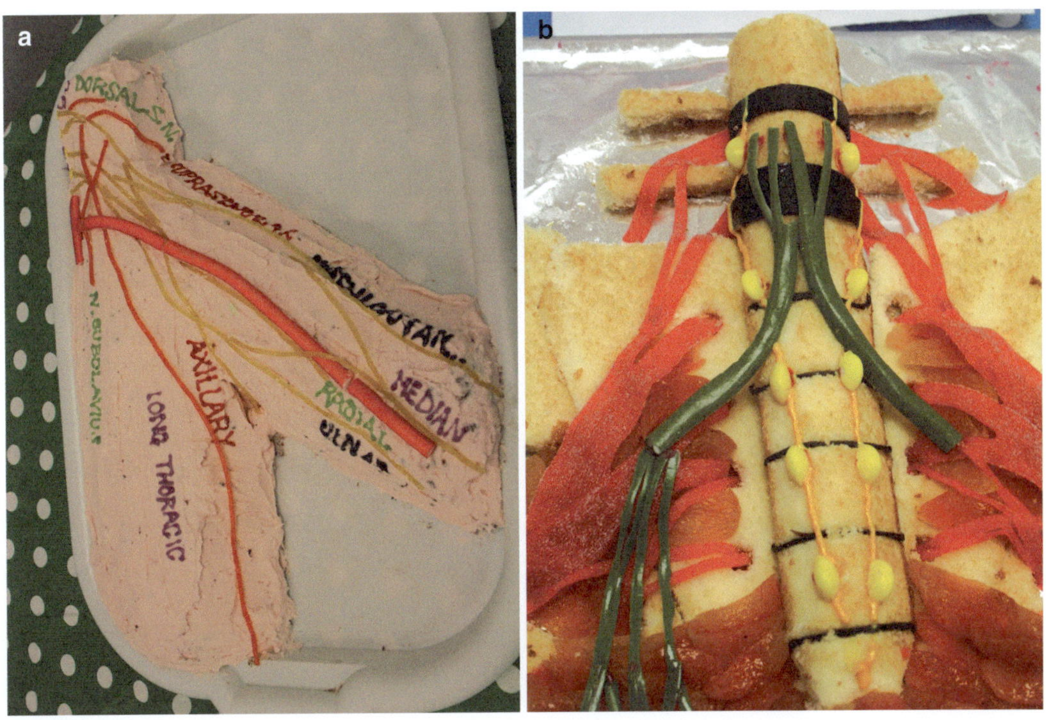

Fig. 20.6 Anatomical cake baking: (**a**) brachial plexus, (**b**) hypogastric plexus

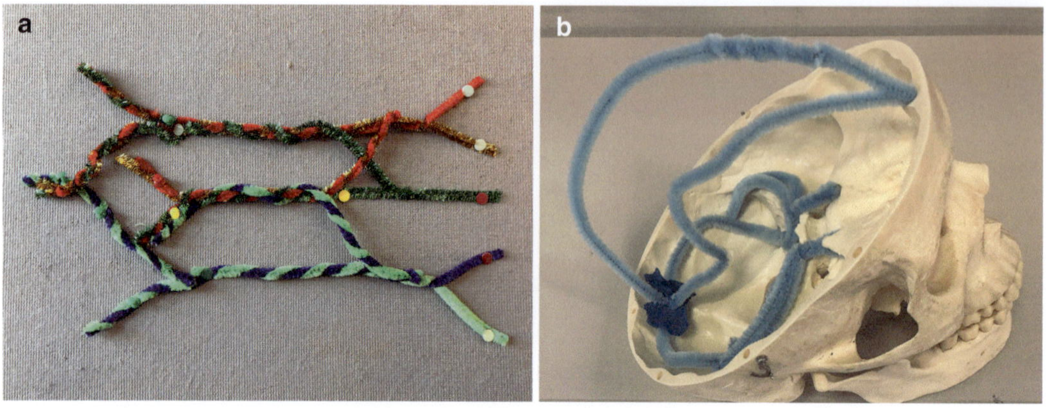

Fig. 20.7 Pipe cleaners in use for (**a**) brachial plexus, (**b**) venous sinuses

Conclusions

Art-based approaches to teaching and learning add new dimensions to the educational environment and to the student experience [4, 5, 8]. Body painting remains one of the most popular art-based modalities employed within anatomy education. Body painting is a useful tool for learning gross and surface anatomy. It actively engages students in the learning process and enables long-term retention of knowledge. The use of bold colour promotes memorability and adds an element of fun. Utilizing body painting encourages students to confront issues associated with clinical examination such as vulnerability, communication, professionalism and body image, in a positive and safe learning environment [4, 5, 8]. Life drawing, clay modelling and other art-based approaches are also educationally beneficial due to their engagement, memorability and context outside of the dissecting room. All of these approaches make excellent tools for outreach and public engagement.

> Body painting is fun, very enjoyable and often a challenge. You do not need to be a good artist, just to be able to follow instructions. Also body painting allows you to realize the situation of organs in the body, often for myself I find it surprising their exact locations and actual size. –A medical student at Durham University, UK

References

1. Op Den Akker JW, Bohnen A, Oudegeest WJ, Hillen B. Giving color to a new curriculum: bodypaint as a tool in medical education. Clin Anat. 2002;15:356–62.
2. Eisendrath D. A text-book of clinical anatomy. 1st ed. Philadelphia: W.B. Saunders; 1904.
3. McMenamin PG. Body painting as a tool in clinical anatomy teaching. Anat Sci Educ. 2008;1(4):139–44.
4. Finn G, McLachlan J. A qualitative study of student responses to body painting. Anat Sci Educ. 2010;3(1):33–8.
5. Finn G. Twelve tips for running a successful body painting teaching session. Med Teach. 2010;32:887.
6. McLachlan JC, Patten D. Anatomy teaching: ghosts of the past, present and future. Med Educ. 2006;40:243–53.
7. Skidmore JR. The case for prosection: comment on R.L.M. Newell's paper. Clin Anat. 1995;8:128–30.
8. Finn G, White P, Abdelbagi I. The impact of color and role on retention of knowledge: a body-painting study within undergraduate medicine. Anat Sci Educ. 2011;4(6):311–7.
9. Cookson NE, Aka JJ, Finn GM. An exploration of anatomists' views toward the use of body painting in anatomical and medical education: an international study. Anat Sci Educ. 2018;11(2):146–54.
10. Aka JJ, Cookson NE, Hafferty F, Finn GM. Teaching by stealth: utilising the hidden curriculum through body painting within anatomy education. Eur J Anat. 2018;22(2):173–82.
11. Parker LM. Anatomical dissection: why are we cutting it out? Dissection in undergraduate teaching. Aust N Z J Surg. 2002;72:910–2.
12. McCormack WTP, Lazarus CMD, Stern DMDP, Small PAJMD. Peer nomination: a tool for identifying medical student exemplars in clinical competence and caring, evaluated at three medical schools. Acad Med. 2007;82(11):1033–9.

13. Granger NA. Dissection laboratory is vital to medical gross anatomy education. Anat Rec B New Anat. 2004;281B:6–8.
14. Pabst R. Gross anatomy: an outdated subject or an essential part of a modern medical curriculum? Results of a questionnaire circulated to final-year medical students. Anat Rec. 1993;237:431–3.
15. Aggarwal R, Brough H, Ellis H. Medical student participation in surface anatomy classes. Clin Anat. 2006;19:627–31.
16. Jariyapong P, Punsawad C, Bunratsami S, Kongthong P. Body painting to promote self-active learning of hand anatomy for preclinical medical students. Med Educ Online. 2016;21(1):30833.
17. Finn GM. Current perspectives on the role of body painting in medical education. Adv Med Educ Pract. 2018;9:701–6.
18. Estai M, Bunt S. Best teaching practices in anatomy education: a critical review. Ann Anat. 2016;208:151–7.
19. Finn GM, Bazira PJ, Bateman J, Sanders KA. Ultraviolet body painting: a new tool in the spectrum of anatomy education. Eur J Anat. 2018;22(6):521–7.
20. Biggs J. Aligning teaching and assessment to curriculum objectives: imaginative curriculum project. LTSN Generic Centre; 2003.
21. Biggs J. Enhancing teaching through constructive alignment. High Educ. 1996;32:347–264.

Evaluating Your Own Performance in Leading a Small Group Discussion

21

Jon Cornwall

Small group discussions can be a valuable tool for aiding education in a tertiary environment. Used successfully, small group discussions can promote and enhance student learning through the implementation and development of discussion skills such as listening, questioning, explaining and responding, skills which form the platform for facilitating discussion and thinking [1, 2]. These skills also provide the basis for the development of teamwork and collaborative learning and allow the exploration of attitudes and the sharing and reflection of experiences [2]. Small groups do have advantages in some areas and have been shown to be better than large groups at promoting thought, developing attitudes and implementing values. However, they are thought not to be as efficient as large group teaching for the purpose of imparting information though this is still debated [2, 3]. They can be challenging, informative, instructive, a lot of fun [1] and can be a successful method of teaching and learning anatomy [1, 4–6].

The teacher has a vital role in the process of the small group discussion as a facilitator, mediator and educator. In order for teachers to continually develop, refine and improve their practice in leading small group discussions, they need to be capable of assessing their own abilities in this skill. This involves being able to measure and evaluate their own performance in leading small group discussions so that future interactions with small groups can be informed and developed on an ongoing basis. In this way, the quality and effectiveness of the small group discussions led by the teacher will continue to be improved.

J. Cornwall (✉)
Centre for Early Learning in Medicine, Otago Medical School, University of Otago, Dunedin, New Zealand
e-mail: jon.cornwall@otago.ac.nz

The Aim of Self-Evaluation for Teachers of Small Group Discussions

Self-evaluation, in most contexts, should have a purpose. Before you undertake any sort of self-evaluation, you should have a clear idea as to why you are performing it. The purpose may be as simple as an answer to one specific question about your performance in a class you have just taught, or as complex as providing information on multiple aspects of the small group discussion. It may be for generating information that is relevant to job promotion or an assessment of teacher performance. In effect, many individuals would like to answer the question: how do you know you have done a good job? This chapter examines how teachers could approach this question in relation to teaching anatomy classes that involve small group discussion.

With any type of evaluation, teachers should have a very clear idea of the framework of the

L. K. Chan, W. Pawlina (eds.), *Teaching Anatomy*, https://doi.org/10.1007/978-3-030-43283-6_21

class they are about to teach and the way they are going to undertake this in the classroom scenario. This may include what they are hoping to achieve (objectives), how they will do this (methods) and how they may measure outcomes satisfactorily (assessment of learning outcomes). They could involve assessing whether specific teaching objectives were met, including working effectively or productively, setting or achieving goals, efficient time management and planning or maintaining group behaviours and balancing contributions from individuals [7]. All these attributes can contribute to good small group teaching and are therefore potential targets for evaluation and may be specifically assessed during the process of self-evaluation.

Tools for Self-Evaluation: Critical Reflection and Feedback

There are many different ways of evaluating performance, including peer review; however, this chapter focuses on two self-evaluation tools that teachers can use to improve their own performance. Broadly speaking, self-evaluation has both individual and social components [8]. The individual component (the "I" or "me") involves critical reflection by the teacher of their own performance, while the social aspect ("they" or "them") utilises feedback from people that attended the small group discussion or class. Both of these components are important tools for teachers as they provide different types of information, and both can be used to improve teaching performance and potentially improve learning outcomes.

Critical Reflection

Critical reflection involves being able to identify, consider and reconstruct the implicit assumptions that underpin an individual's actions to develop a better understanding about the concepts driving their behaviour, thereby enhancing their teaching practice [8, 9]. In essence, it involves the performance of reflecting critically on one's own performance and is a skill that can be improved effectively through training and practice [10]. The terms "reflection" and "critical reflection" are used somewhat interchangeably in education [11, 12], with the key difference between the two being critical reflection supposedly demanding an increased depth of prior knowledge, consideration and focus upon factors such as the underlying assumptions about teaching, or how factors such as student diversity affect teaching style [13]. "Critical reflection" therefore is meant to involve a greater understanding of the factors influencing learning outcomes and it has been suggested that the two processes are different and their definitions require further clarification [14]. Both "reflection" and "critical reflection" involve the same basic processes that are outlined in this chapter [11, 12].

Many teachers indicate that completing self-assessment is a beneficial exercise [15] that has positive effects on teachers' own learning and task orientation [8], and it is therefore important for teachers to be able to critically reflect on their own performance in leading small group discussions. A simple definition of critical reflection is perhaps that it involves "thinking about thinking" [16]. Perhaps more than this, critical reflection also involves knowing how to reflect and analyse thought, draw conclusions from that analysis and put what is learned into practice [9, 16]. It can therefore be framed in terms of *specific tasks* that can enable teachers to evaluate their own teaching. Of course, the term *reflection* has different meanings when used in different contexts; here it is being used to define the high-level analysis that is required for transformative learning, or rather, learning that provides an identifiable transformation in thinking or actions as a result of this process [17].

The process of critical reflection involves an ongoing process of self-assessment and can be summarised as a three-stage model that involves the following components: planning, doing and reviewing [18]. It involves the teacher using this framework on an ongoing basis, using their observations and own knowledge to provide information that guides their teaching. It is a metacognitive process that involves the individual "thinking about thinking" to arrive at a conclusion that informs future practice. Metacognition itself is a self-regulatory process that selects, monitors and evaluates a cognitive process, with the cognitive process in this instance being reflection [18]. This is an important concept as it indicates that critical reflection is a skill that can be informed, controlled and enhanced. In this way critical reflection can be thought of as a framework that allows individuals to critique their own performance and experiences to lead to a new understanding and that this skill can be improved with practice and over time. Such a construct is handy for teachers to enable and facilitate critical reflection, and it is a method that teachers can continually use to evaluate their performance during any stage of a small group discussion, whether this is before, during or after the teaching has taken place. Such evaluation can also be enhanced with the use of simple video or audio recording devices (e.g. a smartphone on a tripod, as a basic tool) so that teachers can reflect on specific elements of their performance in their own time [19], though care should be taken to acquire consent from students prior to making any recordings.

In practice, educators should have a very clear idea about what they are teaching, what their expectations are and how they are going to achieve this. In relation to leading small group discussions and using the model described above, teachers should consider the basic features of critical reflection at each stage of the teaching session: the three steps of planning, doing and reviewing. This basic model of the process utilised to perform critical reflection can be used at any point of small group discussion: before, during and after any class.

Before Small Group Discussions

Undertaking any teaching session involves prior preparation that is initiated sometimes months before the class actually takes place. Evaluation of a teaching session that has not yet happened is perhaps contradictory; however, it may begin with critical reflection of previous classes (the "doing" part of the reflection process) in preparation for optimising performance in the upcoming small group discussion. This may involve such reflections as:

- Were there any problems that I need to address before this next class?
- What outcomes were reached last time I held this class?
- Were the objectives successfully met last time this class was held?
- Were the teaching or learning objectives met in the last class?
- What teaching theory underpins the approach I am taking in this class?
- Based on models or theories relevant to successful small group teaching, such as Tuckman's model of small group development [20], do I need to adjust how I plan to engage the students?

Such critical reflections can help teachers consider their upcoming class and focus their attention on areas that specifically require attention or improvement.

During Small Group Discussions

The act of critical reflection during a small group discussion involves planning, doing and reviewing actions in "real time", as the events occur. Essentially the teacher can be thinking about what they are about to do—including thinking about reasons why and for what purpose—then performing the task or action and then reviewing the response in relation to their expected or anticipated outcome. The teacher may also be critically reflecting on their automatic responses to a given scenario to inform their future actions. Examples of this could include:

- Changing the topic of discussion to direct students to a particular outcome
- Asking a student to perform a specific task or provide input at a particular time
- Suggesting a different seating arrangement in order to facilitate interaction

To do

- Think of a class you have taught that was personally challenging to you.
- Identify three or four individual instances where critical reflection during the class may have altered the outcome that was originally achieved.
- Think about how you may use critical reflection to improve your performance in the future should similar circumstances arise.

- Providing an anecdote at a particular time that is relevant to your class objectives
- Observing group dynamics and implementing an intervention to redirect group focus or steer them toward learning objectives or class outcomes

Each of these examples is different in nature; however, they are similar in that the action was driven by the desire to reach a particular outcome. Critical reflection can be used in each instance to decide what to do (and the basis for this decision), to actually reflect on the doing of the task and to review whether the interaction was successful and provided the intended outcome.

After Small Group Discussions

After a class has taken place, critical reflection enables teachers to be able to assess the preceding class based on their own experience. They can reflect on almost any aspect of the teaching or classroom experience:

- Did students find the objectives to be clear?
- Were the objectives achieved?
- Was the time management appropriate?
- Was the discussion enjoyable?
- Was the class productive?
- Did the group operate as a functional unit?
- Did I communicate effectively?

After the completion of each class it is helpful to critically reflect on the structure and implementation of the small group discussion. What were the reasons for planning the small group discussion? How was each of these achieved? What was the outcome for each of these elements? This circular pattern of questioning and reviewing of performance will help ensure teachers are continually assessing their own actions.

The common theme of action by the teacher before, during and after small group discussions is the process of planning, doing and reviewing that forms the basis of critical reflection. However, in each instance there are subtle differences. Prior to classes teachers are looking to ascertain whether past experiences have been acted upon appropriately and can use this information to "fine-tune" their approach to an upcoming classroom scenario. During classes this process is more rapid and fluid, depending on how the class is progressing. After the class has been completed, the process allows the immediate review of information by the teacher based on their perspective of the completed class. Critical reflection in each instance enables the teacher to be able to identify how their small group discussion skills may be best modified to improve their performance in future classes.

Features of critical reflection

- Employs a framework that requires systematic planning, doing and reviewing of tasks
- Can be used in any stage of teaching to analyse own performance
- Can involve comparing own performance with existing theoretical frameworks such as teaching or learning models
- Can be used on different aspects of performance in "real time"
- Improves with practice

Feedback

Feedback is a beneficial tool that can provide diagnostic evidence of teaching performance [21], and it is also helpful for improving teacher motivation [22]. It is therefore likely to be very useful in terms of improving the teaching of anatomy [1, 23]. The term "feedback" can be used in many different contexts; however, in this instance it is information about how we are doing in our efforts to reach a goal [24]. It should tell the individual something helpful and goal referenced so that the effectiveness of teaching practice can be gauged [25] and subsequent performance can be improved [26]. Feedback is best implemented soon after teaching has taken place for an accurate impression to be given by participants and can be done in public or in private depending on the information or responses you are wishing to elicit from participants. You may wish to obtain feedback from many people or just specific individuals; however, you should have an aim in mind with respect to the type of information that you are wishing to gather. Analysing feedback can be stressful [27] and you should try not to over- or underreact to specific feedback content and think carefully about the feedback before you decide what to do with it [7]. There is a danger of becoming fixated on unimportant problems [28], and therefore gathering opinions on feedback perceived as troubling can be useful. Also remember to thank individuals who provide it, as it can be difficult to receive open and honest feedback from students or individuals who do not feel valued [7].

There are different types of feedback; however, all should provide information that is useful to the individual seeking it [26, 29]. It should be goal referenced, actionable, user-friendly, timely, ongoing and consistent [24]. Goal referenced refers to the feedback being able to be used to improve the task in reference to the student learning outcome or the teacher's objectives. For instance, providing feedback that informs you directly about how good the discussion was about reaching the intended outcome or objective. Actionable feedback is specific and useable and not judgemental. Instances of such feedback could include "I was inattentive until we started discussing a particular topic" as opposed to "I was bored for half the class", which tells you little about how you may teach the class in the future. Feedback that is user-friendly should be easily understood and not overwhelming, such as highly technical comments on performance. It should be in the context of points that are easily understood and that will lead to an immediate improvement in the performance of the teacher. Timely feedback should also occur sooner, rather than later. Teachers do not want to wait for days to hear how their classes were perceived; it is much more useful to get feedback straight away so that improvements can be actioned.

Define your Question

It is helpful to have a specific purpose in mind when trying to decide what sort of feedback you require. This will also determine how you are going to ask for the feedback you require. Making

What will you do?

- You have just completed a small group discussion with students via an online e-platform that allowed you to see and talk with students in real time.
- Several students appeared to be less interested than others during the course of the discussion.
- How are you going to acquire feedback to determine how your online discussions could be improved?
- What sort of questions are you going to ask?
- Once you have the feedback from the students, what are you going to do with it?

Cross-linking

See also Chapter 15 "Facilitating Small Group Learning".

the question useful for the teachers' own practice is essential, so that future practice can be informed and modified accordingly. It also helps to structure your question in relation to the type of information that is required, for instance, relative to objectives, skills or performance. It may also consider the type of learning platform that was being utilised in the classroom (challenge-based, team-based, or problem-based learning).

Points to consider:

- Why do I want feedback?
- What should be the subject of the feedback?
- What specific question(s) am I going to ask?
- How am I going to ask that question to get the information I require?

Time Your Request for Feedback

The timing of feedback can influence the promotion of reflective thinking, and it is therefore an important consideration in planning when and how to gather feedback [30]. It can be important for some feedback to be gathered as teaching is taking place and for other feedback to be gathered after classes have taken place. For instance, feedback on whether anatomical images are better teaching aids than plastinated specimens for learning about specific anatomical relationships does not need to be performed after the class; however, feedback on the format of the class and the questions employed is best suited to after the class has finished.

Points to consider:

- When is the best time for me to ask for feedback?
- Would I get more appropriate feedback if I asked for it at another time?

Modes of Giving Feedback

One should also consider the mode of giving feedback that the feedback providers are comfortable with. In some circumstances it may be more appropriate for feedback to be written and sometimes verbal, or on some occasions individuals could be better served by students being identifiable; sometimes it may be necessary for them to be anonymous. This also takes into account the setting the class was performed in, whether it was laboratory based that involved discussions where gloves and wet prosections were being used, perhaps an e-learning tutorial via live web-link or a group held a few minutes prior to students having to finish for the day. Each of these scenarios may dictate how successful the teacher may be in acquiring appropriate and useful feedback. Examples of different modes include feedback provided by email, via wiki or use of a questionnaire.

Points to consider:

- What is the best mode for me to ask for feedback?
- Would another mode of acquiring feedback be more appropriate for this group of students?

Analyse the Feedback Objectively

Feedback should be judged as positive contributions to improving a teacher's performance; however, this can be difficult. It is easy to take negative comments as personal criticism of how "bad" a teacher or class may be, though it should be realised that if teachers are not receptive to feedback, their efforts in the classroom are unlikely to improve [31]. Even when the feedback is negative rather than positive, it is hopefully a constructive comment that is designed to bring about improvement in the performance of the teacher. Such comments should always be viewed objectively rather than subjectively to assist determining whether comments relate to important or unimportant issues [27] and in a positive way that allows performance to be altered and improved accordingly. It may be useful to go through feedback with a small, trusted group of peers and share feedback, with this supportive environment [32] creating an opportunity to socialise an action plan for future sessions [22, 33]. Feedback should be weighed in context and in reference to other feedback to provide a "picture" of how the class was perceived by those involved.

Points to consider:

- Why were particular comments made?
- What reason did the individual have for making this comment?
- Are these comments valid given other feedback that was gathered?
- Would I feel more comfortable or safer discussing feedback with a small group of peers?

Implement Changes That Consider the Feedback

Once feedback has been appropriately gathered and analysed, changes should be designed and integrated in a way that integrates the feedback that was useful. This should be measured against the goals, outcomes or objectives of the particular course and should take the needs and requirements of students into consideration. Feedback may be general enough to be used by teachers in all their classes, or may be related to a specific mode of teaching or location. It is helpful for teachers to think about where the feedback may most appropriately be utilised.

Points to consider:

- How can I best use this feedback to improve my teaching performance?
- In what setting(s) is the feedback best utilised or implemented?
- How will I monitor whether the changes I make are successful?

Gathering and using feedback

- Define the question you want feedback to answer.
- Time your request for feedback.
- Consider the most appropriate mode for gathering feedback.
- Analyse feedback objectively.
- Implement changes that consider the feedback.

Conclusions

In any aspect of teaching, evaluating one's own performance is important in helping one to understand how to improve the delivery of course material and to better help students achieve learning outcomes in the classroom or laboratory. Critical reflection and feedback are two very important tools for the evaluation of one's own performance in small group teaching, with each providing a unique subset of information that can be used to improve the different aspects of a teachers' performance. Educators should be aware of the process of critical reflection that involves planning, doing and reviewing on an ongoing basis and understanding that this tool can be used to improve performance in all aspects of teaching or educating. Similarly, feedback can be beneficial for teaching performance and gathering feedback can be undertaken in many different ways. However, it must be targeted to specific questions that the educator wishes to answer and used in a way that is appropriate for each teaching environment.

Self-evaluation helps improve a teacher's own performance. Critical reflection and feedback can both be used to provide information that helps refine teaching practice. Both critical reflection and feedback provide useful information but in different ways; understanding how and when to use each of these methods is important for teachers to be able to develop their small group discussion skills.

References

1. Vasan NS, DeFouw DO, Compton S. A survey of student perceptions of team-based learning in anatomy curriculum: favorable views unrelated to grades. Anat Sci Educ. 2009;2:150–5.
2. Edmunds S, Brown G. Effective small group learning: AMEE guide no. 48. Med Teach. 2010;32:715–26.
3. Kooloos JGM, Klaassen T, Vereijken M, Van Kuppeveld S, Bolhuis S, Vorstenbosch M. Collaborative group work: effects of group size and assignment structure on learning gain, student

satisfaction and perceived participation. Med Teach. 2011;33:983–8.
4. Chan LK, Ganguly PK. Evaluation of small-group teaching in human gross anatomy in a Caribbean medical school. Anat Sci Educ. 2008;1:19–22.
5. Kitzes JA, Kalishman S, Kingsley DD, Mines J, Lawrence E. Palliative medicine death rounds: small group learning on a vital subject. Am J Hosp Palliat Med. 2009;25:483–90.
6. Johnson EO, Charchanti AV, Troupis TG. Modernization of an anatomy class: from conceptualization to implementation. A case for integrated multimodal–multidisciplinary teaching. Anat Sci Educ. 2012;5:354–66.
7. Fowler J, Gudmundsson A, Whicker L. Working in groups. In: Fraser K, editor. Studying for continuing professional development in health. New York: Routledge; 2009. p. 114–20.
8. Runhaar P, Sanders K, Yang H. Stimulating teachers' reflection and feedback asking: an interplay of self-efficacy, learning goal orientation, and transformational leadership. Teach Teach Educ. 2010;26:1154–61.
9. Lachman N, Pawlina W. Integrating professionalism in early medical education: the theory and application of reflective practice in the anatomy curriculum. Clin Anat. 2006;19:456–60.
10. Naicker K, van Rensburg GH. Facilitation of reflective learning in nursing: reflective teaching practices of educators. Afr J Nurs Midwifery. 2018;20:1–15.
11. Rogers RR. Reflection in higher education: a concept analysis. Innov High Educ. 2001;26:37–57.
12. Black PE, Plowright D. A multi-dimensional model of reflective learning for professional development. Reflect Pract. 2010;11(2):245–58.
13. Smith E. Teaching critical reflection. Teach High Educ. 2011;16:211–23.
14. Harvey M, Coulson D, Mackaway J, Winchester-Seeto T. Aligning reflection in the cooperative education curriculuml. Asia Pac J Coop Educ. 2010;11:137–52.
15. Borg S, Edmett A. Developing a self-assessment tool for English language teachers. Lang Teach Res. 2018;23:1–25.
16. Downing K, Kwong T, Chan S-W, Lam T-F, Downing W-K. Problem-based learning and the development of metacognition. High Educ. 2009;57:609–21.
17. Aronson L. Twelve tips for teaching reflection at all levels of medical education. Med Teach. 2010;33(3):200–5.
18. Sandars J. The use of reflection in medical education: AMEE guide no. 44. Med Teach. 2009;31:685–95.
19. Tripp T, Rich P. Using video to analyse one's own teaching. Br J Educ Technol. 2012;43:678–704.
20. Bonebright DA. 40 years of storming: a historical review of Tuckman's model of small group development. Hum Resour Dev. 2010;13(1):111–20.
21. Richardson JTE. Instruments for obtaining student feedback: a review of the literature. Assess Eval High Educ. 2005;30:387–415.
22. Tuytens M, Moolenaar N, Daly A, Devos G. Teachers' informal feedback seeking towards the school leadership team. A social network analysis in secondary schools. Res Pap Educ. 2019;34:405–24.
23. Erie AJ, Starkman SJ, Pawlina W, Lachman N. Developing medical students as teachers: an anatomy-based student-as-teacher program with emphasis on core teaching competencies. Anat Sci Educ. 2013;6(6):385–92.
24. Wiggins G. Seven keys to effective feedback. Educ Leadersh. 2012;70:10–6.
25. Mandouit L. Using student feedback to improve teaching. Educ Action Res. 2018;26:755–69.
26. Linet A. From performativity to professionalism: lecturers' responses to student feedback. Teach High Educ. 2009;14:441–54.
27. Lejonberg E, Elstad E, Christopherson KA. Teaching evaluation: antecedents of teachers' perceived usefulness of follow-up sessions and perceived stress related to the evaluation process. Teach Teach Theory Pract. 2018;24:281–96.
28. Moore S, Kuol N. Students evaluating teachers: exploring the importance of faculty reaction to feedback on teaching. Teach High Educ. 2005;10:57–73.
29. Espasa A, Meneses J. Analysing feedback processes in an online teaching and learning environment: an exploratory study. High Educ. 2010;59:277–92.
30. Adadan E, Oner D. Examining preservice teachers' reflective thinking skills in the context of web-based portfolios: the role of metacognitive awareness. Aust J Teach Educ. 2018;43:26–50.
31. Hammer R, Peer E, Babad E. Faculty attitudes about student evaluations and their relations to self-image as teacher. Soc Psychol Educ. 2018;21:517–37.
32. Zhu CJY, White D, Rankin J, Davison CJ. Making meaning from student evaluations of teaching: seeing beyond our own horizons. Teach Learn Inq. 2018;6:127–42.
33. Payette PR, Kendall-Brown M. Gathering mid-semester feedback: three variations to improve instruction. In: IDEA papers; 2018. https://www.ideaedu.org/Portals/0/Uploads/Documents/IDEA%20Papers/IDEA%20Papers/PaperIDEA_67.pdf. Accessed 8 Mar 2019.

22 Establishing and Operating a Body Donation Program

Thomas H. Champney, Douglas C. Broadfield, Evelyn H. Vargas, and David R. Hoodiman

Introduction

Body donation programs for anatomical education have been operating in the United States for over 50 years [1]. These programs were designed to provide a consistent supply of bodies for health-care professionals to learn anatomy by human dissection [2, 3]. The benefits of dissection for learning anatomy as well as developing ethics and professionalism are well documented [4–6]. Each body donor program should be mission-oriented based on the specific needs of the institution and the educational and research goals of the program. Therefore, developing or established programs should align their mission with their institutional mission.

Regulatory and Governance Obligations

To legally implement, develop, and operate a body donation program, the local laws and regulations must be obeyed. In the United States, some states have state-run Anatomical Boards, while other states allow for independent programs to exist (university-based or not-for-profit) that follow the applicable state laws [7, 8]. In order to provide some uniformity to the laws of the individual states, a Uniform Anatomical Gift Act (UAGA) was enacted in 1968 and updated in 2006 [9]. This act established uniform baseline regulations that all states follow. Many states have developed additional regulations above and beyond the UAGA. Internationally, countries can have quite rigorous regulations for body donor programs and these may place limits on the number and types of programs allowed [10–15]. In whatever form they present, all applicable regulations must be followed to legally operate a body donation program.

In addition, a body donation program should establish an institutional oversight committee that is separate from the day-to-day operations of the program [8]. This committee can ensure that the regulatory and ethical obligations are followed while reducing concern about a perceived conflict of interest between the program staff and the program's operations.

T. H. Champney (✉) · D. C. Broadfield
Department of Cell Biology, University of Miami, Miller School of Medicine, Miami, FL, USA
e-mail: tchampney@med.miami.edu; dbroadfield@med.miami.edu

E. H. Vargas · D. R. Hoodiman
Office of Medical Education, University of Miami, Miller School of Medicine, Miami, FL, USA
e-mail: evargas25@med.miami.edu; dhoodiman@med.miami.edu

L. K. Chan, W. Pawlina (eds.), *Teaching Anatomy*, https://doi.org/10.1007/978-3-030-43283-6_22

Ethical Obligations

In addition to legal obligations, there are also ethical considerations when operating a body donation program. There are foundational ethical obligations that have been agreed to by international anatomists [10]. A number of other ethical obligations are currently in a period of transition and body donation programs are being held to higher ethical standards than they were in the past [16–18]. One of the hallmarks of an ethical body donation program is fully informed consent: ensuring that the donor is aware that her body will be used in education or to conduct research and that she consents to this specifically. The donor (or the legally authorized representative) should be provided with as much information as she needs to feel fully informed—especially if there are to be nonmedical, traumatic procedures performed on her body (e.g., crash studies, military defense studies). This consent should also include details about how long the donor program retains the body and when remains may be returned or dispersed.

Other ethical considerations for a body donor program include providing respect and dignity to the donor at all times, establishing transparent and accessible policies, and following the donor's wishes for the use of her body (anonymity, privacy, specific uses, acquisition and use of data or images, as well as transfer and use by other programs in different physical locations). The program should also provide a commemorative service publicly acknowledging the donor's altruistic gift as well as a permanent memorial honoring the donors [19]. The family members of the donor should also be treated respectfully and should be acknowledged for supporting their loved one's unselfish contribution.

Ethical Considerations in Body Donation Programs

1. Fully informed consent
2. Respectful treatment of the donor and the family
3. Following the wishes of the donor
4. Publicly acknowledging the donor's gift

Cross-Linking

See also Chapter 23 "Ethical Considerations of Body Donation".

Budget

When a new body donation program is being established, the proposal should include a budget with salary support for staff sufficient to account for the volume of donations and to allow for separation of duties on specific tasks. Licensed funeral professionals with knowledge of the related regulations, with training in preservation techniques, and with education in grief management are recommended. Budget costs should also include equipment, technology (databases and tracking software), resources for fixatives and other supplies, and support for administration of the program (donation forms, death certificates, memorial service, outreach materials, website, mailing, phone, and Internet). The budget should include how income to support the program will be generated (state support, university support, charges to users for services). Actual costs must be used to determine how a program will be supported and to develop a revenue-neutral budget. User service fees must be based on actual costs and reviewed and adjusted regularly. The budget should be revenue neutral with only enough income generated to support the program including salaries for the staff, equipment, and supplies for the program. The institution that houses the program should have clear and transparent policies on costs borne by the program and costs borne by the institution (space, lighting, security). Income generated by the program should not be used by the institution to offset other costs. The program should not generate profit; it should be self-sustaining and revenue neutral [16].

Facilities

The facilities necessary to operate a body donation program include office space for the staff, a secure morgue with processing stations as well as a storage facility sufficient to accommodate the donors (usually enough space for at least one year's supply). Programs may require cold storage, freezer space, or other specialty storage space depending on their needs and usage. For example, if a donation program will prepare and use fresh frozen tissue, then special freezer space is required. All facilities must be properly secured with limited access. The facilities should have properly monitored and documented air circulation (one pass) to avoid exposure of the staff to formaldehyde and other chemicals that may be used in processing the bodies. The safety requirements for the laboratory and for the preparation and use of human tissues must follow all applicable laws and regulations. The facilities should also include a dedicated secure entrance point, hallway, and elevator to respect the privacy of the donor. Facility costs (including replacement and maintenance costs for equipment) should be factored into the budget.

Standard Operating Procedures

The standard operating procedures (SOP) of a body donation program should provide a detailed description of the program and how it operates, the fee structures associated with the program, and the proper rules and conduct that would be followed by all those who work for the program [7, 8]. Standard operating procedures are an essential part of a governance structure. There is no universal set of SOP and each program may have different procedures, although there should be a basic fundamental set of guidelines followed by all programs. These guidelines include policies for contact with the donors and their families; paperwork for the proper consent forms; protocols for the acceptance, suitability, processing, and storage of the donors after death; guidelines for tracking donors and their parts; the acceptable and unacceptable use of the donors; and the proper handling and return of the donor's remains. The SOP should also contain job descriptions, safety precautions, and training documentation for all personnel associated with the program as well as policies on approving requests for donors from users and approving sites for donor use.

Specifically, the SOP should provide details on the screening of donors for common infectious diseases (hepatitis, human immunodeficiency virus) as well as the criteria for inclusion or exclusion of the donors from the program (weight limits, distance limits, organ transplant, autopsy, traumatic injuries). In addition, all of the methods used to prepare donors should be detailed in the SOP. These can include freezing fresh tissue, standard formaldehyde preparations, and unique embalming procedures such as plastination or Thiel embalming [20, 21].

Local Community Interactions

The majority of established body donation programs rely on "word of mouth" to recruit donors to their programs. When establishing a program, however, there will need to be outreach to the local community to educate the public about donation. This can be accomplished by local media coverage, by speaking to local civic and charity groups, by outreach to the local medical community, and by interacting with local hospice and critical care organizations [22]. Direct advertisement to donors can be ethically challenging and should be avoided especially if the advertising occurs at the place and time where donors and their families are currently dealing with emotional issues (hospices, nursing homes, intensive care units). It is in the best interest of the donation program to maintain contact with the local community and to be fully transparent in its policies and procedures. Body donation programs, like blood donor and organ donor programs, exist due to the good will of the community and the programs should encourage and foster this good will.

Requirements for Operating a Body Donation Program

1. Obey all local laws and regulations.
2. Follow ethical guidelines for donation programs.
3. Develop a budget to cover costs without profit.
4. Provide proper facilities for the processing and storage of donors.
5. Create a set of standard operating procedures (SOP).
6. Interact with the local community to support donation.

The South Florida Body Donation Program

Overview of Program

The South Florida Body Donation Program will be presented as an example of how a body donation program operates. This program is a regional office of the State of Florida Anatomical Board and it reports to the main office of the Anatomical Board in Gainesville, Florida [23]. This regional program covers the majority of Florida south of Lake Okeechobee. The program is housed at the University of Miami Miller School of Medicine and the university provides space and partial funding for the operation of the program. The additional operating costs for the program are obtained by funds supplied to the program from the Anatomical Board by way of user service fees.

Two licensed funeral directors run the day-to-day operations of the donor program (EHV and DRH) and two faculty members oversee the program and coordinate operations with the Anatomical Board (THC and DCB). The program supplies donors to the University of Miami's medical school and health professions classes as well as to seven other academic institutions in South Florida. Approximately 100 donors are utilized each year.

Initial Contact with Donors

After the initial request for forms from potential donors, our licensed funeral directors talk privately with each individual (in English or Spanish). They answer questions and provide guidance about filling out the forms necessary to register as a body donor [24]. These forms include the necessary legal requirements as well as detailed and proper informed consent from the donor (Fig. 22.1). Throughout the conversation, the staff are respectful to the potential donor. They can supply printed materials to the individuals or refer them to the program's website for downloadable forms [24]. When donors register with the program, they are asked to discuss their decision with their family members and to make sure the guidelines for donation are available to the family at the time of their death.

Acceptance and Preparation of Donor's Body

After the donor dies, which can be many years after registration, a local funeral home collects the donor, performs an initial fixation of the donor, and prepares all necessary paperwork (death certificate) for a fixed cost. The donor or their family are required to cover these costs; however, there is a Donor Assistance Fund operated by the Anatomical Board that can reimburse these costs up to $600. The donor is then transported to the medical school by a private transport company utilizing a separate elevator at the university for privacy and respect. A more extensive formaldehyde fixation occurs in the program's facilities and the donor is stored in a refrigerated cold room for eventual use in one of the academic courses at institutions throughout South Florida. The program does not transfer donors to other donation programs outside of Florida unless specifically agreed to by the donor.

UHealth
UNIVERSITY OF MIAMI HEALTH SYSTEM

UNIVERSITY OF MIAMI
MILLER SCHOOL of MEDICINE

SUPPLEMENTAL INFORMATION

The students who will be learning from your donation would like to be able to correlate your previous medical history with what they find during their studies. If you would be willing to provide any information on your past medical history, we would use it to help educate future medical professionals. Any information that you provide is COMPLETELY VOLUNTARY. You are also welcome to have this information removed from your donor file at any time.

<u>MEDICAL HISTORY:</u>

1) Do you have any long-term medical issues? (diabetes, high blood pressure, etc.)

2) Have you ever been hospitalized? (please provide the reason)

3) Have you ever had any surgeries? (please provide the type of procedure)

4) Have you ever smoked tobacco? (please circle one) Yes / No
If yes, how many packs per day, and for how many years?

5) Have you ever drank alcohol? (please circle one) Yes / No
If yes, how many drinks would you consume per week on average, and for how many years?

6) Is there anything else about your lifestyle or personal habits that may be helpful for the students to know? (exercise, diet, etc.)

<u>PERSONAL INFORMATION:</u>

If you would like to provide a short biography of your life and / or a photograph that you would like to share with the students, please feel free to add them. We would request that you not use personally identifying information, since we strive to maintain your anonymity with the students. The students do, however, appreciate knowing about your life and your reasons for donating.

Once per year, we hold a memorial service ("Rose Ceremony") for all those individuals who donated their bodies. Your family members are welcome to attend and can contact our office for the date and time of the ceremony.

Thank you for your willingness to donate your body for the education of future medical professionals. Your altruism, self-sacrifice and generosity are greatly appreciated.

Fig. 22.1 Supplemental information form for donors. This form was developed for donors to provide voluntary information to the students/users, so that the students can have a better understanding of the donor's health and can correlate any findings in the laboratory with the previous health care of the donor. This form also allows donors to provide personal information that they would like to share with the students

Use of Donors in Medical Education

The donors are transported to the academic institution that requests them where they are utilized for anatomical education by medical school or health professional courses. At the University of Miami, each donor is stored in a body bag in a stainless steel tank. Each donor has their head, groin, and hands wrapped in toweling material for protection and privacy. All those utilizing the donors (faculty, students, researchers) are required to treat them with respect and dignity and they sign a Pledge of Respect form stating this commitment. They only uncover the region of the body they are dissecting, they replace all body parts and the skin after they have finished dissecting each day, and they dress and act respectfully at all times. The users also keep all of the donor's tissues in the body bag, so that all of the cremated remains can be returned to the donor's family, if requested.

Fig. 22.2 It is recommended that body donation programs have a permanent memorial to the donors (plaque, display, or garden setting) as well as an annual ceremony to honor the donors and allow the users to reflect on the donation. At the University of Miami Miller School of Medicine, a permanent memorial cabinet contains the plaque (above) along with the metallic rose, video displays of the donors' lives in photographs, and letters from the donor families. The medical school also sponsors a yearly ceremony (Rose Ceremony) where the students reflect on their time in the laboratory with the donors. The family members of the donors are invited to the ceremony

Disposition of Donor Remains

After the anatomy courses are finished, the donors are collected by the program staff and a private transport company takes them to a local crematory to be cremated. The cremated remains are returned to the program office and they are sent to the donor's family (if requested) or are deposited individually at sea. The family receives a thank you letter from the program in both cases and, if deposited at sea, the family can receive a photograph of the actual placement along with the latitude, longitude, and date of the placement.

Donor Commemoration Service

Each year, the University of Miami has a donor memorial service, the Rose Ceremony, to honor the donors used that year. This is a typical memorial service that occurs at most universities where body donor programs are located [19]. The family members of the donors are invited to attend the service where students and faculty reflect on the meaning of the donor's altruism and thoughtfulness.

In addition, a permanent display honoring the body donors is maintained outside of the University of Miami's anatomy laboratories. This includes a plaque thanking all of the donors for their altruism, descriptions of the Rose Ceremony, and the distribution of remains at sea as well as a multimedia tribute to the donors (Fig. 22.2). These are appropriate ways to honor the donors, their altruism, and their willingness to help medical education.

Conclusions and Recommendations

When establishing or modifying a body donor program, the laws and regulations in that jurisdiction must be understood and followed. In addition, to maintain and grow the program, the proper respect and dignity must be afforded to the donors (and their families) at all times [16]. Without these two foundations, a program will not succeed.

When developing a new body donation program, contacting an established program in the region can be very helpful. The established program can provide insight on the local laws regulating body donation and can be helpful with community outreach. Also, examining the websites and forms that other programs use can be valuable (see the associated text box). These can provide templates and background for the new program ("No need to reinvent the wheel"). The authors of this chapter are willing to provide additional information that might be helpful to anyone interested in establishing and operating a body donation program.

Websites of Body Donation Programs

- *South Florida Anatomical Board*: www.sofab.med.miami.edu
- *University of California*: https://www.ucop.edu/ad-program/
- *University of Minnesota*: https://www.med.umn.edu/research/anatomy-bequest-program
- *University of Pittsburgh*: http://www.ooas.pitt.edu
- *McMaster University*: https://fhs.mcmaster.ca/anatomy/bequeathals.html
- *University of British Columbia*: https://cps.med.ubc.ca/bodyprogram/
- *Newcastle University*: https://www.ncl.ac.uk/sme/body/
- *University of Otago*: https://www.otago.ac.nz/anatomy/bequests/index.html
- *Tzu Chi University*: http://www.silent-mentor.tcu.edu.tw/
- *Nanjing Medical University*: http://hdy.njmu.edu.cn
- *Peking Union Medical College*: http://anatomy.sbm.pumc.edu.cn/en/bone.asp

Acknowledgments All donors to body donation programs are gratefully acknowledged for their altruistic and priceless gift. Without their thoughtful donations, body donation programs would not exist. The contributions of the administrators and staff at the University of Miami that support the South Florida program are also acknowledged. Editorial comments on this chapter from Brandi Schmitt (University of California) and Jon Cornwall (University of Otago) are appreciated.

References

1. Garment A, Lederer S, Rogers N, Boult L. Let the dead teach the living: the rise of body bequeathal in 20th-century America. Acad Med. 2009;82:1000–5.
2. Jones DG, Whitaker MI. Speaking for the dead. 2nd ed. Farnham: Ashgate; 2009.
3. Laqueur TW. The work of the dead: a cultural history of mortal remains. Princeton: Princeton University Press; 2015.
4. Korf HW, Wicht H, Snipes RL, Timmermans JP, Paulsen F, Rune G, Baumgart-Vogt E. The dissection course - necessary and indispensable for teaching anatomy to medical students. Ann Anat. 2008;190(1):16–22.
5. ten Brinke B, Klitsie PJ, Timman R, Busschbach JJ, Lange JF, Kleinrensink GJ. Anatomy education and classroom versus laparoscopic dissection-based training: a randomized study at one medical school. Acad Med. 2014;89(5):806–10.
6. Memon I. Cadaver dissection is obsolete in medical training! A misinterpreted notion. Med Princ Pract. 2018;27(3):201–10.
7. American Association of Clinical Anatomists. Best practices guide for donation programs. 2nd ed; 2017. http://clinical-anatomy.org/images/downloads/draft_aaca_bp_asc_.pdf. Accessed 20 Dec 2018.
8. Schmitt B, Wacker C, Ikemoto L, Meyers FJ, Pomeroy C. A transparent oversight policy for human anatomical specimen management: the University of California, Davis experience. Acad Med. 2014;89:410–4.
9. Uniform Law Commission. The National Conference of Commissioners on Uniform State Laws. Uniform Anatomical Gift Act (2006) Summary. 2006. https://www.uniformlaws.org/committees/community-home?communitykey=015e18ad-4806-4dff-b011-8e1ebc0d1d0f&tab=groupdetails. Accessed 19 Feb 2018.
10. Federative International Committee for Humanities and Ethics. Recommendations of good practice for the donation and study of human bodies and tissues for anatomical examination. Plexus, Newsletter of the International Federation of Association of Anatomists. 4–6 Jan 2012. http://www.ifaa.net/wp-content/uploads/2017/09/plexus_jan_2012-screen.pdf. Accessed 31 Mar 2018.

11. Riederer BM. Body donation today and tomorrow: what is best practice and why? Clin Anat. 2016;29:11–8.
12. Riederer BM, Bolt SH, Brenner E, Bueno-López JL, Circulescu AR, Davies DC, De Caro R, Gerrits PO, McHanwell S, Pais D, Paulsen F, Plaisant O, Sendemir E, Stabile I, Moxham BJ. The legal and ethical framework governing body donation in Europe–1st update on current practice. Eur J Anat. 2012;16:1–21.
13. Human Tissue Act. Public Act No 28. Wellington: New Zealand Government. 2008. http://www.legislation.govt.nz/act/public/2008/0028/latest/DLM1152940.html. Accessed 29 Oct 2018.
14. British Medical Association. Human tissue legislation: guidance from the BMA's Medical Ethics Department. 1st ed. London: British Medical Association; 2006.
15. Stimec BV, Drskic M, Fasel JHD. Cadaver procurement for anatomy teaching: legislative challenges in a transition-related environment. Med Sci Law. 2010;50:45–9.
16. Champney TH, Hildebrandt S, Jones DG, Winkelmann A. BODIES R US: ethical views on the commercialization of the dead in medical education and research. Anat Sci Educ. 2019;12(3):317–25.
17. Grow B, Shiffman J. In the US market for human bodies, almost anyone can dissect and sell the dead. Cashing in on the donated dead: The body trade. A Reuters Series. Part 1. Thomson Reuters Corp., Toronto. 2017. https://www.reuters.com/investigates/special-report/usa-bodies-brokers/. Accessed 20 Feb 2018.
18. Winkelmann A. Consent and consensus-ethical perspectives on obtaining bodies for anatomical dissection. Clin Anat. 2016;29:70–7.
19. Štrkalj G, Pather N, editors. Commemorations and memorials: exploring the human face of anatomy. 1st ed. Singapore: World Scientific Publishing Co; 2017.
20. Klaus RM, Royer DF, Stabio ME. Use and perceptions of plastination among medical anatomy educators in the United States. Clin Anat. 2018;31:282–92.
21. Kennel L, Martin DMA, Shaw H, Wilkinson T. Learning anatomy through Thiel- vs. formalin-embalmed cadavers: student perceptions of embalming methods and effect on functional anatomy knowledge. Anat Sci Educ. 2018;11:166–74.
22. Cornwall J. Don't ask, don't tell - who should promote body donation programmes in the public domain? Australas Med J. 2011;4(8):458–9.
23. Anatomical Board of the State of Florida. University of Florida, College of Medicine, Gainesville. https://anatbd.acb.med.ufl.edu/. Accessed 20 May 2018.
24. South Florida Willed Body Program/State of Florida Anatomical Board. http://sofab.med.miami.edu. Accessed 22 Jun 2018.

23 Ethical Considerations of Body Donation

Sabine Hildebrandt and Thomas H. Champney

Introduction

The traditional pedagogical paradigm of anatomy education is that knowledge can be gained through the dissection of dead human bodies. This paradigm itself presents a central ethical quandary for anatomists and their students: by cutting into dead human bodies, they break the taboo of violating the dead body, a taboo shared by many cultures around the world [1, 2]. Thus, anatomists rely on negotiations within their cultures to ensure an adequate supply of bodies for their work, and the results of these negotiations are formalized through laws, including those regulating anatomical body donation. Historically, this has not always been the case, as bodies were obtained through clandestine illegal or legal but questionable mechanisms such as grave robbing, state-sanctioned execution, and the use of unclaimed bodies [2].

"As medical ethics is about things done to the human body" [3], anatomists need to be concerned with how the bodies they use are obtained. Dead human bodies feature prominently in anatomical education, as they not only teach students the structure and function of humans but also provide a rich canvas for other important areas, including ethics, moral decision-making [4] and professionalism. Pertinent questions encountered in the setting of anatomy include reflections on the dignity of the dead, the proper and respectful treatment of the living and the dead, and the experiential learning of how to balance clinical detachment with empathy [5]. Thus, it matters where these bodies come from: whether they were obtained illegally or legally, and whether they were obtained with or without full consent.

After a short overview of the history of anatomical body acquisition, this chapter will describe the meaning of anatomical donation programs for ethics in anatomical education and then will discuss examples of some of the ongoing ethical challenges in body procurement and use.

History of Anatomical Body Procurement

The science and teaching of anatomy are dependent on the availability of dead human bodies for dissection. The legal structure for anatomical body acquisition can be traced in Europe and the United States through various phases and

S. Hildebrandt (✉)
Division of General Pediatrics, Boston Children's Hospital, Harvard Medical School, Boston, MA, USA
e-mail: Sabine.Hildebrandt@childrens.harvard.edu

T. H. Champney
Department of Cell Biology, University of Miami, Miller School of Medicine, Miami, FL, USA
e-mail: tchampney@med.miami.edu

L. K. Chan, W. Pawlina (eds.), *Teaching Anatomy*, https://doi.org/10.1007/978-3-030-43283-6_23

originated from pleas by anatomists to their local governments [6–8]. The first laws on anatomical body procurement were sporadic and often codified the use of bodies of executed persons or others who were deemed to have violated societal norms, such as unwed mothers, persons committing suicide, or duelists [6–9]. However, these legal means of body supply were usually insufficient for medical education in the eighteenth and early nineteenth century, so bodies were also procured by illegal means such as grave robbing and murder [10]. These illegal activities finally led to specific legislation allowing the anatomical use of those who died in public institutions, like prisons, hospitals, and psychiatric wards, and who were unclaimed for burial [11]. Unclaimed bodies often came from economically and racially discriminated groups of the population [11, 12]. While unclaimed bodies made up the majority of the anatomical body supply in Europe and the United States into the twentieth century, body donation did occur but was quite rare [13]. It was only in the middle of the twentieth century that anatomical donation programs were established at university anatomical departments or at state-run anatomical boards in the United States that were loosely regulated by Anatomical Gift Acts [14, 15]. These university-associated or state-run donation programs centre on the altruistic intent of the donor and the not-for-profit status of the educational program using the bodies. The definition of who can donate a body varies between programs. In Germany, Austria, and Switzerland, for example, only prospective donors themselves can donate their bodies [16, 17]. In the United States, however, a family member or the legal representative of the donor is also able to make the donation.

The body donation programs in Europe and the United States became fully functional, i.e. supplied sufficient bodies for medical education, in the second half of the twentieth century. The driving forces behind this increase in body donations are still not fully understood [14]. Some evidence points to government agencies refusing the delivery of unclaimed bodies without voluntary consent by the deceased [18], as well as successful appeals by anatomists to the general public [19]. Other influences that may have led to a change in public opinion are better education, increasing secularism, and the success of modern medicine. In 1963, Jessica Mitford published the first list of United States medical schools that accepted body donations, thus helping to popularize body donation as an alternative to traditional burial [20]. Again, anecdotal evidence shows that, once a body donation program is well established, it attracts further donors through "word of mouth" publicity and "legacy families", that is, families in which it becomes a tradition to donate.

In 1968, the Uniform Anatomical Gift Act (UAGA) was developed in the United States, and it established that the human body was property that could be willed or gifted for organ donation or body donation [21]. It also stated that a donor's wishes for the use of their body superseded those of his or her next of kin. The present, updated UAGA has been adopted by all 50 US states [14]. All US body donor programs accept body donations, but some still supplement their anatomical body supply with unclaimed bodies [22–24], even though this practice is questioned by both the general public and professionals [25]. Worldwide, voluntary body donation is seen by anatomists as the "standard of ethical practice" [26], and many countries have implemented such programs or have plans to do so [27, 28]. Some countries still, however, rely on the use of unclaimed bodies for anatomical education and research, since sufficient numbers of donors are not currently available [17, 29].

The Role of Body Donation for Ethics in Anatomical Education

Ethical anatomical body procurement has a central role in the concept of anatomy as part of a humanistic medical training program [5, 30–32]. Here, the traditional "anatomical cadaver" – a mere object or "specimen" – becomes the "body of the donor" who is seen as a person with human rights. Following from the human right of autonomy is the requirement for fully informed consent for anatomical body donation. Users of the donated bodies are aware of this consent, which helps them – and the staff and faculty – to negotiate the balance in their perception of the donor as both an object to be studied and as an individual with a personal history [33, 34]. Knowing that the donor has consented makes the "violation" of cutting into the donor's body easier for some.

This personalization of the donor, who is often described as "the first patient" or alternatively as a "teacher," illuminates the unique position of anatomy among the other foundational disciplines in the medical curriculum, with anatomy presenting as the "first clinical discipline." Similar to interactions on the clinical wards, the anatomy laboratory allows interactions between student, teacher, and patient, whereby the donor can take on the role of patient or teacher [35, 36]. Therefore anatomical educators should proactively inform their students about the methods and ethics of procurement of their donors to establish the proper respect and dignity for the donors as well as to encourage the students to contemplate the ethics of their actions in the anatomy laboratory.

Not-for-Profit and For-Profit Body Donation

The anatomical donation programs associated with medical schools or anatomical boards are not-for-profit; they are revenue-neutral organizations. However, a new phenomenon has arisen over the last two decades in the United States with the emergence of private body donation companies. With few exceptions, these private companies are based on a for-profit, revenue-generating business model and promote their services through advertising in funeral homes, hospices, nursing facilities, and local newspapers [15]. These companies offer to cover funeral and cremation costs for the family, thus attracting a disproportionate number of economically disadvantaged donors [37]. The donated bodies are dismembered and shipped to national and international clients for research or education purposes, often in postgraduate continuing education courses [37]. From their inception, the for-profit body donation companies, also known as body brokers or non-transplant tissue banks, have come under heavy criticism [38–42]. A lack of specific regulatory guidance in the United States appears to have fostered not only the questionable commercial exploitation of dead human bodies but also transgressions of medical safety and ethics. These include the unsafe handling and shipping of human body parts, mislabelling of the same, and undignified disposal of human remains [37]. Sharp criticism of body brokers is shared not only by individual authors and investigative journalists but also by anatomical associations [26, 43, 44].

Another form of private body donation was established in the 1990s by the anatomist Gunther von Hagens in connection with his business of public exhibits of plastinated bodies, the Body Donation for Plastination program [45]. The von Hagens' enterprises include the Institute for Plastination in Heidelberg and the Gubener Plastinate GmbH [45]. Dr. von Hagens is described as the inventor of a preservation method in which water in human tissues is replaced with silicone. He perfected a previously developed technique [46] and used it to create specimens for public display in travelling exhibits. While popular, these exhibits were soon at the centre of a multidisciplinary ethical debate [47–55]. One of the major criticisms of these exhibits concerned the questionable sources of body procurement [52, 56]. A 2003 audit of von Hagens' plastination facility in Dalian, China, revealed his use of unclaimed bodies, including potential victims of capital punishment [56]. Since that time, von Hagens and his colleagues have stated that the bodies that appear in their travelling exhibits stem from their own private body donation program [57]. This is in contrast to other travelling exhibits of plastinated bodies which display bodies of persons with Asian features and whose origin is still undocumented and presumably includes unclaimed bodies [52, 58]. Because of the undocumented origin of these bodies, the organizer of one of these exhibits "Bodies: The Exhibition" was ordered by the New York Attorney General to provide a refund for tickets and display warning signs that the bodies might include those of Chinese executed prisoners [59].

The concerns raised by for-profit body donation companies should lead anatomical educators to question whether it is ethically appropriate to support these types of companies.

Monetary Value of Donated Human Tissue

The growth of for-profit body brokers has highlighted the economic value of human tissue. Body broker companies have generated millions of dollars in profit [37] and have established prices on whole bodies and body parts. In addition, there are companies that sell human skin, human stem cells, and other human tissues. While for-profit human tissue companies exist, their profit motive can lead to ethically problematic issues. Therefore, ethicists, anatomists, and business researchers have debated the ethical implications of the commercialization and commodification of human tissues [60, 61]. Body donation and human tissue programs should be not-for-profit to avoid these ethical concerns. In addition, they should continually assess their fee structures to determine that they are in proper proportion to their costs. Programs need to be vigilant with their finances and make sure that they do not equate the educational value of the donors with the monetary value of the donor's tissues. The only individuals who should prosper in a human tissue or body donation program are the users who gain knowledge through the invaluable resource inherent in the donors' bodies [32].

Advertising for Body Donation

In order for body donation programs to operate, the local population needs to know that the program exists and that they need donors. However, advertising directly to potential donors – especially just prior to their death – is problematic. These individuals are in a vulnerable state and should not be asked to make this important decision at that time. Most body donation programs reach out to the local news media and the local medical community to make their presence known without direct advertising to donors. For-profit body broker companies, however, are known to reach out to hospices and nursing homes to attract donors (or the families of donors) at or near their death. Most ethicists feel that this is inappropriate and would be preying on individuals at a stressful and vulnerable time [62]. Therefore, all body donation programs are encouraged to thoughtfully consider their outreach methods and to refrain from direct solicitation of donors at or near their death.

Detailed versus Broad Informed Consent

When body donation programs enrol potential donors, they must receive informed consent from the individual who is donating. This is a universally agreed upon principle of anatomical body donation [26]. The type of consent, however, is still debated amongst anatomists, ethicists, and program administrators [63]. Some programs receive broad consent from the donors – giving the programs the ability to use their tissues in any way they feel is appropriate. This consent may consist of a single sentence stating that the donor allows the program to use their body for education or research purposes. Other donation programs have more detailed consent documents that ask the donor to agree to specific uses for their body. For example, the program may ask to be able to retain tissues indefinitely, or to have the ability to transfer the body or tissues to other donation programs, or to use the body in trauma tests. One of the key elements of this debate is the extent of the information offered to potential donors, that is, how much information is necessary to ensure that a donation procedure is ethically appropriate? This question is particularly relevant, as future uses of the donor's tissues may not be known at the time of donation. At a minimum, all programs should truthfully answer any question about the donation program asked by a donor. For living individuals who agree to be research participants, the necessary level of detailed consent is well described and followed by Institutional Review Boards (IRBs). This practice has not, however, been universally accepted for donation programs. Given this uncertainty about the level of consent necessary for body donation, it would be prudent for all programs to obtain detailed informed consent from their donors.

Ethical Considerations in Body Donation Programs

1. Fully informed consent
2. Respectful treatment of the donor and the family
3. Following the wishes of the donor
4. Publicly acknowledging the donor's gift

Conclusion

There are four primary constituents that are involved in body donation: the donors and their families, the donation programs, the users of the donors, and the society at large. Each of these constituents has legal and ethical obligations that are associated with donation. Most importantly for this chapter are the ethical obligations of body donor programs. These include (1) fully informed consent from the donor, (2) respectful treatment of the donor and the family, (3) following the wishes of the donor, and (4) publicly acknowledging the donor's gift. All of these obligations should be followed by an ethically responsible donation program.

In addition, there are guiding principles that should be followed by those utilizing body donors for medical education and research (Fig. 23.1) [61]. The principles that revolve around the donor and the family include detailed informed consent, complete and transparent information, and an understanding for the donor's and family's feelings. The principles that revolve around the donation organization include respect and dignity for the donor, an understanding for the donor's and the family's feelings, an awareness of the cultural and religious background of

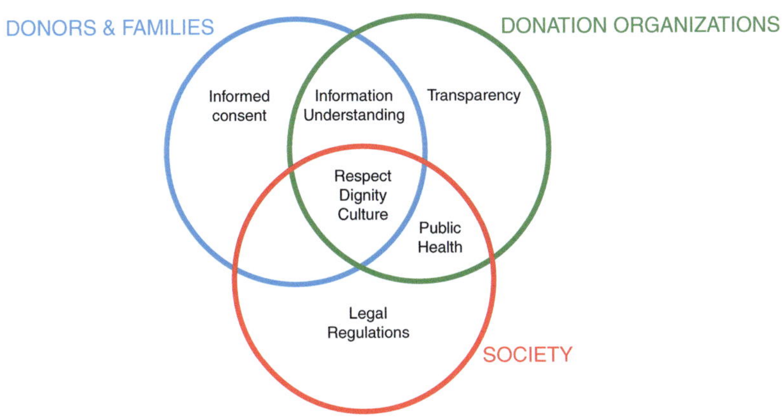

Fig. 23.1 The guiding principles that should be followed by those utilizing willed body donors for medical education and research. The blue circle represents the principles that revolve around the donor and the family: detailed informed consent, complete and transparent information, and an understanding for the donor's and family's feelings. The green circle represents the principles that revolve around the donation organizations: utmost respect and dignity for the donor, an understanding for the donor's and the family's feelings, an awareness of the cultural and religious background of the society, and a complete transparency of the program with the donors and with society. The red circle represents the principles that revolve around the society and the country: the public health concerns of dealing with human tissue, an awareness of the cultural and religious background of the society, and the regulations associated with these practices. (Reproduced with the permission of the publisher of Champney et al. [61])

the society, and a complete transparency of the program with the donors and with society. Finally, there are principles that are societal, including the public health concerns of dealing with human tissue, an awareness of the cultural and religious background of the society, and the legal regulations associated with these practices. All of these principles should be acknowledged and accommodated in body donation and human tissue use.

Medical education, as well as anatomy in particular, is always in transition, as it adapts to the developing values within a society [64]. It is vitally important that body donation programs adapt to changing societal and ethical values, as they may otherwise risk alienating their potential donors. A pertinent example for this need for flexibility is the emerging discussion of the body donors who sought medical aid in dying [65]. Therefore, programs should continually examine their policies and their procedures to ensure the highest level of transparent and ethical behaviour. Transparent and ethical anatomical body procurement is indeed one of the core values of a body donation program and a hallmark of all students' ethical and professional training.

References

1. Laqueur TW. The work of the dead: a cultural history of mortal remains. Princeton: Princeton University Press; 2015.
2. Jones DG, Whitaker MI. Speaking for the dead: cadavers in biology and medicine. 2nd ed. Abingdon, Oxon: Routledge; 2009.
3. Barilan YM. The story of the body and the story of the person: towards an ethics of representing human bodies and bodyparts. Med Health Care Philos. 2005;8:193–205.
4. Dyer GS, Thorndike ME. Quidne mortui vivos docent? The evolving purpose of human dissection in medical education. Acad Med. 2000;75:969–79.
5. Hildebrandt S. The role of history and ethics of anatomy in medical education. Anat Sci Educ. 2019;12:425–31.
6. Stukenbrock K. Der zerstückte Coerper: Zur Sozialgeschichte der anatomischen Sektionen in der frühen Neuzeit (1650–1800). Stuttgart: Franz Steiner Verlag; 2001.
7. Stukenbrock K. Unter dem Primat der Ökonomie? Soziale und wirtschaftliche Randbedingungen der Leichenbeschaffung fuer die Anatomie. In: Helm J, Stukenbrock K, editors. Anatomie: Sektionen einer medizinischen Wissenschaft im 18. Jahrhundert. Stuttgart: Franz Steiner Verlag; 2003. p. 227–39.
8. Sappol M. A traffic of dead bodies. Anatomy and embodied social identity in nineteenth–Century America. 1st ed. Princeton: Princeton University Press; 2002.
9. Hildebrandt S. Capital punishment and anatomy: history and ethics of an ongoing association. Clin Anat. 2008;21:5–14.
10. Ball JM. The body snatchers: doctors, grave robbers and the law. Reprint 1989. New York: Dorset Press; 1928.
11. Richardson R. Death dissection and the destitute. 2nd ed. with a new afterword (2000). Chicago, London: The University of Chicago Press; 1987.
12. Halperin EC. The poor, the black, and the marginalized as the source of cadavers in the United States anatomical education. Clin Anat. 2007;20:489–95.
13. Hildebrandt S. Lessons to be learned from the history of anatomical teaching in the United States of America: the example of the University of Michigan. Anat Sci Educ. 2010;3:202–12.
14. Garment A, Lederer S, Rogers N, Boult L. Let the dead teach the living: the rise of body bequeathal in 20th-century America. Acad Med. 2007;82:1000–5.
15. Champney TH. The business of bodies: ethical perspectives on for-profit body donation companies. Clin Anat. 2016;29:25–9.
16. Riederer BM, Bueno-Lopez J. Anatomy, respect for the body and body donation – a guide for good practices. Eur J Anat. 2014;18:361–8.
17. Riederer BM. Body donation today and tomorrow: what is best practice and why? Clin Anat. 2016;29:11–8.
18. Hildebrandt S. Insights into the Freiburg Anatomical Institute during national socialism. Ann Anat. 2016;205:90–102; here: p. 96.
19. Hildebrandt S. Wolfgang Bargmann (1906-1978) and Heinrich von Hayek (1900-1969): careers in anatomy continuing through German National Socialism to postwar leadership. Ann Anat. 2013;195:283–95.
20. Mitford J. The American way of death. 1st ed. New York: Simon & Schuster; 1963.
21. Uniform Law Commission. The National Conference of Commissioners on Uniform State Laws. Uniform Anatomical Gift Act (2006) Summary 2006. https://www.uniformlaws.org/HigherLogic/System/DownloadDocumentFile.ashx?DocumentFileKey=0e0c3a55-52cf-2082-704f-6185187a21fc&forceDialog=0. Accessed 14 Apr 2019.
22. Jones DG. Resurrecting unclaimed bodies. Clin Anat. 2012;25:674–5.
23. Jones DG. Using and respecting the dead human body: an anatomist's perspective. Clin Anat. 2014;27:839–43.
24. Wilkinson TM. Respect for the dead and the ethics of anatomy. Clin Anat. 2014;27:286–90.

25. Kahn PA, Champney TH, Hildebrandt S. Letter to the editor: the incompatibility of the use of unclaimed bodies with ethical anatomical education in the United States. Anat Sci Educ. 2017;10: 200–1.
26. Federative International Committee for Humanities and Ethics. Recommendations of good practice for the donation and study of human bodies and tissues for anatomical examination. Plexus, Newsletter of the International Federation of Association of Anatomists. 4–6 Jan 2012. http://www.ifaa.net/wp-content/uploads/2017/09/plexus_jan_2012-screen.pdf. Accessed 31 Mar 2018.
27. McHanwell S, Brenner E, Chirculescu ARM, Drukker J, van Mameren H, Mazzotti G, et al. The legal and ethical framework governing body donation in Europe – a review of current practice and recommendations for good practice. Eur J Anat. 2008;1:1–24.
28. Habicht JL, Kiessling C, Winkelmann A. Bodies for anatomy education in medical schools: an overview of the sources of cadavers worldwide. Acad Med. 2018;93(9):1293–300.
29. Stimec BV, Drskic M, Fasel JHD. Cadaver procurement for anatomy teaching: legislative challenges in a transition-related environment. Med Sci Law. 2010;50(1):45–9.
30. Korf HW, Wicht H, Snipes RL, Timmermans JP, Paulsen F, Rune G, Baumgart-Vogt E. The dissection course – necessary and indispensable for teaching anatomy to medical students. Ann Anat. 2008;190(1):16–22.
31. Strkalj G. Humanistic anatomy: a new program for an old discipline. New York: Nova Science Publishers; 2016.
32. Champney T. A bioethos for bodies: respecting a priceless resource. Anat Sci Edu. 2019;12:432–34.
33. Hallam E. Relational anatomy: dissecting and memorializing the dead in medical education. Med Anthropol Theory. 2017;4:99–124.
34. Olejaz M. When the dead teach: exploring the postvital life of cadavers in Danish dissection labs. Med Anthropol Theory. 2017;4:125–49.
35. Bohl M, Bosch P, Hildebrandt S. Medical students' perceptions of the body donor as a "first patient" or "teacher": a pilot study. Anat Sci Educ. 2011;4:208–13.
36. Bohl M, Holman A, Mueller DA, Gruppen LD, Hildebrandt S. The willed donor interview project: medical student and donor expectations. Anat Sci Educ. 2013;6(2):90–100.
37. Grow B, Shiffman J. In the US market for human bodies, almost anyone can dissect and sell the dead. In: Cashing in on the donated dead: The body trade. A Reuters Series. Part 1. Toronto: Thomson Reuters Corp. 24 October 2017. https://www.reuters.com/investigates/special-report/usa-bodies-brokers/. Accessed 20 Feb 2018.
38. Roach M. Stiff—the curious lives of human cadavers. New York: W.W. Norton and Company; 2003.
39. Cheney A. Body brokers—inside America's underground trade in human remains. New York: Broadway Books; 2006.
40. Goodwin M. Black markets—the supply and demand of body parts. Cambridge: Cambridge University Press; 2006.
41. Sharp LA. Bodies, commodities, and biotechnologies—death, mourning, and scientific desire in the realm of human organ transfer. New York: Columbia; 2007.
42. Cantor NL. After we die—the life and times of the human cadaver. Washington DC: Georgetown University Press; 2010.
43. American Association of Clinical Anatomists. Best practices guide for donation programs. 2nd ed; 2017. http://clinical-anatomy.org/images/downloads/draft_aaca_bp_asc_.pdf. Accessed 20 Dec 2018.
44. American Association of Anatomists. A discussion on body donation. No date. https://www.anatomy.org/uploads/4/6/5/1/46517773/aaa-body-donation-article.pdf. Accessed 20 Feb 2018.
45. Institute for Plastination. Body donation for plastination. No date. http://www.bodymobil.de/Downloads/Englisch/download/BD_Broschuere_US_WEB.pdf. Accessed 20 Feb 2018.
46. Von Hagens G. Impregnation of soft biological specimens with thermosetting resins and elastomers. Anat Rec. 1979;194(2):247–55.
47. Wetz FJ, Tag B, Tiedemann K. Schöne Neue Körperwelten. Stuttgart: Klett-Cotta; 2001.
48. Werner J. Körperwelten – endlich unsterblich oder würdelose Leichenshow? Studienarbeit. Norderstedt: Grin Verlag; 2001.
49. Jones DG. Re-inventing anatomy: the impact of plastination on how we see the human body. Clin Anat. 2002;15:436–40.
50. Lozanoff S. Letter to the editor. Re-inventing anatomy: the impact of plastination on how we see the human body. Clin Anat. 2002;15:441–2.
51. Cohn F. Letter to the editor. Re-inventing anatomy: the impact of plastination on how we see the human body. Clin Anat. 2002;15:443–4.
52. Working R. Shock value: Gunther von Hagens has outraged critics with macabre publicity stunts to promote "Body Worlds" and lure 17 million visitors to the exhibits worldwide. It's all in the service of science, he says. Chicago Tribune. 31 July 2005.
53. Barilan YM. Bodyworlds and the ethics of using human remains: a preliminary discussion. Bioethics. 2006;20:233–47.
54. Jespersen TC, Rodriguez A, Starr J. The anatomy of body worlds. Critical essays on the plastinated cadavers of Gunther von Hagens. Jefferson: McFarland and Company; 2009.
55. Lantos JD. Controversial bodies: thoughts on the public display of plastinated corpses. Baltimore: Johns Hopkins University Press; 2011.
56. Peuker T, Schulz C. Der über Leichen geht. Gunther von Hagens und seine Körperwelten. Berlin: Ch.Links Verlag; 2004.

57. Bodyworlds. Questions & answers. No date. https://bodyworlds.com/about/faq/. Accessed 18 Feb 2019.
58. Barboza D. China turns out mummified bodies for displays. New York Times. 8 Aug 2006. http://www.nytimes.com/2006/08/08/business/worldbusiness/08bodies.html. Accessed 23 Feb 2018.
59. Chan S. 'Bodies' show must put up warnings. City room blog. 29 May 2008. https://cityroom.blogs.nytimes.com/2008/05/29/bodies-exhibit-must-put-up-warnings/. Accessed 23 Feb 2018.
60. Anteby M. Markets, morals, and practices of trade: jurisdictional disputes in the U.S. commerce in cadavers. Adm Sci Q. 2010;55:606–38.
61. Champney TH, Hildebrandt S, Jones DG, Winkelmann A. BODIES R US: ethical views on the commercialization of the dead in medical education and research. Anat Sci Educ. 2019;12:317–25.
62. Rady MY, Verheijde JL, McGregor J. Solicitation of deceased and living organ donors. N Engl J Med. 2007;356:2427–9.
63. Winkelmann A. Consent and consensus-ethical perspectives on obtaining bodies for anatomical dissection. Clin Anat. 2016;29:70–7.
64. Jones DG. Human anatomy: a review of the science, ethics, and culture of a discipline in transition. In: Sisu AM, editor. Human anatomy: reviews and medical advances. 1st ed. London: IntechOpen Ltd; 2017. p. 3–20.
65. Wainman B, Cornwall J. Body donation after medically assisted death: An emerging consideration for donor programs. Anat Sci Edu. 2019;12:417–24.

24 A Global Geography of Body Acquisition for Anatomy Education: Issues, Challenges and Prospects

Goran Štrkalj, Joyce El-Haddad, and Anneliese Hulme

Introduction

A majority of anatomy educators maintain that the hallmark characteristic of best practice in anatomy education is the utilisation of human cadavers either through dissection or examination of prosected bodies/body parts [1–3]. In addition to benefits that hands-on study of human tissue provides in learning about the structures of the human body, work with cadavers offers unique opportunities for the development of non-traditional discipline independent skills, such as the demonstration of professionalism and medical ethics [4]. However, to achieve these educational outcomes, human tissue for anatomical examination have to be obtained in an ethical manner and treated with respect in the laboratory.

To facilitate the process of ethical acquisition of human tissue, various national and international groups have issued relevant guidelines and recommendations. The main ideas of these documents were perhaps best explicated in the Recommendations of Good Practice for the Donation and Study of Human Bodies and Tissues for Anatomical Examination [5], by the International Federation of Associations of Anatomists (IFAA). The rationale behind the recommendations and their importance for anatomical education were provided by many anatomists [6–15]. A recent review highlighted three key themes in the IFAA Recommendations: "(1) the consent, understanding and information for the donor and the family; (2) the proper treatment of the deceased by the donation organisation; and (3) the safety, ethical and legal aspects of the proper treatment of the dead by the society" [16].

However, it has also been noted that the recommendations are still a far-off ideal, as they are fully applied in only a small number of countries [13, 14]. Reliable data about body acquisition patterns around the world is therefore needed to further plan local and global strategies to implement the recommendations. In this chapter, we will provide an overview of worldwide practices in obtaining bodies for medical education and the considerable variation within and between different countries. We will also attempt to provide an outline of some factors that influence modes of body acquisition and implementation of body donation programs, together with relevant examples. However, the intricate landscape of obtaining bodies for anatomy laboratories can only be fully comprehended within the context of the complexities of the historical development of the

G. Štrkalj (✉)
School of Medical Sciences, Faculty of Medicine, UNSW Sydney, Sydney, NSW, Australia
e-mail: g.strkalj@unsw.edu.au

J. El-Haddad · A. Hulme
Faculty of Science and Engineering, Macquarie University, Sydney, NSW, Australia
e-mail: joyce.el-haddad@mq.edu.au; anneliese.hulme@mq.edu.au

L. K. Chan, W. Pawlina (eds.), *Teaching Anatomy*, https://doi.org/10.1007/978-3-030-43283-6_24

discipline of anatomy. Therefore, the chapter starts with a brief overview of the history of cadaver acquisition and usage in anatomy laboratories.

History of Cadaver Acquisition for Anatomical Examination

Acquisition of bodies for anatomical dissection has a long, complex and controversial history. Bodies were obtained in a variety of ways, which at times transgressed local cultural norms and were carried out in ethically dubious and even illegal manners [17]. Only practiced for a brief period of time in antiquity, dissections of human cadavers were revived in the late medieval period at the emerging European universities. The cadavers used in dissections were bodies of executed criminals, and condemnation to anatomical dissection was often part of the punishment for the crimes committed. From these early times, anatomists were intermittently faced with a paucity of bodies available for dissection and often reverted to alternative ways of procuring cadavers. Even Andreas Vesalius (1514–1564), the founder of modern, empirically and experimentally oriented anatomy, described how he and his students engaged in stealing human remains from gallows, graveyards and even cadavers of recently deceased from places of mourning [18].

In some countries, mainly in continental Europe, the problem of cadaver paucity was resolved by supplementation or complete replacement of bodies of executed criminals by unclaimed cadavers from hospitals, asylums and prisons. In other places such as Britain, Ireland and the USA, occasional problems with acquisition of cadavers for medical schools turned into an educational crisis in the eighteenth and nineteenth centuries [19]. This came with sociopolitical changes such as population growth, industrialisation, increased travel, trade and protracted wars with more efficient weaponry (e.g. Napoleonic wars), which resulted in enlarged needs for medical care and, consequently, proliferation of medical schools. Most of these schools required a thorough anatomical training through dissection and, hence, cadavers. In many places, anatomists became engaged in shady practices such as grave robbing, either directly or through transactions with their "suppliers"—groups of organised criminals. These practices resulted in profoundly negative attitudes towards the medical profession even causing riots and physical confrontations with anatomists and their students [19, 20]. This strained relationship was somewhat eased with legislations paralleling those in continental Europe, which allowed unclaimed bodies to be sent to the medical schools for anatomical study.

Ethically, this new system was far from perfect as it opened the door for the exploitation of the most vulnerable strata of society, such as the poor and minorities. Whilst the legislations usually allowed for body bequest, donations were quite rare and mostly carried out by enlightened persons, such as the philosopher Jeremy Bentham (1748–1832) [21]. Only in the second half of the twentieth century would body donation become, in some countries, main or even the exclusive source of cadavers for anatomy. At first, this was mainly due to a decrease in unclaimed bodies (caused again by various social factors such as improved social care and standard of living) and, later, the need to obtain bodies in an ethically acceptable way. However, as will be shown below, according to the available data, this transition towards body donation has so far happened only in a rather small number of countries [22].

Body Acquisition Around the World

Systematic mapping of the methods by which cadavers are obtained for medical education has only begun in recent years [10, 23]. In a Herculean effort, Habicht and colleagues produced the first global map of body acquisition patterns in 2018 [22]. Albeit incomplete, this map provides an unprecedented insight into the geography of body donation. Importantly, this map will be continuously updated and presented on the IFAA website [5], as the geography of body donation is not expected to remain static. The continuous updates are thus important to review the changes as data from some countries becomes available, and practices in others are modified [24, 25].

The authors of the body donation map classified the modes of cadaver acquisition into five categories: (1) exclusively body donation (EBD) (the available data, however, did not allow the authors to distinguish between willed body donations and those made by family members), (2) mainly body donation, with less commonly unclaimed bodies (MBD), (3) mainly unclaimed bodies, with less commonly body donation (MUB), (4) exclusively unclaimed bodies (EUB) and (5) unclaimed bodies and other ways (UBO) [22]. The last category, according to existing sources, is found in Nigeria only, where the majority of cadavers (as many as 90% in some medical schools) have been bodies of dead criminals [26, 27]. It appears that these individuals lost their lives in various types of violent confrontations, as many bodies were reported to display horrific injuries [26, 27]. Two MUB and two EUB countries additionally import cadavers from abroad, whilst Libya was reported to rely exclusively on imported bodies. Importing bodies is particularly worrying, especially if private brokers are involved [15, 26]. There have been reports of cases where institutions trying to procure bodies from India have been unsuccessful, due to other institutions offering higher bids [24], essentially dehumanising the human remains to a commercial commodity [24, 28].

It is important to have a distinction between MBD and MUB categories for countries who rely on both body donation and unclaimed bodies. These two categories signify that there is no well-defined, clear-cut boundary between EBD and EUB methods of body acquisition. Analytically, the existence of MBD and MUB categories enable easier capture of possible trends and changes in methods of cadaver acquisition. It would be deemed a positive trend to see countries switching from MUB to MBD with a possible transition to EBD. The reverse would be deemed a negative trend [25, 29].

The most recent published report about body donation patterns worldwide provides data for 71 of 165 countries with medical schools [22]. Of these 71, 68 were reported to use cadavers in anatomy teaching, but only a relatively small proportion, 22 (32%), were found to be EBD countries (Australia, Austria, Canada, Chile, Czechia, Denmark, France, Germany, Ireland, Japan, Malta, Netherlands, New Zealand, Poland, South Korea, Spain, Sri Lanka, Sweden, Switzerland, Thailand, the United Kingdom, Venezuela [22]). Even with incomplete data, it is clear that geographically body donation is restricted, with a disappointingly small number of countries following the IFAA Recommendations. Noteworthy is also the "big unknown" of the other 94 countries where information on body acquisition is still unreported.

Since there are strong arguments that body donation is the only ethical and educationally acceptable manner of obtaining bodies, it is legitimate to question why so few countries rely on this mode of cadaver acquisition. Several studies indicate that the prime motivation for body donation is altruism expressed in various forms such as "helping medical education and research" and "being useful after death" [30–44]. A variety of factors could either facilitate or supress the expression of this altruism. These factors need to be investigated so that anatomy can strive towards a future where body donation within medical education is the norm rather than the exception.

Some Factors Influencing Body Acquisition

Only recently have researchers begun to investigate reasons behind specific modes of body acquisition for anatomy laboratories, and a comprehensive understanding is still in its embryonic stages. However, several factors influencing the way in which cadavers are procured can already be highlighted. These factors differ in nature and are complexly interrelated. They include economic, cultural, sociopolitical factors, as well as modes in which the anatomical communities operate. The role each of these factors play can be fully comprehended only within a specific local context. The consequences of the complexity of interplay between these factors can result in completely different outcomes even between institutions within the same region of a country.

Economic Factors

Even on brief inspection of the map [22] showing body acquisition modes across the globe, it is noticeable that most countries relying on EBD are economically developed. Of the 22 countries utilising EBD, 19 are high income, as defined by the World Bank [45]. Whilst being a high-income country is neither necessary nor a sufficient cause of EBD, as there are high-income countries that do not utilise EBD as well as low-income countries that practice it, it appears that finances strongly impact the implementation of body donation programs. Indeed, organising a body donation program assumes considerable cost in dedicated work hours and substantial infrastructure, including specialised transport, tissue preparation and maintenance and cadaver disposal. Some extra expenditure could be added to this, such as the cost of organisation of commemorations for body donors (see below) [46].

In some places, even basic materials, such as the embalming fluid, are unaffordable [47]. Furthermore, even in high-income countries, budgetary constraints affect the format of body donation programs. Wingfield argued that in the USA, for financial reasons (possible hidden costs), it might be advantageous, especially at smaller institutions, to rely on acquisition of bodies through specialised for-profit organisations [46]. However, these organisations do not seem to always fully comply with the IFAA Recommendations [16].

The thesis of the importance of economic factors is further corroborated by research carried out in regions of different levels of development within the same country. A survey conducted in China revealed significantly higher rates of body donation in economically more developed regions of the country [48].

Cultural Factors

Whilst budgetary issues play an important role, it is clear that even in non-high-income countries, if other factors are in place, body donation could be implemented successfully. Culture is amongst the factors commonly cited as most influential in developing donation programs, which is understandable as culture defines outlook on life, death and handling of human remains [49]. Cultural factors, particularly religion, influence body donation in various ways, both as constraints and as facilitators of the process. Several surveys on body donation, for example, revealed that atheists and agnostics are more inclined to become body donors [41, 49]. At the same time, within other cultural milieus, religious beliefs acted as a prime motivator for body donation [50]. Two of the three non-high-income EBD countries are predominantly Buddhist, namely, Sri Lanka and Thailand. Publications on body donation in both countries reported that Buddhism was the chief motivator for high rates of body donation. However, Buddhism is a complex system of religious beliefs containing various viewpoints, and within certain countries, these variances may impact attitudes towards body donation in different ways [51, 52].

Considering the complexities of cultural differences, it seems important to create body donation programs in accordance with the cultural norms of the society within which it is established. Ethnocentric approach, where other cultures and their practices are viewed from one's own cultural perspective and system of values, can be deleterious. Jones and Nie [53] have, therefore, rightly warned that relying solely on a particular worldview, such as the "Western scientific", and neglecting local cultural context might impede communication with communities and the organisation of body donation programs [53]. Using Confucianism within the Chinese context, as an example, Jones and Nie showed that understanding the cultural background and objections to body donation could allow the development of culturally acceptable practices within the milieu previously thought to be adverse towards it [53]. The actual success stories from Confucian China are in fact the best corroboration of this view [48, 54, 55].

More broadly, it could be noted that there is not a single culture that allows for desecration of the body after death [8, 19, 56–58]. To a layperson, anatomical dissection appears just that—one just has to remember the etymology of the word "anatomy", from the Greek "cutting up". However, if "cutting up" a cadaver is well justified of benefit to the living community and if the process does not violate the dignity of the deceased but rather, following local customs, celebrate it, then within many cultures more positive attitudes towards anatomical dissection are formed.

Although anatomists have often blamed cultural constraint as the most common reason for unsuccessful body donation programs, it may be more a lack of awareness and understanding of the cultural environment surrounding them [24, 27, 53]. It could be suggested that relevant experts, such as cultural/social anthropologists, should be included in designing body donation programs. As most of the anatomy departments are based in university settings, this engagement would be easy to organise and finance [59].

Finally, with full acknowledgement of cultural relativism, it should be noted that certain practices related to body donation created to cater to certain cultural matrices proved to be successfully applied outside that culture. For instance, the "Silent Mentor Program" of body donation organised at the Tzu Chi University in Taiwan to cater specifically within the Buddhist system of values proved successful in other countries with considerably different cultures and religions (see below) [60, 61].

Sociopolitical Factors

In a historical review of body acquisition practices in the USA, Garment and colleagues noted that "during the twentieth century, the United States underwent a population boom, expansion of government, changes in population demographics, developments in science, and proliferation of mass media, all of which affected body acquisition" [62]. Similar complex interplay of societal factors are found in other countries and throughout history [17]. Education, scientific research and medical practice are social activities and could be deeply affected by changes in a social system. Consequently, social policies, both past and present (see section "History of Cadaver Acquisition for Anatomical Examination"), have profoundly influenced the acquisition of bodies for anatomy, including the existence and format of body donation programs.

In countries such as the USA and South Africa, it was noted that people of African ancestry contribute very small numbers to body donations. This reluctance to become a donor is attributed to the trauma stemming from past and present oppressive racial policies and discrimination both explicit/institutionalised and implicit [25, 29, 36, 63, 64]. This trauma makes people of African ancestry feel alienated from society and its medical system. The medical system does not seem to support this group in the same way as the privileged strata of society, creating a distrust which carries forward into body donation.

One key social factor impacting body donation regards the level of legislation governing its regulation and the adequate function of the broader legal system. For example, it was reported that in Serbia, when communists overtook power (in what was then Yugoslavia) soon after the end of the second world war, they "suspended all legal acts" including those related to body acquisition for medical schools [13]. This was followed by several "partial legal acts" on the matter, none of which proved to be satisfactory. Only in 2003, more than a decade after the fall of communist regime, were the comprehensive legal acts regulating body acquisition re-introduced.

Indeed, one important proposition in the IFAA Recommendations is the existence of adequate legislation. It is thus "recommended that legislators in all countries enact laws that are congruent with their societal values and that provide oversight on how willed body programs operate" [16]. Anatomists from several countries have reported that the lack of appropriate legislation presents a key impediment in developing body donation programs [13, 24, 27, 55, 58]. Furthermore, in some places, problems with legislation not only impede the establishment of body donation programs but may actually present a health hazard to students who learn from the donated bodies. In India, for example, there is no requirement to screen donated bodies for infectious diseases, and it was reported that screening is not practiced at several institutions in that country. This is particularly worrying as in some departments, dissection is carried out without gloves, a practice encouraged by their senior anatomists because of the claim that "the feel of the tissues with bare hands is crucial in acquiring anatomical knowledge" [27].

In some countries and states, it appears that the centralisation of body donation programs could be of great benefit, as it would correct the discrepancy between some institutions having an oversupply and institutions in other regions having a shortage [24, 48]. Centralisation would thus enable an even distribution of bodies. Indeed, one ethical imperative is that bodies are adequately and efficiently utilised [27], and this centralised approach could enable better utilisation of donated bodies. One has to be careful with this policy though, as some donors might specify the geographical region where they want their bodies to be examined (e.g. within the local community to which they belonged to in life).

Professional Factors

Body acquisition for medical schools is carried out primarily by anatomy faculty and staff. The way cadavers are obtained, therefore, depends deeply on the organisational and operational capabilities of the specific anatomical communities. Whilst finances play an important role in this respect, other major impacting factors within an organisation include knowledge of relevant processes, alignment with the global bioethical discourse and establishment of good rapport with local communities. A recent survey of medical schools in the USA revealed considerable "lack of awareness of existing practice guidelines" amongst anatomists [65]. Furthermore, it revealed that some members of anatomy faculty were ignorant about their own institution's practices and the important issues pertaining to bodies. For example, it was noted that 6.7% of the survey respondents did not know if their institution had any unclaimed cadavers [65]. The authors of the survey suggested that local experts, legal and other, should be engaged to assist anatomists. This strategy, as mentioned earlier, could be carried out with relative ease within the university environment. However, currently, it is largely local and international professional associations of anatomists that provide key assistance for alignment with guidelines, legislation and demands of the local community.

Community of Practice

Nowadays, anatomists across the globe are organised into professional associations, which often engage in reflecting on their past and present anatomical practices [5, 66, 67]. There is in fact a global community of practice formed through numerous national and international communication channels. Indeed, recommendations issued by various professional bodies are the best examples of positive outcomes stemming from this community of practice. Several discussion fora, some of which are published in anatomical journals or relevant websites, provide a unique opportunity for anatomists from different countries, regions and diverse educational traditions to exchange their experiences, discuss their differences in practices and viewpoints and strive for agreement.

On an individual level, examples of successful collaboration have also started to emerge. Anatomists from the University of Malaya, for example, developed a collaboration with their colleagues from Taiwan's Tzu Chi University, where an elaborate "Silent Mentor Program" with engagement of the families of body donors was developed [60, 61]. The students contact the family of body donors and through personal communication learn about their donors' lives, medical and social background. The program also includes an initiation ceremony at the beginning and a thanksgiving ceremony at the end of the teaching sessions. A similar program, with the help from their more experienced Taiwanese colleagues, was successfully implemented in Malaysia. According to the preliminary report, the program led to increases in body donation and positive feedback from students. It was also reported that the Silent Mentor Program has inspired body donation programs in several institutions in Singapore, Hong Kong and Myanmar [60].

Communication with Community

When all factors outlined above are taken into account, it is apparent that relationships between anatomists and the community within which they practice is of paramount importance. The key step to build good rapport with the community is establishing communication.

In many places, the general public seems to know relatively little about body donation. Several surveys established that one reason for the absence of willingness of people to donate their body for medical education and research is a lack of cognisance of the existence of body donation programs, even amongst medical students [68, 69]. Many anatomists and experts involved in body donation programs have expressed acute awareness of the necessity to improve communication with the community. Consequently, they have suggested strategies to embed the development of communication skills into existing anatomy curricula [42, 70]. Whilst body donation should not be advertised, it does require a carefully conceptualised and implemented public health campaign explaining why body donation is important in provision of better health for the population through adequately resourced medical education.

A body donation program has to be aligned with cultural norms and be transparent by providing information on all stages of the donation process. A program also needs to assure the public that donated bodies will be treated with respect, since concerns about disrespectful treatment of bodies in the anatomy laboratories is often quoted as one reason for unwillingness to donate one's body [62, 71, 72].

In an attempt to raise awareness of the existence of body donation programs, the Federal University of Health Sciences of Porto Alegre, Brazil, has introduced an elaborate outreach program [73]. The program consists of three components: (1) The Dissection Workshop, (2) The Museum of Anatomy and (3) The Ceremony in Honour of Body Donors. The Dissection Workshop is a 40-hour extension course where students are presented with an opportunity to produce anatomical prosections, an educational resource also used in the second component of the outreach program. The Museum of Anatomy is an annual event in which the prosected body parts are exhibited for the public, where students act as museum guides for the visitors. This component of the outreach program is well received amongst the public with a reported 700 visitors in 2012 [73]. The Ceremony in Honour of Body Donors is a commemoration for the body donors, attended by faculty, students and donor families (Fig. 24.1). Whilst the first two components primarily highlight the importance of body donation for medical education, the third focuses on the respectful treatment of donors. According to the 2012 report, since the introduction of the program in 2008, there has been a fivefold increase in the number of people who have signed up for body donation [39, 73].

Commemorations and Memorials

The principal activities to establish a respectful and trustworthy relationship with potential body donors, their families and the community seem to be the establishment of ceremonies (commemorations) and/or creation of objects and spaces (memorials) to honour those who bequeathed their bodies to medical education. Anatomy commemorations usually include academic staff, students and body donors' family members. In many instances, the broader community is also involved with representatives often invited to ceremonies or involved in their organisation. In addition, commemorations and memorials and their significance have been frequently reported in the media (including the social media) [74, 75].

Memorialising the deceased whose bodies were used in dissections could be traced back to the early days of modern anatomy in the late medieval European universities [20]. These early ceremonies were performed in a distinctly different manner compared to contemporary practice and were deeply immersed in the prevailing religious worldview of the time. Lassek noted that "after Mondino [de Luzzi, professor of anatomy at the University of Bologna, thought to be the

Fig. 24.1 The Ceremony in Honour of Body Donors at the Federal University of Health Sciences of Porto Alegre, Brazil. (Photograph courtesy: Prof. da Rocha)

first to incorporate dissection of human cadaver into medical curriculum in the early fourteenth century], dissections became more elaborate and ceremonious. The spiritual factor was not forgotten since special rites were conferred upon the condemned before execution to atone for the deed about to be performed upon his body" [20]. In non-Western educational traditions, similar rituals were carried out in Japan in the Edo period, where memorial addresses for dissected cadavers were given at Buddhist altars. Five texts have been preserved, dating from 1754 to 1861, which record these addresses [76]. In their current form, however, commemorations and memorials emerged only in the twentieth century as body donations for medical education became more common.

Modern commemorations and memorials have multiple functions and address different stakeholders of anatomy education. Most importantly, they enable anatomists to reciprocate the donors and their relatives for their gift – their own bodies or those of their loved ones after death. Bolt investigated anatomy memorials using Marcel Mauss's anthropological analysis of gift giving and thus explored "body donation discourse beyond altruism" [33]. She suggested that there exists a complex network of "gift exchange" between donors, their families and anatomists where the anatomy memorials "arose from desire on part of anatomical professionals to reciprocate body donation, not only to donors but also, in particular, to the donors' relatives" [33].

For the family and friends of body donors, commemorations afford an opportunity to bid a final farewell to the deceased whilst also assisting in the grieving process. A study carried out in Netherlands revealed that the body donors' family members highly valued anatomy commemorations and memorials [33]. Similar results were obtained in a study conducted in China's Jiangsu province, where 64.8% of the surveyed members from the general public expressed the view that memorial practices for body donors were necessary. In addition, 79.98% felt that they would be more willing to donate their bodies if "memorial practices and post-donation services were improved" [77]. For potential donors, donor families and the community, commemorations and memorials also provide reassurance that the bodies in anatomy laboratories are treated with respect and that donation is of great importance in education.

Apart from communicating a positive message to the donors, their families and community in general, commemorations and memorials play an important role in education and development of non-traditional discipline independent skills. The Indiana University School of Medicine – Northwest, for instance, has an elaborate five-principle (the first patient, knowledge, reflection and reflective practice, treating the total patient, professionalism) pedagogy that aims to enhance future physician competency [78]. This approach has incorporated commemorations and promotes a philosophy where students feel privileged that they are given an opportunity and, more importantly, responsibility of working with donated bodies. Just like the body donors' family members, students have highly positive perception of the importance of commemorations and memorials [79]. Indeed, at several institutions, it was students who initiated and often organised memorial services for body donors [75].

Across the world, anatomy commemorations are carried out in different formats, and they tend to be aligned with local cultural norms. Thus, commemorations vary as much as the cultures and societies within which the medical schools are based, even within the same country. For example, the differences in overall academic culture and philosophy between universities in Korea create considerable differences between the modes of commemorations conducted at each institution [80]. Multicultural societies could present a challenge for the organisers of commemorations, and attempts should be made to devise inclusive ceremonies that could cater for attendees belonging to different cultural milieus and social strata. An extensive consultation with the major stakeholders (registered donors, donors' family members, students, community representatives) prior to deciding on the format of the ceremony could be helpful in that regard [81].

Commemorations also follow societal changes. In China, numerous initiatives have been recorded at different institutions to increase body donation through various public campaigns promoting the humanistic approach [55, 77, 82–84]. They include elaborately organised culturally sensitive commemorations and memorials, often presented in an innovative manner. At Nanjing University, a yearly commemoration for body donors is timed to coincide with the Tomb Sweeping Festival, a national holiday that honours the ancestors of the living [55, 77]. The ceremony honouring the body donors is carried out at the memorial garden erected at the university grounds (Fig. 24.2). Body donors, their families and university representatives participate in these well-attended ceremonies. In addition, Nanjing University has introduced new types of memorial activities catering for the sensibilities of the twenty-first century society dominated by digital media. They include the creation of a website where donors have been honoured and memorialised [55, 77]. This website allows donor family members and friends, students and academic staff to login to their individual digital memorial spaces where notes, songs, photos and other modalities for reflection can be electronically uploaded by the user. The report from Nanjing suggests that digital platforms of commemoration have been well received by all stakeholders of anatomy education [55].

Memorials for body donors are also quite diverse. They range from relatively simple memorial plaques to monuments and elaborate places such as parks. Fittingly, in front of the Anatomical and Pathological Museum at the Radboud University in the Netherlands, an old marble dissecting table forms the centre of a monument topped with a sculpture of a phoenix rising from the ashes [33]. The University of the Basque Country created a memorial space in their public grounds, the *Vitae Silva* (Forrest of Life), with weathering-steel stems representing tree trunks connected with steel cables that rustle in the wind, a serene place for remembrance and reflection [85].

The institutions, where needed, also provide the resting place for body donors. Aberdeen's

Fig. 24.2 The 2019 Public Memorial Ceremony in Nanjing, China. This ceremony took place in the memorial garden and the participants included registered organ and body donors, donors' family members, Red Cross and Medical College representatives. (Photograph courtesy: Prof. Luqing Zhang)

Trinity Cemetery has a third of an acre plot reserved for the remains of body donors of the Aberdeen University [86]. The area is marked by a monumental memorial built in 1974, the first in Britain. A dedication to donors overlooks the memorial: "In memory of those who gave their bodies for the increase of knowledge and advance of medicine."

Although commemorations and memorials have marked anatomy education at many institutions for several decades, they are still emerging phenomena. Reports about them, critical investigation and exchange of experiences will undoubtedly help in perfecting their format and implementation.

Conclusions

Based on the incomplete data on the pattern of body acquisition collected so far, only a rather small number of countries have body donation as the sole source of cadavers for anatomical examination. However, there are some signs that this situation is changing in a positive direction. The map of body acquisition practices is constantly updated and is becoming more accurate. Community of practice at a global level, publication of recommendations for the best practice and an ongoing debate about them, development of communication with the local community, together with a growing body of research on the topic seem to be steps in the right direction. Whilst various body donation strategies seem to result in positive outcomes, it must be noted that simple factors, such as budgetary constraints, play an important role. Indeed, without considerable financial investment in anatomy and medical education, improvement to the situation in some places (mainly middle and low-income countries) might be extremely onerous.

Despite difficulties, positive trends in body donation seem to be visible, and one hopes that the ethical way of obtaining bodies for anatomical examination will spread across the globe to more regions

Acknowledgment We thank Professor Andrea Oxley da Rocha for the photographs of the Ceremony in Honour of Body Donors at the Federal University of Health Sciences of Porto Alegre, Brazil, and thank Professor Luqing Zhang for the photograph of the Public Memorial Ceremony in Nanjing, China, and their permission to publish them in this volume.

References

1. Eisma R, Wilkinson T. From "silent teachers" to models. PLoS Biol. 2014;12(10):e1001971.
2. Estai M, Bunt S. Best teaching practices in anatomy education: a critical review. Ann Anat. 2016;208:151–7.
3. Pather N. Teaching anatomy: prosections and dissections. In: Chan LK, Pawlina W, editors. Teaching anatomy: a practical guide. Cham: Springer International Publishing; 2015. p. 213–21.
4. Evans DJ, Pawlina W, Lachman N. Human skills for human[istic] anatomy: an emphasis on nontraditional discipline-independent skills. Anat Sci Educ. 2018;11(3):221–4.
5. Federative International Committee for Humanities and Ethics. Recommendations of good practice for the donation and study of human bodies and tissues for anatomical examination. Plexus, Newsletter of the International Federation of Association of Anatomists. 2012 Jan:4–6. http://www.ifaa.net/wp-content/uploads/2017/09/plexus_jan_2012-screen.pdf. Accessed 1 Mar 2019.
6. Riederer BM, Bueno-López JL. Anatomy, respect for the body and body donation—a guide for good practice. Eur J Anat. 2014;18(4):361–8.
7. Jones DG. Searching for good practice recommendations on body donation across diverse cultures. Clin Anat. 2016;29(1):55–9.
8. Jones DG, Whitaker MI. Speaking for the dead: the human body in biology and medicine. 2nd ed. Farnham: Ashgate; 2009.
9. Hildebrandt S. Thoughts on practical core elements of an ethical anatomical education. Clin Anat. 2016;29(1):37–45.
10. Riederer BM. Body donations today and tomorrow: what is best practice and why? Clin Anat. 2016;29(1):11–8.
11. Štrkalj G. The emergence of humanistic anatomy. Med Teach. 2014;36(10):912–3.
12. Pawlina W, Drake RL. Authentic learning in anatomy: a primer on pragmatism. Anat Sci Educ. 2016;9(1):5–7.
13. McHanwell S, Brenner E, Chirculescu ARM, Drukker J, van Mameren H, Mazzotti G, et al. The legal and ethical framework governing body donation in Europe - a review of current practice and recommendations for good practice. Eur J Anat. 2008;12(1):1–24.
14. Kahn PA, Champney TH, Hildebrandt S. The incompatibility of the use of unclaimed bodies with ethical anatomical education in the United States. Anat Sci Educ. 2017;10(2):200–1.
15. Champney TH. A Bioethos for bodies: respecting a priceless resource. Anat Sci Educ. 2019;12(4):432–4.
16. Champney TH, Hildebrandt S, Jones DG, Winkelmann A. BODIES R US: ethical views on the commercialization of the dead in medical education and research. Anat Sci Educ. 2018;12(3):317–25.
17. Persaud TV, Loukas M, Tubbs RS. A history of human anatomy. 2nd ed. Springfield: Charles C. Thomas; 2014.
18. O'Malley CD. Andreas Vesalius of Brussels, 1514–1564. Berkeley: University of California Press; 1964.
19. Richardson R. Death, dissection and the destitute. Chicago: University of Chicago Press; 2000.
20. Lassek AM. Human dissection: its drama and struggle. Springfield: Charles C. Thomas; 1958.
21. Richardson R, Hurwitz B. Jeremy Bentham's self image: an exemplary bequest for dissection. Br Med J. 1987;295(6591):195.
22. Habicht JL, Kiessling C, Winkelmann A. Bodies for anatomy education in medical schools: an overview of the sources of cadavers worldwide. Acad Med. 2018;93(9):1293–300.
23. Gangata H, Ntaba P, Akol P, Louw G. The reliance on unclaimed cadavers for anatomical teaching by medical schools in Africa. Anat Sci Educ. 2010;3(4):174–83.
24. Biasutto SN, Sharma NA, McBride J, Krishnan S, Vatsalaswamy P, Garud RS, et al. Part II—human bodies to teach anatomy: importance and procurement—experience with cadaver donation. Rev Arg Anat Clin. 2014;6(3):162–75.
25. Kramer B, Hutchinson EF, Brits DM, Billings BK. Making the ethical transition in South Africa: acquiring human bodies for training in anatomy. Anat Sci Educ. 2019;12(3):264–71.
26. Anyanwu GE, Udemezue OO, Obikili EN. Dark age of sourcing cadavers in developing countries: a Nigerian survey. Clin Anat. 2011;24(7):831–6.
27. Biasutto SN, Sharma N, Weiglein AH, Benia FM, McBride J, Bueno-López JL, et al. Human bodies to teach anatomy: importance and procurement—experience with cadaver donation. Rev Arg Anat Clin. 2014;6(2):72–86.
28. Champney TH. The business of bodies: ethical perspectives on for-profit body donation companies. Clin Anat. 2016;29(1):25–9.
29. Kramer B, Hutchinson EF. Transformation of a cadaver population: analysis of a south African cadaver program, 1921–2013. Anat Sci Educ. 2015;8(5):445–51.
30. Anteby M, Garip F, Martorana PV, Lozanoff S. Individuals' decision to co-donate or donate alone: an archival study of married whole body donors in Hawaii. PLoS One. 2012;7(8):e42673.
31. Asad AL, Anteby M, Garip F. Who donates their bodies to science? The combined role of gender and migration status among California whole-body donors. Soc Sci Med. 2014;106:53–8.

32. Bin P, Delbon P, Piras M, Paternoster M, Di Lorenzo P, Conti A. Donation of the body for scientific purposes in Italy: ethical and medico-legal considerations. Open Med (Wars). 2016;11(1):316–20.
33. Bolt S. Dead bodies matter: gift giving and the unveiling of body donor monuments in the Netherlands. Med Anthropol Q. 2012;26(4):613–34.
34. Bolt S, Eisinga R, Altena M, Venbrux E, Gerrits PO. Over my dead body: body donation and the rise in donor registrations in the Netherlands. Omega (Westport). 2012;66(1):57–77.
35. Bolt S, Venbrux E, Eisinga R, Kuks JB, Veening JG, Gerrits PO. Motivation for body donation to science: more than an altruistic act. Ann Anat. 2010;192(2):70–4.
36. Boulware LE, Ratner LE, Cooper LA, LaVeist TA, Powe NR. Whole body donation for medical science: a population-based study. Clin Anat. 2004;17(7):570–7.
37. Cornwall J, Perry GF, Louw G, Stringer MD. Who donates their body to science? An international, multicenter, prospective study. Anat Sci Educ. 2012;5(4):208–16.
38. Cornwall J, Poppelwell Z, McManus R. "Why did you really do it?" A mixed-method analysis of the factors underpinning motivations to register as a body donor. Anat Sci Educ. 2018;11(6):623–31.
39. da Rocha AO, de Campos D, Farina MA, Pacini GS, Girotto MC, Hilbig A. Using body donor demographics to assist the implementation of donation programs in Brazil. Anat Sci Educ. 2017;10(5):475–86.
40. Gurses IA, Ertas A, Gurtekin B, Coskun O, Uzel M, Gayretli O, et al. Profile and motivations of registered whole-body donors in Turkey: Istanbul University experience. Anat Sci Educ. 2019;12(4):370–85.
41. Halou H, Chalkias A, Mystrioti D, Iacovidou N, Vasileiou PV, Xanthos T. Evaluation of the willingness for cadaveric donation in Greece: a population-based study. Anat Sci Educ. 2013;6(1):48–55.
42. Manicom C. A human perspective on body donation: a case study from a psychosocial perspective. S Afr Med J. 2015;105(9):719–20.
43. Rokade SA, Gaikawad AP. Body donation in India: social awareness, willingness, and associated factors. Anat Sci Educ. 2012;5(2):83–9.
44. Saha A, Sarkar A, Mandal S. Body donation after death: the mental setup of educated people. J Clin Diagn Res. 2015;9(6):AC05.
45. The World Bank. World Bank Country and Lending Groups. 2018. https://datahelpdesk.worldbank.org/knowledgebase/articles/906519-world-bank-country-and-lending-groups. Accessed 13 Mar 2019.
46. Wingfield HA. Body donation today: a critical comparison of two current practices, and moving into the future. Clin Anat. 2018;31(1):86–9.
47. Kinfu Y. Spotlight on teaching of anatomy in Ethiopia. Anat Sci Educ. 2008;1(5):230.
48. Chen D, Zhang Q, Deng J, Cai Y, Huang J, Li F, et al. A shortage of cadavers: the predicament of regional anatomy education in mainland China. Anat Sci Educ. 2018;11(4):397–402.
49. Alexander M, Marten M, Stewart E, Serafin S, Štrkalj G. Attitudes of Australian chiropractic students toward whole body donation: a cross-sectional study. Anat Sci Educ. 2014;7(2):117–23.
50. Bajor G, Likus W, Kuszewski P, Kostro K, Los A, Klakus P. "Mortui vivos docent" or who gives his body to science? The analysis of the personal questionnaires of Polish donors in the Conscious Body Donation Program. PLoS One. 2015;10(3):e0121061.
51. Subasinghe SK, Jones DG. Human body donation programs in Sri Lanka: Buddhist perspectives. Anat Sci Educ. 2015;8(5):484–9.
52. Techataweewan N, Panthongviriyakul C, Toomsan Y, Mothong W, Kanla P, Chaichun A, et al. Human body donation in Thailand: donors at Khon Kaen University. Ann Anat. 2018;216:142–51.
53. Jones DG, Nie JB. Does Confucianism allow for body donation? Anat Sci Educ. 2018;11(5):525–31.
54. Shang X, Zhang M. Body and organ donation in Wuhan. China Lancet. 2010;376(9746):1033–4.
55. Zhang L, Xiao M, Gu M, Zhang Y, Jin J, Ding J. An overview of the roles and responsibilities of Chinese medical colleges in body donation programs. Anat Sci Educ. 2014;7(4):312–20.
56. Cantor NL. After we die: the life and times of the human cadaver. Washington: Georgetown University Press; 2010.
57. Carmack HJ, DeGroot JM. Communication apprehension about death, religious group affiliation, and religiosity: predictors of organ and body donation decisions. Omega (Westport). 2020;81:627–47.
58. Rajasekhar SS, Aravindhan K, Gladwin V, Chand P. Body donation—consent from non-related persons: case series, review, and recommendations. J Clin Diagn Res. 2016;10(9):AR01–4.
59. Štrkalj G, Spocter MA, Wilkinson AT. Anatomy, medical education, and human ancestral variation. Anat Sci Educ. 2011;4(6):362–5.
60. Saw A. A new approach to body donation for medical education: the silent mentor programme. Malays Orthop J. 2018;12(2):68–72.
61. Santibañez S, Boudreaux D, Tseng G-F, Konkel K. The Tzu Chi silent mentor program: application of buddhist ethics to teach student physicians empathy, compassion, and self-sacrifice. J Relig Health. 2016;55(5):1483–94.
62. Garment A, Lederer S, Rogers N, Boult L. Let the dead teach the living: the rise of body bequeathal in 20th-century America. Acad Med. 2007;82(10):1000–5.
63. Collins AJ, Smith W, Giannaris EL, Orvek E, Lazar P, Carney JK, et al. Population representation among anatomical donors and the implication for medical student education. Clin Anat. 2018;31(2):250–8.
64. Saucier DA, Miller CT, Doucet N. Differences in helping whites and blacks: a meta-analysis. Personal Soc Psychol Rev. 2005;9(1):2–16.
65. Caplan I, DeCamp M. Of discomfort and disagreement: Unclaimed bodies in anatomy laboratories

at United States medical schools. Anat Sci Educ. 2019;12(4):360–9.

66. Marsland MJ, Lazarus MD. Ask an anatomist: identifying global trends, topics and themes of academic anatomists using twitter. Anat Sci Educ. 2018;11(3):270–81.
67. Sugand K, Abrahams P, Khurana A. The anatomy of anatomy: a review for its modernization. Anat Sci Educ. 2010;3(2):83–93.
68. Kostorrizos A, Koukakis A, Samolis A, Protogerou V, Mariolis-Sapsakos T, Piagkou M, et al. Body donation for research and teaching purposes: the contribution of blood donation units in the progress of anatomical science. Folia Morphol. 2019;78(3):575–81.
69. Abbasi Asl J, Nikzad H, Taherian A, Atlasi MA, Naderian H, Mousavi G, et al. Cultural acceptability and personal willingness of Iranian students toward cadaveric donation. Anat Sci Educ. 2017;10(2):120–6.
70. Evans DJR. The role of the anatomist in communicating anatomy to a lay audience. Eur J Anat. 2007;11(4):79–83.
71. da Rocha AOD, Bonatto-Costa JA, Pedron J, Moraes MPOD, Campos DD. The ceremony to honor the body donor as part of an anatomy outreach program in Brazil. In: Štrkalj G, Pather N, editors. Commemorations and memorials: exploring the face of human anatomy. Singapore: World Scientific; 2017. p. 157–72.
72. Richardson R, Hurwitz B. Donors' attitudes towards body donation for dissection. Lancet. 1995;346(8970):277–9.
73. da Rocha AOD, Tormes DA, Lehmann N, Schwab RS, Canto RT. The body donation program at the Federal University of Health Sciences of Porto Alegre: a successful experience in Brazil. Anat Sci Educ. 2013;6(3):199–204.
74. AbouHashem Y, Brown BT, Štrkalj G. Anatomy commemorations on youtube: a review. In: Štrkalj G, Pather N, editors. Commemorations and memorials: exploring the face of human anatomy. Singapore: World Scientific; 2017. p. 181–93.
75. Štrkalj G, Pather N, editors. Commemorations and memorials: exploring the human face of anatomy. Singapore: World Scientific; 2017.
76. Kozai T. Rituals for the dissected in pre-modern Japan. Nihon Ishigaku Zasshi. 2007;53(4):531–44.
77. Jiang J, Chen Q, Zhang M, Hong T, Huang K, Meng H, et al. Effects of commemorations and postdonation services on public willingness to donate bodies in China. Anat Sci Educ. 2020;13(2):218–29.
78. Talarico EF. A change in paradigm: giving back identity to donors in the anatomy laboratory. Clin Anat. 2013;26(2):161–72.
79. da Rocha AOD, Júnior MAF, Girotto MC, de Moraes MPO, Thomaz GR, de Campos D, et al. The Brazilian Ceremony in Honor of Body Donors: an opportunity to express gratitude and reflect on medical education. Int J Educ Res. 2018;6(2):264–73.
80. Park JT, Jang Y, Park MS, Pae C, Park J, Hu KS, et al. The trend of body donation for education based on Korean social and religious culture. Anat Sci Educ. 2011;4(1):33–8.
81. El-Haddad J. Students' attitudes towards commemorations and memorials in anatomy education. MRes thesis. Sydney: Macquarie University; 2018.
82. Dong S, Sun J, Long Y, Li X, Pi X, Xiao W, et al. Implementation of humanistic education for medical students based on body. J Chin Higher Med Educ. 2009;3:20–2.
83. Zhang L, Ding J. Commemorations and memorials in Chinese body donation programs. In: Štrkalj G, Pather N, editors. Commemorations and memorials: exploring the face of human anatomy. Singapore: World Scientific; 2017. p. 147–56.
84. Zhang L, Wang Y, Xiao M, Han Q, Ding J. An ethical solution to the challenges in teaching anatomy with dissection in the Chinese culture. Anat Sci Educ. 2008;1(2):56–9.
85. Arráez-Aybar L-A, Bueno-López JL, Moxham BJ. Anatomists' views on human body dissection and donation: an international survey. Ann of Anat. 2014;196(6):376–86.
86. Hallam E. The anatomy museum: death and the body displayed. London: Reaktion Books; 2016.

25 Preparing Students Emotionally for the Human Dissection Experience

Anja Böckers

Emotional Reactions of Medical Students to Dissection

For centuries, human dissection has been a well-established teaching method in the gross anatomy laboratory, and it calls for additional professional competencies such as team spirit, self-reflection, or "detached concern," which are also important in novice doctors' later medical practice [1–3]. For many years, these teaching objectives were represented only in the "hidden curriculum." Today's anatomy teaching is guided by the principles of humanism and ethical standards and puts professionalism and reflection in the gross anatomy laboratory into practice [4, 5].

> **Definition**
> The term "detached concern" was introduced by GE Dickinson (1997) and was used in a preclinical teaching context [2]. It describes the effort of medical professionals/students to "care" for the patient/body donor, but yet "not get too close."
>
> Detached concern prevents overly strong emotional reactions which might interfere with the best possible medical treatment and the learning process.

A. Böckers (✉)
Institute of Anatomy and Cell Biology, Ulm University, Ulm, Germany
e-mail: anja.boeckers@uni-ulm.de

In recent decades, teaching time has been markedly shortened in nearly every medical school [6]. As a result, students are subjected to increased stress with regard to learning and examinations. Additionally, students already undergo emotional stress in expectation of the dissection process. This stress might even resemble the symptoms of post-traumatic stress disorder (PTSD), causing somatic symptoms such as decrease in appetite, nausea, or sleeplessness [7, 8].

This emotional impact might have disruptive effects on the students' learning process and might trigger detachment as a coping strategy [9]. An inadequate work up of their emotions in this vulnerable phase might impair the development of medical competencies such as professional empathy for the students' first and future patients [10–12] or could contribute to future mental burnout [13, 14]. Therefore, anatomy staff members should be committed to reducing this mental distress in the anatomy laboratory as much as possible.

Fortunately, "the strongest reactions by medical students to dissection were in anticipation of it" [9, 15–17] (see Fig. 25.1). The nature of these reactions was summarized as follows: "For many, facing the cadaver for the first time elicits a wide range of emotions. These may include thoughts of their own mortality to the sheer admiration of

L. K. Chan, W. Pawlina (eds.), *Teaching Anatomy*, https://doi.org/10.1007/978-3-030-43283-6_25

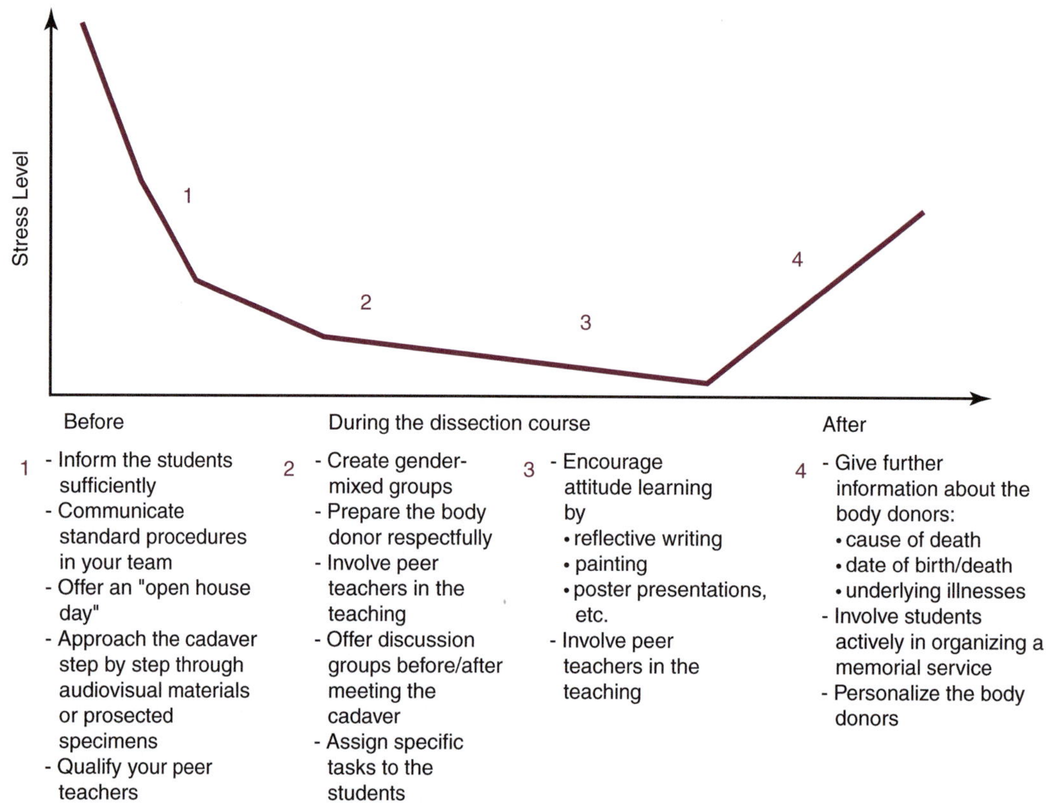

Fig. 25.1 Students' typical stress level before, during, and after the dissection course and an overview of suitable interventions to handle the students' emotional reactions

knowing that someone cared to help others learn about the body, even in death" [18]. Apart from negative emotions a lot of students experience dissection as positive and fascinating [19, 20]. Shortly after the first contact with the cadaver, a habituation process starts in most students, and students' fears reduce significantly.

These findings have been confirmed by many other research groups [21–24]. During the dissection course, students become more aware of mental stress as soon as they have to work on body parts which are intimate or express the human personality such as the face or hands or at times when the cadaver still appears intact [25, 26]. In due course of time, the fear of dissecting the cadaver gives way to professional curiosity, assessment, and occupational stress [19]. However, between 4% and 6% of the students experience difficulties adapting, which is expressed in the form of ongoing nightmares, poor appetite, sleeplessness, and learning difficulties [27]. It is only at the end of the assessment period that students once again focus on the role of the body donor, and this requires further guidance by staff members.

Different strategies were described regarding how students might handle mental distress during the dissection course. Without proper guidance to visualize a cadaver as a learning object and as an individual at the same time, faulty strategies might be learned. If students do not develop this professional ability of "caring for the body and yet not getting too close"—a concept Dickinson et al. [2] labeled "detached concern"—this could in the long run lead to burnout or non-empathetic treatment of patients [14, 24, 28, 29]. Other coping strategies are humor, name-giving, intellectualization, and the application of philosophic or religious attitudes [22, 30]. In addition, the skill of "detached concern" could even be a predictor for assessment results and state examinations [31]. Therefore, emotional distress and its coping strategies demand the faculty's attention.

Reasons Why Anatomists Should Care About Emotional Stress in the Dissection Laboratory

- Interferes negatively with the students' learning process
- Interferes negatively with the development of medical competencies
- Increases the risk of students' burnout
- Reduces students' willingness to donate their own body

Main Factors Causing Students' Emotional Reactions

No Previous Medical Training

Some investigations have shown that students without previous medical training have a higher need for psychological support [24] and take longer to form proper coping strategies than those students who had completed some sort of medical training before entering medical school. But the influence of a previous medical training is still controversial.

Previous Experience with Death and Dying

In general, at the start of dissection, about half of the students have never seen a cadaver before [16]. For these students, the first day of the course is particularly hard. Apart from not being acquainted with the sight of a cadaver, the emotional turmoil caused by a recent death in the family can be connected to strong emotional reactions [32].

While there is no obvious impact of age on the extent of mental distress, there seems to be a relationship between reporting anxiety and personality traits like extraversion or openness measured by the "Big Five" personality inventory [33]. Unfortunately, personality tests are not recommendable as screening instruments due to practical reasons such as anonymity and the considerable expense of performing these tests.

Sex Differences

With regard to the first contact with the body donor, women were shown to start at higher psychological distress levels than their male colleagues [16]. This may be explained by the fact that women have a high body esteem and think more frequently about their own mortality [34]. Subsequently, female students do not get used to the new situation in the dissection laboratory as quickly as male students, and they request introductory courses to get used to the dissection laboratory more often [23, 24, 32].

Strong Emotional Reactions are Likely to be Shown By

- Students with no previous medical training
- Students with recent death experiences in their social environment
- Female students

Recommended Interventions of Psychological Support

If questioned, students wish in particular for adequate preparation before their first contact with the cadaver and dissection—preferably in small groups such as their dissection teams on the first day of the course. Fear of death and additional stress due to dissection are reduced significantly if the students feel well prepared to enter the course [35]. Therefore, mental distress in the dissection course demands anatomists' attention. Practical core elements of an ethical anatomical education that also deal with students' emotions are summarized by Hildebrandt (2006) [4].

In advancing to the hands-on dissection, teachers should emphasize that dissection is a purposeful learning method to acquire the anatomical foundation for clinical practice and that this is in complete accordance with the donors' expectations [36]. Educators have to support the initial habituation process and assist students in developing the professional skill of "detached concern" and encourage students to reflect on their work and emotions.

Preparing students for the dissection experience needs to be implemented with consideration of the local curricular structure. The habituation process needs a preparatory period ahead of the course. Therefore, a dissection course starting in the students' second academic year or second half of their first year is advantageous, because the habituation process can start beforehand. In a modular structured curriculum, which is based on functional body systems and/or uses primarily prosected specimens, the body donor as a human being recedes from being the main focus. In this setting, students might show less emotional reactions, but, on the other hand, indispensable teaching objectives like professionalism and self-reflection are more difficult to address. Thus, the interventions listed below have to be checked in each individual case for their suitability.

Recommended Interventions Before the Dissection Course Starts

1. Create an atmosphere of trust and transparency by passing on comprehensive information to the students about the dissection course itself:
   - Body donation program: From our own experience, it is of particular importance for the students to be told that the cadavers were donated on a voluntary basis during their lifetime according to local regulation and following recommendations for good practice of the IFAA [37]. Body donations are mostly motivated by the donors' own positive experiences as patients during their lifetime and the wish to support young medical students in becoming good doctors. (See also Chap. 22).
   - Techniques of body preservation: Certainly, it would go far beyond the scope of comprehensive information to inform students beforehand about specific body preservation techniques. However, they should be informed about specific features in body appearance after death and preservation such as changes in consistency and color which they have to be aware of on the first day of the dissection course.
   - Counseling services: Inform about possible counseling services in your department, medical school, or university. Many students do not know that these services exist at all or how to contact them.
2. Remember to develop and communicate standard procedures to deal with students' emotional reactions to your staff members. Keep them informed about counseling services too and guide them to an understanding of a uniform role model *you* want them to represent for your students.
3. Offer an "open house day," which allows students to familiarize themselves with the premises, gross anatomy dissection, and learning facilities without yet being in contact with a cadaver.
4. Nowadays, it is easy to get in digital contact with future participants of the dissecting course before the course starts. Besides passing on information (e.g., via websites), it also initiates students' reflection on the expected dissection experience and prepares them for the possible emotional reactions. Students might be invited to express their fears and emotions in any manner they like on appropriate platforms (e.g., internet blogs or forums), thus helping them realize that their fellow students share the same feelings [4].
5. Another possibility for accelerating the students' habituation process is the integration of audiovisual material, which shows the dissection or prosection of human specimens. Audiovisual material might be presented on a single occasion, e.g., during anatomy lectures,

or as web-based presentations for personal usage. It was shown that realistic video presentations on working with human cadavers are able to reduce emotional reactions before the dissection course [38].

The use of audiovisual material should preferably be integrated into the curricular teaching concept. Within this scope, audiovisual material could be part of a preparatory manual for students to work through before actually starting with their first course session, thus utilizing already limited time more effectively. Audiovisual materials might illustrate the preparatory process and its necessary skills and show additional medical images, problem-based case reports, or image-based quizzes.

Perceiving the body donor as a "first patient" could represent important principles to keep the desired balance between the donor as a learning object and a once living being. In this context, students ask for more information about their body donors such as their previous medical history or their motivation to donate their body [39]. In the past, it could be shown that short documentary films illustrating the life and motivation of body donors are very well accepted by students as a preparatory measure to the dissection course. However, a very small number of students are threatened by selected film scenes. Therefore, students' participation in watching such a film with adjoining discussion should preferably be voluntary [40–42].

6. The habituation process for the new situation in the dissection course should allow the students to gradually approach the cadaver. A "step-by-step" approximation can occur through initial demonstrations of prosected specimens such as single organs, progressing to whole body parts, and culminating in the presentation of an intact dead human body. Similarly, the teaching method should be adjusted appropriately with lectures at the beginning, followed by an interactive learning process, and finally active dissection. Depending on their previous knowledge, students should get the chance to follow their natural curiosity to approach the cadaver at their own pace. Several projects adopting such approaches and their positive effects have been described in the past [43, 44].

Recommended Interventions on the First Day of the Dissection Course

1. Female students more frequently experience feelings of fear and disgust than men in expectation of the dissection course. Hence, mixed-sex dissection groups could be advantageous for the purpose of mutual support, a fact that should be considered when organizing the gross anatomy course.
2. It is useful to create a standard operating procedure for the first day in the anatomy laboratory with regard to the students' first confrontation with the body donor, in coordination with your colleagues. Staff members and peer teachers of each dissection group should be adequately briefed beforehand. Hence, preparing students emotionally for the dissection experience should be an explicit learning objective in peer teacher training.
3. The first contact with the body donor can be markedly eased if reverent and respectful preparation and handling of the cadaver is ensured. Students are less emotionally involved if the donor does not appear overly human. Accordingly, the donor's face and genital region should be covered, e.g., with towels. Similarly, emotional reactions are frequently enhanced at the sight of hairy skin regions, therefore requiring a thorough total body shaving of the donor.
4. The majority of students favor emotional preparation immediately on the first day of dissection in a small group setting with peer teachers as their trusted person with whom to share their fears and feelings [24]. This kind of small group setting might occur before and/or after the first contact with the cadaver, and—wherever applicable—this might be supported by audiovisual material (see above). Referring to our personal experiences, quite often, the students' anticipatory fears do not allow a reflective conversation beforehand, yet in some instances, a prolongation of this

tense situation could even increase emotional reactions. Hence, we favor a rather quick guided confrontation with the body donor and sufficient time afterwards for reflection and feedback about one's individual feelings looking back on the first contact with a cadaver.

5. Students might be emotionally relieved to be preoccupied with professional duties. Thus, it could be advisable to have the students perform a physical examination—just as if the donor was their first patient—and document the findings on an admission sheet. Looking at the cadaver in a professional manner diverts the focus from a holistic view toward isolated body parts, regions, or organs.
6. Finally, the practice of personalizing the donor by giving him/her a name offers benefits to medical students as an emotionally focused coping mechanism that takes place at the beginning of the dissection course and that should be guided. Naming seems to aid in increasing students' attention to the humanistic qualities of their cadavers [30].

Recommended Interventions on the First Day of Dissection

- Arrange into mixed-sex dissection groups.
- Create a standard operating procedure for the first day.
- At all times, ensure a respectful handling of the cadaver.
- Arrange a small group setting with peer teachers to reflect on one's fears and feelings before/after the first cadaver contact.
- Engage students with distinct tasks (e.g., admission sheet).

Recommended Interventions During the Dissection Course

Due to stress related to learning and assessments, 75% of the students do not want to participate in extensive programs of psychological support during the time of dissection [44, 45]. Apart from that, once they have distanced themselves from the human being they are dissecting, voluntary measures such as discussion groups are welcomed by the majority of students [46, 47]. Additional consultation with psychosocial services or contact with clergy members might be desirable especially for those 4–6% students who continue to experience emotional distress during their work in the anatomy laboratory.

Most of the students are able to handle their emotions within a short period of time through the mechanisms mentioned above. However, this process might be only short term and superficial, therefore requiring additional interventions. Therefore, the dissection course should offer scheduled time for reflection on the course. Medical humanities projects could be valuable in this process of encouraging students' "self-reflective learning" which in the long term promotes a physician's professional skill in "staying grounded" [47].

1. Students use talking to their peers as their main coping mechanism to overcome their fears about dissection. Implementing peer discussion early in the dissection course by, for example, questionnaires about their feelings concerning dissection is a valuable method of introducing students to an important coping mechanism [24, 46].
2. Previous research suggests that students might be more willing to communicate their feelings associated with death or the dissection experience through *written correspondence* rather than by oral communication. Therefore, memorable experiences are often communicated by reflective writing [18]. Humanistic learning tools such as paintings, vocal performances, writing poetry, interview-projects with the donors' family, and preparing posters or portfolios about the body donor are common instruments for promoting reflection and coping with the new environment of the dissection room [4, 48–53].

3. Peer teachers, specially trained and more experienced students acting as tutors, function as role models and trusted personnel in the dissection room. Therefore, students experience a lower threshold with regard to contacting their peer teacher for any kind of problem compared to contacting staff members. The effectiveness of these peer teaching concepts has been proven in the past [54, 55]. A ratio of one peer teacher assigned to one dissection group around one body donor appears ideal. However, the crucial factor is that qualified peer teachers need in-depth training before entering the dissection course.

Recommended Interventions After the Dissection Course

After course assessments have been completed, students experience a rapid decrease in learning and assessment pressure. Thus, the need to cope with emotions through distancing or depersonalization vanishes. The body donor ceases being only an learning object and turns back into a human being with an individual biography.

Despite the fact that a follow-up meeting might not be explicitly requested by students and that anatomy staff members are not their first choice to discuss their emotions, this is the right time for anatomists to act as role models and to demand students' participation [24]. Generally, at this point in time, the desire to obtain further personal information about the body donor increases. Students seek information such as the donor's cause of death, previous illnesses, and their life or family situation. Likewise, most donors encourage sharing of their information to enhance student education [39].

Anatomists should take the students' desire for a closer personal relationship with the body donor into account in order to help students to reflect on the emotions arising from their dissection experience:

1. Anatomists can meet the desire for information about the donor by disclosing information from the body donor's death certificate about age, date of death, cause of death, and underlying illnesses. An admission sheet documenting the findings before cadaver fixation might be added [15, 56, 57].
2. Most faculties in European or Anglo-American universities conduct a thanksgiving ceremony at the end of the dissection course [58]. This event is a ceremonious occasion of reflection and farewell for all parties concerned. The students "dismiss" their body donor, a person they have not known personally yet who has influenced not only their acquisition of anatomical knowledge but also their personal and professional advancement [59]. Most often, students, tutors, staff members, and relatives participate in this memorial service; students participate actively by expressing their deep gratitude through valuable music and oral contributions. Asian cultures refer to the cadavers as great teachers, and students respect them as highly valued teachers. In some schools, Buddhist ceremonies accompany the dissection course even on the first day in the laboratory, and body donors are personalized by reading their names aloud [60].
3. Nowadays, thanksgiving ceremonies is practiced in western medical schools to set up a personalized connection with the body donor and to possibly strengthen the students' empathic competence. In this context, a paradigm shift away from anonymizing the body donor toward its personalization is discussed as a meaningful measure for a respectful handling of the body donor [58, 61]. A personal relationship between the body donor and the student might be consolidated—among other things—by an informal meeting of the dissection group and family members of the donor or by video-documented interviews of the donor. Up to now, such a concept of de-anonymizing the donor has primarily been offered as a preparatory measure; however, this would also be applicable as a measure at the end of the dissection course to deal with any unaddressed emotional responses [39, 61].

Conclusion

There are many possibilities to structure the dissection course in order to meet the students' different desires to prepare them progressively for the confrontation with the body donor, to appropriately address the variety of feelings arising, and to help students perform the dissection course successfully from an anatomical and also emotional perspective. With the aid of the interventions mentioned here, the dissection course has an eminent potential to help students develop their professional attitudes and competencies.

References

1. Böckers A, Jerg-Bretzke L, Lamp C, Brinkmann A, Traue HC, Böckers TM. The gross anatomy course: an analysis of its importance. Anat Sci Educ. 2010;3(1):3–11.
2. Dickinson GE, Lancaster CJ, Winfield IC, Reece EF, Colthorpe CA. Detached concern and death anxiety of first-year medical students: before and after the gross anatomy course. Clin Anat. 1997;10(3):201–7.
3. Pawlina W. Clinical Anatomy: Professionalism and anatomy. Special issue. Clin Anat. 2006;19:391–479.
4. Hildebrandt S. Thoughts on practical core elements of an ethical anatomical education. Clin Anat. 2016;45:37–45.
5. Winkelmann A. Consent and consensus-ethical perspectives on obtaining bodies for anatomical dissection. Clin Anat. 2016;29(1):70–7.
6. Drake RL, McBride JM, Lachman N, Pawlina W. Medical education in the anatomical sciences: the winds of change continue to blow. Anat Sci Educ. 2009;2(6):253–9.
7. Lee YH, Lee YM, Kwon S, Park SH. Reactions of first-year medical students to cadaver dissection and their perception on learning methods in anatomy. Korean J Med Educ. 2001;23(4):275–83.
8. Getachew D. Reaction of medical students to experiences in dissection room. Ethiop J Health Sci. 2014;4(4):337–42.
9. Tseng WT, Lin YP. "Detached concern" of medical students in a cadaver dissection course: a phenomenological study. Anat Sci Educ. 2016;9:265–71.
10. Hojat M, Gonnella JS, Mangione S, Nasca TJ, Veloski JJ, Erdmann JB. Empathy in medical students as related to academic performance, clinical competence and gender. Med Educ. 2002;36(6):522–7.
11. Brazeau CM, Schroeder R, Rovi S, Boyd L. Relationships between medical student burnout, empathy, and professionalism climate. Acad Med. 2010;85(10 Suppl):S33–6.
12. Rosenthal S, Howard B, Schlussel YR, Herrigel D, Smolarz BG, Gable B. Humanism at heart: preserving empathy in third-year medical students. Acad Med. 2011;86(3):350–8.
13. Dyrbye LN, Harper W, Durning SJ, Moutier C, Thomas MR, Massie FS Jr. Patterns of distress in US medical students. Med Teach. 2011;33(10):834–9.
14. Thomas MR, Dyrbye LN, Huntington JL, Lawson KL, Novotny PJ, Sloan JA. How do distress and well-being relate to medical student empathy? A multicenter study. J Gen Intern Med. 2007;22(2):177–83.
15. Penney JC. Reactions of medical students to dissection. J Med Educ. 1985;60(1):58–60.
16. Romo Barrientos C, José Criado-Álvarez J, González-González J, Ubeda-Bañon I, Saiz-Sanchez D, Flores-Cuadrado A, et al. Anxiety among medical students when faced with the practice of anatomical dissection. Anat Sci Educ. 2019;12(3):300–9.
17. Bob MH, Popescu CA, Suciu ŞM, Buzoianu AD. First year medical students' attitude towards anatomical corpse dissection and its relationship with their personality. Romanian J Morphol Embryol. 2015;56(1):321–4.
18. Wagoner NE, Romero-O'Connell JM. Privileged learning. Anat Sci Educ. 2009;2(1):47–8.
19. McGarvey MA, Farrell T, Conroy RM, Kandiah S, Monkhouse WS. Dissection: a positive experience. Clin Anat. 2001;14(3):227–30.
20. Cahill KC, Ettarh RR. Attitudes to anatomy dissection in an Irish medical school. Clin Anat. 2009;22(3):336–91.
21. Dinsmore CE, Daugherty S, Zeitz HJ. Student responses to the gross anatomy laboratory in a medical curriculum. Clin Anat. 2001;14(3):231–6.
22. Snelling J, Sahai A, Ellis H. Attitudes of medical and dental students to dissection. Clin Anat. 2003;16(2):165–72.
23. Hanock D, Williams M, Taylor A, Dawson B. Impact of cadaver dissection on medical students. NZ J Psychol. 2004;33:17–25.
24. Boeckers A, Brinkmann A, Jerg-Bretzke L, Lamp C, Traue HC, Boeckers TM. How can we deal with mental distress in the dissection room?—an evaluation of the need for psychological support. Ann Anat. 2010;192(6):366–72.
25. Shalev A, Nathan H. Medical students' stress reactions to dissections. Isr J Psychiatry Relat Sci. 1985;22(1–2):121–33.
26. Leboulanger N. First cadaver dissection: stress, preparation, and emotional experience. Eur Ann Otorhinolaryngol Head Neck Dis. 2011;128(4):175–83.
27. Dempster M, Black A, McCorry N, Wilson D. Appraisal and consequences of cadaver dissection. Med Educ Online. 2006;11(1):4592.
28. Lempp HK. Perceptions of dissection by students in one medical school: beyond learning about anatomy. A qualitative study. Med Educ. 2005;39(3):318–25.
29. Robbins BD, Tomaka A, Innus C, Patterson J, Styn G. Lessons from the dead: the experiences of under-

graduates working with cadavers. Omega (Westport). 2008–2009;58(3):177–92.
30. Williams AD, Greenwald EE, Soricelli RL, DePace DM. Medical students' reactions to anatomic dissection and the phenomenon of cadaver naming. Anat Sci Educ. 2014;7(3):169–80.
31. Shapiro J, Nguyen VP, Mourra S, Boker JR, Ross M, Thai TM, et al. Relationship of creative projects in anatomy to medical student professionalism, test performance and stress: an exploratory study. BMC Med Educ. 2009;9:65.
32. Bernhardt V, Rothkotter HJ, Kasten E. Psychological stress in first year medical students in response to the dissection of a human corpse. GMS Z Med Ausbild. 2012;29(1):Doc12.
33. Plaisant O, Courtois R, Toussaint PJ, Mendelsohn GA, John OP, Delmas V. Medical students' attitudes toward the anatomy dissection room in relation to personality. Anat Sci Educ. 2011;4(6):305–10.
34. Hughes BM, Black A. Body esteem as a moderator of cardiovascular stress responses in anatomy students viewing cadaver dissections. J Psychosom Res. 2006;61(4):501–6.
35. Arraez-Aybar LA, Castano-Collado G, Casado-Morales MI. Dissection as a modulator of emotional attitudes and reactions of future health professionals. Med Educ. 2008;42(6):563–71.
36. Chang HJ, Kim HJ, Rhyu IJ, Lee YM, Uhm CS. Emotional experiences of medical students during cadaver dissection and the role of memorial ceremonies: a qualitative study. BMC Med Educ. 2018;18(1):255.
37. Federative International Committee for Humanities and Ethics. Recommendations of good practice for the donation and study of human bodies and tissues for anatomical examination. Plexus, Newsletter of the International Federation of Association of Anatomists. 4–6 Jan 2012. http://www.ifaa.net/recommendations/. Accessed 25 Mar 2019.
38. Casado MI, Castano G, Arraez-Aybar LA. Audiovisual material as educational innovation strategy to reduce anxiety response in students of human anatomy. Adv Health Sci Educ Theory Pract. 2012;17(3):431–40.
39. Bohl M, Bosch P, Hildebrandt S. Medical students' perception of the body donor as a "first patient" or "teacher": a pilot study. Anat Sci Educ. 2011;4(4):208–13.
40. Dosani F, Neuberger L. Anatomy and humanity: examining the effects of a short documentary film and first anatomy laboratory experience on medical students. Anat Sci Educ. 2016;9(1):28–39.
41. Cole TR. Still life: the humanity in anatomy (DVD). Brooklyn: Fanlight Productions; 2001.
42. Trotman P (director/producer). Donated to science (DVD). 2009. Dunedin, New Zealand: PRN Films.
43. Arráez-Aybar L, Casado-Morales MI, Castano-Collado G. Anxiety and dissection of the human cadaver: an unsolvable relationship? Anat Rec. 2004;279B:16–23.
44. Böckers A, Baader C, Fassnacht UK, Ochsner W, Böckers TM. Reduction of mental distress in the dissection course by introducing the body donor experience through anatomical demonstrations of organ systems. Anat Sci Educ. 2012;5(6):321–9.
45. Tschernig T, Schlaud M, Pabst R. Emotional reactions of medical students to dissecting human bodies: a conceptual approach and its evaluation. Anat Rec. 2000;261(1):11–3.
46. Kotzé SH, Mole CG. Talking about death: implementing peer discussion as a coping mechanism to overcome fears about dissection, death, and dying. Anat Sci Educ. 2013;6(3):191–8.
47. Shiozawa T, Griewatz J, Hirt B, Zipfel S, Lammerding-Koeppel M, Herrmann-Werner A. Development of a seminar on medical professionalism accompanying the dissection course. Ann Anat. 2016;208:208–11.
48. Hammer RR. An education that pierces what the knife cannot: a student perspective. Anat Sci Educ. 2010;3(3):151–3.
49. Hammer RR, Jones TW, Hussain FT, Bringe K, Harvey RE, Person-Rennell NH. Students as resurrectionists–a multimodal humanities project in anatomy putting ethics and professionalism in historical context. Anat Sci Educ. 2010;3(5):244–8.
50. Canby CA, Bush TA. Humanities in gross anatomy project: a novel humanistic learning tool at Des Moines University. Anat Sci Educ. 2010;3(2):94–6.
51. Rizzolo LJ. Human dissection: an approach to interweaving the traditional and humanistic goals of medical education. Anat Rec. 2002;269:242–8.
52. Bohl M, Holman A, Mueller DA, Gruppen LD, Hildebrandt S. The willed body donor interview project: medical student and donor expectations. Anat Sci Educ. 2013;6(2):90–100.
53. Dunbar RL, Nichols MD. Fostering empathy in undergraduate health science majors through the reconciliation of objectivity and subjectivity: an integrated approach. Anat Sci Educ. 2012;5(5):301–8.
54. Houwink AP, Kurup AN, Kollars JP, Kral Kollars CA, Carmichael SW, Pawlina W. Help of third-year medical students decreases first-year medical students' negative psychological reactions on the first day of gross anatomy dissection. Clin Anat. 2004;17(4):328–33.
55. Walser J, Horneffer A, Oechsner W, Huber-Lang M, Gerhardt-Szep S, Boeckeers A. Quantitative and qualitative analysis of student tutors as near-peer teachers in the gross anatomy course. Ann Anat. 2017;210:147–54.
56. Gerwer J, Gest TR. Survey of body donation programs in the US concerning the use of donor personal information with medical students. Clin Anat. 2017;30(4):445–9.
57. Hasselblatt F, Messerer DAC, Keis O, Böckers TM, Böckers A. Anonymous body or first patient? A status report and needs assessment regarding the personalization of donors in dissection courses in German, Austrian, and Swiss medical schools. Anat Sci Educ. 2018;11(3):282–93.
58. Jones TW, Lachman N, Pawlina W. Honoring our donors: a survey of memorial ceremonies in United States anatomy programs. Anat Sci Educ. 2013;7:219–23.

59. Elansary M, Goldberg B, Qian T, Rizzolo LJ. The 2008 anatomy ceremony: essays. Yale J Biol Med. 2009;82(1):37–40.
60. Winkelmann A, Guldner FH. Cadavers as teachers: the dissecting room experience in Thailand. BMJ. 2004;329(7480):1455–7.
61. Talarico EF Jr. A change in paradigm: giving back identity to donors in the anatomy laboratory. Clin Anat. 2013;26:161–72.

26 Promoting Active Learning in the Gross Anatomy Laboratory

Lap Ki Chan, Ronnie Homi Shroff, Jian Yang, and Tomasz Cecot

What Is Active Learning?

Although a concise definition of active learning is elusive, different scholars and researchers from a wide range of disciplines have attempted to define the term in different ways. One of the most widely cited definitions is that of Prince: "any instructional method that engages students in the learning process" [1]. Hung et al. posited that active learning is a process by which learners are "actively developing thinking/learning strategies and constantly formulating new ideas and refining them through their conversational exchanges with others" [2]. Bonwell and Eison considered that active learning has the following key attributes [3]:

- Learners doing more than just listening and are engaged in such activities as reading, writing, discussing, or solving problems
- Learners engaging in higher-order cognitive tasks such as problem-solving, knowledge application, and analysis
- Placing less emphasis on the transmission of information and more on the application of knowledge to solve complex problems

There is increasing evidence of the importance of active learning as an instructional approach to facilitate enhanced learner outcomes in terms of both academic performance and the creation of a more positive learning experience and a learner-centered environment. In a study that meta-analyzed 225 studies in undergraduate sciences, technology, engineering, and mathematics (STEM), Freeman et al. compared student scores and failure rates in traditional lecture-based courses and those that used active learning [4]. They demonstrated that in courses that used active learning, student performance was an average of 0.47 standard deviations higher than in those that used traditional lecturing. Moreover, a study conducted by Murdoch and Guy found that when active learning methods were applied in both small and large classes, students from the small classes scored significantly higher on the final examination than did students in the large classes [5].

L. K. Chan (✉)
Department of Biomedical Sciences, Faculty of Medicine, Macau University of Science and Technology, Macao Special Administrative Region, People's Republic of China
e-mail: lapkichan@gmail.com

R. H. Shroff
The Hong Kong Polytechnic University, Hong Kong Special Administrative Region, Hong Kong, People's Republic of China
e-mail: ronnie.shroff@polyu.edu.hk

J. Yang · T. Cecot
School of Biomedical Sciences, Li Ka Shing Faculty of Medicine, The University of Hong Kong, Hong Kong Special Administrative Region, Hong Kong, People's Republic of China
e-mail: jianyang@hku.hk; tscecot@hku.hk

L. K. Chan, W. Pawlina (eds.), *Teaching Anatomy*, https://doi.org/10.1007/978-3-030-43283-6_26

Within a classroom setting, a wide array of student activities is possible [6]. In the gross anatomy laboratory, learning typically occurs through a series of well-structured activities with some degree of student engagement. In this chapter, we aim to examine the various instructional methods that can be employed in the gross anatomy laboratory to facilitate active learning, thus taking full advantage of the small-group and highly interactive learning environment of the gross anatomy laboratory [7–9].

The Types of Active Learning Pedagogies

Active learning is an umbrella term that encapsulates a variety of distinct instructional methods. The proposed table below provides the classification of these approaches, each covering a subset of active learning, and their theoretical underpinnings.

Most, if not all, of these instructional methods are used in medical and health education, with different methods being favored at different stages. Problem-based learning [30, 31], team-based learning [32, 33], and case-based learning [34, 35] are commonly used in the first few years of medical and health programs. Selecting suitable active learning pedagogies for the clinical years and postgraduate training is much more complex, but lifelong learning and self-regulated learning have undeniable roles [36]. Interprofessional education (IPE) is certainly gaining wider acceptance [37]. It is an activity that "occurs when two or more professions learn with, from, and about each other to improve collaboration and the quality of care" [38] and is a combination of at least three active learning pedagogies: collaborative, discovery, and problem-based learning [39]. Although IPE is traditionally used for clinical training, it is increasingly being integrated in basic sciences, including anatomy [40].

In implementing active learning, it cannot be assumed that both teachers and students will naturally embrace it. In fact, both groups may display resistance. A culture of active learning [41] is something that needs deliberate effort to develop. The reasons for teachers' resistance [3] include limited class time, increased preparation time, large classes, the lack of resources, anxiety over loss of control, worry about the lack of necessary skills, and being criticized for unorthodox teaching. But many of these obstacles can be overcome by careful design and planning. Students' resistance to active learning can also be lessened by (1) raising their expectation that they will be able to successfully take part in active learning (e.g., by pitching the activities at the appropriate student level, by consistently using active learning so that students become familiar with how it works), (2) demonstrating the value of active learning (e.g., by establishing explicit links to students' future clinical careers or by creating enjoyable active learning experience), and (3) convincing them that the cost of engaging in active learning is low (e.g., students feel acceptable to make mistakes in engaging in various learning activities if the teacher has established a safe exploratory learning environment) [42].

Examples of Active Learning in the Gross Anatomy Laboratory

Problem-Oriented Dissection (POD)

In traditional cadaveric dissection, students follow step-by-step instructions in a dissection manual. There have been many reports on introducing active learning strategies in dissection classes to better utilize the precious curricular time in the laboratory, with evidence of improved student learning [43–46]. At The University of Hong Kong, the strategy employed is called problem-oriented dissection (POD), in which students need to apply their anatomical knowledge to devise a clinical procedure for a clinical problem artificially created on the cadavers that they are going to dissect, thus stimulating students to integrate anatomy and clinical medicine, and to apply their basic anatomical knowledge in an authentic, and yet safe, environment. Researchers [47–49] pointed out that complex authentic problems may stimulate reflection and integration of the new into the previous learning. Therefore, POD targets the higher levels of the revised Bloom's taxonomy: comprehension, application, analysis, synthesis, and evaluation. POD encompasses a number of instructional methods described in Table 26.1 and therefore has a rich theoretical underpinning.

The essential features of POD include the following:

(i) Student knowledge: The students should have the basic knowledge of the anatomy of the region concerned.
(ii) Small student groups of preferably less than 10 students: Each group of students will collaborate in their learning.
(iii) Clinical case: The case gives the history of a patient for whom a surgical procedure is required, e.g., tracheostomy, joint aspiration, surgical approach to a tumor or fracture, and insertion of catheter. The cadavers on which the students are going to perform the procedure have been prepared to simulate the pathologies described in the clinical case. For example, in the case of septic arthritis, color fluid has been injected into the wrist before the dissection session, so that if the students successfully enter the wrist joint, they will be able to aspirate colored fluid.
(iv) Problem-solving: Students need to work together to apply their basic anatomical knowledge to devise the appropriate way of performing the procedure in the case, to achieve the aims of the procedure without causing unnecessary damage to adjacent structures. The students then perform their devised procedure on the group's cadaver.
(v) Dissection: After the students have performed their devised procedure, they dissect to find out the results of their procedure, e.g., after an aspiration, whether the needle reached the join space and if there are unnecessary damages to adjacent structures. Students then reflect on their self-devised procedure.
(vi) Self-directed learning: After the dissection, the students look up the recommended procedure(s) in the literature and compare theirs to the recommended one(s).
(vii) Sharing: Each group of students presents their devised procedure, the rationale behind it, the results, and how it compares to the recommended one(s). Students evaluate not only their own performance but also those of the other groups and learn from one another. If the number of students is large or if there is not enough time, student groups could report their results on a common platform (a whiteboard or an e-learning platform), and the teacher could then summarize and highlight the learning outcomes.

Although POD is an active learning strategy in which students play a more active role, teachers still have important, though different, roles. Designing the appropriate case is much more demanding than simply asking the students to follow the dissection manual. In addition to helping students to identify the relevant anatomical structures during dissection, teachers also need to help students in evaluating the clinical procedures they have devised, training their problem-solving skills, monitoring the sharing of their learning experience, etc.

Example of a POD Case: Wrist Swelling

A 90-year-old woman who lives in an old age home is admitted to your unit. She has a history of diabetes, hypertension, and stroke and is bedridden. The nurses at the old age home report that she has a fever, is lethargic, and has a very poor appetite. Her right wrist is very red, hot, and swollen. She becomes very irritable when you touch the wrist.

You suspect that she may have septic arthritis, and you want to perform an aspiration of the wrist. Based on your anatomy knowledge, devise an appropriate approach for the aspiration and then perform the aspiration using your suggested approach, followed by a dissection of the wrist to find out what structures the needle has gone through.

Table 26.1 Classification of the different instructional methods of active learning and their theoretical underpinnings

Instructional method	Definition of instructional method	Focus	Theoretical underpinning
Authentic learning	Learners apply knowledge in real-life contexts and situations through problem-solving activities	Extracting meaning and applying knowledge to real-world contexts, issues, and problems Actively participating by considering multiple forms of evidence, weighing ideas, or investigating contradictions	Constructivist learning theory [10, 11] Social constructivism theory [12]
Case-based learning	Learners use case studies within the context of authentic or real-world situations	Examining specific cases of real-world problems or situations Analyzing scenarios that imitate real-life situations	Constructivist learning theory [10, 11] Social constructivism theory [12]
Collaborative learning	Learners actively participate and interact in groups. Knowledge emerges from active dialogue and group work	Autonomous learners participating actively Learner–learner interaction Engaging in a shared learning community Joint construction of meanings Tackling complex problems	Social constructivism theory [12] Socio-cognitive conflict theory [13] Situated cognition [14] Distributed cognition theory [15]
Cooperative learning	Learners work together in groups, exchanging information among themselves and learning from one another, through group-based learning tasks	Engaging with peers Individual and group accountability Learner–learner interaction Selecting/absorbing information Generating ideas and solving problems together through discussion and reflective thought	Social constructivism theory [12] Social interdependence [16] Cognitive development theory [17] Achievement motivation theory [18] Social learning theory [19]
Discovery learning	Learners interact with their environment through investigation and exploration of authentic problems and contexts	Drawing on knowledge, experience, and insights to discover facts, concepts, and connections (i.e., guided discovery) Making connections between concepts and real-world applications Incorporating new information and making connections	Constructivist learning theory [10, 11] Cognitive development theory [17] Social constructivism theory [12]

(continued)

Table 26.1 (continued)

Instructional method	Definition of instructional method	Focus	Theoretical underpinning
Experiential learning	Learners apply their knowledge and conceptual understanding to real-world problems and/or situations	Actively participating and reflecting on experiences Extracting meaning and applying it to real-world problems Making connections between the course material and applying it outside of the classroom	Experiential learning theory [20]; Constructivist learning theory [10, 11] Social constructivism theory [12]
Inquiry-based learning	Learners acquire and construct knowledge through the process of inquiry and discovery	Posing questions, problems, or scenarios Exploring topics, drawing inferences, making connections, and asking questions	Experiential learning theory [20]; Social learning theory [19] Social constructivism theory [12]
Peer-assisted learning	Learners engage with the support of other learners, to help one another to learn more effectively and improve their knowledge in the process	Learners assisting other learners with conceptual understanding and problem-solving Peer tutoring and scaffolding from a more competent peer	Social constructivism theory [12] Cognitive congruence theory [21] Social learning theory [19]
Problem-based learning	Learners work together to apply their knowledge and skills to solve a presented problem through engagement in the problem-solving process	Facilitated problem-solving in which learners work in discussion groups to solve real-world practical problems Applying knowledge to solve the problem and reflecting on what they have learned	Constructivist learning theory [10, 11] Social constructivism theory [12] Theory of multiple intelligences [22]
Self-directed learning	Learners take initiative and assume personal responsibility and autonomous ability to manage their learning process	Taking responsibility for the learning context Taking control of their own learning, their learning goals, and deciding on which learning methods to use and how to evaluate their progress	Metacognition theory [23] Social cognitive theory [24] Constructivist learning theory [10, 11]
Self-regulated learning	Learners apply the necessary strategies to regulate their cognition and exercise control over their learning goals and behavior	Taking control of and responsibility for the learning process Optimizing their learning by planning, monitoring, and evaluating their learning process Emphasizing task mastery and autonomous learning	Self-regulated learning theory [25] Metacognition theory [23] Social cognitive theory [19] Constructivist learning theory [10, 11]

(continued)

Table 26.1 (continued)

Instructional method	Definition of instructional method	Focus	Theoretical underpinning
Situated learning	Learners engage in authentic tasks that take place in real-world contexts	Reflecting on and drawing implications from previous experiences Applying knowledge to real-life problems and complexities in a systematic way	Situated cognition [14] Social constructivism theory [12] Theory of inquiry [26]
Task-based learning	Learners achieve the learning outcomes through performing tasks in solving real-life problems	Working through task-specific learning content, focusing primarily on mastery of tasks	Achievement goal theory [27, 28] Self-determination theory [29] Constructivist learning theory [10, 11]
Team-based learning	Learners engage and interact with each other in class-based teamwork, emphasizing collaboration among group members	Individual and team responsiveness, group engagement, and interaction A "flipped" approach to learning Higher-order cognitive reasoning and thinking skills such as problem-solving, synthesis, application, and evaluation	Social constructivism theory [12] Socio-cognitive conflict theory [13] Situated cognition [14] Distributed cognition theory [15]

Dissection Peer-Support System (DPSS)

Peer and near-peer teaching programs are commonly employed active learning strategies in the gross anatomy laboratory, and numerous reports have documented their benefits to the learners [50–54]. Peer teaching in gross anatomy dissection classes usually involves students taking turn to adopt the roles of tutor and learner. At The University of Hong Kong, a digital system for peer teaching (dissection peer-support system) has been developed so that students can be tutors and learners at the same time in a class.

The dissection peer-support system (DPSS) is a cloud-based mobile platform with multiple functions. The hardware consists of tablet computers, one for each dissection group, connected by Wi-Fi to a secure server in a private local area network (LAN). The iClass Learning Management System (LMS), which was originally developed at The University of Hong Kong to enable interactivity in lectures, was adopted to implement the DPSS on the tablets. Within iClass, teachers create a list of tasks for each dissection session, including mostly the dissection and identification of key anatomical structures but also some clinical problems. When a group of students have successfully identified one of those structures or solved the clinical problems, they can use the high-resolution camera on the tablet to produce a short video (720p or 1080p, up to 60 seconds long) on how they did it. The videos are then uploaded onto the server. All groups can watch, mutually rate, and comment on these videos produced by their peers on their tablets, thus supporting one another in learning to dissect. However, the production, use, and storage of such videos must conform to the local legal regulations and ethical standards.

In producing the videos for the DPSS, students need to concisely describe and demonstrate the relevant anatomy, i.e., they need to teach their peers. This peer teaching increases the students' autonomy in active learning. The DPSS allows teachers to spend more time on guiding students to identify difficult structures, providing more feedback, and clarifying difficult concepts. It also provides concrete evidence (i.e., the videos) on the performance of all the groups. Students' in-app activities (such as view count, peer rating, and peer interactions) provide additional data on their learning. Online assessment and survey can be easily incorporated in the iClass for educational research purposes, such as the Technology-Enabled Active Learning (TEAL), for measuring students' perceptions of active learning in a technology-enabled learning context [55].

Task-Oriented Practical (TOP)

Prosected and plastinated specimens are commonly used in the gross anatomy laboratory in many schools [56]. There are different ways of organizing practical classes using these specimens [57]. In the "loosely-guided, self-directed" instructional method, specimens are spread randomly around the laboratory, and students are free to decide on their own pace and order of studying them. In the "strictly-guided, station-based" instructional method, specimens are organized into stations. Students group into teams and are given a limited time to study each station before they need to move on to the next, a procedure that is similar to an objective structured clinical examination (OSCE). Kooloos and his colleagues [57] have shown that students' factual recall is better after "strictly-guided, station-based" than after "loosely-guided, self-directed." At The University of Hong Kong, we adopt something in between, with learning outcomes higher than the knowledge level in Bloom's taxonomy (i.e., factual recall). Although developing students' ability for self-directed learning is important, we believe that some guidance, if carefully designed, can greatly aid students' learning in prosection classes, especially in an undergraduate-entry medical program such as the one in our universities.

In a task-oriented practical (TOP), which is based on the task-based learning approach [58], the learning materials are organized into stations. These materials can include prosected and plastinated specimens, bones, radiological images (e.g., X-ray, MRI), physical models, two-dimensional images (e.g., anatomical drawings, photos), and digital materials (e.g., videos and three-dimensional digital models).

At each station, students need to complete a task based on the learning materials there. The tasks vary from relatively easy (such as identification), at the level of "knowledge" in the cognitive domain in Bloom's taxonomy, to much more complex and demanding (such as comparing, problem-solving, and procedure simulation), at higher levels in the cognitive domain and even the psychomotor domain in Bloom's taxonomy [58]. Students should discuss how to complete the tasks in their groups, thus developing their communication skill [59]. The role of the teacher is to stimulate students to think and apply what they know to complete the tasks. Therefore, the teachers should either provide short explanations only when needed [60] or adopt the one-minute preceptor technique discussed below [61].

Students should preferably work in small groups of 3–8 people [62] to tackle the stations at their own pace and decide for themselves the order in which they will study the stations. The sequence of stations is not important as each constitutes a single, closed unit focusing on one topic, concept, or problem, which can be completed independent of the other stations. But when there are many students, a more structured approach, by adopting an OSCE structure, can ensure that every student has equal exposure to all of the stations.

Types of Tasks

Discovery: e.g., dissect the heart and find the atrioventricular valves.

Identifying: e.g., identify the three branches of the aortic arch on the specimen.

Ordering and sorting: e.g., trace the pathway of motor impulse to the biceps brachii muscle on the model.

Comparing: e.g., compare the structure of the artery and vein in the slide.

Problem-solving: e.g., how would you approach the pericardial cavity in pericardiocentesis?

Procedure simulation: e.g., intubate the mannequin with the endotracheal (ET) tube, and name the consecutive spaces passed by the tube.

Teacher–Student Interaction: One-Minute Preceptor (OMP)

Across the active learning instructional methods in Table 26.1, there exists a wide spectrum of learner engagement, relative to that of the teacher. Self-directed learning and self-regulated learning occupy one end of the spectrum, where there are only learner activities but no teacher activity. At the other end of the spectrum, some active learning strategies involve much teacher involvement: In POD, the teacher guides the students to evaluate their devised procedures based on the dissection results. Between these two ends of the spectrum, learner activities are more prominent than those of teachers, as in the students-teaching-students process in DPSS. Nevertheless, the role of the teacher in active learning is no less important (except perhaps in self-directed and self-regulated learning). What has changed is the nature of the teacher's involvement: from being a "sage on the stage" to being a "guide on the side" [63].

For a teacher to be a "guide on the side", apart from using the instructional methods in Table 26.1 for organizing learning activities, the teacher in interacting with students should also encourage the students to get engaged and to think and apply their knowledge. A useful model for such teacher behaviors is the one-minute preceptor (OMP), which was originally designed for a busy ambulatory clinical setting, where preceptorial encounters need to be time efficient in order to minimize disruption to clinical service while being effective for the learners [64, 65]. The OMP has been adapted for use in the gross anatomy laboratory, where a teacher oftentimes needs to supervise several groups of students and teaching encounters may also be brief [61, 66, 67].

The Five Microskills of the One-Minute Preceptor

The OMP consists of five microskills, which are not necessarily carried out in the following sequence, since it is a "pliable set of guidelines that can be shuffled and altered as the ever-changing teaching situation warrants" [68]:

(i) Getting a commitment: The teacher needs to respond to the students' question by first inviting the students to reflect on what they already know. The students need to try to answer the question themselves first.
(ii) Probing for supporting evidence: The teacher then further invites the students to articulate the reasons behind the answer, from which the teacher can evaluate how much the students know and how they have applied their prior knowledge to answer the question.
(iii) Reinforcing what was done right: It is important that the teacher gives positive and specific comments so as to indicate to the students the behaviors and competencies that are regarded as desirable and to create a safe environment for students to explore.
(iv) Correcting errors and filling in omissions: The teacher nevertheless also needs to point out the mistakes and omissions in the students' answers, if there are any, in objective, nonjudgmental language.
(v) Teaching a general rule: The teacher should help the students to generalize and transfer what students learn in a specific case or problem to other situations.

The Advantages of the One-Minute Preceptor

The OMP approach is useful in the gross anatomy laboratory because of the following features:

(i) Flexible application: The OMP can be used in virtually any teaching encounter, including the traditional instructional dissection, the problem-oriented dissection, or the task-oriented practical described above.
(ii) Explicit structure: This is useful especially for novice teachers to make sure that their interactions with students can be educationally effective [44]. It provides a framework for them to evaluate their teaching encounters with students and to improve their future teaching.

(iii) Learner-centered environment: The OMP encourages the teacher to adopt the role of a "guide on the side" instead of a "sage on the stage."
(iv) Rich in feedback: The latter three steps of the OMP are feedback that are specific to individual students and questions.
(v) Stimulate reflection: By getting the students' commitment and the reasons behind their answer, the teacher fosters students' ownership of the problem, thus inviting them to reflect on their prior knowledge to think of an answer as best as they can.
(vi) Enhanced teacher immediacy: With appropriate voice, gestures, eye contact, etc., the perceived distance between teacher and students can be reduced, increasing teacher immediacy [69].
(vii) Prepare students for learning in later clinical years: The use of the OMP approach in the early years of the medical curriculum may help students to get used to this type of teacher–student interaction, which they may encounter in their later clinical years.

The Five Microskills of One-Minute Preceptor (OMP)

1. Getting a commitment from the students
2. Probing for supporting evidence
3. Reinforcing what was done right
4. Correcting errors and filling in omissions
5. Teaching a general rule

Summary and Conclusions

Active learning aims to engage learners in higher-order cognitive tasks, such as problem-solving, knowledge application, analysis, synthesis, and evaluation. It is an umbrella term encompassing a wide range of instructional methods with a rich theoretical underpinning. The gross anatomy laboratory, a small-group, highly interactive learning environment, offers rich opportunities for implementing such methods. The few examples discussed are as follows: (1) dissection classes triggered by clinical problems artificially created on the cadavers, to stimulate knowledge application and problem-solving; (2) a cloud-based mobile platform for peer teaching during dissection classes; and (3) a task-based prosection class, wherein students need to complete tasks based on learning materials organized into stations. To further promote student engagement, teachers can adopt the framework of the one-minute preceptor in interacting with students to foster student ownership of a problem and to provide appropriate feedback, even if the preceptorial encounters are brief.

Acknowledgments Parts of the work described in this chapter is supported by the Technology-Enriched Learning Initiative and the iClass team of The University of Hong Kong and also by the project entitled "Developing Active Learning Pedagogies and Mobile Applications in University STEM Education" (PolyU2/T&L/16-19), which is funded by the University Grants Committee of the Hong Kong Special Administrative Region with additional support from the Hong Kong Polytechnic University.

References

1. Prince M. Does active learning work? A review of the research. J Eng Educ. 2004;93(3):223–31.
2. Hung D, Tan SC, Koh TS. Engaged learning: making learning an authentic experience. In: Hung D, Khine MS, editors. Engaged learning with emerging technologies. Dordrecht: Springer; 2006. p. 29–48.
3. Bonwell C, Eison J. Active learning: creating excitement in the classroom. ASHE-ERIC higher education report No. 1. Washington, DC: The George Washington University, School of Education and Human Development; 1991.
4. Freeman S, Eddy SL, McDonough M, Smith MK, Okoroafor N, Jordt H, Wenderoth MP. Active learning increases student performance in science, engineering, and mathematics. Proc Natl Acad Sci U S A. 2014;111(23):8410–5.
5. Murdoch B, Guy PW. Active learning in small and large classes. Acc Educ. 2002;11(3):271–82.
6. Hess GF. Principle 3 good practice encourages active learning. J Legal Educ. 1999;49(3):401–17.
7. Drake R. Anatomy education in a changing medical curriculum. Anat Rec. 1998;253:28–31.
8. Miller SA, Perrotti W, Silverthorn DU, Dalley AF, Rarey KE. From college to clinic: reasoning over memorization is key for understanding anatomy. Anat Rec. 2002;269:69–80.

9. McLachlan JC, Patten D. Anatomy teaching: ghosts of the past, present and future. Med Educ. 2006;40:243–53.
10. Bruner JS. The act of discovery. Harvard Educ Rev. 1961;31:21–32.
11. Piaget J. The child's conception of the world. Littlefield, Adams and Company: Patterson; 1963.
12. Vygotsky LS. Mind in society: the development of higher psychological processes. Cambridge, MA: Harvard University Press; 1980.
13. Doise W, Mugny G, Perret-Clermont AN. Social interaction and the development of cognitive operations. Eur J Soc Psychol. 1975;5(3):367–83.
14. Lave J. Cognition in practice: mind, mathematics and culture in everyday life. New York: Cambridge University Press; 1988.
15. Salomon G. Distributed cognitions: psychological and educational considerations. Cambridge: Cambridge University Press; 1997.
16. Johnson DW, Johnson RT. An educational psychology success story: social interdependence theory and cooperative learning. Educ Res. 2009;38(5):365–79.
17. Piaget J. Part I: cognitive development in children: Piaget development and learning. J Res Sci Teach. 1964;2(3):176–86.
18. Atkinson J. An introduction to motivation. Princeton: Van Nostrand; 1964.
19. Bandura A. Social learning theory of aggression. J Commun. 1978;28(3):12–29.
20. Kolb D. Towards an applied theory of experiential learning. London: John Wiley; 1975.
21. Cornwall M. Students as teachers: peer teaching in higher education. Amsterdam: University of Amsterdam; 1979.
22. Gardner H. Developing the spectrum of human intelligences. Harvard Educ Rev. 1987;57(2):187–93.
23. Flavell JH. Metacognitive aspects of problem solving. In: Resnick LR, editor. The nature of intelligence. Hillsdale: Erlbaum; 1976. p. 231–5.
24. Bandura A. Self-efficacy: toward a unifying theory of behavioral change. Psychol Rev. 1977;84(2):191–215.
25. Zimmerman J. A social cognitive view of self-regulated academic learning. J Educ Psychol. 1989;81(3):329–39.
26. Dewey J. Experience and education. New York: Touchstone; 1938.
27. Dweck CS. Motivational processes affecting learning. Am Psychol. 1986;41(10):1040–8.
28. Locke EA, Latham GP. A theory of goal setting and task performance. Englewood Cliffs: Prentice-Hall; 1990.
29. Ryan RM, Deci EL. Self-determination theory and the facilitation of intrinsic motivation, social development, and Well-being. Am Psychol. 2000;55(1):68–78.
30. Barrows HS. Problem-based learning applied to medical education. Springfield: Southern Illinois University Press; 2000.
31. Bergman EM. Teaching and learning anatomy in a PBL curriculum. In: Chan LK, Pawlina W, editors. Teaching anatomy: a practical guide. Cham: Springer; 2015.
32. Michaelsen LK, Sweet M. The essential elements of team-based learning. New Dir Teach Learn. 2008;116:7–27.
33. Melovitz-Vasan C, Pinhal-Enfield G, DeFouw DO, Vasan N. Team-based learning: an effective pedagogical strategy to teach anatomy. In: Chan LK, Pawlina W, editors. Teaching anatomy: a practical guide. Cham: Springer; 2015.
34. McLean SF. Case-based learning and its application in medical and health-care fields: a review of worldwide literature. J Med Educ Curric Dev. 2016;3:JMECD.S20377.
35. Thistlethwaite JE. Learning and teaching anatomy through case-based learning. In: Chan LK, Pawlina W, editors. Teaching anatomy: a practical guide. Cham: Springer; 2015.
36. Berkhout JJ, Helmich E, Teunissen PW, van der Vleuten CPM, Jaarsma ADC. Context matters when striving to promote active and lifelong learning in medical education. Med Educ. 2018;52(1):34–44.
37. Thistlethwaite JE. Interprofessional education: implications and development for medical education. Educación Médica. 2015;16(1):68–73.
38. Barr H, Low H. Introducing interprofessional education. Centre for the Advancement of Interprofessional Education: Fareham; 2013.
39. O'Halloran C, Hean S, Humphris D, Macleod-Clark J. Developing common learning: the new generation project undergraduate curriculum model. J Interprof Care. 2006;20(1):12–28.
40. Pawlina W, Drake R. Interprofessional education: first steps. Anat Sci Educ. 2015;8(4):289–90.
41. Lachman N, Pawlina W. Drawing from a culture of active learning: student-initiated peer education. Clin Anat. 2016;29:428–9.
42. Cooper KM, Ashley M, Brownell SE. Using expectancy value theory as a framework to reduce student resistance to active learning: a proof of concept. J Microbiol Biol Educ. 2017;18(2):18.2.32.
43. Rizzolo LJ, Stewart WB, O'Brien M, Haims A, Rando W, Abrahams J, et al. Design principles for developing an efficient clinical anatomy course. Med Teach. 2006;28:142–51.
44. Rizzolo LJ, Rando WC, O'Brien MK, Haims AH, Abrahams JJ, Stewart WB. Design, implementation, and evaluation of an innovative anatomy course. Anat Sci Educ. 2010;3(3):109–20.
45. Huitt TW, Killins A, Brooks WS. Team-based learning in the gross anatomy laboratory improves academic performance and students' attitudes toward teamwork. Anat Sci Educ. 2015;8(2):95–103.
46. Fleagle TR, Borcherding NC, Harris J, Hoffmann DS. Application of flipped classroom pedagogy to the human gross anatomy laboratory: student preferences and learning outcomes. Anat Sci Educ. 2018;11(4):385–96.
47. Mamede S, Schmidt HG. The structure of reflective practice in medicine. Med Educ. 2004;38:1302–6.
48. Mann K, Gordon J, MacLeod A. Reflection and reflective practice in health professions education:

a systematic review. Adv in Health Sci Educ. 2009;14:595–621.
49. Moon J. Reflection in learning and professional development. London: Kogan Page; 1999.
50. Evans DJR, Cuffe T. Near-peer teaching in anatomy: an approach for deeper learning. Anat Sci Educ. 2009;2(5):227–33.
51. Duran CEP, Bahena EN, Rodriguez MDG, Baca GJ, Uresti AS, Elizondo-Omana RE, et al. Near-peer teaching in an anatomy course with a low faculty-to-student ratio. Anat Sci Educ. 2012;5(3):171–6.
52. Alvarez S, Schultz JH. Professional and personal competency development in near-peer tutors of gross anatomy: a longitudinal mixed-methods study. Anat Sci Educ. 2019;12(2):129–37.
53. Han ER, Chung EK, Nam KI. Peer-assisted learning in a gross anatomy dissection course. PLoS One. 2015;10(11):e0142988.
54. Manyama M, Stafford R, Mazyala E, Lukanima A, Magele N, Kidenya BR, et al. Improving gross anatomy learning using reciprocal peer teaching. BMC Med Educ. 2016;16:95.
55. Shroff RH, Ting F, Lam WH. Development and validation of an instrument to measure students' perceptions of technology–enabled active learning. Aust J Educ Technol. 2019;35(4):109–27.
56. Cornwall J. The diverse utility of wet prosections and plastinated specimens in teaching gross anatomy in New Zealand. Anat Sci Educ. 2011;4:269–74.
57. Kooloos JG, de Waal Malefijt MC, Ruiter DJ, Vorstenbosch MA. Loosely-guided, self-directed learning versus strictly-guided, station-based learning in gross anatomy laboratory sessions. Anat Sci Educ. 2012;5(6):340–6.
58. Willis J. A framework for task-based learning. Longman handbooks for language teachers. London: Longman; 1996.
59. Channon SB, Davis RC, Goode NT, May SA. What makes a 'good group'? Exploring the characteristics and performance of undergraduate student groups. Adv Health Sci Educ Theory Pract. 2017;22(1):17–41.
60. Phillips NM, Duke MM, Weerasuriya R. Questioning skills of clinical facilitators supporting undergraduate nursing students. J Clin Nurs. 2017;26(23–24):4344–52.
61. Chan LK, Wiseman J. The use of the one-minute preceptor as a teaching tool in the gross-anatomy laboratory. Anat Sci Educ. 2011;4(4):235–8.
62. Wheelan SA. Group size, group development, and group productivity. Small Gr Res. 2009;40(2):247–62.
63. Biggs J, Tang C. Teaching for quality learning at university: what the student does. 3rd ed. Berkshire: The Society for Research into Higher Education and Open University Press; 2007.
64. Neher JO, Gordon KC, Meyer B, Stevens N. A five-step "microskills" model of clinical teaching. J Am Board Fam Med. 1992;5:419–24.
65. Ferenchick G, Simpson D, Blackman J, DaRosa D, Dunnington G. Strategies for efficient and effective teaching in the ambulatory care setting. Acad Med. 1997;72:277–80.
66. Chan LK, Sharma N. Effects of training experienced teachers in the use of the one-minute preceptor technique in the gross anatomy laboratory. Anat Sci Educ. 2014;7(2):124–9.
67. Chan LK, Yang J, Irby D. Application of the one-minute preceptor technique by novice teachers in the gross anatomy laboratory. Anat Sci Educ. 2015;8(6):539–46.
68. Neher JO, Stevens NG. The one minute preceptor: shaping the teaching conversation. Fam Med. 2003;35:391–3.
69. Andersen J, Nussbaum J, Pecchioni L, Grant JA. Interaction skills in instructional settings. In: Vangelisti AL, Daly JA, Friedrich GW, editors. Teaching communication: theory, research, and methods. 2nd ed. Mahwah: Lawrence Erlbaum Associates; 1999. p. 359–74.

Prosection and Dissection Laboratory Sessions: Design and Implement to Enhance Learning

27

Nalini Pather

Introduction

While traditionally dissection has been the gold standard for teaching anatomy, over the last few decades, institutions worldwide have transformed their teaching practice to include the use of varying degrees of prosections as a modality for learning anatomy. At the outset, it is important to acknowledge that an overwhelming number of anatomists concur that using cadaveric material is unsurpassed in ensuring that programs are of high quality and students have sound anatomical knowledge.

The merit of learning by prosection and by dissection has been debated in literature and has largely remained unsettled [1]. Proponents of dissection argue that it is advantageous in acquiring three-dimensional spatial awareness through active learning [2–4], embedding non-technical skills [3, 5], and increasing an appreciation of humanistic patient care [6–9]. In contrast, it has been argued that learning from prosections minimises the emotional anxiety resulting from exposure to cadavers [10, 11]. In light of this, there have been calls for further evidence [12] to support claims that dissection is superior [13–16]. At the same time, others have concluded that the use of prosections to learn anatomy is equally effective [17, 18] or more effective and/or efficient especially in terms of cost and time [5, 19, 20], and that activities can be designed to embed professionalism and the humanities [21–23].

For anatomists, it is important to appreciate the usefulness of each of these modalities in teaching anatomy and, when implementing these, to do so in ways that optimize the student experience and learning, while remaining efficient and feasible. As anatomy teaching morphs with advances in medical education and technology, a growing number of medical schools have embraced integrated curricula with a combination of prosection and dissection teaching modes to serve specific program requirements [24].

This chapter will describe dissection- and prosection-based laboratory sessions and, in particular, key principles in design and implementation, based on a continuous improvement cycle (Fig. 27.1) that will enhance learning through these tools.

Designing Anatomy Laboratory Activities

An initial decision in designing an anatomy program is to decide whether dissection- or prosection-based learning activities would be the most effective pedagogical approach, bearing in mind that these modalities can also be used in combination, and that irrespective of which of these tools are selected, the strategies employed in its implementation will impact its effectiveness.

N. Pather (✉)
School of Medical Sciences, Medicine, University of New South Wales, Sydney, NSW, Australia
e-mail: N.Pather@unsw.edu.au

L. K. Chan, W. Pawlina (eds.), *Teaching Anatomy*, https://doi.org/10.1007/978-3-030-43283-6_27

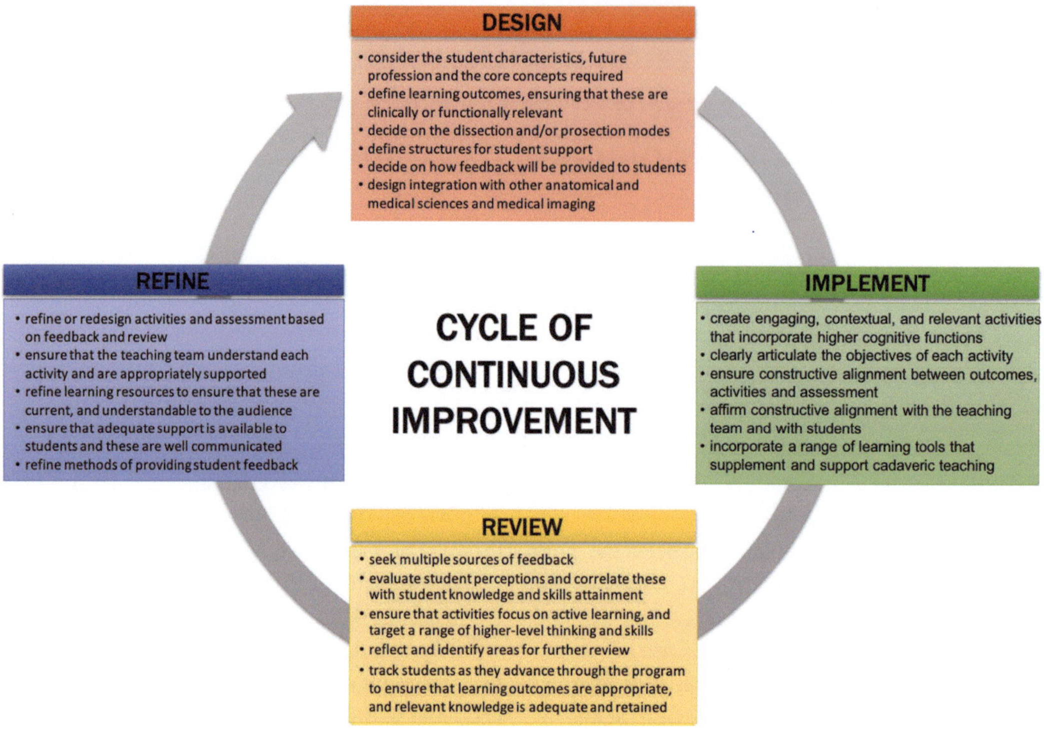

Fig. 27.1 Model of continuous improvement for designing and implementing teaching activities

Definitions

Study by dissection: (etymology: Latin; *dis* "apart" + *sectio* "a cutting, cutting off")

Traditionally, students are assigned to teams and learn by actively undertaking a systematic process of uncovering the body's structure layer by layer in each region. An anatomist is usually present to guide students in this process.

Study by prosection: (etymology: Latin; *pro*- "before" + *sectio* "a cutting, cutting off")

Students learn by observing anatomical structures on specimens that have already been dissected by an experienced dissector. Prosections usually display specific anatomical regions and structures. Study by prosection is flexible enough to utilise teacher-led or student-centred learning approaches.

A growing number of institutions use a combination of dissection and prosection learning activities to suit specific program needs.

Learning Anatomy Using Dissection

Dissection-based laboratory activities afford a number of advantages. The hands-on nature of dissection ensures that students are easily engaged in active learning [25] and thus enhances and promotes deep learning, information retention and confidence building [12, 26–28]. As students dissect and examine relations in situ, they develop clinically relevant visuospatial conceptualisation and appreciation of the multidimensional organisation of the whole human body [5]. Students encounter structures in their anatomical context; for example, the heart is found in the middle mediastinum within the pericardial sac. When comparing the anatomy of neighbouring cadaveric dissections, students encounter anatomical variation and pathology first-hand [5, 29] and begin to understand that human variation is a norm. This awareness of variation can play an important role in ensuring patient safety in future practice [30, 31].

Learning Process in a Traditional Dissection-Based Program

- Typically, anatomy is studied in regions.
- Students are assigned to dissecting groups of usually four to six students per cadaver, dependant on the availability of cadaver resources.
- In order for the team to work optimally and to ensure active participation, roles can be assigned to members (e.g. dissector, assistant, reader/researcher) and rotated over the duration of the dissection course.
- Each group follows a dissection manual or guide to uncover, layer by layer, the regional structures and observe its spatial relationships.
- As the dissection proceeds, teams compare structures and relations in the dissections of other teams. In this way, students become aware of human variation and disease.
- Students use their dissection to review and revise their learning.

Dissection activities allow time to practice important skills. Students develop teamwork skills by working in dissection teams [29] and have opportunity to experience different roles within teams. Mastering a discipline language is an important dimension of learning. During dissection sessions, students are introduced to and have time to practice and use the language of anatomy and of medicine with peers and in teams. At the same time, dissection develops manual dexterity and procedural skills that are important for future practice.

Importantly, for some students, the cadaver is their first close encounter with death and is pivotal to formulating their attitudes towards death and dying [32], and palliative care. Multicultural teams are valuable in exposing students to differences in cultural perceptions and attitudes to these contextual issues and aid in developing attitudes of diversity and inclusivity and of self-awareness.

Things to Consider

- While students place a high value on dissection-based courses [6, 33], there is a high cost-benefit ratio associated with maintaining a cadaveric dissection facility [34]. Modern anatomy laboratories are equipped with expensive computers, monitors and cameras at each dissecting station that can compound the operational cost.
- Dissection-based courses can be labour intensive and require a large amount of time to ensure that the dissections are done appropriately and structures are clearly visible and correctly identified. Sufficiently skilled staff is required to ensure appropriate supervision, guidance and learning. This can be augmented by participation of near-peer tutors and clinical staff and surgeons who will ensure that learning is immediately clinically relevant.
- In some countries, access to a sufficient number of cadavers for teaching by dissection is limited. It is imperative that the use of cadavers complies with the national legal and ethical requirements of the local Anatomy Acts and their associated acts (e.g. The Human Tissue Act in Australia) and their directives.
- While for most students, the dissection experience is positive, a minority would experience negative psychological impact [5, 19, 35, 36]. Appropriate and timely intervention and support for these students should be available to mitigate any risk to self-efficacy.

Advantages of Dissection

- Engages students in active learning
- Enhances deep-learning
- Supports whole-body visuospatial conceptualisation
- Increases awareness of anatomical variation
- Provides opportunities to master discipline language
- Develops manual dexterity and technical skills
- Embeds skills in teamwork
- Develops attitudes to death and dying

Learning Process in Traditional Prosection-Based Courses

- Typically, anatomy is studied in systems.
- Learning activities are flexible in delivery and can be teacher-led in small groups or student-centred in teams.
- If teacher-led, usually, the instructor demonstrates structures on the prosected specimens and facilitates a discussion/tutorial based on these.
- If student-centred, activities can be station-based. Stations can be used to easily integrated with other biomedical and clinical sciences.
- In a station-based session, prosections, models and other learning tools are arranged in a series of stations around the laboratory, with specific learning outcomes and defined activities. Students work in teams progressively through the stations. Students may be guided by a worksheet/guide which includes questions at each station—the questions should include all domains of Bloom's taxonomy.

Learning Anatomy Using Prosection

A prosection-based course can be flexible and has been shown to have a number of advantages. Prosection-based courses are very focussed, contextual and time-efficient [37]. Structures and relations are observed more quickly as these are already dissected and clearly exposed [18, 24, 25]. The flexibility of prosection-based courses allows elegant integration with other biomedical and clinical sciences. Modern anatomy laboratories include technologies that can relate anatomical study to medical imaging, histology and embryology as well as to physiological function and clinical skills. Students can, for example, study the normal anatomy of the respiratory system alongside relevant histological and pathological specimens. Careful planning of prosection resources can also increase awareness of anatomical variation.

Well-thought-out prosection-based courses can be based on collaborative learning pedagogy, e.g. team-based learning, where students are organized into teams to complete laboratory activities. Individual and team readiness tests and checklists can support learning [38, 39] and can reinforce professionalism and the development of teamwork skills.

Prosection-based courses have high self-efficacy and reduce the risk of negative experiences (e.g. failure to dissect specific structures/regions accurately) that may later impact learning in these areas [3, 8]. However, like in dissection, activities can be designed to focus on issues dealing with death and dying, but without the emotional stress of dissecting [37, 40].

Prosection-based courses can be relatively cost-effective [41] with regard to facility and laboratory requirements. Fewer cadavers are required as more than one student cohort can use the prepared prosections. Depending on the relevant legislation, prosections can be retained and used for several years [12, 18, 24]. Prosections can also be plastinated to provide visualisation of anatomical regions while increasing the life cycle of good prosections. These can be used repeatedly and in dry laboratories [42, 43]. When considering using plastinates, it should be noted that evidence suggests that they be used in combination with wet cadaveric material to enrich the learning experience [44].

Things to Consider

- Students place a low effort-high value ratio on prosection-based courses [6, 33]. This is usually because they are able to view and study anatomy in high-quality prosections without investing the time to uncover these by dissection. Student learning and experience is dependent on the quality of prosected specimens. For this reason, it is important to secure the skills of an experienced anatomy dissector. Preparing prosections is labour-intensive, and at all times, care should be taken to comply with national legal requirements regarding how long specimens can be retained.

- Prosections are presented out of their anatomical whole-body context. For example, the heart can be presented as an isolated prosection (i.e. out of the mediastinum and thoracic cage). In these situations, students may have difficulty understanding threshold concepts like, in the case of the thorax, the mediastinum. Care should be taken to design and supplement learning activities that develop students' visuospatial understanding and conceptualisation.
- Anatomy instructors should ideally include prosections that demonstrate anatomical variations so that students develop an understanding of the range of normal. As structures are already dissected and exposed, prosection-based activities must ensure that areas such as surface anatomy are not overlooked [45].

Advantages of Prosections

- Focused, contextual and time-efficient
- Allows flexible use and affords easy cross-discipline integration
- Allows access to a wider range of variations and sections
- Can embed teamwork and professionalism skills
- Develops attitudes to death and dying, with limited social discomfort
- Has less potential for negative experiences that impact self-efficacy
- Extends life cycle of cadaver use and is cost-effective
- Needs fewer cadavers
- Plastinated specimens increase the lifespan of good dissections

Learning Outcomes and Activities

Irrespective of which of the above modes is chosen for specific anatomy sessions, careful planning will ensure a positive learning experience for the students and promote self-efficacy. Important considerations in designing and implementing laboratory activities (Fig. 27.1) include the following:

- *Defining learning outcomes* that are clinically relevant, appropriate and explicit.
- *Developing student-centred, practical and engaging learning activities*. To increase student engagement and enthusiasm, study by dissection and/or prosections should be augmented with models (including ethically sourced 3D-printed models), tactile tools (like "playdough", string, pipe cleaners and felt), body painting and 3D visualisation.
- *Ensuring constructive alignment* between learning outcomes, activities and assessment to promote and reinforce learning.
- *Supporting learning outcomes and activities* with good, relevant resources. Dissection-based activities will require a dissection guide—several good published dissection guides need to be readily available. Prosection-based activities will require a structured learning guide for each session outlining learning outcomes and activities—this can be developed by the course designer. Prosection-based activities should augment sessions by providing a range of anatomical variations (prosections, 3D-printed models, mixed reality experiences).
- Incorporating time for student reflection within dissection and prosection activities to foster professionalism. Feedback from peers and tutors can be valuable in embedding aspects of professionalism [46] and developing humanistic values.
- Strategically designing activities to prevent cognitive overload. Cognitive load theory [47] provides a useful framework to design learning activities that optimize learning. This theory stresses the importance of considering the limited capacity and availability of working memory when designing learning activities, i.e. engaging students' limited working memory will enhance learning and understanding [48].

Implementing Anatomy Laboratory Activities

Once the mode of teaching (dissection- or prosection-based) and learning outcomes have been determined, activities should be implemented in ways that encourage student engagement and promote learning through student-centred approaches [49–53].

Create a Multidimensional Anatomy Laboratory Experience

As the nature of today's student body changes, the anatomy laboratory sessions should include a multifaceted approach to learning. Irrespective of whether a dissection- and/or prosection-based mode is selected, the design should be clinically relevant and engaging. Some techniques to do this include incorporating learning tools such as videos, models, tactile equipment, body painting and 3D virtual models alongside dissection or prosection activities.

Videos, for example, can augment learning in the anatomy laboratory in two ways:

- *Focus students on learning outcomes without cognitive overload.* Tailor-made videos for laboratory sessions can be used to introduce the session's learning outcomes and then comprehensively demonstrate the major anatomical structures, relations and concepts. Depending on the course design, students can view these videos before, during or after the laboratory session. In prosection-based activities, these videos provide an initial quick guide placing organs and regions into the context of the whole body, before students navigate their way through a series of prosections mastering the content along the way. In dissection-based activities, students can refer to the videos as a guide as they proceed through the dissection. As an alternative, dissection videos can alternate with dissection sessions, and this has been demonstrated to significantly increase student performance in examinations [40]. Videos can also be made available to students to use for revision and for consolidating information in their own time and at their own pace.
- *Arouse students' interest and participation in learning activities.* Videos demonstrating surgical and procedural anatomy such as laproscopy, lumbar puncture and cricothyrotomy integrated in the laboratory sessions can ensure that student anatomical learning is contextual and clinically relevant. In a session on hand anatomy, for example, a video on carpal tunnel release can demonstrate application of structural organisation to clinical practice. Similarly, a session on abdominal organ position can incorporate a cholecystectomy procedural video. These videos can be incorporated into laboratory activities, for example, by requiring students to identify anatomical structures, relations and variations demonstrated in the video, thus reinforcing and applying learning.

Videos Can Be Used to

- Reduce cognitive overload and focus on learning outcomes
- Arouse student interest and participation
- Review material before and after a laboratory session, in students' own time and at their own pace
- Introduce the learning activities before students attend the anatomy laboratory and may reduce their stress before entering gross laboratory for the first time
- Implement multidisciplinary integration, e.g. procedural anatomy, and medical imaging

Integrate with Living Anatomy

Anatomy laboratory activities should relate cadaveric anatomy to living anatomy and to clinical practice. Incorporate living anatomy by implementing activities that include a study of surface anatomy and projections, relevant medical imaging and surgical approaches and using techniques such as movement analysis, peer physical examination and body painting.

Medical Imaging The anatomy laboratory provides the ideal platform for side-by-side correlation of the three-dimensional human body with two-dimensional medical imaging [52, 54]. Interpreting medical images provides students with an opportunity to evaluate their anatomical understanding. Ultrasound allows the safe non-invasive visualization of structures in the living and in dynamic movement (see Chap. 37). Portable ultrasound machines can be used in the anatomy laboratory [55] to demonstrate relevant living anatomy, to reinforce clinically relevant learning and to increase student knowledge retention [56, 57], while remaining time-efficient [52]. Similarly, by incorporating virtual reality models (especially those that are segmented from DICOM data), the anatomy laboratory session can relate directly to clinical practice and patient care. In foregut anatomy sessions, for example, the learning outcomes could include correlating cadaveric anatomy with medical images:

- Along with cadaveric specimens, make available barium meal radiographs, angiography of the coeliac trunk and its branches and CT and MRI images of the upper abdomen.
- Additionally, after studying the normal anatomy and medical images, pathological images can be included in problem-solving and reflective exercises.
- Incorporate ultrasound to provide students with dynamic visualisation of anatomy and physiology; for example, echocardiography can demonstrate the opening and closing of the valves as well as cardiac filling during diastole and systole.

Surface Anatomy Surface anatomy is an important part of an anatomy course and is an imperative for patient assessment. Both dissection- and prosection-based sessions should include surface anatomy. Implementing activities involving palpation on peers requires planning and consent from students, with consideration for different cultural attitudes. In dissection sessions, surface anatomy of bony landmarks and the course of nerves and vessels can be outlined on the cadaver before skin incisions are made. Other surface anatomy requiring a living person (like palpation of arterial pulses and organs) also can be easily included by requiring "non-dissecting" team members to demonstrate or paint corresponding surface anatomy as the dissection proceeds. Likewise, in prosection-based sessions, surface anatomy can be studied alongside relevant prosections, for example, when studying a prosection of the axilla, students can palpate the muscular boundaries of the axilla on a peer. There are a few techniques to include surface anatomy, for example, in upper limb sessions:

- Body paint [53] can be used to outline the course and relations of nerves or vessels on the upper limb of the cadaver before dissection, or if using prosections, on peers.
- Body paint can also be used to outline the dermatomes and cutaneous innervation of the upper limb.
- Palpation can be used for arterial pulses (e.g. brachial and radial) and bony and soft tissue landmarks (e.g. the acromioclavicular joint, clavicle, spine of scapula, epicondyles of humerus, boundaries of the cubital fossa) on peers.

Analysis of Movement Analysis of movement provides opportunity for students to correlate cadaveric study of muscle with its actions in the living person, allowing students to understand muscle function and testing, and to extrapolate this to effects of nerve lesions. Analysis of movement can be done in dissection- and prosection-based laboratory activities. For example, in lower limb laboratory sessions, analysis of movement can be incorporated in both dissection- and prosection-based sessions to demonstrate the following:

- muscle action and muscle testing against resistance (palpate muscles as they contract)
- gait analysis with respect to muscle actions in each stage of movement, and relate this to muscle attachments
- joint reflexes in relation to nerve pathways
- changes in anatomy during movement using ultrasound

Integrate with Living Anatomy by Including

- Medical imaging and procedural anatomy
- Surface anatomy
- Analysis of movement

Embed Graduate Attributes

Graduate attributes refer to the qualities or skills that the university or institution seeks to develop in its graduates so that they will be well prepared for successfully engaging in the professional practice of their discipline. In some countries, significant consolidated proposals and arguments have been made to ensure a minimum national core anatomy curriculum for medical programs. The anatomy laboratory presents a unique platform to embed graduate attributes into learning activities. Some techniques to embed common institutional graduate capabilities in dissection- and/or prosection-based activities include the following:

Develop a Knowledge and Skills Base for Professional Practice

Both prosection and dissection teaching modes are based on acquiring a sound three-dimensional knowledge base of anatomy. Laboratory activities enhance conceptual knowledge especially when learning outcomes and activities engage students in finding, evaluating and using information. Care should be taken when developing dissection and prosection activities to ensure that these are clinically relevant. Including clinician and surgeon participation in learning activities can foster direct application of anatomical knowledge to surgical and clinical contexts. This has been shown to increase student engagement and learning in anatomy [58]. In dissection sessions, clinicians and surgeons can be invited to participate and supervise the dissection activities. Similarly, in prosection-based sessions, clinicians and surgeons can be invited to demonstrate relevant surgical approaches to structures in the region or system being studied.

Develop Effective Communication and Team Skills

Team-based learning has been shown to improve student performance in assessment [37]. Both dissection and prosection laboratory sessions lend themselves to team-based learning. This can be fully embedded to include individual readiness tests and team quizzes. Team activities can be designed to develop communication skills. At the end of a dissection session, for example, dissection teams can be rostered to present a "table-side" summary of the session's dissection. Alternately, at the end of a prosection session, teams can present related human variation or solutions to clinical cases.

Critical Thinking and Self-Directed Lifelong Learning

Critical thinking requires students to apply their knowledge and skills to problem-solving. In the anatomy laboratory, this can be done using clinical cases, activities correlating anatomy with medical imaging, implications of human variation, etc. During a dissection- or prosection-based session on thoracic anatomy, for example, students can apply anatomical knowledge learnt

Embed Graduate Attributes

- Develop a sound and relevant knowledge base
- Develop communication and teamwork skills
- Promote critical thinking
- Develop an ethical framework

to a clinical case. The case should require application of knowledge and immediately make the session clinically relevant. It should also allow students to reflect and identify areas that they need to further investigate.

For example, in a thoracic anatomy laboratory session, one can introduce a clinical case of a patient whose chest X-ray reveals fluid accumulation in the thoracic cavity. A thoracocentesis to obtain a sample of pleural fluid is required. There is no ultrasound equipment available at the clinic. Well-designed questions that can trigger student reflection and discussion include the following:

In what situations would fluid accumulate in the thoracic cavity? How would this appear on a radiograph?

Where in the thoracic wall should the needle for the thoracocentesis be inserted? Why?

What are the layers of tissue that the needle passes as it is inserted into the pleural space?

What structures are at risk of damage in this procedure?

Ethical Practice

The use of cadavers provides an opportunity to raise awareness of issues around death and dying with students[23] and with local communities [59]. The increase in student diversity allows natural exploration of cultural differences in attitudes in areas such as medical uncertainty, ethical issues like body and organ donations, the declaration of brain dead and compliance with legal and ethical requirements on cadaveric use. The history of anatomy provides a rich source of stories that can engage students and illustrate complex ethical considerations.

For many students, the exposure to cadavers provides an opportunity to reflect and formulate ideas on death and dying [6–11]. This should be encouraged and opportunities taken to foster respect and dignity for cadaver donors. It is for this reason that many anatomy departments institute a practice of homage to donors and their families, e.g. a dedication or memorial service with or without religious involvement.

Summary of Factors to Consider When Selecting Between Prosections and Dissections

Educational Philosophy and Impact

- Curriculum changes and time available for anatomy
- Opportunity to embed a student-centred approach
- Opportunities to integrate:
  - Anatomical sciences, e.g. histology and embryology
  - Medical sciences, e.g. physiology and pathology
  - Clinical sciences, e.g. surgery
  - Ethics and humanities
  - Non-technical skills
- Opportunities to include a variety of learning experiences:
  - Living anatomy
  - Surface anatomy
  - Clinical imaging and 3D visualization
  - Developing non-technical skills, e.g. teamwork

Assessment methods

- Ensure that assessment is:
  - Reliable
  - Valid
  - Sustainable
- Provide sufficient and timely feedback to enable learning

Faculty and staff

- Develop capacity to support prosection and/or dissection skills
- Enable methods to collect feedback
- Encourage developing skills to support student

Operational

- Access to resources for learning and teaching
- Laboratory facilities and cost of maintenance
- Cost and time to produce prosections
- Access to supporting resources, e.g. medical imaging

Review and Refine Anatomy Laboratory Activities

Course design and implementation needs a process of continuous development (Fig. 27.1) to remain relevant and engaging. Learning activities should be subjected to a process of review, evaluation and refinement. In evaluation, it is ideal to triangulate various sources of information that can inform refinement. Dissection- and prosection-based activities can be reviewed and evaluated using multiple and a variety of ways to obtain student feedback, like perception survey tools and correlation with student performance in assessment. Ideally and if opportunity presents, cross-over studies can provide objective data on innovative techniques. All feedback should then be reviewed and used to refine the course learning activities for the next iteration of the course.

Conclusions

Cadaveric resources provide a rich environment for teaching anatomy using dissection- or prosection-based activities. These should be designed to be engaging, contextual and relevant to professional development, both in cognitive ability and in professional skills. It is not what is used (i.e. the tools) but how it is used (i.e. the pedagogical approach) that makes a difference. As with all educational development, its design and implementation should be a process of continuous development based on robust evaluation and reflection.

References

1. Conwall J. The diverse utility of wet prosections and plastinated specimens in teaching gross anatomy in New Zealand. Anat Sci Educ. 2011;4:269–74.
2. Granger NA. Dissection laboratory is vital to medical gross anatomy education. Anat Rec. 2004;281B:6–8.
3. Lempp HK. Perceptions of dissection by students in one medical school: beyond learning about anatomy. A qualitative study. Med Educ. 2005;39(3):318–25.
4. Aziz MA, McKenzie JC, Wilson JS, Cowie RJ, Ayeni SA, Dunn BK. The human cadaver in the age of biomedical informatics. Anat Rec (New Anat). 2002;269:20–32.
5. Alexander J. Dissection versus prosection in the teaching of anatomy. J Med Educ. 1970;45:600–6.
6. Bertman SL, Marks SC Jr. Humanities in medical education: rationale and resources for the dissection laboratory. Med Educ. 1985;19:374–81.
7. Bertman SL, Marks SC Jr. The dissection experience as a laboratory for self-discovery about death and dying: another side of clinical anatomy. Clin Anat. 1989;2:103–13.
8. Arráez-Aybar LA, Casado-Morales MI, Castaño-Collado G. Anxiety and dissection of the human cadaver: an unsolvable relationship? Anat Rec. 2004;279B:16–23.
9. Biasutto SN, Caussa LI, Criado del Río LE. Teaching anatomy: cadavers vs. computers? Ann Anat. 2006;188:187–90.
10. Arráez-Aybar LA, Castaño-Collado G, Casado-Morales MI. A study of cognitive-affective and physiological-motor reactions to human dissection in Spanish students of human anatomy. Eur J Anat. 2007;11:S67–71.
11. Arráez-Aybar LA, Castaño-Collado G, Casado-Morales MI. Dissection as a modulator of emotional attitudes and reactions of future health professionals. Med Educ. 2008;42:563–71.
12. Yeager VL. Learning gross anatomy: dissection and prosection. Clin Anat. 1996;9:57–9.
13. Patel KM, Moxham BJ. Attitudes of professional anatomists to curricular change. Clin Anat. 2006;19(2):132–41.
14. Rizzolo LJ, Stewart WB, O'Brien M, Halms A, Rando W, Abrahams J, et al. Design principles for developing an efficient clinical anatomy course. Med Teach. 2006;28(2):142–51.
15. Berube D, Murray C, Schultze K. Cadaver and computer use in the teaching of gross anatomy in physical therapy education. J Phys Therapy Educ. 1999;13(2):41–6.
16. Pizzimenti MA, Pantazis N, Sandra A, Hoffmann DS, Lenoch S, Ferguson KJ. 2016. Dissection and dissection-associated required experiences improve student. Anat Sci Educ. 2016;9(3):238–46.
17. Nnodim JO. Learning human anatomy: by dissection or from prosections? Med Educ. 1990;24(4):389–95.
18. Nnodim JO, Ohanaka EC, Osuji CU. A follow-up comparative study of two modes of learning human anatomy: by dissection and from prosections. Clin Anat. 1996;9(4):258–62.
19. Collins J. Modern approaches to teaching and learning anatomy. BMJ. 2008;337:a1310.
20. Winkelmann A. Anatomical dissection as a teaching method in medical school: a review of the evidence. J Med Educ. 2007;41:15–22.
21. Topp KS. Prosection vs. dissection, the debate continues: rebuttal to Granger. Anat Rec. 2004;281B:12–4.
22. Bergman EM, Prince KJ, Drukker J, van der Vleuten CP, Scherpbier AJ. How much anatomy is enough? Anat Sci Educ. 2008;1:184–8.
23. Strkalj G, Pather N. Introduction. In: Strkalj G, Pather N, editors. Commemorations and memorials. Exploring the human face of anatomy. Singapore: World Scientific; 2017. p. 1–14.

24. Drake RL, McBride JM, Lachman N, Pawlina W. Medical education in the anatomical sciences: the winds of change continue to blow. Anat Sci Educ. 2009;2(6):253–9.
25. Turney BW. Anatomy in the modern curriculum. Ann R Coll Surg Engl. 2007;89(2):104–7.
26. Johnson J. Importance of dissection in learning anatomy: personal dissection versus peer teaching. Clin Anat. 2002;15:38–44.
27. Williams SR, Thompson KL, Notebaert AJ, Sinning AR. Prosection or dissection: which is best for teaching the anatomy of the hand and foot. Anat Sci Educ. 2019;12:173–80.
28. Gogalniceanu P, Hardi M, Paraskevas AP, Darzi A. Anatomy teaching in the 21st century—dead cool or cold dead? Anat Sci Educ. 2008;1(3):136–7.
29. McLachlan JC, Bligh J, Bradley P, Searle J. Teaching anatomy without cadavers. Med Educ. 2004;38(4):418–24.
30. Cahill DR, Leonard RJ, Marks SC Jr. Standards in healthcare and medical education. Clin Anat. 2000;13:150.
31. Ahmed K, Rowland S, Patel V, Khan RS, Ashrafian H, Davies DC, et al. Is the structure of the anatomy curriculum adequate for safe medical practice? Surgeon. 2010;8:318–24.
32. Parker LM. Anatomical dissection: why are we cutting it out? Dissection in undergraduate teaching. Aust NZ J Surg. 2002;72(12):910–2.
33. Pabst R, Rothkotter HJ. Retrospective evaluation of undergraduate medical education by doctors at the end of their residency time in hospitals: consequences for anatomical curriculum. Anat Rec. 1997;249:431–4.
34. McLachlan JC. New path for teaching anatomy: living anatomy and medical imaging vs dissection. Anat Rec. 2004;281B:2–14.
35. O'Carroll RE, Whiten S, Jakson D, Sinclair DW. Assessing the emotional impact of cadaver dissection on medical students. Med Educ. 2002;36:550–4.
36. Cahill KC, Ettarh RR. Attitudes to anatomy dissection in an Irish medical school. Clin Anat. 2009;22(3):386–91.
37. Dinsmore CE, Daugherty S, Zeitz HJ. Teaching and learning gross anatomy: dissection, prosection, or both of the above? Clin Anat. 1999;12:111–4.
38. Michaelsen LK, Watson WE, Cragin JP, Fink LD. Team-based learning: a potential solution to the problems of large classes. Exchange Org Behav Teach J. 1982;7(4):18–33.
39. Michaelsen LK, Knight AB, Fink LD. Team-based learning: a transformative use of small groups. New York: Praeger; 2002.
40. Skidmore JR. The case for prosection: comment on R.L.M. Newell's paper. Clin Anat. 1995;8:128–30.
41. Ashdown L, Lewis E, Hincke M, Jalali A. Learning anatomy: can dissection and peer-mediated teaching offer added benefits over prosection alone? ISRN Anatomy. 2013;873825.
42. Lattore RM, García-Sanz MP, Moreno M, Hernández F, Gil F, López O, et al. How useful is plastination in learning anatomy? J Vet Med Educ. 2007;34(2):172–6.
43. Dhingra R, Taranikanti V, Kumar R. Plastination aids in anatomy revisited. Nat Med J India. 2006;19(3):171.
44. Fruhstorfer BH, Palmer J, Brydges S, Abrahams PH. The use of plastinated prosections for teaching anatomy–the view of medical students on the value of this learning resource. Clin Anat. 2011;24(2):246–52.
45. McKeown PP, Heylings DJA, Stevenson M, McKelvey KJ, Nixon JR, McCluskey DR. The impact of curricular change on medical students' knowledge of anatomy. Med Educ. 2003;37(11):954–61.
46. Camp CL, Gregory JK, Lachman N, Chen LP, Juskewitch JE, Pawlina W. Comparative efficacy of group and individual feedback in gross anatomy for promoting medical student professionalism. Anat Sci Educ. 2010;3(2):64–72.
47. Sweller J. Cognitive load theory, learning difficulty and instructional design. Learn Instr. 1994;4:295–312.
48. Khalil MK, Paas F, Johnson TE, Payer AF. Interactive and dynamic visualizations in teaching and learning anatomy: a cognitive load perspective. Anat Rec. 2005;386B:8–14.
49. Nieder GL, Parmelee DX, Stolfi A, Hudes PD. Team-based learning in a medical gross anatomy and embryology course. Clin Anat. 2005;18:56–63.
50. Vasan NS, DeFouw DO, Holland BK. Modified use of team-based learning for effective delivery of medical gross anatomy and embryology. Anat Sci Educ. 2008;1:3–9.
51. Morton DA, Colbert-Getz JM. Measuring the impact of the flipped anatomy classroom: the importance of categorizing an assessment by Bloom's taxonomy. Anat Sci Educ. 2017;10(2):170–5.
52. Griksaitis MJ, Sawdon MA, Finn GM. Ultrasound and cadaveric prosections as methods for teaching cardiac anatomy: a comparative study. Anat Sci Educ. 2012;5:20–6.
53. Finn GM, McLachlan JC. A qualitative study of student responses to body painting. Anat Sci Educ. 2010;3(1):33–8.
54. Ganguly PK, Chan LK. Living anatomy in the 21st century: how far can we go? SE Asian J Med Educ. 2008;2(2):52–7.
55. Fakoya FA. Integrating ultrasound technology into teaching gross anatomy: point of order! Med Educ Online. 2013;18:10.3402/meo.v18i0.20888.
56. Swamy M, Searle RF. Anatomy teaching with portable ultrasound to medical students. BMC Med Educ. 2012;12:99.
57. Smith C, Dilley A, Mitchell B, Drake R. Gray's surface anatomy and ultrasound. Edinburgh: Elsevier; 2018.
58. Böckers A, Mayer C, Böckers TM. Does learning in clinical context in anatomical sciences improve examination results, learning motivation, or learning orientation? Anat Sci Educ. 2014;7(1):3–11.
59. Pather N, Ashwell K. A moment of hope: thanksgiving services for families of donors. In: Strkalj G, Pather N, editors. Commemorations and memorials. Exploring the human face of anatomy. Singapore: World Scientific; 2017. p. 173–80.

Use of Unembalmed/Fresh Cadavers in Anatomy Teaching

28

Jennifer M. McBride and Richard L. Drake

Introduction

Human cadavers have always been an integral part of physician training and medical education [1–4]. Since anatomy courses typically last several weeks to months, they were embalmed or fixed in a process, which involves some combination of ethanol, formaldehyde, glycerin, and phenol, to halt decomposition and extend the usefulness of the cadaver. And while embalming procedures have been modified, the embalmed cadaver still greets most new medical students as they enter their anatomy class. However, an increased role for the cadaver in physician education began as medicine entered the last quarter of the twentieth century.

It was during this time that a new surgical discipline referred to as minimally invasive surgery had its roots. Procedures using these approaches, also referred to as laparoscopic or endoscopic approaches, required unique skills, and a new training paradigm was needed. This led to the development of unembalmed/fresh cadaver/tissue laboratories and training centers [5–8] and an expanded role for the cadaver in physician training and education. However, is there a role for the unembalmed/fresh cadaver in the early education of medical students?

J. M. McBride (✉) · R. L. Drake
Cleveland Clinic Lerner College of Medicine, Case Western Reserve University, Cleveland, OH, USA
e-mail: mcbridj@ccf.org; DRAKER@ccf.org

Why Unembalmed/Fresh Cadavers?

Is it possible, practical, or of any value to use unembalmed/fresh cadavers in an anatomy course for medical students? Some would suggest that for the beginning medical student, gaining basic anatomical knowledge is what's most important. However, what if the learning experience was enhanced in such a way that students felt that knowledge gained was immediately applicable to real clinical situations?

Clearly, this is not the case when learning occurs using an embalmed cadaver which has little in common with the living body. The color is unnatural, and the tissue, organs, and joints are stiff. Add the odor present in most cadaver laboratories, and this is not a very pleasant learning environment. However, would there be a benefit if unembalmed/fresh cadavers were used in this initial exposure to anatomy? One report indicates that using unembalmed/fresh cadavers was found to provide students with more confidence as they proceeded with their training in the operating room [9]. The tissue color and texture of the unembalmed/fresh cadaver are as close as possible to those of the living body, and the young medical student would feel that the experience is like being in an operating room (see Table 28.1). But is this difference worth the added issues of rapid deterioration and possible health hazards?

The individuals involved in planning the unique and innovative educational program at the

L. K. Chan, W. Pawlina (eds.), *Teaching Anatomy*, https://doi.org/10.1007/978-3-030-43283-6_28

Table 28.1 Comparing embalmed and unembalmed/fresh cadavers

Embalmed cadavers	Unembalmed/fresh cadavers
Color is unnatural	Color is natural
Tissue is tough	Tissue is soft
Joints are stiff	Joints are flexible
Strong chemical odor	Natural odors
Possible health hazards from chemical exposure	Possible health hazards – cadavers are tested
Unlike clinical situation	Like clinical situation

Cleveland Clinic Lerner College of Medicine of Case Western Reserve University thought it was. The curriculum at this medical school is a problem-based, organ-systems-oriented approach stressing small-group interactive learning with no lectures and no traditional tests or grades. Each educational offering stresses the integration of basic sciences with clinical medicine which blends nicely with an anatomy program that uses clinical cases to introduce anatomical concepts and facts that are reinforced using unembalmed/fresh cadavers and relevant imaging [10]. Additional factors that made this the perfect setting for a unique laboratory experience was a small class size (32 students), a prosection or previously dissected approach in the laboratory in which residents would be doing the dissections and the teaching, the existence of an unembalmed/fresh cadaver/tissue laboratory that was developed for the practice of minimally invasive surgical procedures, and a strong body donation program (~200 donations/year).

Precautions

The use of unembalmed/fresh cadavers in a dissection laboratory comes with potential health hazards that must be considered and dealt with appropriately. At the Cleveland Clinic, this is handled at four separate and distinct levels: first, when the Body Donation Program is discussed with a potential donor; second, when the Cleveland Clinic is initially notified that a deceased individual's body is being donated; third, when the body arrives at the facility; and finally, when the body is being used in the laboratory.

1. When individuals are considering body donation and contact the Cleveland Clinic Body Donation Program, the first step is to answer any questions they might have and send them a brochure and donation forms. During this initial conversation, it is also indicated to them that we cannot accept a body with an infectious disease such as hepatitis, HIV/AIDS, Creutzfeldt-Jakob disease, methicillin-resistant *Staphylococcus aureus* (MRSA), or *Clostridium difficile (C. diff)*. Additionally, this policy is clearly stated in our brochure.
2. When calls are received that a deceased individual who was either signed up in the Body Donation Program prior to death or always wanted to donate their body to science when they die, but never got around to filling out the form, whoever receives the call asks, "Did this individual have any infectious disease such as hepatitis, HIV/AIDS, Creutzfeldt-Jakob disease, methicillin-resistant *Staphylococcus aureus* (MRSA) or *Clostridium difficile (C. diff)*?" If the answer is yes, the body will not be accepted. Additionally, if there is any indication the individual was using illegal drugs, the body will not be accepted because of the potential for the presence of an infectious disease.
3. When the donated body is received at the Cleveland Clinic, several things occur. First, the body is inspected to make sure that nothing is observed that would suggest the presence of an infectious disease or cause other issues of concern related to potential health hazards. Second, blood samples are taken and sent for serology testing: HIV using an HIV 1 and 2 antibodies by the enzyme immunoassay (EIA), hepatitis C antibody by IA, hepatitis B core antibody total, and hepatitis B surface antigen by EIA. If any tests come back positive, the body is immediately excluded from the Body Donation Program and cremated as soon as possible. Finally, if an individual dies of sepsis, their medical record will be checked and, if possible, the

cause identified. The situation will then be discussed with the Infectious Disease group at the Cleveland Clinic and a decision made as to whether to keep the body or exclude it from use.

4. Finally, it is clearly explained to all participants in any function or activity in the unembalmed/fresh cadaver laboratory that they exercise biohazard safety/universal precautions. A detailed description of biohazardous materials, precautions, blood-borne pathogens, and exposure control plan can be found by reviewing the state Department of Health website. At our institution, they are expected to wear scrubs or a gown, gloves, a mask, and protective eyewear. There are no exceptions to these rules.

Storage

Once a cadaver has been cleared for use in the program, it is stored in a refrigeration unit. Stored in this manner, we have found that cadavers will remain usable for 4–6 weeks. However, once a region of the body is entered, it only remains usable for a few days to 2 weeks and must be covered with a moist towel or sutured closed. But upon entering a new region of the body, the anatomy in that region will appear as if the cadaver was just removed from refrigerated storage for the first time.

The cadaver can be frozen to hold for a later program, but we have found that once it is defrosted, it does not last as long and the quality of the tissue is not as good as cadavers that were only refrigerated and never frozen. We never use a cadaver after a cycle of multiple freeze/defrost patterns as we have found the usefulness of the cadaver after the second freeze/defrost cycle is very limited.

Finally, we try to only have a cadaver out of refrigeration for 2–3 h, with the laboratory temperature maintained at 65 °F, to prevent tissue breakdown. Long exposures at room temperature and constant handling have adverse effects on the length of time a cadaver can be used and tissue quality.

Use in Educational Programs

The majority of unembalmed/fresh cadavers in the Body Donation Program at the Cleveland Clinic are used in educational programs for medical students, residents and fellows, and students in a physician assistant program.

The educational activities for first- and second-year medical students and the physician assistant students are unique and follow a similar format. The anatomy laboratory sessions in these two courses use a prosection or previously dissected approach, and no student dissection occurs. For example, the dissections for the medical student anatomy course are completed by residents from the various surgical disciplines the week before the class. When the class meets, the medical students are divided into groups of six or seven, and they rotate around various dissected cadavers. The anatomy on these cadavers is being demonstrated and discussed by the residents. A similar pattern is followed for the physician assistant anatomy class, but in this case, the dissections are completed by first-year medical students who then do the teaching during each class session. In both of these cases, the instructors, residents, and first-year medical students are able to review their anatomy and gain valuable teaching experience which is expected to be part of the residency program and promoted by the ACGME [11]. Additionally, since unembalmed/fresh cadavers are being used, the students and instructors are learning and teaching in an educational environment that more closely resembles clinical situations.

Use in Research/Clinical Activity

A smaller number of the unembalmed/fresh cadavers in the Body Donation Program at the Cleveland Clinic are used in research projects involving medical students, residents, fellows, and faculty members. These involve a variety of different groups but are especially popular with the ENT, plastics, and dermatology groups. Additionally, requests concerning the availability of an unembalmed/fresh cadaver are made on very short notice when a group may want to evaluate a new or different surgical approach for an upcoming patient.

Conclusion

In conclusion, the Cleveland Clinic has developed a very active program that uses unembalmed/fresh cadavers in a variety of educational and research opportunities. These include a unique anatomy course for first-year medical students where their learning occurs on a cadaver that resembles the living body.

References

1. Aziz MA, McKenzie JC, Wilson JS, Cowie RJ, Sylvanus AA, Dunn BK. The human cadaver in the age of biomedical informatics. Anat Rec. 2002;269:20–32.
2. Cornwall J, Stringer MD. The wider importance of cadavers: education and research diversity from a body bequest program. Anat Sci Educ. 2009;2:234–7.
3. Cornwall J. The diverse utility of wet prosections and plastinated specimens in teaching gross anatomy in New Zealand. Anat Sci Educ. 2011;4:269–74.
4. Ghosh SK. Cadaver dissection as an educational tool for anatomical sciences in the 21st century. Anat Sci Educ. 2017;10:286–99.
5. Human fresh tissue laboratory. Duke University School of Medicine, Durham. 2019. https://surgery.duke.edu/education-and-training/educational-laboratories/human-fresh-tissue-laboratory. Accessed 9 May 2019.
6. Penn human tissue lab. University of Pennsylvania, Philadelphia. 2018. https://www.pennmedicine.org/departments-and-centers/orthopaedic-surgery/education-and-training/human-tissue-lab. Accessed 9 May 2019.
7. Fresh tissue lab. University of Louisville, Louisville. 2019. http://louisville.edu/medicine/departments/anatomy/fresh-tissue-lab. Accessed 9 May 2019.
8. University of California San Francisco. Anatomy Learning Center. San Francisco. 2018. https://meded.ucsf.edu/anatomy-learning-center. Accessed 9 May 2019.
9. Naylor RA, Hollett LA, Castellvi A, Valentine RJ, Scott DJ. Preparing medical students to enter surgery residencies. Am J Surg. 2010;199:105–9.
10. Drake RL. A unique, innovative, and clinically oriented approach to anatomy education. Acad Med. 2007;82:475–8.
11. ACGME Common Program Requirements. Accreditation Council for Graduate Medical Education, Chicago. 2019. https://www.acgme.org/What-We-Do/Accreditation/Common-Program-Requirements. Accessed 9 May 2019.

Essential E-Learning Methods for Evolving Anatomy Laboratories

29

Robert B. Trelease, James Lister, and Stephen Schettler

Decades of advances in computing have promoted the development and pervasive application of new technologies and methods for teaching anatomy [1]. Moreover, ongoing evolution of postsecondary and professional health sciences curricula has favored using newer "adult learning" methods for increased class sizes, with small group activities and problem- and team-based learning (PBL and TBL), and accelerated student progress toward degrees and practice. These implementation challenges are being managed with widespread use of computer-based educational technologies.

Such curricular reforms have also reduced, substituted ("flipped"), or eliminated legacy live anatomy lectures in many medical schools. This has further concentrated remaining teaching contact hours into streamlined laboratory sessions that must integrate cadavers, specimens, models, clinical imaging, and multimedia software in a way appropriate to the specific adult learning curriculum.

Whether readers are early anatomy faculty joining new institutions or seasoned others teaching in new facilities, participating in curricular changes, or planning new multidisciplinary laboratories, all can benefit from understanding the scope and current state of the art in practice in laboratory educational technologies. With anatomy teachers increasingly training different academic-level students for career paths like medicine, dentistry, nursing, and allied health sciences, it is also helpful to be able to repurpose common content and available resources for teaching different types and levels of students. Finally, there is much active research currently on e-learning methods, providing future prospects for rising young faculty.

This chapter appraises practical laboratory e-learning methods in the broader context of different types and levels of undergraduate and health sciences curricula, surveying standard tools that anatomists can use to produce laboratory teaching resources for virtually any curricular model. We then consider whether the current educational use of 3D anatomical models represents a stage in the "maturing" of virtual anatomy tools suitable for undergraduate through medical curricula. We follow with a look at the essentials of creating new 3D digital anatomy models for dual use on PCs and for 3D-printed physical anatomical models. The chapter winds up with a

R. B. Trelease (✉)
Division of Integrative Anatomy, Department of Pathology and Laboratory Medicine, David Geffen School of Medicine at UCLA, University of California, Los Angeles, CA, USA

50-082 UCLA Center for the Health Sciences, Los Angeles, CA, USA
e-mail: trelease@ucla.edu

J. Lister · S. Schettler
Division of Integrative Anatomy, Department of Pathology and Laboratory Medicine, David Geffen School of Medicine at UCLA, University of California, Los Angeles, CA, USA

L. K. Chan, W. Pawlina (eds.), *Teaching Anatomy*, https://doi.org/10.1007/978-3-030-43283-6_29

consideration of curricular changes, the current state of responsive teaching resources development, and the potential benefits and caveats for personally investing time in creating and 3D-printing digital anatomical models.

The Broad View: An Evolving Spectrum of Anatomy Teaching Laboratories

Historically, teaching medical gross anatomy has represented the most comprehensive professional pedagogy, in terms of depth and breadth of clinically relevant anatomical knowledge conveyed. The majority of traditional teaching time was devoted to cadaver dissection laboratories, with remainder given to regional lectures. There is a persisting conviction worldwide that cadaver dissection remains a vital learning experience for medical students, and some schools choose to retain cadaver laboratories despite continuing curricular reforms [2]. This volume also includes excellent chapters on the values of cadaver dissection, the use of prosections, and the design of new generation laboratories.

Current cadaver laboratory facilities should support wired/wireless network access as part of the basic infrastructure [3]. A networked computer with large high-definition (HD) screen at each cadaver tank or group work space is now a mainstay work unit for such legacy "wet" teaching laboratories. Built-in large screen wall displays, HD digital video demonstration cameras, recording, and teleconferencing resources may have more selective use at large schools, depending on program (e.g., for remote-learning demonstrations). Additional adjacent laboratory space(s) can support advanced clinical and imaging activities (e.g., ultrasound and "virtual anatomy" simulation systems; see below). Learning activities with unembalmed cadavers can take place in appropriate controlled-access and universal standard precautions space.

While legacy medical gross anatomy might exemplify the most demanding teaching laboratory setting, our continuing purpose here includes surveying the most useful e-learning methods that can also appropriately support any level of anatomy laboratory. We also recognize a globally diverse "spectrum" of contemporary teaching laboratories, with individual facilities and available resources depending on country, level of education, and discipline(s) served, ranging from undergraduate to research university health sciences professional and postgraduate programs. These respective types of laboratories are likely to use now commonly available base technologies, from physical anatomical models, plastinated specimens, and software learning modules on computer workstations, up to cadaver stations with workstations, digital video, and whole-body digital imaging systems for modern university health sciences.

Career tracks and their associated specialized curricula also influence anatomy laboratory resource availability, with legacy medical, dental, and nursing schools frequently using separate university facilities. In larger health sciences universities, multiple courses including physical therapy, dentistry, nursing, and graduate may use the same physical laboratory space(s). Given continuing demand for interprofessional learning in health sciences, one might expect an increase in the use of shared laboratories [4].

Producing and Managing Core E-Learning Resources for Anatomy Laboratories

Electronic content for anatomy laboratories may include commercially produced dissectors, prosectors, and anatomical visualization software. But primary course faculty at higher levels in health sciences education most often produce the majority of learning resources and documents for anatomy laboratory sessions that integrate with any other lectures, assigned readings, or online (flipped) content defined by the curriculum.

For practical and career development reasons, new anatomy faculty should anticipate progressively repurposing their digital teaching materials (e.g., presentation slides, images, and videos) for laboratory activities. The majority of such content is reasonably produced using the familiar current tools of the trade: word processing software (Microsoft Word or compatible LibreOffice Writer), presentation software (Microsoft PowerPoint or compatible LibreOffice Impress) for creating exercise pages and layout designs in PPT format, and Adobe Acrobat for producing PDF standard format documents (see Chap. 28). With the pervasive use of personal mobile devices, PDF documents have the added advantages of being usable on all PC operating systems, tablets, and smartphones. Digital videos can be incorporated in both PPT and PDF formatted files, or a standalone video player may be used on laboratory workstations.

Design and Use Principles for Multi-exercise Laboratory Sessions

We begin with examples of methods currently used for our medical school's transitional, first-year clinically oriented systems/blocks plus PBL curriculum, with dedicated anatomy laboratory sessions in each block combining small group "exercises" with 14 prosected cadavers, cadaveric specimens, skeletons, traditional anatomical models, embryology stations, and medical images. Due to limited time and complex scheduling constraints, three-hour afternoon laboratory sessions accommodate a large class (180) in thirds, groups of 60 on successive days of the week. The individual regional and integrative anatomy laboratory session activities are organized around assigned small groups of students rotating between prosected cadavers, osteology, models, embryology, and other learning stations. Most often, groups determine their own pace during exercises, but limited occupancy activities (e.g., ultrasound demonstrations) may require regular timing (e.g., 30-minute sessions) to accommodate all.

Current medical curricula require inclusion and integration of medical imaging in preclinical anatomy coursework. In our transitional system/block first-year curriculum, radiology is a "thread" that runs through the year, with relevant medical imaging content included in laboratory exercises and as regional/block radiology presentations before lab sessions.

In our setting, faculty in charge of individual anatomy laboratory sessions create software "modules" for directing group learning activities at each of the rotation stations. Most simply, module can be a short presentation "slide stack" guiding students through a specific exercise, with the PPT or PDF file viewed in presentation mode. For example, a prosection station for the thorax might have three different structural identification exercises (and modules) for the internal and external heart and for the mediastinum in general. An embryology station might have module on the development of the heart, with a narrated video showing the formation and closure of septa. All structures identified in modules are also included in the larger laboratory identification list documents specifying structures that may be tagged or "functionally questioned" on practical examinations.

Creating Laboratory Teaching Modules

- Reduce intrinsic and extrinsic sources of distraction (cognitive load) overall
- Keep each page of a module informative but simple
- Chunk information and highlight what is important
- Simplify images and figures and use color-coded labels to clarify
- List page numbers so students know how far along they are
- Embed questions and answers for students

Faculty developers are urged to make use of multiple adult learning strategies in the design and use of software modules for multiple types of laboratory activities. Modules should be easy to use in small groups and for individual study. As multimedia learning materials, they should adopt page designs that reduce intrinsic and extraneous distractions—cognitive load—and limit important facts and image displays to a few per page. As discussed in the prior chapter, "Role of Image and Cognitive Load in Anatomical Multimedia," best recommendations include the following: (1) pay attention to cognitive load: less is more; (2) whenever possible, use auditory information; (3) chunk, highlight importance, and pace the presentation to avoid overwhelming; and (4) simplify visualizations and color-code labels. Figure 29.1 shows four pages from a hyperlinked PDF module on the anatomy of the heart.

In a laboratory, recommendation 2 (use audio) may be selectively applied to module design. In the multigroup large laboratory setting, student auditory information (and speech) processing systems are used predominantly for vital intragroup communications that direct, reinforce, and correct individual students' ongoing learning. It is most common to encounter one student reading each slide and running the computer while the group identifies and discusses each structure. Any modules with videos and sound are preferentially used at workstations separated from the grouped cadaver tanks, as well as for individual study.

Assessing Student Anatomical Knowledge

Formative and summative assessments (quizzes and examinations) may be the primary responsi-

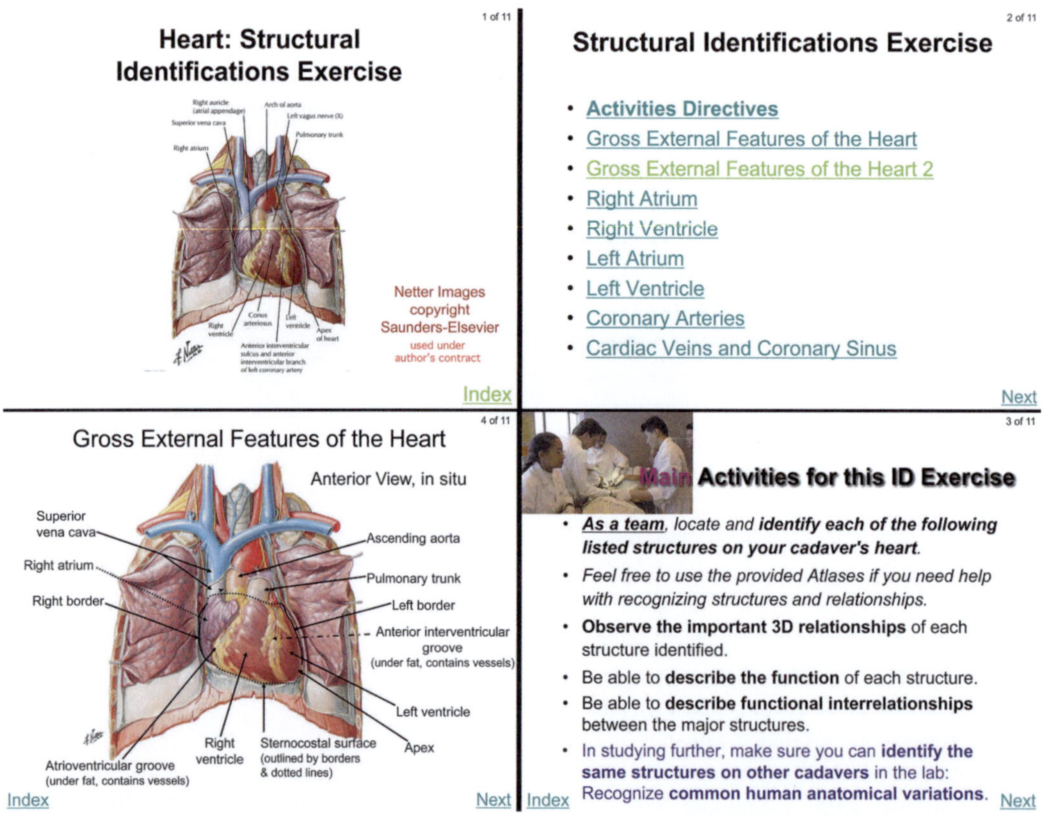

Fig. 29.1 Montage of four screen-capture pages from a heart structure identification module in hyperlinked PDF format, for a cadaver laboratory exercise. The green and aqua texts are hyperlinks that jump to the Index and next pages, just like a Web browser. Upper left, title page of module; upper right, Index page linking to all the slides; lower right, group instructions for the exercise; lower left, first external heart structural identification page

bility of anatomy course faculty, or these functions may be integrated within an online learning management systems (LMS). For example, in our medical system/block curriculum, faculty are required to submit multiple-choice questions (MCQs) in DOC format for use on the LMS-based weekly assessments and summative block examination. Anatomy questions are included in the multidisciplinary content for students' required weekly online assessments, and the LMS automatically calculates weekly and online block final examination scores.

Beyond those curriculum-dictated recorded grade assessments, however, our Anatomy Thread has retained the right to administer a required laboratory practical examination. This contributes a significant fraction of points to the students' final block scores, ultimately "threshold-graded" to pass/fail in accordance with curriculum policy.

Remaining legacy timed practical examinations, with medical students identifying tagged structures on cadavers, specimens, bones, and models, have evolved to include medical images along with new methods for administering and answering questions. Practical enhancements can also include computer grading [5] and MCQ formats [6] as well as collaborative practical examinations [7] compatible with team-based approaches.

Current U.S. standards for accommodating special needs (e.g., identified learning disorders) and remediation can be also supported with computers. In our experience, this has included regular flexible or open timing and separate administration logistics for designated students.

Additional weekly laboratory quizzes may also be administered before laboratory rotations, typically with the use of presentation slides. Properly designed PPTs can be especially effective learning tools with or without audience response systems (ARS), allowing students to see immediately how their anonymous answers to structural questions compared with the rest of the class' response.

For those implementing a team-based learning (TBL) approach to laboratory sessions, students should be presented with online individual and group readiness assurance tests and the application exercise for each laboratory session. Partial implementation in our transitional curriculum has set up selective TBL assessments within the LMS framework. Please see the chapter on team-based learning for more information on such formal assessments.

Laboratory E-Learning Resource Hosting

In a larger school, all courses' digital learning resources and documents are usually made available to students online through the institution's LMS, linked through the individual student's course schedule. By general adult curricular standards, new learning resources might best be posted and available before scheduled laboratory sessions. But for our medical students, required lecture materials (and recordings) are made available the Friday before labs, while the actual laboratory exercise modules are open for download after the last session. This encourages active participation in the laboratory group learning setting while supporting review and outside individual study.

All up-to-date e-learning resources used in each laboratory session are also best stored on a secured and backed-up local web server. This simplifies and organizes multiple scheduled sessions as well as after-hours individual and tutoring use. Other compelling reasons for maintaining a dedicated laboratory server include hosting site-licensed software, the ability to host new software development and other teaching resources not supported by the LMS documents frameworks. Ultimately, simple Web pages can be powerful media for defining overall organization of daily laboratory tasks and small group rotations, as shown in Fig. 29.2.

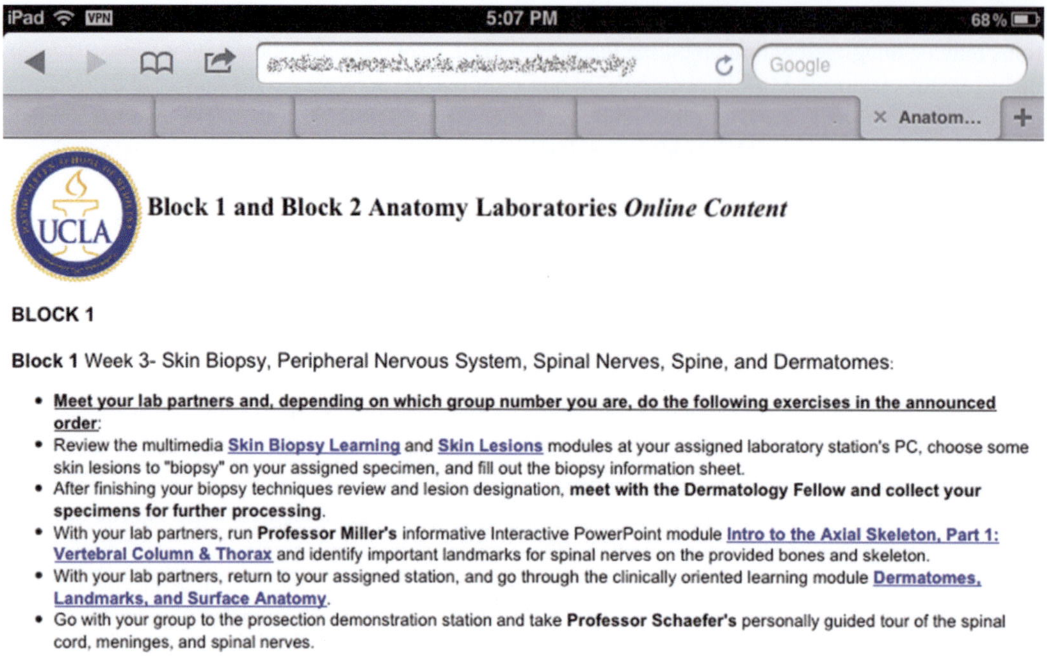

Fig. 29.2 iPad screen capture of a simple HTML Web page demonstrating how a laboratory day's mixed cadaver prosection, demonstration, osteology, and other exercises can be organized. The blue hyperlinks download e-learning materials to the users' devices, whether they are PCs in the anatomy laboratory, personal laptops, or mobile devices

Virtual Anatomy's Third Decade and Practical 3D Models for Laboratories

In the 1990s, it was proposed that computer-based "virtual anatomy"" might be used for helping students to learn by directly perceiving three-dimensional (3D) details and relationships between structures [8, 9]. Digitally reconstructed volumetric models based on the U.S. National Library of Medicine's Visible Human Project cross sections were first conceived for making structurally precise human simulations [10], and other 3D data sources soon became available, based on processing of magnetic resonance (MR) and computed tomography (CT) clinical imaging datasets. It was asserted that a "digital cadaver" might make it possible to learn anatomy without the need for embalmed specimens and dissection, perhaps with the use of stereoscopic three-dimensional (3D) "virtual reality" or holographic display technologies.

How Has the Promise of Virtual Anatomy Technology Worked Out?

From the perspective of this decade's educational efficacy research, several studies have shown little anatomical learning advantage in using 3D digital models versus traditional plastic and wood anatomical models and print images [11–13]. Ongoing reservations were well expressed by Pawlina and Drake, who declared in their 2013 *Anatomical Sciences Education* journal editorial [14] "Anatomical models: Don't banish them from the anatomy laboratory yet."

From the technology access perspective, schools can now purchase a variety of commercial virtual anatomy educational software products for use with flat-screen and projector displays. For example, Anatomy.tv software from Primal Pictures [15] provides interactive flat 2D screen views of a 3D modeled body ("2D/3D") that can be repositioned and revealed in layers, but the tissue realism and precision of the structural renderings are not high. Netter 3D is a comparable online interactive body modeling application based on the legacy Frank Netter anatomical illustrations. The most comprehensive of these products is VH Dissector for Medical Education from Touch of Life Technologies [16], which fulfills most of the educational promises of the Visible Human Project and data. A complex suite of software, VH Dissector, offers multiple interactive 2D/3D viewing functions, along with viewers for rendering models from cross-sectional medical imaging data (e.g., CT).

A few software systems can use electronic viewing glasses to provide stereoscopic (non-holographic) perception of 3D from screen displays. Most of these visually impressive virtual anatomy systems function in large part as interactive digital 3D atlases with freely rotatable perspectives, labeling, removable layers and structures, and a variety of notation features. Currently, the high-end, pool-table-sized touch screen Anatomage Table [17] can render interactive 3D views of a life-sized digital cadaver based on Visible Human or DICOM imaging data, while the VH Dissector software can also be run on a similar-size DICOM radiology workstation if desired.

Commercial virtual anatomy products may be attractive to teachers and supported by administration because they provide comprehensive all-in-one whole-body packages for delivering digital content. They can be useful teaching assets when no cadavers and few other human anatomy laboratory resources are available to the school. However, beyond substantial cost, all share a specific common limitation for current medical and postgraduate curricula: they simultaneously include too much overall content for time-effective goal-directed learning, with too few ready clinical integrators and too much anatomical detail. In the current context of managed, abbreviated basic science instructional time, there may be little opportunity or motivation for all students to make heavy exploratory use of a "virtual cadaver" system as part of scheduled instructional time.

On the other hand, selective use of 3D technology fits very well into multitask laboratory designs like the previously discussed station- and module-based medical anatomy setting. Highly important but difficult structural relationships or vital basics are chosen for effective demonstration with 3D digital models or simulations.

Anatomical modeling and 3D printing (additive manufacturing) are currently active areas for research [18–20], and practical systems are in ongoing use in clinical applications such as producing surgical and dental prostheses and tissue engineering [21]. Early-career anatomists might want to grow their own expertise in 3D modeling and printing, engaging an evolving technology during its continuing digital media-driven growth. Chapter 35 on 3D printing comprehensively examines the different types of printing methods, materials, and processes that can be used for high-precision anatomical models.

Building 3D Digital and Physical Anatomical Models

One of the authors (JL) has begun trials on a new class of mixed virtual anatomy laboratory exercises that combine PC screen-displayed digital 3D models with physical anatomical models built with extruded plastic on a relatively inexpensive (<2500 US$) "high-end" consumer 3D printer. This produces "traditional" physical anatomical models that correspond precisely with interactive, screen-based digital models, providing both media for visual and tactile interactions in learning.

Figure 29.3 schematically depicts the processes, software, file types, and information flows involved in the main steps of creating and preparing 3D anatomical models for such dual use.

Digital 3D anatomical models can be created in several ways. The most useful to anatomists overall may be (1) generating models directly by processing MRI and CT image series data and (2) "hand-building" model geometry using a 3D modeling program or features of digital content creation (DCC) software. In addition to 3D modeling, DCC software usually integrates scene rendering, animation, and digital video recording functions to support cinematography, videography, and computer games design.

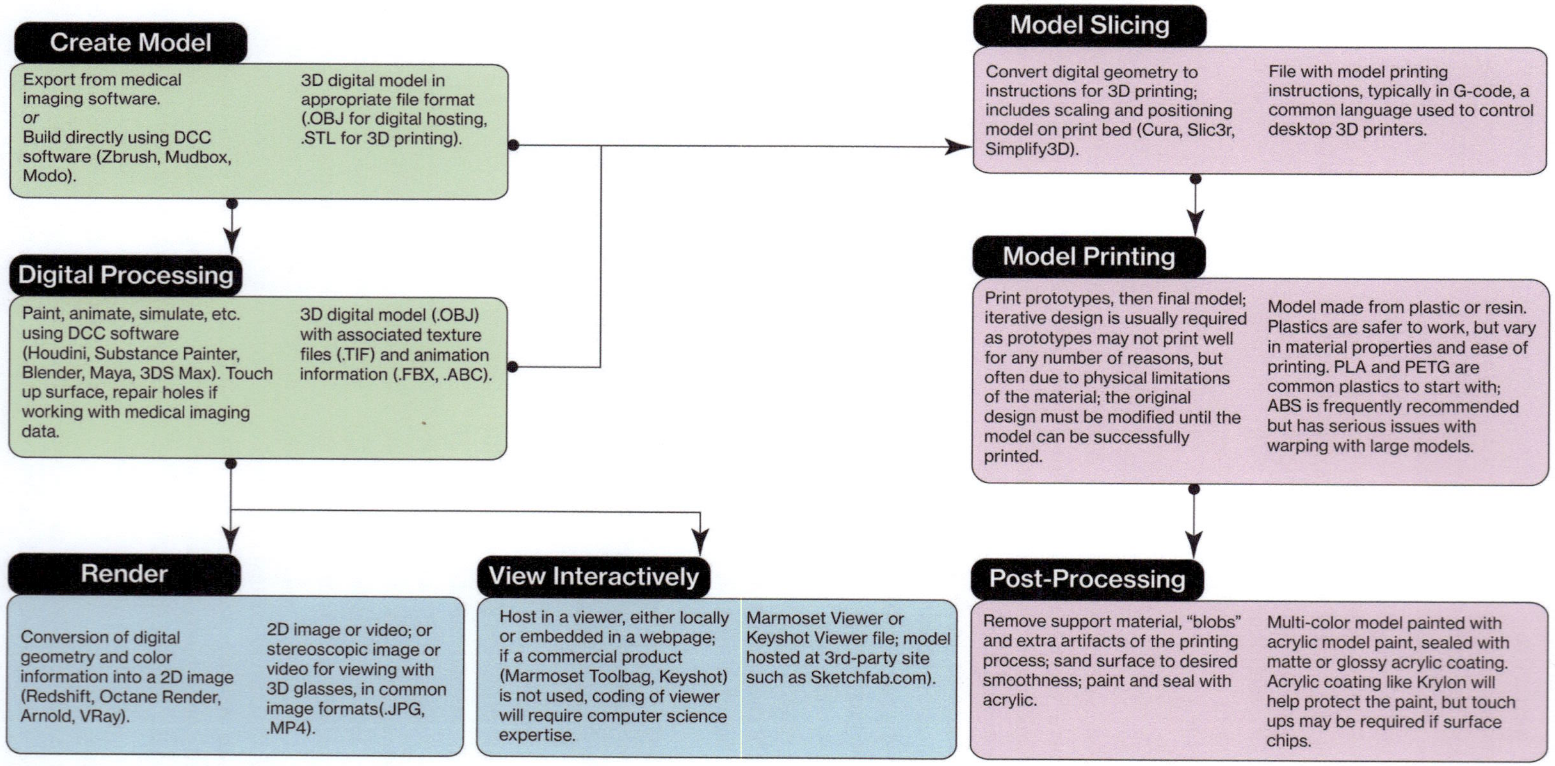

Fig. 29.3 3D Anatomical model development roadmap—Schematic view of the main processes in creating 3D digital anatomical models for use on PCs and for 3D-printing anatomical models. Abbreviations: ABS, acrylonitrile butadiene styrene; DCC, digital content creation; PETG, polyethylene terephthalate glycol; PLA, polylactic acid

Volumetric models have been generated from MRI and CT data segmentation for quite some time now, and these have the benefit of being true to the anatomy. However, the resolution of the image series determines the detail of any exported models. Smaller vital structures like nerves and vessels may be extremely difficult to model effectively from these data.

Generating anatomical models requires a significant investment of the anatomist's time and effort for initially learning the software and then for creating the design, but the benefits can be great. By directly creating the content, the anatomist is not limited to any particular structure list or presentation; the teacher can design exactly the anatomy desired for the model. Content can be modified in response to student feedback or to illustrate anatomical variations/pathological conditions. Furthermore, in developing any anatomical model or illustration, there are cases where "artistic" liberties may need to be taken to improve clarity: for instance, models that maintain the true size and branching patterns of small nerves may be difficult for students to see and learn from. Using 3D modeling software provides complete editorial control over the features in the model, and size and complexity of the model can be adjusted to meet the needs of the class.

Choosing the best 3D modeling or DCC software requires some study beforehand. If the user is only creating models for 3D printing, then modeling software may be all that is needed. For hosting digital models that need to be colored or texturized, another program or additional programs might be required. The model shown in Fig. 29.4 was developed with Pixologic's ZBrush [22].

After creating a suitable digital 3D model, content delivery is the next challenge. Several options exist, including basic 3D model viewers for all operating systems. Currently, the easiest and most informative method may be to use a third-party website to host the displayable models, which takes little effort on the part of the anatomist but may raise concerns over copyright and access. Some DCC packages include their own free viewers for displaying user models, a potentially more secure and reliable option. The model shown in Fig. 29.4 and others have been presented in laboratories and lectures using a web-based player on the Sketchfab.com 3D sharing service [23].

When converting digital models for 3D printing, the complexity of a "raw" model may exceed the limitations of the printing technology. This frequently requires "cleaning up" the original digital model by simplifying its topology (e.g., reducing the number of polygons) or by patching unprintable holes. After the model is fixed, it must be "sliced" by another program into a series of horizontal layers that the printer's computer controller will use in building a physical model from the bottom upward.

There are several different types of nonindustrial 3D printers available for under US$ 3000, the best of which are capable of producing a plastic model large enough to use in an anatomy laboratory. These extrude melted plastic in layers through computer-positioned heated print heads. This project used the LulzBot Taz 5 [24] Cartesian printer with polyethylene terephthalate glycol-modified feedstock, suitable for durable rigid models.

Discussion and Future Prospects

We have presented the scenario of multi-activity anatomy laboratories intercalated into a current clinically oriented system/block plus PBL medical curriculum. We also called it "transitional," with recent introduction of selective block content flipping, introduction of several TBL sessions, and a newer curriculum under development. The current curriculum followed a millennial round of reform after prior cycles of redevelopment, with an overall trend toward continuing reductions in anatomy class time allotted predominantly to mixed resources laboratories.

The pressure of contact-hours reduction and the loss of lecture time for presenting basic anatomy content create the need to deliver important content in alternative formats, such as "self-study modules." For anatomical sciences, this requires that teachers develop adult learning resources

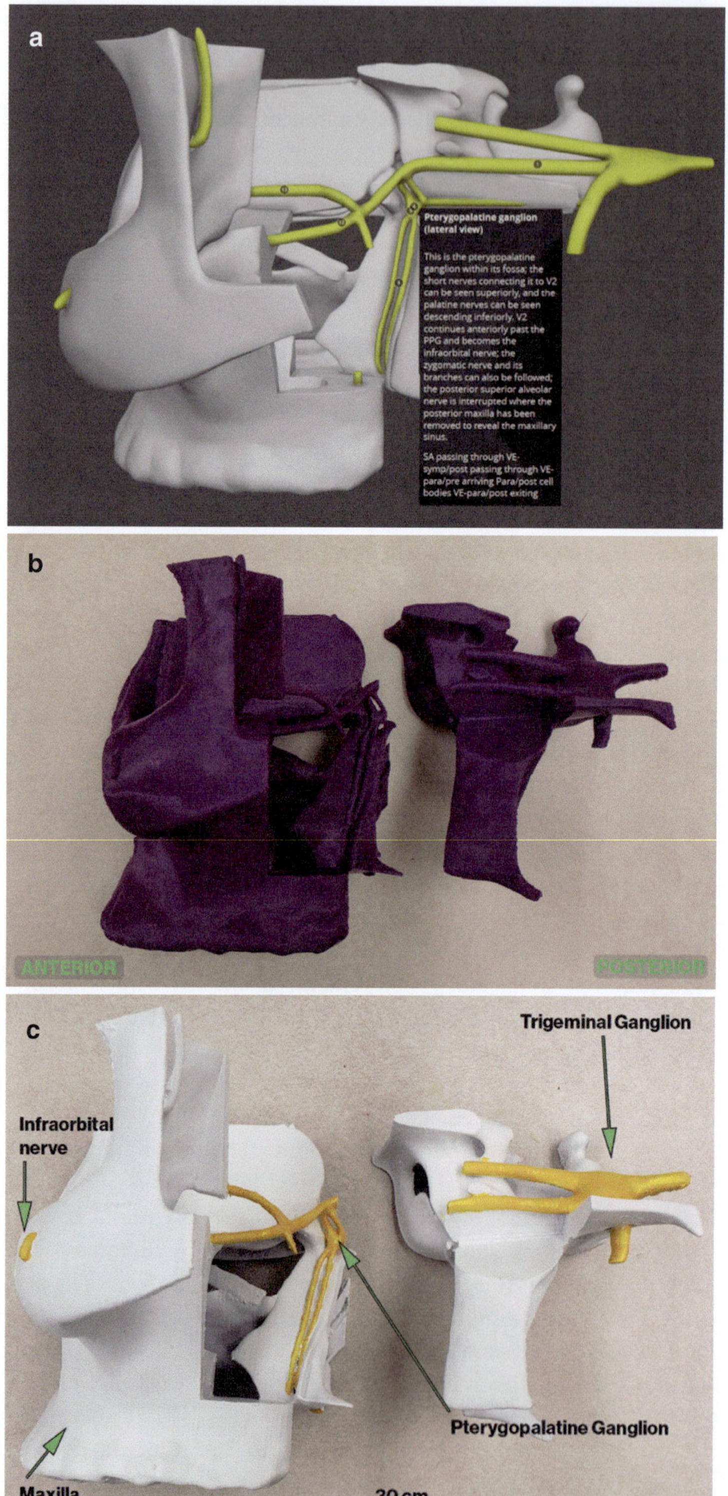

Fig. 29.4 (**a**) Screen capture image of a 3D model of the pterygopalatine (PPG) and trigeminal ganglia in relation to the facial skeleton, created using the digital ZBrush 3D software and displayed by Sketchfab.com webapp. A student has rotated the model to view the ganglia and has selected the PPG annotation tag, popping-up an anatomical information box. (**b**) The same model 3D printed in PETG in two parts. This accommodates printer size limits while allowing student to explore how the articulation of the sphenoid bone with the palatine bone forms the sphenopalatine foramen. (**c**) The plastic model in (**b**), at an intermediate stage in painting. 3D printing is currently limited to just one or two colors on most consumer-level machines, so post-print processing is usually needed to produce the final anatomical model appearance and to enhance structures (e.g., yellow nerves). The model was sanded and primer coated before acrylic painting

and learning activities that manage a high overall density of clinically relevant information in ways that also capture and control the attention of the learner. We have historically addressed this situation in part by providing didactic content online ahead of laboratory time and expecting students to have a basic grasp of this information when they set foot in the laboratory.

Laboratory time is then specifically used for group-based visual recognition and function exercises on cadavers, as well as for smaller laboratory activities such as demonstrations of complex regions, live ultrasound, explorations of osteological materials, and presentations about embryology. These activities provide engaging diversified content learning utilizing small group social interactions (peer group teaching) in addition to other flipped classroom elements. Of course, laboratory time is also used for teachers clarifying questions and assisting individual students with any unique challenges to their mastery of the content.

Managing such e-learning resources entails considerable ongoing costs for equipment support, maintenance, and replacement, and these will vary depending on the organizational structure and scale. In the authors' large research university medical school, network services, security, backup, and class support are provided by a large, budgeted academic information technology (IT) staff. Historically, experienced faculty were also allowed to connect their own laboratory servers securely to the network, an arrangement that smaller schools can make to support their new learning resources development and hosting. Author (RT) has used inexpensive (<800 US$) small PC servers reliably for medical anatomy and embryology laboratories for nearly two decades, with laboratory PCs maintained by an assigned IT worker.

Current U.S. postsecondary schools place emphasis on providing their postmillennial students with extensive mobile learning and campus information resources and media for their ubiquitous smartphones and tablets, so mobile-friendly file formats (e.g., PDF) may reliably be specified for all curricular materials.

Some anatomy laboratories however may proscribe mobile device use, due to regulations which prohibit use of cameras around donor remains, the case at the authors' school. It is generally unwise, if not absolutely forbidden, to post images of cadaver donor remains on laboratory servers, because even with the best IT security and private networks, students may unfortunately choose to share files externally. The student honor code and laboratory regulations should explicitly detail such restrictions, as well as forbidding the outside sharing of learning materials which may include copyrighted and sensitive content.

Since constructivist learning reforms have also continued to pass through primary, secondary, and tertiary schools, career teachers of anatomy at all levels should be prepared to have other opportunities to redevelop their laboratory teaching methods to meet new curricular and student needs in health sciences career tracks. And beyond merely supporting evolution of new curricular content, design of new multimedia learning resources must implement now-familiar adult learning principles, as well as engaging current student mobile technology preferences, with their emphasis on time-efficient, "high-yield" learning.

The limitations of time and other concurrent educational activities in current medical curricula may constrain use of comprehensive virtual anatomy systems (e.g., Anatomage) in primary medical teaching laboratories. Despite recent concerns about educational efficacy, however, multiple introductory undergraduate courses might make better shared use of 3D anatomy software for study in more exploratory laboratories, when use of cadavers and more traditional laboratory resources is not possible, and *especially for distance learning*.

Digitally designing and printing one's own models can be a powerful and still evolving tool in the anatomist's teaching skill set. Operational costs—software licenses, computer, and 3D-printer hardware and printer plastics—may be easily accounted for, but there are very significant costs in terms of personal labor and time to be considered before investing heavy efforts. These personal costs may be prohibitive, depending on the constraints of the academic appointment: If the institution does not credit the development of innovative teaching materials in their promotional process, it is less likely that the anatomist will devote the necessary scholarly time in favor of activities that do lead to promotion.

It takes much time to become proficient with good modeling software and with the processes of 3D printing. Both skills present variably steep learning curves. Once proficiency is gained, the time to build a model from scratch depends on the individual's knowledge of the true proportions and shapes of the anatomy, the efficiency of using the software tools, and the complexity of the model. The time to convert that model into a final physical product depends on mastering the actual printing process, which is governed by a multitude of variables—print speed, print temperature, layer height, etc. These settings will depend not only on the characteristics of the plastic but also on the specific hardware and the ambient environment of the printer.

Tutorials or even paid coursework can help teachers with understanding these modeling and learning processes, and for the more popular 3D programs, there are freely accessible learning materials on the internet. It is also highly recommended that interested anatomists seek out and join appropriate online and local 3D development communities that welcome and provide help for novices.

Any teacher who can afford the research equivalent time and effort to include these technologies in the classroom will have acquired new means for serving the learning needs of their specific student population. Digital modeling grants the educator greater flexibility in responding innovatively to individual and programmatic learning needs, especially when comparable commercial models are unavailable or prohibitively costly. Models can be built or modified based on student feedback, limited only by the speed with which the individual can operate the modeling software. In several cases, one of the authors (JL) has produced quick models overnight to answer questions that arise in laboratory sessions.

In the future, the active pressure from professional film and game studios on responsive software development will make 3D DCC software easier to use, more feature-rich, and more accessible to professionals outside those industries. 3D printing technology is undergoing a similar renaissance as software improves and the variety of available materials broadens. New flexible materials will allow the development of functional models, such as vertebral columns with flexible intervertebral discs, in the range of many possibilities.

References

1. Trelease RB. From chalkboard, slides, and paper to e-learning: how computing technologies have transformed anatomical sciences education. Anat Sci Educ. 2016;9:583–602.
2. Flack NA, Nicholson HD. What do medical students learn from dissection? Anat Sci Educ. 2018;11:325–35.
3. Trelease RB. Anatomy meets architecture: designing new laboratories for new anatomists. Anat Rec B New Anat. 2006;289B:241–51.
4. Cleveland B, Kvan T. Designing learning spaces for interprofessional education in the anatomical sciences. Anat Sci Educ. 2015;8:371–80.
5. Krippendorf BB, Bolender DL, Kolesari GL. Computerized grading of anatomy laboratory practical examinations. Anat Sci Educ. 2008;1:220–3.
6. Shaibah HS, van der Vleuten CP. The validity of multiple choice practical examinations as an alternative to traditional free response examination formats in gross anatomy. Anat Sci Educ. 2013;6:149–56.
7. Green RA, Cates T, White L, Farchione D. Do collaborative practical tests encourage student-centered active learning of gross anatomy? Anat Sci Educ. 2016;9:231–7.
8. Trelease RB. Toward virtual anatomy: a stereoscopic 3D interactive multimedia computer program for cranial osteology. Clin Anat. 1996;9:269–72.
9. Trelease RB. The virtual anatomy practical: a stereoscopic 3D interactive multimedia computer examination program. Clin Anat. 1998;11:89–94.
10. Spitzer VM, Whitlock DG. The visible human dataset: the anatomical platform for human simulation. Anat Rec. 1998;253:49–57.
11. Hoyek N, Collet C, Rienzo F, Almeida M, Guillot A. Effectiveness of three-dimensional digital animation in teaching human anatomy in an authentic classroom context. Anat Sci Educ. 2014;7:430–7.
12. Yammine K, Violato C. A meta-analysis of the educational effectiveness of three-dimensional visualiza-

tion technologies in teaching anatomy. Anat Sci Educ. 2015;8:525–38.

13. Preece D, Williams SB, Lam R, Weller R. "Let's get physical": advantages of a physical model over 3D computer models and textbooks in learning imaging anatomy. Anat Sci Educ. 2013;6:216–24.
14. Pawlina W, Drake R. Anatomical models: don't banish them from the anatomy laboratory yet. Anat Sci Educ. 2013;6:209–10.
15. Anatomy.tv. Powered by Primal Pictures. 2019. http://www.anatomy.tv. Accessed 11 Mar 2019.
16. VH Dissector for Medical Education. 2019. http://www.toltech.net/anatomy-software/solutions/vh-dissector-for-medical-education. Accessed 11 Mar 2019.
17. Anatomage Table 6: magnifying human anatomy to the next level. 2019. https://www.anatomage.com/table6/. Accessed 18 Feb 2019.
18. McMenamin PG, Quayle MR, McHenry CR, Adams JW. The production of anatomical teaching resources using three-dimensional (3D) printing technology. Anat Sci Educ. 2014;7:479–86.
19. Lim KH, Loo ZY, Goldie SJ, Adams JW, McMenamin PG. Use of 3D printed models in medical education: a randomized control trial comparing 3D prints versus cadaveric materials for learning external cardiac anatomy. Anat Sci Educ. 2016;9:213–21.
20. Fredieu JR, Kerbo J, Herron M, Klatte R, Cooke M. Anatomical models: a digital revolution. Med Sci Educ. 2015;25:183–94.
21. Rybicki FJ, Grant GT, editors. 3D printing in medicine: a practical guide for medical professionals. Cham: Springer International; 2017.
22. Zbrush—3D modeling software. Pixologic. 2019. http://pixologic.com/zbrush/features/overview/. Accessed 12 Mar 2019.
23. SketchFab 3D model sharing services. 2019. https://sketchfab.com. Accessed 12 Mar 2019.
24. Introducing the Lulzbot TAZ 5! 2015. https://www.lulzbot.com/learn/announcements/introducing-lulzbot-taz-5. Accessed 13 Mar 2019.

30 Designing Anatomy Teaching Spaces to Meet the Needs of Today's Learner

Quenton Wessels, Adam Michael Taylor, and Christian Jacobson

The importance of space (the physical and material locale) and place (the meaning and value linked to space) in health professions education has gained much attention in recent years [1]. This heterogeneous nature of learning spaces rests upon the alignment of educator mechanisms (e.g., institutional support, the training of academic staff, and the commitment of management) and curricular mechanisms (such as program content, adult learning theories, assessment, and contextual learning) [2]. The use of "learning space/s" in this chapter will refer to the above heterogeneous nature of the educational setting.

The global advancement of technology has had a considerable impact on health professions education, especially anatomy pedagogy, its related facilities, and learning spaces [3]. Classic curricula, those following a traditional sequential examination of preclinical basic science coursework followed by experiential learning in clinical settings, have attempted to move away from pure memorization and didactic teaching. This is, in no small part, in response to changing content, the needs of students, and the adoption by some institutions of the Western Reserve curriculum or derivations thereof. Institutions such as McMaster and Maastricht have, since the 1960s, restructured their medical curricula and associated anatomy courses toward problem-based learning (PBL) [4–6]. Learning in these circumstances is constructive and student-centered and allows active learning [6]. It is practice based and learner driven and as such highlights the importance of the informal and hidden curriculums in medical education. This movement has however been met with some apprehension due to perceived lack of structure and progression, a lack of rigor in specific preclinical disciplines, and pointedly, the taxing resource requirements of these curricula. This becomes most apparent when the student numbers grow in the excess of 100 [6, 7]. Other curricular approaches have evolved from classic and/or PBL origins; these include team-based learning (TBL) [8–10], self-directed learning [11], computer-aided learning (CAL) [12–14], blended learning [15], and hybrid models.

Today, regardless of the formal curriculum, anatomy pedagogy relies strongly on multimedia equipment and prosected specimens. Furthermore, despite the numerous teaching approaches, there appears to be a revival in anatomy pedagogy in medical curricula. This revival

Q. Wessels (✉)
Department of Anatomy, School of Medicine, University of Namibia, Windhoek, Namibia
e-mail: qwessels@unam.na

A. M. Taylor
Lancaster Medical School, Faculty of Health and Medicine, Lancaster University, Lancaster, UK
e-mail: a.m.taylor@lancaster.ac.uk

C. Jacobson
Department of Biology, University of Waterloo, Waterloo, ON, Canada

L. K. Chan, W. Pawlina (eds.), *Teaching Anatomy*, https://doi.org/10.1007/978-3-030-43283-6_30

is occurring in the face of a global reduction in anatomy course content, decreased time spent on cadaver dissections, and a global shortage in anatomy instructors [13, 16–19]. These reductions run counter to an increased emphasis on constructive learning and a holistic approach to medical education but are entirely practical given the availability and associated cost implications of cadavers in most parts of the world. Given these disparate trends, advances in anatomy pedagogy are necessary in modern medical education. The challenge lies in objectively measuring how much anatomy is enough and largely depends on the viewpoints of the traditionalists and the educationalists [13].

Curricula evolve to suit the health-care requirements of patients, and dissection laboratories have similarly adapted to the educational needs of students. Old anatomical theaters paved the way for today's state-of-the-art facilities [20]. The design of anatomy pedagogy facilities for today's student requires an understanding of the current generation and anticipating the needs of the next generation. Today's student has been described as being comfortable with technology and is attracted to the use thereof [21]. From an educational perspective, they have a propensity to prefer a variety of facts that are skillfully and rapidly conveyed [22]. Technological progress and the availability of electronic devices allow today's student to accomplish various tasks simultaneously; this generation has high expectations from technology and expects utility in all situations [21–24]. Furthermore, they are team focused and interdependent with an ability to unify and organize, but they require structure [23, 25]. It is safe to say, given today's trends, that we should anticipate continued adoption of e-learning and a move toward the further integration of mobile devices.

Today's Learner

- Is a team player
- Requires structure and guidance
- Is comfortable with technology
- Relies on fast-paced facts

It is therefore an imperative, from a design perspective, to focus on learning spaces that are flexible and that allow for aspects of TBL and e-learning. The application of TBL in anatomy relies on predetermined reading assignments (pre-class preparation) for the students followed by content-specific in-class discussions. The principle relies in teamwork and the provision of an opportunity to use the assigned reading material and resources to solve problems. TBL encounters are supervised and expect both preparation and attendance by students in order to attain specific competencies and capabilities in the subject matter [8–10]. A flexible learning environment in this instance will allow for easy reconfiguration to suit discussion groups to formal didactic lectures. A major concern for universities is the cost of teaching, the associated space that is required for these activities, and the support services required. In many instances, these support services must allow secure and hygienic accommodation for human remains in accordance with prescribed regulations. If needed, adequate space for the processing of human remains and the preparation of museum specimens should be provided that is not too extravagant or wasteful. It is also vital to consider the university aims beyond that of teaching. These aims typically include research, continued professional development, and institutional cooperation [26].

Obviously, learning spaces are expensive. Long-term resources and careful consideration must thus be taken at the point of investment. Critically, faculty must adapt to and adopt these resources if they are to improve educational outcomes. Lecturers who do not actively use collaborative or cooperative teaching techniques typically do not adopt these practices even if teaching in a space that is conducive to active teaching. Similarly, active lecturers that promote dynamic learning are likely to maintain their particular style of education [27].

Evidence indicates that teaching space, and the implementation of multimedia within that space, has a dramatic effect on learning outcomes [28–31]. Something as simple as applying multimedia design principles to lecture slides significantly improves short- and long-term retention of material [31, 32]. Minimally, new facilities should make some attempt to include student engagement systems. Audience response systems (ARS) or, in the vernacular, "clickers" or "zappers" are electronic voting systems typically deployed in larger lecture environments to increase student participation within traditional, didactic-style, lectures [33, 34]. All modern systems are wireless, but hardware requirements and thus fixed costs vary considerably from one manufacturer to the next. Two basic technologies predominate, however, radiofrequency-based systems that use proprietary hardware built directly into lecture facilities and mobile phone-based systems that append onto and are dependent on Internet connectivity and services. Regardless of platform, data indicates that lecturers often see increased attendance rates and quantifiable improvements in student performance coincident with ARS use [33]. These systems also enjoy a high student satisfaction rate [33–35]. This is not surprising as there is a growing body of evidence that active, "constructivist"-style lectures, as opposed to traditional theory-based "objectivism"-style lectures, are better received by students, and students are more satisfied with the learning experience [30, 31, 36].

Learning in a gross anatomy laboratory can be a function of the various learning activities within a specific community that relates to the subject matter. It is therefore situated as proposed by Jean Lave and Etienne Wenger in 1991 [37]. Generally, two distinct educational settings exist within medical education. The one is where the students learn, such, as the dissection room, and the other where they apply their knowledge. The latter refers to a clinical setting or practice setting, and this is typically separate from the milieu in which students learn anatomy [38]. This division creates a gap between situated learning within a community of practice and needs to be bridged. The environment and context has been suggested to have a positive effect on the recollection of information [38]. For instance, research demonstrated the positive impact of wearing scrubs on contextual learning. Their findings show that those students that were assessed in the same context as they were trained, remember significantly more information [38]. Contextual learning of anatomy sparks ideas such as the incorporation of theater lights, a gowning area, and a scrub room. The reproduction of contextual and environmental factors to match a clinical setting should therefore be considered.

Learning Spaces and Anatomy Pedagogy

Education and the learning space and place are closely intertwined [39]. The conceptual and practical interplay between place, space, and learning is pivotal for the construction and remodeling of learning spaces. The work of Bleakly, Bligh, and Browne refers to these interactions and mentions hospital architectural design and its influence on patient care [40]. The use of place in undergraduate education as well as the influence of vertical hierarchies and horizontal layouts influences interprofessional interplay. Interprofessional education (IPE) relies on such aspects as flexibility, interaction, communication, and student focus [41]. Flexibility in these spaces is pivotal in allowing for the accommodation of current and future technological and pedagogical trends (Fig. 30.1). Future-proofing space is difficult. It is nearly impossible to anticipate the direction of technological advancement; tablets, for instance, comprised a sliver of computer sales until recently. Further, if history is our guide, how anatomy faculty, staff, and students use space may change dramatically.

Medical education during the Renaissance was marked by the study of human anatomy through observation within anatomical theaters [42]. This was a new dimension in medical education as the study of anatomy was previously restricted to the study of ancient texts [43]. The first of these permanent anatomical theaters was completed in Padua, Italy [44], in 1594 and this funnel-shaped wooden construct served as a blueprint for many others [45, 46]. Student involvement or the "Paris method" was brought back to London in 1746 by William Hunter, and cadaveric dissections continued to gain popular-

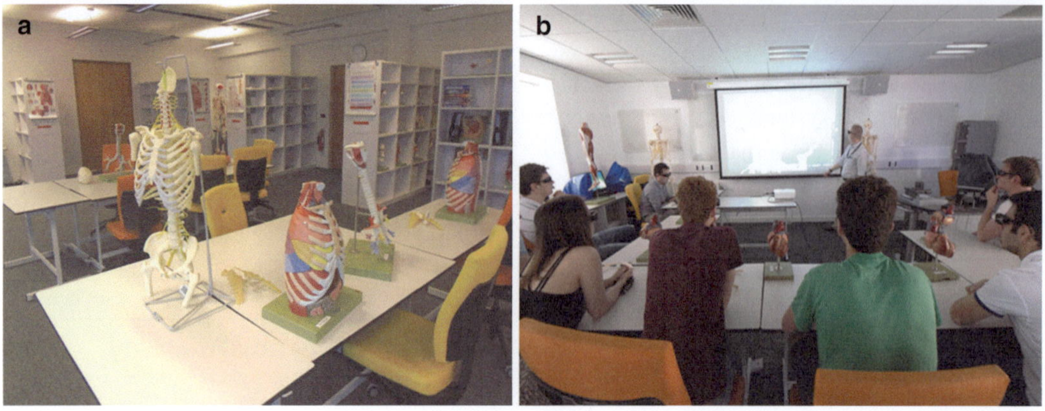

Fig. 30.1 Lancaster University, Medical School's CALC (Clinical Anatomy Learning Centre). (**a**) A combination of anatomical models, digitized medical and histology images, and e-learning resources is used to teach human anatomy. (**b**) The same learning space can be transformed into an interactive platform which permits the use of 3D projections and an Anatomage table. Such a flexible environment can also accommodate augmented reality (AR) teaching modalities

ity in the years that followed [47]. The adoption of PBL curricula by many institutions coincided with the development of lifelike simulators, models, and advanced computer simulations. In many institutions, these developments brought about dramatic changes in the use of anatomy spaces. Certain technologies are likely to play a critical role in future educational space design regardless of the curriculum. Wireless and wired, fixed networks are and will, in one form or another, be critical in future spaces [48].

Key Design Considerations

Defining Your Needs

Modern-day anatomy curricula have become more interactive and clinically orientated and, in contrast to classical didactic lectures, are riddled with detail. The design of a gross anatomy laboratory or appropriate educational or learning spaces depends on the teaching methods employed and with each comes specific challenges. For instance, with a curriculum that focuses on cadaver dissections, there are challenging infrastructural requirements. In general, four broad areas should be established within any anatomy facility: (a) public space, (b) teaching and learning space, (c) practical/simulation space, and (d) related support space. Each of these areas has its own specific requirements as listed below:

- Public space—social space, for leisure and study
- Teaching and learning space—multimedia-ready, multifunctional, reconfigurable
- Practical laboratories/exhibition space—dissection laboratories, simulator and anatomical model space, and anatomy and pathology museum
- Support spaces—offices, cold storage, general storage, locker rooms, embalming facilities, a maceration area, and water purification

These learning spaces correlate with the ideal anatomy learning content, which, as proposed by Sugand and colleagues in 2010, include the following entities: dissection/prosection, anatomical models (Fig. 30.1a), interactive multimedia such as three-dimensional (3D) projections and an Anatomage table (Fig. 30.1b), procedural anatomy, surface and clinical anatomy, and medical imaging [20]. In general, specific design considerations have increased over time beyond the conventional needs of adequate lighting, plumbing and water purification, total laboratory floor space, adequate ventilation in the case of formalin-based embalming techniques, and waste management [18, 45, 49].

Adequate ventilation is also required when formalin-based wet specimens are used for demonstrations or assessment. Air quality, according to the American Society of Heating, Refrigeration, and Air-Conditioning Engineers [50], can be ensured through at least 12 air changes per hour along with a supply of fresh air, a negative pressure, and the expulsion of used air to the outside. Furthermore, an average room temperature of 21 °C should be maintained. The same standards can be applied to other specialized areas such as embalming and maceration rooms [50].

Any formaldehyde-containing wastewater, including water drained from hand basins, should be processed prior to its recirculation into the municipal system (Fig. 30.2 (14)). A water purification plant can accomplish this in conjunction with easily cleanable surfaces and the use of laminated poly-flooring with drains. Sequential processing involves filtration through

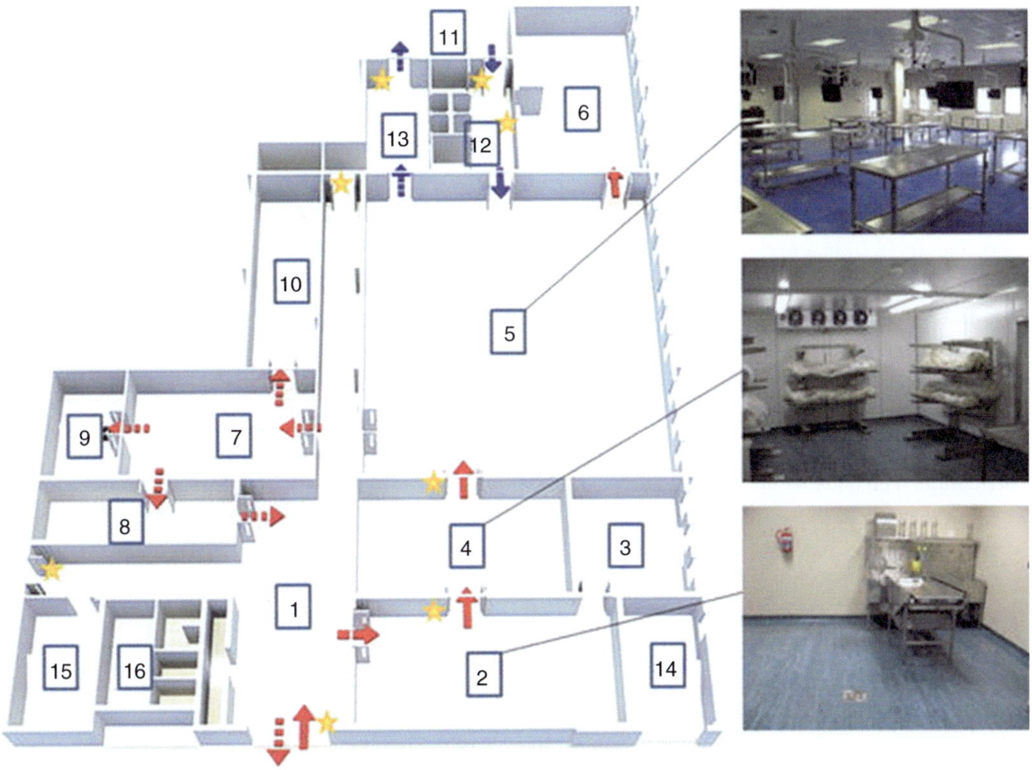

Fig. 30.2 A three-dimensional blueprint of the typical facilities associated with anatomy teaching as well as a selection of photographs. The delivery area (1), embalming facilities (2), mortuary refrigerators (3), and refrigerated storage (4) are separated from the main dissection hall (5) as well as postgraduate dissection hall (6). Students enter the dissection hall from the north (12). Following dissection, the students exit via a second set of doors (13), and soiled coats are dropped off in designated bins (12) toward the atrium of the facility (11). The atrium should be closely associated with resource centers and lecture theaters. Male and female toilets are also provided along with a locker room (not in view). The cadaveric material is processed (7) to either become wet specimens (9) or macerated (8) for osteology material (10). A water purification plant (14) ensures chemical possessing prior to the introduction of the water into the municipal system. Technical staff facilities are adjacent to the embalming and maceration rooms and an office is also provided (15) as well as toilets (16). *Yellow stars* indicate biometric access points and restricted access. The *blue arrows* depict the movement of the students and the *red arrows* the subsequent processing of cadaveric material. *Solid red arrows* represent the sequential movement of cadaveric material into the dissection hall. *Dotted red arrows* point to the movement of cadaveric material away from the dissection hall after completion of the curriculum. *Dotted blue arrows* represent the movement of the students after dissections. (Adapted from Wessels et al. [51])

a polypropylene filter, hydrogen peroxide oxidation and pH correction, and lastly additional filtration through sand and granular activated carbon. From here, the processed water can be introduced into the municipal system as gray water. The specifications vary based on the frequency of embalming, number of workstations, and wet specimen usage [51].

Technological advances include the further integration of audiovisual equipment and associated computer or network support for these facilities. This in turn accommodates teaching modalities such as computer-assisted learning (CAL) and the presentation of medical images, X-rays, and MRIs, in conjunction with dissection or use of prosected specimens [1, 14, 18]. An example of such an application is the work presented by Reeves and colleagues in 2004 that integrated wall-mounted Apple iMac computers at each of their 26 cadaver workstations [52]. In addition, the work of Wessels et al. in 2012 illustrates the incorporation of ceiling- and wall-mounted displays and power points at the cadaver work stations [51]. Wall-mounting preserves floor space, and the anatomy faculty and staff tailored an anatomy software package that complements an integrated systems-based medical curriculum. The package includes a digital dissection guide, medical images (CT scans, X-rays, and MRIs), and cross sections related to the course material. Their work showed that computerization of the workstations, in conjunction with the developed software, promoted autonomy, student proficiency, and the effective use of dissection time. Furthermore, it also provided room for the assessment of specific competencies and the application of anatomical knowledge [52]. However, evidence by McNulty et al. in 2009 emphasizes the significance of understanding student preferences and their learning styles when making use of CAL. Their results show that students do not consistently make use of CAL that relates to the curriculum, and this might be credited to personal partiality [14].

Additional key design considerations have also been highlighted by Van Note Chism [53] and include flexibility that allows for easy reconfiguration and accommodates changing trends in pedagogy, comfortable seating and work surfaces, support for technology and adequate electricity supply, and the concept of the entire campus as a learning space. The latter implies certain "decenteredness" where learning activities occur within the corridors of a building as well as the living spaces of students. This also breaks away from the notion of having a designated front or a privileged space in a classroom. Van Note Chism also recommended the inclusion of sensory stimulation as a design consideration [53].

Anatomy Learning Content That Drives the Design of Space

- Programs offered: forensic medicine, training tomorrow's anatomists, allied health sciences, workshops, and continued professional development
- Dissection/prosection: gross anatomy laboratory/morphology museum
- E-learning: interactive multimedia and wireless technology
- Procedural, surface, and clinical anatomy: the display of 3D digital images and direct link to surgical theaters
- Imaging: C-arm-compatible equipment and visual display of medical images such as X-rays and MRIs
- Context: a gowning area, scrub room, and theatre lights

Choosing the Right Lights

There are two principal characteristics of light that influence perception: the intensity or illuminance of light and its color temperature. The first, intensity, is described as the luminous flux per meter (lux), and the latter, temperature, is related to the principal wavelengths emitted by a light. In general, increased illuminance improves visual acuity [54], and higher illuminance coupled with cooler color temperatures, such as blue-enriched white lights, is stimulating and improves alertness and performance [55]. Lower illuminance with warmer, yellow, color temperatures appears to improve communication and social behavior [56]. As such, lighting in various areas of an anatomy facility should be task specific and in some modifiable to suit various specific uses. For instance, blue-enriched lighting is desirable in practical and simulation areas where a combination of dimmable, warmer lighting and blue-enriched lights might be more appropriate for public spaces where discussions or communication (lower, warmer) or studying (brighter, cooler) might occur. Visual stimulation within the learning environment has the added advantage to reduce monotony and inactivity. Learning spaces should thus incorporate a diversity of colors to combat and reduce boredom while refreshing awareness. Color, in the same way as light and temperature, seems to significantly influence how students learn and their concentration required for a specific task. However, the importance and use of color within the learning environment remains contentious with conflicting results [57].

Reducing Extraneous Noise

Extraneous noise has an effect on cognition, affecting memory and reading comprehension; basically, acoustics influence learning outcomes [58]. It should be noted that most of this research was centered on the performance of primary and secondary school pupils. We might extrapolate and apply these concepts to tertiary institutions. In any new facility, pains must be taken to control sound to improve intelligibility in lecture facilities and reduce background noise in open plan areas to improve concentration on tasks [59].

Sensory Considerations

- Light and color affect mood and behavior.
- Visual stimulation reduces monotony and inactivity.
- Extraneous noise affects cognition and memory.
- The use of irritants in many facilities requires adequate ventilation.

Planning for Assessment

In any design, a critical question will be: Where will assessments take place? Is there a space that is conducive to assessment, and is it suitable to the format of assessment? The Association for Medical Education in Europe (AMEE) Guide No. 25 proposes a multidimensional model of assessment [60]. This guideline suggests selecting suitable assessment tools for the evaluation of a range of learning objectives. Multiple-choice and short-answer questions, oral examinations, and essays are typically used to evaluate knowledge recall as well as applied knowledge. Clinical performance assessment, however, requires more sophisticated methods such as objective structured clinical examination (OSCE), standardized patients (SP), and direct observation of clinical cases [60]. An assessment tool, such as objective structured practical examination (OSPE) (Fig. 30.3), can only be implemented within a suitably designed environment. Flexibility ensures easy transformation of the learning environment for assessment as depicted in Fig. 30.3. There should be ample room for movement in order to allow access to the test material. Figure 30.3 further illustrates that all of the furniture is mobile, creating further flexibility in the environment. Computerization of the stations permits the inclusion of digitized medical images and histology slides. This allows for the employment of various assessment methodologies in a space typically configured for practical sessions.

The Design Process

The design process depends on establishing and building a relationship between the architect, the user client, stakeholders and interest groups, and a professional team of engineers and consultants. The process needs to be interactive, a creative process that is essentially similar to product design. In it, there will be various phases: conceptualization, research, blueprinting, testing, and modification [61]. All will

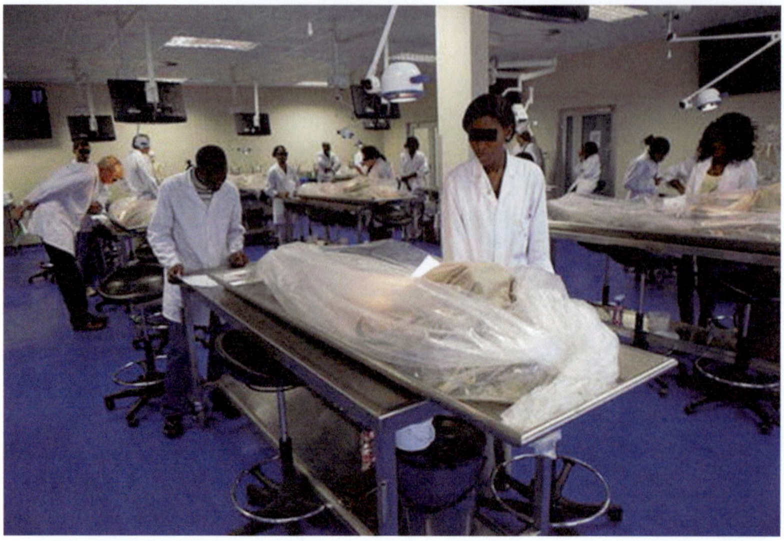

Fig. 30.3 An example of an OSPE assessment process. The practical assessment environment plays an important role. Each station is either located at the head or the toe end of a cadaver and is carefully blueprinted with the learning objectives, and the flow of the students is planned beforehand. The configuration can be changed based on the number of students that will be assessed

occur within a framework provided by the project budget and the interprofessional relationships developed by the design team. With vigilance, the end result should represent the needs of the user client.

> The design process relies on a reflective process of conceptualization, research, blueprinting, testing, and modification. The end result should represent the needs of the user client.

Briefly, the department, or a designated individual from the department, should develop an accommodation list. This list must specify all the departmental requirements for the building—the number and size of the offices, area of public space, specific laboratory requirements, etc., hopefully including everything the department will need over the next 20 years. This list will provide a framework from which the architects and consulting engineers will generate coherent ideas and plans for the facility. This is a dynamic process and a work in progress; it is imperative that faculty and staff play a role in this planning process to ensure alignment with the desired outcomes. Technology will play a critical role in any design. Ensure that all the technology you may need is incorporated early in this process and do not rely on consultants to bring that technology to the table. Research is the key in this regard. During the early planning phase, visit other institutions and ask relevant questions such as: What did they do right? What did they do wrong? Discuss with the architects how dissimilar elements and ideas may be integrated into your design and question how previous designs can be improved.

Someone will need to buy a hardhat and safety shoes. It is pivotal to continue the established relationships after the design is approved. The department needs to play an active role during the construction process. Get faculty involved. Do not rely on individuals who will not be using the space to represent you in the process. Identifying a problem early in construction is significantly cheaper than discovering it after completion. Remember that you will be using a facility for the next 20–30 years, not the contractor, subcontractor, or members of the professional team. Therefore, make sure what you get is what you wanted. In the end, it is never going to be perfect; there will always be some regrets, but these can be minimized by being active in the process from start to finish.

From Design to Commissioning

- Involve all the stakeholders and faculty.
- Continue relationships with the professional team and project manager after the design has been approved.
- Early identification of construction errors and consequent corrections is significantly cheaper than discovering it after completion.
- Participate in every stage until final completion.
- Ensure that you get what you asked for.

Conclusions

Turney aptly pointed out in 2007 that there are three aspects of anatomy pedagogy that need to be resolved: when, how much, and how to teach anatomy [13]. These curricular attributes require an awareness of today's learning as well as the learner's environment. A holistic approach is required in order to enhance teaching, and the system in its entirety should be considered. This includes the methods of assessment. The assessment tools and the assessment environment should be aligned with the learning objectives and teaching methods in order to ensure achievement of outcomes [62, 63]. In creating this learning environment, the following aspects of anatomy teaching space design thus need to be considered: appropriate sensory stimulation, plumbing and electricity, surface area required per student, appropriate assessment space, e-learning capabilities (including 3D projection and virtual dissections), and a dynamic environment that can be suitably reconfigured. The design process relies on adequate research prior to construction and faculty involvement from the conception of the idea, blueprinting, testing, and modification, and finally the commissioning of the facilities.

References

1. Poland B, Lehoux P, Holmes D, Andrews G. How place matters: unpacking technology and power in health and social care. Health Soc Care Community. 2005;13:170–80.
2. Hopkins D, editor. Framework for action on interprofessional education & collaborative practice. 1st ed. Geneva: Health Professions Networks Nursing and Midwifery Human Resources for Health, World Health Organization Press; 2010. http://whqlibdoc.who.int/hq/2010/WHO_HRH_HPN_10.3_eng.pdf
3. Drake RL, McBride JM, Lachman N, Pawlina W. Medical education in the anatomical sciences: the winds of change continue to blow. Anat Sci Educ. 2009;2:253–9.
4. Christopher DF, Harte K, George CF. The implementation of tomorrow's doctors. Med Educ. 2002;36:282–8.
5. Kinkade S. A snapshot of the status of problem-based learning in US medical schools. Acad Med. 2005;80:300–1.
6. Lee RMKW, Kwan C. The use of problem-based learning in medical education. J Med Educ. 1997;1:149–57.
7. Harden RM. Developments in outcome-based education. Med Teach. 2002;24:117–20.
8. Nieder GL, Parmelee DX, Stolfi A, Hudes PD. Team-based learning in a medical gross anatomy and embryology course. Clin Anat. 2005;18:56–63.
9. Vasan NS, DeFouw DO, Compton S. Team-based learning in anatomy: an efficient, effective, and economical strategy. Anat Sci Educ. 2011;4(6):333–9.
10. Vasan NS, DeFouw DO, Compton S. A survey of student perceptions of team-based learning in anatomy curriculum: favorable views unrelated to grades. Anat Sci Educ. 2009;2:150–5.
11. Misch DA. Andragogy and medical education: are medical students internally motivated to learn? Adv Health Sci Educ Theory Pract. 2002;7:153–60.
12. Heylings DJ. Anatomy 1999-2000: the curriculum, who teaches it and how? Med Educ. 2002;36:702–10.
13. Turney BW. Anatomy in a modern medical curriculum. Ann Roy Coll Surg Engl. 2007;89:104–7.
14. McNulty JA, Sonntag B, Sinacore JM. Evaluation of computer-aided instruction in a gross anatomy course: a six-year study. Anat Sci Educ. 2009;2:2–8.
15. Green RA, Whitburn LY. Impact of introduction of blended learning in gross anatomy on student outcomes. Anat Sci Educ. 2016;9(5):422–30.
16. Aziz MA, McKenzie JC, Wilson JS, Cowie RJ, Ayeni SA, Dunn BK. The human cadaver in the age of biomedical informatics. Anat Rec. 2002;269:20–32.
17. Parker LM. Anatomical dissection: why are we cutting it out? Dissection undergraduate teaching. ANZ J Surg. 2002;72:910–2.

18. Trelease RB. Anatomy meets architecture: designing new laboratories for new anatomists. Anat Rec. 2006;289B:241–51.
19. Wilson AB, Notebaert AJ, Schaefer AF, Moxham BJ, Stephens S, Mueller C, et al. A look at the anatomy educator job market: anatomists remain in short supply. Anat Sci Educ. 2020;13:91–101.
20. Sugand K, Abrahams P, Khurana A. The anatomy of anatomy: a review for its modernization. Anat Sci Educ. 2010;3:83–93.
21. Lancaster LC, Stillman D. When generations collide: who they are, why they clash, how to solve the generational puzzle at work. New York: Harper Business; 2003.
22. Murray ND. Welcome to the future: the millennial generation. J Career Plan Employ. 1997;57:36–40.
23. O'Reilly B, Vella-Zarb K. Meet the future. Fortune. 2000;142(3):144–8.
24. Davis DA. Millennial teaching. Academe. 2003;89(1):19–22.
25. Zemke R. Here come the millennials. Training. 2001;38:44–9.
26. Greene JRT. Design and development of a new facility for teaching and research in clinical anatomy. Anat Sci Educ. 2009;2:34–40.
27. Beery TA, Shell D, Gillespie G. Werdman E. The impact of learning space on teaching behaviors. Nurse Educ Pract. 2013;15:382–87.
28. Hunley S, Schaller M. Assessing learning spaces. In: Oblinger DG, editor. Learning spaces. Boulder: EDUCAUSE; 2006. http://net.educause.edu/ir/library/pdf/PUB7102m.pdf.
29. Pascarella ET, Terenzini PT. How college affects students: a third decade of research, vol. 2. San Francisco: Jossey-Bass; 2005.
30. Subramanian A, Timberlake M, Mittakanti H, Lara M, Brandt ML. Novel educational approach for medical students: improved retention rates using interactive medical software compared with traditional lecture-based format. J Surg Educ. 2011;69:253–6.
31. Issa N, Mayer RE, Schuller M, Wang E, Shapiro MB, DaRosa DA. Teaching for understanding in medical classrooms using multimedia design principles. Med Educ. 2013;47:388–96.
32. Issa N, Schuller M, Santacaterina S, Shapiro M, Wang E, Mayer RE, DaRosa DA. Applying multimedia design principles enhances learning in medical education. Med Educ. 2011;45(8):818–26.
33. Caldwell JE. Clickers in the large classroom: current research and best-practice tips. CBE Life Sci Educ. 2007;6:9–20.
34. Mastoridis S, Kladidis S. Coming to a lecture theatre near you: the 'clicker'. Clin Teach. 2010;7:97–101.
35. DiVall MV, Hayney MS, Marsh W, Neville MW, O'Barr S, Sheets ED, Calhoun LD. Perceptions of pharmacy students, faculty members and administrators on the use of technology in the classroom. Am J Pharm Educ. 2013;77:1–7.
36. Prakash ES. Explicit constructivism: a missing link in ineffective lectures? Adv Physiol Educ. 2010;34:93–6.
37. Lave J, Wegner E. Situated learning: legitimate peripheral participation. New York: Cambridge University Press; 1991.
38. Finn GM, Patten D, McLachlan JC. The impact of wearing scrubs on contextual learning. Med Teach. 2010;32(5):381–4.
39. Bleakley A. Broadening conceptions of learning in medical education: the message from teamworking. Med Educ. 2006;40:150–7.
40. Bleakley A, Bligh J, Browne J. Place matters: location in medical education. New York: Springer; 2011.
41. Nordquist J, Kitto S, Peller J, Ygge J, Reeves S. Focusing on future learning environments: exploring the role of space and place for interprofessional education. J Interprof Care. 2011;25(6):391–3.
42. McLachlan JC, Patten D. Anatomy teaching: ghosts of the past, present and future. Med Educ. 2006;40:243–53.
43. Malomo AO, Idowu OE, Osuagwu FC. Lessons from history: human anatomy, from the origin to the renaissance. Int J Morphol. 2006;24(1):99–104.
44. Del Negro P, editor. The University of Padua: eight centuries of history. 1st ed. Padova: Signum; 2003. 296 p
45. Schumacher G-H. Theatrum anatomicum in history and today. Int J Morphol. 2007;25:15–32.
46. Thiene G. Padua University: the role it has played in the history of medicine and cardiology and its position today. Eur Heart J. 2009;30:629–35.
47. Capener N. John Sheldon, F.R.S., and the Exeter medical school. Proc R Soc Med. 1959;52:231–8.
48. JISC. Joint Information System Committee. Designing spaces for effective learning. 2006. http://www.jisc.ac.uk/eli_learningspaces.html.
49. Goldman E. Building a low-cost gross anatomy laboratory: a big step for a small university. Anat Sci Educ. 2010;3:195–201.
50. ASHRAE (American Society of Heating, Refrigerating and Air-Conditioning Engineers, Inc.) ventilation for acceptable indoor air quality. ASHRAE Standard 62–1999. Atlanta, GA: ASHRAE. 1999; 141 p.
51. Wessels Q, Vorster W, Jacobson C. Anatomy education in Namibia: balancing facility design and curriculum development. Anat Sci Educ. 2012;5:41–7.
52. Reeves RE, Aschenbrenner JE, Wordinger RJ, Roque RS, Sheedlo HJ. Improved dissection efficiency in the human gross anatomy laboratory by the integration of computers and modern technology. Clin Anat. 2004;17(4):337–44.
53. Van Note Chism N. Challenging traditional assumptions and rethinking learning spaces. In: Oblinger DG, editor. Learning spaces. Orlando: Educause; 2006. http://net.educause.edu/ir/library/pdf/PUB7102b.pdf.
54. Van Bommel WJM, Van den Beld GJ. Lighting for work: a review of visual and biological effects. Light Res Technol. 2004;36:255–66.
55. Viola AU, James LM, Schlangen LJM, Dijk DJ. Blue-enriched white light in the workplace improves self-reported alertness, performance and sleep quality. Scand J Environ Health. 2008;34:297–306.

56. Baron RA, Rea MS, Daniels SJ. Effects of indoor lighting illuminance and spectral distribution on the performance of cognitive tasks and interpersonal behaviors: the potential mediating role of positive affect. Motiv Emot. 1992;16:1–33.
57. Higgins S, Hall E, Wall K, Woolner P, McCaughey C. The impact of school environments: a literature review. University of Newcastle: The Centre for Learning and Teaching, School of Education, Communication and Language Science; 2005.
58. Clark C, Sorqvist P. A three year update on the influence of noise on performance and behavior. Noise Healt. 2012;14:292–6.
59. Brammer AJ, Laroche C. Noise and communication: a three-year update. Noise Health. 2012;14:281–6.
60. Norcinci J, Burch V. Workplace-based assessment as an educational tool: AMEE guide no. 31. Med Teach. 2007;29:855–71.
61. Ulrich KT, Eppinger SD. Product design and development. 2nd ed. Boston: McGraw-Hill; 2003. p. 384.
62. Miller GE. The assessment of clinical skills/competence/performance. Acad Med. 1990;65(9):S63–7.
63. Johnson EO, Charchanti AV, Troupis TG. Modernization of an anatomy class: from conceptualization to implementation: a case for integrated multimodal-multidisciplinary teaching. Anat Sci Educ. 2012;5(6):354–66.

Role of Image and Cognitive Load in Anatomical Multimedia

31

Timothy D. Wilson

A Brief History

Perhaps the best model in a modern anatomist's teaching arsenal inarguably remains the cadaver [1]. The pedagogic symbiosis of student, facilitator, and cadaver is historically proven and hard to recreate. However, in the last quarter century, multiple iterations of anatomical education reforms [2–5], significant improvements in computing technology (price, ubiquity, software), and an increasing digital revolution in education has created a perfect storm for education. The classic Socratic and/or didactic approach has rapidly changed our "lecture" halls, classrooms, and labs to "shows" which now include visualizations such as diagrams, schematics, computer models, animations, videos, and even simulated environments [6]. Finally, in 2020, COVID-19 caused all formal education environments to pivot further towards digital learning environments. Anatomy departments around the globe that previously used cadaveric materials and small groups for their laboratory experiences now are dealing with government and institutional guidelines that supersede the peer to peer environments to which many of us are familiar [7]. These learning media form the basis of the more general term multimedia instruction, whereby both words and pictures are used to convey an instructional message [8].

Arguably, all anatomy instruction contains multimedia. Multimedia instruction occurs when educational messages are portrayed as a combination of spoken word and pictures [9]. Multimedia instruction should not be regarded solely as a high-tech phenomenon. Richard Mayer suggests that words can be spoken face-to-face or over speaker, speech can be computer generated, or words can exist as text. Pictures, on the other hand, are visual-spatial representations such as illustrations, drawings, schemata, or photographs delivered on a page or screen, while dynamic representations can exist as video or computer-generated animations that are delivered through a screen to an audience. In all cases, motivation for the use of images, and multimedia in general, is to promote learning. The impact of images (and multimedia in general) on learning may not be immediately evident, but over the course of multiple and ranging experiments, Mayer demonstrates that humans learn better from images and words than from the use of words alone. This is known as the multimedia principle [9].

As the reader will see later in this chapter, what and how we "do," "say," and "show" in our classes can have profound effects on learners

T. D. Wilson (✉)
Department of Anatomy and Cell Biology, Schulich School of Medicine and Dentistry, CRIPT – Corps for Research of Instructional and Perceptual Technologies, Western University, London, ON, Canada
e-mail: tim.wilson@uwo.ca

L. K. Chan, W. Pawlina (eds.), *Teaching Anatomy*, https://doi.org/10.1007/978-3-030-43283-6_31

through the multimedia principle. Furthermore, with an underlying social and technological current pushing for increasingly complex multimedia in our lectures, labs, and even assessments, educators may not achieve their objectives despite their best efforts.

The Many Dimensions of Instructional Visualizations

Multimedia instruction utilizing visually rich and interactive computer visualizations is a growing trend in anatomical education. From early computer images from A.D.A.M software (Atlanta, GA) to the newest incarnations of software designed to present human anatomy to a wide audience, the race to make the "best" software has significantly increased complex visualizations to all students and their educators [10, 11]. Regardless of the software manufacturer or approach the software might use, the images and text are the primary methods information is presented to the user; this is multimedia. Originally, the visualizations were termed virtual reality learning objects [12], and these digital anatomical "models" have become increasingly flexible and complex as researchers, educators, and vendors have all progressed with technological advances. There is a growing momentum that computer visualizations, compared to dissection, offer advantages to learners in terms of accessibility convenience, cost, safety, and versatility [1, 13, 14].

Instructional visualizations, whether in your lecture hall or on your laboratory computers, offer users many visual cues beyond the immediate materials demonstrated. It is imperative instructors are aware that in addition to the explicit information images portray to the learner, there are implicit cues therein which may affect the learning outcomes. Some of the visual information may be superfluous to the topic and objectives, and thus, reduce learning. Please see Table 31.1 for an overview of the characteristics and variability instructional software might take as it pertains to visualization media. The reader should consider where these characteristics occur in his/her own pedagogic approach and how they might affect the learner. As Table 31.1 illustrates,

Table 31.1 Characteristics of instructional visualizations and the inherent variability with each characteristic

Visualization characteristic	Variability or type of visualization
Delivery medium	Visualizations can be presented on a page, on a large screen for multiple viewers, or on a portable device (laptop, tablets, or smartphones)
Dimensionality	Visualizations can be presented in 2D (line drawing), as a 3D-rendered image with shading, apposition, and perspective, or may be stereoscopic 3D rendering requiring apparatus to view appropriately
Realism	Visualizations can be highly realistic (photos or videos) or may be of low realism like line drawings or schemata
Interactivity	Images can be a series of drawings paced by the learner, a movie with pause function, and a computer-generated image where users manipulate information within the image or may simple be a noninteractive movie or drawing
Dynamism	Images may be static or involve moving components

visualizations can vary significantly. Some qualities are apparent like the delivery medium or whether components of the visualization move, but there are deeper perceptual cues that permeate the literature of psychometrics and have been linked, with controversy, to intelligence [15] and definitely to learning. These perceptual cues are important features in anatomical education both for the novice learner and the advanced clinician as they aid observers' perception of orientation, location, and identification of the material presented. Once these wayfindings are achieved, the learner must then determine if the presented image is coherent with existing mental models they have formed. Indeed, as digital media steadily creeps into all aspects of modern anatomical education, it is important that the visualizations included in our teaching materials do not undermine good pedagogic intentions.

The viewer's spatial visualization ability may significantly influence the quality of information an image presents to the viewer. Importantly, an individual's spatial ability dictates how difficult it is for the viewer to extract pertinent information from the salient features of the visualization. Spatial ability refers to an individual's ability to search the visual field; apprehend the forms, shapes, and positions of objects as visually perceived; create mental representations of those forms, shapes, and positions; and manipulate such representations "mentally" [15]. In other words, an internal representation of a perceived object or scene must be created and maintained in working memory in such a way that mental manipulations are possible [16]. As the reader can imagine, this is not an easy task for all individuals and it can impact learner comprehension. Garg and colleagues investigated how students performed using a digital model of the carpal bones when asked to study and learn from key views versus interactive or multiple rotational views of the model [17]. Their results did not support the concept that students learn better with rotational virtual reality-type models. Indeed, they "cautioned" educators to carefully consider the inherent characteristics of the models before deciding to use virtual reality-type images in a pedagogical approach [17, 18]. Levinson et al. suggest learner control of multiple views in more complex digital models of a structure, such as the brain, may impede learning, particularly for those with relatively poor spatial ability [19]. Despite caution flags, Brewer et al. suggest that learner control of models might attract novice learners, as the idea of "exploration" of virtual models leads to greater spatial comprehension [20]. Nguyen and colleagues suggest that an individual's spatial ability affects comprehension that involves not only object orientation recognition but also understanding how a 3D structure appears both in shape and locality in cross section [21]. Importantly, their results demonstrate that not all images are created equally as persons with lower spatial ability take longer to complete tests and make more mistakes. Finally, even anatomical testing validity may be inadvertently undermined by incorporating questions that utilize images in their answers as image use significantly influences both test item difficulty and the question's discriminatory power [22]. From this sampling of anatomical pedagogic research, it is clear that the old adage "a picture is worth a thousand words" is indeed true! However, not all viewers will form the same words nor divine the same meaning.

Good teachers are constantly exploring different methods to attract the attention of the entire class. In efforts to mediate perplexed students, technology has led us to believe that dynamic multimedia-embedded "PowerPoint" presentations are the best way to exude information and the learner will simply soak it up. Complex images, movies, 3D digital models, and even personal augmented reality are becoming increasingly accessible to the anatomical landscape of the classroom or lab. Are educators moving in the right direction with good pedagogic principles? Or is the development of tools through technology leading our teaching? It is paramount that educators understand the impacts that these sexy, futuristic, and attractive tools of our educational arsenal have on our students. In some cases, effective teaching may require large powerful tools, while at other times more subtle approaches could be much more useful.

Increased Cognitive Load Diminishes Learning

As the field of research in anatomical pedagogy broadens, new research paradigms are emerging which explore previously ignored facets of our educational practice. The concept of learner's cognitive load capacity, and in particular, the consequences on individual's working memory, is one area of image-coupled pedagogy that is receiving increasing attention. The term "cognitive load" was coined in the late 1970s by John Sweller and has since been the subject of much research and refinement [23, 24]. As digital tools have made their way to the educational mainstream, the theory of cognitive load has undergone changes to incorporate the use of multimedia technologies for learning and instruction.

Richard Mayer is perhaps best known in cognitive psychology for his timely work describing the theory of cognitive load and how it is applied to multimedia intersecting with knowledge translation [25]. Interestingly, even in the early days of multimedia in education, Mayer expresses concern over how much information a learner can process given the novel and increasingly rich methods offered through multimedia channels [26]. Mayer demonstrates that creators and users of multimedia learning should reduce cognitive loads on our learners to enhance learning [27]. But what is cognitive load? Chandler and Sweller [28] suggest that learning can be significantly reduced by multiple drains on the cognitive resources of the learner, hence the electrical analogue of a load. Built on the work of Sweller [29], Mayer [8] suggests a triarchic model of cognitive load highlighting intrinsic, germane, and extraneous cognitive processing, describing how each can separately draw upon the learner's cognitive capacity.

A central tenet of cognitive load theory, as it pertains to multimedia, is that learners engage in different cognitive processing [27, 30, 31] depending on the type and quality of the learning materials presented [32]. Each process draws on the individual's cognitive capacity [25]; thus, some information may be quickly understood and organized into long-term memory for one subset of learners but may be arduous for others. The goal for educators is to adopt proven methodologies and translate these approaches into their own specific style in order to make learning easier—yes easier—for the student! Three types of cognitive load are described below. In each case, the goal is to mediate loads appropriately through the purposeful design or use of our teaching materials and pedagogic approach to enhance learning.

The first form of cognitive load is termed *intrinsic*; it arises from cognitive processing required to apprehend and make sense of novel material. Intrinsic load pertains to the perceived complexity of the "stuff" introduced to the learner. The intrinsic load is described as being an integral part of the learning task that results from interactivity among the elements of to-be-learned materials. This component of cognitive load cannot be easily reduced without impacting the learning objectives [27]. In anatomical education, this "essential" processing is often viewed to be a large hurdle for the novice as the plethora of terms relating not only to the nomenclature but also to the orientation, location, and cross-sectional planes, and even function are often overwhelming. All is not lost, however. Table 31.2 illustrates three principles of essential processing that contribute to intrinsic load, and some general applications educators might follow to mediate intrinsic overload. Readers should recognize that despite using images in their anatomical lessons, the principles of modality and segmentation are important guides to effective presentation and can be achieved with attention to dosing of elements in their visualizations. In the latter part of this chapter, an example of segmentation of images is presented. Lastly, the principle of pretraining is often overlooked in anatomy but can be effectively accomplished utilizing good lecturing principles, whereby the leader of the discussion presents a quick plan for the learning activity incorporating concepts from the previous day.

The next type of cognitive load is termed *germane* or *generative* (Table 31.3), and it is related to cognitive processes dedicated to making sense of the materials, organizing it mentally, and inte-

Table 31.2 Principles of essential processing and applications to improve their effects

Principle	Application
Pretraining	Begin lessons with an overview of important terms, overall relationships, and explain any jargon you might use
Segmentation	Present materials in short sections rather than in one continuous unit
Modality	Use images coupled with spoken words rather than a long string of written words during presentations

Table 31.3 Principles of generative processing and applications to encourage the process

Principle	Application
Multimedia	Present words and images rather than words alone
Personalization	Use conversational speech rather than formal speech when presenting
Voice	Present speech with natural human voice rather than a computerized one in online scenarios
Image	Include speaker's image on the screen from time to time if materials are being presented online

Table 31.4 Principles of extraneous processing and applications to mitigate overload

Principle	Application
Coherence	Delete extraneous words, sounds, or images
Signaling	Highlight important terms and images
Redundancy	Remove redundant captions from narrated animation
Spatial contiguity	Position critical terms next to images
Temporal contiguity	Present corresponding words and images simultaneously

grating it with any prior knowledge and mental schema the student might already possess [33]. Application of germane processing principles in face-to-face scenarios will foster good generative processing in your students. Generative load is often attributed to learner motivation and preference. Therefore, if the educator can facilitate these applications, learner attention will improve, and more information will be appended to the students' long-term memory as knowledge.

The last form of cognitive load is called *extraneous* load (Table 31.4). It refers to the cognitive processing that does not serve the objective of the learning exercise and is mostly imposed by inappropriate approaches that ignore working memory limits. As educator's materials surpass thresholds in students' extraneous load capacity, the effects are manifested by learner confusion and/or frustration and therefore need to be minimized. The overall goal of educators is to manage intrinsic loads, reduce extraneous load, and encourage environments that facilitate germane loads.

Methods to Reduce Cognitive Load Using Visualizations

In order to understand cognitive load as it pertains to lectures, presentations, and multimedia in general, three assumptions are made. These assumptions come from cognitive science and form the basis of how the human mind works with respect to information processing [27]. For a complete review, the reader should see Mayer's cognitive model of multimedia learning [8] and refer to Fig. 31.1 for a pictorial overview. Briefly, the first assumption treats the sensory pathways related to seeing and hearing as dual channels where learners process the inherent sensory information in each channel separately yet simultaneously. The second assumption is that of capacity; learners have very limited capacity to process information in each channel at one time. Finally, the assumption of active processing indicates that learning is not, and cannot, be a passive process. Importantly, learning will not occur unless attention is given to relevant material [34] using vision and audition in what Mayer refers to as sensory memory. Sensory memory is of very short duration and supposed unlimited capacity that attends to all environmental sounds, images, words, etc. Words and pictures are selected from our attention and brought into the limited working memory component of cognition. The working memory of an individual has temporary and limited storage that temporarily holds and manipulates the selected verbal and visual information. The information in working memory must be rapidly organized, contrasted, compared, and worked into mental models [33]. The mental models are either formed anew or may be integrated with schema, already present in the form of prior knowledge. Prior knowledge of a subject differentiates novices from experts. The learner appends this newly constructed information to their long-term memory, which is thought to be of unlimited capacity [27, 30].

In order to demonstrate methods aiming to decrease cognitive overload, an example of an introductory lesson on the heart is employed. In the accompanying figures, the instructor is giving an overview of the heart chambers and the valves separating them. In Fig. 31.2, examples of potential cognitive overloads are presented, while in Fig. 31.3, the multimedia approach was altered to reduce cognitive load. The following paragraphs outline what the educator can do mechanically with the visualization during a demonstration or lecture. Realize that this is a static picture; the importance of the following paragraphs will be heightened should the educator utilize one of the

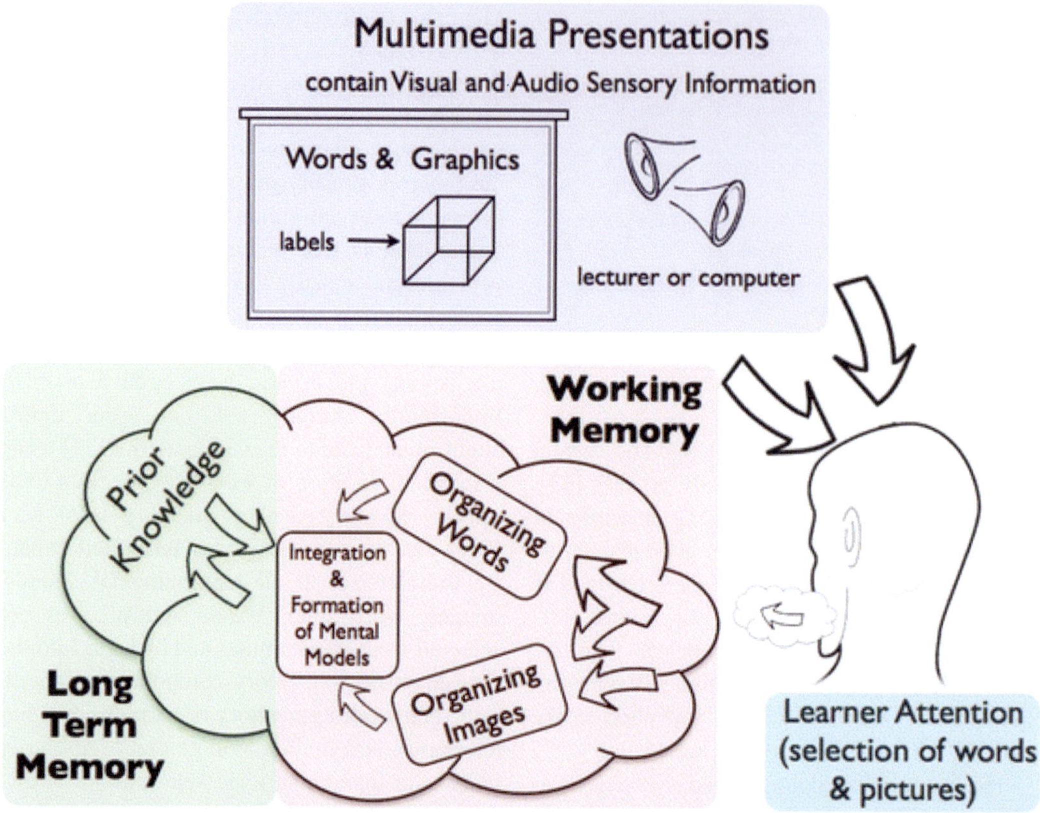

Fig. 31.1 Illustration of the cognitive theory of multimedia learning. The learner (*bottom right*) participates through apprehension of multimedia where dual sensory channels bring sensory information to the working memory of the learner. Information in the working memory must quickly be organized and encoded into mental models for organization in long-term memory

many software programs available on the market that are dynamic in the sense that they can move or can strip layers off the anatomy, in essence, change in front of the learner.

In Fig. 31.2, the instructor utilizes an actual photograph of the mammalian heart with numbers close to pertinent, but somewhat ambiguous, locations. The numbers then refer to a lengthy legend below where information is presented. It may be tempting in your slides or demonstrations to include dense information but doing so will only serve to increase cognitive load on the learner. The increased extraneous load of Fig. 31.2 will lead to decreased learning as the multimedia principles illustrate low coherence as there is too much information, lack of signaling as the numbers are next to ambiguous structures, and poor spatial contiguity characteristics as explicatory text is distant from the picture (see Table 31.4). The inclusion of excessive information on one's slides is a common mistake and should be avoided at all costs. Instead, the instructor might commence with the photograph to demonstrate how complicated the actual structures appear in vivo (Table 31.2—overview). The instructor would then transition to simplified diagrams (increasing coherence) in a very prescribed and stepwise fashion giving only parts of the diagram sequentially (Table 31.2—segmentation), using good oration (Table 31.2—modality), and present the diagram and perhaps a few key visual examples to learners as demonstrated in Fig. 31.3a, b. Utilizing a "less is more" approach with multimedia takes advantage of both the visual and auditory (dual) channels in the learner and enables better cognitive processing, thus

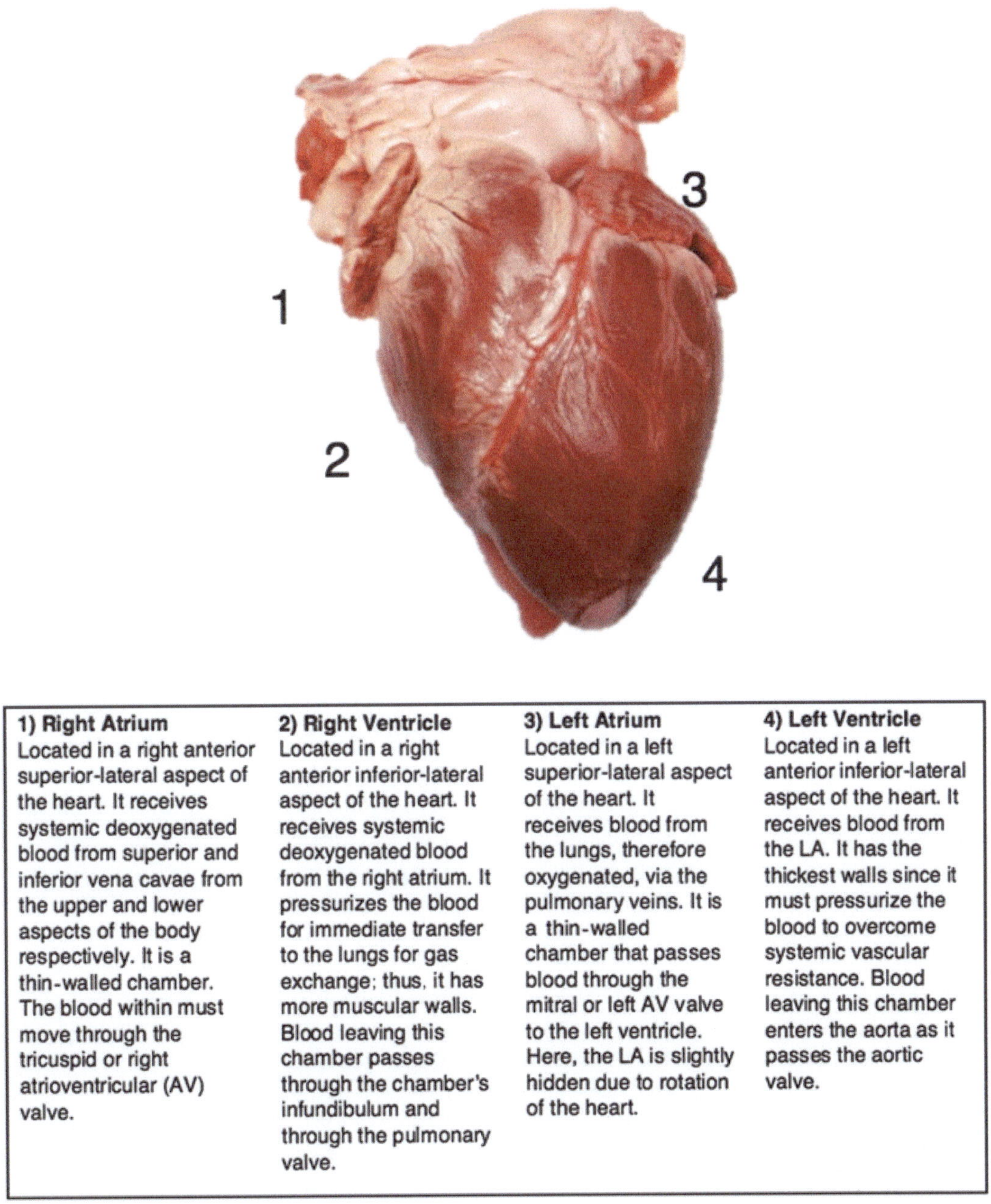

Fig. 31.2 Example of a multimedia slide incurring increased extraneous loads. The use of ambiguous color photographs, lack of cues for the viewer, and much interesting, but superfluous, information will all detract the learner from essential processing through increased extraneous load

cognitive load reduction. Taking advantage of maximizing essential processing principles (Table 31.2), the learner can then better attend to key information, rather than trying to decipher the message. Utilizing a simplified visualization while making use of the spoken word accompanying these images effectively achieves the effect of modality in a lecture scenario.

In concert with the modality effect, good instructors also know how to pace their messages

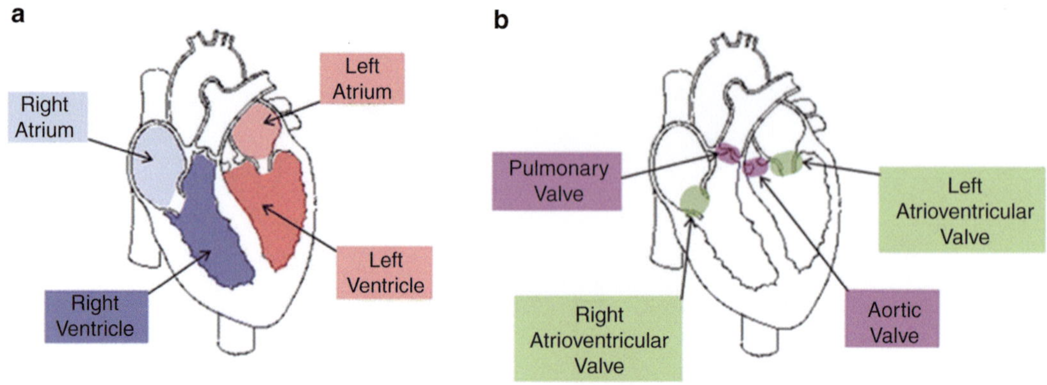

Fig. 31.3 (**a**, **b**) Depiction of a simple heart illustration that might be used in multimedia environment, including lectures. These diagrams incorporate multiple aspects of cognitive load-reducing techniques that affect extraneous processing. This simple set of diagrams would be part of a series of diagrams that are "built" in a stepwise fashion for students that introduce one element (label) at a time. They would be used in concert with good presenter narration with meaningful repetition of the important terms geared to the objectives of the lecture or multimedia exercise. The use of color and transparency should also be incorporated to make borders of adjoining anatomy clearer for the learners. This approach incorporates coherence, signaling, and spatial and temporal contiguity (Table 31.4), enabling students to chunk information and link increasing preknowledge with new information

and visual materials. The first approach affecting pace is the segmentation principle (Table 31.2). Segmentation, sometimes called chunking, can be a temporal factor where the instructor simply needs to slow down to employ purposeful repetition of important information (verbal signaling), highlighting important structures (verbally and visually) or giving pertinent analogy for the students to both see and hear (modality) information presented in chunks. Instead of an overwhelming image (Fig. 31.2), educators should consider simplifying visuals and using sequential figure components that parse the learning materials into a step-by-step fashion (Fig. 31.3a, b) matching visual representation with auditory explanation (spatial and temporal contiguity). By combining the principles of segmentation, modality, and contiguity, educators enable learners to "chunk" information more efficiently [6]. For the learner, chunking doses the amount of information present in working memory, aiding information transition to long-term memory (Fig. 31.1).

Often, to reduce cognitive load, multimedia images and accompanying oration need to be simplified and streamlined in a process Mayer and Moreno describe as weeding [27]. The goal of weeding is to eliminate potentially interesting but extraneous material to reduce the processing of the ill-fated extraneous overload. By weeding out many of the words in Fig. 31.2, or potentially distracting materials such as background noises, or poor video clips, the coherence principle (Table 31.4) is applied and the learner can move more information from working memory to long-term memory. Contrast the image and accompanying information in Fig. 31.2 with that of the images in Fig. 31.3. This process may be difficult for educators, for as we become proficient in a discipline, the discipline and its intricacies tend to take on an ego of its own, and we convince ourselves that students need every individual detail. This simply is not true and Mayer describes this information as "seductive details" [35], and these details should be removed to enhance learning.

Finally, as many courses are moving to online formats where video and narration is often used, educators need to keep narration in time with the presentation and to avoid overly repetitive delivery. Mayer refers to this approach as the temporal contiguity principle [36] (Table 31.4). Although elaborate digital demonstrations may be tempting, they hold limited quantitative pedagogical support, for by their nature, they introduce additional and different cognitive demands on the novel learner.

Students with more advanced base knowledge will not fall victim of high cognitive loads for they already have prior knowledge and append new information to schema in long-term memory [37]. Lowe questions the widely assumed learning edge animated graphics have over static graphics in a learning environment. Lowe suggests that cueing, in the form of "specific strategies and targets," need to be in place in order for learners to create accurate mental representations from interactive animations [38]. If the use of complex visuals or software is deemed necessary with novel anatomists, these visualizations should be well explained within the context of the lesson or used to provide a summary of the material in question rather than a vehicle by which the information is introduced to the students.

Summary

To consider multimedia visualizations as ubiquitous tools for successful pedagogy in anatomical education is overly simplistic and incorrect. Much hinges on the base knowledge of the learners, and the goal of the educator is to meet the objectives of the course or activity while challenging the learner. Uses of multimedia that are too wide in the spectrum between the basal needs or challenges to the learner will result in boredom or confusion, respectively. In either case, learning will decrease. Addressing the pertinent anatomical details in accordance with the objectives is truly important, but it is equally important to apply strategies that will enhance learning. The use of multimedia has complicated the challenge to educators, but enhanced learning can be achieved by careful material preparation and presentation with cognitive load in mind. Good instructional methods can be successful across multiple forms of media (face-to-face and online) and any discipline [35]. These methods can be summarized to four short rules that maximize learning:

1. Pay attention to the cognitive loads generated with your learners. The tools (programs, presentations, pictures, and words) will affect cognitive load. *Less is always more*. That is, less "cognitive load" will yield more "learning."
2. Orate to the ears of the learners and demonstrate to the eyes. Use our two main sensory channels (dual channels) to your pedagogic advantage.
3. Chunk, highlight importance, and pace your presentations using multimedia so the learner is not overwhelmed.
4. Simple visualizations (non-dynamic line drawings) with proximal labels that are color-coded are the best place start with new learners. As the base knowledge of your learners rises, so too can the complexity of your demonstrative tools.

Educators should strive to simplify all aspects of their multimedia use to reduce cognitive load, in particular extraneous load [8] as it is counterproductive to the objectives of their overall mission as a teacher. Multimedia design principles are dynamic and will be heavily influenced by the experience of the learners [39] as advanced learners can draw upon previous knowledge more readily [37], which the instructor should be aware in order to adjust the visualization, language, and pace accordingly. Just as lecturers attend to the verbal detail given to our class, care must be also exercised as to what and, importantly, how we demonstrate to the class as well. By paying attention to the aforementioned principles of multimedia learning, and by organizing visualizations accordingly, regardless of the tool involved, educators will find student satisfaction and success to rise.

As educators in an increasingly fast-paced and digital society, it is important to stay abreast of novel and evidence-driven pedagogic principles and practices. We should constantly weigh ongoing practice with our previous experience in the face of ever-changing student expectations and technology. Until technology is nimble and adaptive to the multitude of learner styles [40], we as educators are the only intelligent buffer that is able to modify our approach to support deep learning in all our learners.

References

1. McLachlan JC, Bligh J, Bradley P, Searle J. Teaching anatomy without cadavers. Med Educ. 2004;38(4):418–24.
2. Drake RL. Anatomy education in a changing medical curriculum. Anat Rec. 1998;253(1):28–31.
3. Drake RL, McBride JM, Lachman N, Pawlina W. Medical education in the anatomical sciences: the winds of change continue to blow. Anat Sci Educ. 2009;2(6):253–9.
4. Collins TJ, Given RL, Hulsebosch CE, Miller BT. Status of gross anatomy in the US and Canada: dilemma for the 21st century. Clin Anat. 1994;7(5):275–96.
5. Stull AT, Hegarty M, Mayer RE. Getting a handle on learning anatomy with interactive three-dimensional graphics. J Educ Psychol. 2009;101(4):803–16.
6. Mathewson JH. Visual-spatial thinking: an aspect of science overlooked by educators. Sci Educ. 1999;83(1):33–54.
7. Evans DJR, Bay BH, Wilson TD, Smith CF, Lachman N, Pawlina W. Going virtual to support anatomy education: a STOPGAP in the midst of the Covid-19 pandemic. Anat. Sci. Educ. 2020;13:279–83.
8. Mayer RE. Instruction based on visualization. In: Mayer RE, Alexander PA, editors. Handbook of research on learning and instruction. New York: Routledge; 2011. p. 427–45.
9. Mayer RE. Multimedia learning. New York: Cambridge University Press; 2009.
10. Attardi SM, Rogers KA. Design and implementation of an online systemic human anatomy course with laboratory. Anat Sci Educ. 2015;8:53–62.
11. Trelease RB. From chalkboard, slides, and paper to e-learning: How computing technologies have transformed anatomical sciences education. Anat. Sci. Educ. 2016;602:583–602.
12. Trelease RB, Rosset A. Transforming clinical imaging data for virtual reality learning objects. Anat Sci Educ. 2008;1(2):50–5.
13. McLachlan JC, Patten D. Anatomy teaching: ghosts of the past, present and future. Med Educ. 2006;40(3):243–53.
14. Aziz MA, McKenzie JC, Wilson JS, Cowie RJ, Ayeni SA, Dunn BK. The human cadaver in the age of biomedical informatics. Anat Rec. 2002;269(1):20–32.
15. Carroll JB. Abilities in the domain of visual perception. In: Carroll JB, editor. Human cognitive abilities: a survey of factor-analytic studies. 1st ed. New York: Cambridge University Press; 1993. p. 304–63.
16. Nguyen N, Mulla A, Nelson AJ, Wilson TD. Visuospatial anatomy comprehension: the role of spatial visualization and problem solving strategies. Anat Sci Educ. 2014;7(4):280–8.
17. Garg AX, Norman G, Spero L, Taylor I. Learning anatomy: do new computer models improve spatial understanding? Med Teach. 1999;21(5):519–22.
18. Garg A, Norman GR, Spero L, Maheshwari P. Do virtual computer models hinder anatomy learning? Acad Med. 1999;74(10 Suppl):S87–9.
19. Levinson AJ, Weaver B, Garside S, McGinn H, Norman GR. Virtual reality and brain anatomy: a randomised trial of e-learning instructional designs. Med Educ. 2007;41(5):495–501.
20. Brewer DN, Wilson TD, Eagleson R, de Ribaupierre S. Evaluation of neuroanatomical training using a 3D visual reality model. Stud Health Technol Inform. 2012;173:85–91.
21. Nguyen N, Nelson AJ, Wilson TD. Computer visualizations: factors that influence spatial anatomy comprehension. Anat Sci Educ. 2012;5(2):98–108.
22. Vorstenbosch MA, Klaassen TP, Kooloos JG, Bolhuis SM, Laan RF. Do images influence assessment in anatomy? Exploring the effect of images on item difficulty and item discrimination. Anat Sci Educ. 2013;6(1):29–41.
23. Ozcinar Z. The topic of instructional design in research journals: a citation analysis for the years 1980–2008. Australas J Educ Technol. 2009;25(4):559–80.
24. Sweller J. Cognitive load during problem-solving – effects on learning. Cogn Sci. 1988;12(2):257–85.
25. Mayer RE. The science of learning: determining how multimedia learning works. In: Mayer RE, editor. Multi-media learning. 2nd ed. New York: Cambridge University Press; 2009.
26. Mayer RE, Heiser J, Lonn S. Cognitive constraints on multimedia learning: when presenting more material results in less understanding. J Educ Psychol. 2001;93(1):187–98.
27. Mayer RE, Moreno R. Nine ways to reduce cognitive load in multimedia learning. Educ Psychol. 2003;38(1):43–52.
28. Chandler P, Sweller J. Cognitive load theory and the format of instruction. Cogn Instr. 1991;8(4):293–332.
29. Sweller J. Instructional design in technical areas. Camberwell: ACER Press; 1999.
30. Mayer RE. Cognitive theory of multimedia learning. In: Mayer RE, editor. Cambridge handbook of multimedia learning. New York: Cambridge University Press; 2005. p. 31–48.
31. DeLeeuw KE, Mayer RE. A comparison of three measures of cognitive load: evidence for separable measures of intrinsic, extraneous, and germane load. J Educ Psychol. 2008;100(1):223–34.
32. Leacock TL, Nesbit JC. A framework for evaluating the quality of multimedia learning resources. Educ Technol Soc. 2007;10(2):44–59.

33. Kirschner PA. Cognitive load theory: implications of cognitive load theory on the design of learning. Learn Instr. 2002;12(1):1–10.
34. Khalil MK, Paas F, Johnson TE, Payer AF. Interactive and dynamic anatomical visualizations: the implication of cognitive load theory. Anat Rec. 2005;286B(1):15–20.
35. Hegarty M, Just MA. Understanding machines from text and diagrams. In: Mandl H, Levin JR, editors. Knowledge acquisition from text and pictures. Amsterdam: Elsevier; 1989. p. 171–94.
36. Lowe RK. Animation and learning: selective processing of information in dynamic graphics. Learn Instr. 2003;13(2):157–76.
37. Sweller J. Implications of cognitive load theory for multimedia learning. In: Mayer RE, editor. The Cambridge handbook of multimedia learning. New York: Cambridge University Press; 2005. p. 19–48.
38. Lowe R. Interrogation of a dynamic visualization during learning. Learn Instr. 2004;14:257–74.
39. Kalyuga S. Prior knowledge principle in multimedia learning. In: Mayer RE, editor. Cambridge handbook of multimedia learning. New York: Cambridge University Press; 2005. p. 325–38.
40. Kalyuga S. Adapting levels of instructional support to optimize learning complex cognitive skills. In: Kalyuga S, editor. Managing cognitive load in adaptive multimedia learning. Hershey: IGI Global. p. 246–71.

Essential E-Learning and M-Learning Methods for Teaching Anatomy

32

Robert B. Trelease

Computer-enhanced learning has evolved rapidly over the past three decades, along with the development and explosive growth of the Internet-based World Wide Web and its associated hyperlinked multimedia methodologies [1]. Early work on "electronic learning" ("e-learning" [2]) focused primarily on establishing effective methods for using educational multimedia on personal computers (PCs) and via the Web. More recent refinements in basic e-learning concepts have included mobile ("m-learning" [3]) and ubiquitous learning ("u-learning" [4]).

M-learning and u-learning methods acknowledge that current-generation students (especially in health sciences) are generally very mobile and relatively experienced in using the Web and self-adopting personal technology, such as smartphones and tablet computers. The u-learning concept recognizes that with ubiquitous computing technologies, students can acquire knowledge virtually anywhere with well-designed online educational resources and mobile applications. This chapter will most frequently refer simply to *e-learning* when discussing computer-based methods for teaching, with the understanding that learning resources should ideally be equally usable on PCs and mobile devices, in-classroom and without.

With continuing curricular evolution, innovation, research, and development with new e-learning resources may be keys to successful implementation of new curricula and classes. For the new anatomist, understanding and mastering e-learning methods can provide the basis for career-long innovation in education, as well as a historical share in the progressive application of useful new technology. And although this chapter will refer to anatomy frequently in the context of medical gross anatomy, these methods and strategies apply equally well to teaching histology, histopathology, neuroanatomy, and embryology at multiple educational levels.

Types of Curricula, Methods, and E-Learning Resources

Whether a new anatomist is entering a long-established school with mature instructional resources management or a newer program still building online learning capabilities, it is helpful to understand the context of anatomical e-learning development relative to changing health sciences curricula. Although e-learning is not that old globally, having appeared and evolved rapidly during the careers of senior anatomists, it will

R. B. Trelease (✉)
Division of Integrative Anatomy, Department of Pathology and Laboratory Medicine, David Geffen School of Medicine at UCLA, University of California, Los Angeles, CA, USA
e-mail: trelease@ucla.edu

L. K. Chan, W. Pawlina (eds.), *Teaching Anatomy*, https://doi.org/10.1007/978-3-030-43283-6_32

continue to evolve with progressively changing curricula and availability of new technologies. *In pragmatic terms, it is easier to figure how to influence change, when you understand sources and drivers of change.*

Anatomy E-Learning: Legacy and Curricular Change

Legacy basic medical anatomy instruction in the latter twentieth century typically centered around daily and weekly serial lectures on regional topics, coupled with progressive student dissection (and microscopy) laboratories. As PCs became more multimedia-capable and widely available in the early 1990s, the first e-learning resources were developed. These variably included desktop-published lecture notes, early presentation software, computer-based laboratory instructions, and the first, non-networked PCs in anatomy laboratories, equipped with videodisc and CD-ROM players and early structural image archives. These disc media were transitory: Other types of technology will also have limited lifespans for various reasons.

Digital video became a practical educational tool in the mid-1990s, supporting the recording of lectures and laboratory demonstrations. The ongoing revolution in computer-based clinical imaging modalities (computed tomography, CT; magnetic resonance imaging, MR; and ultrasonography, US) also facilitated increased integration of digital radiology with anatomy instruction. With the advent and rapidly spreading usage of the Web, all these digital media became distributable via networked computers, institutional Web pages, and servers. By the end of the twentieth century, with the stationing of PC workstations at individual dissection stations, what we now recognize as basic e-learning resources became an integral part of classical "lectures plus laboratories" instruction.

As Western medical school curricular reforms continued to spread, increased time was reallocated for multidisciplinary problem-based learning, early clinical experience components, and patient simulations. Reduced time was allotted for conventional basic sciences classes, especially in the United States, and overall anatomy class time decreased [5–7]. Hours available for lectures and laboratory dissection and microscopy were drastically reduced, forcing greater reliance on prosections, anatomical models, and increasingly capable multimedia software. E-learning resources became essential for the delivery of anatomy instructional content in these new curricular regimes.

Effective in-class implementation of e-learning methods came to depend on the support and daily use of institutional instructional technology infrastructure, including lecture hall projection systems, network connectivity, and laboratory-based PCs with large-screen displays. Earlier classes of students may have had computer ownership requirements as a condition of admission; current-generation students typically own laptops and carry mobile devices with broadband access, although this may not be the case uniformly throughout the world. Current e-learning resources should make heavy use of omnipresent student-owned mobile technologies [8].

Types of Curricula and E-Learning Methods

New anatomists beginning their first teaching assignment in a health sciences school will find themselves involved in an evolving educational model. The eventual long-term task could be to transform or to replace content from existing lectures and laboratories via some new, e-learning technology-enhanced regime. In that context, it is useful to consider briefly some more recent online learning models that have been defined by universities for online instruction or e-learning. The anatomy components of different types of curricula are examined in greater detail in Chap. 44.

A given school may have a spectrum of e-learning or online course methodologies, ranging between the extremes of conventional lectures and laboratories supported by discrete computer-based media to fully online (classroom-free) courses dominated by an integrated digital learning environment.

"Online lecture" can describe a range of content available for Web access, within a continuum extending from text files with simple audio or video-enhanced presentations, to captured lecture videos, to professionally produced studio recordings. Lecture capture records an instructor's presentation as it is delivered live in the classroom. When used for a well-spoken live presentation, lecture capture can be an effective review tool for students, as well as providing media content for transitioning to other types of instruction (see below). Production studio recordings are typically created outside the classroom by multimedia developers and may include features of a highly polished professional video, such as split screens, custom animations, and transitions. Online lectures can play an important part in a course, but they alone usually do not constitute the entirety of a class. Chapter 12 covers preparation and recording of lectures for online delivery.

In "blended," "hybrid," or "mixed-mode" instruction, critical elements of course content may be delivered online, while student-faculty contact is largely shifted to digital means, with other possible forms of contact [9]. For example, all lectures could be delivered online or in-person, while student interactions might be conducted through discussion boards or chat rooms. Perceived institutional benefits of blended instruction might be that physical classroom space and/or faculty and student time can be freed for other uses (e.g., in "asynchronous learning," wherein lectures can be accessed online at virtually any time).

In "inverted" or "flipped" instruction, lectures are prerecorded and provided as online videos, while on-campus class time is spent discussing lecture subjects and other course materials in greater depth and scope. It generally differs from blended instruction primarily in that there is no intended saving of physical classroom time: Instructional time is repurposed rather than reduced. Inverted or blended instruction may also employ integrated self-learning modules (see next section).

Other curricular models include problem-based, case-based, and team-based learning. Although these methods are often used in discrete, multidisciplinary class sessions separate from those dedicated primarily to anatomy learning, case- and team-based methods can also be used as a focus for anatomy laboratory exercises. Although e-learning methods can be used to support such small-group learning experiences, they will not be covered here (see Chaps. 16–18).

Learning Theories and Instructional Design Principles

Design and use of e-learning resources are best guided by established principles of *instructional design*, rooted in cognitive and behavioral research applied to adult learning. Such principles apply to the organization of specific courses and to delivery methods and tools, including e-learning. Although there are several different models for instructional design [10], perhaps the most simply compelling is that of Gagné and colleagues, which predates the rise of computer-enhanced learning and has since proven sound for e-learning [11].

Gagné's theories distinguish nine different instructional events and the associated cognitive processes involved in learning (parenthesized). These are as follows: (1) gaining student attention (reception), (2) informing students of the learning objective(s) (expectancy), (3) stimulating recall of prior learning (retrieval), (4) presenting the stimulus/lesson content (selective perception), (5) providing learning guidance (semantic encoding), (6) eliciting student performance (responding), (7) providing feedback (reinforcement), (8) assessing performance (retrieval), and (9) enhancing retention and transfer (generalization).

Gagné further defines several additional learning principles. In a nutshell, instructional design principles dictate developing and using educational resources appropriate to the type of learning and desired outcomes, with a hierarchical approach to skills acquisition in sequence.

Comprehensively supporting these processes is the foundation of sound instructional design and effective learning media development. For example, attention-getting e-learning resources should clearly define and focus on discrete learning objectives, recapitulate earlier knowledge, provide self-assessment with feedback, and encourage application of knowledge with clinical content.

In contemporary education, Gagné's familiar taxonomies are considered to apply most directly to "transmissive learning," wherein the teacher is viewed as primarily responsible for conveying knowledge directly to students (e.g., in lectures and laboratories). Over the last two decades, transmissive approaches have been depreciated in medical school curricular changes, and other theory-based learning methods have been preferentially promoted.

With *constructivist* approaches to learning [8], the meaning of knowledge is assumed to be personally defined ("constructed") by individual students, frequently in groups. Rather than being a knowledge transmitter, an instructor is viewed as a facilitator. Sharing some common principles with *constructive learning*, *situated learning* [12] depends on knowledge and skills being learned in social interactions in practical working contexts. In modern anatomy curricula, problem-based learning and anatomy laboratories can be viewed as instances of both constructive and situated learning. Anatomy learning during clerkship and residency training would be considered situated.

When designing e-learning resources, it is best to support these aforementioned basic adult learning processes, while implementing what *instructional research* has shown works best in practice. Mayer [13] has provided a useful perceptual/cognitive framework practically integrating additional multimedia learning principles with instructional software design objectives. Multimedia learning requires processing both verbal and visual information via separate, limited-capacity cognitive channels, progressively through sensory memory and short-term into long-term memory.

In designing instructional multimedia, it is thus important to avoid presentations that interfere with students "*selecting*" relevant verbal and visual information, cognitively *organizing* and *integrating* it, and processing it through successive, limited short-term and long-term memory channels. For example, placing descriptive text on separate pages from associated images should be avoided because this demands "*extraneous*" processing that interferes with *essential* (representative) and *generative* (integrative/sense-making) cognitive processing. Designs should be avoided that overload short-term memory with too much information per page.

Methods, Media, and Software Tools for E-Learning

The beginning teacher of anatomy requires a basic set of tools—software—and methods for creating and using e-learning content, whether intended for live lectures, online presentations, laboratory applications, or integrated self-learning modules. One must also understand the best methods for deploying teaching resource files for specific classes, either by direct loading onto PCs or distribution via the Web. Fortunately, with prior graduate, medical education, or teaching experience, there is a good chance you have mastered many of the basic programs and methods that you need.

The most essential e-learning production software collection for anatomy faculty includes programs for word processing, image processing, presentation, production of files in portable document format (PDF), video production, Web development, and Web hosting . With some appropriate planning, the anatomical e-learning resources developed should be usable on legacy PC systems, lecture hall, and laboratory presentation systems, as well as on popular mobile devices like tablets and smartphones (m-learning) for functionally ubiquitous learning (u-learning) that is not tied only to classroom settings.

Word and Image Processing for E-Documents

Word processing is a foundational part of e-learning, for preparation of distributable lecture notes and learning objectives, laboratory exercise instructions, structural identification lists, and examinations. Well-crafted lecture outlines can be imported into presentation software, to serve as the organizing content for lecture slide production. Microsoft Word by default saves files in the .doc or .docx (XML) formats, which are editable (unless locked) and may not be read uniformly on all mobile devices and PCs due to font differences and other platform-specific restrictions. Word and other Open Source word processing programs are also capable of exporting and importing HTML (hypertext markup language) files, which can assist with developing e-learning Web pages. Furthermore, Word documents can be exported as PDF files or serve as another source for developing e-books (see below).

Image processing is also a necessary part of producing multimedia documents, presentations, and Web pages with embedded pictures. Whether images are obtained from a published collection, digital photos, or Web-distributed images (e.g., from a Google image search), some rudimentary image processing will usually be necessary, such as cropping, color balancing, captioning, or file-type conversion. All these functions are easily and routinely performed with Photoshop or other comparable programs.

In general, even if large TIFF (tagged image file format) images are the starting point, e-learning content typically works most efficiently with files output in compressed JPEG (Joint Photographic Experts Group) format. For some Web and mobile applications, the PNG (Portable Network Graphics) files may be preferred, even if they use more memory than JPEG files.

Digital Presentations for Lectures and Other Uses

Since the mid-1990s, the de facto standard for presentation software on Windows and Apple PCs has been Microsoft PowerPoint (PPT), still useful for developing and giving lectures, for creating laboratory learning module applications, and for designing Web resources. In fact, astute development of PPT and other presentation files will allow the same content to be used in lectures and in associated laboratory exercises. PowerPoint presentations can also incorporate internal hyperlinks to create more interactive content that allows jumping between pages and linking to external Web pages on networked computers. Lecture presentations can be enhanced by embedding videos, using audience response system (ARS) technology [14], or in-line quiz pages and slide-based quiz games (e.g., Jeopardy [15]).

Lecture content can also serve as the starting point for "inverting" or "blending" instruction, or it can serve as core content in online self-learning modules [16] that can complement existing lectures and laboratories (see integrated self-learning modules, below). As an alternative for adventurous anatomists, presentations can be developed and delivered as hyperlinked Web pages and embedded media. Apple iPad tablets and AirPlay wireless streaming have been used with HDTV monitors to present faculty explanations of dissections in the author's anatomy laboratories.

Portable Document Format

Portable document format (PDF) is a multimedia file standard developed by Adobe, based on the PostScript page composition and printing language. At the basic level, it supports formatted text files with embedded images that will display consistently (WYSIWYG) on different PC platforms as well as most mobile devices (e.g., iPad and Android tablets). It can also support internal navigation hyperlinks and select video file and 3D object embedding, although these advanced features are not generally supported in standard mobile device PDF apps. PDF is the most commonly used, externally loadable e-book format, so basic text + image PDFs can be read on most popular mobile readers (e.g., iBooks, Kindle, and Nook). PDFs can be directly exported from virtually any PC application (e.g., PowerPoint, Word, or Web browser). Specialized features (e.g., enabling "User Extended" note taking for students in Adobe Reader) require using Acrobat Pro, part of the Adobe Creative Suite that includes PhotoShop and the DreamWeaver HTML editor.

Advantages of exporting lecture slides and notes as PDFs

- Compact, standardized file format for what you see is what you get (WYSIWYG) display on any PC or mobile device.
- Most commonly used and supported e-book/e-reader format.
- Provides a degree of protection against easy, uncredited appropriation and reediting of author's content.
- Usable on mobile devices (e.g., iPads and Android tablets).
- Students can make notes with PCs and laptops when document is saved as Reader Extended PDF (Export setting in Acrobat Pro).
- Students can make notes with iPads and Android tablets using selected PDF apps.

E-Books

Although books reproduced in digital format—e-books—have been around in some form for nearly two decades, the popularity of e-books began to surge in 2010 with the availability of smartphones and tablets. A Pew Internet and American Life survey in 2016 found that about 29% of the American public had read an e-book in the preceding year [17]. Although it has been popularly asserted that millennial students read less than their predecessors, given a range of individual differences in learning styles [18], the current widespread availability of popular anatomy texts as e-books offers another e-learning option for students with mobile devices.

In addition to collecting all their curricular content in PDF format as previously mentioned, students have additional options for building personal e-learning libraries with open EPUB standard or proprietary iBooks formats. The EPUB format is an international file standard based on document encoding in XHTML with cascading style sheets defined formats, table of contents, and indexing. Apple iBooks format is an extension of the latest EPUB3 standard. Major biomedical publishers make many of their textbooks available in EPUB and PDF formats, and many US library systems are now loaning such e-books freely, including textbooks.

Useful free programs for creating these e-book format files include calibre (multiplatform EPUB and other formats) and iBooks Author (Apple proprietary EPUB3 format). Unlocked PDF and EPUB files can be "side-loaded" (i.e., directly downloaded outside of Apple or Android bookstores) onto popular mobile devices, e-readers, and PC reader programs.

Video Production and Acquisition

Videos can contribute to e-learning resources in many ways, and digital video has improved greatly since the first Web streaming formats became available in the 1990s. Currently the most widely used high-quality formats are based on the Motion Picture Experts Group (MPEG)-4 or h.264 standards, which are compatible with PCs and most mobile devices. There are several major methods for producing curricular MPEG-4 video recordings, and perhaps the easiest is the previously mentioned lecture capture, which saves all presenting computer screen displays (including pointers) and audio. Screen capture programs, available for most computing platforms, can easily be used by faculty to record their own live PowerPoint presentations. In more elaborate solutions, lecture capture files can be automatically recorded through a dedicated hardware server integrated with the digital projection system (e.g., as at the author's school).

Offline or production studio video can be created with collected presentation slides, camcorder recordings, audio and video clips, images, and animations using programs like iMovie, Premiere, or Avidemux. Such videos can be extremely sophisticated, presenting visual information in a more dynamic way than captured lectures. Furthermore, with proper permissions, clinical vignettes can be integrated.

"Podcasting" is a more recent, mobile-centric method for online distribution of multimedia, especially video and audio. The essential elements of a podcast are prerecorded media files (which can include documents), an associated descriptive Web page with media links, and an XML file which sets up a "Really Simple Syndication" (RSS) feed with metadata that define and control automatic distribution of media files to subscribers. Lecture podcasts can be linked to a school's learning management system (see Online Hosting Methods below).

Before beginning an elaborate video production, it is usually worth doing an online search for freely available content. Among other sources, Google's Web-based YouTube video service can serve as a source for usable files, and many donated educational videos can be found there [19]. It is important to list full presentation and copyright credits for any such freely accessible and distributable video.

Web Design

The Web has become the essential online infrastructure that integrates all methods of e-learning, just as it integrates most aspects of modern urban life. The Web also provides the means by which integrated course management systems organize curricula and operations for whole schools (see next section). But even if a course management system is available to organize major functional services and generic learning resources for all classes in a curriculum, having stand-alone Web services for anatomy instruction can be invaluable for supporting laboratories, self-learning modules, and hosting specialized anatomical media distribution.

Although many postmillennial students and teachers in postsecondary institutions may have experience that can be drawn on for educational Web resources design, it is extremely helpful for a new anatomist to learn the essentials of Web development and hosting, including HTML editing. Learning fundamental (or even advanced) Web design should not be an overwhelming challenge: There are numerous high-quality texts available, and free online development courses can be found with a bit of Web searching. Overall, the best teacher is personal experience gained by learning to use a HTML editor. See section "Online Hosting Methods for E-Learning and M-Learning Resources" for additional information about Web hosting and anatomy learning resources.

Creating Mobile/Ubiquitous E-Learning Web Resources

- Appropriate HTML5 or XHTML coding together with cascading style sheets (CSS) can be used to create adaptive Web pages that display optimally on PCs and mobile devices.
- (X)HTML "meta" statements scale Web page display size appropriately for the screen of the specific connecting device.
- Different CSS parameter sets for PCs and mobile devices set the font and other layout characteristics appropriate for each device.
- Use MPEG-4 (not Flash or other plug-in formats) for video.
- JavaScript programming can be used to replace Flash animation programming and complex 3D media formats.

Virtual Anatomy, Augmented Reality, and Other Simulation Methods

Different types of 3D image-based anatomical simulations or "virtual anatomy" have been employed for teaching for over two decades [20–21]. Although immersive, stereoscopic 3D anatomical resources have not been widely used, screen-based photogrammetric 3D simulations have been practical resources for online and laboratory learning [22]. Earlier screen simulations used proprietary software standards [23–24], but current HTML5 programming methods make "virtual reality learning objects," derived from clinical imaging data, usable on mobile devices [25] (Fig. 32.1). Similar methods have been applied to producing virtual anatomical models and panoramic virtual microscopy slides.

Other virtual anatomy simulations and commercial learning resources have been success-

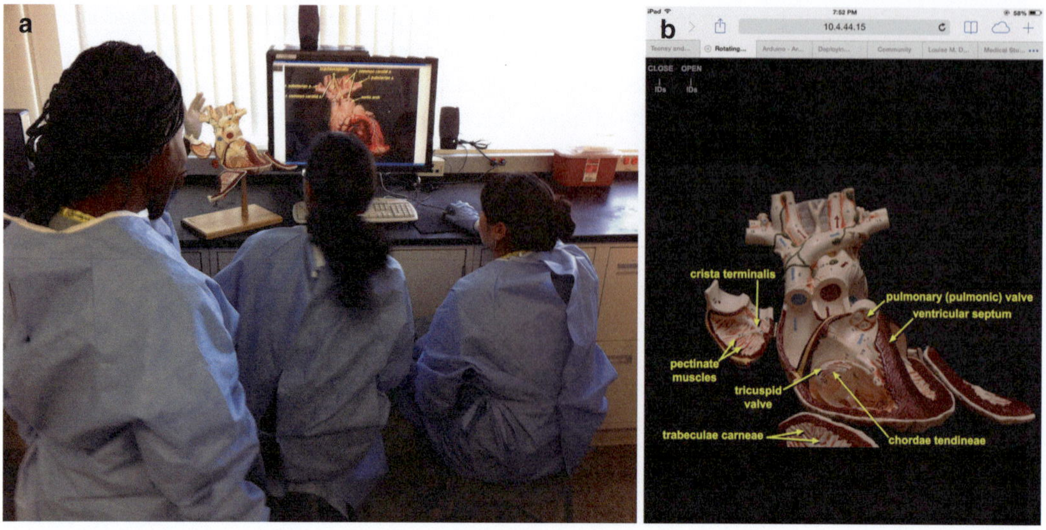

Fig. 32.1 In-class "mixed reality": (**a**) Medical students learning from a classical heart anatomy model while using a Web-based virtual reality learning object produced photogrammetrically from that model. (**b**) Apple iPad screen capture of the same interactive 3D heart model, demonstrating mobile device compatibility of this HTML5 format 3D object

fully developed from the Visible Human dataset [26], which was specifically conceived as a foundation for human simulation efforts [27]. One of the most recent developments of such learning technology has been the Anatomage Table system [28], a touch screen computer table that supports "virtual dissection" of digital human models, along with integrated medical imaging. Photogrammetric 3D models of cadavers and organ specimens also have been introduced for use with Anatomage [29]. Chapter 29 considers whether and how such virtual anatomy methods can supplant or complement current laboratory exercises using cadavers, models, and a mix of learning tools.

More recently, the original "virtual reality" concept has been expanded to include "augmented" and "mixed reality" (AR and MR) applications. These allow mobile devices with cameras to present additional data and images on top of real object views, such as models and specimens [30]. Other mobile-related augmentation of anatomical learning has involved scanning QR (quick response) codes that link to information [31] to specimens and models.

Quizzes and Self-Assessment

Self-testing modules can play a vital role in motivating students to desired levels of performance and in allowing them to gauge their knowledge and progress [32]. Frequently, these functions will be integrated into a course management system (see below), and development consists of creating assessment questions that fit the system's question-and-answer boilerplate requirements. Separately, self-assessments can be implemented in lecture presentations using ARS technology [14] or embedded Jeopardy quiz game slides [15], associated with laboratory exercises hosted on an anatomy Web server or integrated in self-learning modules [16].

Integrated Self-Learning Modules

Self-learning modules [16] are comprised of a varying number of linked and interactive software components that students can progress through individually to learn specific content. These typically include a presentation divided into segments, with illustrative animations and videos, simulations or game components, self-evaluation, and clinical case or problem-based learning components.

Adobe's (Shockwave) Flash multimedia programming environment had been a popular tool for producing such self-contained modules, since they could be downloaded from websites. With the widespread use of mobile devices not supporting Flash, Web-based multimedia alternatives became much more popular. Adobe recently ended their support of the product.

Current HTML5 Web browser standards support animation and play common media files without proprietary plug-ins. A special advantage of HTML5 and other Web-programmed approaches to self-learning modules is that they may also be set up to use server databases, supporting the collection of assessment data and other learning performance measures. Implementation tools for Web-based self-learning modules include HTML [5] or XHTML, JavaScript, PHP (hypertext programming language) or Java (programming language), and MySQL database server.

Online Hosting Methods for E-Learning and M-Learning Resources

As the prime enabling technology for the "revolution" in effective e-learning methods, the Web remains the dominant communications channel or "medium" for sharing educational resources. It is unusual to encounter a major health sciences school that does not currently use some type of Web-based course or learning management system (CMS or LMS, virtual learning environment) to organize the distribution of required content and information in current "clinically integrated" systems or multidisciplinary preclinical curricula.

LMSes are the overall unifying software for online learning resource access in contemporary schools. They integrate password-secured access for students and faculty with database-driven hierarchical organization of course- and year-based curricular content, date-sensitive calendaring, news distribution, small group communication channels, global/group email capabilities, and other socially networked information exchange in a combined large and small group educational environment.

How new anatomy teachers can make use of LMS or Web services

- Find out how an available LMS is being managed and familiarize yourself with how to use it.
- Identify the file standards and online media types supported by the LMS and be prepared to lobby management if you have special format or media needs.
- If LMS and/or anatomy Web services are unavailable, investigate departmental and administrative support and seek other faculty collaborators for establishing services.
- Learn the basics of Web design and multimedia production via books and online courses.
- If a curricular development group is not available, investigate support for Web-savvy student workers.
- Learn about available laboratory computing and network infrastructure, management, and planning.

Although a LMS may be used to distribute anatomy learning resources (e.g., lecture notes, slides, videos, and podcasts) uniformly with other programmatic curricular content, a separate Web server is typically needed for running computing-enhanced anatomy laboratories and for supporting other multimedia resources that cannot be directly hosted by the LMS.

It is worth learning how to run a Web server, in order to facilitate hosting and control of specialized anatomy resources (e.g., virtual anatomy and online simulation archives) and commercial instructional software, as well as to support the development and testing of new visualization and teaching methods. This need not be a forbidding or daunting challenge for scientifically trained anatomists. Web services are readily managed on the most popular Apache HTTP Server software [33], freely installable on networked Windows PCs and preinstalled on Apple Macintoshes. For distribution, HTML and other media files are simply loaded into Apache's "home folder." The author has used a number of Web servers extensively and continuously over the last 25 years to provide daily anatomical educational resource access, including integrated embryology simulations and laboratories, basic integrated gross anatomy laboratory session software, and online laboratory prosection guides for surgery and obstetrics/gynecology clerkship reviews.

A dedicated Web server can also support the development and use of more complex learning resources linked to databases, such as the previously mentioned self-learning modules, that integrate presentation components with self-assessment components collecting student performance data. Such Web servers must be secured on a virtual private network or behind a firewall, in order to protect student data, faculty intellectual property, copyrights, and any copyrighted commercial media used.

Perspectives for Continuing E-Learning Development and Research

It been recognized that variations in individual student capabilities and learning habits affect the implementation and individual usage of e-learning methods. Despite common aggregate perceptions of "millennial" students, research evidence demonstrates their considerable diversity in backgrounds, personalities, and learning styles, encompassing differences in technological predilection, multitasking, reading, and professional behaviors [16]. Chapter 2 further examines the characteristics of millennial generation students.

Nguyen et al. [34] have shown that students demonstrate individual differences in visualization ability that influence the effects of dynamism and interactivity on learning task performance with virtual anatomical models. Furthermore, in studying more conventional e-learning media, McNulty et al. [35, 36] have demonstrated that, although most may be familiar with Web-based educational resources, medical student usage of Web-based computer-aided instruction is complexly related to gender, individual learning styles, personality variables, and preferences. For example, one might expect greater elective use of certain kinds of interactive e-learning resources by students with more exploratory learning styles and less by those with pragmatic behaviors more focused on acquiring itemized core knowledge needed for passing examinations.

Other findings about relevant behavioral and student attributes have been reported by Nieder and Borges [37] in an eight-year study of student use of online lectures. Predictable trends were seen in higher usage related to examinations, but there was also a high degree of individual variability, partially attributed to gender and differences in academic ability. In a more limited study, Nieder et al. [38] also reported correlations between online lecture use and achievement motivation and VARK learning styles. These lines of evidence indicate that it is important to consider a range of individual students' behavioral and cognitive attributes in planning how to use e-learning resources in a specific curriculum.

There are other practical caveats for the development and implementation of new e-learning resources. Faculty members need to consider protecting their intellectual property rights in producing e-learning media that might potentially be expropriated with free public distribution. They could also be exposed to copyright violation challenges for embedded media that are shared externally by students. In many US schools, individual faculty own the copyrights to their lectures, and e-learning materials should show copyright notices appropriate to their institution, with acknowledgement credits for other included materials. Also, to reduce the risks of expropriation and other copyright violation, online learning materials should be secured on a password-protected website and/or on a virtual private network not accessible to the public.

Concluding Remarks

With continuing waves of curricular change, anatomy teachers' concerns about intensive use of e-learning may have been amplified by public political claims that online lectures and discussions can satisfactorily substitute for in-person classes. Such concerns might be further reinforced by continuing promotion of "massive" open online courses (MOOCs) for higher education. Certainly, the Association of American Medical Colleges welcomed MOOC progenitor Salman Khan as plenary speaker at their 2012 annual meeting [39]. But MOOCs largely engage entirely online learning near the low-yield "lecture tip" of the "interactive learning pyramid," with limited network-based student-faculty interaction. Whatever the case, the increase and spread of "flipped" basic sciences courses in medical schools suggests that mindful anatomists should expect their legacy teaching roles will be further deconstructed, along with the continuing growth of curricular e-learning and asserted institutional ownership of its content.

Fortunately, educational evolution will provide many new opportunities. Mastery of humanistic instructional design and multimedia software methods can facilitate lifelong innovation, research in education, and publication, as well as provide creative satisfaction in developing new practical ways to enhance learning with technology. Chapter 52 provides more insight into educational research opportunities in anatomy.

References

1. Trelease RB. From chalkboard, slides, and paper to e-learning: how computing technologies have transformed anatomical sciences education. Anat Sci Educ. 2016;9:583–602.
2. Ruiz JG, Mintzer MJ, Leipzig RM. The impact of e-learning in medical education. Acad Med. 2006;81:207–12.
3. Walton G, Childs S, Blenkinsopp E. Using mobile technologies to give health students access to learning resources in the UK community setting. Health Inform Libr J. 2005;2:51–65.
4. Yahya S, Ahmad EA, Jalil KA. The definition and characteristics of ubiquitous learning: a discussion. Int J Educ Devel Inform Communic Tech. 2010;6:117–27.
5. Drake RL. Anatomy education in a changing medical curriculum. Anat Rec. 1998;253:28–31.
6. Drake RL, Lowrie DJ, Prewitt CM. Survey of gross anatomy, microscopic anatomy, neuroscience, and embryology courses in medical school curricula in the United States. Anat Rec. 2002;269:118–22.
7. Drake RL, McBride JM, Lachman N, Pawlina W. Medical education in the anatomical sciences: the winds of change continue to blow. Anat Sci Educ. 2009;2:253–9.
8. Chakraborty TR, Cooperstein DF. Exploring anatomy and physiology using iPad applications. Anat Sci Educ. 2018;11:336–45.
9. Green RA, Hughes DL. Student outcomes associated with use of asynchronous online discussion forums in gross anatomy teaching. Anat Sci Educ. 2013;6:101–6.
10. Reigeluth CM, Carr-Chellman AA. Instructional-design theories and models, volume iii: building a common knowledge base. New York: Routledge; 2009.
11. Gagné RM, Wager WW, Golas K, Keller JM. Principles of instructional design. 5th ed. Belmont: Wadsworth Publishing; 2004.
12. Barab SA, Duffy T. From practice fields to communities of practice. In: Jonassen D, Land S, editors. Theoretical foundations of learning environments. 2nd ed. New York: Routledge; 2012.
13. Mayer RE. Applying the science of learning to medical education. Med Educ. 2010;44:543–9.
14. Alexander CJ, Crescini WM, Juskewitch JE, Lachman N, Pawlina W. Assessing the integration of audience response system technology in teaching of anatomical sciences. Anat Sci Educ. 2009;2:160–6.
15. Webb TP, Simpson D, Denson S, Duthie E Jr. Gaming used as an informal instructional technique: effects on learner knowledge and satisfaction. J Surg Educ. 2012;69:330–4.
16. Khalil MK, Nelson LD, Kibble JD. The use of self-learning modules to facilitate learning of basic science concepts in an integrated medical curriculum. Anat Sci Educ. 2010:219–26.
17. Perrin A. Book Reading 2016. In: Pew Research Center. 2016. http://www.pewinternet.org/2016/09/01/book-reading-2016/. Accessed 21 Feb 2019.
18. DiLullo C, McGee P, Kriebel RM. Demystifying the millennial student: a reassessment in measures of character and engagement in professional education. Anat Sci Educ. 2011;4:214–26.
19. Jaffar AA. YouTube: an emerging tool in anatomy education. Anat Sci Educ. 2012;5:158–64.
20. Trelease RB. Toward virtual anatomy: a stereoscopic 3-D interactive multimedia computer program for cranial osteology. Clin Anat. 1996;9:269–72.
21. Trelease RB. The virtual anatomy practical: a stereoscopic 3D interactive multimedia computer examination program. Clin Anat. 1998;11:89–94.

22. Attardi SM, Rogers KA. Design and implementation of an online systemic human anatomy course with laboratory. Anat Sci Educ. 2015;8:53–62.
23. Trelease RB, Nieder GL, Dørup J, Hansen MS. Going virtual with QuickTime VR: new methods and standardized tools for interactive dynamic visualization of anatomical structures. Anat Rec. 2000;261:64–77.
24. Trelease RB. Diffusion of innovations: smartphones and wireless anatomy learning resources. Anat Sci Educ. 2008;1:233–9.
25. Trelease RB, Nieder GL. Transforming clinical imaging and 3D data for virtual reality learning objects: HTML5 and mobile devices implementation. Anat Sci Educ. 2013;6:263–70.
26. Spitzer VM, Scherzinger AL. Virtual anatomy: an anatomist's playground. Clin Anat. 2006;19:192–203.
27. Spitzer VM, Whitlock DG. The visible human dataset: the anatomical platform for human simulation. Anat Rec. 1998;253:49–57.
28. Anatomage Table 6: magnifying human anatomy to the next level. 2018. https://www.anatomage.com/table6/. Accessed 18 Feb 2019.
29. Petriceks AH, Peterson AS, Angeles M, Brown WP, Srivastava S. Photogrammetry of human specimens: an innovation in anatomy education. J Med Educ Curric Dev. 2018;5:2382120518799356.
30. Küçük S, Kapakin S, Göktaş Y. Learning anatomy via mobile augmented reality: effects on achievement and cognitive load. Anat Sci Educ. 2016;9:411–21.
31. Mogali SR, Vallabhajosyula R, Ng CH, Lim D, Ang ET, Abrahams P. Scan and learn: quick response code enabled museum for mobile learning of anatomy and pathology. Anat Sci Educ. 2019;12:664–72.
32. Logan JM, Thompson AJ, Marshak DW. Testing to enhance retention in human anatomy. Anat Sci Educ. 2011;4:243–8.
33. Welcome! – The Apache HTTP server project. http://httpd.apache.org/. No date. Accessed 20 Feb 2019.
34. Nguyen N, Nelson AJ, Wilson TD. Computer visualizations: factors that influence spatial anatomy comprehension. Anat Sci Educ. 2012;5:98–108.
35. McNulty JA, Espiritu B, Halsey M, Mendez M. Personality preference influences medical student use of specific computer-aided instruction (CAI). BMC Med Educ. 2006;6:7.
36. McNulty JA, Sonntag B, Sinacore JM. Evaluation of computer-aided instruction in a gross anatomy course: a six-year study. Anat Sci Educ. 2009;2:2–8.
37. Nieder GL, Borges NJ. An eight-year study of online lecture use in a medical gross anatomy and embryology course. Anat Sci Educ. 2012;5:311–20.
38. Nieder GL, Borges NJ, Pearson JC. Medical student use of online lectures: exam performance, learning styles, achievement motivation, gender. Med Sci Educ. 2011;21:222–8.
39. 2012 AAMC Annual meeting – final program. 2012. https://www.aamc.org/download/370200/data/2012aamcannualmeetingprogram.pdf. Accessed 25 Feb 2019.

The Use of Low-Fidelity Models to Enhance the Learning of Anatomy

33

Lap Ki Chan and Rocky Chun Chung Cheung

Anatomical models have been used for a long time to facilitate the teaching and learning of human anatomy. Earlier models were made of wax, wood, ivory, and papier-mache [1]. When plastic was invented in the early twentieth century, it allowed the production of a lot of the modern commercialized models that we now see in anatomy laboratories.

The newest kinds of anatomical models are the highly accurate three-dimensional (3D) digital models of the human body, reconstructed using photos of actual sections of the human body or radiological images [2–7]. These models allow scaling, coloring, manipulation, and dissection and have been used extensively in the teaching and learning of anatomy [8–12], including the creation of animation, lecture slides, and interactive online materials [13]. Moreover, advances in 3D printing technology also see the possibility of turning digital three-dimensional models into physical models, offering even more exciting possibilities for teaching and learning [14–16].

Despite the popularity of digital models, there is some evidence that students may learn better using physical three-dimensional models than using digital three-dimensional models [17, 18]. Less technologically advanced physical models are still being invented and effectively used in anatomical education and even the training of advanced surgical skills [19–21]. These physical anatomical models have unique educational value despite their low fidelity [22]. Learning of 3D information requires both perception and imagery [23]. Perception is the extraction of information from pattern and form, and imagery is the reconstruction of the perception in one's own mind. Since complexity is a function of the organization of pattern and form [24], the low-fidelity models present a less complex 3D image and therefore the time needed for reconstruction is also less. This means students learning with low-fidelity models would learn quicker as there is lower associated cognitive load. When used appropriately, they can also create a learning environment that enables the teacher and learners to interact, the learners to reflect on their knowledge, and the teacher to provide appropriate feedback. This chapter describes the features and construction of these physical models and how they can be used effectively and enjoyably in the teaching and learning of anatomy.

L. K. Chan (✉)
Department of Biomedical Sciences, Faculty of Medicine, Macau University of Science and Technology, Macao Special Administrative Region, People's Republic of China
e-mail: lapkichan@gmail.com

R. C. C. Cheung
School of Biomedical Sciences, Li Ka Shing Faculty of Medicine, The University of Hong Kong, Hong Kong Special Administrative Region, Hong Kong, SAR, People's Republic of China
e-mail: u3005043@connect.hku.hk

L. K. Chan, W. Pawlina (eds.), *Teaching Anatomy*, https://doi.org/10.1007/978-3-030-43283-6_33

Examples of Low-Fidelity Anatomical Models

The Midgut Rotation Apron

Chan [25] developed a model in the form of an apron for the learning of midgut rotation (Fig. 33.1). When the teacher puts on the apron, he or she represents the developing embryo. On the front of the apron are three tubes made of fabric, which are aligned end to end, representing the foregut, midgut, and hindgut. The midgut tube can be pulled out from behind the apron, representing the elongation of the midgut during embryological development. The teacher then rotates the midgut loop in a counterclockwise direction (when observed from the front), so that the distal limb of the loop lies anterior to the proximal limb, thus demonstrating to the students why the transverse colon lies anterior to the duodenum in adult humans. Another such wearable anatomical model is a T-shirt for demonstrating the organization of the peritoneal cavity [26].

The Perineal Space Model

This is a model invented by the first author for teaching the spatial relationship between the pelvic cavity and the superficial and deep perineal spaces (Fig. 33.2). On a hung skeleton, the pelvic diaphragm, the perineal membrane, and Scarpa's/Colles' fascia are represented by three pieces of paper of different colors and cut to different shapes to fit their anatomical positions. Therefore, the paper representing the pelvic diaphragm is cut and folded into a cone that fits inside the bony pelvis, and the one representing the perineal member is cut into a triangle to fit between the two ischiopubic rami. The remaining piece of paper represents only the inferior portion of Scarpa's fascia, which stops at the inguinal ligaments on each side but flows over the pubis into the perineum, turns into Colles' fascia, and eventually attaches to the posterior margin of the paper representing the perineal membrane. The space above the pelvic diaphragm is the pelvic cavity, and the one between the pelvic diaphragm and the perineal membrane is the deep perineal

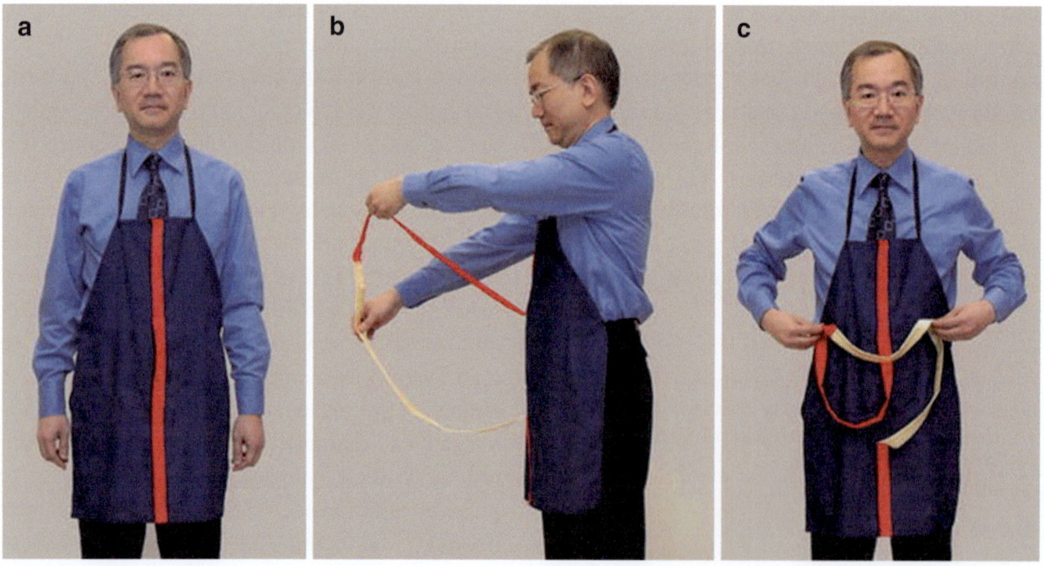

Fig. 33.1 The midgut rotation apron. (**a**) Foregut, midgut, and hindgut aligned. (**b**) Protrusion of the midgut. (**c**) After rotation of the midgut. (Based on Fig. 1 in Chan [25])

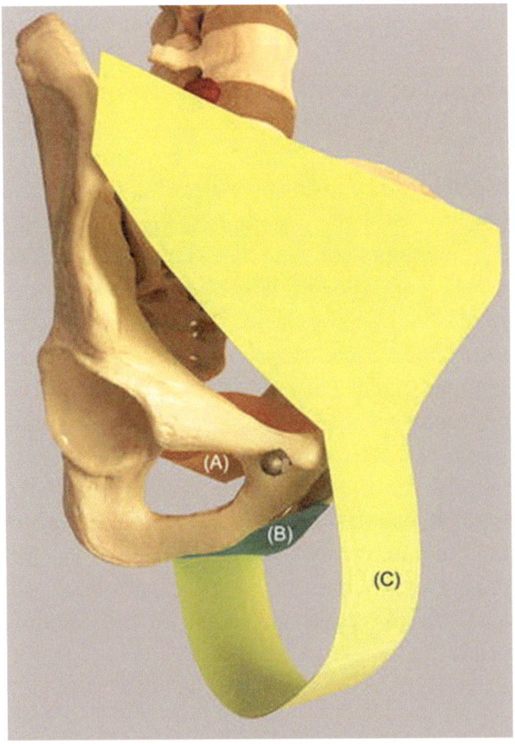

Fig. 33.2 The perineal space model. (**a**) The paper cone representing the pelvic diaphragm. (**b**) The paper representing the perineal membrane. (**c**) The paper representing Scarpa's and Colles' fasciae

space, while the one between the perineal membrane and Colles' fascia is the superficial perineal space. The model can be constructed easily with students, with plenty of opportunities for interaction, e.g., when creating the perineal membrane, the students can be asked where it is attached to.

The Hair-Band Model of the Digital Extensor Mechanism

Cloud et al. [27] described an ingenious model of the digital extensor mechanism for demonstrating the anatomy and function of the intrinsic muscles (Fig. 33.3). Two hair bands are knotted together to form a figure eight. But to complete the model, the hair bands need to be put onto a finger, with the finger going through both loops of the figure eight and the knot of the figure eight staying on the dorsal side of the proximal phalanx. While the proximal loop is pulled toward the palm, the distal loop is caught just under the fingernail. This completes the model, since the hair bands are now dorsal to both the distal and proximal interphalangeal joints, but palmar to the metacarpophalangeal joint. When the proximal loop of the hair band is pulled

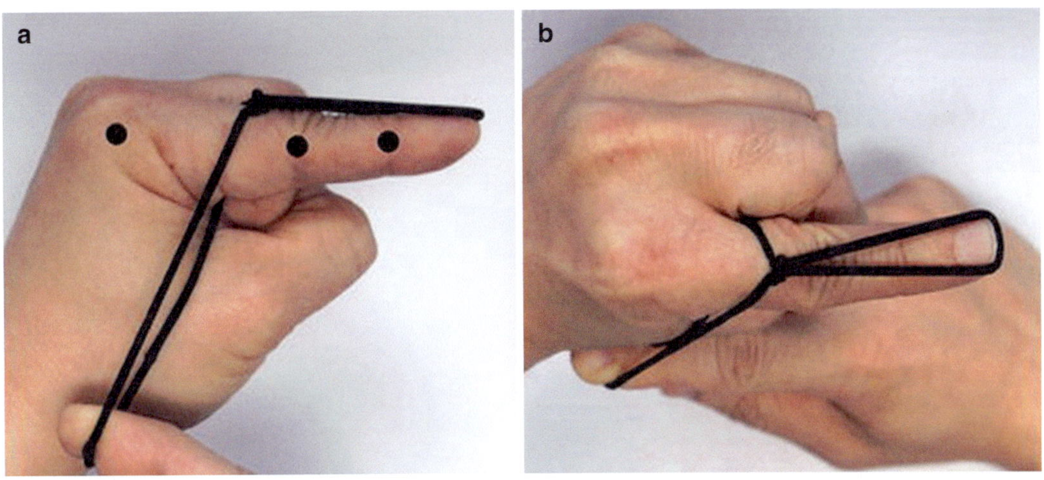

Fig. 33.3 The hair-band model of the digital extensor mechanism. (**a**) *Side view*. The *dark circles* represent the axes of the metacarpophalangeal, proximal interphalangeal, and distal interphalangeal joints. The hair bands pass palmar to the metacarpophalangeal joint and dorsal to the proximal interphalangeal and distal interphalangeal joints. (**b**) *Dorsal view*. (Based on Cloud [27])

toward the palm (representing the contraction of the intrinsic muscles), it causes extension of the interphalangeal joints and flexion of the metacarpophalangeal joint.

The Paper Model of the Muscles

This is a model described by Gangata [28] for students to learn the three-dimensional anatomy of muscles in the human body. Outlines of muscles were printed on paper and then cut out. Students then need to attach the paper muscles onto a human skeleton.

The Inguinal Canal Model

This model invented by the first author describes the formation of the inguinal canal in relation to the muscular layers of the anterior abdominal wall (Fig. 33.4). The external oblique (EO), internal oblique (IO), and transversus abdominis (TA) are represented by three pieces of paper of different colors. They are stacked together and the lower margin of the stacked paper is folded up obliquely to represent the inguinal ligament. One hole is cut in each of the three layers, eccentrically aligned, just superior to the "inguinal ligament." The canal formed by these three eccentric holes thus represents the inguinal canal. The holes in the EO and TA layers represent the superficial and deep inguinal rings, respectively. This model is particularly helpful for illustrating the boundaries of the inguinal canal. A teacher can insert a finger into the "inguinal canal" and identify with the students the structures forming the roof, floor, and anterior and posterior walls of the inguinal canal.

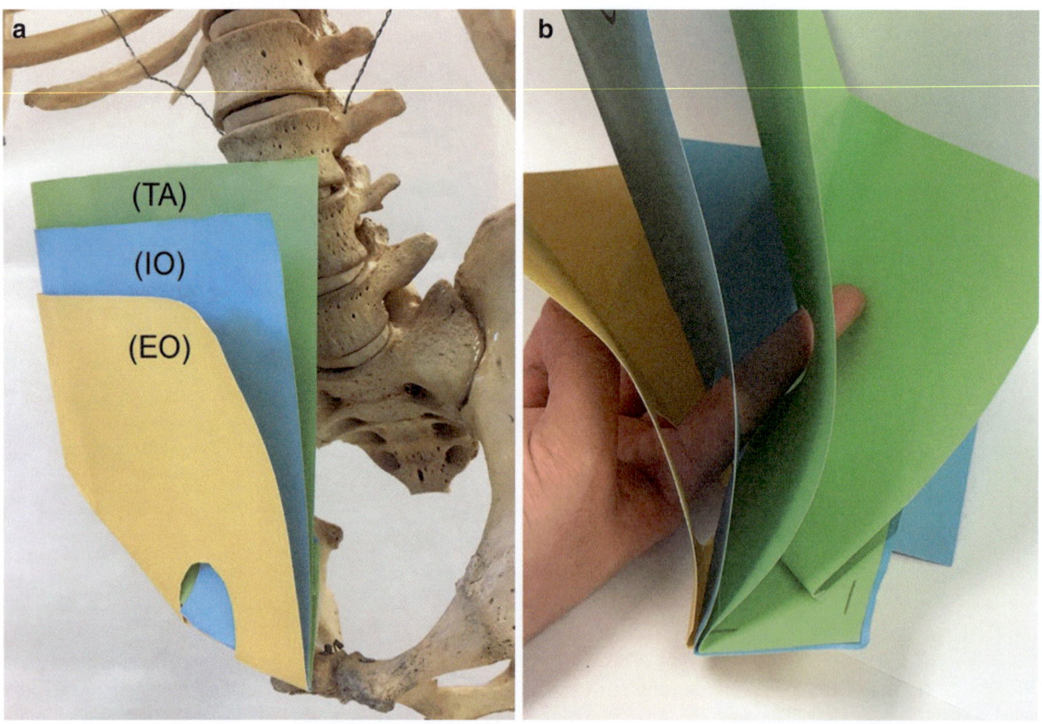

Fig. 33.4 The inguinal canal (right side) model. (**a**) *Left anterior view*. The three pieces of paper of different colors represent the three muscles forming the anterolateral abdominal wall: EO external oblique, IO internal oblique, TA transversus abdominis. (**b**) *Superior view*. A finger is inserted through the "inguinal canal," represented by the eccentric holes cut in the three pieces of paper

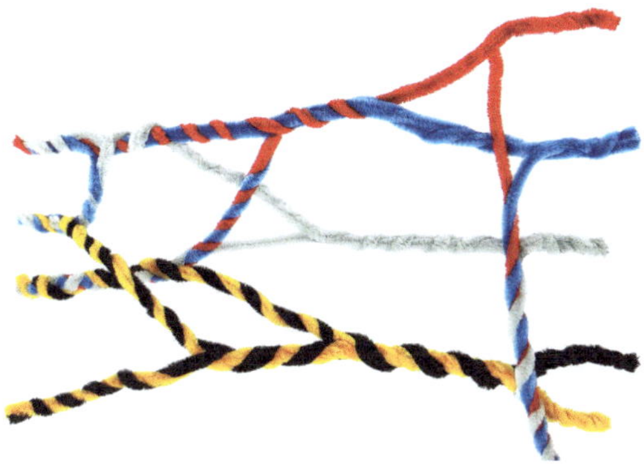

Fig. 33.5 A pipe-cleaner brachial plexus model. (Based on Lefroy et al. [29])

The Pipe-Cleaner Brachial Plexus Model

It is not often easy to teach or learn the anatomy of the brachial plexus using human specimens because of the complexity of the plexus itself and its surrounding structures. The task may be made easier by using the model described by Lefroy et al. [29], wherein the five nerve roots of the plexus are represented by pipe cleaners of five different colors and the branches are formed by the intertwining of these "nerve roots" (Fig. 33.5).

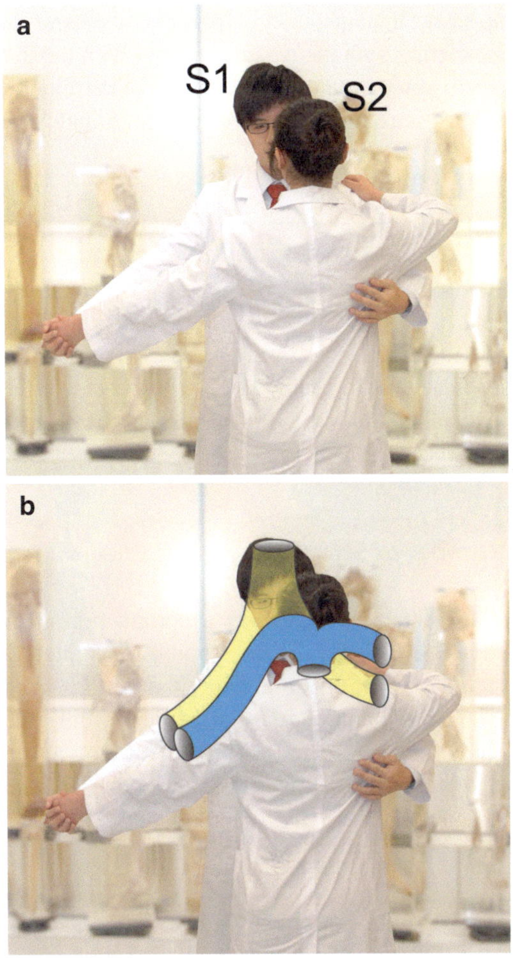

Fig. 33.6 (a, b) The mediastinal waltz model of the asymmetrical anatomical relationship between the main bronchi and pulmonary arteries. (Reproduced with permission from the publisher of Chan [30])

The Mediastinal Waltz

The anatomical structures in the mediastinum are arranged asymmetrically and understanding of this relationship can be challenging to students. Chan [30] described a representation of this relationship by the arms and heads of a waltzing couple (Fig. 33.6). The head of student one (S1) represents the trachea and the abducted arms represent the main bronchi, with the left one more so than the right to demonstrate the more horizontal position of the left main bronchus. The direction that S1 faces thus represents the anterior aspect of the mediastinum. Student two (S2) stands with the back facing anteriorly. The head of S2 represents the pulmonary trunk, while the right arm and left arm represent the left and right pulmonary arteries, respectively. S2 stands slightly to the left of S1 (i.e., they are not standing exactly face to face) because the bifurcation of the pulmonary trunk is not anterior to the bifurcation of the trachea but to the left main bronchus. The right hand of S2 is positioned on the left shoulder of S1 so that the right arm of S2 (left pulmonary artery) is superior to the left arm of S1 (left main bronchus). The left hand of S2 holds the right hand of S1 so that the left arm of S2 (right pulmonary artery) is anterior to the right arm of S1 (right main bronchus). The relative positions of the arms of the two students mimic those found between the pulmonary arteries and the main bronchi at the lung roots. The teacher can build up such a human model in front of students while interacting with them, perhaps even with the use of prosected specimens, but needs to be mindful of such sensitive issues as gender roles and body contact.

Features of Low-Fidelity Physical Anatomical Models

The low-fidelity physical models used in anatomical teaching and learning may not resemble the human body they are representing in the following aspects [22].

Number of Structures

When a model represents a certain region or structure of the human body, it typically includes only a small number of structures from the actual human body. For example, only the gut is represented in the midgut rotation apron [25], while the blood vessels, mesenteries, and ligaments and all the other abdominal organs are left out.

Shape and Surface Details of the Structures

Details are typically lacking in low-fidelity physical anatomical models. For example, in the hair-band model of the digital extensor mechanism [27], the hair bands do not resemble the extensor expansion or the intrinsic muscles in shape, texture, color, etc.

Dimensionality of the Structures

Sometimes low-fidelity physical models use two-dimensional structures to represent the three-dimensional human body. For example, Gangata [28] used paper (two-dimensional) cutouts to represent muscles (three-dimensional).

Size of the Structures

The absolute and relative sizes of the structures in a physical model may not follow those of the structures in the human body they are representing. For example, in Chan's [25] "human" model of the uterus, the torso of the person (representing the body of the uterus) wearing the poncho (representing the peritoneum around the uterus) is much bigger than the actual uterus.

Anatomical Relationship

The only aspect in which low-fidelity physical anatomical models always closely resemble the human body is the anatomical relationship. For example, the hair bands in the model of the digital extensor mechanism [27] closely mimic the extensor mechanism in terms of their relationship to the finger joints. The paper model of the inguinal canal also accurately reproduces the spatial relationships of the openings in the three muscular layers, thus allowing students to identify, on the model, the boundaries of the inguinal canal.

Features of low-fidelity physical anatomical models

- Include only a small number of structures
- Lack shape and surface details
- Can be two-dimensional
- Can have structures of different absolute and relative sizes
- Reproduce anatomical relationships closely

Advantages of Using Low-Fidelity Physical Anatomical Models

Despite the apparent lack of resemblance between the low-fidelity physical anatomical models and the human body, these models are welcomed enthusiastically by teachers and students. Some of the advantages of using low-fidelity models are as follows:

Serve as Memory Aids

By depicting a smaller number of structures and by ignoring many structural details, low-fidelity physical models simplify complex anatomical regions into abstract symbolic representations [24] that can be more easily remembered and manipulated mentally [23]. These models contain just enough information to help students solve problems that involve mental manipulation of three-dimensional anatomy, without being distracted by irrelevant information [23]. This means students learning with low-fidelity models would learn quicker as there is lower associated cognitive load. With reduced cognitive load, students can focus on accomplishing specific learning goals, such as deducing cross-sectional anatomy from three-dimensional anatomy [31] or understanding the anatomical basis of some clinical procedures [32]. Another example is the midgut rotation apron, which simplifies the developing gut into a simple tube, ignoring all other details. This allows the students to focus on the manipulation of the rotating gut and how this rotation results in the adult anatomy.

Arouse Students' Enthusiasm and Participation

Students enjoy the simplicity and ingenuity of low-fidelity physical models [33–35], but this simplicity also means that the models require greater involvement of the teacher (e.g., to establish the resemblance between the model and the parts of the body they are representing). This need for greater teacher involvement can be seen as an advantage of physical models as it provides opportunities for teachers and students to interact, for students to actively participate and reflect, and for teachers to aim teaching at the right level and give feedback; all of these characteristics are important for students to learn effectively and enjoyably.

Easy and Inexpensive to Produce

The low-fidelity physical models presented here are easy to produce, at very low cost, with materials that are easily obtained. Since damaging these models is of little consequence, students are not afraid to participate in manipulating them in order to learn from them.

Advantages of low-fidelity physical anatomical models

- Serve as memory aids
- Arouse students' enthusiasm and participation
- Easy and inexpensive to produce

Constructing Low-Fidelity Physical Anatomical Models

The construction of a good low-fidelity anatomical model involves the four steps in a plan, do, study, and act cycle (or the Deming cycle):

Plan: Identify Learning Outcomes and Design the Model

The low-fidelity physical models are very specific in that one model aims at one, or at most just a few, learning outcomes, which usually involve some anatomical relationships. It helps to be as specific as possible in identifying the learning outcomes of a model. For example, the intended learning outcome of the hair-band model of the digital extensor mechanism is for students to explain the actions of the intrinsic muscles of the hand through an analysis of their spatial relationships with the finger joints. It is not about the precise attachments of these muscles, their detailed anatomical features, their neurovascular supplies, or their anatomical relationships with the adjacent structures. The model is focused on helping the students to achieve just one outcome.

The design of low-fidelity physical models is only limited by the creativity of the designer. One important consideration is that the model should be easily contextualized; that is, it should allow students to easily establish the relationship between the model and the human body. For example, the models described by Zumwalt et al. [32] were built on a human skeleton. Such a design allows the students to position the models in the human body, which would otherwise be difficult because of the non-resemblance of the models to the real structures. The midgut rotation apron [25] is worn by the demonstrator, so that the students know the fabric tube being pulled out comes from the abdomen and represents the midgut.

To Do

- Think of an anatomical relationship that students usually find difficult to comprehend.
- Identify the outcomes you want your students to learn in that anatomical relationship (be specific).
- Think of a simple model to target these outcomes.

Do: Use the Model

This refers to the actual use of the models in helping students to learn. Some tips are included in the next section.

Study: Evaluate the Use of the Model

After putting a physical model to use in the classroom, the teacher needs to evaluate whether the design of the model aids in the achievement of the intended learning outcomes. Reflection is the starting point of such an evaluation. Useful questions to ask oneself after inventing and using a low-fidelity physical model in the classroom can include the following: What went well? Did the students learn what I would like them to learn using the model? Did I establish the resemblance between the model and the human body? Did the students enjoy learning with the model? Did I show the right immediacy behaviors, direct the teaching at the right level, give appropriate feedback, etc.? What can I do differently next time to improve the learning experience for my students?

Act: Improve the Model

Suitable modifications to the model should be made to facilitate the achievement of the intended outcomes. For example, the original design of the midgut rotation apron required the teacher to continue to hold the two ends of the tube representing the transverse colon, which prevented the use of the hands to explain the model and to interact with the students. This can easily be solved by the use of Velcro to attach the two ends of the tube, freeing the hands of the teacher.

Using Low-Fidelity Physical Anatomical Models for Teaching and Learning

Use Models in Both Large and Small Group Settings

The use of these models is not limited to small group settings. Even in a lecture, the use of these models can also improve the interaction between the teacher and students and maintain the attention of the students.

Establish the Correlation Between the Model and the Human Body

Many of these models are so simple that it may be difficult for students to realize how they represent the human body. Therefore, the first step in

What Will You Do?

- You have invented a model for demonstrating the anatomy of the inguinal canal. It consists of just three pieces of paper, one on top of another, representing the three muscles of the anterior abdominal wall of the inguinal region. Each piece of paper has a hole cut into it, being most lateral in the deepest layer and most medial in the most superficial layer. The holes thus represent the inguinal canal.
- Once you start your demonstration, you try to establish the correspondence between the three pieces of paper and the three muscles of the anterior abdominal wall. But instead of telling your students, you ask "Which muscle of the anterior abdominal wall does this most superficial piece of paper represent?" One of the students says transversus abdominis, while the others remain quiet or look blank. What will you do?

using these models is to establish the correspondence between the model and the human body, including clarifying orientation (e.g., anterior/posterior, medial/lateral, etc.) and describing the region and specific anatomical structures represented in the model.

Involve the Students

These models should not be used for a one-way demonstration, in which the teacher demonstrates without involving the students. Instead the teacher should make full use of the opportunity for two-way interactions with students, which make it possible for the teacher to give feedback on the students' learning, promote reflection by the students, motivate the students, etc.

Pitch at the Right Level

From interactions with the students, the teacher should be able to gauge the knowledge level of the class. The teacher should then adjust the level of subsequent interactions to the level of the students. Teaching at either too high or too low a level can decrease students' learning.

Give Feedback

During interactions with students, the teachers can observe the performance of individual students and appropriate feedback can be given with sensitivity and encouragement.

Promote Student Reflection

During the model demonstration, the teacher should ask questions that stimulate the students to think, reflect on, and integrate what they have learned about the anatomy of the region being explored.

Motivate the Students with Appropriate Immediacy Behaviors

Immediacy behaviors are those shown by the teachers which help to shorten the physical and

Tips on using low-fidelity physical anatomical models

- Use them in both large and small group settings.
- Establish the correspondence between the model and the human body.
- Involve the students.
- Pitch at the right level.
- Give feedback.
- Promote reflection.
- Motivate with appropriate immediacy behavior.

psychological distance between the teacher and the students [36]. Appropriate immediacy behaviors enable students to feel the enthusiasm of the teacher, which can in turn motivate the students to learn.

Conclusion

Low-fidelity physical anatomical models are useful tools for teaching and learning anatomy. Despite their apparent non-resemblance to the human body, they can accurately depict a targeted anatomical relationship. Low-fidelity physical models help students to learn anatomy by serving as memory aids, helping them solve three-dimensional anatomical problems, and, best of all, providing opportunities to draw students and teachers closer together for further interactions. To make the best educational use of these interactions, the teacher needs to involve the students, motivate them with the appropriate immediacy behaviors, pitch at the right level, give appropriate feedback, and promote reflection.

References

1. Marković D, Marković B. Development of anatomical models – chronology. Acta Med Median. 2010;49(2):56–62.
2. Spitzer VM, Whitlock DG. The visible human dataset: the anatomical platform for human simulation. Anat Rec. 1998;253:49–57.
3. Park JS, Chung MS, Hwang SB, Shin BS, Park HS. Visible Korean human: its techniques and applications. Clin Anat. 2006;19:216–24.
4. Zhang SX, Heng PA, Liu ZJ. Chinese visible human project. Clin Anat. 2006;19:204–15.
5. Nguyen N, Wilson TD. A head in virtual reality: development of a dynamic head and neck model. Anat Sci Educ. 2009;2:294–301.
6. Spitzer VM, Scherzinger AL. Virtual anatomy: an anatomist's playground. Clin Anat. 2006;19:192–203.
7. Sergovich A, Johnson M, Wilson TD. Explorable three-dimensional digital model of the female pelvis, pelvic contents, and perineum for anatomical education. Anat Sci Educ. 2010;3:127–33.
8. Rosse C. The potential of computerized representations of anatomy in the training of health care providers. Acad Med. 1995;70:499.
9. Jastrow H, Vollrath L. Teaching and learning gross anatomy using modern electronic media based on the visible human project. Clin Anat. 2003;16:44–54.
10. Hisley KC, Anderson LD, Smith SE, Kavic SM, Tracy JK. Coupled physical and digital cadaver dissection followed by a visual test protocol provides insights into the nature of anatomical knowledge and its evaluation. Anat Sci Educ. 2007;1:27–40.
11. Peterson CA, Tucker RP. Undergraduate coursework in anatomy as a predictor of performance: comparison between students taking a medical gross anatomy course of average length and a course shortened by curriculum reform. Clin Anat. 2005;18:540–7.
12. Nguyen N, Nelson AJ, Wilson TD. Computer visualizations: factors that influence spatial anatomy comprehension. Anat Sci Educ. 2012;5(2):98–108.
13. Fredieu JR, Kerbo J, Herron M, Klatte R, Cooke M. Anatomical models: a digital revolution. Med Sci Educ. 2015;25:183–94.
14. McMenamin PG, Quayle MR, McHenry CR, Adams JW. The production of anatomical teaching resources using three-dimensional (3D) printing technology. Anat Sci Educ. 2014;7(6):479–86.
15. Lim KH, Loo ZY, Goldie SJ, Adams JW, McMenamin PG. Use of 3D printed models in medical education: a randomized control trial comparing 3D prints versus cadaveric materials for learning external cardiac anatomy. Anat Sci Educ. 2016;9(3):213–21.
16. Ratinam R, Quayle M, Crock J, Lazarus M, Fogg Q, McMenamin P. Challenges in creating dissectible anatomical 3D prints for surgical teaching. J Anat. 2019;234:419-37.
17. Khot Z, Quinlan K, Norman GR, Wainman B. The relative effectiveness of computer-based and traditional resources for education in anatomy. Anat Sci Educ. 2013;6:211–5.
18. Preece D, Williams SB, Lam R, Weller R. "Let's get physical": advantages of a physical model over 3D computer models and textbooks in learning imaging anatomy. Anat Sci Educ. 2013;6:216–24.
19. Nicholson RA, Crofton M. Training phantom for ultrasound guided biopsy. Br J Radiol. 1997;70:192–4.
20. Schreuder HWR, Van de Berg CB, Hazebroek EJ, Verheijen RHM, Schijven MP. Laparoscopic skills training using inexpensive box trainers: which exercises to choose when constructing a validated training course. BJOG. 2011;118:1576–84.
21. Bastos EM, Silva RDP. Proposal of a synthetic ethylene-vinyl acetate bench model for surgical foundations learning: suture training. Acta Cir Bras. 2011;26(2):149–52.
22. Chan LK, Cheng MMW. An analysis of the educational value of low-fidelity anatomy models as external representations. Anat Sci Educ. 2011;4:256–63.
23. Marks SC. The role of three-dimensional information in health care and medical education: the implications for anatomy and dissection. Clin Anat. 2000;13:448–52.

24. Miller R. Approaches to learning spatial relationships in gross anatomy: perspective from wider principles of learning. Clin Anat. 2000;13:439–43.
25. Chan LK. Pulling my gut out—simple tools for engaging students in gross anatomy lectures. Anat Sci Educ. 2010;3:148–50.
26. Noël GPJC. A novel patchwork model used in lecture and laboratory to teach the three-dimensional organization of mesenteries. Anat Sci Educ. 2013;6(1):67–71.
27. Cloud BA, Youdas JW, Hellyer NJ, Krause DA. A functional model of the digital extensor mechanism: demonstrating biomechanics with hair bands. Anat Sci Educ. 2010;3:144–7.
28. Gangata H. An innovative approach to supplement the teaching of the spatial gross anatomy relationships of muscles to undergraduates in health sciences. Clin Anat. 2008;21:339–47.
29. Lefroy H, Burdon-Bailey V, Bhangu A, Abrahams P. A novel technique for teaching the brachial plexus. Clin Teach. 2011;8:196–9.
30. Chan LK. The mediastinal waltz – a representation of asymmetrical mediastinal anatomy. Anat Sci Educ. 2011;4(3):174–5.
31. Oh CS, Kim JY, Choe YH. Learning of cross-sectional anatomy using clay models. Anat Sci Educ. 2009;2:156–9.
32. Zumwalt AC, Lufler RS, Monteiro J, Shaffer K. Building the body: active learning laboratories that emphasize practical aspects of anatomy and integration with radiology. Anat Sci Educ. 2010;3:134–40.
33. Op Den Akker J, Bohnen A, Oudegeest W, Hillen B. Giving color to a new curriculum: bodypaint as a tool in medical education. Clin Anat. 2002;15:356–62.
34. McMenamin PG. Body painting as a tool in clinical anatomy teaching. Anat Sci Educ. 2008;1:139–44.
35. Finn GM, McLachlan JC. A qualitative study of student responses to body painting. Anat Sci Educ. 2009;3:33–8.
36. Andersen J, Nussbaum J, Pecchioni L, Grant J. Interaction skills in instructional settings. In: Vangelisti AL, Daly JA, Friedrich GW, editors. Teaching communication theory and research methods. Hillsdale: Lawrence Erlbaum; 1999. p. 359–74.

The Use of Augmented Reality Technology in Medical Museums

34

Atsushi Sugiura, Toshihiro Kitama, Masahiro Toyoura, and Xiaoyang Mao

Introduction and Historical Background

Medical museums focusing on anatomical and pathological specimens appear to have originated in the sixteenth century [1, 2] and were further developed in the seventeenth century. One of the most famous examples, the "Hunterian Museum," belonging to the Royal College of Surgeons in England [3], was founded during this era. Since the early nineteenth century, these museums have played an active part in medical education. Indeed, there are many notable examples of success in their contribution to modern medical education [4–8]. Historically, another important method in medical education was the use of anatomical wax models. Such models, which flourished in Florence, Italy, in the eighteenth century, were exalted as virtual anatomy education tools by the Austrian Emperor Joseph II (exhibited in the Josephinum Wax Models Museum in Austria [9]). They were in use until at least the end of the nineteenth century as visual aids in the education of surgeons and midwives. The quality and accuracy of these reproductions can be considered as the pioneer work for the virtual anatomical education tools of today. Recently, in contrast to their former popularity, the use of these time-honored institutions and methods, with the exception of a few world-famous museums, appears to have declined. Several causes for this can be identified. One is the increase in universities' need for more space for biological and biomedical research, which has progressed rapidly. Another is the difficulty of enhancing and updating the exhibits and models in conformance with the continuous development of knowledge and technology in biomedical science [8, 10–12]. In spite of these circumstances, various information technology (IT) tools offer the opportunity to revolutionize medical museums to act as a new educational tool [4, 5, 8]. For instance, the medical museum of Leiden University has been renovated based on the concept that medical museums can play an irreplaceable role in motivating self-learning by means of various interactive digital tools [5]. The medical museum itself is involved in the anatomy and pathology curriculum of medical students. Diaz-Perez et al. [4] reported on an evaluation of a pathology teaching strategy that combines medical museums with clinical problems, virtual autopsy, and digital microscopy. Moreover, other examples of significant influence on teaching and learning techniques in

A. Sugiura · T. Kitama (✉)
Center for Life Science Research, University of Yamanashi, Yamanashi, Japan
e-mail: asugiura@yamanashi.ac.jp; tkitama@yamanashi.ac.jp

M. Toyoura · X. Mao
Interdisciplinary Graduate School, University of Yamanashi, Yamanashi, Japan
e-mail: mtoyoura@yamanashi.ac.jp; mao@yamanashi.ac.jp

L. K. Chan, W. Pawlina (eds.), *Teaching Anatomy*, https://doi.org/10.1007/978-3-030-43283-6_34

medical education through developments in IT, such as three-dimensional printing [13–17], audiovisual techniques [18–23], and computer-generated imagery including virtual reality (VR) combined with a mobile device [24–30], have been demonstrated. In addition to these relevant examples, a brief description of the augmented reality (AR) technology and our AR-based tour support systems for our medical museum will be introduced in the following sections.

Impact of Medical Museum

- A tour of the medical museum can provide an invaluable learning experience to beginners in medicine.
- Each specimen provides visitors an opportunity to deepen their knowledge and satisfy their curiosity.
- The tour also engenders thoughts regarding the dignity of life and respect for organ donation.

Augmented Reality Technology

Figure 34.1 shows a schematized image of VR and AR. VR is a computer technology that can create a simulated environment (virtual world); it enables users to experience anything they want, anywhere, and at any time. A head-mounted display (HMD) is the most commonly used wearable device to realize the VR environment. Instead of viewing a personal computer (PC) monitor in front of users, they wear an HMD and are then immersed in and able to interact with the three-dimensional (3D) virtual world. By contrast, AR is a technology, using either an HMD or mobile device, that combines real world with virtual objects or information and gives the users the impression that the virtual objects are part of the real world. To achieve these aims, an AR system needs to be able to adjust the image of the virtual object that the user sees according to the real-time 3D position of the user relative to the object. To meet this technical requirement, two methods are available: sensor-based AR and marker-based AR.

Sensor-based AR accurately identifies a position in the real-world environment using a sensor such as a global positioning system (GPS) or infrared radiation sensor. AR systems using handheld GPS devices have previously been used by field workers [31–33]. An outstanding recent example of GPS-based AR is the mobile game Pokémon GO (Niantic Inc., San Francisco, CA), although various kinds of positive and negative effects of its use have been reported [34–37]. Despite the effectiveness of position detection in outdoor areas, the technique is limited indoors

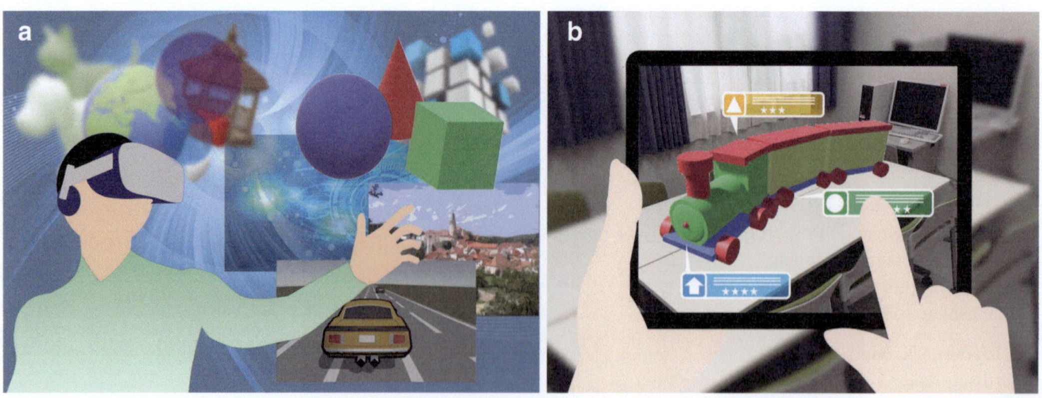

Fig. 34.1 A schematized image of the VR and AR. (**a**) VR is a simulated environment (virtual world) created with software and presented to the user. HMD are used to provide an immersive experience. (**b**) AR is the integration of virtual information with the user's environment in real time. Tablets, smartphones, and wearable devices support AR

due to poor GPS signal reception inside buildings. As a solution, a technique using infrared radiation sensors is proposed for tour guidance systems in museums to specify a visitor's location [38]. Regarding an anatomy learning system, there is a magic mirror-like system that uses a Kinect sensor (Microsoft Corp, Redmond, WA) comprising an infrared radiation sensor and a video camera [39]. The sensor can overlay an anatomical image on the user's body. In addition, when the user moves around, the system can track the user and move the projected image accordingly, so that the projected image remains on the user. Another example of a sensor-based AR is combining a virtual image with haptic feedback using a sensor glove or a 3D stylus device, which provides users with a tactile sensory image of the virtual object. The sensor glove system (CyberTouch, CyberGlove Systems LLC., San Jose, CA) has small vibrotactile stimulators and motion sensors on each finger, which provide an impression of the user touching or grasping real objects when the user's virtual hand handles virtual objects [40]. The stylus device (Phantom Omni, SensAble Technologies Inc., Wilmington, MA) consists of an articulated arm and a stylus pen at the end, attached to a base. The arm joints allow for free 3D manipulation of the stylus. When the cursor on a PC monitor—indicating the stylus-tip position—does not touch any virtual object, the user feels as if he/she is holding only weight of the stylus pen. When the cursor encounters a virtual object, the user can feel the forces that push back on the stylus, simulating touch and interaction with virtual objects [41, 42]. Each of the systems is constructed using AR techniques to enable 3D structural comprehension of anatomy or surgical skill training.

Marker-based AR estimates the position and orientation of a characteristic marker in real time from a camera; it displays virtual objects over the position of the marker found in the real world. Two popular kinds of markers with the different design are used. The first kind consists of figures with a thick bold square border, within which is a symbol for identification (referred to as an AR marker hereafter) (Fig. 34.2). Interactive learning systems using an AR marker have been proposed [43, 44]. These systems recognize an AR marker, and then, the virtual anatomical figure (e.g., a lower limb or skull) appears in the computer screen. The user can modify the position and orientation of the anatomical structure by manipulating the AR marker or by moving the tablet that the user is holding. Such intuitive operation with visual feedback facilitates students' comprehension of the complex anatomy structure. The second kind of markers uses the whole or a part of an

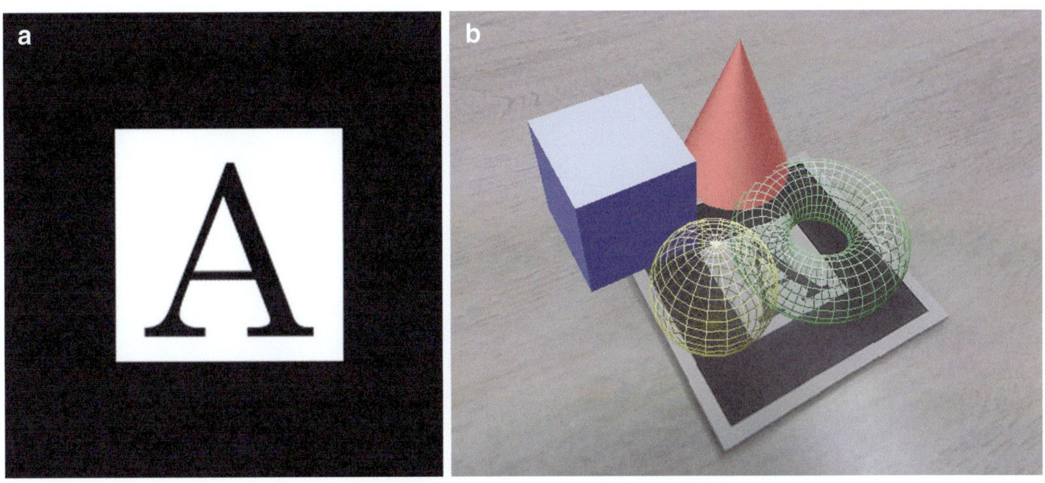

Fig. 34.2 AR marker and an example of displayed virtual objects. (**a**) Example of an AR marker that has a thick square frame and a pattern to distinguish it in the center. (**b**) The virtual objects are displayed on the AR marker when it is detected by the AR system

object image with a characteristic feature as a marker (referred to as image marker). Various types of image marker-based AR system have also been developed for exhibition facilities. For example, in an Acropolis museum, a mobile device with an AR system can allow visitors to see mythological modifications or restored images superimposed on archaeological artefacts [26]. It has also been demonstrated that an AR guide, which presents the contextual content for an artwork, clearly increases the time spent on the exhibition tour, which may reflect an increase in visitors' interest [45]. These examples illustrate the unique applications of AR systems in museum exhibitions and how they can enhance visitors' learning experiences. AR technology appears to have great potential for various exhibition facilities. Moreover, image marker-based AR has also been applied for anatomy education [6, 12]. The tablet screen displays a virtual anatomy model, which is presented over a picture in a textbook. When the user touches the virtual model displayed on the screen, the touched area is highlighted, and the name of the body part is displayed. In addition, a two-finger touch enables the user to remove the part of an object, while a pinching gesture allows zooming in and out [6]. Real-time interaction with the anatomy model could potentially increase user satisfaction.

AR Support Systems for Tours in Medical Museums

The medical museum at the University of Yamanashi's Faculty of Medicine (UYFM) is relatively small compared to the prominent museums cited earlier. The authors manage the museum and act as docents in presenting exhibitions to visitors (i.e., students of medical, nursing, or paramedic courses). Medical students at the UYFM may have limited need to visit the museum, as their curriculums contain numerous lectures and trainings regarding anatomy and pathology. In contrast, as nursing and paramedical students learn anatomy and pathology through a limited number of lectures, visiting a medical museum can provide them with an invaluable experience for their education. In addition, explanations offered by a docent can help promote the students' understanding, depending on their knowledge or interests. When visitors browse at their own pace, however, traditional description panels, even those with illustrations, are unable to satisfactorily fulfill the educational aims. We believe that advanced technology could provide more engaging descriptions and may help with self-learning. The following three methods were devised: a guidance system using quick response (QR) codes with network connections to digital contents on a website, a virtual medical museum using VR technology, and a guidance system using AR technology. We regarded an AR-based description as appropriate, because it offers many advantages for guiding learning in medical museums (see text box). In the following sections, we introduce three examples of recently developed AR support systems [46]. The first two systems consist of a tablet device (referred to as tablet AR) and a simple development environment using two marker types: AR marker and image marker, respectively. The third system is a relatively advanced version with a combination of an HMD in place of a tablet (referred to as HMD AR) and a natural-click interface.

Advantages of Using AR for Educational Tours in Medical Museums

- Visitors only need to capture the specimen with a device's camera to immediately obtain supplementary information.
- Combining actual specimens with virtual information can produce a synergistic effect, resulting in efficient acquisition of important knowledge.
- The AR system can improve the descriptions of exhibits in traditional museums and thus raise visitors' learning motivation.

Tablet AR System (AR Marker and Image Marker)

Several kinds of free AR software libraries allow the construction of a simple AR environment (Table 34.1). These libraries have four elemental functions: marker image registration, AR information setup, marker tracking, and AR information presentation. We develop the tablet AR system using ARToolKit [47] as the AR software library and Visual Studio [48] as the integration software (Fig. 34.3a). For a tablet device, Surface Pro 3 (Microsoft Corp., Redmond, WA) is used to display AR information. For constructing the AR marker system, first, AR marker images corresponding to each specimen are registered using an ARToolKit

Table 34.1 Software libraries for construction of AR environment (in alphabetical order) and their websites

Library name	Website
AR Studio	https://sparkar.com/ar-studio
ARCore	https://developers.google.com/ar/
ARKit	https://developer.apple.com/arkit/
ARToolKit	http://www.hitl.washington.edu/artoolkit/
DeepAR	https://www.deepar.ai/
EasyAR	https://www.easyar.com/
Kudan	https://www.kudan.eu/
Lens Studio	https://lensstudio.snapchat.com/
MAXST	http://maxst.com/#/
Vuforia	https://www.vuforia.com/
Wikitude	https://www.wikitude.com/
XZIMG	https://www.xzimg.com/

They can be used for free, unless the user plans to develop a commercial product

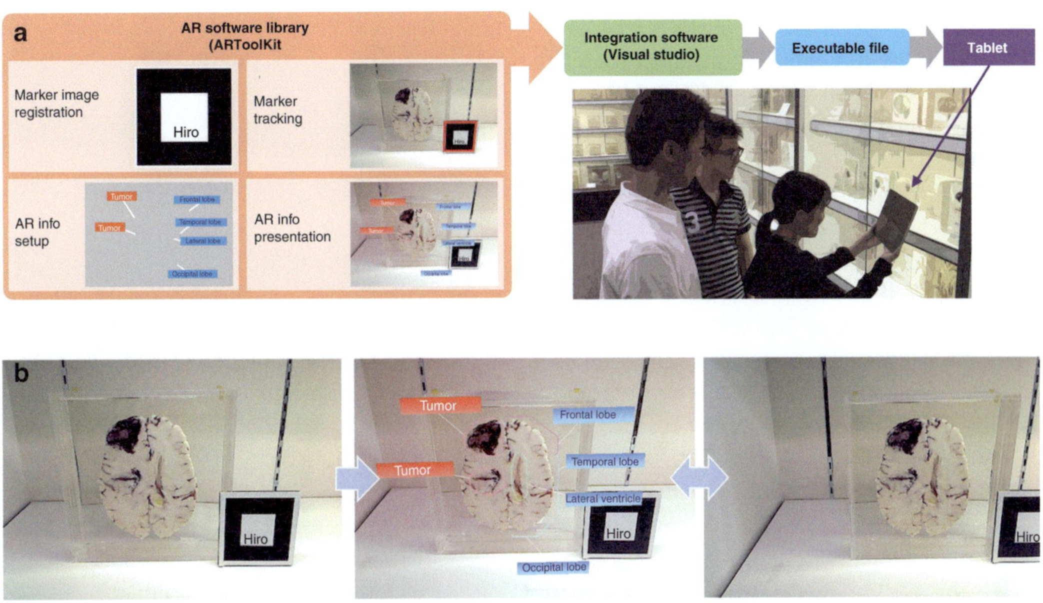

Fig. 34.3 Overview of a tablet AR system using an AR marker. (**a**) A schematic image of the system configuration. The AR software library (ARToolKit) features four elemental functions (marker image registration, AR information setup, marker tracking, and AR information presentation), which are integrated and compiled by an integration software (Visual Studio) to create an executable file. The file is installed on the user's tablet. (**b**) An example to show how the system works. Left: Plain view of an AR marker placed beside a specimen (not viewed through the AR system). Middle: Display of virtual labels on the specimen on the user's tablet when the system detects the AR marker. Right: Virtual labels are not displayed when the AR marker is outside the video frame. (Modified from Sugiura et al. [46])

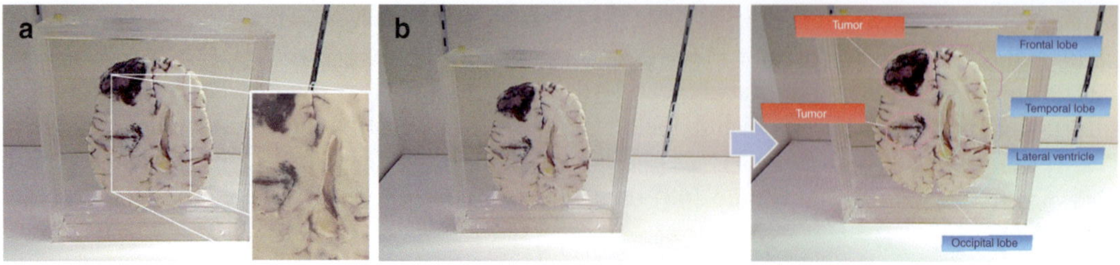

Fig. 34.4 Overview of a tablet AR system using an image marker. (**a**) Part of the overall cross-sectional image is registered as the marker. (**b**) An example of how the system works. Left: Plain view of a specimen (not viewed through the AR system). Right: The system detects the marker and displays the virtual labels. (Modified from Sugiura et al. [46])

function, and the printed AR marker is set on the side of each specimen (Fig. 34.3b left). Second, the textual contents, the size of the virtual label, and the boundary and leader line positions are set up in the coordinate system using the center of the AR marker as the center. Care was taken that the number of virtual labels and the color of the leader lines do not conceal the specimen image (Fig. 34.3b middle). Third, the AR marker position is tracked in real time from the video image using a tablet camera. Fourth, while the marker is being tracked, virtual objects are presented on the tablet screen to align them with real specimen images. All four elemental functions of the ARToolKit are integrated and compiled into an executable file, which is installed on the user's tablet for practical use. When the AR marker is out of the video frame, the virtual objects are not displayed (Fig. 34.3b right). The exhibit cannot fully occupy the screen of the tablet since the AR marker must always be included. The total cost of developing the tablet AR system is about $1500 (e.g., standard-performance PC: about $1000; tablet device: $500–$700; application software: free).

As for an image marker system, the specimen image itself is used as a marker (Fig. 34.4a). Since the UYFM medical museum has many slice-shaped specimens, a rectangular part with distinctive characteristic features is clipped from each overall cross-sectional image as an image marker. Users can obtain AR information relating to each exhibit without being aware of the presence of a marker, because it is not an artificial add-on placed next to the specimen but is in fact part of the specimen (Fig. 34.4b). Medical students at the UYFM usually visit the museum individually or in a small group. The tablets are lent to individuals or to the group. In case that a certain number of nursing and paramedical students visit the museum, they are divided into small groups, and the tablet is lent to each group. In spite of students' positive perceptions for the image marker system, they experienced some inconvenience in holding the tablet (see text box). Therefore, we improve the

Students' Perceptions

Advantages of a Tablet AR System

- The system provides a simple, low-cost AR environment.
- Visitors can obtain AR information without being aware of the marker.
- The system is favorably accepted by students as a useful tool.
- AR guide seems to be effective in helping student learn.

Disadvantages of a Tablet AR System

- Visitors must continuously capture the specimen via the tablet's camera.
- It may be difficult to assign part of a specimen with few distinctive features as an image marker.
- Holding a tablet might be a burden to users.

tablet AR for more convenient and sophisticated learning system. The system will be introduced in the following sections. For the two tablet ARs described thus far, the developer needs some programming skills with Visual Studio to integrate the AR software library. There are other all-in-one software products, such as InstantAR [49], that do not need the users to have the programming skills, in constructing AR system.

Head-Mounted Display (HMD) AR System

An HMD (Wrap920AR, Vuzix Corp., West Henrietta, NY) is used as a hands-free terminal device to relieve the physical burden of holding a tablet. The HMD has a display very close to the front of each eye of the user and is connected to a laptop PC (Think Pad, Lenovo Corp., Hong Kong, China). The system employs one of the two built-in USB cameras with a horizontal viewing angle of 31° and presents what it captures to both displays for the left and right eyes. Notably, although the HMD device has the advantage that users do not need to hold the tablet to point its camera toward the exhibits, users cannot perform touch-screen operations, and operating a computer keyboard to interact with the system is bothersome. To solve this problem, a natural-click-interface technique [50] is incorporated into the system by using a computer vision library (OpenCV [51]), in addition to the same software used for developing the tablet AR systems (Fig. 34.5a). The OpenCV has three additional functions: hand movement tracking, virtual button setup, and virtual button presentation. The hand movement tracking serves to recognize fingertip movements—so-called click gestures—through real-time detection of skin-colored areas in a video image. In the real world, a click means pressing a button with a finger, while in the AR environment, it means pressing a virtual button with a virtual finger that moves according to the real finger's motion (Fig. 34.5b). The second function is for the setting up virtual buttons on the screen at specified positions in the HMD coordinate system. Our system has two virtual buttons—INFO and AR (Fig. 34.5b left). The third function concerns the presentation of further information upon clicking the virtual buttons. In our system, a click on the AR button presents virtual labels for the different parts of the specimen, while a click on the INFO button presents the additional information about the specimen. These labels and information will disappear when the virtual buttons are clicked again. The opacity level of the information label is set at 0.5 (i.e., semitransparent, with 0 being completely transparent and 1 completely opaque) so that the specimen image is still visible through the label. All elemental functions (four functions of ARToolKit and three functions of OpenCV) can be properly executed through integration and compilation using a Visual Studio software. The total cost of developing the HMD AR system is about $2600 (e.g., about $1000 for a standard-performance desktop PC; over $1000 for a HMD unit; about $600 for a laptop PC; application software, free). HMD can free the user's hand from holding a tablet but may have shortcomings such as poor visibility due to small display size with low resolution. The natural-click gesture would facilitate active operating behavior for more effective learning. Recently appeared HMD devices, like HoloLens (Microsoft Corp., Redmond, WA), could provide a simple environment for the development of AR system with a hand-gesture operation.

Students' Perceptions

Advantages of an HMD AR System

- Visitors do not need to hold the device when viewing the exhibits.
- Visitors can make notes and sketches while wearing the HMD.
- The system provides an inspiring experience.

Disadvantages of an HMD AR System

- The HMD has a small screen and low resolution.
- Using an HMD may cause cybersickness.

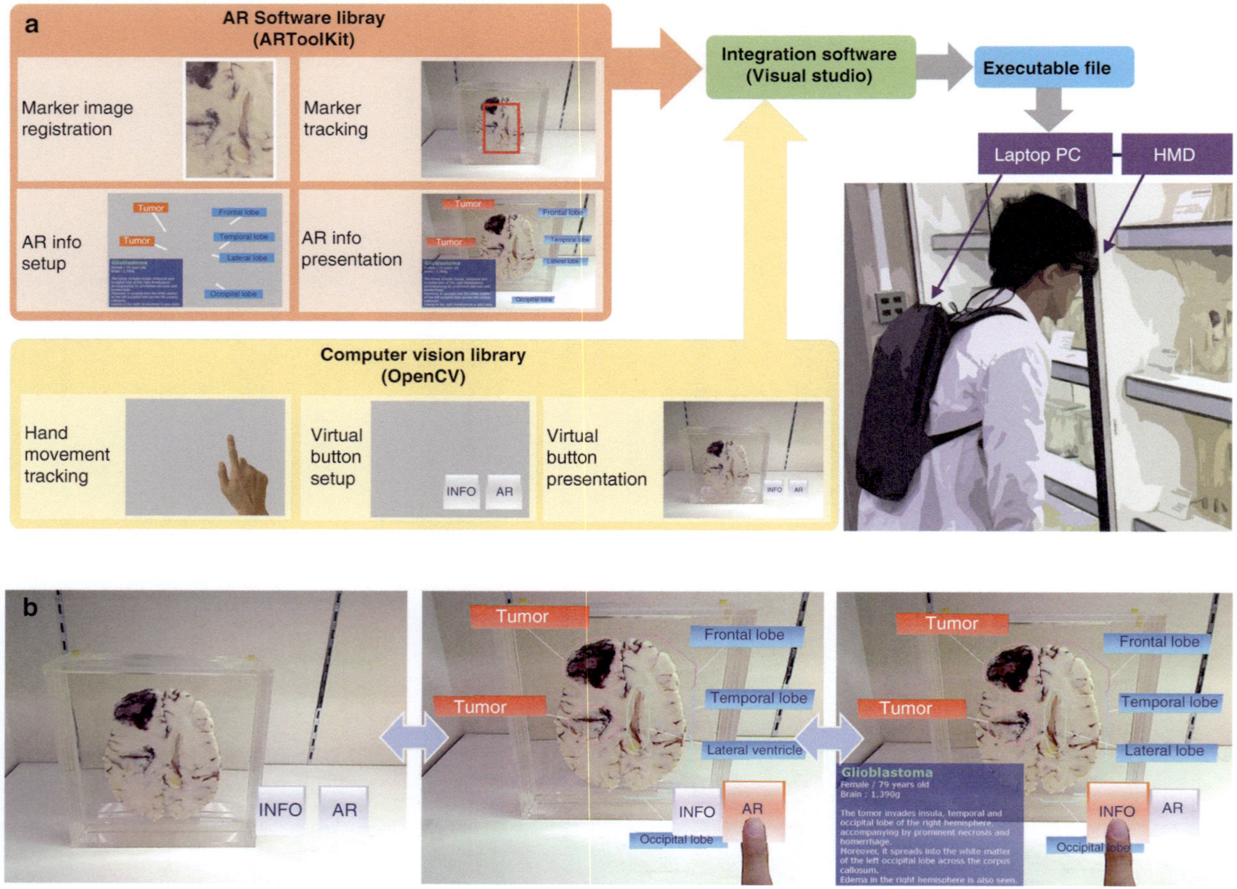

Fig. 34.5 Overview of a head-mounted display (HMD) AR system. (**a**) A schematic image of the system configuration. In addition to the same ARToolKit software as the tablet AR system (Fig. 34.3), computer vision library (OpenCV) is used to provide three additional functions (hand movement tracking, virtual button setup, and virtual button presentation). All elemental functions are integrated and compiled by Visual Studio software to create an executable file. The file is installed on the user's laptop PC connected to the HMD. (**b**) An example to show how the system works. Left: Two virtual buttons overlaying a specimen image are displayed on the HMD display. Middle: Virtual labels are displayed when the user presses the AR button using a click gesture. Right: The information about the specimen is displayed when the user presses the INFO button. (Modified from Sugiura et al. [46])

Limitations and Scalability

The AR system's implementation as a learning tool in medical museums should be considered carefully. Because medical museums house sensitive human remains, taking photos and videos is generally not allowed. For this purpose, devices used in the present study were loaned to visitors during tours inside the museum. Software applications that can be installed on visitors' mobile devices would be another solution, provided that they are used exclusively during the tour and no recording can occur. Despite their usefulness, AR technologies, especially those that use HMD, can cause an affliction called cybersickness. An optical see-through HMD could create a more comfortable environment without blurred vision or dizziness in the future. Another limitation is that our AR system works better on sliced specimens than on three-dimensional ones, because the present system uses part of the specimen image as the marker and three-dimensional specimens will only yield the same image when viewed from a specific angle. The use of multiple marker images from different angles for a single specimen would help in such recognition.

In this chapter, we introduced three examples of using AR technology as a framework for designing tour support systems. These systems are highly scalable. For instance, in the present system, only some features on the museum specimens bear virtual labels and explanations. The knowledge content can be expanded by (1) labelling additional features or adding supplementary information, such as 3D computer graphics to demonstrate the structural details of specimens, and (2) adding a function that can provide the AR information on the basis of a visitor's level of knowledge or interest and can even suggest optimal tour routes. AR systems, in combination with other technologies such as haptic devices or holography, for the use of group learning activities and so forth, would contribute to advancing the teaching of anatomy and pathology.

Conclusion

This chapter introduced AR-based support systems to establish new learning modalities in medical museums. The tablet AR system enables visitors to capture images of specimens and obtain AR information without being aware of markers. The HMD AR system with a natural-click interface employs virtual buttons that facilitate intuitive hand-gesture operation by users and enables them to obtain AR information. According to students' perceptions, both tablet and HMD AR systems were favorably accepted as a useful tool, which may also conform to the concept of system development with a user-centered design. Tours of the medical museum that combine actual specimens with AR information may produce a synergistic effect, resulting in the efficient acquisition of important knowledge. They contribute not only to raising students' learning motivation but also to improving the descriptions of exhibits in traditional medical museums. The application of AR technology to medical museums may even be a paradigm shift that could be expanded throughout the world and be shaping the future of anatomical exhibitions.

References

1. Cole FJ. A history of comparative anatomy: from Aristotle to the eighteenth century. 1st ed. London: Macmillan & Co. Ltd.; 1949.
2. Bates AW. "Indecent and demoralising representations": public anatomy museums in mid-Victorian England. Med Hist. 2008;52:1–22.
3. Turk JL. The medical museum and its relevance to modern medicine. J R Soc Med. 1994;87:40–2.
4. Diaz-Perez JA, Raju S, Echeverri JH. Evaluation of a teaching strategy based on integration of clinical subjects, virtual autopsy, pathology museum, and digital microscopy for medical students. J Path Inform. 2014;5:25.
5. Marreez YM, Willems LN, Wells MR. The role of medical museums in contemporary medical education. Anat Sci Educ. 2010;3:249–53.
6. Moro C, Štromberga Z, Raikos A, Stirling A. The effectiveness of virtual and augmented reality in health sciences and medical anatomy. Anat Sci Educ. 2017;10:549–59.

7. Shibata S, Manabe T, Yamashita K, Kajita H. Role of the medical museum in teaching medical students. Arch Pathol Lab Med. 1991;115:539–43.
8. Wakefield D. The future of medical museums: threatened but not extinct. Med J Aust. 2007;187:380–1.
9. Lukić I, Vicko Glunćić V, Ivkić G, Hubenstorf M, Marusić A. Virtual dissection: a lesson from the 18th century. Lancet. 2003;362:2110–3.
10. Ferrari L, Coda R, Fulcheri E, Bussolati G. The role of the pathological anatomy museum: past glory, present crisis, and future prospects. Pathologica. 2001;93:196–200.
11. Fulcheri E. The museums of pathologic anatomy: a very much neglected sector of scientific museology, worth of reconsideration. Pathologica. 1996;88:291–6.
12. Küçük S, Kapakin S, Göktaş Y. Learning anatomy via mobile augmented reality: effects on achievement and cognitive load. Anat Sci Educ. 2016;9:411–21.
13. Preece D, Williams SB, Lam R, Weller R. "Let's get physical": advantages of a physical model over 3D computer models and textbooks in learning imaging anatomy. Anat Sci Educ. 2013;6:216–24.
14. McMenamin PG, Quayle MR, McHenry CR, Adams JW. The production of anatomical teaching resources using three-dimensional (3D) printing technology. Anat Sci Educ. 2014;7:479–86.
15. Watson RA. A low-cost surgical application of additive fabrication. J Surg Educ. 2014;71:14–7.
16. Jones DB, Sung R, Weinberg C, Korelitz T, Andrews R. Three-dimensional modeling may improve surgical education and clinical practice. Surg Innov. 2016;23:189–95.
17. Yao R, Xu G, Mao SS, Yang HY, Sang XT, Sun W, et al. Three-dimensional printing: review of application in medicine and hepatic surgery. Cancer Biol Med. 2016;13:443–51.
18. Pavese P, Coulouma M, Sellier E, Stahl JP, Wintenberger C, François P. CD-ROM continuous medical education model for the management of urinary tract infections in family practice. Med Mal Infect. 2012;42:321–6.
19. Rössler B, Lahner D, Schebesta K, Chiari A, Plöchl W. Medical information on the internet: quality assessment of lumbar puncture and neuroaxial block techniques on YouTube. Clin Neurol Neurosurg. 2012;114:655–8.
20. Benninger B, Matsler N, Delamarter T. Classic versus millennial medical lab anatomy. Clin Anat. 2014;27:988–93.
21. Blake T. Teaching musculoskeletal examination skills to UK medical students: a comparative survey of rheumatology and orthopaedic education practice. BMC Med Educ. 2014;14:62.
22. Dash S, Kamath U, Rao G, Prakash J, Mishra S. Audio-visual aid in teaching "fatty liver". Biochem Mol Biol Educ. 2015;44:241–5.
23. Rinaldi VD, Lorr NA, Williams K. Evaluating a technology supported interactive response system during the laboratory section of a histology course. Anat Sci Educ. 2017;10:328–38.
24. Luursema JM, Verwey WB, Kommers PAM, Geelkerken RH, Vos HJ. Optimizing conditions for computer-assisted anatomical learning. Interact Comput. 2006;18:1123–38.
25. McGhee J. 3-D visualization and animation technologies in anatomical imaging. J Anat. 2010;216:264–70.
26. Engelke T, Becker M, Wuest H, Keil J, Kuijper A. MobileAR browser – a generic architecture for rapid AR-multi-level development. Expert Syst Appl. 2013;40:2704–14.
27. Liu K, Fang B, Wu Y, Li Y, Jin J, Tan L, et al. Anatomical education and surgical simulation based on the Chinese visible human: a three-dimensional virtual model of the larynx region. Anat Sci Int. 2013;88:254–8.
28. Kockro RA, Amaxopoulou C, Killeen T, Wagner W, Reisch R, Schwandt E, et al. Stereoscopic neuroanatomy lectures using a three-dimensional virtual reality environment. Ann Anat. 2015;201:91–8.
29. Miki T, Iwai T, Kotani K, Dang J, Sawada H, Miyake M. Development of a virtual reality training system for endoscope-assisted submandibular gland removal. J Craniomaxillofac Surg. 2016;44:1800–5.
30. Yeo CT, MacDonald A, Ungi T, Lasso A, Jalink D, Zevin B, et al. Utility of 3D reconstruction of 2D liver computed tomography/magnetic resonance images as a surgical planning tool for residents in liver resection surgery. J Surg Educ. 2018;75(3):792–7.
31. Kamat VR, El-Tawil S. Evaluation of augmented reality for rapid assessment of earthquake-induced building damage. J Comput Civ Eng. 2007;21:303–10.
32. Odashima T, Kanbara M, Yokoya N. An outdoor wearable annotation presentation system using augmented reality: technical Report of IEICE, PRMU2002–181. IIEEJ Trans Image Electron Vis Comput. 2003;32(6):832–40.
33. Schall G, Mendez E, Kruijff E, Veas E, Junghanns S, Reitinger B, et al. Handheld augmented reality for underground infrastructure visualization. Pers Ubiquit Comput. 2009;13:281–91.
34. Althoff T, White RW, Horvitz E. Influence of Pokémon Go on physical activity: study and implications. J Med Internet Res. 2016;18:e315.
35. Nigg CR, Mateo DJ, An J. Pokémon Go may increase physical activity and decrease sedentary behaviors. Am J Public Health. 2017;107:37–8.
36. Wagner-Greene VR, Wotring AJ, Castor T, Kruger J, Mortemore S, Dake JA. Pokémon Go: healthy or harmful? Am J Public Health. 2017;107:35–6.
37. Xian Y, Xu H, Xu H, Liang L, Hernandez AF, Wang TY, et al. An initial evaluation of the impact of Pokémon Go on physical activity. J Am Heart Assoc. 2017;6:e005341.
38. Bombara M, Cali D, Santoro C. KORE: a multi-agent system to assist museum visitors. In: Proceedings of the workshop on objects and agents (WOA 2003): intelligent systems and pervasive computing. Villasimius (CA); 10–11 Sept 2003. p. 175–8.
39. Meng M, Fallavollita P, Blum T, Eck U, Sandor C, Weidert S, et al. Kinect for interactive AR anatomy

learning. In: Proceedings of the 2013 IEEE international symposium on mixed and augmented reality. Adelaide, SA; 1–4 Oct 2013. p. 277–8.
40. Sakellariou S, Ward BM, Charissis V, Chanock D, Anderson P. Design and implementation of augmented reality environment for complex anatomy training: inguinal canal case study. In: Proceedings international conference on virtual and mixed reality (VMR 2009). San Diego, CA; 19–24 July 2009. p. 605–14.
41. Luciano CJ, Banerjee PP, Bellotte B, Oh GM, Lemole M Jr, Charbel FT, et al. Learning retention of thoracic pedicle screw placement using a high-resolution augmented reality simulator with haptic feedback. Neurosurgery. 2011;69:ons14–9.
42. Yeom SJ. Augmented reality for learning anatomy. In: Williams G, Statham P, Brown N, Cleland B, editors. Proceedings of annual Australasian society for computers in learning in tertiary education (ASCILITE 2011) conference: changing demands, changing directions; 2011; Hobart, TAS; 4–7 Dec 2011. p. 1377–83.
43. Ferrer-Torregrosa J, Torralba J, Jimenez MA, Garcia S, Barcia JM. ARBOOK: development and assessment of a tool based on augmented reality for anatomy. J Sci Educ Technol. 2015;24:119–24.
44. Chien CH, Chen CH, Jeng TS. An interactive augmented reality system for learning anatomy structure. In: Proceeding of the international multiconference of engineers and computer scientists (IMECS 2010). Hong Kong; 17–19 Mar 2010. p. 370–5.
45. Tilon AB, Marchal I, Houlier P. Mobile augmented reality in the museum: can a lace-like technology take you closer to works of art? In: Proceedings of 2011 IEEE international symposium on mixed and augmented reality - arts, media, and humanities (ISMAR-AMH). Basel; 26–29 Oct 2011. p. 41–7.
46. Sugiura A, Kitama T, Toyoura M, Mao X. The use of augmented reality technology in medical specimen museum tours. Anat Sci Educ. 2019;12:561–71.
47. Kato H, Billinghurst M. Marker tracking and HMD calibration for a video-based augmented reality conferencing system. In: Proceedings 2nd IEEE and ACM international workshop on augmented reality. San Francisco, CA; 20–21 Oct 1999. p. 85–94.
48. Visual Studio. 2019. https://www.microsoft.com/en-us/store/b/visualstudio/. Accessed 15 Aug 2019.
49. InstantAR. 2019. https://www.praxik.com/instant-ar/. Accessed 15 Aug 2019.
50. Sugiura A, Toyoura M, Mao X. A natural click interface for AR systems with a single camera. In: Proceedings of graphics interface 2014 (GI '14). Montreal, QC; 7–9 May 2014. p. 67–75.
51. OpenCV. 2019. https://opencv.org/. Accessed 15 Aug 2019.

3D Printing in Anatomy

35

James F. X. Jones, Conan McCaul, Laura Gorman, Thomas Campbell, and Mark Pickering

Introduction

Basics of Additive Manufacturing or 3D Printing

The principle that underpins 3D printing is simple. If a digital object is sliced from top to bottom, then the whole object can be reconstructed by printing the first (most caudal) slice and then adding each subsequent slice on top (i.e., additive manufacture) until the final (most rostral) slice is completed. In order for printers to render these digital slices into physical objects, there must be a *Z*-axis controller which regulates the slice thickness and linear movement from object bottom to object top. This latter movement involves either a print head rise or a printing platform descent. The printing platform is the solid surface which accepts the first slice. The *X–Y* control which creates the detail of each individual slice involves either a moving extruder connected to two stepper motors (one for each axis) or in the case of digital light projection stereolithography a single black and white whole slice projection of UV light onto a photocurable liquid resin.

Since the expiry of patent restrictions in photopolymerization and powder bed fusion (2013–2015), among others, there has been an explosion in papers relating to 3D printing. These methods involve the use of liquid and powder, respectively, in additive manufacturing. This expiry of patents continued into 2018. As the graph shows (Fig. 35.1), this has also been reflected in papers that relate to printing in anatomy.

The value of 3D printing to anatomy has been highlighted recently [1–5], and the range of applications involve many topographical regions, cross-sectional anatomy [6], and functional anatomy [7, 8]. The anatomical dividends of this rich new medium affect students [1], physicians [9], surgeons [10], and patients [11].

J. F. X. Jones (✉) · L. Gorman · T. Campbell · M. Pickering
Discipline of Anatomy, Biomedical Section, School of Medicine, University College Dublin, Dublin, Ireland
e-mail: James.Jones@ucd.ie; laura.gorman@ucdconnect.ie; thomas.campbell@ucd.ie; mark.pickering@ucd.ie

C. McCaul
Department of Anaesthesia, Rotunda Hospital Dublin and Mater Misericordiae University Hospital, Dublin, Ireland
e-mail: cmccaul@rotunda.ie

Four Steps to 3D Printing

STEP 1: Choose an Anatomical Model

There is a growing number of Web-based archives of free printable anatomical files. Of particular note, the most complete collection can be found on BodyParts3D. Any anatomist new to this field may begin at the website of BodyParts3D. There

L. K. Chan, W. Pawlina (eds.), *Teaching Anatomy*, https://doi.org/10.1007/978-3-030-43283-6_35

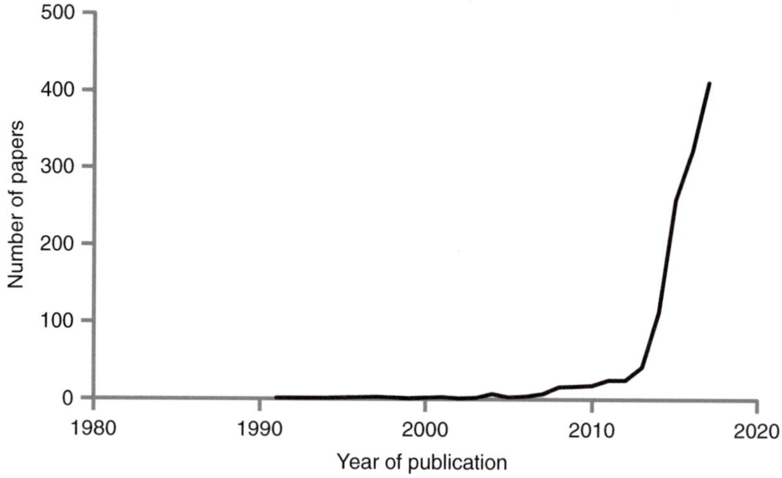

Fig. 35.1 Graph showing increase in publications relating to keywords 3D printing and anatomy in PubMed. (This graph was constructed by authors using PubMed CSV data imported into Microsoft Office Excel software)

are just a few simple steps to follow in order to select the anatomical part(s) one wishes to model and then print.

Websites offering free anatomical models:

BodyParts3D	http://lifesciencedb.jp/bp3d/
NIH 3D Print Exchange	https://3dprint.nih.gov/
GrabCAD Community	https://grabcad.com/library/tag/anatomy
MyMiniFactory	https://www.myminifactory.com/
Thingiverse	https://www.thingiverse.com/

STEP 2: Refine the Model

The common file format for printable files contains the suffix .stl (stereolithography or standard tessellation language). The next step involves importation of selected digital files into the free software Meshmixer. This allows for some manipulation and sectioning of models if required. There are also interesting Boolean subtractive and additive manipulations that can be executed in order to increase the educational impact of the model. In other words, individual anatomical parts can be conjoined or subtracted according to educational need. A Boolean subtraction of an object from a solid block results in a mold when printed which can then be filled with soft silicone material to represent muscle, for example.

Websites offering free 3D printing software:

Meshmixer	http://www.meshmixer.com/
Blender	https://www.blender.org/
Slicer 4.0	https://www.slicer.org/

STEP 3: Slice the Model

As 3D printing involves printing one thin layer at a time in sequential slices, the final digital model must be “sliced.” In the authors’ opinion, the best slicing software for fused deposition modelling (FDM) (extrusion of molten filaments) is Simplify3D although the free software Cura is a close second. Many slicing engines have similar features, but the useful features for anatomical model building include variable density and pattern infill (which can create a multi-textural feel to the printed part) and variable slice thickness to create smooth gradients. When a printing nozzle traces a closed path, the infill percentage of the region within the loop can range from 0% to 100%. The greater the density of infill, the harder the resulting model. Various infill patterns are also possible such as a simple rectilinear grid or honeycomb. A variable slice thickness regulates smoothness because many thin slices are smoother that fewer thicker slices. An unsightly staircase effect of an object’s edge can ensue if this is not optimized.

Table 35.1 Corresponding basic tissues with material choices

	3D printing process		
Tissue	FDM	Binder jetting	Photopolymers
Bone	ABS, PLA	Calcium sulfate	Hard rigid resins
Muscle	Filaflex, Ninjaflex, porous filaments, TPU		Soft flexible resins
Cartilage	Flexible PLA		
Skin	Flesh-colored Filaflex		

Note that although FDM is the cheapest option for 3D printing, it now offers the greatest range in material choice. Post processing of porous filaments can increase biomechanical properties according to which solvent is chosen to wash out the polyvinyl alcohol

FDM fused deposition modelling, *ABS* acrylonitrile butadiene styrene, *PLA* polylactic acid, *TPU* thermoplastic urethane

Slicing software:

Simplify3D	https://www.simplify3d.com/
Cura	https://ultimaker.com/en/products/ultimaker-cura-software

The final file is called a G-code, and it contains all the instructions to operate the X, Y, and Z stepper motors as well as heaters and extruders.

STEP 4: Choose a 3D Printer

Choosing a printer depends on choosing the material(s) for model building (see Table 35.1) and affordability. The key generic material categories include thermoplastic, powder, or liquid resin. Each approach offers certain advantages but is also restricted by certain limitations. The cost of FDM printers ranges from €2000 to €5000 (about US$2200–5600), and consumables are cheap (filaments typically cost €30 per kg [about US$34 per kg]). The Z Corp binder jet printer costs approximately €30,000 (about US$36,000), and consumables (powder and binder) are expensive, €1000 (about US$1100) to fill and run the machine. The stereolithography Ember printer costs €10,000 (about US$11,000), and resin is approximately €120 per liter (about US$130 per liter). No special engineering support is required, and even undergraduate students quickly become expert in their use. This probably reflects the recent advances in internet educational resources and technological assistance.

Thermoplastic Material: Fused Deposition Modelling (FDM)

This is the most affordable option for anatomy departments, and recent advances have greatly extended the capabilities of this method and material choice. The principle of this technique involves the controlled melting and extrusion of thermoplastic filaments through a regulated moving hot end. An ideal printable filament has an appropriate melting point, passes easily without friction through various conduits from reel to hot end, and solidifies quickly. Realistic anatomical models require multiple materials and colors; hence, multi-material extrusion machines are recommended (Fig. 35.2). In addition, it is very important that for flexible filaments, the extruder is close to the hot end so that coiling and buckling of filament are minimized. Machines with Bowden tubes (such as Ultimaker machines) do not work well with this method, whereas machines with direct drives (e.g., Builder 3D Printers) do (Figs. 35.3 and 35.4).

Powder Material: Powder Binder Jet Printing

This technique works by harnessing the ejected droplet control found in ordinary inkjet printers and replacing the black ink with adhesive or binder. In place of paper, the printer uses 0.1 mm layers of calcium sulfate powder spread on a descending platform (the *Z*-axis control) (Fig. 35.5). The advantage of this method is that no model supports are required as the whole platform is covered in powder during the forma-

Fig. 35.2 FDM printer with a single dual extrusion hot end (Builder Premium) which can mix filaments. Since there is a single hot end (and therefore operates at a single temperature), only thermocompatible filaments can be co-extruded

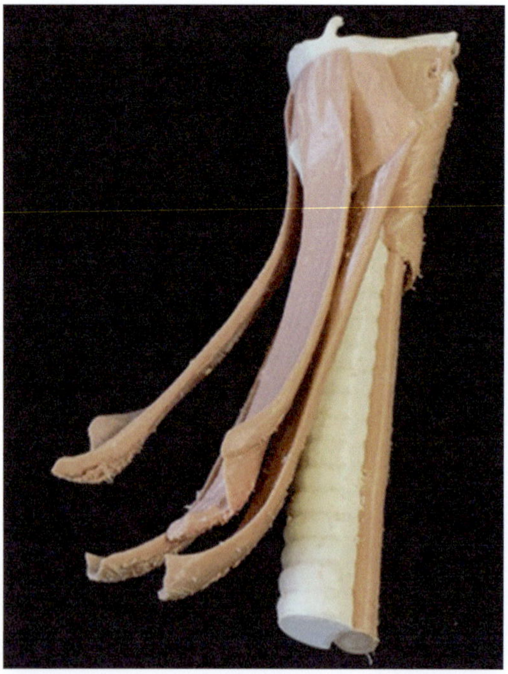

Fig. 35.3 A two-material (flesh Filaflex and ivory PLA) model of upper airway manufactured in a single print. This was created using the machine shown in Fig. 35.2. The cervical strap muscles are very flexible and can be dissected

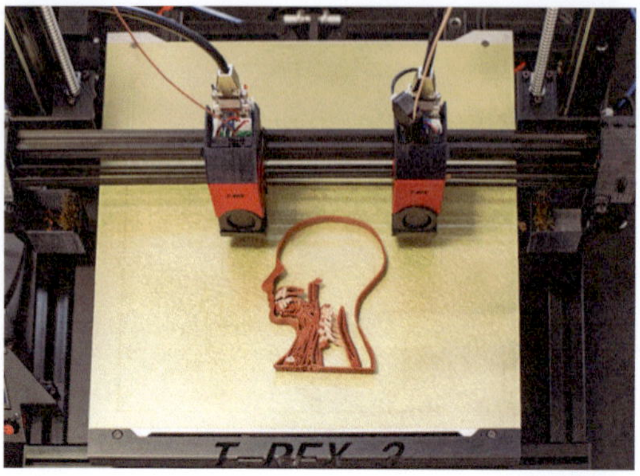

Fig. 35.4 An FDM 3D printer (T-Rex3) with independent moving hot ends and extruders. This permits the use of very different filaments melted at different temperatures and printed at different speeds. If identical filaments are passed into both extruders, a duplication mode can be selected to increase manufacture rate

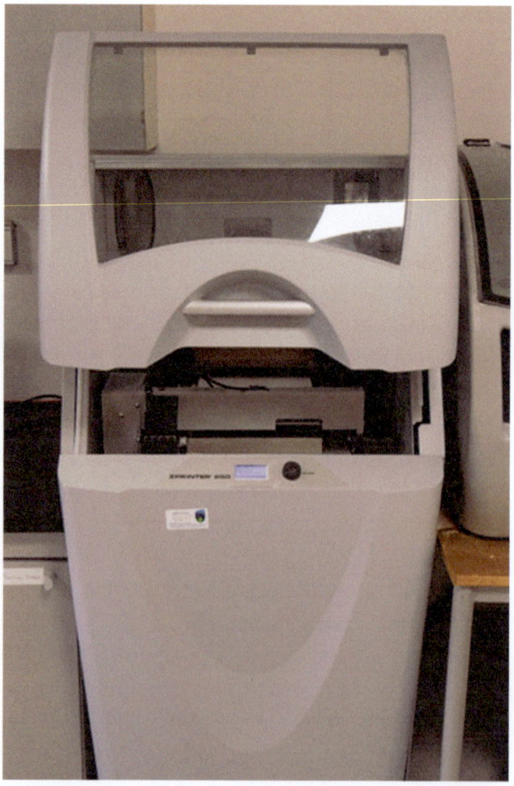

Fig. 35.5 A Z Corp powder binder jetting machine. The parts are supported by powder which does not receive the liquid binder

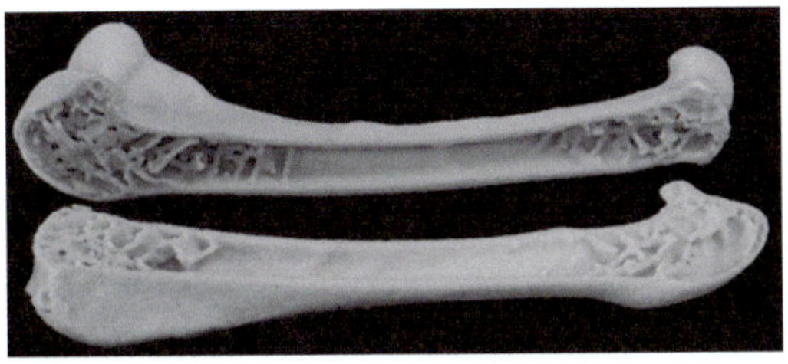

Fig. 35.6 A vertically sectioned femur from a swan which had been imaged with MRI. The bone was segmented out selectively from the DICOM stack and transformed to printable format (stl file) using the free software 3D slicer. The print was rendered using calcium sulfate and binder in the machine shown in Fig. 35.5. Note that the trabecular detail is apparent within the medullary cavity and this internal complexity is not possible to attain by traditional injection molding methods

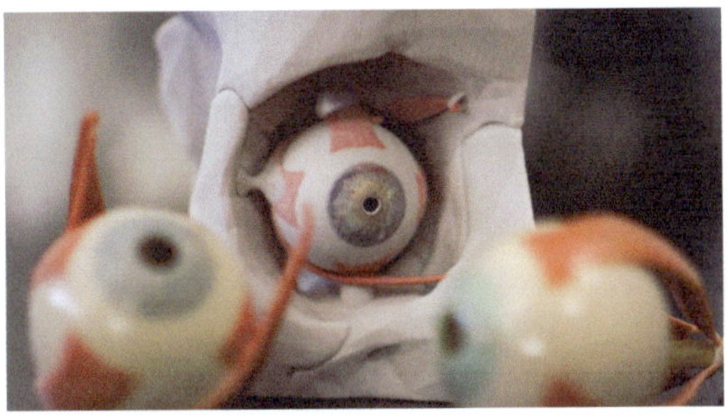

Fig. 35.7 Examples of color prints of the orbit and some extraocular muscles. These were made with the machine shown in Fig. 35.5

tion of each layer (Fig. 35.6) and a color inkjet cartridge can add colors to the part (Fig. 35.7). When the part is printed, the object is fragile and needs to be post processed with cyanoacrylates or infused with wax.

Liquid Resin Material: Stereolithography

Stereolithography uses a UV light source and digital projector to cure thin layers of photopolymer (Fig. 35.8). An ordinary projector transmits each layer as a black and white image. Wherever there are white pixels, UV light can penetrate and solidify the resin; conversely, black regions stop transmission, and these regions remain liquid. Thus, a constantly updating UV mask is applied to the whole model. The advantage of this technique is that there are no moving parts in the *X* and *Y* direction as the whole slice is transmitted

Fig. 35.8 A stereolithography Ember printer which has a 50 μm spatial resolution in *X*, *Y*, and *Z* directions. This printer uses light to cure photopolymers and results in detailed prints

at once as a binary mask (just a *Z* stepper motor is required). Various resins are commercially available which produce hard to flexible prints, and it is also relatively straight forward to create novel recipes which are biocompatible. The advantage of this approach is that the spatial resolution is quite fine. However, the build volume tends to be small as it is harder to regulate homogenous photocures over large surface areas.

Model Making: Old and New

In addition to direct printing of anatomical parts (Fig. 35.9), some ingenious applications have produced models with "a page turning style" such as a paranasal sinus-skull base anatomical model that vividly presents sectional anatomy [6]. Embedded magnets for tarsal [7] and carpal bones [12] contribute to the interactivity of models. In addition to directly printing anatomical parts, some old techniques such as mold

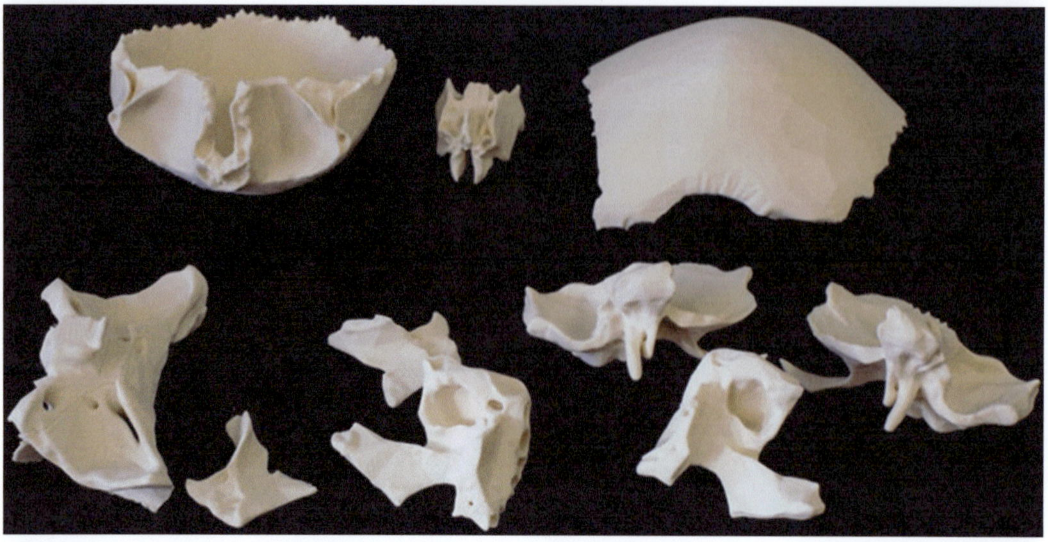

Fig. 35.9 A collection of cranial bones printed with FDM using PLA filament. An ivory-colored PLA filament is superior to white PLA as the resultant prints are more visually realistic. This method augments osteology classes as ethmoidal bones in particular are perishable due to the thinness of the lamina papyracea but are easy to print in large numbers. In addition, classic fracture patterns can be created digitally and then printed rather than directly inflicting the stress on a physical part which can yield unpredictable results

making and the lost-wax technique can be integrated with this new technology. A soft Sylgard model of the upper airway has been constructed using a negative dissolvable mold [13]. This model was ventilated with high nasal flow to study the effects of upper airway resistance and resistive work of breathing. As some filaments can be dissolved when exposed to the appropriate solvent, hollow blood vessels can be produced by coating polyvinyl alcohol (water soluble) with silicone [14] or acrylonitrile butadiene styrene (acetone soluble) with silicone [15]. It is also possible to print with wax and by heating the finished part to melt and lose the wax core. This has been applied to a 3D-printed wax pelvicalyceal system which was coated with a soft silicone (Ecoflex) in a mold [16]. A mock pulmonary circulatory system incorporating a compliant 3D-printed model has been constructed with PLA and latex [17].

In addition to the lost-wax technique, 3D printing can be readily employed for mold making. Meshmixer can be used to perform a Boolean subtraction of a model or block, and the resultant negative can be printed and used repeatedly as a mold.

Comparison of 3D Printing to Previous Model-Making Methods

The principal advantage of 3D printing (additive manufacturing) over older techniques such as injection molding is that internal complexity can be captured in unprecedented detail. It is not an uncommon experience for anatomy educators when dealing with traditional plastic skulls to attempt to pass a probe through a small cranial foramen and find an impasse (e.g., no internal canal). This is not the case for 3D prints (Figs. 35.9), and 35.6 shows how internal trabeculae can be present internally but cut off from view unless the model is split open. In this manner, the question of dissectibility arises with a 3D printing approach. As the number of printing materials continues to evolve, the anatomical models of the future may attain physiological biomechanical realism.

The advantages of 3D printing compared to older techniques are as follows:

- Fast production compared to hand-painted models.
- Scalable models.
- Rapid prototyping of many model versions.
- Easily sectioned from any angle.
- When combined with CT, laser scanning or photogrammetry archival material can be preserved [18].
- Multi-material components can add functionality.
- Internal complexity is preserved compared to injection molding or subtractive manufacturing methods.
- Models can be made dissectible and disposable.

Figure 35.10 shows an example of a brainstem model which was constructed by printing nuclei directly and inserting them in translucent resin held in a 3D-printed mold of the rhombencephalon. Figure 35.11 illustrates how 3D printing can be used to create a functional skull/brain model of penetrating trauma.

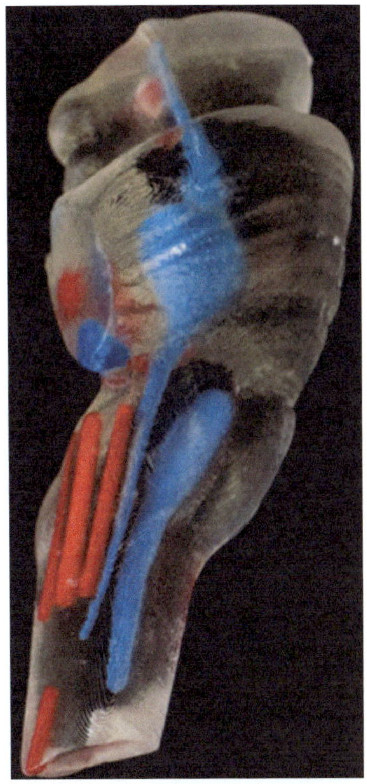

Fig. 35.10 Model of brainstem showing principal motor (red) and sensory nuclei (blue). The nuclei were printed directly with colored PLA, and a PLA mold was filled with translucent acrylic

Educational Applications and Benefits

It has been reported that some 3D-printed models yield superior score results to wet or plastinated specimens after tests of anatomical knowledge [1]. Randomized control trials comparing 3D prints and cadaveric materials have shown some benefit for learning cardiac anatomy through 3D prints [19] and brain ventricular systems [20]. However, it is acknowledged that the limited spatial resolution of 3D prints results in models that are not as realistic as plastinated specimens [3]. The range of applications of 3D printing to anatomy and surgical education are impressive and expanding. As 3D models can be digitally sectioned easily, sectional anatomy benefits from physical slices of CT, MRI, or sonographic data sets. This is particularly useful for interpretation of echocardiography and cerebral MRI. Osteology probably benefits the most as high-density bony tissues are easily segmented from radiological data sets, and this is also true when radiodense high-contrast agents are injected to perform angiography. Soft tissue such as muscle is more difficult to delineate and usually requires customized spin-labelling protocols with MRI.

3D-printed models are used in our institution for undergraduate education of medical and allied health students. They have proven particularly useful in teaching neuroanatomy as illustrated in Figs. 35.10 and 35.11 and cranial bone anatomy in Figs. 35.7 and 35.9. At the postgraduate level, an airway model is being perfected in collaboration with the Department of Anaesthesia to teach emergency cricothyroidotomy and airway intubation.

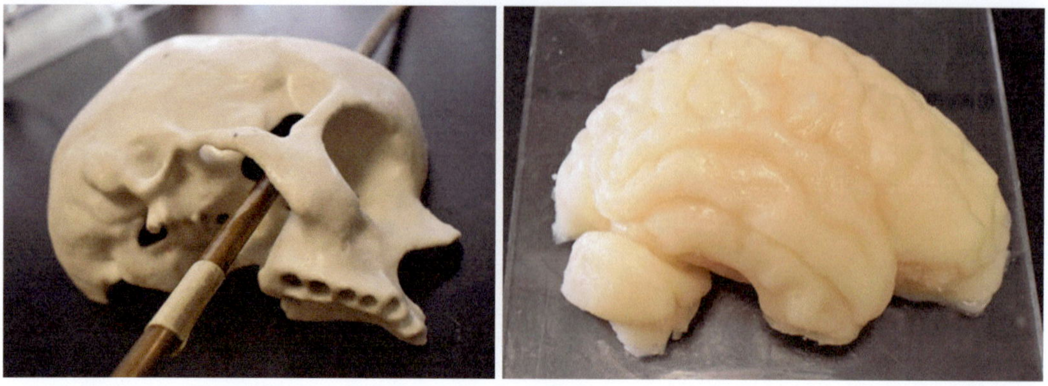

Fig. 35.11 Model of a "Phineas Gage type injury" by a tamping iron. Left panel shows the calcium sulfate bone model created by binder jetting, and right panel is a soft gelatin model of the right cerebrum created with a 3D-printed mold. By placing the soft brain inside the cranium, students can simulate the rod's trauma and establish which cerebral regions are affected

At present, the range of materials and control of material properties limit the degree to which 3D printing can compete realistically with cadavers [21]. But this is a rapidly evolving field, and the recent demonstration of the use of light to selectively cure and inhibit photopolymers [22] and also regulate material properties [23] suggests that the problem may be tractable and reducible to one of polymer chemistry.

Conclusion

Anatomists should embrace the new technology of 3D printing because it offers so much to them in their educational enterprise. But the field of 3D printing also needs the expertise of anatomists to ensure accuracy and relevance of printed models. The authors predict a renaissance will soon envelop anatomical research which has been considered moribund. An artificial myology will be required to create anatomically accurate robots of the future driven by 3D-printed artificial muscles.

Acknowledgments Special thanks to our photographer Owen Humphreys. The financial support of the School of Medicine at University College Dublin is gratefully acknowledged. Some of the printers were purchased with funding received from the College of Anaesthesiologists (Ireland) and the Difficult Airway Society.

References

1. Garas M, Vaccarezza M, Newland G, McVay-Doornbusch K, Hasani J. 3D-printed specimens as a valuable tool in anatomy education: a pilot study. Ann Anat. 2018;219:57–64.
2. Wu AM, Wang K, Wang JS, Chen CH, Yang XD, Ni WF, et al. The addition of 3D printed models to enhance the teaching and learning of bone spatial anatomy and fractures for undergraduate students: a randomized controlled study. Ann Transl Med. 2018;6(20):403.
3. Mogali SR, Yeong WY, Tan HKJ, Tan GJS, Abrahams PH, Zary N, et al. Evaluation by medical students of the educational value of multi-material and multi-colored three-dimensional printed models of the upper limb for anatomical education. Anat Sci Educ. 2018;11(1):54–64.
4. McMenamin PG, Quayle MR, McHenry CR, Adams JW. The production of anatomical teaching resources using three-dimensional (3D) printing technology. Anat Sci Educ. 2014;7(6):479–86.
5. Cai B, Kanagasuntheram R, Bay BH, Lee J, Yen CC. The effects of a functional three-dimensional (3D) printed knee joint simulator in improving anatomical spatial knowledge. Anat Sci Educ. 2019;12:610–18.
6. Zhang XD, Li ZH, Wu ZS, Lin W, Lin WJ, Lin JC, et al. A novel three-dimensional-printed paranasal sinus-skull base anatomical model. Eur Arch Otorhinolaryngol. 2018;275(8):2045–9.
7. O'Reilly MK, Reese S, Herlihy T, Geoghegan T, Cantwell CP, Feeney RN, et al. Fabrication and assessment of 3D printed anatomical models of the lower limb for anatomical teaching and femoral

vessel access training in medicine. Anat Sci Educ. 2016;9(1):71–9.

8. Smith ML, Jones JFX. Dual-extrusion 3D printing of anatomical models for education. Anat Sci Educ. 2018;11(1):65–72.
9. Ghazy A, Chaban R, Vahl CF, Dorweiler B. Development and evaluation of 3-dimensional printed models of the human tracheobronchial system for training in flexible bronchoscopy. Interact Cardiovasc Thorac Surg. 2019;28(1):137–43.
10. Yang T, Lin S, Xie Q, Ouyang W, Tan T, Li J, et al. Impact of 3D printing technology on the comprehension of surgical liver anatomy. Surg Endosc. 2019;33(2):411–7.
11. Yang T, Tan T, Yang J, Pan J, Hu C, Li J, et al. The impact of using three-dimensional printed liver models for patient education. J Int Med Res. 2018;46(4):1570–8.
12. Kanagasuntheram R, Geh NKT, Yen CC, Dheen ST, Bay BH. A composite 3D printed model of the midcarpal joint. Anat Sci Int. 2019;94(1):158–62.
13. Adams CF, Geoghegan PH, Spence CJ, Jermy MC. Modelling nasal high flow therapy effects on upper airway resistance and resistive work of breathing. Respir Physiol Neurobiol. 2018;254:23–9.
14. Liu Y, Gao Q, Du S, Chen Z, Fu J, Chen B, et al. Fabrication of cerebral aneurysm simulator with a desktop 3D printer. Sci Rep. 2017;7:44301.
15. Kaneko N, Mashiko T, Ohnishi T, Ohta M, Namba K, Watanabe E, et al. Manufacture of patient-specific vascular replicas for endovascular simulation using fast, low-cost method. Sci Rep. 2016;6:39168.
16. Adams F, Qiu T, Mark A, Fritz B, Kramer L, Schlager D, et al. Soft 3D-printed phantom of the human kidney with collecting system. Ann Biomed Eng. 2017;45(4):963–72.
17. Knoops PGM, Biglino G, Hughes AD, Parker KH, Xu L, Schievano S, et al. A mock circulatory system incorporating a compliant 3D-printed anatomical model to investigate pulmonary hemodynamics. Artif Organs. 2017;41(7):637–46.
18. Young JC, Quayle MR, Adams JW, Bertram JF, McMenamin PG. Three-dimensional printing of archived human fetal material for teaching purposes. Anat Sci Educ. 2019;12(1):90–6.
19. Lim KH, Loo ZY, Goldie SJ, Adams JW, McMenamin PG. Use of 3D printed models in medical education: a randomized control trial comparing 3D prints versus cadaveric materials for learning external cardiac anatomy. Anat Sci Educ. 2016;9(3):213–21.
20. Yi X, Ding C, Xu H, Huang T, Kang D, Wang D. Three-dimensional printed models in anatomy education of the ventricular system: a randomized controlled study. World Neurosurg. 2019;125:e891–e901.
21. Ratinam R, Quayle M, Crock J, Lazarus M, Fogg Q, McMenamin P. Challenges in creating dissectible anatomical 3D prints for surgical teaching. J Anat. 2019;234:419–37.
22. de Beer MP, van der Laan HL, Cole MA, Whelan RJ, Burns MA, Scott TF. Rapid, continuous additive manufacturing by volumetric polymerization inhibition patterning. Sci Adv. 2019;5:eaau8723.
23. Schwartz JJ, Boydston AJ. Multimaterial actinic spatial control 3D and 4D printing. Nat Commun. 2019;10:791.

36 The Use of Virtual, Augmented and Mixed Reality in Anatomy Education

Allan C. Stirling and Christian Moro

Introduction

Modern advances in technology have allowed educators to increasingly utilise a variety of devices and digital tools within their curricula. Modern-day anatomy students now have access to smartphones and tablets with faster processing power and larger memory. This allows them access to digital anatomy textbooks, anatomical movies, virtual dissection applications and medical imaging datasets, to name a few [1].

Many of these technologies are now ubiquitous in a student's daily study practice. More recently, aspects of the traditional 2D anatomy practice have been reconfigured to make use of newer devices that allow for 3D visualisations. The main modalities that are now making their way into modern-day anatomy laboratories are virtual reality (VR), augmented reality (AR) and mixed reality (MR). Implementing these as educational tools can increase engagement and enhance learner experiences [2]. The recently released consumer-level virtual reality devices and the accessibility of smartphones allow these modes of teaching to be surprisingly user-friendly to all educators looking to provide unique and engaging experiences.

For new educators entering the field of anatomy education or, indeed, for experienced anatomists who are less confident with technology, the idea of using virtual reality headsets or augmented reality devices can seem quite daunting. A modern anatomy curriculum like any other requires continuous refinement of the content and tools used in its teaching [3]. This curricular redesign is complicated by the increasingly discussed issue of reduced allocation of time to the basic sciences in a modern medical programme [4].

The challenge presents itself, therefore, to supplement your existing anatomy curricula with technology that is engaging to the learner and adds to their knowledge acquisition and retainment all the while managing the limitations placed on anatomy teaching time.

A common misbelief is that 3D anatomy sets out to replace traditional anatomy teaching methods such as dissection- or prosection-based practical classes. This is not necessarily the case, and the early research into the use of three-dimensional technologies in anatomy curricula indicates it as a useful addition to traditional methods to supplement existing methods [2, 5].

In some studies, computer-based learning was shown to be less effective in certain aspects of anatomy study [6]. More recently with newer devices that generate high-quality, high-fidelity visualisations, researchers are demonstrating their benefit in a variety of anatomy-related areas such as radiology [7] and musculoskeletal domains [8].

A. C. Stirling (✉) · C. Moro
Faculty of Health Sciences and Medicine, Bond University, Gold Coast, QLD, Australia
e-mail: astirlin@bond.edu.au; cmoro@bond.edu.au

L. K. Chan, W. Pawlina (eds.), *Teaching Anatomy*, https://doi.org/10.1007/978-3-030-43283-6_36

The adoption of augmented or virtual reality need not necessarily mean the discontinuation of cadaver-based education unless this is the ultimate endpoint for say ethical, financial or epistemological reasons. The educator should view the use of virtual or augmented reality from a pedagogical perspective much in the same way you would consider implementing any new teaching method into your existing curricula; these technologies have their advantages in different situations but equally can be counter-constructive if used improperly.

Advantages of Virtual and Augmented Reality in the Classroom

- Generates interest in the subject matter
- Leads to greater understanding of anatomical relationships when returning to same structures on the cadaver or model
- Allows for groups of students to all view the same anatomical structure at the same time that is physically not possible on a single cadaver or model

Anatomy by its very nature is a visual subject, and it, therefore, stands to reason that an additional means of visualising the human body can have profound learning benefits if incorporated correctly. Your student cohort will have a mix of learner types, and it has been shown that low spatial ability students, when provided with additional views of a structure (with reference points), elevate their learning to a level near that of high-spatial ability students [9].

Understanding Virtual, Augmented and Mixed Reality in Anatomical Teaching

Virtual Reality

Within virtual reality (VR), the user's senses (sight, hearing and haptics) are fully immersed in a synthetic environment that mimics the properties of the real world through high-resolution, high refresh rate, head-mounted displays, stereo headphones and motion-tracking systems [10]. This technology enables an individualised learning experience, even in a busy or noisy laboratory or teaching environment. The cost of these devices although having reduced substantially over the last few years remains one of the limiting factors for many potential adopters. For a VR headset that can run anatomy applications and generate high-quality images, the price point is between $300 and $500 USD depending on the model chosen. Devices that incorporate the user's phone such as the Gear VR cost about $130 USD but only works with Samsung phones.

One of virtual reality's strengths as a teaching tool is that it allows the learner to visualise, navigate and explore the anatomy of the body in ways not possible with traditional methods [2]. Users can travel through the layers of the body and enter inside structures getting a unique viewpoint from inside anatomical structures such as the heart or the pelvis (Fig. 36.1).

The nature of most virtual reality anatomy applications is that they are immersive and therefore less collaborative than other learning experiences. The question then arises of what the educator can do while the student is wearing the headset or goggles. Most systems allow for an additional computer monitor to also display the users' viewpoint, and this allows the teacher and potentially other learners also to comment and discuss the anatomical structures being viewed (Fig. 36.1).

One continuing issue when using VR applications is the motion sickness it can generate with the user [2, 10]. This is certainly less evident in anatomy applications than say in computer games where the user is moving at high speed and making multiple turns or loops. However, some people still feel nauseous after using VR for learning. The effects of this can be mitigated by only spending short periods in VR (under 10 minutes), by sitting down, and some users have reported the use of a fan helps. Our own experience has shown that using VR regularly builds up something of a tolerance.

In *virtual reality* the user is fully immersed in a 3D environment that mimics the properties of the real world. Typically, this involves the user wearing a headset or goggles. Audio narration via headphones or inbuilt speakers in the device allows for audio narration or a guided lesson to also enhance the student experience.

Recommendations for the Implementation of Virtual Reality

1. Create lessons and modules that run no longer than 10-minutes at a time. After this timeframe, many learners become slightly motion sick or distracted from the virtual environment.
2. Virtual reality has the added benefit of placing the learner inside an enclosed space, completely immersed in the environment. This removes background noises, lights, clutter and other distractions, making VR particularly useful for individual learning while in a busy teaching laboratory, hospital or clinical environment.
3. Motion sickness is primarily prevalent in VR and can be an inhibitor for the use of this technology. This is commonly induced if the application takes control over the movement or navigation. It is recommended that VR application design allows the user to hold complete control over all navigation. As an alternative, augmented reality does not induce motion sickness.
4. Within VR, it is difficult for users to write notes. The provision of additional resources containing written information eliminates the need to take notes and is therefore advised.
5. Incorporate VR within existing laboratory practical classes to supplement other methods of learning. Don't assume that all of the learning objectives can be achieved using the technology alone. Get students to return to the cadavers following their VR learning to relate the 3D with the 'real life'.

Augmented Reality

Through the use of a camera and screen, digital models are superimposed onto the real world when viewed on the user's screen. As such, devices such as student's smartphones or tablets can be utilised for AR [11]. The user can interact with both the real and virtual elements of their surrounding environment. Implementing AR as a

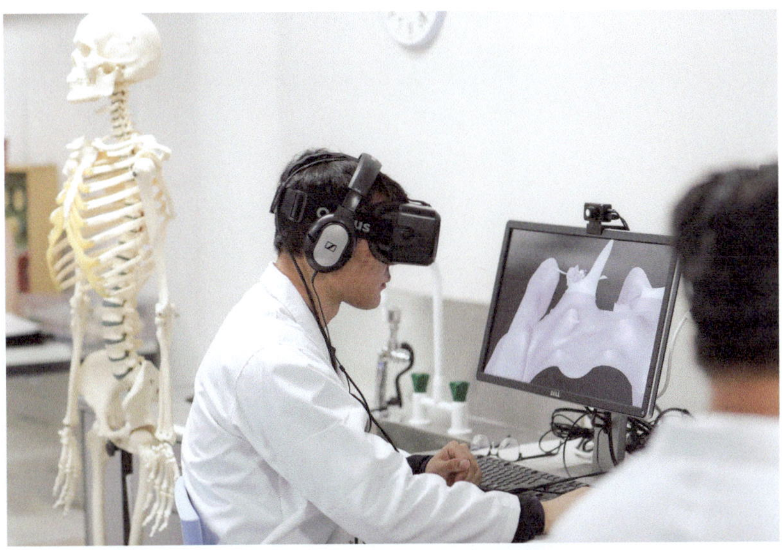

Fig. 36.1 A student in an anatomy laboratory undergoes a 6-minute lesson on spinal and pelvic skeletal anatomy using the Oculus Rift, a desktop-based virtual reality device

mode of learning can increase engagement and enhance the learner experience while being shown in research to also reduce the cognitive load on the learner [12].

One great strength of this form of visualisation technology is that it allows far more collaboration than other modalities such as VR (Fig. 36.2). Learners can work together through a worksheet or session guide and gain instant feedback from their interaction with the screen. In this way much like VR they control the pace of the learning [2]. Another benefit to AR is that it can run with smartphones or tablets and doesn't require the expenditure of VR or MR to start using it in the classroom. The open format of augmented reality, where students remain within the physical learning environment, also allows peer assistance and for students to work in groups. In the modern medical classroom, for example, where there is often a mix of undergraduate and postgraduate students [13], student collaborative exercises can be of great benefit to learning.

In *augmented reality* digital models or computer-generated images are superimposed on a user's view of the surrounding area, thus providing a composite view. The user is then able to interact with both the real and virtual elements as represented on the screen display.

Recommendations for the Implementation of Augmented Reality

1. Augmented reality is the preferred choice for surface anatomy, with VR being particularly effective at viewing inner structures.
2. Augmented reality is best used as a supplement to support other teaching methods, rather than being considered a complete replacement for prosection-based learning or dissection.

Fig. 36.2 Anatomy students utilise augmented reality to explore the human brain. This uses the tablet's camera to render a 3D model on the screen in place of the colourful cube marker that they hold

3. Different students prefer different speeds, and so it is best for the presenter not rush the content or force slow audio narrations.
4. When adding pre-recorded narration to accompany the AR learning, a less formal style of narration (i.e. not robotic) is preferred by the students. The narrator should aim to speak as naturally as possible.
5. Use natural colours for the 3D models and features.
6. Student fingers can get in the way of augmented reality models and become distracting. A staff member should be available and aware of this to assist the learner in becoming used to interacting with the devices.

Mixed Reality

Mixed reality (MR), a continuum of these innovative technologies, combines real and virtual worlds through a head-mounted, see-through display. This subsection of modality type includes the Microsoft HoloLens 2. Objects can be projected as holograms in front of the user's eyes in 3D space reducing the need for holding a screen or camera (Fig. 36.3). Objects can be manipulated using gesture controls with the hand and fingers or with a more traditional keyboard or handset. Mixed reality (also called Hybrid Reality) is still a developing technology and is not necessarily consumer-ready at this time. It is, however, being researched in the educational setting and shows great promise in its potential to allow multiple users to visualise the same structure if the headsets communicate with each other. This may lead to small-group-based anatomy teaching sessions being carried out utilising holograms with an instructor guiding the learning and all users able to see what the instructor is 'pointing' to in 3D space. As the price point of these devices diminishes (development kits for the HoloLens 2 currently cost $3500 USD) we may see more of these in the anatomy classrooms in the coming years.

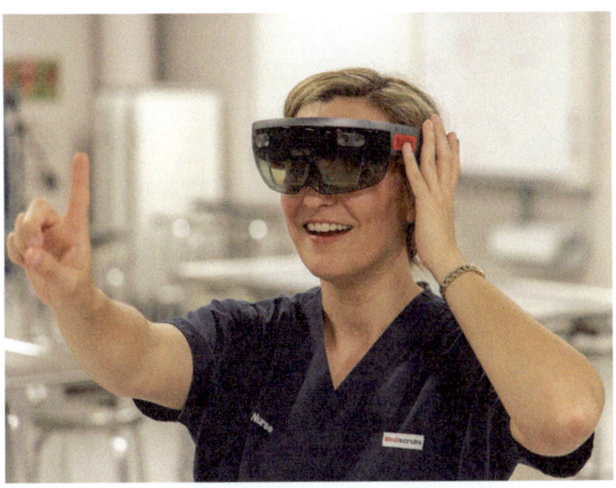

Fig. 36.3 An anatomy student completes a 6-minute guided holographic lesson on the human skeletal system using the HoloLens

> In *mixed reality*, the real and virtual worlds are merged to create elements of new environments and visualisations in front of the user. Physical and digital objects coexist as the user looks through a transparent head-mounted display.

Recommendations for the Implementation of Mixed Reality

1. Be aware that this is a developing technology. At best, current kits are prototypes with better quality, wider field of view and more powerful devices set to enter the consumer market soon.
2. Mixed reality allows the user to have their hands completely free. This allows the learner to take notes, discuss concepts with the educator or interact during the session. This is difficult in virtual or augmented realities.
3. The learner will need time to adapt to this new technology. Provide introductory activities to practice motion gestures, voice commands and other interactivities before commencing the lesson.
4. Maintain a well-lit and large area free of obstacles, for the user to walk around the hologram and interact with all sides and perspectives of the models.

Setting Up an Anatomy Laboratory for Virtual, Augmented or Mixed Reality

Deciding on Which Devices to Procure

There is excitement and awe when navigating through a digital human body in virtual and augmented reality, such as within the applications described by Kuehn (2018) [14]. However, while these devices are now widely available to consumers, their deployment remains niche in medical and clinical education, primarily utilised by tech-savvy universities, clinical institutions, hospitals and allied health schools. There are intrinsic benefits to learning when utilising VR and AR for studying anatomy, surgical skills or health education, and it is recommended that educators consider adopting this technology [2]. However, the decision of *which* device to incorporate within an anatomical teaching program can be challenging. The primary consideration revolves around the content wishing to be described.

In the case of virtual reality, this is best suited to visualising the visceral aspects of the body and allowing the students to gain insights into anatomical relationships and spatial understanding. Supplementing the dissection of the abdomen, for instance, with VR would allow students to view complex regions such as the epiploic foramen and entrance to the lesser sac in 3D from any angle before dissecting it on a cadaver or viewing it on a model or prosection. At the time of writing the major companies allowing consumer-based access to virtual reality are Oculus Rift, HTC Vive and Sony PlayStation VR with the Oculus and HTC being more applicable to classroom educational use (Table 36.1).

A low-cost way of bringing anatomy into the classroom can be achieved with a mobile phone and Google Cardboard or Samsung Gear VR [10]. At the more expensive end of the spectrum, the HTC Vive or Oculus Rift require a good-quality computer to run them with a graphics card capable of displaying the models at high frames per second.

For surface anatomy or study of discrete, individual organs, augmented reality allows students the ability to scan a region and often, with the use of markers or interaction with the screen, highlight the structure or bring up an annotation.

Cost and Funding Considerations for Laboratory Setups

There can be a considerable setup costs in integrating virtual, augmented or mixed reality into a laboratory. This includes the desktop computer, software and head-mounted display device. However, VR laboratories do not need that many devices to run successful teaching sessions. One effective method of teaching is through a rotational system with students moving between technology and traditional learning modes. This could mean that a group of 5–10 students spend 20 minutes on a worksheet, models and dissection/prosection activity and then rotate between these and the virtual reality devices. This means that even large cohorts can take part in technology-enhanced virtual and augmented reality sessions with only a few VR devices physically available. Other factors to consider include physical space, Internet connectivity and ability to update and maintain the technology.

Table 36.1 Main hardware devices used for virtual and augmented reality and their benefits in visualising anatomical structures

Virtual reality devices	Augmented and mixed reality devices
Oculus Rift Oculus Go HTC Vive/Vive Pro Samsung Gear VR Sony PlayStation VR Google Cardboard	Any smartphone or tablet (AR) Microsoft HoloLens 2 (MR) Magic leap one (AR) Meta 2 (MR)
Advantages: Good for visualising visceral anatomical structures and navigating around deeper structures to gain an understanding of anatomical relationships	Advantages: Good for surface anatomy or virtual dissection of organs, layer by layer

Building your Own Modules

Educators may wish to embark upon developing their own modules or teaming up with other departments within their institution to develop custom in-house anatomy modules and teaching material. Although a more time-consuming undertaking this route has the advantage of potentially producing a more authentic and learner-centred experience that is aligned with your particular educational needs and curriculum. It also allows the decision between exactly which technology to use, such as desktop or mobile mixed reality [15], and to maximise the benefits of these. This approach does not necessarily require 'reinventing the wheel' as there are repositories that exist with very high-quality 3D models of the human body that can be purchased and incorporated into your application. One such website that exists is Turbosquid (https://www.turbosquid.com/), a marketplace for designers and artists to distribute their 3D designs and models. Other archives exist, and many of these have been designed with input from anatomists and content experts.

Pitfalls to Avoid

As with any new teaching method, the educator must decide if it will benefit the learner with its introduction. In the example of technology-enhanced learning and in particular VR, AR and MR, you must ensure your adoption is grounded strongly by the pedagogy. There may be a temptation to use all modalities because it is current, cutting-edge or even 'cool'. The intended learning outcomes must always be front of mind and not lost with the technology.

Equally important is ensuring you have a robust method of evaluation to ensure the addition of virtual, augmented or mixed reality is improving the students' learning experience. If there is no feedback loop or discussion with the learners, you run the risk of adding something to your practical classes that, while being fun or enjoyable, does not result in enhanced learning or improved learning outcomes.

Another potential pitfall when technology is added to a practical class is for the students to see it as a separate part of the class. The design and physical placement of the technology should be considered so as to ensure the students see this as an integral and embedded part of the laboratory session.

The authors redesigned many of their practical anatomy classes to incorporate questions relating to the digital content and an experienced tutor was assigned to the computer stations to facilitate discussion about what was being observed. The augmented reality and virtual reality learning stations were embedded within our class structure, so the students didn't see this as a separate part of the practical laboratory session to be completed after the 'proper anatomy with cadavers' was finished.

An example lesson plan for a technology-enhanced anatomy lesson:

- Take an anatomical structure that is difficult to visualise or where you lack a good physical representation of the structure in model form (e.g. the inner ear)
- Strip away all the detail until you are left with the just the surrounding bony structure
- Slowly, layer by layer allow the students to add in more detail all the while letting them control the learning experience and moving around the virtual dissection
- Get them to narrate or list the anatomical relationships
- Return to the textbook, model or cadaver and relate it to what they have just visualised

Conclusion

Incorporation of three-dimensional visualisations into your anatomy curriculum is an effective way of engaging students irrespective of which modality or device you use. The actual hardware and number of physical devices available will be dictated by the curricula and the number of students in the cohort. It is important from the outset to recognise that virtual or augmented reality is greatly beneficial as an additional method of instruction and can be woven into anatomy practicals with the other forms of learning such as models and dissection-based activities. Clear learning objectives that incorporate digital content, as well as student feedback on which elements are most beneficial to them, can greatly enhance teaching and learning in this space.

References

1. Trelease RB. From chalkboard, slides, and paper to e-learning: how computing technologies have transformed anatomical sciences education. Anat Sci Educ. 2016;9(6):583–602.
2. Moro C, Štromberga Z, Raikos A, Stirling A. The effectiveness of virtual and augmented reality in health sciences and medical anatomy. Anat Sci Educ. 2017;10(6):549–59.
3. Moxham BJ, Plaisant O. Perception of medical students towards the clinical relevance of anatomy. Clin Anat. 2007;20(5):560–4.
4. Collins JP. Modern approaches to teaching and learning anatomy. BMJ. 2008;337:a1310.
5. Peterson DC, Mlynarczyk GS. Analysis of traditional versus three-dimensional augmented curriculum on anatomical learning outcome measures. Anat Sci Educ. 2016;9(6):529–36.
6. Khot Z, Quinlan K, Norman GR, Wainman B. The relative effectiveness of computer-based and traditional resources for education in anatomy. Anat Sci Educ. 2013;6(4):211–5.
7. Bork F, Stratmann L, Enssle S, Eck U, Navab N, Waschke J, Kugelmann D. The benefits of an augmented reality magic mirror system for integrated radiology teaching in gross anatomy. Anat Sci Educ. 2019;12:585–98.
8. Barmaki R, Yu K, Pearlman R, Shingles R, Bork F, Osgood GM, Navab N. Enhancement of anatomical education using augmented reality: An empirical study of body painting. Anat Sci Educ. 2019;12:599–609.
9. Stull AT, Hegarty M, Mayer RE. Getting a handle on learning anatomy with interactive three-dimensional graphics. Educ Psychol. 2009;101(4):803–16.
10. Moro C, Stromberga Z, Stirling A. Virtualisation devices for student learning: comparison between desktop-based (Oculus Rift) and mobile-based (Gear VR) virtual reality in medical and health science education. Australas J Educ Technol. 2017;33(6)1–10
11. Birt J, Stromberga Z, Cowling M, Moro C. Mobile mixed reality for experiential learning and simulation in medical and health sciences education. Information. 2018;9(2):31.
12. Küçük S, Kapakin S, Göktaş Y. Learning anatomy via mobile augmented reality: effects on achievement and cognitive load. Anat Sci Educ. 2016;9(5):411–21.
13. Moro C, McLean M. Supporting students' transition to university and problem-based learning. Med Sci Educ. 2017;27(2):353–61.
14. Kuehn BM. Virtual and augmented reality put a twist on medical education. JAMA. 2018;319(8):756–8.
15. Stretton T, Cochrane T, Narayan V. Exploring mobile mixed reality in healthcare higher education: a systematic review. RLT. 2018;26:2131.

The Use of Ultrasound in the Teaching and Learning of Anatomy

37

Danielle F. Royer and Cory A. Buenting Gritton

Introduction

Ultrasound (US) is an increasingly utilized imaging modality within point-of-care clinical settings including emergency medicine, anesthesiology, internal medicine, and diverse surgical specialties and is safe because it does not employ ionizing radiation [1, 2]. US imaging is relatively low cost compared to other medical imaging modalities, and with portable equipment that has decreased drastically in size, bedside imaging is now more accessible than ever [3]. Additionally, US has gained recognition not only as a supplement to enhance existing medical curricula, but as a clinical skill that can be introduced in the preclinical years [4].

National surveys in the United States suggest that US scanning is an important skill in medical education and graduate anatomy training [5, 6]. In an increasing number of medical schools, US has been introduced in the preclinical years as electives or adjunct sessions, mainly within gross anatomy and physical examination courses, while clinical years have incorporated US into various specialty rotations as well as stand-alone electives [7, 8]. Other institutions have longitudinally incorporated US across all 4 years of undergraduate medical education (UME) [9–11]. The Alliance of Medical School Educators in Radiology (AMSER) has proposed a standardized US curriculum for UME, and a survey of UME ultrasound directors in the United States has produced a list of core clinical milestones for US in medical school [12, 13]. These proposals, along with existing integrated four-year approaches, have been the model for most US curricula development, as no nationwide US curriculum currently exists. In contrast, specific US training guidelines, which vary by specialty, have been adopted for board-certified clinicians and residency programs [14, 15]. At the UME level, US curricula have typically included hands-on training, where students manipulate the transducer themselves to obtain live views inside the body, including normal anatomical structures or pathologies. However, some programs opt to incorporate US demonstrations, where expert sonographers manipulate the transducer while explaining image acquisition and on-screen views to a student audience [16, 17].

The cost and specifications of US machines vary depending on the needs of the user. For UME, both the laptop-based US machines and handheld US devices have been successfully used [18]. Based on the AMSER recommendations and core clinical US milestones, the key ultrasound functions needed for UME teaching include

D. F. Royer (✉)
Department of Cell & Developmental Biology, University of Colorado School of Medicine, Aurora, CO, USA
e-mail: Danielle.Royer@cuanschutz.edu

C. A. Buenting Gritton
Cell & Developmental Biology, University of Colorado School of Medicine, Aurora, CO, USA
e-mail: Cory.Gritton@cuanschutz.edu

L. K. Chan, W. Pawlina (eds.), *Teaching Anatomy*, https://doi.org/10.1007/978-3-030-43283-6_37

B-mode, M-mode, depth adjustment, gain adjustment, freeze, measurement calipers, color Doppler, image save, and cine loop. Additionally, transducer type depends on the intended scope of US scanning. The higher frequency linear transducer is ideal for superficial scanning such as neck or limbs, while the lower frequency curvilinear transducer is suited for deep structures found within the abdomen, and the phased array transducer is best for imaging the beating heart. The curvilinear, linear, and phased array transducers are sufficient for the surface scanning approaches typically used in UME, while more advanced coursework and US-guided procedures may require specialized equipment such as transesophageal and endocavity transducers [12, 13].

This chapter will focus on the incorporation of hands-on US scanning of living volunteers into an existing gross anatomy course, as a supplement to classroom activities and cadaveric dissections or prosections. However, we acknowledge there are other ways to include US in an anatomy curriculum, including static images or live demonstrations and video, either displayed during live US scanning or prerecorded. Static US images could be appropriate and sufficient in a stand-alone anatomy course; however, in the context of clinical or preclinical courses where there is the added goal of learning US scanning as a skill set, live hands-on US scanning sessions are recommended, as they offer the most benefits to the learner.

Advantages of Ultrasound in Anatomy

US is well suited to supplementing an anatomy course because the dynamic nature of this imaging modality allows students to see inside the living body in real time, the hands-on nature of scanning engages students with active learning and recall of key anatomical concepts, and it readily provides clinical relevance to anatomy.

Noninvasive and Structurally Intact

There are many methods of image acquisition (e.g., x-ray, CT, MR), yet most are time-consuming and expensive, carry a risk of ionizing radiation, and cannot be performed by students directly. Surface US is noninvasive and safe and allows for real-time acquisition of anatomical images by students in a classroom setting (Fig. 37.1). Moreover, medical students are often learning palpation and physical examination alongside gross anatomy during UME, which pairs well with US, as palpation is often necessary during scanning. US allows for real-time viewing of dynamic anatomical relationships, such as the changes in shape of the diaphragm and the corresponding movement of the liver and lungs during respiration. The nondestructive and noninvasive nature of surface US allows for the visualization of structures, spaces, and relationships, which are otherwise challenging to view during dissection and in prosected cadavers, or using static models and images.

Active Learning: Application and Review

Active learning is the process of using mental or physical activities that force students to recall, apply, and assess their knowledge [19]. By directly manipulating the transducer, students are applying their knowledge to interrogate the anatomy in a living body, in order to acquire views of dynamic regions, recognize structures, and make correlations with surface anatomy. While scanning, in-depth discussion of concepts such as neuronal pathways, functions, or variations can occur after identification of anatomical structures. For example, Fig. 37.1 shows the relationship of the left renal vein crossing superficial to the abdominal aorta and deep to the superior mesenteric artery.

Clinically Relevant

Early exposure to clinically relevant imaging modalities has become increasingly important in health professions education. Today, there are many examples of the use of US for diagnosing medical conditions at the bedside (i.e., point-of-care ultrasound) across various medical

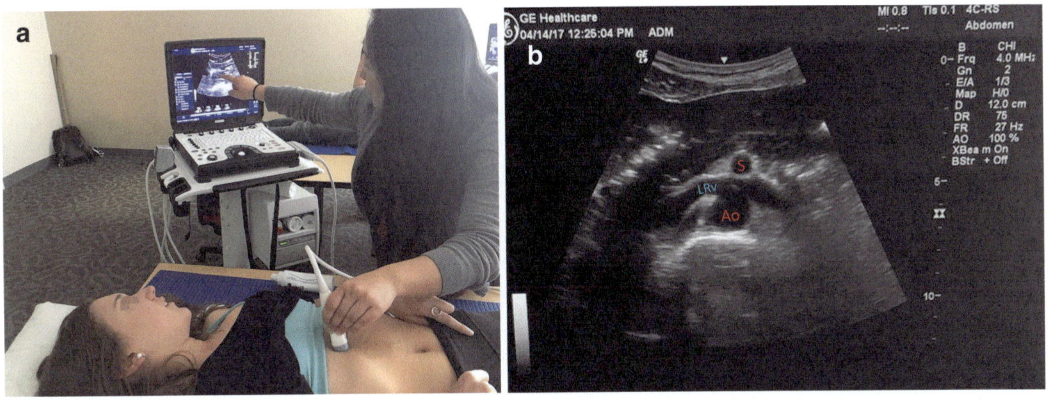

Fig. 37.1 Image (**a**) shows student performing live abdominal ultrasound on a peer in a classroom setting to obtain real-time images of abdominal vessels. Image (**b**) shows a transverse US image of abdominal vascular relationships (Ao indicates abdominal aorta, S indicates superior mesenteric artery, and LRv indicates left renal vein)

specialties. Similarly, US is commonly used in procedures including ultrasound-guided peripheral nerve blocks and central line placement. The addition of US early during preclinical training is an opportunity to demonstrate the clinical relevance of gross anatomy and build the foundation for clinically important skills. Outside of career relevance for healthcare students, the inclusion of concrete or real-life examples has been proven to help students learn and retain information [20].

Challenges of Ultrasound in Anatomy

The inclusion of US into a course will present challenges, as would any course addition. The main challenges typically faced when integrating US into an anatomy course are reviewed below. Recommendations on how to address these challenges are provided in the section on the development of US sessions.

Access to Resources

Multiple resources are required for hands-on US sessions, including equipment and suitable space. US machines are becoming increasingly compact and affordable; however, the initial purchase can be substantial, especially if many machines are needed. New machines range from around $2000+ USD for handheld devices (usually with more limited functionality and small screens which may be unsuitable for group teaching) to $30,000 USD or more for hospital-grade laptop systems with the full range of imaging and diagnostic capabilities.

Hands-on US sessions require adequate and flexible small group space. For the most part, US sessions do not work well in a gross anatomy dissection laboratory environment. The small group sessions need to comfortably accommodate a living scanning model in a reclined position, typically on a table or examination bed, and offer an option to dim the lighting for ideal viewing of US images.

Student volunteers can be utilized as scanning models to enable live US viewing, or other live models can be recruited, either as volunteers or with compensation. Traditional embalming limits the ability to scan cadavers, but soft embalming methods (e.g., Thiel) may permit scanning of a cadaver. However, the dynamic aspect of US (e.g., heart contractions, effects of respiration, blood flow) will be lost when scanning a cadaver, even using soft embalming techniques [21].

Facilitators

Small group, hands-on US sessions within a gross anatomy course require clear guidance and accessible help for the novice learner. Facilitators such as physicians and sonographers have training in both US and anatomy, but may find it challenging to work within the schedule of an existing course. For anatomy faculty, who are likely already involved in the course, specific training in US may be needed, as they often lack experience with this modality. However, anatomy faculty and students can be readily trained to facilitate live US to sufficiently meet the goals of supplementing a gross anatomy course [22–24]. A single facilitator leading 1–2 small groups at a time provides the most direct guidance for novices engaged in hands-on live scanning. Large group sizes or multiple groups per session may require a greater number of facilitators, which may be problematic. Alternatively, a combination of a large group live demonstration followed by small group hands-on practice can ease the burden on facilitators.

Large Class Sizes

Small groups are ideal for US sessions as only one transducer can be used at a time, and only one model can be scanned per machine. Ideally, groups should consist of five or fewer students per machine, to maximize the amount of active scanning for each student and to allow all participants to readily view both the probe placement and on-screen images. If students serve as both model and operator during a small group session, they should be encouraged to take turns using the transducer and as the scan model. For large classes, achieving this optimal small group size can be a challenge depending on available resources. However, a combination of increased number of sessions, machines, and/or facilitators can help overcome this barrier.

Ethical and Cultural Sensitivity

Another challenge of live peer scanning in US sessions can be the ethical and cultural barriers of baring skin in front of friends and classmates or palpating (touching) peers. Generally, this can be overcome by asking for student volunteers, as many students are excited to see their own anatomy. However, institutional-specific guidelines on student volunteers should be consulted prior to sessions. The comfort, privacy, and respect of volunteer scan models should always be a priority. No student should be forced or coerced to serve as a scan model, and a student who volunteers for one region (e.g., neck) should not be assumed to automatically serve as a volunteer for another region (e.g., pelvis). Statements regarding scan model privacy, professionalism, and the ability of students to refuse to serve as scan models should be discussed and addressed before each US session. These statements can be verbal, in writing, or a combination of both and may include waivers. Faculty are strongly advised to research the policies and guidelines at their home institution prior to implementing a hands-on live US scanning curriculum using student volunteers.

Pre-scanning volunteers or recruiting standardized patients (who have also been pre-scanned) prior to US sessions can overcome these barriers as well, though it increases the burden on session preparation and may incur additional costs. When student volunteers are recruited and not pre-scanned, it is important to issue a disclaimer before each session that acknowledges that the US scanning in the course is performed for educational purposes only. That is, volunteers are not receiving a formal diagnostic scan or medical screening, and pathologies will not be deliberately sought out. The statement should specify that while incidental findings (incidentaloma) may be observed, the focus of the scan is on normal anatomy, and as such students and facilitators are not responsible for any missed pathologies. Moreover, upon observation of suspected incidental findings (e.g., variations, pathologies such as cysts, tumors, heart valve dysfunction), a scan volunteer should be advised to follow-up with their medical provider on their own. Again, the statement can be verbal, written, or in the form of a waiver; faculty must inform themselves on relevant institutional policies.

In training US teaching assistants and facilitators, scan model comfort, privacy, and professionalism should also be addressed, and all facilitators should be familiar with the required scan model disclaimers or waivers. Training should include identification of some key pathologies that may appear in each region, along with the appropriate steps to inform scan models of incidental findings. It is especially critical to emphasize this during the facilitator training prior to surface scanning female reproductive anatomy, in light of possible unknown pregnancy findings.

Examples of Ultrasound in the Teaching and Learning of Anatomy

Specific examples of ultrasound within an anatomy course are discussed below, to highlight different regions and illustrate how US can complement a course. Suggestions for how to implement US sessions into a course will be discussed later.

Wrist

The wrist, a relatively small region, contains many structures continuing from the forearm into the hand, which are easily visible using US. For example, the tightly packed structures of the carpal tunnel, often disturbed through dissection, are easily visible in situ (Fig. 37.2). With live scanning, the individual being scanned can move muscles, such as flexor pollicis longus and flexor digitorum superficialis, in real time, to help identify structures passing through the carpal tunnel and promote discussions of structure and function. The relationships of the median nerve within the carpal tunnel can be observed while carpal tunnel boundaries and contents remain intact. The median nerve can be traced proximally into the forearm to view its relationship between flexor digitorum superficialis and flexor digitorum profundus, leading to discussion of anterior forearm muscle innervations. Nearby, the radial artery can be viewed lateral to the carpal tunnel, along with the ulnar nerve and ulnar artery medially. Both neurovascular bundles can be traced proximally and linked with clinical procedures or palpation of peripheral pulses.

Hepatorenal Recess (Morison's Pouch)

The hepatorenal recess, a space between the liver and the right kidney, can be difficult to visualize in the cadaver once the abdominal organs are mobilized through dissection. Through US, this space is easily visible as the interface between the intact liver and right kidney (Fig. 37.3). The relationship between the thorax and abdomen during respira-

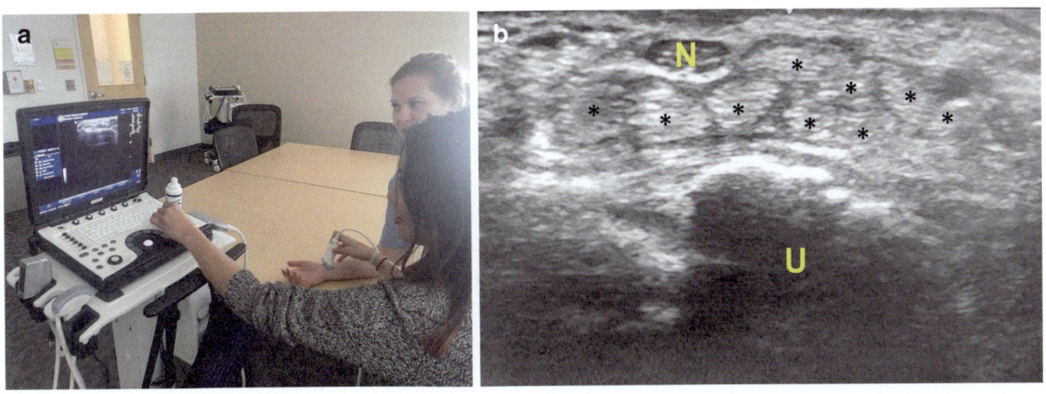

Fig. 37.2 Image (**a**) shows transducer placement to scan the carpal tunnel. Image (**b**) shows the on-screen US image (N indicates median nerve, U indicates ulna, and asterisks indicate each of the nine tendons that pass through the carpal tunnel)

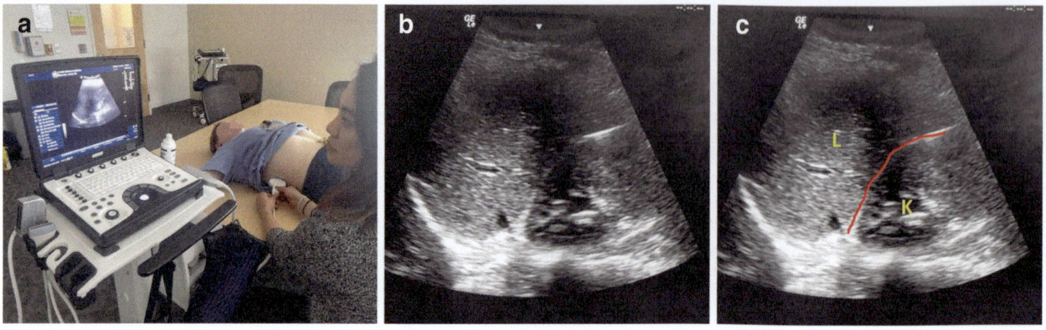

Fig. 37.3 Image (**a**) shows transducer placement to scan the hepatorenal recess. Image (**b**) shows the on-screen image of the hepatorenal recess. Image (**c**) shows the hepatorenal recess (red line) between the liver (L) and kidney (K)

tion can be discussed while the dynamic change in the shape of the diaphragm and costodiaphragmatic recess are visualized on US. Although lung parenchyma is difficult to view, signs of respiration are visible with in sync diaphragm movement and excursion of the liver. The clinical relevance of the hepatorenal recess as a location for free fluid accumulation in a supine patient can be discussed. This can set the stage for continued dialogue about US examination of trauma-related injuries and the importance of patient body position in the location and diagnosis of free fluid accumulation in the peritoneal cavity.

Eye

The eye is a small, 3D structure with many components not easily visible after embalming and macroscopic dissection. The structure and relationship of surrounding muscles are not easily visible in laboratory, due to the small space within the bony orbit. While the extraocular eye muscles cannot be directly visualized in US, the position and dynamic movement of the eye within the orbit can be viewed. Specifically, the model can be instructed to move their eye in a simulated eye examination, prompting a review of extraocular muscles, actions, and innervations (Fig. 37.4). Although the nerves involved in pupillary dilation and contraction are not visible, observing the model's pupillary light reflex can prompt discussion of intraocular muscles, actions, innervations, and autonomic pathways.

Heart

The heart is a complex three-dimensional structure that can be difficult to view intact and in situ in the cadaver. Often, two-dimensional images fail to fully convey the organization of the heart and its orientation within the thoracic cavity. Due to the destructive nature of dissection, the anatomical position of the heart within the mediastinum is often lost and, even after dissecting the heart itself, the relationship of the cardiac chambers and internal anatomy can be difficult for students to relate to surface anatomy. Utilizing US, multiple views of the heart can be obtained, fostering discussions of the orientation of the heart within the thorax, as well as the relationship of the heart's own chambers (Fig. 37.5). The dynamic nature of the heart functioning in situ can be appreciated during live scanning and can be used to frame a discussion of blood flow through cardiac chambers and valves, as well as cardiac pathologies, congenital defects, and related cardiac physiology. Given the complexity of the heart, two-dimensional images and sectioned plastinated hearts or heart models can be useful to assist students in understanding the orientation of the heart and its common planes of section (i.e., parasternal long axis, parasternal short axis, subxiphoid, apical) in US.

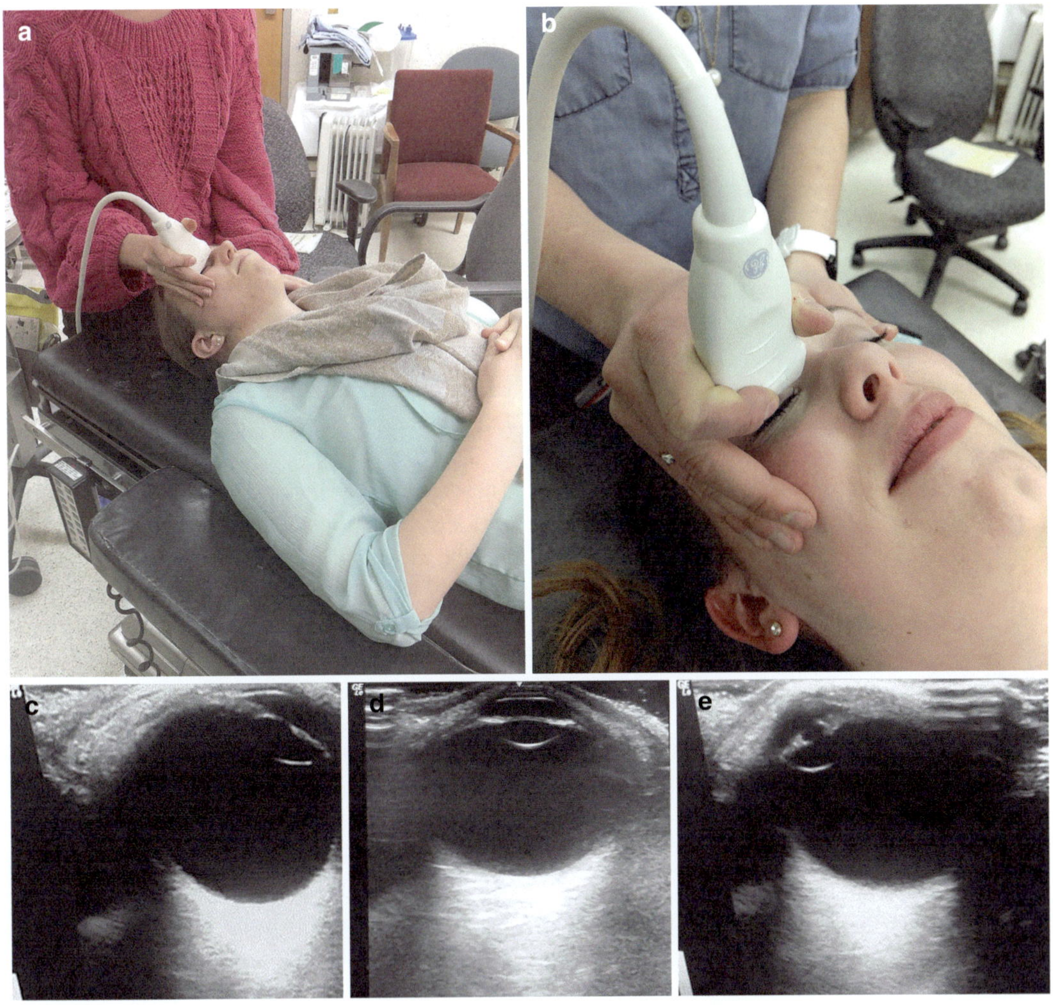

Fig. 37.4 Images (**a**, **b**) indicate the transducer placement for scanning the eye. The dynamic nature of US is demonstrated when the model looks laterally (**c**), centrally (**d**), and medially (**e**)

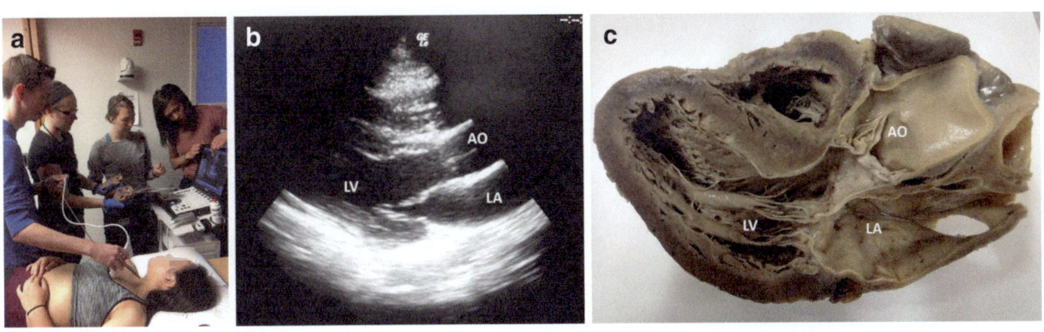

Fig. 37.5 Image (**a**) demonstrates transducer placement to capture a parasternal long axis view of the heart. The relationship and internal anatomy of the cardiac chambers are easily visible using US (**b**) and can be compared to plastinated hearts that have been sectioned along the same parasternal long axis (**c**). In images (**b**, **c**), the left atrium is indicated as LA, left ventricle as LV, and aortic outflow as AO

Other Anatomical Structures That Lend Themselves to Ultrasound

- *Carotid sheath structures:*
  - Common carotid artery bifurcation
  - Carotid sheath content relationships
  - Valsalva maneuver to visualize venous distention
- *Anterior neck:*
  - Thyroid lobes and isthmus
  - Sternocleidomastoid and infrahyoid muscles
  - Larynx and trachea
- *Abdominal organs and spaces:*
  - Peritoneal recesses
  - Right and left upper quadrant contents and relationships
- *Liver, associated vasculature and biliary tree:*
  - Inferior vena cava, portal veins, and hepatic veins
  - Portal triad structures
  - Gall bladder
  - Costodiaphragmatic recess and diaphragm
- *Abdominal vasculature:*
  - Abdominal aorta
    - Celiac trunk: splenic artery and common hepatic artery
    - Superior mesenteric artery
  - Inferior vena cava
    - Left renal vein
  - Vasculature relationships
- *Urogenital structures (viewed through abdominal surface scanning):*
  - Urinary bladder and ureteric jets
  - Female reproductive structures
    - Uterus (fundus, body, cervix)
    - Uterine orientation (e.g., anteverted, anteflexed)
    - Vaginal canal
  - Male reproductive structures
    - Prostate
    - Seminal vesicles
  - Relationships between urogenital structures and spaces

Developing Ultrasound Sessions in an Anatomy Curriculum

The integration of US into an existing anatomy curriculum depends on many factors. This section discusses aspects to consider while developing sessions and suggestions for overcoming specific challenges related to the creation of US sessions in an anatomy course.

Identify Ultrasound Session Learning Objectives

There are many anatomical areas that can be viewed with US; however, it is not necessary or feasible to include them all when US is an adjunct to gross anatomy coursework. Similarly, it is important to recognize that anatomy students are not going to be trained sonographers by course end, and while they will gain sonography skills from hands-on sessions, the focus should be on using US as a supplement to understanding and visualizing anatomical structures and relationships. If an anatomy course already exists, it is important to assess and identify what regions would benefit most from the inclusion of US sessions. One method to consider is identifying existing course concepts or objectives which overlap with anatomical structures that lend themselves well to ultrasound; a few are listed in this chapter. A foundation of US knowledge is required in order to utilize this modality for viewing live anatomy. In addition to objectives for anatomy, it is important to develop objectives regarding basic ultrasound scanning techniques, image orientation, image quality, common artifacts, ultrasound physics, knobology, and proper handling of US equipment. Deliberate curriculum addressing such objectives should be scheduled early on in an ultrasound curriculum [25].

Session Scheduling

Considering when to implement US sessions into an existing course is important. It is beneficial to consider where time is available, how the sessions will align with the anatomy curriculum, as well as how much time is available for each session. Before exploring anatomical structures with US, an introductory session is necessary to familiarize students with the imaging modality. It is important in the introductory session to focus on US machine functions, on-screen orientation, transducer functions, and basic understanding of US physics. Static models like a training phantom can be considered, so as not to overwhelm the learner with the anatomy they might not yet know, the dynamic aspects of live scanning, or the need to palpate another person. For regional anatomy US sessions, it is recommended for corresponding cadaver laboratories to occur first, followed by US sessions, because students have seen the structures once already and are better primed to make the most of the hands-on US session.

The amount of time designated for each session needs to be planned with the number of students in mind, as well as the specific goals for the session. For example, with one US machine, five students can complete a hands-on regional anatomy US session (e.g., neck, abdomen) in about 1 hour. However, if group sizes are larger, more time may be required to allow all students active scanning time. Presession work such as readings, PowerPoints, or online resources can be used to introduce regional US prior to sessions and maximize hands-on US scanning during the session itself. Presession materials can reduce the need for intense direct facilitator interactions during each session, which can be important if time or facilitators are limited.

Facilitator and Equipment Resources

If you are at a medical school or health sciences center, speak with clinicians to see if there is used equipment or clinic equipment that can be borrowed. This can minimize the start-up cost of implementing US sessions into an anatomy curriculum. Consider the number of small groups at each session as this will dictate the number of US machines required. Alternatively, the number of machines accessible can dictate group size and number. Both handheld and laptop US machines have been used successfully in undergraduate medical US education; however, key machine features needed include B-mode, M-mode, color Doppler, calipers to measure, gain adjustment, depth adjustment, freeze, image storage, and cine loop [12, 13]. The small screens of handheld devices may limit the ability of all group members to view the ultrasound scans, which makes them potentially challenging for use in group settings of more than 2–3 people.

Facilitators are an important resource as well. A single facilitator can successfully lead 1–2 hands-on groups at a time; thus the number of groups and sessions will determine the number of facilitators needed. Alternatively, the number of facilitators can dictate the number of small groups and sessions. Both clinicians and anatomists, with some US training, can successfully lead US sessions. In fact, this can be a wonderful opportunity for students to see clinicians and basic scientists working together. Additionally, graduate teaching assistants have successfully been trained to serve as US facilitators [23]. All facilitators must be proficient in US and anatomy as well as understand the goals of using US in the course; otherwise, there is risk of malalignment of the US activities and the learning outcomes of the course.

Assessment

US can be integrated into an anatomy course with or without an assessment component. If assessments are included in the course, they must be tightly aligned with US session content and objectives. Both knowledge and skill in US can be assessed, but different assessment tools will be required in order to measure these different domains. For example, image-based fill-in-the blank or multiple-choice questions may be sufficient in written or laboratory practical examinations for assessing anatomical knowledge as captured by US images. However, assessment of US skills requires different assessments, in which students are asked to acquire predetermined US views showing specific structures or relationships. There has been limited use of Objective Structured Clinical Examinations (OSCE) in US curricula integrated at the UME level; however, OSCEs have been used to evaluate US competencies [10,

26–28]. Live assessments can be undertaken using a predetermined checklist or rubric, but this type of assessment will require additional time, scheduling, machines, and a skilled proctor. Alternatively, if assessments are not included, the incorporation of US into an anatomy course can be an opportunity to actively review anatomy in an interactive, low-stakes, formative environment.

Student Feedback

It is important to acquire student feedback about alignment of the US sessions with the gross anatomy curriculum. A benefit of the small group work during these sessions is that facilitators can solicit feedback informally while the sessions are happening. Formally, targeted questions should be included in the course evaluations to address the newly implemented curriculum. Student feedback is critical in guiding and directing the future structure, organization, and alignment of the US sessions within the gross anatomy course. A common student request is for more time to practice scanning, which often increases when US is added to assessments. This request can be accommodated by open scan times, where US equipment is available and students are free to scan without facilitation.

Building an Ultrasound Curriculum for Anatomy: To-Do List

- Compare anatomical structure/content with structures and regions that lend themselves well to ultrasound.
- Assess the number of sessions that can be incorporated into the course.
- Evaluate the amount of time which can be allotted for each session.
- Identify ideal group size and number of groups.
- Evaluate US machine access and quantity.
- Identify available space for US sessions that accommodates all groups.
- Identify the number of facilitators needed and whether additional facilitators can be trained.

Conclusion

There are many curricular and logistical aspects to consider when using US for the teaching and learning of anatomy. This dynamic imaging modality is safe, noninvasive, and hands-on, allowing students to actively review anatomy content while viewing real-time anatomy in situ. Creation of US sessions within anatomy can be constrained by large class sizes, resources, and the need for trained facilitators, but these challenges can be overcome through alignment, assessment of resources, and flexibility during the development of US sessions. Students generally welcome the addition of hands-on US into an anatomy course, and it can be a fun and engaging way to bring life back into anatomy [29].

References

1. So S, Patel RM, Orebaugh SL. Ultrasound imaging in medical student education: impact on learning anatomy and physical diagnosis. Anat Sci Educ. 2017;10(2):176–89.
2. Shriki J. Ultrasound physics. Crit Care Clin. 2014;30(1):1–24.
3. Hoppmann R, Blaivas M, Elbarbary M. Better medical education and health care through point-of-care ultrasound. Acad Med. 2012;87(2):134.
4. Tarique U, Tang B, Singh M, Kulasegaram KM, Ailon J. Ultrasound curricula in undergraduate medical education: a scoping review. J Ultrasound Med. 2018;37(1):69–82.
5. Bahner DP, Goldman E, Way D, Royall NA, Liu YT. The state of ultrasound education in U.S. medical schools: results of a national survey. Acad Med. 2014;89(12):1681–6.
6. Royer DF. The role of ultrasound in graduate anatomy education: current state of integration in the United States and faculty perceptions. Anat Sci Educ. 2016;9(5):453–67.
7. Rempell JS, Saldana F, DiSalvo D, Kumar N, Stone MB, Chan W, et al. Pilot point-of-care ultrasound curriculum at Harvard medical school: early experience. West J Emrg Med. 2016;17(6):734–40.
8. Prats MI, Royall NA, Panchal AR, Way DP, Bahner DP. Outcomes of an advanced ultrasound elective: preparing medical students for residency and practice. J Ultrasound Med. 2016;35(5):975–82.
9. Hoppman RA, Rao VV, Poston MB, Howe DB, Hunt PS, Fowler SD, et al. An integrated ultrasound curriculum (iUSC) for medical students: 4 year experience. Crit Ultrasound J. 2011;3(1):1–12.
10. Hoppman RA, Rao VV, Bell F, Poston MB, Howe DB, Riffle S, et al. The evolution of an integrated ultra-

sound curriculum (iUSC) for medical students: 9-year experience. Crit Ultrasound J. 2015;7(18):1–12.
11. Bahner DP, Adkins EJ, Hughes D, Barrie M, Boulger CT, Royall NA. Integrated medical school ultrasound: development of an ultrasound vertical curriculum. Crit Ultrasound J. 2013;5(1):6.
12. Baltarowich OH, Di Salvo DN, Scoutt LM, Brown DL, Cox CW, DiPietro MA, et al. National Ultrasound Curriculum for medical students. Ultrasound Q. 2014;30(1):13–9.
13. Dinh VA, Lakoff D, Hess J, Bahner DP, Hoppmann R, Blaivas M, et al. Medical student core clinical ultrasound milestones: a consensus among directors in the United States. J Ultrasound Med. 2016;35(2):421–34.
14. Accreditation Council for Graduate Medical Education. https://www.acgme.org (no date). Accessed 14 Apr 2019.
15. American Institute of Ultrasound in Medicine. https://www.aium.org (no date). Accessed 14 Apr 2019.
16. Stringer MD, Duncan LJ, Samalia L. Using real-time ultrasound to teach living anatomy: an alternative model for large classes. NZ Med J. 2012;125:37–45.
17. Paganini M, Rubini A. Chest ultrasound integrated teaching of respiratory system physiology to medical students: a first experience. Adv Physiol Educ. 2015;39(1):129–30.
18. Galusko V, Khanji MY, Bodger O, Weston C, Chambers J, Ionescu A. Hand-held ultrasound scanners in medical education: a systemic review. J Cardiovasc Ultrasound. 2017;25(3):75–83.
19. Michael J. Where's the evidence that active learning works? Adv Physiol Educ. 2006;30(4):159–67.
20. Sumeracki MA, Weinstein Y. Six strategies for effective learning. Acad Med. 2018;93(4):666.
21. Munirama S, Eisma R, Columb M, Corner GA, McLeod GA. Physical properties and functional alignment of soft-embalmed Thiel human cadaver when used as a simulator for ultrasound-guided regional anaesthesia. Br J Anaesth. 2016;116(5):699–707.
22. Celebi N, Zwirner K, Lischner U, Bauder M, Ditthard K, Schurger S, et al. Student tutors are able to teach basic sonographic anatomy effectively – a prospective randomized controlled trial. Ultraschall Med. 2012;33(2):141–5.
23. Smith JP, Kendall JL, Royer DF. Improved medical student perception of ultrasound using a paired anatomy teaching assistant and clinician teaching model. Anat Sci Educ. 2018;11(2):175–84.
24. Jurjus RA, Dimorier K, Brown K, Slaby F, Shokoohi H, Boniface K, et al. Can anatomists teach living anatomy using ultrasound as a teaching tool? Anat Sci Educ. 2014;7(5):340–9.
25. Jamniczky HA, McLaughlin K, Kaminska ME, Raman M, Somayaji R, Wright B, et al. Cognitive load imposed by knobology may adversely affect learners' perception of utility in using ultrasonography to learn physical examination skills, but not anatomy. Anat Sci Educ. 2015;8(3):197–204.
26. Sissley AC, Johnson SB, Erickson W, Fortune JB. Use of an objective structures clinical examination (OSCE) for the assessment of physician performance in the ultrasound evaluation of trauma. J Trauma. 1999;47(4):627–31.
27. Rebel A, Srour H, DiLorenzo A, Nguyen D, Ferrell S, Dwarakanatli S, et al. Ultrasound skill and application of knowledge assement using an innovative OSCE competition-based simulation approach. J Educ Perioper Med. 2016;18(1):E404.
28. Dinh VA, Dukes WS, Prigge J, Avila M. Ultrasound integration in undergraduate medical education: comparison of ultrasound proficiency between trained and untrained medical students. J Ultrasound Med. 2015;34(10):1819–24.
29. Royer DF, Kessler R, Stowell JR. Evaluation of an innovative hands-on anatomy-centered ultrasound curriculum to supplement graduate gross anatomy education. Anat Sci Educ. 2017;10(4):348–62.

The Roles of Radiology in Anatomy Education

38

Patrick Schiller, Andrew Phillips,
and Christopher Straus

Introduction

Medical imaging has been used in formal anatomy instruction internationally since at least the 1960s [1–3] and has been growing in prevalence continually [4], evolving in line with advances in both technology and educational theories [5, 6]. Trainees, teaching faculty, physicians, and medical education experts have embraced medical imaging in anatomy instruction [7–14], as does much of the anatomy literature [15]. Radiology provides a distinct medium with which to understand anatomical structure and concepts [8, 16–26]. The incorporation of medical imaging has also objectively improved student performance in gross anatomy, both in short-term and long-term retention [8, 22, 27, 28]. Furthermore, curated radiological examples can supplement gross normal variation and pathology seen in the anatomy laboratory [29–32].

P. Schiller (✉)
University of Chicago, Department of Medicine, Ellis Ave, IL, USA
e-mail: patrick.schiller@uchospitals.edu

A. Phillips
University of North Carolina Chapel Hill, Department of Emergency Medicine, Chapel Hill, NC, USA

C. Straus
University of Chicago, Department of Radiology, Ellis Ave, IL, USA
e-mail: cstraus@uchicago.edu

Additionally, physicians currently report insufficient anatomy and medical imaging knowledge [33–37]. Incorporating radiologic imaging into anatomy courses allows future physicians to better apply basic anatomy to their future clinical setting; this leads to improvements in physical examinations, differential diagnoses, and patient treatment [17, 27, 38–47]. Of course, being exposed to medical imaging during anatomy also increases the future physician's comfort in both using and interpreting medical imaging [48], one of the primary ways anatomy is utilized in daily practice [49].

This chapter provides practical advice to determine what images are needed, how to obtain them, and how to implement them into a gross anatomy course.

Determining Course Objectives

Before incorporating radiology into an anatomy curriculum, instructors must consider the course objectives. There is a massive amount of information available in the field of radiology (enough to require 4 years of dedicated radiology residency for burgeoning radiologists), so it is essential to prioritize the knowledge, concepts, and skills to be covered. Some of these goals will be impacted by practical considerations, including access to images, software, hardware, and imaging experts, all discussed below; others will be

L. K. Chan, W. Pawlina (eds.), *Teaching Anatomy*, https://doi.org/10.1007/978-3-030-43283-6_38

impacted by instructor preference for important anatomical structures. Published goals and objectives exist [14, 29, 30, 49–51] and can be used as templates and starting points for content. In general, we recommend that radiological anatomy be used to supplement gross anatomy instruction as an additional tool for conveying and understanding concepts, just as textbooks or models are used as adjuncts to the course. Furthermore, incorporating multiple and varied imaging modalities (discussed below) has been shown to maximize learning of and appreciation for anatomy [52].

Overview of Image Types

Table 38.1 describes the most common types of medical imaging and what structures and/or tissue types they best describe. For example, X-rays are good choices to teach bone anatomy, but poorly suited for soft tissues like the abdominal organs, which are best seen using axial imaging such as computed tomography (CT, or "CAT scan") or magnetic resonance imaging (MRI).

Sometimes colloquialisms are used by physicians that may be helpful for the anatomist to know. "Plain films" refer to X-rays. "Axial imag-

Table 38.1 Common medical imaging modalities, a brief description of the physics used in their production, examples of tissue types they best describe, and medical indications for their use in clinical practice

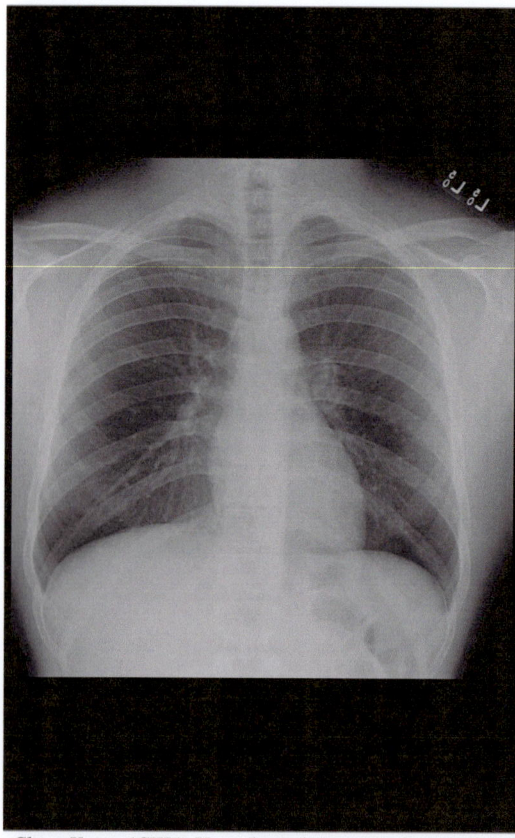	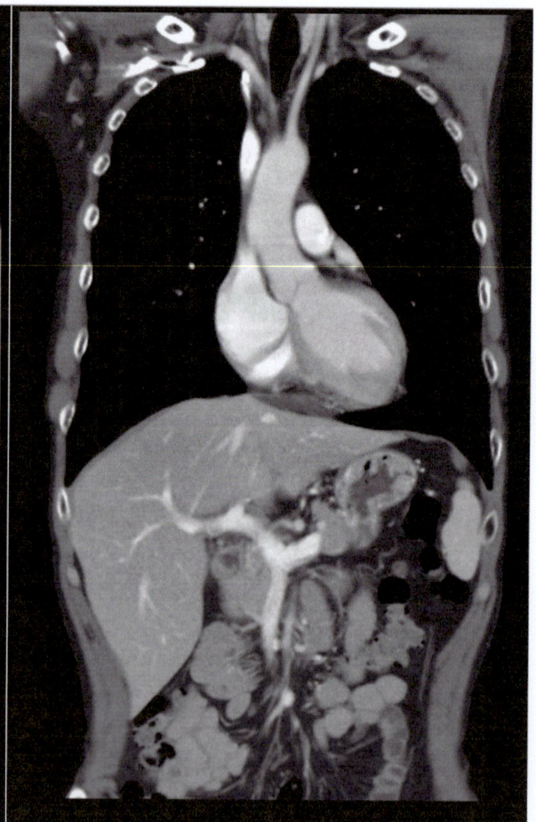
Chest X-ray (CXR): Uses through-transmission X-rays. When shot through a patient, different tissues absorb these x-rays in varying amounts, creating a 2D image or pattern *Best for*: Bones, some tumors, and other dense matter *Indication*: Bones, screening	*Computer tomography (CT)*: Uses directed transmission of X-rays, similar to 100 s of X-rays in rapid sequence from all directions. Computers reconstitute these data into 3D images *Best for*: Bones, organs, tissues, and tumors *Indication*: Serial sectioning of any body part, can be viewed in axial, coronal, sagittal, or 3D reconstruction

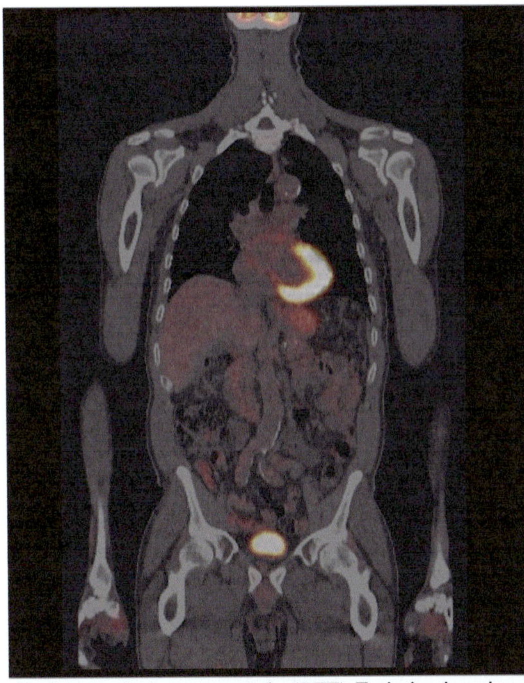

Positron-emission tomography (PET): Emission imaging using radioactive molecules, most commonly glucose. The radioactive molecule is injected into a patient and tissues with greater uptake of the molecule are captured on imaging (in the case of glucose, this tends to be tumors, cardiac, and brain tissues)
Best for: Bones, organs, tumors, and brain activity
Indication: Tumor detection

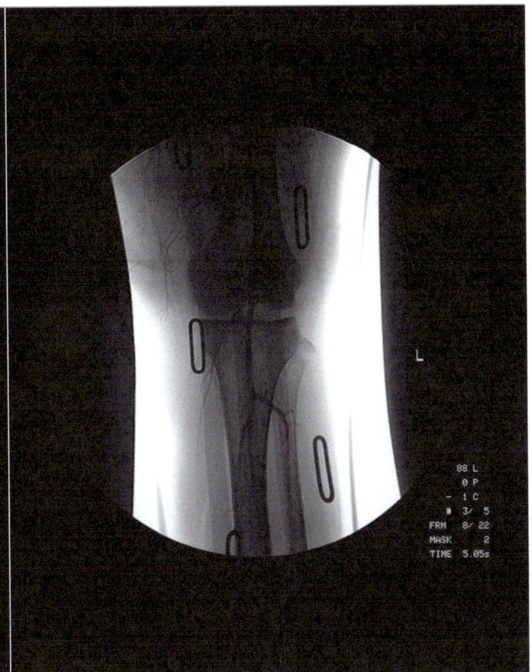

Angiography: Use of through-transmission X-rays in a constant stream (not a single momentary picture), allowing cine loops of imaging. Images are best with added contrasting materials within spaces not otherwise highlighted, like injecting iodine-based contrast in the vessels
Best for: Blood vessels
Indication: Arteries of any location

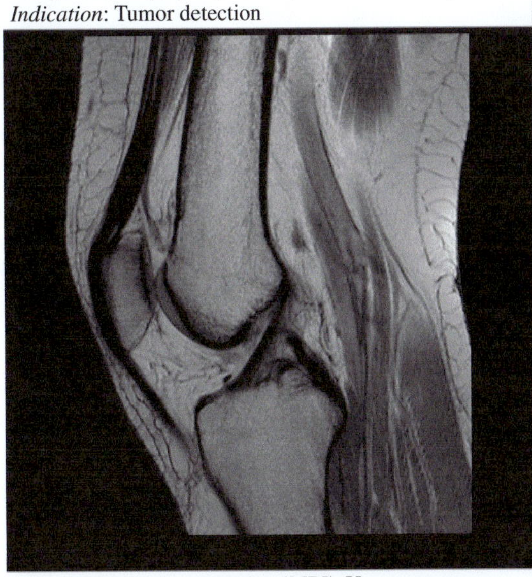

Magnetic resonance imaging (MRI): Uses protons or water in tissues to send a signal generated by a magnet.
Best for: Soft tissue differentiation—brain, spinal cord, ligaments, cartilage; *not* cortical bone or air-filled structures
Indication: Best option for detecting differences in soft tissues such as the brain and joints

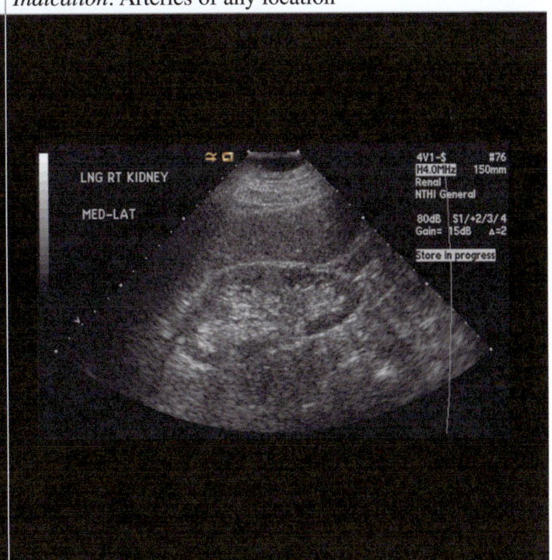

Ultrasound (US): Uses high-frequency sound waves to map out tissue differences based on echo characteristics, similar to using sonar to map the floor of the ocean
Best for: Soft tissue, blood vessels; *not* bone or air-filled structures (needs fluid to transmit sound)
Indication: Cardiac motion, fetal activity; can show live motion

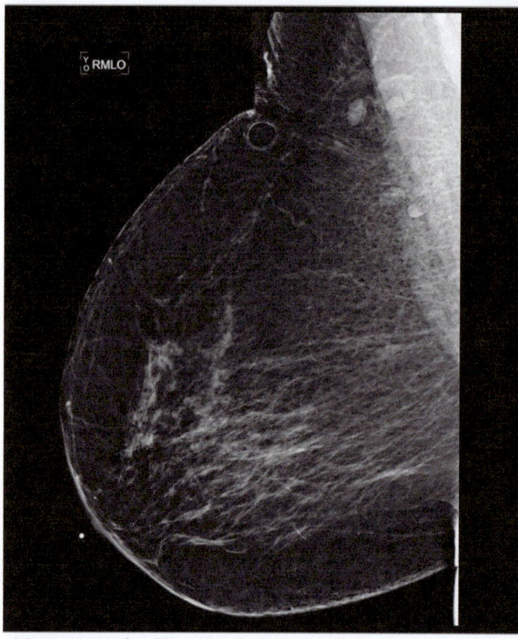

Mammography: Uses through transmission to image breast tissue
Best for: Breast tissue
Indication: Breast cancer screening and diagnosis

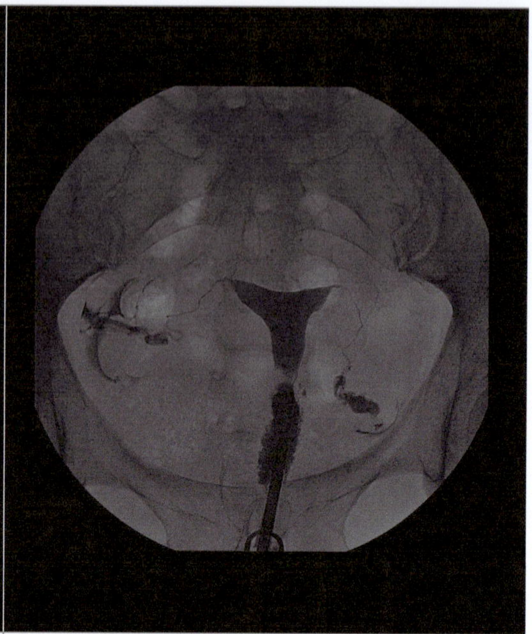

Hysterosalpingogram: A radiopaque contrast solution is injected through the patient's cervix, allowing for crisp imaging of the soft tissues of the uterus and fallopian tubes using X-rays. These X-rays can be taken in rapid succession to allow for real-time imaging to follow the flow of the contrast
Best for: Fallopian tubes and uterine cavity
Indication: Checking for scarring in infertile patients

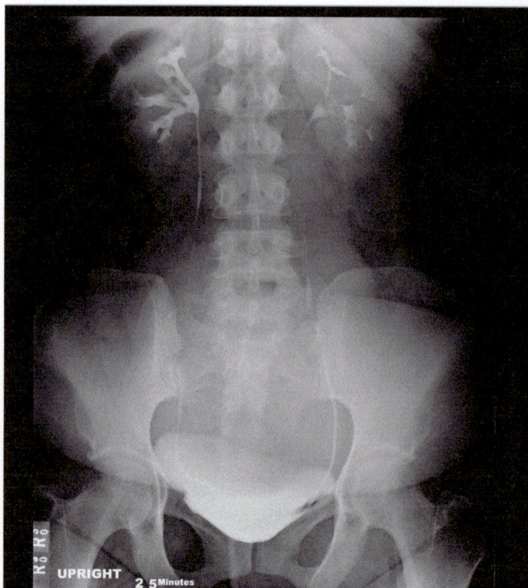

Ureterogram: A renally excreted radiopaque contrast solution is injected into the patient, allowing for improved imaging of the soft tissues of the urinary tract using X-rays. These X-rays can be taken in rapid succession to allow for real-time imaging
Best for: Urogenital system
Indication: Checking for scarring/abnormalities of the kidney and ureters/urethra

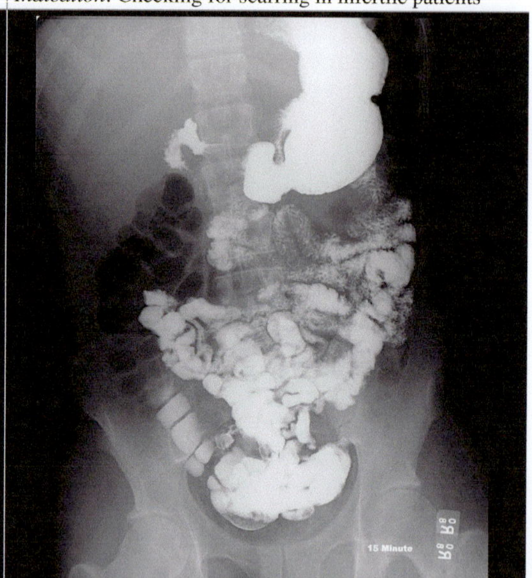

Barium imaging: A radiopaque contrast solution is consumed by the patient, allowing for improved imaging of the soft tissues of the GI tract using X-rays. These X-rays can be taken in rapid succession to allow for real-time imaging. This image was taken in the prone position
Best for: Alimentary tract—augmented with barium
Indication: Evaluating the oral cavity to the rectum

ing" refers to imaging methods that take slices in series, specifically CT and MRI. POCUS stands for point-of-care ultrasound which is when a small, portable ultrasound is used at the point of patient care, which is in contrast to the larger formal ultrasound machines that are used for formal imaging assessments.

Obtaining Images

The best way to obtain images is to utilize your home institution's imaging experts (information on this collaboration is detailed below). Radiologists and radiology departments often have collections of medical images used for teaching, reference, and personal review stored in a picture archiving and communication system (PACS). These images often depict classic normal anatomy useful for anatomy instruction. On a practical note, any patient information attached to these images must be removed prior to incorporation into the course [48].

Anatomy instructors can also find extensive imaging resources online, many of which are free for instructional use, but often require some medical imaging knowledge to use effectively. As such, radiologist involvement is strongly encouraged. Table 38.2 shows a variety of resources, notes about their specific features and functionality, as well as their respective fees. McEntee [53] has also created a useful summary of online radiological anatomy resources, distilled from over 100 explored websites.

Cadaveric CT Imaging

Some anatomy courses have opted to directly image course cadavers prior to dissection, especially with CT scans. The use of cadaveric CTs has been argued to pique student interest in both dissection and image interpretation while also increasing their ability to map structures from imaging to gross anatomy since the gross anatomy structures have been directly imaged [54]. However, imaging every cadaver each time the course is offered costs more time, money, and resources than using the same medical images every year. Several studies have shown that adding cadaveric CT images to anatomy courses increases student learning and appreciation of anatomy [55–58], but none of them involved comparisons to the use of CT scans obtained from living patients. Furthermore, Lufler et al. [28] found that there was no significant difference in course performance between students who used cadaveric CT scans of the body they used for dissection compared to students who used cadaveric CT scans of other bodies. Additionally, the embalming and preservation process of cadavers alters the way tissues appear on CT imaging (both due to gross physical changes in the structures and the way altered tissues interact with the underlying physics of CT scanners), degrading image quality and making many soft tissues appear more similar than they do in living patients [59]. Some structures can be seen almost as well on cadaveric CTs, such as the thoracic wall and vertebral structures, but others, including large bony joints and cardiac structures, have significantly less definition [60]. As such, using CT scans obtained from living patients will offer higher-quality images with significantly greater clinical relevance and will also be easier to obtain while still offering improvements in anatomy appreciation, understanding, and learning. We do not recommend cadaveric CTs based on cost/benefit considerations, especially for initial incorporation of medical imaging into anatomy courses.

Manipulating Images

Once images have been acquired, they need to be edited for teaching purposes. Edits may include adding labels, arrows, stars, or colored shading to aid in indicating structures. It is a good idea to keep images in pairs—with and without editing—so students can test themselves with unedited images. Tam [61] has general tips and approaches to manipulating images, and the NIH has created a resource for processing and analyzing medical images using Java [62]. Thapa et al. [63] created a tutorial for converting medical images to line art (also called vector graphics), which can simplify images, allowing for clearer understanding of anatomical features for those new to radiology, and also reduce the digital memory needed to store converted images.

Storing Images

High-quality images require large amounts of storage space, and it can take time to transfer from one location to another. Many institutions have access to software programs used by the radiology department to store, transfer, and view images. May et al. [30] provide a brief overview of what to look for in radiology imaging display software if course instructors are able to obtain images, but cannot gain access to institutional software programs. Simpler alternatives include online image resources described in Table 38.2, cloud-based storage systems (many of which have free accounts for limited data storage), or uploading images to course websites, all of which allow students remote access. Lectures with high-resolution images, especially serial imaging, are best transported on high-capacity thumb (flash) drives.

Displaying Images

Digital displays are strongly preferred for image instruction as they tend to offer higher resolution, are easily updated, and are the method used in clinical practice (opposed to long obsolete physical films). Projected images from lecture slides are usually sufficient to convey introductory and broad topics, but computer monitors offer better definition for optimal viewing of subtle details.

Implementation

There are many ways in which medical imaging can be incorporated into anatomy courses, depending on the overall goals, available images, and comfort of instructors with imaging material.

Lectures

The most traditional method of incorporating radiology into anatomy, lectures, can include medical images interspersed alongside illustrations and gross anatomy photos in existing lecture slides, or exist as free-standing radiology lectures devoted to medical imaging. For example, the use of text, drawings, and gross dissection can be used to explain the branching schematic of arteries. Then medical imaging (e.g., CT angiogram, which isolates the arteries from surrounding structures) can be used to further depict this topic while providing a clear clinical correlation. Lectures can also take place in the anatomy laboratory (although this requires screens in the laboratory). These targeted lectures can focus on imaging concepts related to the dissection of the day, rather than global concepts which can be harder for students to contextually incorporate. These shorter, more focused lectures can be given to smaller groups of students, inviting questions and active learning. Although traditional large group format lectures have been described as the least useful form of learning radiology by recent medical school graduates [16], they remain a straightforward and efficient means to cover important information.

Online Lectures

The same lectures that would otherwise be given in person can be recorded and uploaded in a format accessible to students [42, 64–69]. This method is optimal if high-quality radiology images are available for students to download or access electronically, either piecemeal or organized into a slide deck. Ideally, these slide decks have both unlabeled and labeled images, so students can practice identifying structures without the immediate aid of an instructor [17, 70]. Webb and Choi [70] provide some general tips for organizing these lectures. Prerecorded lectures are also available through resources listed in Table 38.2.

Table 38.2 Online resources for acquiring, storing, editing, and teaching with medical imaging

Resource name and access	Cost	Features/functionality
Diagnostic imaging pathways http://www.imagingpathways.health.wa.gov.au	Free	Labeled and unlabeled searchable images, basics of imaging modalities
JET-EM https://jetem.org	Free	Peer-reviewed online journal of emergency medicine curricula
MedEdPORTAL https://www.mededportal.org	Free	Peer-reviewed online journal of curricula, organized by topic, sponsored by the AAMC. Requires free login credentials. Contains imaging tutorials, cases, and online lectures
MedPix https://medpix.nlm.nih.gov/home	Free	Database of medical images, teaching cases, and clinical topics, searchable by structure, organ, or disease
NetAnatomy http://www.netanatomy.com	Subscription	Learning software combining gross and radiological anatomy
Openi https://openi.nlm.nih.gov	Free	Searchable database of images and figures from all open access journals on PubMed
OsiriX https://www.osirix-viewer.com	Subscription	Software for accessing, storing, and viewing images in both 2D and 3D
Radiopedia https://radiopaedia.org	Free	Online wiki for all things radiology including plethora of images
RSNA teaching file system http://mirc.rsna.org/query	Free	Approachable case-based images to allow for clear clinical correlates between anatomy and radiology; the RSNA website has further information for teaching a radiology curriculum (https://www.rsna.org/education/educator-resources)
STATDx https://www.statdx.com	Subscription	Labeled and unlabeled searchable images, cases, and anatomy modules
Smart Sparrow https://www.smartsparrow.com	Subscription	E-modules including lectures, images, and cases

Small Groups

In general, interactive small groups are favored over lectures by both instructors and students [16, 71, 72]. These can occur during dissection laboratory, with instructors or teaching assistants going from table to table, pulling up images relevant to the current stage of dissection. In this way, students can manipulate their dissections while directly drawing correlates to imaging. Alternatively, small group learning can take place outside of dissection laboratory, with instructors guiding students through cases, problem-based learning, or flipped classrooms, all of which can allow for a deeper and more individualized appreciation for the relationship between medical imaging and gross anatomy [29, 71–75].

Cases can be completed by students either individually or in small groups (the latter being a learning strategy highly recommended by recent medical school graduates [16]), with groups ideally moderated by both anatomists and physicians. These clinical cases would include basic clinical symptoms, leading to an investigation of how surface and internal anatomy play significant roles in both the physical examination and symptomatology. Medical imaging can serve as the link between gross anatomical concepts and how they're used in the clinical context. Table 38.2 identifies some online sources where cases can be found, and McEntee [53] summarizes some useful online resources as well.

E-Modules

Several courses have had success with online e-modules [42, 64, 70]. These modules can combine lectures, case studies, and questions in a prescribed order and format. Completed either individually or in groups, these e-modules can be an effective, flexible, broad way for students to engage with desired material. Trelease and Rosset [76] have created a guide with suggestions of how to develop these interactive online modules for radiology.

Virtual Dissection

"Virtual dissection" uses three-dimensional renditions of computed anatomy to simulate gross dissection. These renditions can be based off of medical imaging techniques, making them hyper-realistic, allowing students to dissect renditions of cadavers at their own pace. This has the potential to avoid the practical limitations and learning curves associated with traditional dissection, yet still gaining many of the same conceptual benefits of cadaveric dissection [48, 57, 76–80]. Some virtual dissection programs even have haptic and tactile feedback for learners [4].

Testing

In order to be most useful, incorporating medical imaging in anatomy courses should extend to the inclusion of imaging into student assessment [48, 81, 82].

Creating Examination Questions

Questions can ask students to identify structures on images themselves (e.g., have a star or arrow designating the descending aorta on a chest X-ray), either by multiple choice or written response. Building from that, direct integration can be tested by asking students to identify the correct pin from several placed in the cadaver with a medical image being the question stem. For example, the inferior vena cava (IVC) can be starred on an image, and students are asked to identify the corresponding pin from several adjacent structures in the thorax of a cadaver, such as the ascending aorta, IVC, and esophagus, testing students' ability to correlate two-dimensional image interpretation with the physical three-dimensional cadaver. Combining two different orientations is especially effective in testing a deeper level of understanding (e.g., using axial CT slices compared to a supine cadaver), limiting the ability to use short-term recall [83]. Furthermore, students can be asked to give short imaging-related presentations if preferred, covering concepts taught in the course applied to normal variation or pathology found in their cadavers.

The ability to identify structures on imaging has been deemed an incredibly useful skill by young physicians [16], yet it is important to realize that simple recitation of "what is this" fails to test deep understanding or appreciation of imaging's relationship to gross anatomy. For example, asking a question about the aortic arch using a sagittal or axial medical image will require more than simply recognizing the walking stick appearance appreciated on a supine cadaver. This is a variant of Bloom's taxonomy and allows course directors to utilize imaging to its greatest advantage by requiring students to function beyond the simplest of levels of recognition [16]. By asking higher-order questions which have been shown to be linked to higher integrated learning and retention of knowledge, students are required to operate at a deeper understanding of anatomy, improving their comprehension and long-term retention.

Resources for finding example questions and writing your own can be found in Table 38.2, with a brief overview of writing effective multiple-choice questions found in JET-EM [84].

Administering the Examination

There is no empirical evidence for the best testing practices specific to medical imaging in anatomy. As such, testing can follow the format of the course as a whole (e.g., administered on paper, electronically, or orally). However, based solely on experience, we recommend presenting test images on screens, whether desktop monitors for small groups or lecture screens with lights dimmed for large groups, in as high quality of a view as possible. Paper printouts are difficult to interpret and introduce an element of ambiguity, and students are more likely to have studied the material using electronic displays. As such, the radiology portion of the examination may need to be administered in a separate location from other examination portions in order to use projectors or screens as needed.

Grading

Past experience has shown that if radiology material is a small portion of the assessed material, students choose not to study imaging materials, reducing the benefits of combining radiology with anatomy. Learner motivation can be improved by having radiology occupy a larger portion of the examination and/or requiring students to separately pass the radiology portion of the examination, independent of their performance on other sections.

Formative Assessment

Prior to testing, many instructors and students find it useful to complete imaging-related practice questions. These are most effective if their format resembles how they will be tested and the depth of content aligns with their academic and career interests. Ideally, these practice questions will involve different images from those used on examinations, minimizing the use of short-term memorization.

Practical Notes for Implementation

In-Person Assistance

The in-person teaching described above does not need to be done solely by the existing course instructors. Students who have taken the course in prior years, either MD or PhD anatomy graduate students, can be irreplaceable assets as teaching assistants [34], as can radiologists and other specialists (e.g., cardiologists, surgeons, etc.) [39]. Physicians in the dissection laboratory or small groups can answer ad hoc questions as they arise. In our experience, teaching assistants (TAs) perform best when they receive preparatory lessons before each module, especially if they are upper-level preclinical students. TAs can also be involved in actively creating, updating, and maintaining course content, either through their roles as TAs or through education-centered research projects [48, 85]. See Chaps. 5 (Engaging Residents and Clinical Faculty in Anatomy Education) and 6 (A Significant Role for Sessional Teachers in the Anatomy Education Landscape) (tentative titles) for engaging students, residents, and physicians as teachers.

Access to Images in the Dissection Laboratory

The use of radiology images in the anatomy laboratory can enhance the cadaveric dissection experience, and vice versa. Having students examine images prior to starting a dissection (e.g., cross-sectional abdominal CTs prior to the abdominal dissection) gives them the opportunity to think about and predict how the gross anatomy will be arranged, as well as what anatomical relationships to expect, a skill required by surgeons and other proceduralists [58, 60]. Additionally, having images tableside, or even imbedded into an electronic dissector manual, can boost learning of gross anatomy by allowing students to directly compare two-dimensional imaging to the three-dimensional anatomy they are dissecting [49, 86, 87]. Furthermore, should the anatomy laboratory have access to fixed cross-sectional slices of cadavers, direct correlations can be made between these cross-sectional slices and corresponding CT cross sections [17].

All of these comparisons are most successful if there are electronic displays (e.g., monitors, tablets, projectors) close to the dissection table that allow students to scroll through images as needed. A number of papers have overviews of hardware logistics, including the use of computers, tablets, and phones to view imaging, in case these devices are not currently available in the dissection laboratory [15, 48, 57, 88].

Integration into Subsequent Courses

The best way to achieve long-term retention of anatomy concepts is to integrate new information into familiar knowledge. In the case of anatomy, medical imaging is the link which can extend anatomy well beyond its designated course [34, 46, 47, 51]. Radiology images have a natural place in nearly any use of case- or problem-based learning in a medical curriculum [73–75] and can allow students to apply their anatomy knowledge subsequently in new avenues. Additionally, a radiology-based anatomy course can be integrated into the clinical years, allowing students to return to anatomy concepts as they begin to apply them to clinical contexts.

Economic Costs

There is an upfront investment to newly incorporating radiology images into an established anatomy course. At the very least, time must be devoted to developing course materials. An investigation done in 1994 estimates that 200 hours of work went into creating and supporting a new radiology course for third-year medical students, split among instructors, administrators, and ancillary staff; this time dropped considerably during subsequent offerings of the class, as updating materials requires less time than building them from scratch [89]. Furthermore, anatomy instructors will not need to develop an entirely new course; adding radiology materials to an existing course will be much more efficient. Even more efficiency can be achieved by collaborating with radiologists, who are likely to have existing resources that will not need to be newly developed.

Depending on existing institutional resources, instructors may need to purchase access to images, viewing software, online resources, and/or hardware (e.g., computers or tablets that allow for viewing images in dissection laboratory). The costs of these vary widely depending on the existing home institution resources. Many of these resources must be purchased only once before lasting years (or in the case of images, essentially indefinitely), requiring little or no maintenance. As such, the largest economic costs should be expected while developing the medical imaging materials, with significantly lower costs in subsequent years. Furthermore, the costs to developing a medical imaging curriculum in anatomy are similar to general course costs within medical school [90].

It is also possible to incorporate medical imaging into the anatomy course with only the expense of time, using the free resources aforementioned and having students use their own electronic devices to view the images. It is reasonable to start with free images in lectures and free at-home study modules and build from there if the course does not currently have any imaging. Doing so can also make the case to include medical imaging in the budget for future years.

Although institutions are often initially resistant to change, sometimes providing short-term pushback before objective improvements in test scores or other metrics has been achieved [30], the benefits of incorporating radiology into anatomy should outweigh the costs. Every institution is unique, as are their financial arrangements, making it difficult to provide general recommendations to address these costs. However, we hope that the benefits of incorporating radiology into anatomy courses allow for the procurement of pilot grants, resource reallocation, or other forms of funding.

Collaborating with Radiologists and Other Experts

Of course, the more familiar course instructors are with medical imaging, the easier it will be to incorporate imaging into an anatomy course and, in general, the higher quality the content will be [48]. In fact, recent graduates tend to prefer medical imaging to be taught by practicing radiologists [91], and collaborations between anatomists and radiologists have been shown to be mutually beneficial and optimal for teaching both anatomy and radiology [5, 47]. Although there have been recent trends of decreasing rates of collaboration with radiologists [10, 92], there exist national movements in the field of radiology to increase the presence of radiologists in anatomy courses [93]. Ideally, a radiologist will be able to join the course faculty, creating a vertical integration of radiology throughout the course. Their expertise and imaging resources (e.g., image banks, software, questions, cases, lectures, and other collections through professional society memberships) are likely invaluable to anatomy courses.

One persistent barrier to radiologist collaboration with anatomists is the lack of financial, promotional, and academic credit for teaching anatomy at many institutions [5, 92]. Although this is unlikely to be changed by an anatomy course instructor, contacting radiology departments can help aid in national movements to address this shortcoming [93]. In the absence of a designated medical education liaison in the radiology department, do not hesitate to reach out to the departmental chairperson; the lack of a formal education position is often the result of a lack of awareness for the need of preclinical curricular collaboration with radiologists. Many radiologists remain unaware of shifts in expectations within education programs and the ability to use resources shared among radiologists, available through their professional societies, for the education of broader audiences. If needed, medical school curricular leaders associated with anatomy (e.g., Chairs of Anatomy, Medical School Deans) can be effective in championing radiologists' or other clinical physicians' recruitment on instructors' behalf.

It should be noted that active collaboration with the radiology department is more important than access to images alone, since this collaboration can be immensely helpful to the anatomy course moving forward. Although direct institutional support is useful in developing this collaboration, it is not an outright necessity for all course components. As such, believing a change on this level is an all-or-none concept would be a disservice.

Conclusion

Medical imaging is an incredibly powerful tool that can enhance nearly every aspect of an anatomy course. It provides an additional medium with which to convey topics, requires a different type of spatial manipulation that deepens learning, offers a direct clinical correlation that can increase motivation for learning anatomy, and improves short-term and long-term learning and utilization of anatomical knowledge. Medical images can be obtained through collaborations with institutional radiologists or through online resources and can be incorporated into anatomy courses in a wide variety of ways. These collaborations with radiologists can prove to be the most meaningful, as their radiological and anatomical expertise is unique and irreplaceable. We highly recommend formally including medical imaging in anatomy courses based on extensive literature findings and personal experience.

References

1. Johnson TH. Medical school radiology teaching and examination methods. Radiology. 1969;93(2):443–6.
2. Steele JP, Homan JM, American College of Radiology Committee on Audio Visual Media. Roentgen anatomy of the normal bones and joints. JAMA. 1971;218:1579.
3. Stassa G, Evans JA. Radiology, anatomy, and the medical student. Radiology. 1969;92(7):1562–3.
4. Shaffer K. Teaching anatomy in the digital world. N Engl J Med. 2004;351(13):1279–81.
5. Phillips AW, Smith SG, Straus CM. The role of radiology in preclinical anatomy: a critical review of the past, present, and future. Acad Radiol. 2013;20(3):297–304.e1.
6. Rahim S, Ros P. Moving away from spoon-feeding as a teaching style in radiology. AJR Am J Roentgenol. 2016;207(6):1232–8.
7. Schober A, Pieper C, Schmidt R, Wittkowski W. "Anatomy and imaging": 10 years of experience with an interdisciplinary teaching project in preclinical medical education – from an elective to a curricular course. Rofo-Fortschr Rontg. 2013;186(05):458–65.
8. Murphy KP, Crush L, O'Malley E, Daly FE, Twomey M, O'Tuathaigh CMP, et al. Medical student perceptions of radiology use in anatomy teaching. Anat Sci Educ. 2015;8(6):510–7.
9. Korner N. Diagnostic radiology in the medical curriculum. Med J Aust. 1973;1(12):605–11.
10. Bassett LW, Squire LF. Anatomy instruction by radiologists. Investig Radiol. 1985;20(9):1008–10.
11. Squire LF, Novelline RA. Radiology should be a required part of the medical school curriculum. Radiology. 1985;156(1):243–4.
12. Novelline RA, du Cret RP, Weinberg EJ, Sellers TA, Seybolt LM, Kuni CC, et al. Role of radiology in medical education: perspective of nonradiologists. Acad Radiol. 1994;1(1):70–4.
13. Holt NF. Medical students need more radiology education. Acad Med. 2001;76(1):1.
14. Mirsadraee S, Mankad K, McCoubrie P, Roberts T, Kessel D. Radiology curriculum for undergraduate medical studies – a consensus survey. Clin Radiol. 2012;67(12):1155–61.
15. Vuchkova J, Maybury TS, Farah CS. Testing the educational potential of 3D visualization software in oral radiographic interpretation. J Dent Educ. 2011;75(11):1417–25.
16. Phillips AW, Eason H, Straus CM. Student and recent graduate perspectives on radiological imaging instruction during basic anatomy courses. Anat Sci Educ. 2018;11(1):25–31.
17. Moscova M, Bryce DA, Sindhusake D, Young N. Integration of medical imaging including ultrasound into a new clinical anatomy curriculum. Anat Sci Educ. 2015;8(3):205–20.
18. Bleakley A. Broadening conceptions of learning in medical education: the message from teamworking. Med Educ. 2006;40(2):150–7.
19. Croskerry P. Context is everything or how could I have been that stupid? Healthc Q. 2009;12(sp):e171–6.
20. Shapiro L. In: Shapiro L, editor. The Routledge handbook of embodied cognition. New York: Routledge/Taylor & Francis Group; 2014.
21. Robbins P, Aydede M. In: Robbins P, Aydede M, editors. The Cambridge handbook of situated cognition. Cambridge: Cambridge University Press; 2009.
22. Stanford W, Erkonen WE, Cassell MD, Moran BD, Easley G, Carris RL, et al. Evaluation of a computer-based program for teaching cardiac anatomy. Investig Radiol. 1994;29(2):248–52.
23. McLachlan JC, Bligh J, Bradley P, Searle J. Teaching anatomy without cadavers. Med Educ. 2004;38(4):418–24.
24. McLachlan JC, Patted D. Anatomy teaching: ghosts of the past, present and future. Med Educ. 2006;40(3):243–53.
25. Temkin B, Acosta E, Malvankar A, Vaidyanath S. An interactive three-dimensional virtual body structures system for anatomical training over the internet. Clin Anat. 2006;19(3):267–74.
26. Tam MDBS, Hart AR, Williams S, Heylings D, Leinster S. Is learning anatomy facilitated by computer-aided learning? A review of the literature. Medl Teach. 2009;31(9):e393–6.
27. Phillips AW, Smith SG, Ross CF, Straus CM. Improved understanding of human anatomy through self-guided radiological anatomy modules. Acad Radiol. 2012;19(7):902–7.
28. Lufler RS, Zumwalt AC, Romney CA, Hoagland TM. Incorporating radiology into medical gross anatomy: does the use of cadaver CT scans improve students' academic performance in anatomy? Anat Sci Educ. 2010;3(2):56–63.
29. Louw G, Eizenberg N, Carmichael SW. The place of anatomy in medical education: AMEE guide no 41. Med Teach. 2009;31(5):373–86.
30. May H, Cohen H, Medlej B, Kornreich L, Peled N, Hershkovitz I. Computed tomography-enhanced anatomy course using enterprise visualization. Anat Sci Educ. 2013;6(5):332–41.
31. Shaffer K. Teaching anatomy in the digital world. N Engl J Med. 2004;351(13):1279–81.
32. The Royal College of Radiologists. Structured Training Curriculum for Clinical Radiology. 2007. https://www.rcr.ac.uk/sites/default/files/docs/radiology/pdf/CURRICULUM%20CR%20FINAL%20January%202009%20incorporationg%20revised%20SI%20curricula.pdf. Accessed 15 Mar 2019.

33. Bergman EM, Vleuten CPMVD, Scherpbier AJJA. Why don't they know enough about anatomy? A narrative review. Med Teach. 2011;33(5):403–9.
34. Fillmore EP, Brokaw JJ, Kochhar K, Nalin PM. Understanding the current anatomical competence landscape: comparing perceptions of program directors, residents, and fourth-year medical students. Anat Sci Educ. 2016;9(4):307–18.
35. Prince KJAH, Scherpbier AJJA, Mameren HV, Drukker J, Vleuten CPMVD. Do students have sufficient knowledge of clinical anatomy? Med Educ. 2005;39(3):326–32.
36. Schiller PT, Phillips AW, Straus CM. Radiology education in medical school and residency: the views and needs of program directors. Acad Radiol. 2018;25(10):1333–43.
37. Subramaniam RM, Hall T, Chou T, Sheehan D. Radiology knowledge in new medical graduates in New Zealand. N Z Med J. 2005;118:U1699.
38. Erkonen WE, Albanese MA, Smith WL, Pantazis NJ. Gross anatomy instruction with diagnostic images. Investig Radiol. 1990;25(3):292–4.
39. Erkonen WE, Albanese MA, Smith WL, Pantazis NJ. Effectiveness of teaching radiologic image interpretation in gross anatomy. A long-term follow-up. Investig Radiol. 1992;27(3):264–6.
40. Erkonen WE, Vydareny KH, Sandra A, Ferguson KA, Kreiter CD. Interinstitutional study to compare the effectiveness of a radiology-anatomy module of instruction. Acad Radiol. 2000;7(9):700–4.
41. Glinkowski W, Ciszek B. WWW-based e-teaching of normal anatomy as an introduction to telemedicine and e-health. Telemed J E Health. 2007;13(5):535–44.
42. Ketelsen D, Schrödl F, Knickenberg I, Heckemann RA, Hothorn T, Neuhuber WL, et al. Modes of information delivery in radiologic anatomy education: impact on student performance. Acad Radiol. 2007;14(1):93–9.
43. Marker DR, Bansal AK, Juluru K, Magid D. Developing a radiology-based teaching approach for gross anatomy in the digital era. Acad Radiol. 2010;17(8):1057–65.
44. Rizzolo LJ, Rando WC, O'Brien MK, Haims AH, Abrahams JJ, Stewart WB. Design, implementation, and evaluation of an innovative anatomy course. Anat Sci Educ. 2010;3(3):109–20.
45. Shaffer K, Small JE. Blended learning in medical education: use of an integrated approach with web-based small group modules and didactic instruction for teaching radiologic anatomy. Acad Radiol. 2004;11(9):1059–70.
46. Heptonstall NB, Ali T, Mankad K. Integrating radiology and anatomy teaching in medical education in the UK–the evidence, current trends, and future scope. Acad Radiol. 2016;23(4):521–6.
47. Lazarus MD, Chinchilli VM, Leong SL, Kauffman GL. Perceptions of anatomy: critical components in the clinical setting. Anat Sci Educ. 2012;5(4):187–99.
48. Caswell FR, Venkatesh A, Denison AR. Twelve tips for enhancing anatomy teaching and learning using radiology. Med Teach. 2015;37(12):1067–71.
49. McHanwell S, Davies DC, Morris J, Parkin I, Whiten S, Atkinson M, et al. A core syllabus in anatomy for medical students - adding common sense to need to know. Eur J Anat. 2007;11(S1):3–18.
50. The Royal College of Radiologists. The undergraduate curriculum. 2012. https://www.rcr.ac.uk/clinical-radiology/specialty- training/curriculum/clinical-radiology-curriculum. Accessed 5 Feb 2015.
51. Naeger DM, Webb EM, Zimmerman L, Elicker BM. Strategies for incorporating radiology into early medical school curricula. J Am Coll Radiol. 2014;11(1):74–9.
52. Ward PJ, Walker JJ. The influence of study methods and knowledge processing on academic success and long-term recall of anatomy learning by first-year veterinary students. Anat Sci Educ. 2008;1(2):68–74.
53. McEntee MF. Web watch: radiographic anatomy on the web. Radiography. 2004;10(4):293–8.
54. Slon V, Hershkovitz I, May H. The value of cadaver CT scans in gross anatomy laboratory. Anat Sci Educ. 2014;7(1):80–2.
55. Buenting M, Mueller T, Raupach T, Luers G, Wehrenberg U, Gehl A, et al. Post mortem CT scans as a supplementary teaching method in gross anatomy. Ann Anat. 2016;208:165–9.
56. Bohl M, Francois W, Gest T. Self-guided clinical cases for medical students based on postmortem CT scans of cadavers. Clin Anat. 2011;24(5):655–63.
57. Murakami T, Tajika Y, Ueno H, Awata S, Hirasawa S, Sugimoto M, et al. An integrated teaching method of gross anatomy and computed tomography radiology. Anat Sci Educ. 2014;7(6):438–49.
58. Paech D, Giesel FL, Unterhinninghofen R, Schlemmer H-P, Kuner T, Doll S. Cadaver-specific CT scans visualized at the dissection table combined with virtual dissection tables improve learning performance in general gross anatomy. Eur Radiol. 2016;27(5):1–8.
59. Chew FS, Relyea-Chew A, Ochoa ERJ. Postmortem computed tomography of cadavers embalmed for use in teaching gross anatomy. J Comput Assist Tomogr. 2006;30(6):949.
60. Schramek GGR, Stoevesandt D, Reising A, Kielstein JT, Hiss M, Kielstein H. Imaging in anatomy: a comparison of imaging techniques in embalmed human cadavers. BMC Med Educ. 2013;13(1):143.
61. Tam MDBS. Building virtual models by postprocessing radiology images: a guide for anatomy faculty. Anat Sci Educ. 2010;3(5):261–6.
62. Rasband WS. ImageJ. Bethesda: US National Institutes of Health; 1997. https://imagej.nih.gov/ij/, 1997–2018. Accessed 26 Apr 2019.
63. Thapa M, Vaidya S, Dighe M, Richardson M. Creating radiology line art with photoshop tools and filters. In: MedEdPORTAL. Association of American Medical Colleges. 2011. https://www.mededportal.org/publication/8296/. Accessed 17 Mar 2019.

64. Khalil MK, Paas F, Johnson TE, Su YK, Payer AF. Effects of instructional strategies using cross sections on the recognition of anatomical structures in correlated CT and MR images. Anat Sci Educ. 2008;1(2):75–83.
65. Carriero A, Bonomo L, Calliada F, Campioni P, Colosimo C, Cotroneo A, et al. E-learning in radiology: an Italian multicentre experience. Eur J Radiol. 2012;81(12):3936–41.
66. Salajegheh A, Jahangiri A, Dolan-Evans E, Pakneshan S. A combination of traditional learning and e-learning can be more effective on radiological interpretation skills in medical students: a pre- and post-intervention study. BMC Med Educ. 2016;16(1):46.
67. Pinto A, Brunese L, Pinto F, Acampora C, Romano L. E-learning and education in radiology. Eur J Radiol. 2011;78(3):368–71.
68. Scalese RJ, Obeso VT, Issenberg SB. Simulation technology for skills training and competency assessment in medical education. J Gen Intern Med. 2008;23(1):46–9.
69. Sendra-Portero F, Torales-Chaparro OE, Ruiz-Gómez MJ. Medical students' skills in image interpretation before and after training: a comparison between 3rd-year and 6th-year students from two different medical curricula. Eur J Radiol. 2012;81(12):3931–5.
70. Webb AL, Choi S. Interactive radiological anatomy eLearning solution for first year medical students: development, integration, and impact on learning. Anat Sci Educ. 2014;7(5):350–60.
71. Belfi LM, Bartolotta RJ, Giambrone AE, Davi C, Min RJ. "Flipping" the introductory clerkship in radiology: impact on medical student performance and perceptions. Acad Radiol. 2015;22(6):794–801.
72. O'Connor EE, Fried J, McNulty N, Shah P, Hogg JP, Lewis P, et al. Flipping radiology education right side up. Acad Radiol. 2016;23(7):810–22.
73. Navsa N, Boon JM, L'Abbé LN, Greyling LM, Meiring JH. Evaluation of clinical relevance of a problem-orientated head and neck module. SADJ. 2004;59(3):113–7.
74. Subramaniam RM. Problem-based learning: concept, theories, effectiveness and application to radiology teaching. Australas Radiol. 2006;50(4):339–41.
75. Zumwalt AC, Lufler RS, Monteiro J, Shaffer K. Building the body: active learning laboratories that emphasize practical aspects of anatomy and integration with radiology. Anat Sci Educ. 2010;3(3):134–40.
76. Trelease RB, Rosset A. Transforming clinical imaging data for virtual reality learning objects. Anat Sci Educ. 2008;1(2):50–5.
77. Hisley KC, Anderson LD, Smith SE, Kavic SM, Tracy JK. Coupled physical and digital cadaver dissection followed by a visual test protocol provides insights into the nature of anatomical knowledge and its evaluation. Anat Sci Educ. 2008;1(1):27–40.
78. Petersson H, Sinkvist D, Wang C, Smedby Ö. Web-based interactive 3D visualization as a tool for improved anatomy learning. Anat Sci Educ. 2009;2(2):61–8.
79. Tam MDBS. Building virtual models by postprocessing radiology images: a guide for anatomy faculty. Anat Sci Educ. 2010;3(5):261–6.
80. Turmezei TD, Tam MDBS, Loughna S. A survey of medical students on the impact of a new digital imaging library in the dissection room. Clin Anat. 2009;22(6):761–9.
81. Raupach T, Brown J, Anders S, Hasenfuss G, Harendza S. Summative assessments are more powerful drivers of student learning than resource intensive teaching formats. BMC Med. 2013;11(1):61.
82. Wormald BW, Schoeman S, Somasunderam A, Penn M. Assessment drives learning: an unavoidable truth? Anat Sci Educ. 2009;2(5):199–204.
83. Phillips AW, Smith SG, Straus CM. Driving deeper learning by assessment: an adaptation of the revised Bloom's taxonomy for medical imaging in gross anatomy. Acad Radiol. 2013;20(6):784–9.
84. Jones JS, Phillips AW, King AM, Estes MK, Conlon LW, Scott KR. A Brief Didactic Intervention to Improve Multiple-Choice Item-Writing Quality. JETem. 2018;3(1). Retrieved from https://escholarship.org/uc/item/9z22m7d7.
85. Furmedge DS, Iwata K, Gill D. Peer-assisted learning – beyond teaching: how can medical students contribute to the undergraduate curriculum? Med Teach. 2014;36(9):812–7.
86. The General Medical Council. The good medical practice. 2013. http://www.gmc-uk.org/guidance/good_medical_practice.asp. Accessed 28 Jan 2017.
87. The General Medical Council. The trainee doctor. 2011. http://www.gmc-uk.org/Trainee_Doctor.pdf_39274940.pdf. Accessed 29 Jan 2017.
88. Gregory JK, Lachman N, Camp CL, Chen LP, Pawlina W. Restructuring a basic science course for core competencies: an example from anatomy teaching. Med Teach. 2009;31(9):855–61.
89. Smith W, Erkonen W, Hough D. Calculating the costs of undergraduate medical education in radiology. Investig Radiol. 1994;29:856–8.
90. Jones RF, Korn D. On the cost of educating a medical student. Acad Med. 1997;72(3):200–10.
91. Branstetter BF, Humphrey AL, Schumann JB. The long-term impact of preclinical education on medical students' opinions about radiology. Acad Radiol. 2008;15(10):1331–9.
92. Samuel S, Shaffer K. Profile of medical student teaching in radiology: teaching methods, staff participation, and rewards. Acad Radiol. 2000;7(10):868–74.
93. Chowdhury R, Wilson IDC, Oeppen RS. The departments of radiology and anatomy: new symbiotic relations? Clin Radiol. 2008;63(8):918–20.

39 Assessing Anatomy as a Basic Medical Science

Dujeepa D. Samarasekera, Eng-Tat Ang, and Matthew C. E. Gwee

> Assessment Drives Student Learning.
> –Author unknown

Introduction

The practice of medicine is undoubtedly driven by scientific knowledge [1]. Basic sciences (e.g., anatomy, biochemistry, pharmacology, and microbiology) are heavy in terms of content, and learners need to be properly assessed to ensure effectiveness.

D. D. Samarasekera (✉)
Centre for Medical Education (CenMED) Unit, Yong Loo Lin School of Medicine, National University of Singapore, National University Health System, Singapore
e-mail: dujeepa@nus.edu.sg

E.-T. Ang
Department of Anatomy, Yong Loo Lin School of Medicine, National University of Singapore, National University Health System, Singapore

M. C. E. Gwee
Centre for Medical Education (CenMED) Unit, Dean's Office, Yong Loo Lin School of Medicine, National University of Singapore, National University Health System, Singapore

What Is Assessment?

Assessment is essentially a systematic process (i.e., a strategy) to obtain valid and reliable evidence of student acquisition of learning outcomes, i.e., the knowledge (K), skills (S), and attitudes (A) in various domains of learning. Current thinking on assessment puts major emphasis on assessment that facilitates student learning, i.e., from the assessment of learning to the assessment for learning.

Why Is There a Need to Assess Student Learning?

The primary goal of assessment is twofold: firstly, assessment serves to maintain academic standards (achieved mainly through marking and grading), as commonly practiced in most institutions of higher learning. In such a context then, assessment will determine whether a student has acquired the relevant learning outcomes (K, S, A) to progress to the next phase of learning or clini-

> Students can, with difficulty, escape from the effects of poor teaching, they cannot escape the effects of poor assessment.
> –Boud [2].

L. K. Chan, W. Pawlina (eds.), *Teaching Anatomy*, https://doi.org/10.1007/978-3-030-43283-6_39

cal practice. However, assessment of students in medical and health professional education is also aimed at ensuring public safety. Therefore, such assessments also have national implications since the acquisition of K, S, and A should equip the graduates with necessary professional competencies that enable them to deliver healthcare as the new generation of competent practitioners (who must also be accredited by the national body responsible for maintaining practice standards).

Focus of Traditional Assessment Strategies

In a traditional learning environment, assessment strategies commonly designed by the disciplines focused more on testing *factual content* that imposed on students the need to "memorize, recall, and regurgitate" in examinations. Thus, assessments tested mainly the acquisition of *lower-order* learning outcomes through mere rote memorization and recall. Moreover, assessment scores were used primarily as final (summative) tests without considering their potential as a learning tool. A major concern of such assessment practices is the *negative steering effect* of assessment on student learning; i.e., students tended to undertake *superficial* or *rote learning* rather than *deep* learning with enhanced conceptual understanding. Thus, students are likely to become *knowledge-rich* but *application-poor* with respect to medical problem-solving and critical reasoning. Myers and Jones [3] have clearly expressed that "What matters…is not what students know but what they can do with what they know. What's at stake is the capacity to perform, to put what one knows into practice."

Assessment Strategies for Twenty-First-Century Medical Education

Today, however, there is clear recognition that assessments have excellent potential as learning tools since they will drive how students learn: "…assessment has a major impact on students' learning behaviour [and it is]…for the test maker to capitalize on this behaviour. …the driving influence of assessment is a powerful tool to ensure that students learn what, and how, teachers want them to learn" [4].

Consequently, there is now a major paradigm shift from the assessment of learning to assessment for learning. The latter is best achieved through providing students with *feedback* to explain the "rights" and "wrongs" of test items, preferably, on an individual basis in a reflective dialogue. Feedback is further enhanced through the use of *formative* assessments as *diagnostic* tools rather than as test instruments, which is consistent with the paradigm shift to drive student learning behavior through assessments [5].

Thus, today, assessments should not focus predominantly on student acquisition of *lower-order* learning outcomes based primarily on *rote memorization* of factual content knowledge. Instead, assessments should now focus more on the acquisition of *higher-order* learning outcomes to help students develop their intellectual skills, i.e., their ability to *analyze*, *integrate*, *evaluate*, and *apply* the foundational knowledge they have acquired. Myers and Jones [3] have stated that "Students learn not by just absorbing content (taking copious notes and studying for exams), but by critically analyzing, discussing and using content in meaningful ways."

Thus, major paradigm shifts in assessment practices have made it imperative to strategize the design of test instruments used in the assessment of students in health professional education. Instead of using instruments to test a student's ability to remember a large amount of factual content, testing should now be aimed at how students can *apply* the knowledge and skills which they have acquired to their next level of learning or to problem-solve in future professional practice [6, 7], including the relevant *domain-independent* skills such as communication, teamwork, and professionalism.

Rigorous Assessment in Anatomy Knowledge Is Essential

> Each department is responsible for some part of the education of a medical student, but no department should forget that it is no more than a part of the whole school which is responsible for the education of the whole student and the fulfilment of the overall objective. –Miller, 1961

The status and relevance of anatomy in twenty-first-century education to medical students have recently been reaffirmed in authoritative reports such as the General Medical Council UK Tomorrow's Doctors, Joint Publication of the Howard Hughes Medical Institute and the American Association of Medical Colleges, and the Association of Faculties of Medicine, Canada.

Anatomy is the study of the structure of the human body and associated body functions and, therefore, provides fundamental core knowledge in medical, nursing, and other health professional training programs. Thus, students need to acquire a solid *foundational knowledge* of anatomy as a BMSc which would subsequently enable them to apply the knowledge they acquired to what they will learn about associated structural abnormalities occurring in disease processes and also to patient management. In view of the highly critical role of assessment in medical and health professional education, especially the public safety aspects, knowledge of anatomy must also contribute to the education, training, and transformation of today's students to become tomorrow's practitioners who are fit to deliver twenty-first-century healthcare. In other words, anatomy must be considered an essential educational "building block," together with many from other disciplines, required for the transformation of students to become competent healthcare practitioners. Thus, a rigorous assessment process in anatomy will optimize its contribution to the overall educational "building blocks" which the students must acquire and be equipped with to become competent practitioners

Recommended Best Practices for Anatomy Assessment

> Everything that can be counted does not necessarily count; everything that counts cannot necessarily be counted. –Albert Einstein

Constructive Alignment in Anatomy Assessment

For any assessment process to be successful, there is a need to ensure that test items are closely aligned to the expected learning outcomes [8] as well as the need to have a clear understanding of the primary goal(s) of assessment.

Anatomy is taught mainly in the early part of the medical course. The assessment modalities used in anatomy examinations must, therefore, be constructively aligned to the expected course outcomes, i.e., for learning to be effective [9]. Figure 39.1 outlines how best to achieve this close alignment through proper planning in the development of an assessment process.

The close alignment of assessment to learning outcomes (i.e., to requisite future practice competencies) will have a strong *positive* steering effect on students' learning behaviors [10]. The best practice in constructive alignment is to develop an *assessment blueprint* (see Table 39.1) which allows the faculty to have a bird's-eye view of the discipline-specific outcomes. It also serves to inform test developers, teachers, and students how different assessment instruments test the overall learning outcomes. Furthermore, it helps to conceptualize the *level* of testing in the cognitive (knowledge) domain using Bloom's taxonomy. A blueprint will facilitate the development of items that test *higher-order* cognitive skills and hence steering students away from *rote learning* [11].

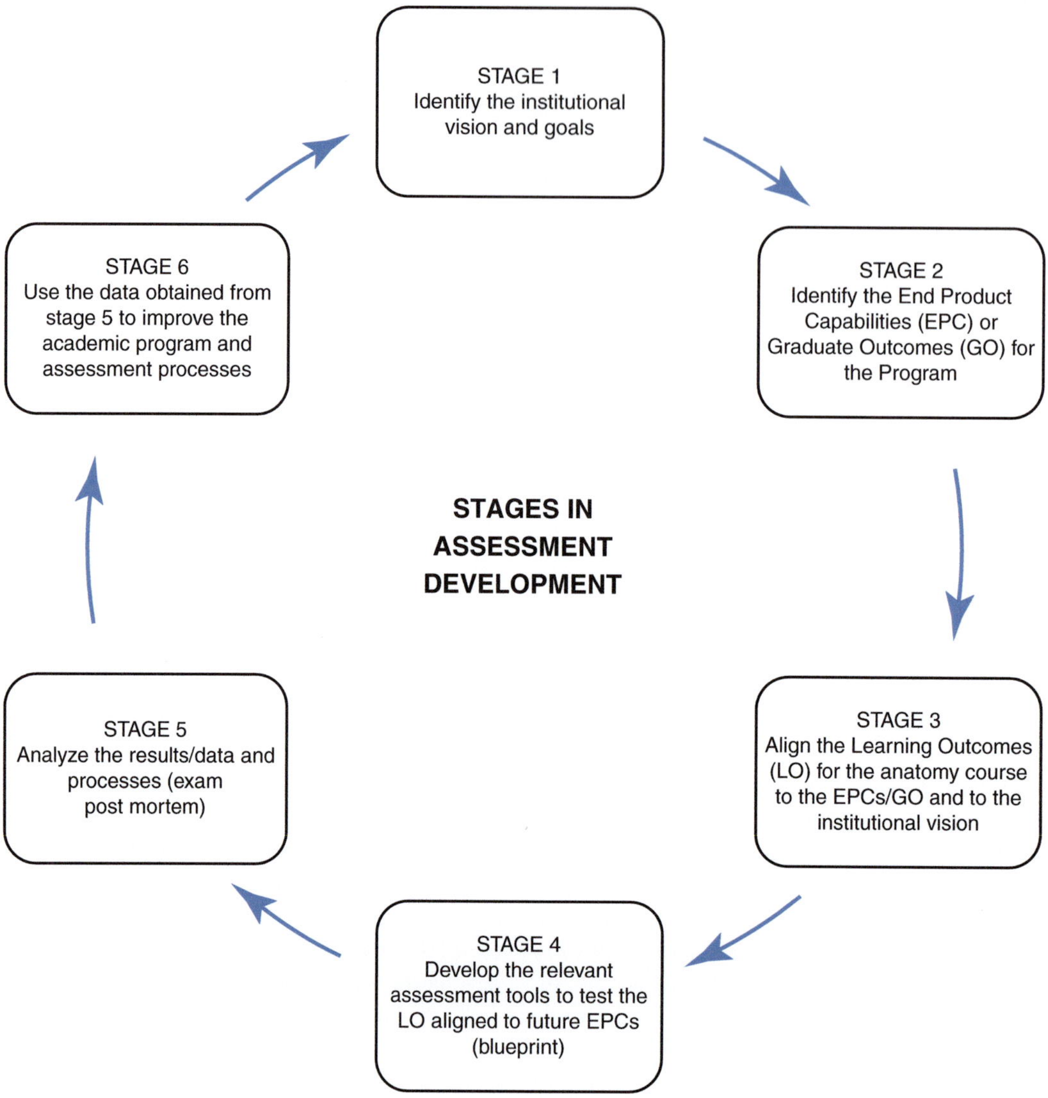

Fig. 39.1 Stages in the development of an assessment process

However, anatomy, like the other biomedical sciences, is a highly content-based discipline. Thus, assessment has often focused on the acquisition of factual content knowledge in the cognitive (i.e., knowledge) domain of learning. Today, assessment of anatomy in the cognitive domain must also assess the intellectual skills of students, which is often achieved through regular practice in knowledge processing (i.e., how students analyze, integrate, evaluate, and apply the knowledge acquired to resolve a problem or issue posed). This would further facilitate the constructive alignment of anatomy learning to program/course outcomes. Thus, the design of such assessments should be based on a future practice or a clinical scenario that poses a problem, as the stem in multiple-choice questions (MCQs). Regular practice in knowledge processing can also be expected to facilitate knowledge retrieval, knowledge transfer to various new and novel situations (as in many clinical contexts), as well as to the progressive development of students' intellectual skills—aimed primarily at enhancing the development of critical thinking and clinical rea-

Table 39.1 Example of a final assessment blueprint for an undergraduate course in anatomy

Outcome capabilities modules	Knowledge				Skills (basic clinical skills/ anatomical specimens)	Professionalism	Teamwork	Communication
	recall/ comprehension	Application	Analysis	Evaluate				
Musculoskeletal	MCQ	MCQ		MEQ	MSE		MSE	
Respiratory		MCQ	MCQ/ MEQ					
Cardiovascular		MCQ	MEQ		MSE			MSE
GIT	MCQ	MCQ/ MEQ	MCQ		MSE	MSE		
Neuro/CNS	MCQ	MCQ	MCQ	MEQ	MSE			
Genitourinary		MCQ	MCQ					
Reproductive		MCQ						
Special senses	MCQ	MCQ			MSE			

MCQ multiple-choice questions, *MEQ* modified essay questions, *MSE* multi-station examination, e.g., objective structured practical/clinical examination (OSP/CE)

soning skills which will, ultimately, facilitate the application of knowledge acquired in resolving patient problems that will be encountered in future practice. The regular assessment of knowledge-processing skills will, therefore, serve as educational preparation of the students for future medical problem-solving and medical decision-making.

The design of assessment strategies in anatomy should therefore focus less on testing factual content knowledge and more on testing the intellectual skills (ability) of students to process the knowledge acquired.

Utility Index of an Assessment System

The *utility index*, described by Cees van der Vleuten in 1996, provides a good foundation to choose and incorporate different assessment instruments when assessing students [12]. The formula is more conceptual than mathematical, highlighting the fact that no single assessment instrument can achieve high utility if used alone. This is because each instrument has its own limitations and strengths with regard to validity, reliability, educational impact, cost (and feasibility), and acceptability by the stakeholders. However, if the assessment planning is done carefully and incorporates multiple instruments strategically into the curriculum at different time points of student learning, the data gathered from the instruments will provide useful information about the learner. This will assist to build a profile of the student's abilities and for examiners to decide on whether the learner is ready to progress to the next level of learning [13].

Assessment Utility Index

U = Utility index
R = Reliability
V = Validity
E = Educational impact
A = Acceptability
C = Cost
W = Weight

$$U =_{w_r} R \times_{w_v} V \times_{w_e} E \times_{w_a} A \times_{w_c} C$$

Validity

Validity is defined as the extent to which the instrument measures the competency which it is supposed to measure [14]. Validity can be affected by several factors, including: poor alignment of test instruments to the competencies and domains assessed, confusing instructions to assessors and students, poor test development leading to low-quality test items/questions, developed test items/questions too difficult or too easy for the students, and poor sampling of testable content/domain areas.

The *validity* of an assessment can be described under different categories or types, including:

- *Content Validity*: This refers to whether the assessment actually tests the intended subject area.
- *Construct Validity*: This refers to the extent to which the assessment instrument measures a particular behavior or trait (e.g., 360° evaluation instrument can measure traits such as *teamwork* and *professionalism*).
- *Predictive Validity*: This refers to whether student performance on an assessment instrument could predict his/her future performance.
- *Concurrent Validity*: This refers to how well the performance in one test correlates with other validated tests administered at the same time.
- *Face Validity*: This refers to stakeholders' views and perceptions on whether the test measures what it is supposed to measure and the fairness of the assessment.

Reliability

Reliability refers to the consistency or reproducibility of student performance. If the assessment is unreliable, then the student's score may vary widely, depending on whether other factors are also aligned to the intent of the examination. The reliability can also be affected by several factors such as efficiency of the test administration, adequacy of the sampling of the content/domain areas tested, and objectivity of scoring by assessors as a consequence of a lack of or poor assessor training.

Reliability also consists of a few categories as outlined below:

- *Test-Retest Reliability*: This refers to whether a student's performance in an examination will be similar if the test is administered repeatedly at different times.
- *Inter-rater Reliability:* This refers to the concurrence of scoring among assessors on a single performance.
- *Split-Half Reliability*: This refers to the internal consistency of an assessment. The process involves dividing the examination questions/items into two halves which assess the same knowledge or domain skills. The examination is then administered and the scores for each half are obtained and correlated. The *split-half correlation* is considered reliable if the scores of the two sets correlate positively.

Educational Impact

One of the most important aspects of assessment is its *relevance* to and its *impact* on student learning. Assessments could be used in a strategic way to drive students to learn what is important in the curriculum and for their future learning and practice. Incorporating this constructive alignment of learning relevance and future practice relevance to assessment is also called the *consequential validity* of assessment [15].

If the test item in the anatomy MEQ examination uses contextualized future practice clinical scenarios, then the students, during their learning, will focus on applying knowledge of the gross anatomical structures and the associated functions which they learned to applied clinical anatomy. The teachers can also be expected to emphasize knowledge application to students during their teaching-learning activities.

Acceptability

Acceptability and the perceived fairness by the stakeholders, i.e., teachers, students, administrators, professional and employing bodies, patients, and the communities, regarding the robustness of the assessment and its processes are also important. This forms the basis of the trust placed by these stakeholders with regard to the graduate's effectiveness as a practitioner.

Cost/Feasibility

Another factor that needs careful attention when developing an assessment instrument or a process is the cost of development or the overall feasibility of employing the instrument or the process.

Effective Use of Formative and Summative Assessments

As discussed, assessment should drive student learning behavior, not only to develop higher-order discipline-specific cognitive skills but also to prepare students to acquire competencies for future professional practice. However, this cannot be achieved using a single assessment instrument. The students need to be closely supervised and guided on how to develop their skills and gain knowledge and also to be given feedback on how best to further improve/develop through the learning program. This is best achieved using formative assessments (FAs). See the relevant chapter on FA in this book. On the other hand, summative assessments (SAs) are used to evaluate whether students have reached the required level of competency, so that they will be effective learners at the next level of learning, safe to patients, and contributing to team care when applying their acquired skills in practice as healthcare professionals.

Effective Use of Feedback in Assessment

Feedback is the cornerstone of assessment, and in contemporary health professional education, it is becoming one of the critical components of learning. Providing relevant, focused, and immediate feedback on student performance assists learners to identify areas for improvement and enhances their individual areas of strength.

A typical feedback setting involves a face-to-face verbal discourse between the teacher and student. Students usually get feedback on their skills assessments, whereas only a score is often provided for written assessments. However, increasingly more elaborate feedback is also provided on written assessments, especially for MCQs through the use of technology.

The computer-based online feedback can provide more granular feedback regarding the students' performance by comparing the individual's score relative to the cohort as well as for each content area tested. It can further benchmark the student performance to local or international cohorts taking the same examinations.

For face-to-face feedback, the most commonly used format is the modified *Pendleton's* feedback model. The model not only identifies areas of strength and areas for improvement but also improves students' *metacognition* and *empowers* learners to be more self-directed in their learning [16].

Modified Pendleton's Feedback Model [16]

- Ask the learner what went well and why.
- The teacher says what went well and why.
- Ask the learner what can be done better and how.
- The teacher says what can be done better and how.
  - Summarize strengths and list three things to concentrate on.
  - How you (or your colleagues) could help the learner in achieving the above.

Assessment Formats in Anatomy

Written Assessment

Written assessments are used mainly to assess in-depth knowledge acquisition: the lower-order outcomes of cognitive or knowledge-processing skills of students, relating to the "knows" and "knows how" levels in the modified Miller's pyramid of competence shown in Fig. 39.2 [17].

Multiple-Choice Question

MCQs are the most common assessment used in medical and health professional programs for anatomy [18]. If the test items are developed properly (see section on various developmental stages) and constructed using blueprinting, MCQ can achieve high psychometric qualities such as high test-retest reliability, content validity, and educational impact [19]. MCQs which are context-rich with a single best response are widely used today. The other formats (e.g., true/false) have now been largely phased out, whereas the extended matching items are increasingly gaining popularity.

Modified Essay Questions

Modified (or structured) essays are commonly used to assess the students' *knows how* level of learning. In most instances, these are best used to assess in-depth knowledge of a particular content area in anatomy. The disadvantage of this format is that it cannot sample widely, and hence, the reliability is generally low.

Long Essays

Long essay questions are no longer used widely today, due to the expected low reliability when testing a subject area like anatomy. The long essays are now limited for use only in special circumstances where students need to develop or synthesize a response to a question relating to a given situation or condition.

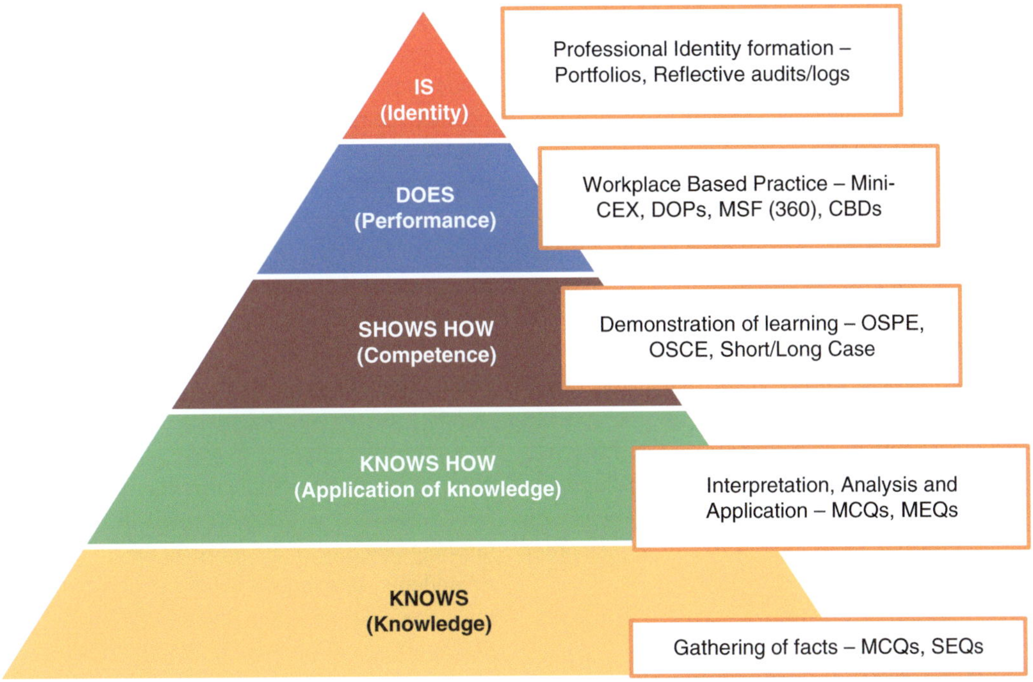

Fig. 39.2 An assessment framework using modified Miller's pyramid of competence [17]

Assessing Skills

Multi-station Examinations (MSE, e.g., Objective Structured Practical/Clinical Examination, OSPE/OSCE)

The most common format used to assess skills is the MSE such as the OSPE/OSCE. The use of MSE to assess relevant skills acquired using task trainers and simulated participants (SPs) with checklist-based scoring is a popular strategy used in most medical and health professional training programs. The advantage of this format is that it allows assessment of multiple domains in a single test. The domains could be applied knowledge—radiological images or anatomical specimens, clinical case vignettes, audiovisual materials, communication skills stations using SPs, etc. See blueprint shown in Table 39.1. Therefore, the MSE, if developed properly and based according to the overall assessment blueprint, can have high validity and educational impact. However, the main disadvantages are that it requires a large amount of resources to develop as well as to conduct and poor reliability due to low sampling of subject content area.

Spot Tests

Anatomy *spot* tests are used for quick identification of anatomical structures. The students are presented with several prosected specimens bearing anatomical structures such as muscles, nerves, blood vessels, etc., which are highlighted for students to identify. The identification of these structures may lead to further questions about their functions or the clinical presentations of their dysfunctions. The major disadvantage of this assessment strategy is that it promotes a negative steering effect (i.e., *rote memorization* and *recall* of knowledge), and it has largely been replaced by MSE.

Oral Examination

Oral (or viva voce) examinations were used extensively in the past in medical and health professional training programs. Oral examinations (OEs) have been phased out in many assessment settings due to their apparently poor reliability. However, recent studies show that if the OE is well structured (using a blueprint for specific item development), examiners are briefed and trained, and scoring is based on a marking template; the OE can achieve good reliability as well as identify the learner's abilities in higher-order cognitive skills [20].

Portfolio Assessment

The use of *portfolios* to assess student learning during early years is now gaining in popularity. The portfolio not only assesses different *cognitive* skills such as reflective learning but also evaluates some areas of affective skills [21]. The students need to show achievement of required outcomes in these domains and provide evidence of their learning at multiple time points of the program. Feedback regarding their performance is given by faculty who have been trained to evaluate portfolios. The main advantage of using this modality of assessment in the early part of the curriculum is that the faculty can evaluate, over different phases, how the learners are developing in a discipline and, consequently, can provide immediate feedback for remediation whenever required. Portfolios are also used as summative assessments, especially to make value judgments and high-stakes decisions on a learner's ability to move to the next level of learning or to certify that a learner is fit for practice.

Assessments must also test, not only what students learn but also how they learn: this is now considered an important aspect in assessment because the assessment of factual content knowledge regularly and predominantly will lead to student development of learning habits focused on fact memorization, similar to the traditional practice in which students "… sit, listen, take notes, and then memorize, recall and regurgitate in examinations (followed quickly by knowledge fading)." Such a practice is likely to encourage rote learning habits (i.e., superficial learning) in which learners will find difficulty in connecting related ideas and facts into conceptual frameworks that promote deep learning with understanding of the subject matter to be learned. It becomes imperative then that, today, the assessment of anatomy should focus less on fact memorization (which is likely to result in lower-order learning outcomes) and more on the development of students' intellectual skills to become competent medical problem-solvers and decision-makers. Use of a portfolio allows the learner to document and reflect one's own learning and also for the teacher to identify learning gaps in K-S-A domains so that proper guidance can be given to improve student performance. The process contributes to and enhances student's professional identity formation.

Programmatic Assessment

Programmatic assessment has recently been advocated [22]. This involves a paradigm shift from just one (single point) summative assessment to a program of several continuous (multi-point) assessments which can include testing in other domains of learning, such as "psychomotor skills" as well as "attitudes." A student's competence is determined, not from a single test, but by an expert panel going through all assessment points and making a value judgment on the student's performance. Such a paradigm shift can be expected to enhance the reliability of assessments since a large sampling of test items is involved. This approach leverages on student portfolios to document and track examination performance.

Standard Setting in Assessment

Deciding on whether a learner has achieved the required competency, based on his/her performance in tests, is an important process in deciding whether or not a learner is ready for the next level of learning or professional practice. The score deciding whether a student can pass or fail should be based on relevant and context-specific educational rationale [23]. According to Norcini "A standard is a special score that serves as a boundary between those who perform well enough and those who do not" [24]. Broadly speaking, there are two types of assessment standards: relative (also known as norm referenced) and absolute (fixed or criterion referenced) [24]. In *relative* standards, a student's performance is compared with the performance of other students in the cohort, and the pass/fail mark is then set accordingly. For example, this method is used widely on entry tests to programs where the numbers of student vacancies are limited. The advantage is that you can limit the number of passes and failures based on the requirements of the school's program. However, the main disadvantage is that setting pass/fail standard scores will differ from one cohort to the other, since it is based mainly on student performance.

When deciding on a pass/fail score using *absolute* standards, the passing standard is set before the examination is conducted. The standard set will not, therefore, differ from cohort to cohort. Consequently, an entire cohort can pass or fail the assessment, depending on whether the minimum passing score based on the standard has or has not been achieved. The score is set using a *criterion-based* standard setting process in which experts first define the acceptable minimum passing standard before students sit for the assessment. The major advantage of this method is that the passing score is based on a minimum competency standard and will not vary according to the performance of the cohort. It has also been shown that criterion-based standards promote teamwork and collaborative learning as the passing grade is not set based on cohort performance. The main disadvantage is that it is a time-consuming process as well as being resource intensive. There are a few well-established methods of criterion-based standard setting, such as Angoff, Ebel, and Nedelsky [25–27].

Conclusion

The chapter describes the need to revamp the assessment techniques for anatomical knowledge among the students. This is done to ensure that learning has taken place beyond the rote memorization of factual materials, through a proper conceptual framework for assessment and feedback that is acceptable and cost effective to all stakeholders. The ultimate aim would be that the students will be equipped with the knowledge, skills, and the right attitude to practice competently in their respective fields. This will no doubt be impactful for medical education.

References

1. Association of American Medical Colleges-Howard Hughes Medical Institute Committee. Scientific foundations for future physicians. Report of the AAMC-HHMI Committee. 2009. https://www.aamc.org/download/271072/data/scientificfoundationsforfuturephysicians.pdf. Accessed 26 Feb 2019.
2. Boud D. Assessment and learning: contradictory or complementary? In: Knight P, editor. Assessment for learning in higher education. London: Kogan Page; 1995.
3. Myers C, Jones TB. Promoting active learning: strategies for the college classroom. San Francisco: Jossey-Bass; 1993.
4. Dijkstra J, van der Vleuten CP, Schuwirth LW. A new framework for designing programmes of assessment. Adv Health Sci Educ Theory Pract. 2010;15:379–93.
5. Van der Vleuten C, Schuwirth L, Driessen E, Dijkstra J, Tigelaar D, Baartman L, et al. A model for programmatic assessment fit for purpose. Med Teach. 2012;34:205–14.
6. Woods NN, Brooks LR, Norman GR. The role of biomedical knowledge in diagnosis of difficult clinical cases. Adv Health Sci Educ Theory Pract. 2007;12:417–26.
7. Larsen DP, Butler AC, Roediger HL III. Test-enhanced learning in medical education. Med Educ. 2008;42:959–66.
8. Biggs J. Teaching for quality learning at university: what the student does. Buckingham/Philadelphia: The

Society for Research into Higher Education and Open University Press; 1999.
9. Ramsden P. Learning to teach in higher education. London/New York: Routledge Falmer; 1992.
10. Brown SA, Knight P. Assessing learners in higher education. London: Kogan Page; 1994.
11. Bloom BS, Englehart MB, Furst EJ, Hill WH, Krathwohl DR. Taxonomy of educational objectives, the classification of educational goals – handbook I: cognitive domain. New York: McKay; 1956.
12. Van der Vleuten CPM. The assessment of professional competence: developments, research and practical implications. Adv Health Sci Educ. 1996;1:41–67.
13. Van der Vleuten CP, Schuwirth LW. Assessing professional competence: from methods to programmes. Med Educ. 2005;39:309–17.
14. Moskal BM, Leydens JA. Scoring rubric development: validity and reliability. Pract Assess Res Evaluat. 2000;7(10):2–6.
15. Paul R, Elder L. Consequential validity: using assessment to drive instruction. In: The foundation for critical thinking. 2007. http://www.criticalthinking.org/files/White%20PaperAssessmentSept2007.pdf. Accessed 26 Feb 2019.
16. Tan CH, Lee SS, Yeo SP, Ashokka B, Samarasekera DD. Developing metacognition through effective feedback. Med Teach. 2016;38:959.
17. Cruess RL, Cruess SR, Steinert Y. Amending Miller's pyramid to include professional identity formation. Acad Med. 2016;91(2):180–5.
18. Craig S, Tait N, Boers D, Mcandrew D. Review of anatomy education in Australian and New Zealand medical schools. ANZ J Surg. 2010;80:212–6.
19. Case SM, Swanson DB. Constructing written test questions for basic and clinical sciences. 3rd ed. National Board of Medical Examiners; 2002.
20. Kearney RA, Puchalski SA, Yang HY, Skakun EN. The inter-rater and intra-rater reliability of a new Canadian oral examination format in anesthesia is fair to good. Can J Anaesth. 2002;49:232–6.
21. Driessen E, Van Tartwijk J, Vermunt JD, van der Vleuten CP. Use of portfolios in early undergraduate medical training. Med Teach. 2003;25:18–23.
22. Schuwirth LWT, Van der Vleuten CP. Programmatic assessment: from assessment of learning to assessment for learning. Med Teach. 2011;33(6):478–85.
23. Wass V, van der Vleuten C, Shatzer J, Jones R. Assessment of clinical competence. Lancet. 2001;357:945–9.
24. Norcini JJ. Setting standards on educational tests. Med Educ. 2003;37:464–9.
25. Ebel RL. Essentials of educational measurement. 3rd ed. Upper Saddle River: Prentice Hall Inc.; 1979.
26. Angoff WH. Scales, norms and equivalent scores. In: Thorndike RL, editor. Educational measurement. 2nd ed. Washington, DC: American Council on Education; 1971.
27. Nedelsky L. Absolute grading standards for objective tests. Educ Psychol Measur. 1954;14:3–19.

40 Developing Multiple-Choice Questions for Anatomy Examinations

Andrew R. Thompson and Polly R. Husmann

Introduction

Multiple-choice questions (MCQs) have become a standard form of assessment at levels ranging from grade school to licensing examinations for advanced degrees. This is primarily due to the fact that MCQs are generally objective, have a high level of reliability and validity, and are easy to score [1, 2]. In settings such as medical school, examinations utilizing MCQs are often the primary method of assessing student performance. Thus, it is vital that MCQs are constructed to reflect the understanding and abilities of the students in a challenging, yet clearly written fashion. Therein lies the problem: while it is not difficult to develop an MCQ, it takes skill and thoughtfulness to write a *good* MCQ.

While the general guidelines related to constructing MCQs [3] apply to constructing anatomy questions, there are also some unique considerations that anatomists must face. As a starting point, it is important to decide *what* to assess. While this question is somewhat dependent on the goals of each course and/or institution, it is critical that the content being assessed can be mapped to specific learning objectives. This concept is referred to as constructive alignment, which indicates that course learning objectives should be aligned with the learning activities and assessments. Determining what to assess in anatomy largely depends on the learning objectives specific to the target audience. One of the principles of adult learning theory (andragogy) is that learning is optimal when the information has relevance to learners' professional and/or personal lives [4]. This concept applies to constructing MCQs; the content being assessed should reflect what anatomy is important to a particular field. For anatomists teaching students entering healthcare fields, there is general agreement that the focus should be on clinically relevant information. This might differ slightly for an anatomy course training medical illustrators or doctoral students. It has been shown numerous times that assessment drives learning and students will adapt their study strategies to what and how information is tested [5–9]. Thus, it is important that assessments are designed with the learning objectives in mind.

Anatomy courses typically utilize both laboratory and lecture components. As a result, testing often occurs in both written (lecture) and practical (laboratory) formats. Historically, MCQs have been used almost exclusively on

A. R. Thompson (✉)
Department of Medical Education, University of Cincinnati College of Medicine, Cincinnati, OH, USA
e-mail: Thomp3ar@ucmail.uc.edu

P. R. Husmann
Anatomy and Cell Biology, Indiana University School of Medicine, Bloomington, IN, USA
e-mail: Phusmann@indiana.edu

L. K. Chan, W. Pawlina (eds.), *Teaching Anatomy*, https://doi.org/10.1007/978-3-030-43283-6_40

written examinations, whereas practical examinations were largely comprised of identification style questions using tagged structures on human cadavers and/or models. However, as will be discussed in a later section, practical examinations offer a great opportunity to reinforce lecture content by utilizing multiple-choice style questions that target higher cognitive levels [10].

Test Item Construction

As many authors have pointed out, writing a good test question is not a simple task [2, 11]. This section will begin by defining the components of a question and then discuss the key points of constructing a good MCQ by outlining some of the common flaws. Note that in most educational fields, the word "item" is used as a synonym for "question." These terms will be used interchangeably for the remainder of this chapter.

Typically, MCQs have two main components: a stem and options. The portion of an item that contains the background (often a clinical scenario) and states a question is referred to as the stem. The stem can be further subdivided into two parts: the vignette, which outlines the relevant history and problem, and the lead-in, which states the actual question or prompt. The stem is followed by a list of answer options, usually labeled A–E, that include at least one correct (keyed) answer and typically four incorrect, but plausible, options known as distractors. Figure 40.1 shows an example item with these components labeled.

Depending on the course objectives, an item may not include all of the components described above. For example, in an undergraduate anatomy course, it might not be appropriate to include a clinical vignette if this type of information is not emphasized in the course. However, in the context of most healthcare-related fields (and especially in medical school), structuring the item around a clinical vignette is recommended since that is the context in which students will use the information professionally.

Developing quality distractors and a plausible stem can be a very challenging task. This is especially true for those teaching in healthcare fields with limited clinical experience since distractors need to be as conceivable as the keyed option. Trying to view a question through the lens of a student is a good starting point. If you can determine the most common misconceptions or points of confusion about a particular topic, those can often serve as good distractors. Peer-reviewed online resources such as www.uptodate.com, www.statpearls.com, or www.radiopaedia.org can be used to better understand the clinical context of certain pathologies, which can help gener-

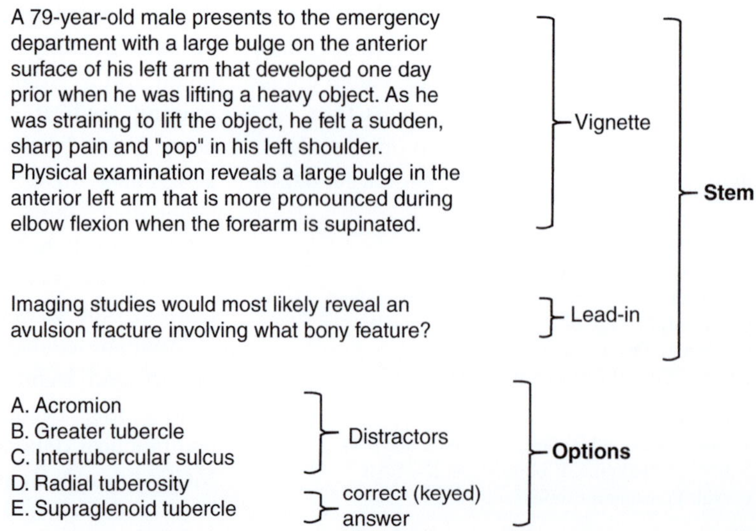

Fig. 40.1 Example MCQ with components labeled

ate quality distractors and create a realistic vignette in the stem. Having a clinical faculty member help write or edit questions is a great way to ensure your distractors and stem make sense from a clinical perspective.

There is debate as to the optimal number of answer options. Some research suggests that as few as three options provides similar results to questions with four or five options [12, 13]. For a standard-format MCQ, the National Board of Medical Examiners (NBME) recommends three to five options [3]. The quality of an item does not improve by adding additional distractors just for the sake of reaching five answer options.

MCQ Types

MCQs can be classified into two broadly defined formats: true-false or one-best-answer. There are several different types of true-false question, but all of these share the common feature of the potential for more than one correct answer. K-type true-false items include a list of statements or options (labeled by numbers or letters), and the student must determine which are true. Answer options include various combinations of the labeled statements. K-type items were commonly used in the past but have fallen out of favor because they can be confusing.

Example: K-Type Question
Which of the following nerves arise from anterior divisions of the brachial plexus?(*)

1. Musculocutaneous
2. Ulnar
3. Radial
4. Axillary

A. 1, 2, and 3
B. 2, 3, and 4
C. 1 and 2*
D. 3 and 4
E. All of the above

An X-type item is similar to a K-type, but instead of providing a series of potential answer combinations, the student must individually select all the correct answers.

Example: X-Type Question
A lesion in the posterior limb of the internal capsule would interrupt which of the following pathways?

A. Lateral corticospinal*
B. Dorsal column-medial lemniscus*
C. Corticobulbar
D. Spinothalamic*
E. Vestibulocerebellar

While true-false type items can be constructed to be technically acceptable, this format is no longer used on most standardized examinations. This is mainly due to the fact that creating answer options that are 100% true or false is challenging and, even when it is done well, students are often only required to recall facts [3].

As the name implies, one-best-answer items, referred to as A-type, include a stem and series of answer options, one of which is considered the best answer. The concept of a single "best" answer is what sets these apart from their true-false counterpart. In true-false items, each option must be either 100% true or 100% false. Developing options that meet this criteria, and are still plausible, can be challenging. Conversely, one-best-answer items may have several plausible options that can be viewed on a continuum of least correct to most correct. While certain material may lend itself better to the true-false format, one-best-answer type items are generally preferred because they are less problematic, more closely mimic reality, and are able to test higher-order thinking, such as the application or analysis of information [3].

Example: A-Type Question

A 48-year-old male presents to the emergency department with weakness and tingling in his right upper limb that began after a recent surgery where a tumor was removed from the patient's axillary region. Physical examination demonstrates weakness in shoulder abduction and elbow and wrist extension. Decreased sensation is observed on the superolateral shoulder, posterior arm and forearm, and dorsolateral hand. The most likely cause for the patient's symptoms is damage to the

A. axillary nerve.
B. posterior cord.*
C. radial nerve.
D. superior trunk.
E. lateral cord.

Common Flaws in Item Writing

Perhaps the best way to learn how to construct a good MCQ is to appreciate the most common item writing flaws. Item flaws are a serious concern because they can frustrate test-takers and compromise the reliability and validity of an examination. In a study of 10 high-stakes nursing examinations, Tarrant and Ware [14] found that up to 75% of questions on each examination had flaws. Flawed items were more difficult and less discriminating than their well-written counterparts. The most concerning finding of this study was that borderline students tended to benefit from flawed items, indicating that these students could have passed an examination when, in reality, they did not demonstrate requisite understanding of the material. This finding can have serious implications in healthcare fields with regard to licensing and future patient outcomes. Tarrant and Ware cite a lack of faculty training as the main reason for poor item writing. In a similar study, Vahalia et al. [2] analyzed over 2000 anatomy questions from 90 in-house medical school examinations and found that over one-fourth of the questions had errors that could be attributed to negligence on the part of the item writer. These two examples demonstrate the importance of proper item writing training.

There are two broad categories of item flaws: irrelevant difficulty and testwiseness. The discussion below highlights the types of flaws within each of these categories that are most likely to impact anatomists. For a more comprehensive review of item flaws, refer to Paniagua and Swygert [3].

Irrelevant Difficulty

I. *Answer options or stems that are long and/or overly complicated.* Answer options should be concise and to the point. In general, each option should be restricted to a single sentence. If the options have to be read through multiple times to understand what is being stated, this takes time away from the examination and creates an unnecessary burden on the student. Similarly, a stem should be brief and to the point but include background information that is relevant to answering the item. In the following example, there are several extraneous pieces of information in the stem, and the answer options are overly detailed. If you are unsure whether an option or stem is clear and concise, ask a colleague to review the item.

Example: Irrelevant Difficulty

A 30-year-old male presents to the emergency department with sudden vision loss. He is a construction worker and drinks alcohol regularly. He fractured a carpal bone in his right wrist one year ago, but it healed normally with no loss of function. Visual field testing reveals a right homonymous hemianopia. The patient most likely has a lesion affecting the

A. right optic nerve past the point where it emerges from the optic canal of the sphenoid bone.
B. optic chiasm medial to the anterior clinoid processes of the sphenoid bone.
C. optic tract near the superior margin of the left crus cerebri of the midbrain.*
D. fibers in Meyer's loop traveling through the right temporal lobe.

II. *Using "none of the above" as an answer option.* Avoid the temptation to include "none of the above" when it is a struggle to think of an additional plausible answer. Providing this as an option makes even the knowledgeable students question whether there is a subtle point that negates a seemingly correct answer option. In some instances, replacing "none of the above" with an option that is more clearly worded to be either correct or incorrect may be a good solution. In the example question below, "none of the above" could easily be replaced with "there would be no sensory deficits." The knowledgeable student would quickly realize this was incorrect and eliminate it as a potential correct option.

Example: None of the Above

A 25-year-old female presents with a fracture of the right medial epicondyle of the humerus, which resulted in damage to the ulnar nerve. Damage at this location would most likely result in paresthesia in which of the following regions?

A. Thenar eminence
B. Anterolateral forearm
C. Hypothenar eminence*
D. Dorsolateral hand
E. None of the above

III. *Using stems that are negatively phrased.* Another flaw that is all too common is having the word "except" or "not" added to a lead-in. This requires the student to use the process of elimination to determine which option is the least accurate. In addition to taking longer to answer, this flaw can result in a student getting an item wrong simply because they failed to notice the word "except" or "not" in the stem. Often, a negatively phrased question can be revised to ask for the most likely, rather than the least likely, option.

Example: Negative Phrasing

A 29-year-old female presents to the emergency department with a neck laceration that occurred during a motor vehicle accident. She is rushed to surgery where it is noted that the plexus of nerves arising from the posterior border of the sternocleidomastoid muscle was damaged. All of the following nerves would potentially be affected except the

A. great auricular.
B. recurrent laryngeal.*
C. supraclavicular.
D. transverse cervical.
E. lesser occipital.

Testwiseness

I. *The presence of grammatical cues.* The lead-in should not allow the student to automatically exclude any options simply due to nonsensical grammar or poor phrasing. This flaw can be avoided by careful examination of an item by the item writer and a reviewer. The example below has several flaws, including answer options that do not make sense because of the phrasing and incorrect grammar.

Example: Grammatical Cues

A 15-year-old male presents with right arm pain after falling off his bike. There is swelling, and light touch to the area is painful. The next best step to evaluate this patient would be to order a

A. radiograph.*
B. lumbar puncture.
C. arteriogram.
D. rest and revaluate after one week.
E. arthroscopy.

II. *Use of absolute terms.* In most instances, absolute terms such as "always" or "never" should be avoided in answer options or stems because almost nothing is absolute. In the example below, an absolute term is used in the stem. This is problematic because it could be argued there is no correct answer since not every patient will exhibit sensory deficits with an ulnar nerve injury, especially when the location of the damage is not described.

Example: Absolute Terms

Fertilization of an oocyte always occurs in which organ?

A. Ovary.
B. Uterine tube.*
C. Uterus.
D. Vagina.

III. *Unequal answer options.* Answer options should be similar in length and in detail. Frequently, item writers spend so much time crafting the correct option that they fail to make the distractors similar in detail. This discrepancy can cue students to the correct answer without requiring any knowledge of the subject. Similarly, explanations should not be used anywhere in an item. If something has to be explained in an option or stem, the item should be revised.

Example: Unequal Answer Options

A 45-year-old male presents with severe weakness in his left upper and lower limbs. When asked to stick his tongue out, it protrudes to the right side. The patient most likely has a lesion in the

A. thalamus.
B. posterior limb.
C. ventral area of the medulla, near the preolivary sulcus.*
D. precentral gyrus.
E. hypoglossal canal.

Cognitive Levels of MCQs

Historically, MCQs have been viewed somewhat negatively since many believed they could only assess straightforward recall of information [1, 15]. However, when carefully constructed, most educators now agree that MCQs can assess higher cognitive levels [10, 16–19]. Bloom's taxonomy is the most commonly used framework to evaluate cognitive levels. It was initially developed to standardize the language used in educational objectives in order to facilitate the exchange of information related to curricula and assessments across institutions [20, 21]. Three educational domains were identified: cognitive, affective, and psychomotor. In the seminal publication, *The Taxonomy of Educational Objectives* [22], the cognitive domain was elaborated and described. Most notably, a six-category hierarchy that can be used to classify student behaviors was outlined. These six categories represent the intended outcomes of the educational processes.

A former student of Bloom later revised the taxonomy by including a more detailed description of different types and levels of knowledge, as well as reorganizing the hierarchy and rephrasing the terms associated with each taxonomic level [20]. Of the six revised categories, MCQs can target the first four levels: remember, understand, apply, and analyze (Table 40.1). Because Bloom's taxonomy was designed for a general audience, its interpretation within a particular discipline may be subjective. Thompson and O'Loughlin [19] developed a rubric known as the Blooming Anatomy Tool (BAT) that provides guidelines tailored to the discipline of anatomy that can be used to construct and classify items according to cognitive level. This tool has been shown to improve consistency in determining the cognitive level of anatomy examination items, regardless of previous experience with either the discipline or Bloom's taxonomy.

When constructing an MCQ, it is important to align the cognitive level of the question with the level of the student and the learning objectives. If the learning objectives are written in such a way that recall is emphasized, constructing an examination with a large number of items that require

Table 40.1 Examples of examination MCQs arranged by Bloom's taxonomy

Cognitive level	Example question
Remember	The common fibular nerve is a branch of which of the following nerves? A. Femoral B. Deep fibular C. Sciatic* D. Obturator E. Tibial
Understand	Which of the following structures would be most vulnerable to damage following a fracture to the neck of the fibula? A. Common fibular nerve* B. Deep femoral artery C. Lateral plantar nerve D. Femoral nerve E. Anterior tibial artery
Apply	A 25-year-old female presents to the emergency department after being struck by a car while crossing the street. Imaging reveals a fracture to the neck of the fibula. If the nerve traveling near the fracture was damaged, which of the following would most likely be observed? A. Weakness in plantarflexion B. Sensory loss on the medial leg C. Sensory loss over the heel D. Weakness in foot eversion* E. Weakness in toe flexion
Analyze	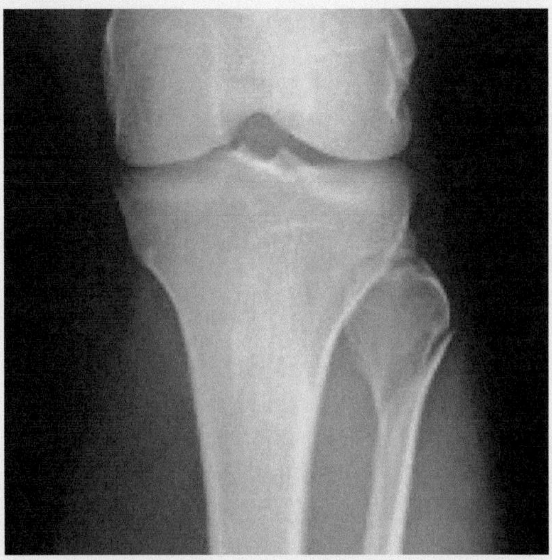 A 25-year-old female presents to the emergency department after being struck by a car while crossing the street. Imaging studies are provided. If the nerve traveling near the injured area was damaged, which of the following would most likely be observed? A. Weakness in plantarflexion B. Sensory loss on the medial leg C. Sensory loss over the heel D. Weakness in foot eversion* E. Weakness in toe flexion

analysis or application of information could lead to poor overall student performance and is unfair to students. For example, in a study of gross anatomy practical examination questions, Thompson et al. [10] found that higher-order questions tended to be more difficult and more discriminating compared to lower-order questions. The authors describe a situation in which a particular examination had a disproportionally high number of higher-order questions, which resulted in a low class average and a high number of examination failures. Nonetheless, Burns [23] argues that testing higher-order learning skills in anatomy will make students in healthcare-related fields better equipped to handle complex clinical situations. One way to ensure an examination is testing the appropriate level is to assign a cognitive level to each examination item based on a standardized system (e.g., the Blooming Anatomy Tool). The entire examination can then be reviewed based on frequencies of questions targeting each cognitive level to ensure the overall structure of the examination is commensurate with the learning objectives.

Interpreting Examination Results

After an examination is constructed and administered comes the critical step of evaluating the results. It is likely that many readers already utilize testing software (e.g., ExamSoft®) that automatically calculates the relevant psychometric data. If not, interpreting examination results can be challenging since these values would need to be calculated by hand or the raw examination data has to be formatted and then imported into a program capable of computing the desired statistics. An additional benefit of most testing software is that the performance history of each item is recorded. This allows for examinations to be constructed using items that have proven to be reliable on previous assessments. Over time, an entire examination can be constructed using items with proven performance history to target a desired class average. This may help avoid making the mistake of creating an examination that is too difficult or too easy. It is important to keep in mind that many of the item statistics described below are influenced by the sample size, so caution must be used when interpreting results from an examination with a small number (less than approximately 30) of students and/or items.

When interpreting results for individual items on an examination, there are a few important psychometric values to consider. The first is referred to as item difficulty and is reported as the proportion of students who answered the item correctly, or the p-value (not to be confused with the p-value associated with statistical significance). Item difficulty, or p-values, range from 0 to 1. For example, a p-value of 0.67 indicates that 67% of students answered the item correctly.

In addition to difficulty, most testing software provides at least one measure of item discrimination. In simple terms, item discrimination indicates how well an item separates the high- and low-performing students. The two most commonly used measures of item discrimination are point biserial (PBS) and discrimination index. PBS is a simple correlation between a dichotomous variable and a continuous one. In the case of item statistics, PBS represents the relationship between getting a single item correct/incorrect (dichotomous variable) and a student's overall average on the examination (continuous variable). PBS values range from −1 to +1. A large, positive PBS value indicates that students who did well on the examination tended to answer the item correctly, whereas students who did poorly on the examination tended to get the item incorrect. A negative PBS value indicates that students who did poorly on the examination tended to get the item correct, whereas students who did well on the examination got the item incorrect. A negative PBS value can be a red flag that something is wrong with the item.

The other commonly reported type of item discrimination statistic is referred to as the discrimination index (DI). While the exact formula used to calculate DI varies, it is generally a ratio that compares the performance on an individual item between students that scored in the upper and lower quartiles on the total examination. Similar to PBS, DI values range from −1 to +1, and the higher the value, the better that item discriminated between higher- and lower-performing students.

While the interpretation of each of these item statistics in isolation is relatively straight forward, they are most informative when used together. One of the first things to note is that item difficulty and discrimination are related. Extremely easy and extremely difficult items are more likely to have poor item discrimination. If an item is well written and difficult, it should have a relatively high PBS or DI. For example, an item with a difficulty of 0.5 and a PBS of 0.1 is concerning and should be reviewed for flaws because many of the higher-performing students answered it incorrectly. Conversely, an item with a difficulty of 0.3 and a PBS of 0.6 suggests the item is very challenging, but many of the higher-performing students got it correct. Unfortunately, there is no established ideal figure for the relationship between difficulty and discrimination. LeBlanc and Cox [24] suggest that an examination composed of items ranging in difficulty from 0.2 to 0.8 would provide a compromise between a reasonable class average and discrimination. As a general rule, a PBS value of <0.1 is poor and >0.3 is good. However, focusing on the exact value can be misleading. Instead, item discrimination and difficultly should be used as screening tools to help highlight problematic items or confirm well-written ones.

When provided both the DI and PBS, PBS is generally preferred because it is calculated using all students instead of just the upper and lower quartiles. An important underlying assumption to keep in mind, especially for those teaching in integrated curricula, is that item discrimination measures the degree to which an individual examination item and the entire examination are measuring a specific ability or knowledge base. When an examination evaluates a wide range of content from different disciplines, discrimination values tend to be lower. It might be possible to circumvent this issue by calculating discrimination values from a subset of items that can be classified to a specific discipline. However, this functionality is likely missing from most testing software, and in a truly integrated curriculum and assessment, it can be challenging to classify an item that requires knowledge from multiple disciplines.

In addition to looking at each individual item, it is important to consider the reliability of the examination as a whole. The most commonly used measure of reliability in this context is the Kuder-Richardson Formula 20 (KR-20), which is very similar to the more general Cronbach's alpha. KR-20 is a measure of reliability that uses inter-item consistency to demonstrate that students performed similarly on items that assess the same constructs. Values range from 0 to 1. Higher values indicate that if the same set of students were to retake the same examination, they would likely receive a similar score. A KR-20 value of at least 0.70 is considered desirable on instructor-written, in-house examinations. However, it is important to realize that KR-20 is influenced by a number of factors. Similar to the item discrimination statistics described above, KR-20 is designed to measure how well an examination measures a single construct. Thus, integrated examinations may result in lower reliability coefficients [25]. KR-20 is also impacted by the number of items on an examination. The higher the number of items, the more likely that an acceptable KR-20 will be achieved. To preempt problems with KR-20 results, a formula called the Spearman-Brown prophecy can be used to predict reliability based on the number of examination items or to determine the approximate number of examination items needed to achieve a specified level of reliability.

Other Considerations

Using MCQs on Anatomy Practical Examinations

While anatomy practical examinations have historically used free response "identification" style items, the laboratory environment is a great place to utilize MCQs. Generally speaking, there are two common ways to utilize MCQs on practical examinations (Table 40.2). The first can be referred to as a single-structure MCQ and involves having a single structure tagged on a cadaver or model that is referred to in the item stem. The answer options are provided in written format. The second way to introduce MCQs is to have a question prompt with a stem, but in this case, there are several tagged structures on the cadaver or model from which the student must select their answer. This can be referred to as a multiple-structure MCQ. As a general guideline, MCQs utilized in a laboratory setting should target higher cognitive levels and focus on the appli-

Table 40.2 Examples of anatomy laboratory practical examination MCQs

Item type	Tagged structure(s)	Station prompt
Single-structure MCQ	Paracentral lobule on a half-brain	The tagged structure receives its blood supply from which of the following arteries? A. Anterior cerebral* B. Middle cerebral C. Posterior cerebral D. Anterior choroidal E. Posterior communicating
Multiple-structure MCQ	A. Anterior cerebral a.* B. Middle cerebral a. C. Posterior cerebral a. D. Anterior choroidal a. E. Posterior communicating a.	A 35-year-old male presents to the emergency department with sudden onset of hemiparesis and hemisensory deficits of the lower limb below the level of the knee. Occlusion of which of the tagged vessels would most likely account for these findings?

cation of clinically relevant anatomy. However, it is important to again consider the learning objectives and prepare students to answer these types of items since they tend to be more difficult than the traditional identification items [10]. Additionally, the length of timed examinations may need to be increased since these types of items usually take longer to answer compared to strict identification.

Using Images in MCQs

Due to the visual nature of anatomy, items on written examinations often include an image in the stem. Previous research has shown that many students prefer learning with visual aids [26] and that visual aids enhance learning [27]. However, it is also possible that visual sources can increase cognitive load [28], which could hamper learning. Given this concern, it is important to utilize images on examinations in a deliberate manner since they could have an unintended effect on student performance. In a study looking at the effect of images on item statistics in an undergraduate anatomy course, Notebaert [29] found no significant difference in item discrimination or difficulty between items with or without images, suggesting the inclusion of images did not have an appreciable influence on examination outcomes. Vorstenbosch et al. [30] also looked at the influence of images and found that the type and context in which an image is used impacted student performance. The authors attribute this finding to the fact that some images require interpretation, whereas others have a cueing effect. When deciding whether to include an image in an item, it is important to consider *why* it is being used. In particular, is interpreting something on the image necessary to determine the correct answer option? If this is the case, using an image can potentially increase the cognitive level assessed by the item. For example, Table 40.1 demonstrates how the cognitive level of an item can increase from apply to analyze simply by adding an image and slightly adjusting the text.

Conclusion

The amount of thought and skill that goes into writing a quality MCQ can be somewhat overwhelming to novice item writers. Even after reading this chapter, some may still find themselves wondering, well now what? *How* do I write a good MCQ? Unfortunately, becoming a good item writer is not something that can be learned by simply reading a set of instructions. It takes time, practice, and experience. The following points summarize the key concepts that should be kept in mind while crafting MCQs:

I. Test from learning objectives. Students will adapt their study habits based on what is assessed, so assessments should reflect learning objectives. If you are teaching students in a healthcare field, construct the items in a way that mimics real-world clinical scenarios. Students are more likely to be motivated to learn when the focus is on content that is important and applicable to their future careers.

II. Avoid item flaws. There are many opportunities to make mistakes when writing examination items. Flaws, such as overly complicated options, negatively phrased stems, or poor grammar, can compromise the quality of the examination and reward testwise students who might not actually understand the material.
III. Distractors should be as plausible as the correct answer. After spending the time to craft a good stem, do not ruin a question by choosing distractors that can be eliminated without much thought or understanding. If only one or two good distractors can be identified, the stem may need to be revised to make other options more plausible. Typically, MCQs should have three to four distractors. Avoid adding irrelevant distractors simply to increase the total number of answer options.
IV. Review items multiple times before putting them on an examination. Ideally, have a colleague (or two) review new questions to make sure there are no issues. You do not need a content expert to identify general item writing flaws, so do not be afraid to ask a colleague who teaches a different discipline.
V. Lastly, use psychometric data to flag and improve items. Even if you think your item is well crafted, item statistics can highlight flaws that you may have overlooked. Assuming the information was taught well and the examination items were constructed from learning objectives, poor item statistics are a good indication that something is wrong with the question. If you do not currently utilize any psychometric data to review and construct examinations, you are missing a critical piece of information that can improve assessment quality, validity, and reliability.

References

1. Levine HG, McGuire CH, Nattress LW. The validity of multiple choice achievement tests as measures of competence in medicine. Am Educ Res J. 1970;7(1):69–82.
2. Vahalia KV, Subramaniam K, Marks SC, De Souza EJ. The use of multiple-choice tests in anatomy: common pitfalls and how to avoid them. Clin Anat. 1995;8(1):61–5.
3. Paniagua MA, Swygert KA. Constructing written test questions for the basic and clinical sciences. Philadelphia: National Board of Medical Examiners; 2016.
4. Knowles MS. Andragogy in action. 1st ed. San Francisco: Jossey-Bass; 1984.
5. Newble D, Entwistle N. Learning styles and approaches: implications for medical education. Med Educ. 1986;20:162–75.
6. Stanger-Hall KF. Multiple-choice exams: an obstacle for higher-level thinking in introductory science classes. CBE-Life Sci Educ. 2012;11(3):294–306.
7. Wood T. Assessment not only drives learning, it may also help learning. Med Educ. 2009;43(1):5–6.
8. Tan CM, Thanaraj K. Influence of context and preferred learning environments: approaches to studying physiology. Med Educ. 1993;27(2):143–59.
9. Ross ME, Green SB, Salisbury-Glennon JD, Tollefson N. College students' study strategies as a function testing: an investigation into metacognitive self-regulation. Innov High Educ. 2006;30(5):361–75.
10. Thompson AR, Kelso RS, Ward PJ, Wines K, Hanna JB. Assessment driven learning: the use of higher-order and discipline-integrated questions on gross anatomy practical examinations. Med Sci Educ. 2016;26(4):587–96.
11. Nnodim JO. Multiple-choice testing in anatomy. Med Educ. 1992;26(4):301–9.
12. Vyas R, Supe A. Multiple choice questions: a literature review on the optimal number of options. Natl Med J India. 2008;21(3):130–3.
13. Tarrant M, Ware J. A comparison of the psychometric properties of three- and four-option multiple-choice questions in nursing assessments. Nurse Educ Today. 2010;30(6):539–43.
14. Tarrant M, Ware J. Impact of item-writing flaws in multiple-choice questions on student achievement in high-stakes nursing assessments. Med Educ. 2008;42(2):198–206.
15. Wood EJ. What are extended matching sets questions? Biosci Educ. 2003;1(1):1–8.
16. Epstein R. Assessment in medical education. N Engl J Med. 2007;356(4):387–96.
17. Palmer EJ, Devitt PG. Assessment of higher order cognitive skills in undergraduate education: modified essay or multiple choice questions? Research paper. BMC Med Educ. 2007;7(1):1–7.
18. Clifton S, Schriner C. Assessing the quality of multiple choice test items. Nurse Educ. 2010;35(1):12–6.
19. Thompson AR, O'Loughlin VD. The blooming anatomy tool (BAT): a discipline-specific rubric for utilizing Bloom's taxonomy in the design and evaluation of assessments in the anatomical sciences. Anat Sci Educ. 2015;8(6):493–501.

20. Anderson L, Krathwohl D, Airasian P, Cruikshank K, Mayer R, Pintrich P, et al. A taxonomy for learning, teaching and assessing: a revision of Bloom's taxonomy of educational objectives. New York: Longman; 2001.
21. Anderson L, Sosniak L, editors. Bloom's taxonomy: a forty-year retrospective. Chicago: University of Chicago Press; 1994.
22. Bloom B, Englehard M, Furst E, Hill W, Karathwohl D. Taxonomy of educational objectives: cognitive domain. New York: McKay; 1956.
23. Burns ER. "Anatomizing" reversed: use of examination questions that foster use of higher order learning skills by students. Anat Sci Educ. 2010;3(6):330–4.
24. LeBlanc V, Cox MA. Interpretation of the point-biserial correlation coefficient in the context of a school examination. The Quant Meth Psych. 2017;13(1):46–56.
25. McGahee TW, Ball J. How to read and really use an item analysis. Nurse Educ. 2009;34(4):166–71.
26. Jastrow H, Hollinderbäumer A. On the use and value of new media and how medical students assess their effectiveness in learning anatomy. Anat Rec B: The New Anatomist. 2004;280B(1):20–9.
27. Gross MM, Wright MC, Anderson OS. Effects of image-based and text-based active learning exercises on student examination performance in a musculoskeletal anatomy course. Anat Sci Educ. 2017;10(5):444–55.
28. Mayer RE. The Cambridge handbook of multimedia learning. 2nd ed. New York: Cambridge University Press; 2014.
29. Notebaert AJ. The effect of images on item statistics in multiple choice anatomy examinations. Anat Sci Educ. 2017;10(1):68–78.
30. Vorstenbosch MATM, Klaassen TPFM, Kooloos JGM, Bolhuis SM, Laan RFJM. Do images influence assessment in anatomy? Exploring the effect of images on item difficulty and item discrimination. Anat Sci Educ. 2013;6(1):29–41.

41 Peer and Faculty Assessment of Nontraditional Discipline-Independent Skills in Gross Anatomy

Nirusha Lachman and Wojciech Pawlina

Introduction

Quality improvement and patient safety remain at the forefront of healthcare delivery. As management systems recognize the impact that poorly developed nontechnical skills may have on patient safety, efforts to strengthen leadership and communication skills, situational awareness, teamwork, and professionalism among practicing healthcare workers have increased. The importance of nontechnical skills in healthcare is discussed in the literature [1–4], and improvement in their use requires practice and repetition [5]. In addition, several studies in surgical environment showed that nontechnical skills are linked to surgical skills (technical skills) especially when evaluated during intraoperative crisis management situation [6]. Especially, surgical residents and trainees could benefit from formative assessments to appraise their levels of nontechnical skills during surgical procedures [7, 8]. In general, it has been concluded that surgeons who possessed good technical skills also tended to show similar levels of performance in the nontechnical skills domains. The opposite relationship is also observed where poor performers in nontechnical skills domains were often deficient in technical skills [9, 10]. While organizations continue to seek opportunities to build on these skills through formal training sessions [11, 12], medical educators share the responsibility by finding opportunities within medical curricula to promote early development of such *nontraditional discipline-independent skill (NTDIS)* [13, 14]. It is evident that health profession educators must begin to embrace teaching of nontechnical skills (NTS) or nontraditional discipline-independent skills (NTDIS) both in their medical and allied health curricula and throughout the continuum of professional career [4].

In this chapter, we provide insight into understanding nontraditional discipline-independent skills and how it relates to anatomy education. We highlight opportunities for integration of nontraditional discipline-independent skills within the anatomy curriculum and ways by which these skills may be evaluated in a team-based learning environment and contribute to overall student performance.

In the global context of healthcare delivery, improvement of patient treatment and safety through multidisciplinary teamwork has been strongly recognized. In medical education, teamwork and the skill of working effectively within a professional team have become a valued component of the curriculum embedded within the core competencies and milestones of the Accreditation Council for Graduate Medical Education

N. Lachman (✉) · W. Pawlina
Department of Clinical Anatomy, Mayo Clinic College of Medicine and Science, Mayo Clinic, Rochester, MN, USA
e-mail: Lachman.nirusha@mayo.edu; pawlina.wojciech@mayo.edu

L. K. Chan, W. Pawlina (eds.), *Teaching Anatomy*, https://doi.org/10.1007/978-3-030-43283-6_41

(ACGME) in the USA [15, 16]. Milestones describe performance levels residents and fellows are expected to demonstrate for skills, knowledge, and behaviors in the six clinical competency domains of ACGME [17].

Over the years, many anatomy curricula have evolved to incorporate objectives that allow for the development of skills not directly related to the acquisition of anatomical knowledge. These *nontraditional discipline-independent skills (NTDIS)* focus on team interaction, communication, leadership, and other aspects of professionalism that are required for delivery of quality healthcare.

The process of educating medical students however is as intuitive as it is fundamental. Anatomy courses that include laboratory experiences are never designed for secluded/individual learning. Just as dissection of the human body is a shared experience, learning anatomy in a curriculum that aligns with principles of healthcare delivery cannot be chalked up to basic factual assimilation of knowledge by an individual student. In today's work culture, collaboration and teamwork underpin a successful work environment. In the current healthcare environment, team-based models in the teaching and practice of medicine have become more a norm than a preference [18]. Successful teamwork, however, involves more than just a group of random individuals working together toward a common goal. It involves an understanding of individuals' strengths and weaknesses and of professionalism and openness to sharing and gaining from others' perspectives and experiences [18].

While incorporating teamwork in the classroom is often challenging for most medical school courses, team-based exercises are central to the learning of anatomy. Lerner and colleagues [19] define teamwork as the "ability of team members to [20] work together [5], communicate effectively [21], anticipate and meet one another's demands, and inspire confidence" which is said to result in effective collective action toward the achievement of a common goal [22].

Within the context of anatomy courses, students may be provided with opportunities to demonstrate these skills by being formally assigned to teams (within the anatomy laboratory and classroom) [23] with rotational leadership roles within their team and attaching value to peer and faculty assessments by having them contribute toward the final grade. Providing clear objectives and expectations at the beginning of the course with regular short reflective presentations (one to two slides showing examples of good and poor practices through teaching team observations or simply reminding students to be reflective) throughout the course encourages student awareness and subsequent development of nontraditional discipline-independent skills.

> Today's physicians must demonstrate both professionalism and leadership skills in order to succeed in largely team-based healthcare environments. –Pawlina et al. [20]

Assessing Nontraditional Discipline-Independent Skills

Peer Assessment

The learning process in the anatomy laboratory largely depends on the presence of cohesive learning teams. To achieve that goal, every member of the team needs to be accountable for his or her individual contribution to the team. Dissection teams are in the best position to recognize multiple and different kinds of contributions from their respective members. Some contribute a wealth of creative ideas for the team to consider, others are stronger in analyzing and assessing those ideas, and yet other students are very helpful in managing the group work. Hence, the team members develop a clear understanding of the varied ways in which individual members contribute to the work of the group. Peer assessment is essential because team members are typically the ones who have enough interaction and information to accurately assess one another's contributions.

Demonstrating a Good Work Ethic in the Laboratory

Students working with a dissection team:

- Are prepared for the day's dissection.
- Are reliable and attend laboratory and briefing sessions/classroom activities.
- Attend team meetings that occur outside of formal laboratory or class time.
- Make positive contributions to team discussions and assignments.
- Value and encourage input from fellow team members.
- Are respectful in considering cultural differences and diversity within the team.

It is important to remember that most students if not all have little or no experience evaluating peers. For most of them, the task is daunting and often associated with the fear of social repercussion among fellow students [24–26] (see Chap. 19). As faculty continues to keep in mind that inasmuch as peer evaluations provide information on student demonstration of nontraditional discipline-independent skills, for students, the process of providing the assessment is still a learning activity. Student learning and ability to conduct an evaluation may be enhanced by clarifying goals and objectives for providing feedback, explaining to them how the feedback will be used, setting clear criteria for assessment, and paying attention to the anonymity and timing of the assessment [27, 28].

As a form of best practice, peer evaluations should be conducted in the middle of the course/block as well as at its conclusion [15]. In addition to assigning grades, students should be asked to provide written comments on the positive and negative aspects of their interactions within the dissection group.

How Do Students Benefit from Performing Peer Assessment?

The benefits of performing peer evaluation are twofold. On the one hand, students are given the opportunity to learn how to critically judge attributes that are demonstrated in the process of fulfilling course objectives, and on the other, students are provided with valuable feedback that directs focus on their own self-improvement [29].

Peer evaluation:

- *Is engaging for students*: Students are held responsible for their contribution toward the success of the team. It creates an awareness for behaviors and characteristics that would otherwise not be considered.
- *Is a good source of feedback*: Enables faculty to provide anonymous peer-generated constructive commentary to students so that they may improve team performance.
- *Promotes student reflection and collaborative learning*: Helps identify areas of weakness in conceptual understanding. Allows students to understand course objectives better and improve knowledge and overall performance.
- *Helps develop interpersonal skills*: Allows students the opportunity to work with individuals who may have varying learning styles, cultural backgrounds, and personality traits different from their own.
- *Promotes professionalism*: Peer assessment is designed around aspects of professionalism. It promotes increased awareness of expectations and required competencies.

Guiding Students Through the Peer Evaluation Process

- Introduce the concept of peer evaluation.
- Relate to core objectives and curriculum requirements.
- Teach how to provide useful constructive feedback using examples.
- Demonstrate good examples of student-generated feedback and allow students to analyze and revise poorly constructed comments.

- *Promotes accountability*: Student participation becomes incentive based. It encourages fair and equal participation and contribution to the group, thereby limiting "social loafing" [30, 31] of team members.

> When students develop the capacity to judge their peers' performance, they might improve their capacity to examine their own clinical performance. –Hulsman et al. [21]

Considering that every educational environment may be different, there is no definitive way to conduct peer evaluation. On the spectrum of methods discussed by Michaelsen and colleagues [31], the advantages and disadvantages of these are clearly recognized. The common principle, however, is based on allocation of points through differentiation among team members. In our 10-year experience, we have found the use of a modified version of the Michaelsen's method (Fig. 41.1) [31] in conjunction with formal orientation and dedicated time for teaching students how to recognize positive and poor nontraditional discipline-independent behaviors effective and well accepted among students (Mayo Medical School 2007–2016) [25, 31–33].

We provide an interpretation for a set of nontraditional discipline-independent skills as they relate to observable behaviors within the anatomy teaching and learning environment (Table 41.1) [32].

One of the strengths of designing the anatomy curriculum within a team-based environment is its potential for emphasizing leadership and the importance of developing leadership skills. Again, students in the early stages of professional development cannot be expected to show natural leadership abilities. While strong leadership may be intuitive, basic leadership skills require effort, reflection, and a commitment to taking on greater responsibility for the purpose of team success. Instituting leadership evaluation following outline of clear expectations for the leadership role is recommended as an adjunct to peer evaluation. We have adapted a student leader evaluation form [34] that is completed for each rotating leader. The evaluation (Fig. 41.2) uses a Likert scale to rate the leader on five of nontraditional discipline-independent skills presented in Table 41.1 in addition to rating the student leader's problem-solving ability and overall professionalism. Written feedback regarding the leader's strengths and areas for improvement is also required and shared with the leader during formal feedback sessions with faculty (see Chap. 19). Student leadership evaluation also provides opportunity for self-reflection on leadership experience and skills. Unlike with the peer evaluation, the student being evaluated in this case is also required to complete the evaluation form. Scores are then analyzed and presented as a bar graph comparison of peer and self-ratings. The learning impact is significant in the sense that it provides insightful information on the student's ability to critically assess their own strengths and weaknesses and compare evaluation of themselves with the perception of others (Fig. 41.3).

> Early introduction of peer-assessment in medical education would facilitate early acceptance of this mode of evaluation and would promote early on the habit of critical evaluation of professional clinical performance and acceptance of being evaluated critically by peers. –Hulsman et al. [21]

Faculty Evaluation of Nontraditional Discipline-Independent Skills

Objective assessment of nontraditional discipline-independent skills is challenging, and little can be done to limit subjectivity when assigning formal grades. One method to consider may be to categorize evaluating skills under two broad categories: (1) *cognitive skills*, relating to a student's capacity to think (reason clinically), make decisions, and formulate actions, and (2) *social (interpersonal) skills*, relating to how a student interacts with other individuals within the course (peers, near-peers, and faculty) to convey information (Table 41.2) [35]. Faculty should:

GROSS ANATOMY PEER EVALUATION FORM

NAME ____________ TEAM/TABLE #____________

At the end of the Block III, it is necessary for all members of the dissection group to assess the contributions that each member of the group made to the work of the group. This contribution sholud reflect your judgment on:

- Preparation for the laboratory dissection and other team assignments
- Contribution to the dissection process, group discussions and assignments
- Reliable attendance in the laboratory and other team-based activities
- Flexibility and ability to resolve disagreements
- Personal commitment to honor the choices and rights of other members
- Responsibility for own actions and decisions
- Overall professionalism as demostrated in the commitment to demonstrate values and attributes that

constitute professionalism

Please assign scores that reflect how you really feel about the extent to which the other members of your team contributed to your learning and team performance. This will be your only opportunity to reward the members of your dissection team who worked hard on your behalf. (Note: If you give everyone the same score, you will be hurting those who did the most and helping those who did the least.)

Instructions:

In the space below please rate each of the other members of your team except yourself by distributing 60 points. Each member's peer evaluation score will be the average of the points they receive from the other members of the team. To complete the evaluation you should: 1) List the name of each member of your team except yourself, and 2) assign an average of 20 points to the members of your team (you should assign a total of 60 points in a four member team), and 3) differentiate some in your ratings; eg. You must give at least one score of 21 or higher (maximum = 30) and one score of 19 or lower.

Team Members Scores

1) ____________________

2) ____________________

3) ____________________

Additional Feedback:

In the space below briefly describe your reasons for your highest and lowest ratings. These comments - but not information about who provided them - will be used to provide feedback to students during formal feedback sessions with faculty

Reason(s) for your highest rating(s). (Use back if necessary.)

Reason(s) for your lowest rating(s). (Use back if necessary.)

Fig. 41.1 Gross anatomy peer evaluation form

Table 41.1 Observable behaviors related to nontraditional discipline-independent skills in anatomy

1. **Respect** *Shows personal commitment to honoring the choices and rights of others*	Acknowledges the rights of others in an inclusive and culturally sensitive way Works consciously with donor bodies to ensure careful and sensitive handling of cadaveric material Conforms with dress code within the laboratory Acknowledges and makes accommodation for team members with learning difficulties and physical challenges
2. **Integrity** *Shows commitment to honesty and trustworthiness in evaluating and demonstrating own skills and abilities*	Assigns realistic grades to peers and provides honest evaluations based on peers' contribution to students own learning Recognizes strengths and weaknesses when contributing to achievement of team outcomes
3. **Responsibility** *Accepts responsibility for own actions and decisions without blaming others*	Admits to shortcomings in team activities (e.g., dissection quality, team assignments) Shows commitment to attendance of lectures and laboratory sessions Keeps team and faculty updated on absences and personal challenges that may affect individual and team performance
4. **Compassion** *Shows adequate appreciation of other person's special needs for comfort and provides support without overt emotional involvement*	Supports team members by making allowances for unexpected personal challenges during course Shows understanding when one or more team members are experiencing difficulty with course material Aware of the importance of wellness for individual team members
5. **Commitment to excellence** *Shows adequate commitment to the pursuit of excellence and continuous quality improvement*	Providing high-quality dissections for evaluation Striving for excellence in team presentations (verbal, laboratory "bedside" presentations; written, embryology pamphlets, autopsy reports, etc.) Ensuring reading assignments are completed on a regular basis to ensure valuable contribution in team-based learning (e.g., ARS, dissection, etc.)

1. Interact with students in the lab as much as possible (moving around to tables to ensure equal interaction, questioning, and guiding through more difficult dissections).
2. Communicate with students during course.
3. Observe students during class (attentiveness to lectures, focus, punctuality, interaction with team members during ARS discussion sessions).
4. Monitor student progress (ARS scores, follow up with students when performance declines).
5. Provide feedback whenever possible (formal sessions mid-block and on the job feedback).
6. Maintain paper trails and records of student interaction with more challenging students.

For a total of 10% contribution toward the final grade, assign 5% to cognitive skills and 5% to social skills. Assuming that all students in general were able to meet expectations, allocate 4/5 to each category, and reserve a 5/5 for students who have demonstrated exceptional skills, and score 3/5 and below for students who need improvement. In addition, use the same categories to provide written feedback to qualify faculty-generated scores. In addition, faculty evaluations should include submission of a written evaluation of each student at the conclusion of the course (Fig. 41.4). Included in these evaluations are observations on students' professional attitudes, respect for others, preparation, contributions to team learning, leadership experience, decorum, responsibility, and interest and participation in a variety of course activities (see Chap. 19).

Peer and Faculty Evaluation of Nontechnical Competencies of Teamwork as It Relates to the Authentic Learning Environment

In the past decade, nontechnical competencies have become a crucial element and determinant

STUDENT LEADER EVALUATION FORM

Please rate the student leader below with whom you have worked. Circle one response per Item. Circle the appropriate number between 1 and 9 where 1 is the lowest rating and 9 is the highest rating. If you had Insufficient contact to evaluate this student/leader on a particular characteristic, circle UA (unable to evaluate).

Name of Leader ____________________

Please check one of the following: ☐ I am the leader being assessed (i.e. I am assessing myself)
☐ I am a team member of the leader being assessed

	1	2	3	4	5	6	7	8	9	UA
Respect	Shows inadequate personal commitment to honoring the choices and rights of other persons.						Always shows exceptional personal commitment to honoring the choices & rights of other persons.			
Integrity	Shows inadequate commitment to honesty and trustworthiness in evaluating and demonstrating own skills and abilities.						Always shows exceptional commitment to honesty and trustworthiness in evaluating and demonstrating own skills and abilities.			
Responsibility	Does not accept responsibility for own actions and decisions; blames other professionals.						Fully accepts responsibility for own actions and decisions; does not blame other professionals.			
Compassion	Shows inadequate appreciation of other person's special needs for comfort and help, or develops inappropriate emotional involvement.						Always appreciates other person's special needs for comfort and help, but avoids inappropriate emotional involvement.			
Problem-Solving	Fails to critically assess information, risks, and benefits; does not identify major issues or make timely decisions.						Critically assesses information, risks, and benefits; identifies major issues and makes timely decisions.			
Commitment to Excellence	Shows inadequate commitment to the pursuit of excellence and continuous quality improvement.						Always demonstrates a commitment to the pursuit or excellence and continuous quality improvement.			
Overall Professionalism	Shows Inadequate commitment to the characteristics and attributes that constitute professionalism.						Always demonstrates a commitment to the characteristics and attributes that constitute professionalism.			

ADDITIONAL COMMENTS FOR LEADERS

Leadership Strengths

Suggestions for Improvement

Fig. 41.2 Student leadership evaluation form

of success in the healthcare practice environment [36]. Where safety and efficacy are key, the expectation that individuals will demonstrate behaviors consistent with that of a high-performing professional team is unequivocal. More specifically, within the surgical division, several tools have been implemented to assess nontechnical skills both within and outside the operating room environment [8, 36]. With a focus on teamwork, collaboration, and leadership, the Oxford NOTECHS II assessment tool [37, 38] was designed to evaluate behaviors associated with

MAYO CLINIC

Mr.
Group Leader

Dear Block III Leader,

I would like to take this opportunity to thank you for being a leader during the Human structure block. Please find below tabulated results of your self-evaluation and compare them with the results of the evaluation from your peers. I think these evaluations are very valuable exercises in assessment of leadership, teamwork, and professionalism at this stage of your medical career. They provide you with thoughts for self-reflection and with an opportunity for recognizing directions for further self-development.

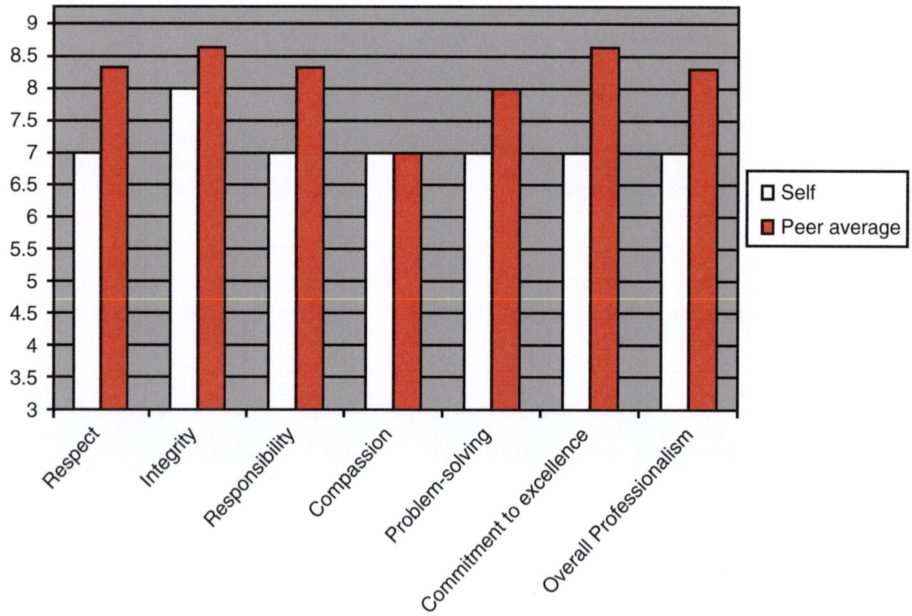

If you have any comments, please do not hesitate to contact me.

Thanks,

Wojciech Pawlina M.D.

WP:drf

Fig. 41.3 Graphic comparison of peer vs. self-evaluation

Table 41.2 Opportunities for assessment of cognitive and social skills

Cognitive skills	Opportunities for assessment
Anticipation of problems	During dissection and laboratory sessions Near-peer feedback through faculty debriefing sessions Faculty interaction
Decision-making	Peer feedback Leadership evaluation
Situation awareness	Faculty and near-peer observations during class and laboratory sessions
Problem-solving ability, knowledge retention, and conceptual ability	ARS performance Practical tests Overall performance on knowledge-based evaluations
Flexibility	Faculty interaction Peer and leadership evaluations
Adaptive strategies and workload distribution	Leadership evaluations
Social skills	**Opportunities for assessment**
Communication	Written communications Verbal communication Faculty and near-peer observations of overall disposition Peer and leadership evaluations Team presentations (oral and written, e.g., patient communication brochures, autopsy reports, bedside presentations)
Leadership	Leadership evaluations Faculty and near-peer observations
Teamwork	Peer evaluations Faculty and near-peer observations
Commitment to self-improvement (seeking feedback/advice/learning opportunities)	Peer evaluations Faculty and near-peer interactions Self-evaluations
Commitment to personal well-being	Multifaceted input including feedback from medical school administration
Resource management	Leadership evaluation

(1) leadership and management, (2) teamwork and cooperation, (3) problem-solving/decision-making, and (4) situational awareness.

In an effort to create curricular alignment within an authentic learning setting [39] and reduce the learning curve during transition into residency, we adapted the validated Oxford NOTECHS II assessment tool [37, 38] to evaluate student behaviors as they relate to effective teamwork within the anatomy course. The learning environment is designed to provide multiple opportunities for peer-peer and faculty-student interaction where nontechnical skills can be observed and identified against a list of demonstrable skills representative of a team-based anatomy course for medical and allied health curricula. Effective utilization of the "anatomy team-adapted nontechnical skills (NOTECHS II)" assessment tool requires student and faculty orientation before the start of the course that heightens awareness of the required competencies and their demonstrable skills [36–38] (Table 41.3).

The assessment should be supported by additional feedback that provides examples of behaviors that offer opportunity for improvement. As with the leadership and peer evaluation assessments previously described, the "anatomy team-adapted nontechnical skills (NOTECHS)" assessment tool can be used as supporting evi-

MAYO CLINIC

Memo

Wojciech Pawlina, M.D.
Professor and Chair
Department of Anatomy
Stabile 9-38

DATE: December 3, 2012

TO:

Mayo Medical School

FROM: Wojciech Pawlina, M.D.
Chair of Block III Human Structure

RE: Block III Final Course grade

Dear

The purpose of this memo is to inform you of your performance in Block III Human Structure.

The final grade for Block III Human Structure
80.07% (Pass)

Specific components of your final grade are listed in the table and graph below:

Objective Components:	Your Score	Class Average	Percent of Final Grade	Subjective Components:	Your Score	Percent of Final Grade
Written Examination	79.66%	76.61% (SD±6.298)	30%	Laboratory Peer-Assessment Score[1]	95.00%	10%
Practical Examination	72.73%	76.62% (SD±12.92)	20%	Faculty and TAs Evaluation Score	100.00%	10%
Audience Response System Questions	68.23%	74.50% (SD±5.04)	10%	Anatomy bed-side Presentation	94.00%	5%
Dissection Quality	64.15%	74.24% (SD±4.58)	2.5%	Embryology brochure	80.00%	2.5%
NBME Examination Gross Anatomy + Embryology	77.47% 580[2]	77.62% (SD±10.49)	10%	Your NBME US rank	77 percentile rank	
NBME Examination Gross Anatomy only	77.15% 570[2]	77.62% (SD±10.49)	N/A	Your NBME US rank	68 percentile rank	

[1]This score represents the avearge percent value calculated from the sum of points assigned by your team members. Twenty (20) points equates to 100% of the laboratory peer-assessment score.

[2]This is your original test score. For example, if your score in Gross Anatomy + Embryology is 500 with the corresponding percentile rank of 53, it indicates that 53% of the national group of examinees who took the Gross Anatomy + Embryology Subject Examination as an end-of-course or end-of-year assessment had scores at or below 500. For detailed explanation see next page.

Fig. 41.4 Student communication of performance

Here is the graph of your performance (in percentage) on individual components of the Human Structure Block.

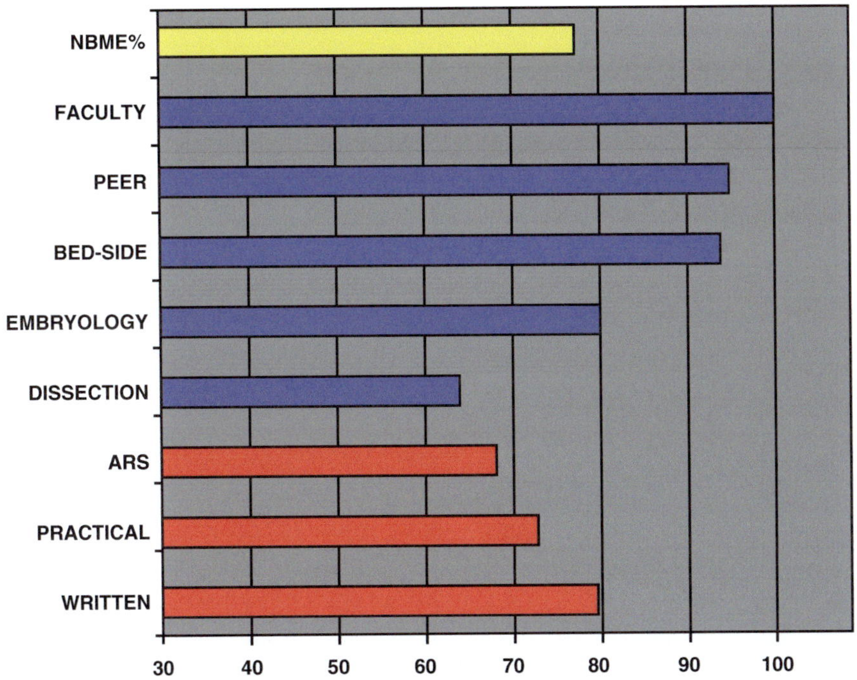

The NBME score scale is defined to have *a mean of 500 and a standard deviation of 100* for a group of 4.285 first-time takers from 36 U.S. LCME-accrediated medical schools who took the Gross Anatomy and Embryology Subject Examination or a modular component of the exam as an end-of-course exam under standard testing conditions during the 2002-03 and 2003-04 academic years. The vast majority of scores range from 200 to 800. Two sets of norms have been developed to aid in the interpretation of examinee performance: 1) the Gross Anatomy + Embeoyology score and 2) the Gross Anatomy score.

The mean score for MMS Class 2016 on the NBME Gross Anatomy + Embryology Subject Examination was 581.1 points (SD±78.5) which corresponds to 77 US percentile rank.
The mean score for MMS Class 2016 on the NBME Gross Anatomy Subject Examination was 573.4 points (SD±77.5) which corresponds to 69 US percentile rank.

Only your original test score of the NBME Gross Anatomy + Embryology Subject Examination Subject examination was used for calculating 10% of your grade and was normalized to the class average of both the written and practical examinations (77.62%).

If you have any questions or comments regarding your performance, please do not hesitate to contact me.

Thanks,

Wojciech Pawlina M.D.

WP:drf

Fig. 41.4 (continued)

Table 41.3 Anatomy team-adapted nontechnical skills (NOTECHS) assessment tool

Nontechnical competencies (NTC)	Demonstrable skills (DS)	Behavior			
		Directly compromises effective teamwork	*In other conditions could directly compromise effective teamwork*	*Maintains an effective level of teamwork*	*Enhances teamwork; a model for all other teams*
		1 Below standard	2 Basic standard	3 Standard	4 Excellent
Leadership and management	• Assumes responsibility for assigned leadership in team (authoritative, confident, assertive) • Encourages group interaction and effective communication • Directs activities in order to meet specific learning outcomes • Places value on team members contribution to team success (presentations, activities, team assessments, dissection) • Demonstrates emotional intelligence to harness the best from team members • Considers cultural diversity among team members • Holds self and team members accountable to team charter (norms and behaviors identified as important for team) • Demonstrates critical thinking and reflection, receptive to feedback • Demonstrates effective time management skills (prioritizes deliverables)				
Teamwork and cooperation	• Demonstrates commitment to team success (team learning as opposed to individualistic approach, supportive of team goals, helps team members achieve success) • Demonstrates sensitivity to team members challenges (uses humor appropriately, shows empathy, inclusive behaviors, non-competitive) • Fosters a healthy collegial working environment (open communication, relaxed, friendly, polite, culturally aware) • Holds self, accountable (reflective, admits to error, comfortable with uncertainty, accepts and applies meaningful feedback) • Demonstrates good communication skills (verbal, written, minds body language, honest and professional, listens to views of others) • Collaborative (holds the needs of the team above own, gives off best, finds joy in success of others, steps back when necessary) • Demonstrates commitment to conflict resolution (focusses on what is factual, reflects on what is the right course of action, suggests solutions, discourages ongoing conflict)				
Problem-solving and decision-making	• Uses knowledge to help guide discussions and find answers during team discussions (comes prepared to class) • Actively contributes to team discussions when there are conflicting answers • Demonstrates ability to use appropriate resources (consult faculty, consult text, use course resources)				
Situational awareness	• Option generation (suggests alternative options, reviews answers, critically evaluate options (e.g., multiple choice questions) • Seeks feedback when uncertain • Present and engaged in team work, laboratory, and classroom activities • Considers multiple factors before acting (inappropriate humor, lack of sensitivity to team members, dominating discussions, lack of patience if understanding exceeds rest of the team) • Vigilant and aware of perceptions and actions that impact team/self • Forward thinking (identifies potential challenge, considers contingencies, anticipates requirements)				

dence for providing feedback to students both during verbal one-one sessions and in final course written evaluations.

Conclusion

As we move forward in the teaching and learning of anatomy, the dynamic continuum between tradition and change commits educators to incorporate aspects of learning not traditionally associated with its basic science. As the expectations for healthcare professionals continue to evolve, medical educators are obliged to find opportunities within their curricula for the early development of nontraditional discipline-independent skill sets. Applied to this setting, these nontechnical skills translate as nontraditional discipline-independent skills that are recognizable through student interaction, communication, execution of tasks, and overall display of professionalism [33]. Because of the importance of these skills, their impact on student learning is best achieved through formal assessment. In this chapter, we have drawn from 10-year experience of assessing nontraditional discipline-independent skills. While assessment of nontraditional discipline-independent skills may be best achieved in a team-based learning environment, peer- and faculty-generated feedback may be equally effective in larger group settings. Early exposure to systems of peer evaluation [24] during the medical curriculum provides an effective means of developing students' confidence in evaluating their peers, while repeated exposure enhances the quality of the evaluations they will be able to provide [40].

> I liked the emphasis on professionalism, teamwork, and peer evaluation which will be so important down the road! –Year 1 student, Mayo Medical School, Class 2017

References

1. Higham H, Greig PR, Rutherford J, Vincent L, Young D, Vincent C. Observer-based tools for non-technical skills assessment in simulated and real clinical environments in healthcare: a systematic review. BMJ Qual Saf. 2019;28(8):672–86.
2. Flin R, Maran N. Identifying and training non-technical skills for teams in acute medicine. Qual Saf Health Care. 2004;13(1):i80–4.
3. Yule S, Flin R, Paterson-Brown S, Maran N. Non-technical skills for surgeons in the operating room: a review of the literature. Surgery. 2006;139(2):140–9.
4. White N. Understanding the role of non-technical skills in patient safety. Nurs Stand. 2012;26(26):43–8.
5. Glavin RJ, Maran NJ. Integrating human factors into the medical curriculum. Med Educ. 2003;37(1):59–64.
6. Bierer J, Memu E, Leeper WR, Fortin D, Fréchette E, Inculet R, et al. Development of an in situ thoracic surgery crisis simulation focused on nontechnical skill training. Ann Thor Surg. 2018;106:287–92.
7. Spanager L, Konge L, Dieckmann P, Beier-Holgersen R, Rosenberg J, Oestergaard D. Assessing trainee surgeons' nontechnical skills: five cases are sufficient for reliable assessments. J Surg Educ. 2015;72(1): 16–22.
8. Gostlow H, Marlow N, Thomas MJ, Hewett PJ, Kiermeier A, Babidge W, et al. Non-technical skills of surgical trainees and experienced surgeons. Br J Surg. 2017;104(6):777–85.
9. Riem N, Boet S, Bould MD, Tavares W, Naik VN. Do technical skills correlate with non-technical skills in crisis resource management: a simulation study. Br J Anaesth. 2012;109(5):723–8.
10. Almutairi AA. Do technical skills correlate with non-technical skills during crisis resource management? Health Sci J. 2017;11(6):1–5.
11. Cheraghi-Sohi S, Bower P. Can the feedback of patient assessments, brief training, or their combination, improve the interpersonal skills of primary care physicians? A systematic review. BMC Health Serv Res. 2008;8:179.
12. Norgaard B. Communication with patients and colleagues. Dan Med Bull. 2011;58(12):B4359.
13. Darbyshire D, Gordon M, Baker P. Teaching handover of care to medical students. Clin Teach. 2013;10(1):32–7.
14. Gordon M. Non-technical skills training to enhance patient safety. Clin Teach. 2013;10(3):170–5.
15. Camp CL, Gregory JK, Lachman N, Chen LP, Juskewitch JE, Pawlina W. Comparative efficacy of group and individual feedback in gross anatomy for promoting medical student professionalism. Anat Sci Educ. 2010;3(2):64–72.

16. Swing SR. The ACGME outcome project: retrospective and prospective. Med Teach. 2007;29(7):648–54.
17. Holmboe ES, Edgar L, Hamstra S. The milestones guidebook. In: Accreditation Council for Graduate Medical Education. 2016. https://www.acgme.org/Portals/0/MilestonesGuidebook.pdf. Accessed 26 June 2019.
18. Dougherty RW, Wyles CC, Pawlina W, Lachman N. "The team is more than the sum of its parts": implementation of charters to improve team dynamics in an anatomy course. TAPS. 2018;3(1):6–14.
19. Lerner S, Magrane D, Friedman E. Teaching teamwork in medical education. Mt Sinai J Med. 2009;76(4):318–29.
20. Pawlina W, Hromanik MJ, Milanese TR, Dierkhising R, Viggiano TR, Carmichael SW. Leadership and professionalism curriculum in the Gross Anatomy course. Ann Acad Med Singap. 2006;35(9):609–14.
21. Hulsman RL, Peters JF, Fabriek M. Peer-assessment of medical communication skills: the impact of students' personality, academic and social reputation on behavioural assessment. Patient Educ Couns. 2013;92(3):346–54.
22. Sasou K, Reason J. Team errors: definition and taxonomy. Reliab Eng Syst Safe. 1999;65(1):1–9.
23. Swartz WJ. Using gross anatomy to teach and assess professionalism in the first year of medical school. Clin Anat. 2006;19(5):437–41.
24. Arnold L, Shue CK, Kalishman S, Prislin M, Pohl C, Pohl H, et al. Can there be a single system for peer assessment of professionalism among medical students? A multi-institutional study. Acad Med. 2007;82(6):578–86.
25. Arnold L, Shue CK, Kritt B, Ginsburg S, Stern DT. Medical students' views on peer assessment of professionalism. J Gen Intern Med. 2005;20(9):819–24.
26. Spampinato CM, Wittich CM, Beckman TJ, Cha SS, Pawlina W. "Safe Harbor": evaluation of a professionalism case discussion intervention for the gross anatomy course. Anat Sci Educ. 2014;7(3):191–8.
27. Peer NF. Feedback marking: developing peer assessment. Innovat Educ Train Int. 1995;32(1):175–85.
28. Panadero E, Alqassab M. An empirical review of anonymity effects in peer assessment, peer feedback, peer review, peer evaluation and peer grading. Assess Eval High Educ. 2019;44(8):1253–78.
29. Spandorfer J, Puklus T, Rose V, Vahedi M, Collins L, Giordano C, et al. Peer assessment among first year medical students in anatomy. Anat Sci Educ. 2014;7(2):144–52.
30. Tan HH, Tan ML. Organizational citizenship behavior and social loafing: the role of personality, motives, and contextual factors. J Psychol. 2008;142(1):89–108.
31. Michaelsen LK. Team-based learning for health professions education: a guide to using small groups for improving learning. 1st ed. Sterling: Stylus; 2008.
32. Escobar-Poni B, Poni ES. The role of gross anatomy in promoting professionalism: a neglected opportunity! Clin Anat. 2006;19(5):461–7.
33. Bryan RE, Krych AJ, Carmichael SW, Viggiano TR, Pawlina W. Assessing professionalism in early medical education: experience with peer evaluation and self-evaluation in the gross anatomy course. Ann Acad Med Singap. 2005;34(8):486–91.
34. American Board of Internal Medicine Foundation. ABIM Project Professionalism. 1st ed. Philadelphia: ABIM Communications; 1999.
35. Lachman N, Christensen KN, Pawlina W. Anatomy teaching assistants: facilitating teaching skills for medical students through apprenticeship and mentoring. Med Teach. 2013;35(1):e919–25.
36. Hull L, Sevdalis N. Advances in teaching and assessing nontechnical skills. Surg Clin North Am. 2015;95(4):869–84.
37. Robertson ER, Hadi M, Morgan LJ, Pickering SP, Collins G, New S, et al. Oxford NOTECHS II: a modified theatre team non-technical skills scoring system. PLoS One. 2014;9(3):e90320.
38. Mishra A, Catchpole K, McCulloch P. The Oxford NOTECHS system: reliability and validity of a tool for measuring teamwork behaviour in the operating theatre. Qual Saf Health Care. 2009;18(2):104–8.
39. Pawlina W, Drake RL. Authentic learning in anatomy: a primer on pragmatism. Anat Sci Educ. 2016;9(1):5–7.
40. Brutus SD, Donia MBL, Ronen S. Can business students learn to evaluate better? Evidence from repeated exposure to a peer-evaluation system. Acad Manage Learn Educ. 2013;12(1):18–31.

Designing and Using Formative Assessment in Anatomy

42

Darrell J. R. Evans

The 'Whats and Whys' of Formative Assessment

The ability of students to understand where their learning is at and for educators to be able to adjust their learning design and provide appropriate feedback requires curriculum designers to have a deliberate focus on the inclusion of formative assessment opportunities as an education program is constructed. However for many educators, formative assessments have been more of an afterthought, an added extra or an element that is not really part of the overall curriculum design, with most of the attention given to summative assessment. The word summative implies that students must reach the summit, but of course the learning journey is developmental and often requires educators to facilitate, guide and help students to reach that summit. Formative assessment provides one of the most important and impactful tools educators can use to achieve this.

Formative assessment is not new, and educators have been using such an approach in teaching and learning for centuries [1]. Some argue that philosophers such as Socrates practiced formative assessment to assess where his students were in their learning and subsequently adapted his teaching to guide and develop their understanding. However formative assessment as a recognised pedagogical term became part of the conventional nomenclature in the 1960s with Michael Scriven describing the different roles evaluations can play within a curriculum [2]. It was advanced that during the development of a curriculum, evaluation could be used to guide necessary changes and refinement ("formative"), whereas at the end of the finished curriculum, evaluation was used to determine whether the desired goal had been met ("summative"). Others then took this approach further, applying it to educational assessment recognising that assessing students should not only be used to make summative judgements. This became part of the foundations of the mastery of learning concept [3].

Since then, formative assessment has become a well-recognised part of learning theories, and a variety of approaches have been designed and used at all levels of education, including the tertiary sector. Formative assessment has been defined in a variety of ways [4, 5], but in general it is accepted that formative assessment includes tasks that provide feedback for both students and educators, with students able to develop and adjust their learning as a result and educators to adapt their teaching approach, both with the aim of enabling student achievement of intended

D. J. R. Evans (✉)
Faculty of Health and Medicine, University of Newcastle, Callaghan, NSW, Australia

Faculty of Medicine, Nursing and Health Sciences, Monash University, Clayton, VIC, Australia
e-mail: darrell.evans@newcastle.edu.au

L. K. Chan, W. Pawlina (eds.), *Teaching Anatomy*, https://doi.org/10.1007/978-3-030-43283-6_42

learning outcomes. Used appropriately formative assessment should be designed to close the gap between current learning and desired goals. As such, formative assessment can be best defined as a feature of "assessment for learning".

Why Formative Assessment?
Formative assessment provides feedback for both students and educators, with students able to develop and adjust their learning as a result and educators to adapt their teaching approach, both with the aim of enabling student achievement of intended learning outcomes.

The place of formative assessment within a constructive alignment approach to learning is clear. However, for educators to spend valuable time designing formative assessment opportunities that are appropriately entwined and matched with learning outcomes, it is necessary to provide evidence that such assessments have a positive impact on student learning. Black and Wiliam designed a comprehensive meta-analysis of over 250 research studies focused on formative assessment and demonstrated that such an approach within the learning design can make a difference in learning gain at all levels [6]. Importantly they also concluded that successful formative assessment was related to active student involvement, enhanced feedback, self-assessment and the ability for educators to adapt their teaching during the learning cycle.

In anatomy, formative assessment has been used for many years, often focusing on "mock" assessments such as prosection "spotter" tests or in the form of quizzes and case studies. However students can often use these approaches in a superficial way to focus on rote learning of anatomical facts and with the aim of passing a summative assessment. Approaches that enable the retention of knowledge and particularly understanding beyond the learning period is therefore critical so that the students instead develop deeper learning, which enables effective application within their practice [7]. This relies on an active learning process delivered in context and where assessment opportunities are spaced and repeated over a defined period and designed for ongoing learning and not final examination cramming [8–11].

Principles of Formative Assessment

The principles of formative assessment relate specifically to the intended benefits of using such approaches.

Student-Focused Principles

The inclusion of formative assessment within the curriculum should be aimed at providing students with opportunities to practice and evaluate their learning as the program proceeds, supporting a developmental and progressive approach to learning. Placing formative assessments at defined intervals, perhaps when difficult anatomical concepts have been introduced or when a change of topic has occurred, allows a student to identify gaps in their learning and provides opportunity to make adjustments to fill those gaps. This relies on the inclusion of high-quality, effective and immediate feedback within formative assessments. Feedback should be designed to allow the student to determine both what they have learned and also where their misunderstandings and gaps are and preferably some guidance on what to do to improve, although deeper learning is likely to come from learners self-correcting. Such an approach allows the student to identify their strengths and weaknesses without incurring any academic penalty. Formative assessment can assist students to define the learning expectations including the level of competence required at different stages of the program, which when used appropriately can alleviate the compulsion for students to ask "what do I need to know for the examination?", a clear indication of a superficial approach. Ultimately formative assessment should enable students to engage with learning during the learning delivery phase and not at the end of the process and strengthen the student's ability to self-regulate their performance and progress [12].

Educator-Focused Principles

Formative assessment is also important for the educator and it provides them with opportunities to evaluate and modify their teaching during the learning cycle, including the approaches used, the emphasis on particular subject areas or concepts and the pace of teaching and learning. Educators are able to gather information about their students' progress and learning needs at times when changes can be made with the results of some assessments acting as early warning mechanisms [13]. Educators can ensure even small learning successes can be celebrated, building confidence and self-esteem in the learners. Because formative assessment is most effective when the emphasis is on an active approach, such opportunities also promote staff-student interaction.

Formative Assessment Principles

- Incorporating practice and evaluation opportunities
- Establishing learning expectations
- Identifying learning strengths, weaknesses and gaps
- Providing high-quality feedback
- Supporting progressive learning
- Guiding educators to adjust teaching approach
- Encouraging active and motivational learning

Learning to Learn

In many healthcare programs, anatomy is taught quite early to ensure students have developed an appropriate foundation of knowledge and understanding. However we now recognise that anatomy provides the opportunities to develop a range of other skills and attributes often referred to as non-traditional discipline-independent skills (NTDIS) [14]. One attribute that could be included in these is the ability to learn, probably the most important skill our students will develop as this will underpin their ability to engage with lifelong learning, a critical attribute for all healthcare professions. The principles behind formative assessment demonstrate how important this approach is to supporting the ability for students to learn and the strategies they employ.

Designing and Delivering Formative Assessment

Curriculum Design

When developing the anatomy curriculum, whether as a stand-alone program or part of an integrated program, it is essential to ensure assessment components, both formative and summative, are inherent to the design of the curriculum and program, enabling focused alignment to the student learning outcomes [15]. Assessment, especially formative assessment, should not be an afterthought of the design and just incorporated in an uncoordinated way, but rather an integrated thread. Formative assessments provide the opportunity to help students engage with anatomical concepts, knowledge and understanding as well as sets of skills and practices, but because of this, they need to be varied in nature (aligning to required focus) and be meaningful to learning. They should not be an extra load on students or those designing them.

Successful formative assessment strategies need appropriate planning, and a number of steps should be considered during the design of the curriculum.

When in the Curriculum to Offer Formative Assessment?

Deciding when to offer formative assessment activities is dependent on the program design but should be placed throughout the learning cycle, creating a spaced approach to assessment [8–10].

This allows for perhaps an early focus on testing anatomical concepts or terminology, getting students comfortable with subject matter and the program approach, whilst later having a greater emphasis on understanding, application and further exploration. A spaced approach also encourages students to act on feedback and apply ongoing learning and thus maximises the potential for deeper learning and can have a positive impact on summative assessment outcomes [16]. Therefore, including a factual recall approach might have a place early in the journey, but progressive learning requires other assessment types to be created and aligned within the learning design which lead up to any summative or competence-based assessment. Inclusion of "just-in-time" formative assessments in preparation for summative assessments is popular in some programs, but the educator should be aware not to overly focus on a rote-learning support mechanism but encourage students to think metacognitively about their learning [17].

Where in the Curriculum to Offer Formative Assessment?

The anatomy curriculum can feature as a standalone course or part of an integrated program, and as such some of the decisions about where formative assessments can take place will be part of a wider design process. However formative assessments can be easily incorporated into a range of curriculum designs and settings. When designing a particular anatomy teaching session or group of sessions, consideration should be made about whether an assessment activity is offered before a class as part of preparation. Using this approach, the educator must make sure that the subsequent session uses the intended outcomes of the assessment so the students see the benefit of the assessment in their learning and are therefore more likely to keep engaging. Assessments can be incorporated during the class either as part of the learning activities that make up the session or perhaps as an extra, voluntary task, but clearly aligned to the learning outcomes of that session. Post-class formative assessments may prove the most difficult to elicit engagement from students and rely on a particular set of attributes (see section "Creating Engagement Around Formative Assessment") but can be the most powerful by providing opportunities to reinforce and review what has been learned, providing an opportunity for practice.

How to Offer Formative Assessments?

Incorporating formative assessment into a curriculum is most effective when students are active in the process. This can materialise with formative assessments integrated within active learning tasks in face-to-face sessions such as a tutorial or practical class. Such an approach places formative assessment directly within the most focused part of the learning process and at a time when the educator can be present and facilitate. Using such an approach often means the student may be unaware that formative assessment is being used. Where assessment is provided outside the classroom, technology-enabled tasks might be most appropriate and can be provided through learning management systems, apps or bespoke systems/platforms. Where possible tasks should always be related to the specific learning outcomes, and therefore when a generic commercial assessment approach or product is used, it is important that students can associate the tasks to the learning outcomes at hand, and the educator ideally needs to guide and curate this. Many of the commercial products available are based around quizzes, and whilst these can provide an engaging approach, often the number of question types is restricted. Quizzes should not be the sole focus of any formative assessment regime but can be used successfully early within the learning cycle and as ways to encourage interaction when delivered in a face-to-face environment. Quizzes are often used by students as "just-in-time" exercises prior to summative examinations [18]. When looking at the types of assessment that might be used, it is important for the educator to use an approach that recognises that students are developing their learning progressively; therefore the approach, level and content will need to change over time. It is also important that a formative assessment regime accommodates different learner types and may require a varied set of formats. The connection of formative assessment to the approach used for summative assessment is an important consideration as one outcome of formative assessment is to practice the skills required in the summative assessment. Reports show that including even one formative assessment can have a positive effect on summative outcomes [13, 19, 20]. However this does not mean copying the type of assessment format; rather the focus should be on the learning outcomes.

How Will Feedback Be Provided?

One of the inherent principles of effective formative assessment design is the focus on providing feedback to the student that they can subsequently use to benefit future learning. Therefore when designing a formative assessment task, ensuring attention has been provided to how feedback will be delivered is crucial and must be aligned to the type of assessment used. If the assessment task is delivered as a self-contained face-to-face activity, feedback should also be preferably delivered at the same face-to-face session and as close to the assessment as is practical. The advantage of many anatomy courses is that educators get the opportunity to interact with students in practical-based classes providing lots of opportunity for engagement and feedback. In other types of formative assessment, feedback can take the form of written feedback, audio or video feedback or peer-led feedback and can use the benefit of technology through automation and more recently artificial intelligence (AI). In all cases feedback should focus on areas of accomplishment, where there are gaps or misunderstanding in the required learning and with guidance on how to improve so the student is able to action appropriately. The educator must facilitate and fully support this process, requiring ongoing interaction and dialogue with students.

Providing Guidance and Expectations to Students

As with the design of any teaching and learning activity, students need clear and effective instructions for any formative assessment activity that is being provided, and this can be the key to whether a student will engage or not. The purpose and approach of the assessment need to be defined up front, informing the students what the task will focus on, why and how this will aid their learning. The inclusion of criteria and rubrics where appropriate will provide students with a clear understanding of what is expected of them and can later act as part of the feedback process. When introducing a new or different type of formative assessment, including a worked example can help students identify the level of learning required and the expected detail or approach. Where a formative assessment is built into an active learning task in the face-to-face environment, additional instruction about the assessment element may not be required.

Avoiding the Dangers of Over-Packing the Curriculum

The inclusion of formative assessment within the curriculum must always be designed to assist student learning and carefully aligned to the learning outcomes. Therefore a well-spaced approach to formative assessment where tasks are deliberately designed and incorporated either directly into learning activities or throughout the learning cycle at points of most benefit should be effective. However there can be a tendency to incorporate too many assessment tasks or to keep adding further tasks as new ones are thought of. This can have the disadvantage of students perceiving assessment overload and can lead to a disengagement with the tasks at best and with the actual learning at worst [16]. For the educator, designing new assessment tasks can be time-consuming, especially when linked to the associated feedback that needs to be provided, and therefore consideration of educator workload is also necessary. Therefore creating an appropriate balance is important, evaluating what is working and what is not so that tasks can be removed and replaced only as appropriate.

Incorporating Formative Assessment into Current Curricula

Where a curriculum is not undergoing redesign, it should still be a goal to include formative assessment where this has not been a previous practice.

Incorporating formative assessment into current curricula is possible but must avoid the appearance of an add-on; otherwise, students will be less likely to engage with it. Therefore educators need to focus on the intended learning outcomes and design the most appropriate tasks and activities that align with the delivery approach for the course/sessions. This might include changing or adapting an in-class activity to be a formative assessment-focused activity promoting interaction and collaboration. Alternatively it could be the result of speaking with previous students and understanding what types of formative assessment would have helped their learning and when this should have been offered during the program. Such an approach provides the opportunity to refresh the look and feel of the curriculum with a continued emphasis on student learning, whilst not having to launch a full curriculum overhaul.

What Will You Do?

You have a limited set of resources and materials within the anatomy laboratory, yet you wish to provide a set of active learning-based formative assessment for all your students, focusing on collaboration and occurring at different times during the course. How might you go about designing such a set of activities so that they do not rely on technology but do provide immediate feedback?

Creating Engagement Around Formative Assessment

Without applying credit to an assessment, many argue that students fail to engage, and therefore spending time designing and developing formative assessment is wasted. However the basis of such an argument suggests that such educators have not bought into the benefits of formative assessment, not set expectations appropriately with their students and just as importantly not designed assessments that students would wish to engage with.

Importance of Educator Commitment

For formative assessments to be an effective learning tool, it is critical that educators not only focus on good curriculum design as described elsewhere in this chapter but also demonstrate commitment to and the value and emphasis of using these tools. If the educator demonstrates the belief in these activities, there is more opportunity for the approach to succeed and to become a mainstream learning mechanism. This commitment needs to be clear with students. It is unhelpful, for instance, for an educator to imply "have a go if you want to, but there are no marks up for grabs". Instead there should be an emphasis on setting clear positive expectations from the outset which show students that this is part of the holistic learning frame. If students can see how this enables their learning, how they gain from the activity/task and ultimately how this connects to the summative assessment or competency required, there is likely to be more engagement. Some educators may not use the term formative in case that it is seen as less important and therefore not something to engage with, and instead they may focus on "progressive learning tasks".

Making Formative Assessment Fun

Designing formative assessment provides an opportunity for educators to be creative and innovative in their approach and to focus on stimulating and motivating learners [21, 22]. Whilst some may prefer to use commercially available products (where effective curation by the educator is required for the best outcomes), developing in-house approaches or adapting what others have used can be more effective, by connecting directly to the learning outcomes and tying more closely to the educator's student cohorts. Recently educators have begun to use approaches that focus on more entertaining approaches including competition as a way to engage students with formative assessments. One such approach is to use TV quiz and game show formats or a focus on games and puzzles [10, 23–26]. These have been used in a variety of scientific and health-related curricula, including anatomy, and report effective outcomes for many students. The reason for their success seems to be related to the familiarity of the approach (because students have seen it before), the entertainment value and the variety of questions or approaches so students do not become trained on particular question types. Instead the content, the level and the complexity become the focus. These types of approaches can help reduce the anxiety felt by students and convey an air of fun to the process, which may subsequently help when it gets to the summative assessment. A cultural shift in the way in which learners see assessment as part of the learning cycle is an important step, and including conduits where there might be some familiarity can facilitate this.

Involving Students in Creating Formative Assessment

One way to become more student/learner focused in assessment design or to provide the opening for more creative approaches is to involve students in the design and development [27]. Students are more likely to come up with approaches that will engage fellow learners; they know what interests them, they will act as champions of the approach and thus increase the chance of the tasks being taken more seriously by other students. In turn the student developers learn about assessment design; if they are near-peers, they become refocused on the learning outcomes, and overall they are provided with opportunities to express their creativity and enterprising skills.

What Will You Do?

You have introduced an online quiz as part of the locomotor anatomy module. However you have followed the activity on the learning management system and detected that whilst there was some early engagement this has tailed off and only a few students are now using the quiz. How would you discover why students were no longer engaging and what adaptations might you make to either change the assessment itself or the expectations you set?

Types of Formative Assessment

A wide variety of assessment approaches and types have been incorporated into anatomy curricula over many years, with some deliberately designed with anatomy in mind whilst others have been adapted to be effective for students learning anatomy [10, 13, 18, 23, 28–33]. Although it is not possible to list all of these assessment types, a selection is provided which can be used as a guide for anyone looking to start exploring opportunities (Table 42.1). Some of these assessment types can also be used as summative approaches too and in many cases should be part of an active learning-based curriculum. In determining which formative assessments to use, the educator should make a number of considerations including:

- Delivery mode (face-to-face or online)
- Delivery focus (individuals or groups)
- Class size (large class, small group)
- Session type (laboratory class, lecture, tutorial)
- Level of the learner (nascent, advanced)
- Number of assessments (one-off or multiple)
- When within the program (early, mid, just-in-time)

Tips on Incorporating Formative Assessment

- Introduce appropriately spaced assessment opportunities.
- Build assessments into classroom activities.
- Use a variety of assessment types but avoid overloading.
- Include assessments other than just quizzes.
- Do not only focus on digital assessment solutions.
- Ensure feedback is actionable by the student.
- Consider reducing some summative assessment.
- Make assessments fun and engaging.

It is important when introducing new formative assessment opportunities that time is spent subsequently on evaluating the efficacy and outcomes of those assessments. This provides an opportunity to assess which activities are working well, which ones students are engaging with (when and how often), the student feedback and whether it may be having an effect on outcomes. Evaluation should lead to evolving of the assessment which may range from small tweaks to major changes to removal.

Challenges of Designing and Delivering Formative Assessment

Introducing the concept of formative assessment into curricula in a meaningful, integrated and deliberate way comes with challenges that should be appreciated by educators, program leaders, those making the decisions with regard to overall curriculum content and, of course, those controlling the resources. The main challenge has been mentioned earlier but revolves around the change

Table 42.1 Types of formative assessment that can be used within anatomy programs

Assessment	Assessment description
Audience response systems (ARS)	Assessment approach where combinations of software and hardware present a variety of anatomy questions, record responses and provide feedback, usually using "clickers" or mobile devices. Non-technology-based approaches can substitute such as coloured card
Puzzles including word searches and crosswords	*Word search*—a word game where the objective is to find and mark all the anatomical words hidden inside the box *Crossword*—a word puzzle where the objective is to fill a grid of squares forming anatomical words or phrases by solving clues
Prosection spotter tests (steeplechase)	Assessment where wet anatomical specimens, plastic models, 3D specimens, bones, histological slides and imaging are used which are pre-pinned with labels/flags with questions ranging from basic identification of structures to ones of application and function
Integrated anatomy practical paper	Assessment combining spotter test and elements of an objective structure clinical examination (OSCE). This can focus on higher levels of taxonomy
Autopsy reports	Assessment approach where students work in teams to provide a report detailing pathologies identified during dissection. This can be used to assess anatomical knowledge, visual cues, reasoning and interpretation
Illustration tasks	Assessment where students demonstrate their anatomical knowledge and understanding by making illustrations and models. This may extend to body painting exercises
Concept maps	Students use a graphical organiser approach to represent relationships between concepts to demonstrate anatomical knowledge and applied understanding
Oral assessments	Educators or peers pose questions to students in spoken form. The student has to answer the question to demonstrate sufficient knowledge and understanding. This can be used to discuss complex applied anatomical and clinical problems and scenarios
Online quizzes	Technology-based approaches to assess anatomical knowledge, understanding and application using multiple choice, true/false, extended matching, identification, short answer and other question types. Successful approaches include interactive quiz approaches and those that have a fun or competitive perspective
Online forums	Students divided into groups and prompted into anatomical discussions through forums working through conceptual problems and case studies. This can also be used to discuss key program topics and provide safe environment for students to bring to light misunderstandings and provide feedback to each other
One-minute papers	Educator asks questions at the end of session about main points of the session, and students have 1 minute to submit an answer (on paper or through technology). Educator can show summary of main points then or assess student responses after class and addresses misunderstandings in the next session
3-2-1 summariser	Educator asks students to write 3 concepts learned in the session, 2 questions about the topic and 1 misunderstanding or misconception

Table 42.1 (continued)

Assessment	Assessment description
Sticky note challenge	Educator provides a variety of questions about the topic on sticky notes stuck around the room in three colours denoting easy, medium and difficult. Students must select one of each colour and answer the question when the educator asks
Interactive tasks	*Think-pair-share*—Educator asks a question, students write down the answer, and then students discuss their answers in pairs *Peer interviews*—Students interview each other about what they have learned on a particular topic. This can be used to demonstrate understanding from previous sessions *Incremental case study*—Educator provides different information on an applied problem to different groups of students. Groups interview other groups to gain additional information until they can solve the problem *Numbered heads*—Each student in a group is assigned a number. Groups are provided with mini-problems and case studies and agree on answers. The educator randomly selects a number and the student with that number answers for the group *Elbow partners*—During a session students turn to person at left or right elbow to talk about the topic or ask question about the topic
Learning logs/journals	Piece of writing that summarises a student's critical reflection or main learning on a subject. This provides for a later study review guide
Peer demos	Educator assigns particular students to demonstrate learned anatomical concepts to small groups of students and answer questions
Revision bingo	Instead of numbers being called out, educators give students a verbal clue that fits with one of the anatomical answers on their playing grid. Get five correct answers in any straight line and they win the game
Diagnostic tests (are included here as a type of formative assessments)	Variety of approaches are used including quizzes, polling and scenarios undertaken at the beginning of a unit of study to assess the skills, abilities, interests, experiences, levels of achievement or difficulties of an individual student or a whole class

in practice and attitude that is needed by many to ensure formative assessments are an integral part of the learning process. This takes understanding, commitment and leadership so that formative assessment is not seen as of secondary importance to summative assessment. Piloting some approaches with purposeful evaluation will provide evidence of success which can be used in any subsequent curriculum planning debates and when determining resource allocation.

Developing effective formative assessments takes the time, patience and commitment of the educator and can therefore be seen by some as a less important part of the curriculum design process. Where time to develop formative assessments is short, incorporating readily available commercial products is a possibility (resource permitting). However, these must still be entwined appropriately within the program design, and there is an additional expectation on the educator to curate the content and approach to ensure the desired learning outcomes are those that are addressed. Alternatively, starting off with small-scale interventions built within the curriculum that are not resource or time intensive can provide students with some opportunities. Over time additional elements can be added so that the program develops further. In this way the development of a formative assessment portfolio is gradual and more manageable.

Getting the right balance between formative and summative assessment is difficult, and naturally, although perhaps unfortunately, the emphasis has usually been on summative assessment. Scaife and Wellington [34] have argued that by spending more time on formative assessment and less on summative assessment, the educator's energy and time are used in a way that can make a meaningful difference. However, letting go of some summative assessment is difficult for educators because of tradition and sometimes because of restrictions in the program curriculum that are out of the control of anatomy educators. Where possible, opportunities to reduce summative assessment in favour of introducing some formative assessment should be encouraged.

Introducing formative assessment does not just rely on time but also capability, and it can be challenging for educators to think differently about assessment and have the ability to design new approaches or use technology to realise ideas. Therefore it is important that curriculum design help is available through training opportunities, the appointment of instructional designers focused on transforming the approach to curriculum development and other facilities and resources that will enable new assessment approaches including technologies. If this is combined with opportunities to recognise educators for their subsequent creativity through a variety of means, this will help encourage further curriculum change, ultimately benefiting student learning.

The opportunity to provide more creative approaches to assessment comes with risks that the approach may not be as effective as planned. This should not be used as a barrier to avoid innovation as formative assessments can be more easily adapted, changed or removed from the curriculum than the summative assessment, which may be a more regulated element of the program. However, this does not minimise the potential disruption to the students learning though, and so setting expectations with students when first introducing a new formative assessment is important. Allowing the students to work with you and to provide the feedback on the assessment will be beneficial.

Challenges in Introducing Formative Assessment

- Available time in the curriculum
- Design and development time of educators
- Capability of educators
- Commitment of educators and departments
- Overt focus on summative assessment
- Risk of ineffective approaches

Conclusion

Formative assessment can be a powerful tool in enabling students to understand where their anatomical learning is at and for educators to be able to adjust their learning design and provide appropriate feedback. Formative assessment must be a deliberate part of the curriculum design with a focus on developing deeper learning and not used purely as a mechanism for testing rote learning of anatomical facts. When designing formative assessments, it is important for the educator to consider when, where and how the formative assessment activities will be included and the approach for providing effective feedback and guidance in order to maximise the learning opportunities for their students.

References

1. Greenstein L. What teachers really know about formative assessment. Alexandria/Virginia: Association for supervision and curriculum development; 2010.
2. Scriven M. The methodology of evaluation. In: Tyler RW, Gagné RM, Scriven M, editors. Perspectives of curriculum evaluation. Chicago: Rand McNally; 1967. p. 39–83.
3. Bloom BS, Hastings JT, Madeus GF. Handbook on formative and summative evaluation of student learning. New York: McGraw-Hill Book Co.; 1971.
4. Popham WJ. Transformative assessment. Alexandria/Virginia: Association for supervision and curriculum development; 2008.
5. Hegarty M. Formative assessment: making it happen in the classroom. Thousand Oaks: Corwin Press; 2010.
6. Black P, Wiliam D. Assessment and classroom learning. Assess Educ. 1998;5:7–74.
7. Rolfe I, McPherson J. Formative assessment: how am I doing? Lancet. 1995;345:837–9.
8. Krasne S, Wimmers PF, Relan A, Drake TA. Differential effects of two types of formative assessment in predicting performance of first-year medical students. Adv Health Sci Educ Theory Pract. 2006;11:155–71.
9. Kerfoot BP, Baker HE, Koch MO, Connelly D, Joseph DB, Richey ML. Randomized, controlled trial of spaced education to urology residents in the United States and Canada. J Urol. 2007;177:1481–7.
10. Evans DJR, Zeun P, Stanier RA. Motivating student learning using a formative assessment journey. J Anat. 2014;224:296–303.

11. Konopasek L, Norcini J, Krupat E. Focusing on the formative: building an assessment system aimed at student growth and development. Acad Med. 2016;91:1492–7.
12. Nicol DJ, Macfarlane-Dick D. Formative assessment and self- regulated learning: a model and seven principles of good feedback practice. Stud High Educ. 2006;31:199–218.
13. Azzi AJ, Ramnanan CJ, Smith J, Dionne E, Jalali A. To quiz or not to quiz: formative tests help detect students at risk of failing the clinical anatomy course. Anat Sci Educ. 2015;8:413–20.
14. Evans DJR, Pawlina W, Lachman N. Human skills for human[istic] anatomy: an emphasis on non-traditional discipline-independent skills. Anat Sci Educ. 2018;6:221–4.
15. Vanderlelie JJ, Alexander HG. Learning-oriented assessment increases performance and written skills in a second year metabolic biochemistry course. Biochem Mol Biol Educ. 2016;44:412–20.
16. Das S, Alsalhanie KM, Nauhria S, Joshi VR, Khan S, Surender V. Impact of formative assessment on the outcome of summative assessment – a feedback based cross sectional study conducted among basic science medical students enrolled in MD program. Asian J Med Sci. 2017;8:38–43.
17. Brady M, Forest CP. Metacognition, formative assessment and student perspective: learning about metacognition through in-class comparison of response systems. J Physician Assist Educ. 2018;29:104–8.
18. McNulty JA, Espiritu BR, Hoyt AE, Ensminger DC, Chandrasekhar AJ. Associations between formative practice quizzes and summative examination outcomes in a medical anatomy course. Anat Sci Educ. 2015;8:37–44.
19. Carrillo-de-la-Pena MT, Bailles E, Caseras X, Martinez A, Ortet G, Perez J. Formative assessment and academic achievement in pre-graduate students of health sciences. Adv Health Sc Educ Theory Pract. 2009;14:61–7.
20. Bekkink MO, Donders R, van Muijen GNP, Ruiter DJ. Challenging medical students with an interim assessment: a positive effect on formal examination score in a randomized controlled study. Adv Health Sci Educ Theory Pract. 2012;17:27–37.
21. Harlen W, James M. Assessment and learning: differences and relationships between formative and summative assessment. Assess Educ. 1997;4:365–79.
22. Hudson JN, Bristow DR. Formative assessment can be fun as well as educational. Adv Physiol Educ. 2006;30:33–7.
23. Willmott CJR. Revision bingo. Biochem Mol Biol Educ. 2001;29:193–5.
24. Howard MG, Collins HL, DiCarlo SE. "Survivor" torches "who wants to be a physician?" in the educational games ratings war. Adv Physiol Educ. 2002;26:30–6.
25. Zakaryan V, Bliss R, Sarvazyan N. Non-trivial pursuit of physiology. Adv Physiol Educ. 2005;29:11–4.
26. Shah S, Lynch LMJ, Macias-Moriarity LZ. Crossword puzzles as a tool to enhance learning about anti-ulcer agents. Am J Pharm Educ. 2010;74:117.
27. Poot R, de Kleijn RA, van Rijen HV, van Tartwijk J. Students generate items for an online formative assessment: is it motivating? Med Teach. 2017;39:315–20.
28. Lukic IK, Gluncic V, Katavic V, Petanjek Z, Jalsovec D, Marusic A. Weekly quizzes in extended-matching format as a means of monitoring students' progress in gross anatomy. Ann Anat. 2001;183:575–9.
29. Alexander CJ, Crescini WM, Juskewitch JE, Lachman N, Pawlina W. Assessing the integration of audience response system technology in teaching of anatomical sciences. Anat Sci Educ. 2009;2:160–6.
30. Ranaweera SP, Montplaisir LM. Students' illustrations of the human nervous system as a formative assessment tool. Anat Sci Educ. 2010;3:227–33.
31. Smith CF, McManus B. The integrated anatomy practical paper: a robust assessment method for anatomy education today. Anat Sci Educ. 2015;8:65–73.
32. Choudhury B, Gouldsborough I, Shaw FL. The intelligent anatomy spotter: a new approach to incorporate higher levels of Bloom's taxonomy. Anat Sci Educ. 2016;9:440–5.
33. Choudhury B, Freemont A. Assessment of anatomical knowledge: approaches taken by higher education institutions. Clin Anat. 2017;30:290–9.
34. Scaife J, Wellington J. Varying perspectives and practices in formative and diagnostic assessment: as case study. J Educ Teach. 2010;36:137–51.

43 Core Syllabi in Anatomy

Claire F. Smith, Gabrielle M. Finn, Jane Holland, Jane Stewart, Siobhan A. Connolly, Catherine M. Hennessy, and Stephen McHanwell

Introduction

Health profession education is undergoing a period of rapid change. The prime impetus for these changes is the continuing advancement of our knowledge in both basic and clinical sciences. Together with the reassessment of what should be taught major changes have occured in medical, dental, and allied health profession syllabi and curricula. Advances in genetics and molecular biology are giving rise to new treatments, and an increasing emphasis on primary care and prevention has seen these disciplines grow in importance. In addition, there is increasing recognition of the importance in the early years of programs of teaching clinical, communication, and professionalism skills. No course can just expand indefinitely, and so, as new content is included, other parts of the curriculum have to be reduced. The reduction in the time available for teaching has been felt keenly across all the basic sciences [1, 2], none more so than anatomy with perceived negative consequences for student learning [3–5]. One response to the need to reduce the time spent on learning and teaching in anatomy has been a focus on creating core syllabi, however defined, focusing on topics deemed to be of the greatest clinical relevance [6–9].

Anatomy comprises of a large factual base with its own specialized language, a large fraction of which has to be mastered before that knowledge can be deployed effectively [10, 11].

C. F. Smith (✉)
Brighton and Sussex Medical School, University of Sussex, Brighton, East Sussex, UK
e-mail: c.smith@bsms.ac.uk

G. M. Finn
Faculty of Biology, Medicine and Health, University of Manchester, Manchester, UK
e-mail: gabrielle.finn@manchester.ac.uk

J. Holland
Royal College Surgeon Ireland, Dublin, Ireland
e-mail: jholland@rcsi.ie

J. Stewart
School of Medical Education, Faculty of Medical Sciences, Newcastle University, Newcastle upon Tyne, UK
e-mail: jane.stewart@newcastle.ac.uk

S. A. Connolly
Department of Anatomy, Edinburgh Medical School: Biomedical Sciences, College of Medicine and Veterinary Medicine, University of Edinburgh, Edinburgh, UK
e-mail: s.connolly@ed.ac.uk

C. M. Hennessy
Brighton and Sussex Medical School, University of Sussex, Brighton, East Sussex, UK
e-mail: c.hennessy@bsms.ac.uk

S. McHanwell
School of Medical Education and School of Dental Sciences, Faculty of Medical Sciences, Newcastle University, Newcastle upon Tyne, UK
e-mail: stephen.mchanwell@newcastle.ac.uk

L. K. Chan, W. Pawlina (eds.), *Teaching Anatomy*, https://doi.org/10.1007/978-3-030-43283-6_43

Anatomy is a core subject for students on a wide range of healthcare programs. The expectations about levels of anatomical knowledge and topics to be covered and the competencies to be demonstrated will vary between professions.

Defining Terms

The terms "syllabus" and "curriculum" are often used interchangeably even though they mean different things. The term "syllabus" is a list of topics defined as content items. The term "curriculum" describes all the elements of a program and not simply the content of the course but also the course structure, aims and objectives, teaching methods, assessment processes, and philosophical approaches to learning. Learning outcomes capture what a student is expected to know by the end of a program and can be used to detail a syllabus by expressing the cognitive, psychomotor, or affective level of competence that students must demonstrate to pass the program. It is important to note that a syllabus will not express the mode of delivery of a program (didactic format, through problem-based, scenario-based, or discovery learning). That information will be held within the other components of the curriculum.

Developing a Core Syllabus

How does one set about developing a core syllabus? Knowledge of what constitutes a subject is often held tacitly by professionals and shared in collegial, informal, and anecdotal ways. At its simplest, one way of uncovering this knowledge is to gather a group of senior professionals together to reach agreement on a syllabus which then might be consulted upon. This informal and unstructured method can be very helpful at defining a local need and should involve a wide range of stakeholders, for example, teaching faculty, course director, administration, and educationalists. An informal method can be subject to bias and thus undermine its validity. Alternatively, the Delphi method [12] can be adopted as it is a methodology designed specifically to identify such phenomenon of expert opinion. Delphi methods provide a means by which the opinion of a group can be sought on an issue, a series of issues, or simply a series of statements to arrive at a consensus.

In the healthcare arena, Delphi methods have been used in a variety of situations, in emergency medicine [13–15] and in nursing education [12, 16–18] to establish consensus for diagnostic criteria [19]. They have been used in general practice information requirements [20] and for identifying applicable skills, attitudes, and practices in clinical teachers [21]. More specifically, as illustrated within this chapter, they have been used in curriculum development [22] for musculoskeletal anatomy for physical medicine and rehabilitation residents [23]. More recently, they have been used in anatomy education to provide a clear steer for regulated professions as to the breadth and level of anatomical knowledge required by graduates [24].

There is no generally accepted and standardized way of implementing a Delphi study [12], but if the necessary degree of rigor is to be achieved, key decisions about methodology need to be resolved before the study is undertaken. The key decisions include how to decide upon the composition of the expert Delphi panel, the nature of the statements or content put to the panel, and the degree of agreement that would constitute consensus. These decisions will be taken by the researcher or research panel managing the study.

Composition of Delphi Panels

The first decision is deciding on what defines an expert and converting this into inclusion and exclusion criteria. In doing so the researcher(s) identifies who holds the collegial knowledge that the study wishes to uncover. So, in the case of a Delphi panel looking at a core syllabus, this might be on the basis of identifying staff with appropriate discipline knowledge—specific qualifications and a minimum number of years of teaching in a relevant discipline to a particular group or groups(s) of students. The aim being also to select staff with a range of experience and a range of roles, from staff directly involved in teaching to course leaders and curriculum planners. Panel members need to understand the knowledge required at the end goal. The key is who holds the knowledge that the process is trying to extract. The size of the panel needs also to be considered. Larger panels will, of necessity, create additional work, but the panel does need to be large enough to capture the completeness of the views and to understand variations in perspectives.

Designing the Study

The first stage in a core syllabus study is deciding upon what will constitute the content. This might be on the basis of a preexisting set of learning outcomes, creating learning outcomes from scratch, or deciding content by some other means such as devising a topic list from, for example, one or more standard textbooks in a particular discipline [9].

There then follows a series of stages in which the content is put to the panel and their input sought. Where the items are in the form of statements or outcomes, then a first stage might be to invite panel members to accept or reject a particular item or to accept but with suggestions for modification. Modifications would be captured in free text boxes. The failure to achieve a preset level of consensus (normally between 70% and 90%) would trigger the rejection of the statement. Statements where modifications were suggested are redrafted based on the comments made by the panel. A degree of adjudication by the research panel might be required if modifications contradict one another. In the second stage, the statements would be presented again, including the modified ones as well as the ones accepted by all panel members. This time the statements that were previously accepted can be seen, but depending on the methodology adopted, panel members may or may not be permitted to judge them again. This second stage could seek further modifications which would be dealt with in the same way. A third stage would then ask the panel members to simply accept or reject each statement. The process is essentially similar where topic lists, rather than statements, are used except that panel members may be asked to grade their response from essential to nonessential.

From a methodological point, it is a matter of good practice, and arguably essential, that the researcher must be as objective as possible, adopting a post-positivist approach. Decisions are made based on staying true to the expert opinion and not on introducing their own. The researcher must be able to show a robust audit trail of how decisions were made. Part of the robustness is to set a required consensus level at the outset and retain those throughout the study. Consensus levels vary, as with many aspects of a Delphi study, but values between 70% and 90% are common. After the Delphi is complete, there is further work to integrate it into local curricula.

Current Core Syllabi

The section will now outline current core syllabi that exist in anatomy (refer to Table 43.1), the methods adopted, and the involvement of professional bodies and learned societies.

Medicine

Anatomy has been described as the cornerstone of medicine for centuries [1], and hence it is not surprising that there is a need to define what is considered core knowledge for a doctor to know. There have been a number of attempts to develop a core syllabus in anatomy. The purposes for which these syllabi were devised do vary, and consequently the level of detail they aspire to varies significantly between them [25–29]. The core regional anatomy syllabus for undergraduate medicine developed by the Anatomical Society first disseminated in 2003 and later refined and published in 2007 [29] had a very clear aim. This was to produce a syllabus that steered a commonsense course toward what could be considered as

Table 43.1 An overview of existing syllabi, including the number of learning outcomes and year of publication

Professional area	Published by	Number of learning outcomes or topics	Format	Year published
Gross anatomy medicine	Anatomical Society *Journal of Anatomy* (Smith et al.) [30]	156	Regional	2016
Gross anatomy medicine	Anatomical Society (McHanwell et al.) [29]	163	Regional	2007
Gross anatomy medicine	American Association of Clinical Anatomists (Leonard et al.) [27]	~ 500	Regional	1996
Gross anatomy medicine	Dutch Anatomical Society (Griffieon et al.) [26]	1000 plus	Systems and regional	1999
Head and neck medicine	International Federation of Associations of Anatomists (Tubbs et al.) [28]	646	Regional	2014
Musculoskeletal anatomy of the vertebral column and limb medicine	International Federation of Associations of Anatomists (Webb et al.) [50]	1112	Regional	2018
Embryology medicine	Anatomical Society (Holland et al.) [49]	61	Systems	2019
Embryology medicine	Fayoka et al. [42]	276 core or recommended	Systems	2017
Embryology medicine	Leonard et al. 2000 [41]	~700	Regional and systems	2000
Embryology	Liaison Committee for Medical Education (LCME) and the Commission on Osteopathic College Accreditation (COCA) (Das et al.) [43]	221 primary or secondary level	Competencies	2018
Neuroanatomy medicine	International Federation of Associations of Anatomists (Moxham et al.) [25]	209 core	Regional	2015
Gross anatomy nursing	Anatomical Society (Connolly et al.) [37]	64	Systems	2018
Gross anatomy pharmacy	Anatomical Society (Finn et al.) [39]	49	Systems	2018
Oral anatomy histology and embryology dentistry	International Federation of Associations of Anatomists (Moxham et al.) [51]	54	Regional	2017

necessary for a medical student to know upon graduation and away from the highly detailed courses sometimes taught in the past ("adding common sense to the need to know"). The Anatomical Society core syllabus created in 2007 comprised of 182 learning outcomes that were grouped regionally [29]. This syllabus was subsequently referenced in the UK *General Medical Council's (GMC) Teaching Tomorrow's Doctors*. This first Anatomical Society syllabus was generated as a consensus document and so was open to criticism on the grounds that the learning outcomes had been generated by a relatively small group of anatomists, albeit some of whom were clinically qualified, and lacked broader validity. Consequently, a second study was undertaken using a modified Delphi technique research tool to seek a wider consensus involving a broader range of expert's knowledge ($n = 39$). The revised core syllabus showed a number of small but significant changes including a reduction in the number of learning outcomes to 156. It was published in 2016 and has also been taken up by the GMC [30].

In parallel other syllabi for medicine have been produced by the International Federation of Associations of Anatomists (IFAA), again using a Delphi method, employing a content list rather than a set of predefined learning outcomes [9]. This approach has been used to develop a head and neck syllabi [28]. Twenty experts were asked to rank a total of 646 structures on a scale ranging from essential to not necessary. The result is thus a list of how important the panel found these structures, not a list of learning outcomes to be imported into a curriculum. This approach is clearly very helpful in understanding the depth of a subject that a student may be expected to go into but requires a further stage to develop these into learning outcomes. A further approach has been developed by the American Association for Anatomy (AAA) who has published a competency-based syllabus linking competencies to clinical relevance and also clearly defining the limits to detail in key topics [31]. The approaches of AS, IFAA, and AAA are clearly complementary, and the extent to which they might converge or not will be helpful in promoting further discussion about the extent of necessary anatomical knowledge at any particular stage [9].

Nursing

Historically, medicine employed the "medical model" that primarily focused on pathophysiology, diagnostics, and therapeutics. In contrast nursing advocated for a holistic "biopsychosocial model" of health that focuses on pathophysiology, psychology, health determinants, and social considerations aimed at improving patient's quality of life and overall health where possible, through shared goals and advocacy [32]. However, in recent years, all health professionals have adopted the biopsychosocial model as knowledge of health and safety within health systems has grown increasingly complex. This has created a seismic shift toward team-based decision-making that is diminishing outdated stereotypes [33], maximizing on nurses' clinical acumen by advancing their roles into preliminary diagnostics, prescribing and performing enhanced technical interventions to provide continuity of care as medical teams rotate around practice settings [34–36]. Anatomy is a fundamental element for team-based decision-making, shared understanding, and patient safety, and yet there has been no evidence-based consensus on the level and depth of anatomy required within undergraduate nursing education. Using the Anatomical Society's core gross anatomy syllabus for medical students as a foundation, a modified Delphi technique was used to develop discipline-specific outcomes to nursing graduates. The output from this study was 64 nursing-specific anatomy learning outcomes that are applicable to all undergraduate (preregistration) programs [37]. The new core anatomy syllabus for undergraduate nursing can be used internationally alongside local regulatory standards and guidelines offering a standardized framework upon which nurse educators, clinical mentors, and nursing students can use to build a core knowledge that provides a foundation for specialist knowledge.

Pharmacy

The need for anatomical knowledge is not limited to the doctor or nurse. Given the interprofessional nature of healthcare delivery, it is becoming increasingly important for healthcare professionals from all disciplines to possess a significant grounding in the anatomical sciences. One such discipline is pharmacy; it is worth noting that in the United Kingdom a pharmacist holds a masters' degree and is a licensed practitioner under the General Pharmaceutical Council (GPhC), who is the independent regulator for pharmacists, pharmacy technicians, and pharmacy premises in Great Britain. The role of the pharmacist has changed from one that was traditionally based on dispensing to one that encompasses treatment and diagnosis and acting as the first port of call for patients [38–40]. Given this considerable shift in duties, it is of paramount importance for pharmacy graduates to have a solid foundation upon which to build their pharmacological knowledge—specifically including anatomy.

Examples of the need for a pharmacy graduate to understand anatomy are abundant, but one favored example from the United Kingdom is the increasing use of community-based pharmacies as a point of delivery for influenza vaccines. In such a scenario, the administering pharmacist needs an awareness of basic surface anatomy, musculature, and neurovasculature of the upper limb. Similarly, drug metabolism cannot be fully understood without at least a basic understanding of the anatomy of the liver.

A modified Delphi technique was used to develop outcomes specific to pharmacy graduates [39]. A Delphi panel consisting of medical practitioners, pharmacists, and anatomists (n = 39) was created. The output from this study was 49 pharmacy-specific learning outcomes that are applicable to all pharmacy programs. The syllabus offers a basic anatomical framework upon which pharmacy educators can build the necessary clinical practice and knowledge [39]. Unlike medicine, anatomy is not always explicitly taught within the pharmacy curriculum. For this reason, the syllabus also included a mapping document which highlighted potential clinical scenarios that would be common to most curricula, in which anatomy could be integrated. An illustrative example "describes the anatomy, histology, and function of the different structures of the gastrointestinal tract: esophagus, stomach, duodenum, ileum, jejunum, colon, rectum, and anal canal," which was aligned to the following: ulcerative colitis, disease, peptic ulcers, drug absorption, and delivery. Furthermore, within the M. Pharm syllabus, due to the mechanisms of drug delivery and absorption, histology was integrated where appropriate, thus showing the flexibility that can be afforded when creating a syllabus, so long as a robust methodology is adopted.

Embryology

While previous developmental or embryological syllabi have been published [41–43], there was criticism that they were not suited to the local environment, and hence the syllabus developed by the Anatomical Society combines an outcomes-based approach [44, 45] with the methodological rigor of the Delphi process. Embryology is perhaps often considered primarily of interest to those practicing as obstetricians, pediatricians, or within some surgical specialties, but an understanding of developmental anatomy and teratology has a core role in multiple additional specialties [46, 47]. Teaching time for embryology has reduced within the last 50 years and varies considerably from institution to institution but typically averages at around 14 hours in undergraduate courses (range 0–50 hours) [48]. Given these time constraints, it is essential that educators prioritize what core embryological concepts require inclusion within their curriculum and recognize what content may be appropriately omitted and addressed within specialized postgraduate training programs [48]. This is perhaps why some embryology syllabi, for example, Fayoka et al. [42] and Das et al. [43], have not been widely adopted, being considered too in depth for the time allocated to teaching, or, in the example of the syllabus proposed by Das et al. [43], not based on methodological rigor. The Anatomical Society embryology syllabus sets to address both of these issues and comprises of 61 core learning outcome statements [49], linked to optional examples of conditions providing clinical context and created by a Delphi panel of 18 members. The syllabus provides flexibility for educators to integrate and adapt this within existing institutional curricula, while still providing a common core level of knowledge of embryology essential to all newly qualified doctors.

Impact

As discussed at the beginning of the chapter, core syllabi have to be usable and used. Only one core syllabus in medicine, by Smith et al. [24], has been evaluated on the impact its creation has had on medical students and anatomy educators. The study by Smith et al. (2019) [52] reported that 46% (n = 75) of medical students at Brighton and Sussex Medical School estimated that they knew

Tips on Using Core Syllabi

- If a regulatory body governs your course, check its syllabus requirements.
- Speak to everyone delivering anatomy or anatomy-related material to see what is already taught.
- Create a matrix of the learning outcomes in the core syllabi to cross-reference where and when the teaching of anatomy occurs, who is responsible for it, and if it is assessed.
- Core syllabi only specify what needs to be taught, not how. So give thought as to how best to deliver the content in your curriculum.
- Where integrated curricula exist, map the outcomes of the core syllabus to individual cases, scenarios, problems, or conditions to facilitate horizontal and vertical integration.

over 50% (n = 78) of the learning outcomes. In addition, all anatomy educators surveyed in the United Kingdom were aware of the core syllabus, and 48% (n = 24) had checked the core syllabus against their own institutional learning outcomes. Sixty-four percent (n = 32) of anatomy educators had shared the core syllabus with colleagues, and 34% (n = 17) had shared it with students. Forty-six percent (n = 23) of anatomy educators reported changing their teaching in some way as a result of reading the syllabus. Anatomy educators reported valuing the syllabus because it offered a standard of anatomy knowledge which allowed them to validate their own teaching content and also the syllabus offered leverage to include or remove content where appropriate [52].

Implementing Core Syllabi

The step from reading a core syllabus in a journal or website to fully implementing it may not be an easy one. There are a number of key stakeholders in any curriculum: students, teachers, curriculum leads or planners, administrators, and, of course, the relevant professional, statutory, regulatory bodies. The two cases below highlight different approaches.

Case 1

A system-based curriculum for medicine where students study anatomy in year 1 and 2 predominantly uses a mix of dissection and prosection. A new head of anatomy starts and decides to ensure alignment to a core syllabus. The first step involved creating a matrix of the learning outcomes of the syllabi, which are then matched to their own learning outcomes. The words may have varied but if the content matched a box was ticked. It became very clear that teaching in two areas was missing: the breast and ear. The head also felt that teaching of the pelvis was not to the level of depth required. They met with the curriculum lead and showed them the syllabus and the areas of concern. It was discussed that teaching "the breast" could easily be introduced into a session that covered the thoracic wall and later revisited in a breast cancer lecture given by a clinician. The module where content on "the ear" was best aligned was fiercely protected by the module leader who felt the ear was not neuroscience. It was decided that this material in its very basic form would be added to a clinical case that involved anatomy as a prosection session in year 2 and that the more complex anatomy of the ear would be picked up by ear, nose, and throat specialists in a lecture and a prosection session in year 4. The head met with the teacher of the pelvic region and asked them, giving examples of where some further detail needed to be added. This was completed for the next teaching session.

Case 2

In the United Kingdom, the majority of nursing programs adopts a didactically delivered, integrated approach to teaching and learning anatomy and physiology (A&P) as one module. There was initial resistance toward a stand-alone core anatomy syllabus. Once the A&P component is complete, students progress onto pathophysiology and clinical simulation as a separate component. This approach is favored to contextualize the material and to compensate for the curricular time restrictions that have been placed on the biological sciences that range from 6 up to 300 hours. Due to this disparity, some institutions are in a better position to adopt the syllabus in comparison to others, who may favor adaption. Nonetheless, the time limitation for teaching and learning is a point of contention. It has drawn widespread criticism from students, registered nurses, and academics, who have accused program planners of bias toward biological science in favor of nursing's predominant research base, the social sciences. As a result, there is a disequilibrium between science and social science within curriculum, despite the advocated 50-50 division of theory. Despite this reproach, numerous institutions have condensed A&P even further, delivering the content in one class rather than as paralleling subjects. As a result, science lecturers cannot distinguish how many teaching hours are allotted to each distinct subject. Consequently, there is little differentiation between the disciplines of anatomy and physiology. This led to calls for concomitant learning outcomes in physiology. The time-pressured, didactic teaching, passive learning approach coupled with a lack of access to anatomical resources that have led to widespread integration of anatomy and physiology has created several issues for consideration:

(i) In the short term, a core physiology syllabus is required to compliment and merge with that of anatomy to enable more widespread curricular mapping. This is currently underway.

(ii) In the medium term, an urgent dialogue is needed between anatomy, nursing, and senior faculty about the local factors that limit accessibility to resources, thus creating educational inequality in clinical education.

(iii) In the long term, how will the adoption of recommended syllabi be supported? Will they be adapted? If so, how? Is it a local issue? Should there be workshops? If so, delivered by whom? How do we ensure that the guidance stays relevant and up to date?

(iv) Ultimately, who is responsible for overseeing the quality and relevancy of the recommended syllabi after they are created and disseminated?

Tips for Implementing Core Syllabi

- Core syllabi can be implanted at any time, although curriculum redesign often offers a good opportunity.
- Look at student feedback for trends as to areas of anatomy students are finding difficulty. Consider if this is due to those areas not being taught.
- Create a master sheet as the means to check those learning outcomes you/your department already covers.
- Identify the areas where change is needed.
- Decide where and how these learning outcomes can be included.
- Get "buy in" from other teachers and curriculum planners to bring about change—this can take time and patience.
- Prepare to remove areas that you/your department overteach(es) to allow inclusion of important, omitted content, especially if your curriculum has a "one in, one out" rule.
- Map the core syllabi to the assessment blueprint.

Conclusion

Core syllabi in anatomy are essential as they define the "what do I need to know." There is no need for there to be just "one" global core syllabi. Each syllabus should be defined and robustly created to represent local or national needs based on the populations they serve.

References

1. Heylings DJA. Anatomy 1999-2000: the curriculum, who teaches it and how? Med Educ. 2002;36:702–10.
2. Craig S, Tait N, Boers D, McAndrew D. Review of anatomy education in Australian and New Zealand medical schools. ANZ J Surg. 2010;80:212–6.
3. Goodwin H. Litigation and surgical practice in the UK. Br J Surg. 2000;87:977–9.
4. Ellis H. Medico-legal litigation and its links with surgical anatomy. Surgery. 2002;20:i–ii.
5. Lynn-Macrae AG, Lynn-Macrae RA, Emani J, Kern RC, Conley DB. Medicolegal analysis of injury during endoscopic sinus surgery. Laryngoscope. 2004;114:1492–5.
6. Bergman EM, Verheijen IWH, Scherpbier AJJA, Van der Vleuten CPM, De Bruin ABH. Influences on anatomical knowledge: the complete arguments. Clin Anat. 2014;27:296–303.
7. Older J. Anatomy: a must for teaching the next generation. Surg J R Coll Surg Edinb Irel. 2004;2:79–90.
8. Sugand K, Abrahams P, Khurana A. The anatomy of anatomy: a review for its modernization. Anat Sci Educ. 2010;3:83–93.
9. Moxham BJ, Plaisant O, Smith CF, Pawlina W, McHanwell S. An approach toward the development of core syllabuses for the anatomical sciences. Anat Sci Educ. 2014;7:302–11.
10. Regan de Bere S, Mattick K. From anatomical 'competence' to complex capability. The views and experiences of UK tutors on how we should teach anatomy to medical students. Adv Health Sci Educ. 2010;15:573–85.
11. Young MFD. Bringing knowledge back in: from social constructivism to social realism in the sociology of education. Abingdon: Routledge; 2008.
12. Keeney S, Hasson F, McKenna H. The Delphi technique in nursing and health research. 1st ed. Oxford: Wiley-Blackwell Publishing; 2011.
13. Kilroy D, Driscoll P. Determination of required anatomical knowledge for clinical practice in emergency medicine: national curriculum planning using a modified Delphi technique. Emerg Med J. 2006;23:369–6.
14. Kilroy DA, Mooney JS. Determination of required pharmacological knowledge for clinical practice in emergency using a modified Delphi technique. Emerg Med J. 2007;24:645–7.
15. Penciner R, Langhan T, Lee R, McEwen J, Woods RA, Bandiera G. Using a Delphi process to establish consensus on emergency medicine clerkship competencies. Med Teach. 2011;33:e333–9.
16. McKenna HP. The Delphi technique: a worthwhile research approach for nursing? J Adv Nurs. 1994;19:1221–5.
17. Williams PL, Webb C. The Delphi technique: a methodological discussion. J Adv Nurs. 1994;19:180–6.
18. Sumsion T. The Delphi technique: an adaptive research tool. Br J Occup Ther. 1998;61:153–6.
19. Graham B, Regehr G, Wright JG. Delphi as a method to establish consensus for diagnostic criteria. J Clin Epidemiol. 2003;56:1150–6.
20. Green B, Jones M, Hughes D, Williams A. Applying the Delphi technique in a study of GP's information requirements. Health Soc Care Comm. 1999;7:198–205.
21. Yeates PJ, Stewart J, Barton JR. What can we expect of clinical teachers? Establishing consensus on applicable skills, attitudes and practices. Med Educ. 2008;42:134–42.
22. Stritter FT, Tresolini CP, Reeb KG. The Delphi technique in curriculum development. Teach Learn Med. 1994;6:136–41.
23. Lisk J, Flannery J, Loh E, Richardson D, Agur A, Woods N. Determination of clinically relevant content for a musculoskeletal anatomy curriculum for physical medicine and rehabilitation residents. Anat Sci Educ. 2014;7:135–43.
24. Smith CF, Finn GM, Stewart J, McHanwell S. Anatomical Society core regional anatomy syllabus for undergraduate medicine: the Delphi process. J Anat. 2016;228:2–14.
25. Moxham B, McHanwell S, Plaisant O, Pais D. A core syllabus for the teaching of neuroanatomy to medical students. Clin Anat. 2015;28:706–16.
26. Griffioen FM, Drukker J, Hoogland PV, Godschalk M. General plan anatomy objectives of the teaching of anatomy/embryology in medical curricula in the Netherlands. Eur J Morphol. 1999;37:228–325.
27. Leonard RJ, Acland RD, Agur A, Blevins CE, Cahill DR, Collins JD, et al. A clinical anatomy curriculum for the medical student of the 21st century: gross anatomy. Clin Anat. 1996;9:71–99.
28. Tubbs RS, Sorenson EP, Sharma A, Benninger B, Norton N, Loukas M, et al. The development of a core syllabus for the teaching of head and neck anatomy to medical students. Clin Anat. 2014;27:321–30.
29. McHanwell S, Davies DC, Morris J, Parkin I, Whiten S, Atkinson M, et al. A core syllabus in anatomy for medical students—adding common sense to need to know. Eur J Anat. 2007;11:3–18.
30. Smith CF, Finn GM, Stewart J, Atkinson M, Davies DC, Dyball R, et al. The anatomical society core regional anatomy syllabus for undergraduate medicine. J Anat. 2016;228:15–23.
31. American Association of Anatomists. Anatomical Competencies. 2018. https://www.anatomy.org/anatomy-education-resources.html. Accessed 23 Feb 2019.
32. Kitson A, Marshall A, Bassett K, Zeitz K. What are the core elements of patient-centred care? A narrative review and synthesis of the literature from health policy, medicine and nursing. J Adv Nurs. 2013;69:4–15.
33. Joseph-Williams N, Lloyd A, Edwards A, Stobbart L, Tomson D, Macphail S, et al. Implementing shared decision making in the NHS: lessons from the MAGIC programme. BMJ. 2017;357:1744.

34. Griffiths P, Maruotti A, Recio Saucedo A, Redfern O, Ball J, Briggs J, et al. Nurse staffing, nursing assistants and hospital mortality: Retrospective longitudinal cohort study. BMJ Qual Saf. 2019;28:609–17.
35. Scottish Government. Transforming nursing, midwifery and health professions roles: advanced nursing practice. 2017. https://www.gov.scot/publications/transforming-nursing-midwifery-health-professions-roles-advance-nursing-practice/. Accessed 14 Jan 2017.
36. McDonnell A, Goodwin E, Kennedy F, Hawley K, Gerrish K, Smith C. An evaluation of the implementation of advanced nurse practitioner (ANP) roles in an acute hospital setting. J Adv Nurs. 2015;71:789–99.
37. Connolly SA, Gillingwater TH, Chandler C, Grant A, Greig J, Meskell M, et al. The anatomical Society's core anatomy syllabus for undergraduate nursing. J Anat. 2018;232:721–8.
38. Ridge K. Role of pharmacists is set to grow and grow. 2015. https://www.england.nhs.uk/blog/keith-ridge-2/. Accessed 9 Oct 2017.
39. Finn GM, Hitch G, Apampa B, Hennessy CM, Smith CF, Stewart J, et al. The anatomical society core anatomy syllabus for pharmacists: outcomes to create a foundation for practice. J Anat. 2018;232:729–38.
40. Finn GM, Gillingwater T, Connolly S, Smith CF. Putting gross anatomy into the curriculum: new anatomy syllabi for nursing and pharmacy students. Anat Sci Edu. 2018;11:427–8.
41. Leonard RJ, Hoos PC, Agud R, Gilroy AM, Lozanoff S, Nelson ML, et al. A clinical anatomy curriculum for the medical student of the 21st century: developmental anatomy. Clin Anat. 2000;13:17–35.
42. Fakoya FA, Emmanouil-Nikoloussi E, Sharma D, Moxham BJ. A core syllabus for the teaching of embryology and teratology to medical students. Clin Anat. 2017;30:159–67.
43. Das M, Claeson KM, Lowrie DJ, Bolender DL, Fredieu JR, Guttmann G, et al. A guide to competencies, educational goals, and learning objectives for teaching human embryology in an undergraduate medical education setting. Med Sci Educ. 2018;28:417–28.
44. Harden RM. AMEE guide no. 14: outcome-based education: part 1-an introduction to outcome-based education. Med Teach. 1999;21:7–14.
45. Kennedy D. Writing and using learning outcomes: a practical guide. In: Cork Open Research Archive, University College Cork. 2007. https://www.cmepius.si/wp-content/uploads/2015/06/A-Learning-Outcomes-Book-D-Kennedy.pdf. Accessed 4 Jan 2019.
46. Mascio CE, Pasquali SK, Jacobs JP, Jacobs ML, Austin EH. Outcomes in adult congenital heart surgery: analysis of the Society of Thoracic Surgeons database. The J Thorac Cardiov Sur. 2011;142:1090–7.
47. Lee NK, Kim S, Jeon TY, Kim HS, Kim DH, Seo HI, et al. Complications of congenital and developmental abnormalities of the gastrointestinal tract in adolescents and adults: evaluation with multimodality imaging. Radiographics. 2010;30:1489–507.
48. McBride JM, Drake RL. National survey on anatomical sciences in medical education. Anat Sci Educ. 2018;11:7–14.
49. Holland JC, Smith CF, O'Shea M, Stewart J, Ockleford C, Finn G. The anatomical society Core embryology syllabus for undergraduate medicine. J Anat; In press. 2019;235:847–60.
50. Webb AL, Green RA, Woodley SJ. The development of a core syllabus for teaching musculoskeletal anatomy of the vertebral column and limbs to medical students. Clin Anat. 2019;32:974–1007.
51. Moxham B, McHanwell S, Berkovitz B. The development of a core syllabus for the teaching of oral anatomy, histology, and embryology to dental students via an international 'Delphi Panel'. Clin Anat. 2017;31:231–49.
52. Smith CF, Finn GM, Hennessy C, Luscombe C, Stewart J, McHanwell S. The initial impact of the Anatomical Society gross anatomy core syllabus for medicine in the United Kingdom: student and teacher perspectives. Anat Sci Edu. 2019;12:494–506.

Anatomy Education in an Innovative Medical School Curriculum

44

Jennifer M. McBride and Richard L. Drake

You have just accepted the position as director of anatomy at a new medical school. The new school has decided on an integrated organ system-oriented curriculum using a problem-based format interspersed with additional classroom activities. Your challenge is to develop an innovative approach to anatomy education at this forward-thinking institution with limited contact hours.

First Considerations

Whether you find yourself in the situation described in the first paragraph or you have to significantly modify an existing course due to a major curriculum revision, there are three basic guidelines that must be followed.

Match the Curriculum with the Academic Program's Foundation

When a major curricular reform occurs, there is usually a specific curriculum that someone or some group has chosen to institute. Examples would be a discipline-based curriculum, an integrated curriculum, or a mixture of the two. Similarly, when a new institution is being created, there is usually a specific educational philosophy that has been chosen as the basis for the curriculum that will be designed. An example might be student-centered learning versus teacher-centered learning. Whatever the case, these factors are the foundation for the academic program, and the anatomy course that is being developed must seamlessly fit into this environment. If it does not, the success of the program will likely be limited.

Incorporate Faculty into the Planning Process

As a new course is developed, the skills, abilities, and interests of the faculty who will teach this course must be foremost in the considerations of the planners. This also applies when a major curricular reform is being planned. The faculty are the individuals who will have the responsibility of instilling confidence in the students that the course they are taking or the curriculum they are following will meet their needs as they continue their medical training. If a course, or a curriculum, is developed by a small group of individuals who have a vision but they do not evaluate whether the faculty have the skills, abilities, and much less the interest to carry out the new program, the planning process is severely flawed.

J. M. McBride · R. L. Drake (✉)
Cleveland Clinic Lerner College of Medicine, Case Western Reserve University, Cleveland, OH, USA
e-mail: mcbridj@ccf.org; draker@ccf.org

L. K. Chan, W. Pawlina (eds.), *Teaching Anatomy*, https://doi.org/10.1007/978-3-030-43283-6_44

The capabilities of the faculty must be considered during the entire planning process and their opinions obtained to provide the best opportunity for a successful outcome. A feeling of ownership in the final product is absolutely essential.

Know Your Student Base

The outcome that demonstrates the true success of a new course, or a major curricular reform, is whether student learning has been enhanced. All planning must take into consideration (1) the number of students who will be involved in the new course or curriculum, (2) the academic background of these students, and (3) their capabilities, i.e., can they accomplish what we are asking them to do? Educational approaches and activities that are successful with 30 students may not realize the same success with 130 students. What works at one institution may not be successful at another simply because of the population of students involved. So design a course/curriculum with your consumer in mind.

Basic Guidelines of Course Development

- The course being developed must match the curriculum or philosophy of education at the institution.
- The course must complement the group of faculty involved in the teaching.
- The course must be appropriate for the type of students that will be participating.

Current Trends

During the past 15 years, curricular reform has been a major event especially in medical schools in the USA. Driving forces behind these reforms include, but are not limited to, the desire to decrease the number of contact hours and increase the time available for self-directed learning, reducing lectures and increasing the number of activities that involve interactive learning, reducing unnecessary redundancy between courses and increasing the integration of various subjects, increasing the opportunities for early student contact with clinical faculty, and increasing the use of electronic resources when the opportunity presents itself. These issues have led course developers in all subjects, including anatomy, to consider the following guiding principles when designing/planning a course in a modern medical school curriculum.

Use Active Versus Passive Learning

It is well established that learning and retention are improved through the use of more interactive activities in an educational program [1–3]. Examples of interactive approaches include problem-based learning, team-based learning, small-group interactive sessions, peer-facilitated workshops, the use of audience response systems, incorporation of web-based instructional materials, and simulations. While lectures can still play an important role in education, the current trend in course design has been to decrease the number of lectures and increase the use of interactive educational approaches.

Provide Efficient Laboratories

One of the biggest changes has been a decrease in the available contact hours for all courses. This has had a significant impact on anatomy courses since they are nearly the only course in a modern medical school curriculum that maintains a laboratory. Furthermore, this laboratory activity is regarded by many as the most important component of an anatomy course. So in order to maintain this valuable educational experience, it has to become more efficient. Examples of some of the types of changes that have been introduced to facilitate a more streamlined experience in the anatomy laboratory include the increased use of prosections or previously dissected specimens, peer-teaching-peer laboratory exercises, and the use of plastinated demonstration specimens. These approaches have resulted in a reduction in course hours but allowed the unique learning opportunity in the anatomy laboratory to remain a major component of the modern medical school curriculum.

Vary the Educational Experience

Individuals learn in different ways [4], and to take advantage of this situation, current suggestions are to use a variety of educational experiences to enhance and stimulate learning. This multimodality approach has been used in anatomy for many years (lectures and laboratories), but the current trend is to expand these offerings. In anatomy this is easily accomplished by adding imaging, interactive sessions in the classroom, body painting, and a variety of approaches in the laboratory as mentioned in the preceding section [5–7].

Integrate Anatomy with Clinical Medicine

Integrating anatomy with clinical medicine is not new to anatomy courses. Learning in context has been occurring in these programs for several decades. However, the current trend is to increase this activity throughout the course which gives students a reason to learn and understand the subject [8, 9].

Establish Opportunities for Longitudinal Learning

Learning should be longitudinal. Studies have shown that if information is revisited, retention is improved [10]. Make an effort to establish some type of unique educational activity in every aspect of the curriculum where it is appropriate, including upper-level clinical rotations. But always demonstrate how your contribution is an additional opportunity for students to reinforce their understanding of the anatomical sciences, has a positive impact on student learning, is a great example of longitudinal integration, and enhances their training as a future physician. The amount of time you get is not as important as is your presence in other parts of the curriculum [9].

Five Guiding Principles of Course Development

- Use active versus passive learning.
- Time is limited so the laboratory must be efficient.
- Use a variety of educational experiences.
- Integrate anatomy with clinical medicine.
- Learning should be longitudinal.

Example

The best way to indicate how a course can be designed using the basic guidelines and guiding principles described in this chapter is to provide an example. In this case, the three basic guidelines:

- The course being developed must match the curriculum or the philosophy of education of the institution.
- The course must complement/fit the group of faculty involved in the teaching.
- The course must fit/be appropriate for the type of students that will be participating.

and the five guiding principles:

- Use active versus passive learning.
- Time is limited so laboratory must be efficient.
- Use a variety of educational experiences – a multimodality approach.
- Integrate anatomy with clinical medicine – learning in context.
- Learning should be longitudinal.

were followed to develop a new program in anatomy education at the Cleveland Clinic Lerner College of Medicine of Case Western Reserve

University. The challenge was to create an anatomy course in the context of a problem-based organ system-oriented curriculum that stressed small-group interactive learning, with no lecture and no traditional tests or grades. The time available for this class in the first year would be a 2 hour time slot every Monday morning for approximately 30 weeks.

Keeping the three basic guidelines and five guiding principles in mind, what emerged from the course development process was an education program in anatomy that was unique in its approach and innovative in its design. The basics of this approach were to use clinical cases to introduce anatomical concepts and facts that would be reinforced using cadavers and imaging. Additionally, this program would be implemented in a longitudinal continuum with anatomy reviews occurring in the second year and the third year prior to the rotation in surgery [11].

Fourteen years later, the anatomy faculty at this institution have a new challenge. Next year, the entire medical school program is moving to a new building that does not have any facilities for the use of cadavers. Anatomy instruction in this new facility will consist of a more digital approach. New teaching options will include the 3D4Medical Complete Anatomy app, Touch of Life Technologies Sectra Table, U/S Mentor simulator, and virtual reality modules [12–14].

As the anatomy faculty began to consider their options, in the context of the previously described course development guidelines and principles, they focused on how all of the educational modalities at their disposal – including unembalmed cadavers – could be woven into a unique approach that would enhance learning and meet the needs of medical students. The solution was a blended approach providing learners with a mixture of educational tools that would benefit the learning styles of all students.

To accomplish this goal, education needs to occur in two facilities – the new building and the existing cadaver laboratory – a mixed location approach. How is this going to occur? Using as an example the Cardiovascular and Respiratory Sciences 1 course for first-year medical students that lasts 6 weeks, students would have class in the new building weeks 1, 3, and 5, using the new teaching options, and class in the existing anatomy laboratory weeks 2, 4, and 6, using unembalmed cadavers. This new approach meets the educational needs of medical students, provides learners with a multimodality approach that should enhance learning, and consists of a variety of educational tools capable of matching the learning styles of any student. Additionally, it still follows the three basic guidelines and five guiding principles stated earlier.

Conclusion

In this day of curricular reform and the development of new educational programs, it is important to remember a successful anatomy course should incorporate active versus passive learning; have an efficient laboratory since time is limited; use a variety of educational experiences – a multimodality approach; integrate anatomy with clinical medicine, learning in context; and be longitudinal in design.

References

1. Louw G, Eizenberg N, Carmichael SW. The place of anatomy in medical education: AMEE guide no. 41. Med Teach. 2009;31:373–86.
2. Sugand K, Abrahams P, Khurana A. The anatomy of anatomy. A review for its modernization. Anat Sci Educ. 2010;3:83–93.
3. Zumwalt AC, Lufler RS, Monteiro J, Shaffer K. Building the body: active learning laboratories that emphasize practical aspects of anatomy and integration with radiology. Anat Sci Educ. 2010;3:134–40.
4. Fleming ND. I'm different; not dumb. Modes of presentation (VARK) in the tertiary classroom. In: Zelmer A, editor. Research and development in higher education: proceedings of the 1995 annual conference of the higher education and research development society of Australasia. HERDSA, vol. 18; 1995. p. 308–13.
5. Korf HW, Wicht H, Snipes RL, Timmermans JP, Paulsen F, Rune G, Baumgart-Vogt E. The dissection

course – necessary and indispensable for teaching anatomy to medical students. Ann Anat. 2008;190:16–22.
6. Finn GM, McLachlan JC. A qualitative study of student responses to body painting. Anat Sci Educ. 2010;3:33–8.
7. Lufler RS, Zumwalt AC, Romney CA, Hoagland TM. Incorporating radiology into medical gross anatomy: does the use of cadaver CT scans improve students' academic performance in anatomy? Anat Sci Educ. 2010;3:56–63.
8. Pabst R, Westermann J, Lippert H. Integration of clinical problems in teaching gross anatomy: living anatomy, X-ray anatomy, patient presentations, and film depicting clinical problems. Anat Rec. 1986;215:92–4.
9. Rizzolo LJ, Rando WC, O'Brien MK, Haims AH, Abrahams JJ, Stewart WB. Design, implementation, and evaluation of an innovative anatomy course. Anat Sci Educ. 2010;3:109–20.
10. Custers EJFM. Long-term retention of basic science knowledge: a review study. Adv Health Sci Educ. 2010;15:109–28.
11. Drake RL. A unique, innovative, and clinically oriented approach to anatomy education. Acad Med. 2007;82:475–8.
12. Moro C, Stromberga Z, Raikos A, Stirling A. The effectiveness of virtual and augmented reality in health sciences and medical anatomy. Anat Sci Educ. 2017;10:549–59.
13. Erolin C, Reid L, McDougall S. Using virtual reality to complement and enhance anatomy education. J Vis Commun Med. 2019;6:1–9.
14. Bairamain D, Liu S, Eftekhar B. Virtual reality angiogram vs 3-dimensional printed angiogram as an educational tool-a comparative study. Neurosurgery. 2019;85:E343–49.

The Role of the Anatomist in Teaching of Nontraditional Discipline-Independent Skills

45

Darrell J. R. Evans and Wojciech Pawlina

Anatomy as a basic science has been at the heart of many healthcare curricula including medicine, dentistry, and physical therapy for centuries. However in the last decade, the role of anatomy as a basic science course in medical curriculum has undergone significant change that has challenged the very existence of anatomy departments and their educational philosophy [1, 2]. It has been shown that a traditional delivery and assessment of anatomy no longer have a place within the modern medical curriculum.

Many medical schools implement competency-based educational systems which assess observable outcomes to judge student competence based on defined course criteria [3]. Whereas course instructors have been historically more accustomed to using learning objectives as the defining criteria for anatomy knowledge in written and practical examinations, many are less familiar with implementing and evaluating competencies not related to concrete anatomical knowledge and thus have been neglected in curricula.

As healthcare curricula evolve, conventional courses need to adopt a new approach to basic science instruction which extends beyond the expected discipline-based skills and enables the integration of sets of human skills. Because anatomy courses are often placed early in most medical and related curricula, they can provide an ideal forum for introducing a range of human skills and attributes important for subsequent practice. These skills are often referred to as nontraditional discipline-independent skills (NTDIS). Teaching NTDIS in anatomy should be viewed as a bridge between anatomy and clinical disciplines. The overarching objective of the anatomy course has been to provide discipline-specific fundamental scientific theories and concepts for clinical application. However, in the new environment, anatomy courses should incorporate additional objectives and outcomes that would be revisited in clinical courses. These include subjects such as leadership, teamwork, professionalism, and effective communication as well as other less tangible skills and attributes including resilience and empathy [4].

Upon entering a healthcare profession course, students need to begin the process of professional socialization. This process is contingent upon the development of discipline-independent skills related to their chosen career. As a result,

D. J. R. Evans (✉)
Faculty of Health and Medicine, University of Newcastle, Callaghan, NSW, Australia

Faculty of Medicine, Nursing and Health Sciences, Monash University, Clayton, VIC, Australia
e-mail: darrell.evans@newcastle.edu.au

W. Pawlina
Department of Clinical Anatomy, Mayo Clinic College of Medicine and Science, Mayo Clinic, Rochester, MN, USA
e-mail: pawlina.wojciech@mayo.edu

L. K. Chan, W. Pawlina (eds.), *Teaching Anatomy*, https://doi.org/10.1007/978-3-030-43283-6_45

curricula have become more holistic in approach, increasingly focused on the patient and providing opportunities for students to develop professionally. Consequentially, anatomists are finding that they are being asked to contribute to professional development aspects of the curriculum in collaboration with colleagues from other disciplines. In more recent years, students have also recognized that anatomy provides effective opportunities for developing many of the attributes and competencies demanded by professional bodies and future employers. Students themselves therefore see anatomy as one of the cornerstones of their medical education. This is important as healthcare professionals are entering health systems that are undergoing disruptive change through technology and innovation, which will have knock-on effects to the roles they will need to play.

The emphasis on developing professional identity formation (PIF) through medical education has intensified in recent years [5, 6]. PIF is a central milestone for every person who aspires to become a professional. There is growing recognition that the transformation to a healthcare professional involves the complex integration of core values, moral principles, and a self-awareness/belief [7]. An active and constructive approach to incorporating these into the curriculum will facilitate the development of professional identity. Anatomy as a discipline and anatomy educators are well positioned through effective curriculum design and pedagogy to introduce and promote PIF through the incorporation of a range of skills, experiences, and opportunities for social interaction into the learning journey [8]. Students must be made aware that they are not only learning anatomy for their own sake in terms of a grade or higher honors status but that their investment in learning will directly affect the quality and safety of their future patients [8]. Anatomy sessions should therefore expose students to an array of NTDIS through deliberate design and intervention: dexterity, as students learn and practice the art of dissection and palpation; communication, as students develop interpersonal skills within teams; leadership; and the art of effective teamwork. Development of professional identity is influenced by the hidden curriculum; thus, modeling professional behavior by anatomy educators is an essential aspect of the implicit education at work in the anatomy course [9, 10]. Anatomy educators should be role models for those students with diverse educational backgrounds, cultures, and experiences and be counsellor and mentor to students in a journey to develop their professional identity which for many starts in anatomy [8].

The use of cadavers provides an early focus for examining legal and ethical issues. Cadavers are also an emotional nidus for many students who have an opportunity to learn coping skills and healthy detachment. Students should learn the professional behavior expected within the healthcare professions early in the anatomy lab, with teachers and supervisors serving as role models [3, 8]. The focus on early professionalism has become increasingly prevalent in anatomical education, which adds further onus upon anatomy educators to polish their own skills in leadership, communication, and human sensitivity. This chapter will outline a number of the skills, competencies, and attributes we expect of our healthcare students. We shall describe how these skills can be accommodated and enhanced by the anatomist.

Nontraditional Discipline-Independent Skills, Competencies, and Attributes

The ability to incorporate nontraditional discipline-independent skills, competencies, and attributes in anatomy teaching will depend on the nature of the particular curriculum including the style of the teaching/learning activities, the level and focus of the content, the assessment opportunities, and the time available. In general, skills and competencies in the clinical discipline are referred to as nontechnical skills. They can be divided into those that refer to a person's clinical reasoning and the ability to make decisions and act upon them. Such skills include situational awareness, mental readiness, assessment of risk, anticipation of problems, decision-making, flexibility, and workload distribution. In contrast social or interpersonal skills relate to how a person interacts with other people to convey information. Examples of these skills include communication, leadership, teamwork, resource management, willingness to seek advice and feedback, and coping under pressure with stress [11]. Many NTDIS are naturally embedded as co-elements in the activities that anatomical learning activities feature and therefore may only need appropriate exposure and emphasis.

Nontraditional discipline-independent skills and attributes which anatomical learning experiences can provide within a healthcare curriculum:

- Practical skills
- Interpersonal and communication skills
- Teamwork and leadership
- Professionalism including ethical and legal issues
- Teaching skills
- Multidisciplinary interaction
- Additional transferrable skills including situational awareness, resilience, and empathy

Practical Skills

Whatever the design approach, anatomy courses often include practical-based elements as key features. Dissection has been an integral part of anatomical curricula for centuries, and while dissection is geared toward an exploration of the body in situ, it also provides the opportunity for the development of dexterity with an array of different tools for dissecting tissues [12, 13]. Although not all students will decide to enter into surgical or pathology professions where dissection skills are routine practice, most students will still need to be proficiently dexterous to use medical tools and equipment.

If dissection is incorporated as an element in a particular program, it must be accompanied by appropriate training and guidance to ensure hands-on skills are developed appropriately and that additional learning outcomes are achieved. Assessing the quality of students' laboratory work should also be incorporated into the anatomy course evaluation. It helps students focus on their laboratory objectives more efficiently and emphasizes the importance of quality handiwork.

If dissection is not an available element within a program, there are other ways of incorporating laboratory practical skills such as the use of surface anatomy sessions or living anatomy sessions. Students can be given the opportunity to palpate each other and models; they may draw and paint anatomical features [14–16]. In addition to developing practical skills, these methods also introduce students to elements of patient physical examination [17].

Interpersonal and Communication Skills

The success of most anatomy courses relies on students' abilities to communicate and work with others. If classroom activities are designed with communication as an outcome in mind, anatomy sessions can provide many opportunities for students to assess and develop appropriate interpersonal skills. In a century where communication has become increasingly global, the imperative to develop our students' skills is changing. Not only must students be able to communicate with cultures and people familiar to themselves, students must gain the skills to interconnect appropriately with widely variable audiences in the global community [18].

Many healthcare-associated professional bodies (e.g., the General Medical Council (GMC) in the UK or the Accreditation Council for Graduate Medical Education (ACGME) in the USA) have also recognized the importance of incorporating directed communication skills training within the medical curriculum and have made this a distinct outcome competency [19, 20]. There is a range of approaches that can be used to provide opportunities for developing different communication skills, and in most cases these approaches can be appropriately interwoven into existing activities so that the course is not overburdened with content but enhanced by additional outcomes [3, 21, 22]. One example might be to have students adapt an oral presentation to a patient audience rather than a presentation directed to student peers [18].

Most anatomy classroom activities naturally involve communication and related skills, but these are often informal or "hidden" aspects of a course, and as a result, students may be unaware that communication skills are included as learning outcomes of the sessions. It is therefore essential to set appropriate expectations at the outset of the anatomy course and to highlight opportunities for the development of such skills.

To enhance opportunities for communication skills practice and assessment, practical sessions can be designed so as to include material for discussion. Case studies, checklists, and formative quizzes are all topics that may be assigned as required discussion topics [23, 24]. Students should be assessed for how well they can discuss with others class content such as the identification of specific features through morphology or understanding how anatomy relates to function and what constitutes normal versus abnormal anatomy [3]. Furthermore, interactions with teachers and teaching assistants during anatomy sessions are ideal opportunities for students to debate and discuss, thereby continually developing the interpersonal and communication skills necessary for anyone entering into a healthcare-related profession. The longitudinal nature of the anatomy course is conducive to designing regular cycles of assessment and feedback regarding student communication skills, and students should have a sense that they are able to improve over time.

There are a range of other communication skills training activities that can be incorporated into an anatomy course. Activities emphasizing written communication skills include the creation of patient-focused information, online material (e.g., a student-designed teaching package), popular media-focused material (e.g., news stories, opinion editorials), discussion boards and diary writing, as well as mock paper or grant writing [22, 25–29]. Activities emphasizing oral communication skills and interpersonal opportunities might include using mock interviews, peer observation, simulated interactions, and the creation of audio/visual casts [22, 28, 30–34].

Teamwork and Leadership Skills

Teamwork is a major focus of healthcare. Many people are involved in making the decisions that affect patients. It is hoped that a more progressively holistic attitude toward the teams directing patient care will lead to more optimal patient-centered management. Accordingly, most medical schools have recognized that teamwork and leadership are essential professional competencies that need to be highlighted and developed during training. These leadership skills can be fostered through classroom team-based learning activities where groups of students are able to achieve sets of educational objectives under direction of a designated student team leader. Students are actively immersed in collaborative learning in team settings and, as a team, are responsible for obtaining, sharing, and presenting knowledge to one another. Recently many curricula have been improved to provide more opportunities for students to hone the skills of working within a team or taking on the role of team leader. Given the nature of the activities that occur within the anatomy practical class, opportunities for collaborative learning are already firmly embedded [12, 35]. Therefore, in recent years, anatomists have looked at ways to introduce and develop approaches that extend further the leadership opportunities in the team-based elements of the anatomy course. The focus on teaching and practicing leadership within teams in anatomy is dramatically increasing. Anatomy will remain a vanguard course in the medical curriculum where teamwork and team-based education are essential [36, 37].

There are a number of ways that leadership and teamwork skills can be developed: (1) the incorporation of reciprocal peer teaching where students alternate roles as teachers and learners [38, 39], (2) team-based learning with an emphasis on group problem-solving using clinical case scenarios and objective-oriented assignments [40, 41], (3) rotating appointments of student-leaders for each laboratory team [3], or (4) teamwork encouraged through administering group quizzes, audience response system activities, and team presentations [42]. These strategies all provide opportunities for individuals to be accountable for their contributions to a collective body [43, 44]. The development of teamwork and leadership skills is not just for face-to-face classroom activities but can also be encouraged in extracurricular time via electronic means such as through the provision of online discussion boards and chat rooms for groups of students to work collaboratively on team projects. To encourage teams to "play nice," it is often helpful to explicitly include group peer assessment in students' course grades [24].

Professionalism

Professionalism has been identified as a core competency of medical and other healthcare training and one of the six areas of competency defined by the ACGME [19]. Curriculum developers have been conscientious to design opportunities for students to develop the appropriate skills and attributes needed not only so that they can meet accreditation requirements but also to develop professional competencies that strengthen students' abilities to uphold their commitment to patient care and public trust [45].

Anatomy has played a central role in educating medical students early on perceived values of professionalism. The dissection laboratory and the use of cadaveric material in particular provide a focus for activities directed at teaching, practicing, and rewarding professional behavior and associated skills including respect, compassion, altruism, autonomy, and ethical principles [21, 35, 43, 44, 46, 47].

The use of donors within anatomy classes provides an early opportunity for students to develop and reinforce attitudes of respect and compassion. Increasingly, body donors are recognized as a student's first patient [44, 46, 48–50]. The growing emphasis on the donor as patient seems to increase students' curiosity as to the background of their donors including the names, life histories, and the causes of death. Course leaders may desire to disclose such information to students [46, 50]; however, this decision may be complicated by local and national regulations protecting confidentiality. Therefore, when deciding what information to disclose, the consenting wishes and rights of the donor must be protected. Such a situation provides students with insight into the ethical and legal issues of patient confidentiality, even after the patient is no longer living. These issues can be further explored using reflective exercises that present students with cases of compromised donor confidentiality. In some schools, the reflective exercises include meeting the families of donors prior to the dissection or following dissection in student-led convocations [51–53].

Constituent components of the ACGME professionalism competencies:

- Demonstrate respect, compassion, and integrity.
- Demonstrate responsiveness to patient needs that supersedes self-interest.
- Demonstrate accountability to patients, society, and the profession.
- Demonstrate excellence and ongoing professional development.
- Demonstrate adherence to ethical principles.
- Demonstrate sensitivity and responsiveness to diverse patient populations.
- Demonstrate respect for patient privacy and autonomy.

While the use of cadaveric material is key in providing the scientific basis of student knowledge, human bodies also expose students to questions and reflections on human values. For many students, anatomy may be the first time they have had to confront the issues of death and dying, and this can be a particularly challenging experience

[54–57]. A number of approaches can be incorporated in anatomy courses to provide students with opportunities to explore their emotions, to be introspective, and to cope with related stress. These include exercises such as group discussions, orientation and exploratory seminars, elective courses, debriefing sessions, peer evaluation, portfolios/journals, and creative projects [43, 46, 48, 55–64]. It is important that students are able to cope with the anxiety that they might feel within the dissecting room; however, it is also critical to limit the extent to which students become desensitized over time [51]. Discussion on ethics-related issues concerning dissection and the anatomical donor provides students with opportunities to develop the skills they need to interact with colleagues, teachers, and where appropriate, donor families, on sensitive issues [21, 44, 65].

Students increasingly appreciate the opportunity to formally thank their donor and their families. Many medical schools around the world hold services of thanksgiving, often inviting donor friends and relatives as well as students and staff [48, 49]. When devising such services, depending on whether a program decides to have a more formal (lecture theaters, events halls, and places of religious worship) or less formal event (held in classrooms and laboratories) [52], students should be allowed to decide how they wish to contribute to the proceedings (e.g., simple attendance versus event organization, the reading of poems and reflections, etc.) [41, 49, 54, 55, 66]. Whatever the format, the event affords students an opportunity to bring a sense of closure to an emotionally intense period of training. Convocations of thanks reinforce respectful and compassionate attitudes among students, who will have an opportunity to interact with donor families and acknowledge the human gift they have been given [43, 44, 46, 67].

Some anatomy students inordinately focus too much on knowledge accumulation for forthcoming examinations, in so doing, neglecting the deeper learning objectives and the discipline-independent competencies that an anatomy course presents. It is the responsibility of the course designers to ensure that the relevance of discipline-independent learning activities to the anatomy course is clearly articulated and that careful attention to professionalism is expected. Other students might focus too little on anatomical content. One way of emphasizing the importance of anatomical knowledge is to demonstrate how poor knowledge of anatomy can lead to serious mistakes and the potential for medical malpractice law suits. In the surgical specialties, for instance, this has been termed "anatomical ignorance," and recently there has been a considerable increase in cases of malpractice and litigation as a result [68]. It is important for students to realize that the consequence of such cases leads to a weakening of public trust on the healthcare system. Students, through reflection, must be able to link their efforts on anatomy practical sessions and examinations to the success of their future practice and the safety of their future patients.

Inclusion of opportunities to develop reflective practice and critical thinking allows students to explore experiences and develop new understanding and appreciation [60]. These most commonly include personal reflections on a dissecting experience, extracurricular collaborative learning reflection, online discussion boards and chat rooms, and investigative assignments such as the autopsy report [3, 29, 40, 58, 60, 69]. These additional approaches allow for the integration of theory into practice, for the acceptance of professional responsibility, and for the expansion of confidence and self-esteem. Validation instruments have been developed for measuring student reflection on gross anatomy [70].

Teaching Skills

Teaching skills have become increasingly important in the healthcare professions. While healthcare professionals are often experts in their fields, they may not have the ability to share their knowledge and teach their patients effectively. Therefore, appropriate skills training is needed. Until recently, few curricula included any formalized training opportunities for students to develop their own teaching skills. However, formalized teacher training is becoming more prevalent and, in some programs, is a required component of course accreditation.

Several reasons why teaching skills are important for medical students to acquire early in their education are as follows: (1) as future residents and faculty members, students will have teaching roles; (2) teaching is an essential aspect of physician-patient interaction; and (3) medical students who understand teaching and learning principles will become better learners themselves [71].

Anatomists have for a number of years played a key role in developing teaching skills within their students and have used a variety of means to provide students with opportunities to have face-to-face contact as teachers and facilitators of anatomical learning. Course leaders can choose strategies such as directed peer teaching, reciprocal peer teaching, near-peer teaching, and team-based learning [24, 38, 40, 72]. Through these strategies, a range of primary teaching skills can be included such as gauging the knowledge level of the learner group, elucidating key information for the learner, demonstrating techniques effectively, providing useful feedback, encouraging, and motivating student learning [73–75]. Providing opportunities for students to act as teachers can result in improvements in anatomical knowledge and deeper understanding for the students who assume the teacher role [24].

The provision of student-oriented teaching opportunities results in additional outcomes that relate well to developing the attributes of professionalism as described previously in this chapter. Requiring students to adopt the role of near-peer teaching assistants gives students the opportunity to hone their skills of mentoring. As mentors, students can identify the desirable attributes associated with professional interactions. Both students and near-peer teaching assistants appear to believe that role modeling is an appropriate feature of such programs [76]. Students learn what a good role model is both by seeing one and by being one.

Promoting Multidisciplinary and Interprofessional Interactions

A holistic approach to managing the care of patients requires healthcare workers to demonstrate effective teamwork. Teams most often involve practitioners from a range of specialties coming together to assess, treat, and care for the patient. The teaching and learning of anatomy can also be enhanced by adopting a multidisciplinary approach [77]. Anatomy can be taught by an interprofessional team of radiologists, physical therapists, surgeons, speech therapists, and others [78–80] each bringing a different perspective to the subject area and demonstrating the importance of collaboration between disciplines [1, 2].

Students from different disciplines can also learn together within the anatomy classroom. Dissection sessions can be shared at the same time by students studying medicine, biomedical sciences, nursing, and physical therapy. Intermixing student bodies is an effective method to teach skills in interprofessional learning and collaboration [77, 81].

Bringing together students from different healthcare backgrounds or with different career aspirations is quite common. For instance, some anatomy courses have utilized interactions between physical therapy students, physician assistant students, and medical students as a platform for interprofessional learning. Mixed student cohorts benefit from teaching one another the anatomy of regions receiving different emphasis depending on their program of study. For example, physical therapy students may feel more comfortable teaching musculoskeletal areas of the back and extremities, while medical students may be able to offer better teaching on the hepatobiliary anatomy. Students also have an opportunity to learn about the strengths and limitations of other healthcare specialties, while sharing time around the laboratory table.

Anatomy course directors can also consider bringing even more disparate student bodies together in order to develop specific skills and attributes. For example, some programs find it useful to pair medical students with art students to work on specific projects and analyze common and differing perspectives. This approach has been shown to develop humanistic sensitivities and more reflection into the classroom [82].

Effective Incorporation of Nontraditional Discipline-Independent Outcomes into an Anatomy Course

The design and implementation of learning and teaching sessions that incorporate nonanatomical skills, competencies, and attributes involve a number of considerations.

What to consider when incorporating nontraditional discipline-independent skills and attributes:

- Identifying outcomes and setting clear expectations
- Highlighting relevance
- Avoiding curriculum overload
- Establishing need and avoiding repetition
- Ensuring appropriate staffing skills
- Analysis, reflection, and improvement

Identifying Outcomes and Setting Clear Expectations

The identification of learning outcomes is a key element of curriculum design and, in anatomy, should include aspects directly related to the discipline itself. Further, as this chapter demonstrates, these learning outcomes should also include nontraditional discipline-independent skills, competencies, and attributes. As has been shown, many sessions can provide ideal opportunities for developing practical, communication, and teamwork skills as well as developing a functional awareness of ethics, law, and professionalism; however, these can be hidden within the curriculum and are not always obvious to students. It is therefore important that clear expectations are made evident at the beginning of the course and are emphasized on a regular basis and that any "additional" expected outcomes are defined and well signposted. When additional outcomes are to be assessed, students must be provided with the training and practice to achieve the measurable outcome.

Highlighting Relevance of Skills and Attributes

For students to become engaged in developing nontraditional skills in the context of an anatomy course, it is essential that the relevance to anatomy and to their ultimate educational endpoint is made apparent through learning NTDIS. This might be demonstrated, for example, by showing how communication skills exercises are applicable to their future patient interactions or how an ethical debate will likely influence decisions that students will find themselves having to make as they progress in their careers. Therefore, while the anatomy course provides the conduit for development of NTDIS, appropriate context needs to be set so that students can recognize how the skills they are learning in their teams, through oral presentations, and in their written assignments will apply more broadly to the work they will do for the rest of their lives.

Ensuring the Course Is Not Overloaded

It is easy over a period of time to continually add or extend elements to a teaching session or course of learning activities. However, this can lead to "curriculum creep" – a tendency to overburden the course or program with copious and extraneous material. More is not necessarily always "better." It is therefore important when considering the inclusion of additional skills or new material that course leaders do not overload their courses with too much content or assessment. A careful balance must be struck to ensure students gain the necessary knowledge and understanding of the subject, while also developing other essential associated skills and competencies. In many cases these elements are already natural parts of an anatomical course, and therefore additional time or focus may not be necessary, and instead careful curation and signposting of material and opportunities are required. Even a brief seminar highlighting NTDIS can have positive effects on a student's nontechnical skills [83]. Integrating both the discipline and non-discipline elements together provides opportunities to assess more holistically the knowledge, skills, and attributes a student has acquired and minimizes the need for space in the curriculum.

What will you do?

- You are designing opportunities for students to demonstrate anatomical knowledge using oral communication skills. How do you balance the training and feedback students receive in terms of learning basic skills of communication versus assessing proficiency on anatomical information?
- You provide a forum for discussing the ethics of dissection as an important tool for learning anatomy, but you become concerned by some of the student views expressed as they appear to be disrespectful to donors. How will you engage with students on this issue without dictating your own ethical position?

Cross-Linking

- See also Chapter 41 "Peer and faculty assessment of nontraditional discipline-independent skills in gross anatomy" and 47 "Exploring the Hidden Curriculum in Anatomy Education".

Establishing Need and Avoiding Repetition

While there are plenty of opportunities to incorporate new dimensions into anatomy activities that emphasize additional learning outcomes, it is important that these additions are viewed in the context of the whole curriculum. Inclusion of training that includes development of a particular skill or attribute should have a sound pedagogical basis and should map well to the curriculum for a particular healthcare specialty. While one skill might be important for a particular career route, it may not be relevant for another, and/or a different emphasis on that skill or attribute may be required. The anatomy team may identify additional skills on which to focus, but it is important to check to make sure these skills are not already covered elsewhere in the curriculum. Anatomy teams must therefore engage with colleagues across a range of disciplines so that initiatives are not implemented in isolation, redundancies are minimized, and so the potential for successful outcome can be fully maximized.

Ensuring Appropriate Skills for Teaching Faculty

While the anatomy environment provides some great opportunities for developing additional skills and focusing on different attributes, the anatomist, without appropriate training, is not necessarily the best person to lead, develop, or assess such approaches. It is therefore important that anatomy instructors collaborate with colleagues across the curriculum working together to ensure appropriate inculcation of NTDIS. Taking this a step further, appropriate colleagues from outside of anatomy but with expertise regarding NTDIS should not only be involved in the process of such curriculum development but also welcomed into the anatomical environment to facilitate within the course. In addition, there should also be opportunities for anatomy teaching faculty to receive training through relevant faculty development programs.

Analyzing, Reflecting, and Improving

The inclusion of additional skills, competencies, and attributes into anatomy curricula must be accompanied by appropriate analysis to ensure that the approaches used for the development or assessment of new elements are effective for the learners, the new elements fit well with the expected course outcomes, and the outcomes align well with the entire healthcare training program. This analysis should include informal observation and feedback from students and other stakeholders. Additionally, a more formalized approach including qualitative and quantitative staff and student surveys, as well as a review of student demonstrable attainment through formative and summative assessments, should be obtained. The anatomy teaching staff should reflect on the success of new or redefined initiatives as it is important that all garnered analysis of new initiatives is viewed in light of the whole curriculum to determine whether the additions help to meet the overarching outcomes of the healthcare program and whether further adaptations are necessary.

Conclusion

The anatomy curriculum provides the opportunity for students to develop and demonstrate a range of skills, competencies, and attributes, which may at first glance not be directly related to the discipline of anatomy but which can help lay a foundation for effective healthcare training. These additional competencies, NTDIS, include aspects related to practical skills, interpersonal and communication skills, teamwork and leadership, and professionalism. When devising a course to include such elements, it is important that all opportunities for development are outcomes-based and relevant to the particular profession and well within the wider context of the healthcare training program. The decision to incorporate a new and distinctive activity within an anatomy course must also be balanced against the tendency to overload the curriculum and should avoid repetition.

Successful inculcation of NTDIS development requires a paradigm shift in thinking and action, whereby educators and students internalize a change in philosophy [4]. Such action focuses on educators committing to facilitating NTDIS development within the curriculum, students recognizing the importance of NTDIS in their professional socialization and future success, and patient care becoming the focus of authentic learning.

When NTDIS are appropriately integrated within the curriculum, students should be able to recognize that anatomy not only serves as a fundamental building block for their knowledge and understanding of medical science but also is a fundamental opportunity for developing the interpersonal skills and attributes that are expected of them as professionals.

> Neither economic incentives nor technology nor administrative control has proved an effective surrogate for the commitment to integrity evoked in the ideal of professionalism. [84]

References

1. Sugand K, Abrahams P, Khurana A. The anatomy of anatomy: a review for its modernization. Anat Sci Educ. 2010;3:83–93.
2. Estai M, Bunt S. Best teaching practices in anatomy education: a critical review. Ann Anat. 2016;208:151–7.
3. Gregory JK, Lachman N, Camp CL, Chen LP, Pawlina W. Restructuring a basic science course for core competencies: an example from anatomy teaching. Med Teach. 2009;31:855–61.
4. Evans DJR, Pawlina P, Lachman N. Human skills for human[istic] anatomy: an emphasis on nontraditional discipline-independent skills. Anat Sci Educ. 2018;11:221–4.
5. Hafferty FW. Professionalism and the socialization of medical students. In: Cruess RL, Cruess SR, Steinert Y, editors. Teaching medical professionalism. 1st ed. New York: Cambridge University Press; 2009. p. 53–73.
6. Cruess RL, Cruess SR, Steinert Y. Amending Miller's pyramid to include professional identity formation. Acad Med. 2016;91:180–5.
7. Cruess RL, Cruess SR, Boudreau JD, Snell L, Steinert Y. Reframing medical education to sup-

port professional identity formation. Acad Med. 2014;89:1446–51.
8. Pawlina W. Not "How should I learn?" or "How should I act?" but, "Who shall I become?": a précis on the roots of early professional identity formation in the anatomy course. Anat Sci Educ. 2019;12:465–7.
9. Goldie J. The formation of professional identity in medical students: considerations for educators. Med Teach. 2012;34:e641–8.
10. Hafferty FW, O'Donnell JF, editors. The hidden curriculum in health professional education. 1st ed. Lebanon: Dartmouth College Press; 2015.
11. Yule S, Flin R, Paterson-Brown S, Maran N, Rowley D. Development of a rating system for surgeons' non-technical skills. Med Educ. 2006;40:1098–104.
12. Older J. Anatomy: a must for teaching the next generation. Surgeon. 2004;2:79–90.
13. Chambers J, Emlyn-Jones D. Keeping dissection alive for medical students. Anat Sci Educ. 2009;2:302–3.
14. McMenamin PG. Body painting as a tool in clinical anatomy teaching. Anat Sci Educ. 2008;1:139–44.
15. Collett T, Kirvell D, Nakorn A, McLachlan JC. The role of living models in the teaching of surface anatomy: some experiences from a UK medical school. Med Teach. 2009;31:e90–6.
16. Nanjundaiah K, Chowdapurkar S. Body-painting: a tool which can be used to teach surface anatomy. J Clin Diagn Res. 2012;6:1405–8.
17. Chinnah TI, Regan de Bere S, Collett T. Students' views on the impact of peer physical examination and palpation as a pedagogic tool for teaching and learning living human anatomy. Med Teach. 2011;33:e27–36.
18. Evans DJR. Connecting with different audiences: the anatomy of communication is essential. Anat Sci Educ. 2013;6:134–7.
19. Swing SR. The ACGME outcome project: retrospective and prospective. Med Teach. 2007;29:648–54.
20. General Medical Council. Outcomes for graduates 2018. London: General Medical Council; 2018.
21. Pawlina W. Professionalism and anatomy: how do these two terms define our role? Clin Anat. 2006;19:391–2.
22. Evans DJR. Promoting knowledge and understanding in society. Training our students for effective communication. J Appl Res High Educ. 2011;3:35–46.
23. Hofer RE, Nikolaus OB, Pawlina W. Using checklists in a gross anatomy laboratory improves learning outcomes and dissection quality. Anat Sci Educ. 2011;4:249–55.
24. Evans DJR, Cuffe T. Near-peer teaching in anatomy: an approach for deeper learning. Anat Sci Educ. 2009;2:227–33.
25. Reif-Lehrer L. Teaching good communication/proposal-writing skill: overcoming one deficit of our educational system. J Sci Educ Technol. 1992;1:211–9.
26. Bregman BM, Irvine C. Subjectifying the patient: creative writing and the clinical encounter. Fam Med. 2004;36:400–1.
27. Rees C, Sheard C. Undergraduate medical students' views about a reflective portfolio assessment of their communication skills learning. Med Educ. 2004;38:125–8.
28. Pearce M, Evans DJR. Developing multimedia resources: for students, by students. Clin Teach. 2012;9:178–82.
29. Choudhury B, Gouldsborough I. The use of electronic media to develop transferable skills in science students studying anatomy. Anat Sci Educ. 2012;5:125–31.
30. Van Dalen J, Bartholomeus P, Kerkhofs E, Lulofs R, Van Thiel J, Rethans JJ, et al. Teaching and assessing communication skills in Maastricht: the first twenty years. Med Teach. 2001;23:245–51.
31. Losh DP, Mauksch LB, Arnold RW, Maresca TM, Storck MG, Maestas RR, et al. Teaching inpatient communication skills to medical students: an innovative strategy. Acad Med. 2005;80:118–24.
32. Windish DM, Price EG, Clever SL, Magaziner JL, Thomas PA. Teaching medical students the important connection between communication and clinical reasoning. J Gen Intern Med. 2005;20:1108–13.
33. Nuovo J, Bertakis KD, Azari R. Assessing resident's knowledge and communication skills using four different evaluation tools. Med Educ. 2006;40:630–6.
34. Koenigsfeld CF, Wall GC, Miesner AR, Schmidt G, Haack SL, Eastman DK, et al. A faculty-led mock residency interview exercise for fourth-year doctor of pharmacy students. J Pharm Pract. 2012;25:101–7.
35. Böckers A, Jerg-Bretzke L, Lamp C, Brinkmann A, Traue HC, Böckers TM. The gross anatomy course: an analysis of its importance. Anat Sci Educ. 2010;3:3–11.
36. Lerner S, Magrane D, Friedman E. Teaching teamwork in medical education. Mt Sinai J Med. 2009;76:318–29.
37. Vasan NS, DeFouw DO, Compton S. Team-based learning in anatomy: an efficient, effective, and economical strategy. Anat Sci Educ. 2011;4:333–9.
38. Krych AJ, March CN, Bryan RE, Peake BJ, Pawlina W, Carmichael SW. Reciprocal peer teaching: students teaching students in the gross anatomy laboratory. Clin Anat. 2005;18:296–301.
39. Youdas JW, Krause DA, Hellyer NJ, Hollman JH, Rindflesch AB. Perceived usefulness of reciprocal peer teaching among doctor of physical therapy students in the gross anatomy laboratory. J Phys Ther Educ. 2007;21:30–8.
40. Nieder GL, Parmelee DX, Stolfi A, Hudes PD. Team-based learning in a medical gross anatomy and embryology course. Clin Anat. 2005;18:56–63.
41. Huitt TW, Killins A, Brooks WS. Team-based learning in the gross anatomy laboratory improves academic performance and students' attitudes toward teamwork. Anat Sci Educ. 2015;8:95–103.
42. Chen LP, Gregory JK, Camp CL, Juskewitch JE, Pawlina W, Lachman N. Learning to lead: self- and peer evaluation of team leaders in the human structure didactic block. Anat Sci Educ. 2009;2:210–7.

43. Escobar-Poni B, Poni ES. The role of gross anatomy in promoting professionalism: a neglected opportunity. Clin Anat. 2006;19:461–7.
44. Swartz WJ. Using gross anatomy to teach and assess professionalism in the first year of medical school. Clin Anat. 2006;19:437–41.
45. Epstein RM, Hundert EM. Defining and assessing professional competence. JAMA. 2002;287:226–35.
46. Weeks SE, Harris EE, Kinzey WG. Human gross anatomy: a crucial time to encourage respect and compassion in students. Clin Anat. 1995;8:69–79.
47. Pearson WG, Hoagland TM. Measuring change in professionalism attitudes during the gross anatomy course. Anat Sci Educ. 2010;3:12–6.
48. Evans DJR, Fossey S. Perspectives on anatomical donation and holding services of thanksgiving. Clin Ethics. 2011;6:195–9.
49. Jones TW, Lachman N, Pawlina P. Honoring our donors: a survey of memorial ceremonies in United States anatomy programs. Anat Sci Educ. 2014;7:219–33.
50. Williams AD, Greenwald EE, Soricelli RL, DePace DM. Medical student' reactions to anatomic dissection and the phenomenon of cadaver naming. Anat Sci Educ. 2014;7:169–80.
51. Kostas TR, Jones DB, Schiefer TK, Geske JB, Carmichael SW, Pawlina W. The use of video interview to enhance gross anatomy students' understanding of professionalism. Med Teach. 2007;29:264–6.
52. Crow SM, O'Donoghue D, Vannatta JB, Thompson BM. Meeting the family: promoting humanism in gross anatomy. Teach Learn Med. 2012;24:49–54.
53. Halliday NL, Moon MB, O'Donoghue DL, Thompson BM, Crow SM. Transformation and closure for anatomical donor families that meet medical students. Anat Sci Educ. 2019;12:399–406.
54. Hull FM. Death, dying and the medical student. Med Educ. 1991;25:491–6.
55. Marks SC Jr, Bertman SL, Penney JC. Human anatomy: a foundation for education about death, dying in medicine. Clin Anat. 1997;10:118–22.
56. Tschernig T, Schlaud M, Pabst R. Emotional reactions of medical students to dissecting human bodies: a conceptual approach and is evaluation. Anat Rec. 2000;261:11–2.
57. Kotze SH, Mole CG. Talking about death: implementing peer discussion as a coping mechanism to overcome fears about dissection, death and dying. Anat Sci Educ. 2013;6:191–8.
58. Coulehan JL, Williams PC, Landis D, Naser C. The first patient: reflections and stories about the anatomy cadaver. Teach Learn Med. 1995;7:61–6.
59. Tschernig T, Pabst R. Services of thanksgiving at the end of gross anatomy courses: a unique task for anatomists? Anat Rec. 2001;265:204–5.
60. Lachman N, Pawlina W. Integrating professionalism in early medical education: the theory and application of reflective practice in the anatomy curriculum. Clin Anat. 2006;19:456–60.
61. Shapiro J, Nguyen VP, Mourra S, Boker JR, Ross M, Thai TM, et al. Relationship of creative projects in anatomy to medical student professionalism, test performance and stress: an exploratory study. BMC Med Educ. 2009;9:65.
62. Camp CL, Gregory JK, Lachman N, Chen LP, Juskewitch JE, Pawlina W. Comparative efficacy of group and individual feedback in gross anatomy for promoting medical student professionalism. Anat Sci Educ. 2010;3:64–72.
63. Alt-Epping B, Lohse C, Viebahn C, von Steinbuchel N, Benze G, Nauck F. On death and dying – an exploratory and evaluative study of a reflective, interdisciplinary course element in undergraduate anatomy teaching. BMC Med Educ. 2014;14:15–21.
64. Weyers S, Noack T, Rehkamper G, Wenzel-Mayburg U, Schulz C, Schmitz A. "I just realized that something happened to me today" – a innovative educational approach to support students in dealing with the dissection course, death and dying. Ann Anat. 2010;210:160–3.
65. Hildebrandt S. Thoughts on practical core elements of an ethical anatomical dissection. Clin Anat. 2016;29:37–45.
66. Rizzolo LJ. Human dissection: an approach to interweaving the traditional and humanistic goals of medical education. Anat Rec. 2002;269:242–8.
67. Pawlina W, Lachman N. Dissection in learning and teaching gross anatomy: rebuttal to McLachlan. Anat Rec. 2004;281B:9–11.
68. Waterston SW, Stewart IJ. Survey of clinicians' attitudes to the anatomical teaching and knowledge of medical students. Clin Anat. 1995;18:380–4.
69. Wisner BP, Cravens GT, Holley KE, Carmichael SW, Pawlina W. Autopsy report: a valuable tool for developing observational skills. Clin Anat. 2004;17:197.
70. Wittich CM, Pawlina W, Drake RL, Szosteck JH, Reed DA, Lachman N, et al. Validation of a method for measuring medical students' critical reflections on professionalism in gross anatomy. Anat Sci Educ. 2013;6:232–8.
71. Dandavino M, Snell L, Wiseman J. Why medical students should learn how to teach. Med Teach. 2007;29:558–65.
72. Youdas JW, Hoffarth BL, Kohlwey SR, Krammer CM, Petro JL. Peer teaching among physical therapy students during gross anatomy: perceptions of peer teachers and students. Anat Sci Educ. 2008;1:199–206.
73. Ocel JJ, Palmer BA, Wittich CM, Carmichale SW, Pawlina W. Outcomes of the gross and developmental anatomy teachin assistant experience. Clin Anat. 2003;16:526–30.
74. Lockwood AM, Roberts AM. The anatomy demonstrator of the future: an examination of the role of the medically qualified anatomy demonstrator in the context of tomorrow's doctors and modernizing medical careers. Clin Anat. 2007;20:455–9.
75. Peluso MJ, Hafler JP. Medical students as medical educators: opportunities for skill development in the

absence of formal training programs. Yale J Biol Med. 2011;84:205–9.
76. Bulte C, Betts A, Garner K, Durning S. Student teaching: views of student near-peer teachers and learners. Med Teach. 2007;29:583–90.
77. Smith CF, Hall S, Border S, Adds PJ, Finn GM. Interprofessional anatomy education in the United Kingdom and Ireland: perspectives from students and teachers. Anat Sci Educ. 2015;8:360–70.
78. Purkayastha S, Paraskevas P, Darzi A. Make surgeons more active in teaching anatomy at all levels. BMJ. 2007;334:110.
79. Raftery AT. Anatomy teaching in the UK. Surgery. 2007;25:1–2.
80. Collins JP. Modern approaches to teaching and learning anatomy. BMJ. 2008;337:a1310.
81. Mitchell BS, McCrorie P, Sedgwick P. Student attitudes towards anatomy teaching and learning in a multiprofessional context. Med Educ. 2004;38:737–48.
82. Moore CM, Lowe C, Lawrence J, Borchers P. Developing observational skills and knowledge of anatomical relationships in an art and anatomy workshop using plastinated specimens. Anat Sci Educ. 2001;4:294–301.
83. Hagemann V, Herbstreit F, Kehren C, Chittamadathil J, Wolfertz S, Dirkmann D, et al. Does teaching nontechnical skills to medical students improve those skills and simulated patient outcome? Int J Med Educ. 2017;8:101–13.
84. Sullivan W. Work and integrity: the crisis and promise of professionalism in North America. New York: Harper Collins; 1995.

Role of Anatomists in Building an Integrated Medical Curriculum

46

Douglas F. Paulsen, Brenda J. Klement, and Lawrence E. Wineski

Introduction

Following the *Flexner Report* in 1910 [1], most US medical schools adopted a standard curricular model encompassing 4 years of medical education. The first 2 (preclinical) years included basic science courses, each administered by an academic department of the same name (e.g., anatomy, biochemistry, etc.). The third and fourth (clinical) years then focused on training in discipline-based clerkships (e.g., medicine, pediatrics, etc.). A shift away from the traditional model and toward an integrated approach arguably began in 1952 with the cross-disciplinary, organ system-based curriculum at Western Reserve University (now Case Western Reserve University) [2]. This model was adopted, to varying degrees, by other schools in the following decades. Another shift occurred in 1975 when McMaster University introduced its problem-based learning model, which promoted further integration of basic and clinical concepts in service to understanding and solving clinical problems [3]. Since then, the uniformity of traditional medical curricula has given way to greater diversity in which each medical school has adopted its own blend of traditional, integrated, and problem-based approaches. Most schools have adopted, modified, and, in some cases, abandoned one or more of these approaches along the way. The result is a rich diversity of approaches that have succeeded in preparing students to participate in a rapidly changing professional landscape.

In recent years, a reconvergence has occurred around the concept of integration, although again in an institution-specific manner. Integration is a complex and dynamic process [4] and is accomplished through a series of steps in order to help students see the big picture landscape and to integrate concepts [5]. Many medical schools have undertaken curricular modifications to enhance topic integration along both horizontal (e.g., among preclinical subjects) and vertical (preclinical with clinical subjects) lines [6–10]. The rationale for change has varied but often included enhancing students' appreciation of the need for and their skills in (1) integrating diverse and complex concepts in support of critical thinking and clinical problem-solving, (2) preparing for integrative questions on the United States Medical Licensing Examination (USMLE) Step examinations, and (3) eliminating unnecessary redundancy to provide more independent study time in a crowded curriculum.

Although the need for integration is increasingly accepted, the process for achieving it has

D. F. Paulsen · L. E. Wineski
Department of Pathology and Anatomy, Morehouse School of Medicine, Atlanta, GA, USA
e-mail: dpaulsen@msm.edu; lwineski@msm.edu

B. J. Klement (✉)
Department of Medical Education, Morehouse School of Medicine, Atlanta, GA, USA
e-mail: bklement@msm.edu

L. K. Chan, W. Pawlina (eds.), *Teaching Anatomy*, https://doi.org/10.1007/978-3-030-43283-6_46

proven a more fractious problem. In early integrated approaches, because a system-based approach to dissection is impractical, anatomy courses were conducted alongside rather than as components of integrated curricula. Finding a place for dissection in integrated curricula remains challenging, but an increasing number of programs now incorporate anatomy teaching into an integrated curriculum [11], and cadaver dissection continues to be included in medical training throughout the world [12]. Institutions seek to establish innovative approaches of incorporating anatomy into their curriculum to meet the needs of future physicians [13].

This chapter describes an approach to curricular integration involving a sequence of changes which, instead of marginalizing human dissection, incorporate it as a cornerstone of integration. The approach described is noteworthy because it was initiated by anatomists who led the way by first integrating regional dissection of the body with system-based approaches to histology and embryology.

Human Morphology: Integrating the Anatomical Sciences

In 1995, the first-year curriculum at Morehouse School of Medicine (MSM) focused on normal human structure and function with courses in gross anatomy and embryology, histology and cell biology, neurobiology, biochemistry, and physiology. The second-year curriculum focused on disease processes and treatments with courses in pathology, pathophysiology, microbiology and immunology, and pharmacology. At that time, anatomy faculty teaching in the "gross anatomy and embryology" and "histology and cell biology" courses chose to integrate their courses by combining the allotted time and adjusting the topic sequence. The aims were to reduce redundancy and encourage integration of knowledge of human structure from the electron-microscopic to the gross-anatomical level and from fertilization to old age. The result was a course entitled "Human Morphology" [14] that began a transition to curricular integration.

Initiating Integration

- Take the lead. Begin at the departmental level integrating the anatomical sciences.
- Combine curriculum time allotted, and adjust presentation sequence to reduce redundancy.
- Integrate cell biology with early embryology and histology of basic tissue types.
- Integrate regional dissection with organ system histology and embryology.

At the time, all courses were administered by the academic departments. Beginning curricular integration within one department sidestepped the turf battles that plague many attempts at curricular integration. In the anatomy department, there was a common appreciation that despite the system-based approach common in teaching histology and embryology, course integration would require maintaining regional dissection as a foundation.

Topic Sequencing in Human Morphology

- Cell biology, early embryology, and basic tissue types
- Blood and bone marrow, vascular, lymphoid, and skin histology as transitional topics with back dissection
- Limb embryology with upper limb dissection
- Cardiovascular and pulmonary histology and embryology with thoracic cavity dissection
- Digestive, urinary, endocrine, and reproductive system histology and embryology with abdominal and pelvic cavity dissections
- Lower limb dissection
- Head and neck dissection with neurobiology

Regional dissection traditionally divides the body into seven regions: the back, upper limb, lower limb, thorax, abdomen, pelvis and perineum, and head and neck. Overlaying those regions with the body systems best represented in each provided an initial basis for interlacing the topics of gross anatomy, histology, and embryology. The order in which those regions are dissected, and how they are grouped for examinations, varies among institutions. Flexibility in ordering the regions facilitated adjustments from year to year before arriving at the most appropriate sequence. Other principles important in arranging the order included the need to introduce themes and basic principles prior to delving into variations or more complex concepts, the skill required to complete the dissections, and whether key principles were better observed in cadavers, models, or other teaching aids.

Thus, for example, cell biology, early embryology, and basic tissue types (epithelium, connective tissue, nerve, and muscle) and their histogenesis were introduced prior to beginning dissection. Dissection began with the back, where students could remove skin and define muscles with little damage from unrefined dissecting skills. The blood, bone marrow, skin, and vascular and lymphoid system histology were covered concurrently as a transition from tissues to organ systems.

Because the upper limb has key elements associated with both the back and the thoracic wall and because thoracic wall structure is appropriately dealt with before entering the thoracic cavity, the upper limb dissection served as an appropriate bridge between the back and the thoracic cavity.

Entering the thoracic cavity after the upper limb provided a welcome change of pace and brought a new level of excitement to the laboratory, refreshing the students' enthusiasm and curiosity. The thoracic unit included the histology and embryology of the cardiovascular and respiratory systems. Although the esophagus is located here, its uncomplicated histology was reserved until the abdominal cavity dissection, where most digestive organs are located and can be presented as a system.

Progressive dissection through the abdominopelvic cavity and perineum dovetailed with the histology and embryology of the digestive, urinary, endocrine, and reproductive systems. This required the students to divide their time nearly equally between gross and microscopic anatomy topics and provided important opportunities to reinforce concept integration. Pituitary histology was covered with the endocrine system during the pelvic cavity unit. This illustrates the need for flexibility in blending system-based and region-based approaches to course integration.

Study of the lower limb followed that of the pelvis and perineum. These dissections are highly complementary because of the neuromuscular and vascular continuities between these regions. At this point, the histology coursework was complete. The students were able to focus again on their dissections and were happy to return to what they now perceived as a straightforward limb dissection.

The head and neck dissection was reserved for last for two reasons. First, it requires refined dissecting skills to demonstrate the smaller, more delicate, and closely apposed structures. The students were now better prepared to perform this challenging dissection. Second, this sequence provided an excellent opportunity to blend topics with our neuroscience curriculum, facilitating further concept integration in coverage of the brain, cranial nerves, and special senses.

With the integrated human morphology course well established, supported by the faculty and showing clear evidence of positive impact on student performance, the institution began pursuing a more extensive curricular integration in 2006. At this point, because of their extensive experience in subject integration, the anatomy faculty were in an optimal position to assume a leadership role in the larger-scale curricular integration process.

Institutional Governance and Curricular Change

Curricular integration across departmental boundaries is challenging. It requires a clear institutional commitment to change and support by the academic leadership. Ideally, that leadership will provide the requisite mechanisms and resources and a clear rationale for the institution's commitment in ways that encourage commitment by the faculty charged with implementation. At MSM, steps leading to this change included external encouragement from the Liaison Committee on Medical Education (LCME) to shift course governance from the academic departments to a central institutional body. This is reflected in the 2019 LCME publication (standard 8.3) that requires responsibility by a faculty committee for the components of a medical education program including development, design, sequencing, review, and evaluation [15]. MSM thus placed control for the design, implementation, and evaluation of the medical curriculum in the hands of the institutional Curriculum and Evaluation Committee (CEC), composed of faculty from most academic departments, academic administrators, and student representatives. The CEC determined that increased integration of the first-year curriculum was needed to accomplish the following objectives:

- Reduce the number of contact hours in the preclinical curriculum to provide more time for self-directed learning by eliminating unnecessary redundancy.
- Increase the students' capacity for cross-disciplinary concept integration to facilitate critical thinking and clinical problem-solving.
- Better prepare students for the growing number of integrative questions on USMLE Step and National Board of Medical Examiners (NBME) shelf exams.

The individual serving as both chair of the Department of Medical Education and senior associate dean for educational affairs was charged with facilitating the first-year medical curriculum integration and convened a faculty working group to undertake the integration process.

Institutional Requirements for Curricular Change

- Institutional commitment to change
- Clear rationale and objectives for the change
- Support by top-level academic leadership
- Central rather than departmental control over the curriculum
- Focus on curricular change as a continuing process rather than a one-time event
- Ongoing course director and faculty engagement in design and implementation
- Decrease in the number of contact hours in the preclinical curriculum to provide more time for self-directed learning by eliminating unnecessary redundancy
- Enhancement of the students' capacity for cross-disciplinary concept integration to facilitate critical thinking and clinical problem-solving
- Preparation of students for the growing number of integrative questions on USMLE Step and National Board of Medical Examiners (NBME) shelf exams

Human Morphology as a Foundation for Integrating the First-Year Basic Science Courses

Human morphology, which integrated regional gross anatomy dissection with system-based histology and embryology, accounted for approximately half of the first-year curriculum and provided a foundation on which to build the larger curricular integration process. Because the existing first-year curriculum had been successful, there was some initial faculty resistance to changes involving other courses. As a compromise, the faculty working group began full curricular integration by reordering existing lectures to complement the order of system presentation developed for human morphology. Because there was room for everything in the existing curriculum, there would be room for everything in one that was simply reordered. In the initial version of the integrated curriculum, although the schedule was rearranged, the individual courses remained intact as independent graded entities, and the original course directors retained control of their subject areas.

Certainly, rearranging the order of subject presentation alone does not constitute integration, but when a culture change is called for, it is a useful place to start.

Scheduling class sessions was achieved by creating a magnetic calendar grid on which each curricular element (e.g., class or laboratory) could be moved about [9]. With the course faculty gathered, each course director took a turn at the calendar grid proposing and justifying rearrangements to assure appropriate concept sequencing for their discipline and integration with the others. Such scheduling meetings were held weekly during the noon hour and lunch was provided. This added to the collaborative atmosphere needed to overcome a prior culture of separation and turf guarding. As the group worked toward its common goal, defensiveness declined, and the faculty developed a remarkable camaraderie and investment in the new class sequence. Further, an unprecedented understanding of the content and educational philosophy underlying the prior strengths and weaknesses of each other's courses developed, leading to a broader and more distributed understanding of the curriculum as a whole and of the student experience. This was a valuable resource in moving toward a culture of integration.

Interlacing the biochemistry, physiology, and neurobiology courses with human morphology spread the original one-semester human morphology course across two semesters and created some logistical challenges. For example, associating the many cardiovascular and respiratory physiology lectures with the thoracic cavity dissection and the relevant anatomy, histology, and embryology lectures involved a significant break in the dissection schedule to accommodate the related topics. This created challenges associated with longer cadaver preservation that were eventually resolved.

Integrating organ system physiology with histology proved straightforward. However, overlapping the biochemistry topics with only cell and tissue histology provided insufficient time for the students to process that material, and the duration of biochemistry coverage was later expanded. The experience demonstrated that providing adequate time for students to process new information is an important consideration in an effective plan for integration.

Achieving Faculty Acceptance

- Culture shifts require incremental change.
- Begin by reordering existing curriculum components without challenging course autonomy.
- Use a scheduling method that allows iterative changes to be proposed, discussed, and revisited as necessary.
- Use working lunches to create a congenial atmosphere and a sense of common purpose.

Establishing New Integrated Courses

In the second phase of the integrated curriculum, the department-based course designations were dropped, and the first-year curriculum was divided into four half-semester courses, entitled Basic Principles of Human Biology, and Organ Systems 1, 2, and 3 [10]. The anatomy content continued to form the backbone for each new course structure, and logical breaks in the anatomy content provided the markers to form dividing lines between new courses and between examinations within courses (Fig. 46.1). The subject matter of each new course was integrated and replaced the previous subject-oriented courses.

A new curriculum management system was installed which included choosing a course director for each course from among the faculty working group. The course directors represented each of the first-year curriculum academic departments and were charged with oversight of course operations, scheduling adjustment, and examination assembly. A new position of first-year curriculum manager was created to oversee all curriculum operations and support the course

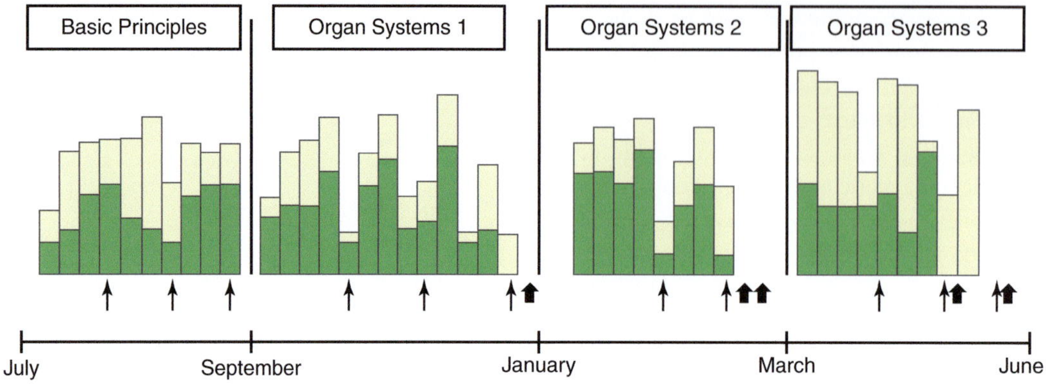

Fig. 46.1 Outline of the new integrated curriculum structure. Vertical lines designate the break between course boundaries, and the name of each course is listed at the top. The width of each bar represents 1 week in the curriculum, and the height of each bar represents the relative amount of time spent during that week on course content. The topics from the previous human morphology course (including anatomy, histology, cell biology, and embryology) are represented by the green bars. Other topic areas such as biochemistry, physiology, and neurobiology are represented by the yellow bars. The thin black arrows represent locations within each course where in-course examinations are given. The thick arrows represent the locations within the courses where NBME subject examinations are given

directors. Prior course directors and other key faculty were appointed as content experts for their respective fields and continued to participate in the curriculum operation meetings. This first-year curriculum operation group still meets monthly to review course operations, student performance, and opportunities for further integration.

In subsequent years, system-specific subgroups of content experts began meeting separately to assess and eliminate unnecessary content overlaps and to refine concept sequencing for each system. Thus, for instance, physiology faculty who had traditionally reviewed the anatomy of an organ system prior to covering system function were relieved of that duty by having physiology lectures follow the lectures and laboratory sessions conducted by the anatomy and histology faculty. In some cases, this led to the discovery and filling of gaps in coverage that one group had assumed that the other had been covering. Again, this facilitated faculty understanding of information covered in each other's lectures, enhanced integration and efficiency, and reduced contact hours. These efforts continue with an eye toward further consolidation and horizontal integration as well as providing time for vertical integration with pathology, pathophysiology, and clinical topics.

The new integrated courses include flexibility for utilizing multiple teaching modalities depending on curriculum needs and faculty preferences. Most lecture sessions are audio recorded to allow student review of the content and for distance education. It also forms the foundation for selected sessions that are presented in virtual fashion instead of in-person, enabling students to work independently at the time of their choosing. Designated curriculum time also incorporates clinical learning activities that take advantage of the integrative nature of the courses. These activities include specialized laboratory sessions, small group case study discussions, clinical demonstrations, basic physical examinations, and radiology reviews [16].

Curriculum modification is an evolutionary process, and its structure and organization should remain flexible enough to allow for small changes every year and larger, more substantial changes when needed. For example, since the implementation of the second phase of our curriculum, entire topic coverage for some subjects has been moved from one course to another, the practical examinations are now administered the day before the multiple-choice examinations, and the order of certain topic coverage has been revised to enhance the integrative nature of the curriculum content. To better enable curriculum modification, it is important to conduct yearly evaluations of the curriculum, particularly the topics taught, their allocated coverage within the curriculum, as well as faculty teaching expertise.

Student Assessment

Consolidating the examinations for the various courses was a major step forward in curricular integration and significantly reduced curriculum hours. Previously, each of the multiple courses running concurrently had its own 2-hour multiple-choice examination approximately every 3–6 weeks. In the new approach, one examination covering all topics was administered at the same frequency, where logical content breaks occurred. Instead of nearly a full week for examinations for each examination period, a reading day and an examination day sufficed. The examination time released was returned to the students for independent study.

To provide examinations of reasonable length, the number of test items for each topic was reduced. Four questions were included per class session (with some exceptions when justified). This provided equitable coverage of each topic. The Department of Medical Education coordinated all examination-related tasks. Examination software (ExamSoft Worldwide) was used to deliver the examinations, analyze question performance, and bank questions. The software was also used to monitor and track student performance and to identify subjects needing improved coverage. Gross anatomy laboratory (practical) examinations were initially administered on the same day as the multiple-choice examination but were later changed to the day before the multiple-choice component. These examinations involve fill-in-the-blank questions for tagged cadavers and prosections with four practical questions per scheduled laboratory session.

Five National Board of Medical Examiners (NBME) subject examinations (in biochemistry, gross anatomy and embryology, histology and cell biology, neurobiology, and physiology) were also given each year, each administered after the relevant subject material had been covered. Scores on the subject examinations were incorporated into the grade for the course in which each was given, and students were required to achieve a predetermined minimum score on each subject examination.

Curriculum Integration Considerations

- Establish clear goals and outcomes.
- Establish course and curriculum management team.
- Determine in-course examination scheme (frequency, number of items, delivery process).
- Determine NBME subject examination scheme (when to administer, minimum score, grade).
- Determine course grading structure.
- Evaluate student performance (compared to pre-integration).

Student Performance in the Integrated First-Year Curriculum

Student performance on the in-course examinations was monitored by the course operation group and compared to performances prior to integration. Overall, the students performed as well as or better than they did prior to integration, and the percent of students performing in the A range has steadily increased since the start of the new curriculum.

Curricular integration has correlated with progressively higher NBME subject examination scores. The class average for all five NBME examinations improved in the first 3 years of the new curriculum and has remained steady ever since. The histology and cell biology subject examination average increased by 7%, and the gross anatomy and embryology subject examination average increased by 9% in the first year of integration. The class average for each examination increased by smaller increments in subsequent years. In addition, the number of students scoring in the upper range on these examinations also increased annually. Since the move to curricular integration, the institution has maintained a high USMLE Step 1 pass rate, as well as increased Step 1 examination averages.

Results of Curriculum Integration on Students and Faculty

Despite the number of anatomy hours within a curriculum, students strongly view anatomy content as a foundation for building further knowledge [17]. Students learn to integrate course information through a developmental process that improves as students gain maturity and experience [18]. One component of this maturity is the development of learning strategies to assist the student in course performance [19]. To gain maturity, students need to adjust to the pace, information volume, and pressures of medical school. The MSM curriculum was intentionally structured to allow for this adjustment by first presenting foundational information delivered at a relatively moderate pace, then gradually increasing the rigors of content volume and rate of delivery throughout the year. Students have expressed positive feedback on the structure and function of the curriculum and have consistently given all courses high evaluation marks.

Despite the maturation opportunities inherently built into the curriculum, students that do not perform well on examinations or have difficulty developing new learning strategies are placed in an academic support program. This structured program is similar to others [20] and includes faculty-led tutorials, peer tutoring, self-evaluation, and advisory conferences. Students that need additional time in the basic sciences (for personal or academic reasons) can enroll in a decelerated program during the first or second year. This entire academic support program has progressively gotten stronger over the years so that the amount of support and opportunities provided to students is quite robust.

Faculty have also experienced growth and maturation of their professional abilities during the curriculum integration process. Outcomes that we and others [21] have seen include enhanced awareness of what others are teaching, increased communication between faculty, increased teaching collaborations, and improvement in the teaching process. After some initial reluctance to make a curriculum change, MSM faculty have been very eager and agreeable to teach and participate in the revised curriculum. To help faculty improve their teaching abilities, the institution has created a faculty development program that also includes a student, peer, and course director evaluation and feedback system as well as acknowledgment of exceptional teaching through student- and dean-nominated teaching awards.

Comparison of the Anatomy Teaching Structure in the MSM Curriculum to Other Programs

Anatomy content can be integrated within a medical curriculum in several ways. Decreasing the number of course hours by condensing anatomy, histology, and embryology into a short course can be as effective as a traditional gross anatomy course [22]. This demonstrates that some medical programs can have an effective curriculum with fewer hours of anatomy teaching and still provide students with enough anatomical training for their continued medical studies and successful board results.

In addition to MSM, other institutions have created an integrated curriculum that includes anatomy teaching throughout multiple courses and interwoven into an organ system-based structure. The curriculum design for the University of New Mexico School of Medicine [17] includes an initial course that covers the anatomy and pathology of the musculoskeletal system, skin, and connective tissue, while the anatomy of the other organ systems is taught during the first week of the relevant organ system block. The anatomy content is taught over the course of the 18-month preclinical curriculum and is always near organ physiology and pathophysiology so that the relationships of structure, function, and clinical relevance can be taught within the same time span. The curriculum revision for the University of Alabama School of Medicine [23] includes two sequential courses taught in the first semester, the first of which covers the basic information of cell biology, genetics, biochemistry, histology, and gross anatomy. The second course covers the basic concepts of microbiology, immunology, pathology, and pharmacology. The remaining three semesters of the preclinical curriculum contain sequential organ system modules, each integrating the anatomy, histology, embryology, physiology, pathology, and pharmacology topics. Therefore, the anatomy content in this curriculum is also spread out over most of the preclinical years.

Summary

Curricular integration at MSM has been relatively unusual among medical curricula. The process was initiated by anatomy faculty and then followed by broader faculty input building on the template of a dissection-oriented, anatomy-centric curriculum. Elsewhere, system-based approaches have led to significant difficulties in fitting active dissection by students into an integrated curriculum. Our experience shows that it is not only possible but desirable to blend regional dissection with a system-based approach to integrating teaching of human structure and function. The entire process was carefully organized and phased in over several years, with a focus on developing cross-departmental participation and collegiality. The four half-semester integrated courses that emerged from this first-year curriculum integration process are now well established. Faculty acceptance has been exceptional, with most now embracing the changes and fully supportive of the new courses. Examination scores have improved and student satisfaction is high. Overall, the new curriculum is more efficient, has better topic alignment, and is better organized than the previous curriculum.

References

1. Flexner A. Medical education in the United States and Canada. New York: Carnegie Foundation for the Advancement of Teaching; 1910. http://www.carnegiefoundation.org/sites/default/files/elibrary/Carnegie_Flexner_Report.pdf. Accessed 26 Apr 2019.
2. Ham TH. Medical education at Western Reserve University. A progress report for sixteen years, 1946–1962. N Engl J Med. 1962;267:868–74.
3. Sweeney GD, Mitchell DL. An introduction to the study of medicine: phase I of the McMaster M.D. program. J Med Educ. 1975;50:70–7.
4. Vergel J, Stentoft D, Montoya J. Extending the theoretical framework for curriculum integration in pre-clinical medical education. Perspect Med Educ. 2017;6(4):246–55.
5. Alsharif NZ. Curriculum integration: a self-driven continuum. Am J Pharm Educ. 2014;78(1):2.
6. Dahle LO, Brynhildsen J, Behrbohm Fallsberg M, Rundquist I, Hammar M. Pros and cons of vertical integration between clinical medicine and basic science within a problem-based undergraduate medical curriculum: examples and experiences from Linköping, Sweden. Med Teach. 2002;24(3):280–5.
7. Vidic B, Weitlauf HM. Horizontal and vertical integration of academic disciplines in the medical school curriculum. Clin Anat. 2002;15(3):233–5.
8. Muller JH, Jain S, Loeser H, Irby DM. Lessons learned about integrating a medical school curriculum: perceptions of students, faculty and curriculum leaders. Med Educ. 2008;42(8):778–85.
9. Klement BJ, Paulsen DF, Wineski LE. Anatomy as the backbone of an integrated first year medical curriculum: design and implementation. Anat Sci Educ. 2011;4:157–69.
10. Klement BJ, Paulsen DF, Wineski LE. Implementation and modification of an anatomy-based integrated curriculum. Anat Sci Educ. 2017;10:262–75.
11. McBride JM, Drake RL. National survey on anatomical sciences in medical education. Anat Sci Educ. 2018;11(1):7–14.
12. Memon I. Cadaver dissection is obsolete in medical training! A misinterpreted notion. Med Princ Pract. 2018;27(3):201–10.
13. Yaqinuddin A, Ikram MF, Zafar M, Eldin NS, Mazhar MA, Qazi S, et al. The integrated clinical anatomy program at Alfaisal University: an innovative model of teaching clinically applied functional anatomy in a hybrid curriculum. Adv Physiol Educ. 2016;40(1):56–63.
14. Wineski LE, Paulsen DF. Course syllabus for human morphology: gross anatomy, embryology, histology, cytology. Atlanta: Department of Anatomy and Neurobiology, Morehouse School of Medicine; 2005. https://www.coursehero.com/file/6698820/The-Human-Morphology-Course-Syllabus/. Accessed 26 Apr 2019.

15. Liaison Committee on Medical Education. Functions and Structure of a Medical School – (contains the LCME Standards). 2019. http://lcme.org/publications/#standards. Accessed 26 Apr 2019.
16. Klement BJ, Paulsen DF, Wineski LE. Clinical correlations as a tool in basic science medical education. J Med Educ Curric Dev. 2016;3:179–85.
17. Hartley RS, Smith GA, Rosenberg MJ. Anatomy integration: effective change or change of affect? Anat Sci Educ. 2018;11(6):535–46.
18. Chipamaunga S, Prozesky D. How students experience integration and perceive development of the ability to integrate learning. Adv Health Sci Educ Theory Pract. 2019;24(1):65–84.
19. Hernandez-Torrano D, Ali S, Chan CK. First year medical students' learning style preferences and their correlation with performance in different subjects within the medical course. BMC Med Educ. 2017;17(1):131.
20. Hortsch M, Mangrulkar RS. When students struggle with gross anatomy and histology: a strategy for monitoring, reviewing, and promoting student academic success in an integrated preclinical medical curriculum. Anat Sci Educ. 2015;8(5):478–83.
21. Khalil MK, Kibble JD. Faculty reflections on the process of building an integrated preclerkship curriculum: a new school perspective. Adv Physiol Educ. 2014;38(3):199–209.
22. Halliday N, O'Donoghue D, Klump KE, Thompson B. Human structure in six and one-half weeks: one approach to providing foundational anatomical competency in an era of compressed medical school anatomy curricula. Anat Sci Educ. 2015;8(2):149–57.
23. Brooks WS, Woodley KT, Jackson JR, Hoesley CJ. Integration of gross anatomy in an organ system-based medical curriculum: strategies and challenges. Anat Sci Educ. 2015;8(3):266–74.

Exploring the Hidden Curriculum in Anatomy Education

47

Gabrielle M. Finn and Frederic W. Hafferty

Social scientists have long been interested in the impact of anatomy and the dissection experience on the professional formation of medical students [1]. Two of the earliest and most famous studies of undergraduate medical education, Robert K. Merton and colleagues' 1957 study of medical training at Cornell (*The Student-Physician*) and Howard Becker and colleagues' 1961 examination of the medical school experience at the University of Kansas (*Boys in White*), highlight anatomy as a formative experience—albeit from contrasting theoretical perspectives. A third publication, *Report of the Fifth Teaching Institute, Association of American Medical Colleges, Atlantic City, New Jersey, October 15–19, 1957*, while not mentioning anatomy,[1] took an ecological and system approach to the undergraduate training experience along with a focus on how students can be quite sensitive to the difference between what faculty preach versus what faculty practice. Although this report did not deploy the literal term "hidden curriculum," its focus on the potential for discordances between the intended and formal curriculum versus the variety of other-than-formal learning that takes place for students did anticipate, by 40 years, the rise of a hidden curriculum literature within the fields of medicine and medical education.

> **Definition**
> The hidden curriculum refers to the tacit, implied, unwritten, unofficial, and often unintended behaviors, lessons, values, and perspectives that students learn during their education.

In this chapter, we will examine the function and framing of anatomy laboratory and cadaver dissection as a pedagogical space (for faculty) and learning environment (for students)—and do so using the analytical lens of the hidden curriculum. We begin with a brief look at how laboratory session has been characterized by both social scientists and medical trainees using a variety of resources including academic reference materials and physician autobiographies. With this as our foundation, we then will introduce the hidden curriculum (HC) as an analytical tool to broaden and deepen our understandings of the various types of learning that take place in anatomy. Within this context, we will highlight anatomy as

[1]Anatomy was featured in an earlier report (*Third Teaching Institute; October 18–22, 1955*) by the same sponsoring organizations.

G. M. Finn (✉)
Faculty of Biology, Medicine and Health, University of Manchester, Manchester, UK
e-mail: gabrielle.finn@manchester.ac.uk

F. W. Hafferty
Division of General Internal Medicine, Program in Professionalism and Values, College of Medicine and Science, Mayo Clinic, Rochester, MN, USA
e-mail: hafferty.frederic@mayo.edu

L. K. Chan, W. Pawlina (eds.), *Teaching Anatomy*, https://doi.org/10.1007/978-3-030-43283-6_47

a source of occupational socialization and professional formation. We also will reference, but not review, the spate of new teaching tools currently being used in anatomy teaching in order to link these tools to hidden curriculum issues. We conclude with a challenge for anatomy educators as to how they might encourage other basic science courses to build upon their innovations.

In high-stakes environments such as medical education:
Students constantly are on the lookout for message gaps between what faculty formally tell them about course/learning standards and what students come to learn (also from faculty) about what they should "really be doing to pass the course."

Case Examples of the Hidden Curriculum: Laboratory Sessions and Beyond

Anatomy and the Importance (or Not) of Dissection

At Institution X, students repeatedly are told by faculty that it is "critically important" for students to attend laboratory sessions to dissect and to engage in self-directed study on their cadavers—all because this "hands-on" and "tactile" experience is the best way for them to "really learn" their anatomy. However, students also hear, via the student grapevine, that they can fail the practical examination and still pass the course based solely upon their written examination performance. As a result, many students "get the message" and do not bother to attend laboratory sessions, even though they also admit that knowledge of anatomy will be quite important once they begin clinical practice.

Think of the hidden curriculum:

- As a heuristic device for exploring the difference between what people and/or organizations say… and what they actually do
- As a tool for reconciling the formal versus the other-than-formal dimensions of organizational life (including anatomy education)
- As a form of "gap analysis"
- Not as a teaching tool that can be deliberately exploited to "teach by stealth"

In this example, we can see where the long-term benefits of dissection seen by both faculty and students (e.g., clinical expertise) have given way to short-term student goals of "passing the test." Extending these rationalizations even further, students at X also believe that if dissection is all that important, then the skills and knowledge it provides will be resupplied at some future time(s). Students also believe that faculty actually know they are passing (grade wise) students who did not dissect and therefore passing students who, according to those same faculty, "really do not know their anatomy." In short, students feel they are being told one thing (e.g., "dissection is critical to learning") yet seeing another (e.g., grading policies) all of which lead students, via the tacit participation of anatomy faculty, that the "really" most important thing is to "pass the course" rather than "learn the anatomy."

Conflict-of-Interest (COI) Disclosure Statements

Many medical schools now require their faculty, and/or visiting speakers, to include a statement at the beginning of their presentation referencing "outside" financial interests and/or whether the speaker will be referring to "off-label" prescribing. Most often, this requirement is linked to some formal policy covering a range of conflict-of-interest (COI) issues. Similar to our previous WCC and anatomy examples, the formal curriculum is the requirement and/or formal policy—including supporting rationales for why the policy exists. However, and once again referring to our previous examples, there can be rifts or gaps between policy and practice. In other words, how disclosure statements *actually are presented* to an audience represents another set of messages to organizational members such as students and clinical faculty about the importance of COI issues. Over the past 2 years, one of the chapter authors has witnessed disclosures ranging from one that flashed across the screen at the speed of light, accompanied by a mutter of "moving right along," to a faculty member who showed his COI slide, read through it—twice—and then announced that this was the most important slide of his talk. Each COI behavior delivered a distinctly different message to the audience. The former was dismissive, the latter affirming. In both instances and for all the other instances that fell in between, community members were treated to a cascade of messaging about how one should think about COI—regardless of the actual policies that lay in the background. In short, while policies/formal curricula may be important, they hardly are the entire story.

Hidden Curriculum and Anatomy Education: A Brief History

Anatomy, particularly with its classic foundation in cadaver dissection, has long been considered *the* iconic first-year medical school experience. Although no longer reserved exclusively for medical trainees and no longer ubiquitously built around the dissection of human cadavers, it still is considered an important point of transition in the professional formation of trainees as they move from laypeople into a different and distinctive culture of medicine. Outsiders, including the public and social scientists, harbor a certain fascination with anatomy, including its social practices (e.g., what actually does go on in laboratory session), as well as the impact of dissection on trainees and faculty [2].

Since the 1950s, a number of books and journal articles have given witness to this fascination. For the purposes of this chapter, we group these materials into three overarching categories: (1) trainee–faculty reactions to laboratory session, (2) new approaches to teaching anatomy, and (3) the role of dissection as a necessary (or not) part of the anatomy learning environment. Publications have ranged from observational studies and fieldwork to the administration of various psychometric and learning assessment tools. As perhaps befitting anatomy's enduringly iconic status, conclusions have ranged from (1) dissection as having no emotional impact to laboratory sessions as having a "profound effect" on the socioemotional lives of faculty and students, (2) new teaching tools and modalities as having no benefit in stimulating anatomical learning to these tools having a substantive impact, and (3) dissection as essential or superfluous or anachronistic to learning human anatomy.

In certain instances, this chaos of conclusions can be viewed as rooted in the methodological approaches used and the questions being asked. Thus, and broadly speaking, observational and field-based studies have tended to emphasize the impact (positive and negative) of laboratory session/dissection, while instrument-based studies have tended to advance more of a "no impact" conclusion—whether impact was viewed on a social–psychological plane or in terms of subject matter mastery. The role and place of anatomy—particularly human dissection—remain hotly contested.

This conundrum brings us to a point fundamental to the integrity of this chapter. As humans move into, and then through, new, unique, and/or novel situations, they move from being acutely aware of differences, nuances, or "unusualities" to reframing what is now *not new* as routine and commonplace. In short, humans go from being hyperaware to hypoaware as the startling quickly becomes taken for granted. In this way, learning moves more from more intentional, deliberate, and even strategic to a domain that functions more at the tacit and unconscious levels. For these reasons, asking "old timers" to explain "what is happening" or "how things work around here" generates one set of responses, while asking newcomers the same questions often produces a totally different set of responses. It is not that one is correct and the other wrong (particularly to those providing the responses) but rather the existence of two potentially contrasting sets of social realities.

The difference between insider and outsider views on the "nature of things" is of particular relevance to this chapter. With anatomy, we have one group (e.g., faculty) who are considered, by insiders and most outsiders alike, as having institutionally sanctioned authority and expertise. Meanwhile, other groups, such as trainees, lack not only that authority but also are supposed to adopt the knowledge base, skill sets, and value orientations of those insiders. In short, neophytes are supposed to want to become "just like" insiders. This shift from outsider to "true insider" status is captured in Robert K. Merton's classic definition of socialization in which "…socialization designates the process by which people selectively acquire the values and attitudes, the interests and knowledge—in short the culture—current in the groups of which they are, or seek to become a member" [3], p. 287, as well as Lave and Wenger's concept of legitimate peripheral participation [4].

There is, however, a problem with classic views of socialization. They tend to frame learners as passive and empty vessels waiting to be filled by the wisdom and expertise of their insider-elders (e.g., faculty). In its extreme form, such a framing tends to discount as callow or naive the views and/or experience of learners. In turn, failures to learn (however this is established) often are attributed to the learners themselves—as opposed to the learning environment. Similarly, solutions thus target the student side of the equation by advocating the selection of "better" recruits or by ramping up teaching hours or the addition of more "rigorous" material. Finally, and as a consequence of such marginalizing strategies, students may construct their own learning environment and do so in ways that negate what they see as the mixed messages of faculty or contradictory teaching practices being delivered by faculty.

As faculty:

Do not be surprised when the learning environment experienced by your students is at some odds with the anatomy you intended when constructing your course.

Consider how the hidden curriculum in anatomy may impact upon your students' professional identity formation.

The Language and Structure of the Hidden Curriculum

Over the years and across the fields of education (general), sociology, and medical education, a number of terms have been used to capture the distinctions between the formal and the other-than-formal aspects of the learning environment. Some of these terms are depicted in Fig. 47.1.

Cross-Linking:

- Chapter 1 "Elements of Successful Adult Learning"
- Chapter 13 "Evaluating Your Own Performance in a Lecture"
- Chapter 19 "Giving Feedback to Students"
- Chapter 20 "Using Body Painting and Other Art-based Approaches for the Teaching of Anatomy and for Public Engagement"
- Chapter 25 "Preparing Students Emotionally for the Human Dissection Experience"
- Chapter 39 "Assessing Anatomy as a Basic Medical Science"
- Chapter 41 "Peer and Faculty Assessment of Nontraditional Discipline-Independent Skills in Gross Anatomy"
- Chapter 45 "The Role of the Anatomist in Teaching of Nontraditional Discipline-Independent Skills"

As captured in this figure, the basic distinction worth remembering is between the formal versus the other-than-formal aspects of learning. On the left side of Fig. 47.1, we list several terms that have been used to capture this side of the pedagogical coin. For the most part, these terms stand as synonyms. On the right side are examples of other-than-formal types of curricula. Here, the terms are not synonymous. Although the term "hidden curriculum" often is used as a catchall

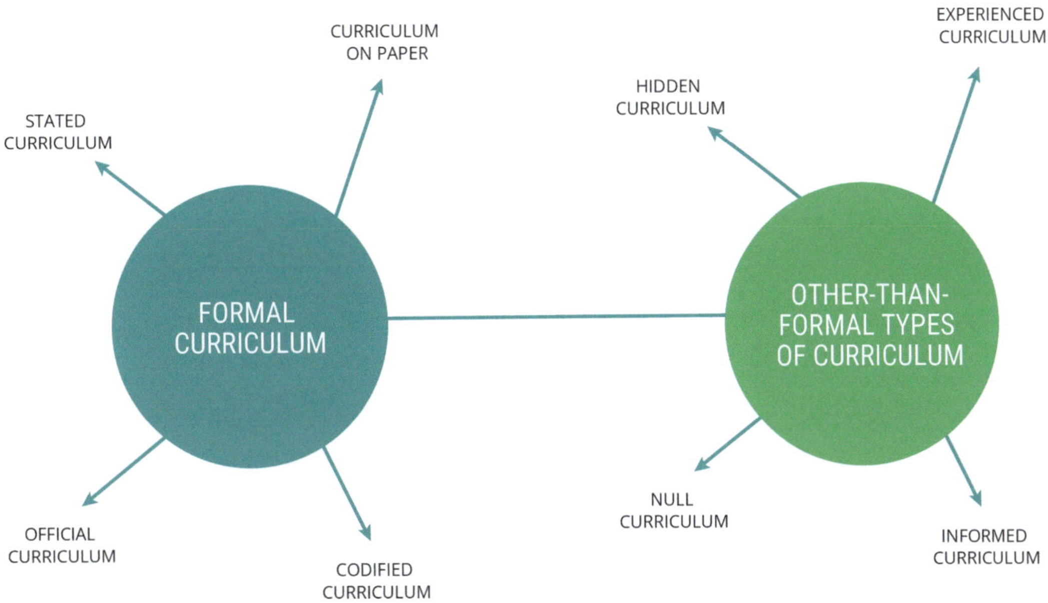

Fig. 47.1 Types of curriculum: a lexicon

descriptor for everything that might fall within the other-than-formal camp, this actually is not the case. Nor is it good analytic practice. For example, there are lots of ways that work gets done in clinical medicine that are quite well understood by all of the participants but where those practices are quite different from those specified within an organization's formal policies (e.g., the formal curriculum). One term often used to capture these more informal understandings of how work "really happens" versus "how things are supposed to take place according to the rules" is work-around [5]. There is nothing "hidden" here. Everyone knows these "rules of the road"—including management. It is just that everyone agrees that doing it "the way we do it" is better for the sake of efficiency or good patient care, etc., than following the letter of formal policies. The term in Fig. 47.1 that best fits these kinds of open and shared understandings is the *informal curriculum*.

In turn, the *null curriculum* is what gets taught by what is *not* said or done [6]. Once again, students are relentless, at least early in their training, to make sense of all the newness that surrounds them. If within a given anatomy course certain topics are never mentioned or certain cadaver body parts are never dissected, then students may well reason that these topics or body parts are not important. The same is true in clinical settings. If members of the clinical staff never sit on the side of a patient's bed, then students may well reason that this is not what clinicians do at the X medical center. They may even go one step further and assume that sitting on a patient's bed is something they should never do (e.g., "This just isn't done at X!"). Remember, no one directly has told students anything—pro or con. Instead, the absence of something (sitting) is imbued with meaning—and often without faculty having any clue about what is going on.

Finally, the *hidden curriculum* is, technically speaking, hidden. Most often, the HC functions at the level of organizational culture and therefore largely operates at a tacit or unconscious level. The HC largely deals with implicit understandings of "how things are done around here." As such, the HC operates within domains that are taken for granted and thus often fly beneath the reflective radar of both faculty and students. Finally, one underappreciated feature of the HC is that while it need not either be negative or bad, it is primed (as a methodological approach) to identify alternative ways of looking at things—at least alternative to the official or the formal account of what is going on.

In summary, while there is nothing particularly wrong with trying to get a handle on the HC by simply differentiating between the formal versus the other-than-formal aspects of learning, if one wishes to move beyond a simple either–or dichotomy, then one needs to be at least somewhat familiar with some of the various subtypes of other-than-formal learning.

Where to Find the Hidden Curriculum

While it may be reassuring to think of the HC as a singular overarching and essentially homogeneous domain of learning—and thus a domain that stands in singular contrast to *the* (again, singular) formal curriculum—in point of fact, the other-than-formal curriculum is awash with a complex array of different types of curricula: informal, hidden, null, etc., along with different shades of these subtypes. In this respect, the HC is quite similar to the formal curriculum, which itself is made up of all kinds for formal learning experiences. From this vantage point, it essentially is meaningless—as well as analytically misleading—to list an array of educational practices or events and go down that list, as a checklist, to indicate whether each event falls into the formal, informal, hidden, null, etc., camp. Instead, it is more fruitful to identify a particular pedagogical issue—say patient centeredness, or professionalism, or learning surface anatomy—and at this point begin to explore how different aspects of the formal and the other-than-formal curricula intersect to convey particular messages about the "real importance" of these learning objectives. "Respect for the cadaver" may be formally listed in course materials as the "number one value in laboratory session," but if students are free (e.g., unchecked by faculty) to act however they want at night when instructors are not around or if instructors routinely refer to the bodies as "learning tools" or as an "it," then students may be receiving another set of messages—no matter what formally is stated in the course description.

In Fig. 47.2, we seek to capture some of the places one might look for other-than-formal messages about a particular issue or concern. References to the HC in the medical literature often refer to the "classroom" versus the "clinic" or the classroom versus the "wards" when differentiating between locales of formal versus the hidden curriculum. While this may be a rough guideline, like most guidelines, it is incorrect in its particulars. "The classroom," or a particular course, usually has all kinds of curricula coursing through it. A given classroom not only will contain formal curricula but also a variety of hidden, informal, null, etc., messaging. The first case example that appears at the beginning of this chapter covering faculty exhortations encouraging students to dissect, but with those urgings countered by the course examination policies, is a case in point. There is no "classroom versus clinic" here. Everything is taking place "in class."

One influence on the HC that is more difficult to capture is that of the HC on professional identity formation, especially within the anatomy context. Examples, which can be extrapolated from Fig. 47.2, could include the influence of black humor negative role modeling or the impact of the learning environment culture on students' dignity. Physical gestures, such as the way a cadaver is handled, all signpost negative behaviors, which although seemingly distal to professional practice when anatomy is typically studied during the early years of medical school contribute to professional identity formation and proto-professionalism.

How to Explore the Hidden Curriculum: Three Steps

As summarized below, whenever one wants to explore a particular leaning environment or issue, one begins with the formal. It is not that the formal curriculum is a more important source of

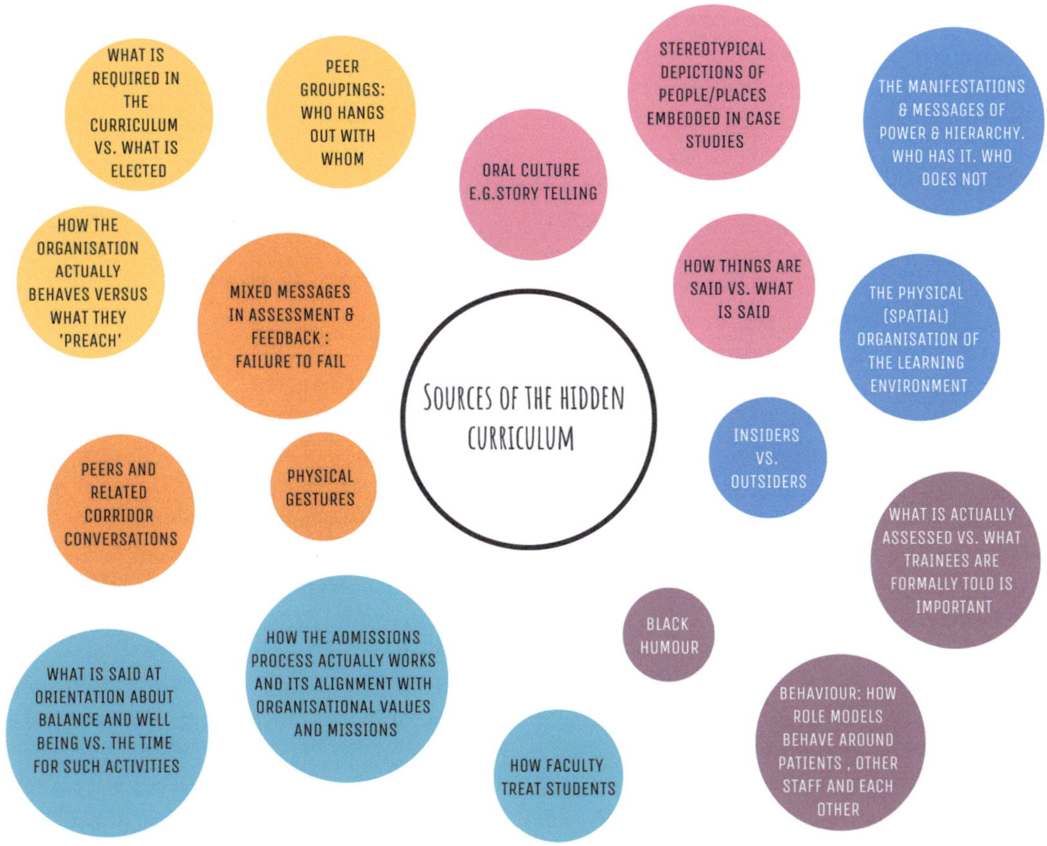

Fig. 47.2 Sources of the hidden curriculum

learning—it actually may constitute a minority, percent wise, of the overall learning taking place—but whatever the issue at hand (e.g., patient-centered care, professionalism, learning surface anatomy, etc.), we need to begin with what the powers-to-be say they are doing to promote that issue. Furthermore, we need to be aggressive and exhaustive in pursuing the formal side of this equation. Whatever the official explanation or account offered, we must follow those initial declarations with an array of supplemental questions (e.g., "And what else do you do?" "Why are you doing these things?" "Can you give me another example?"). Thus, when you hear WCC organizers claiming that HTE emphasizes patient-centered care, one also needs to scour the school's website, its student and faculty handbooks, course syllabi, etc., so that when one concludes "*this* is how HTE says it emphasizes patient-centered care in its curriculum," you now are certain you have the formal side of the equation well in hand.

The second step, and a corollary of the first, is not to leave the formal side of things too quickly. Most organizations have a vast array of formal policies and practices, and not all of them, when taken as a whole, send a consistent set of messages to members of the community. In short, we may find mixed messages without ever leaving the formal curriculum. For example, most courses are built around two core activities—the delivery of content and the assessment of that content in terms of learner mastery. However, what faculty teach and what faculty assess may be two different things. Moreover, and as we saw in our first case example above, we not only should look at the content of examinations but also at how points are distributed across the examination or how much emphasis is given to one topic area rather than another. The difference between what is taught and what is assessed can be rife with contradictions.

Finally, only when we have exhausted the formal side of our equation do we move into exploring the other-than-formal aspects of the learning environment as our third step. Here, there are endless venues to be explored, and investigators would do well to restrict the number of avenues they might venture down. For example, early in his fieldwork and interviews, Hafferty [7] found that laboratory session contained a certain oral culture and that students were telling stories to each other, particularly early on in laboratory session, about pranks that students played on each other in laboratory or on laypeople using body parts, usually told as happening at some other medical school or at some other time. The issue is not whether these stories depicted actual events or even that students today tell similar "cadaver stories" but rather that in the laboratory session being studied, and at the time of that study (the early 1970s), students told each other a lot of stories about laboratory session and these stories carried important messages (often tacit) about things students were worried or concerned about. Studying the oral culture of students thus became an important part of this study. In another laboratory session or at another time, such "tellings" may not be a part of the environment although it is hard to imagine any laboratory setting devoid of student storytelling and thus devoid of an oral cultural dimension. In short, the fundamental strategy is to be sensitive to context and to let the setting and/or situation help to dictate the lines of inquiry you wish to take.

The three steps to biopsying the hidden curriculum:

- Begin with the formal curriculum.
- Next, be exhaustive in exploring this side of the formal/other-than-formal equation.
- Only then, begin to explore the other-than-formal aspects of learning.

I plan to will my body to science. I have wondered on occasion what the med students carving on me would be thinking when cutting into me. Would they make comments about my tattoos? What will they think and say about the various scars I have collected over the yrs from injuries as a motorcycle racer? –A reader's comment to a *Time* article on anatomy training [8]

Future Challenges

Over the years, anatomy has become much more deliberate and reflective in defining and expanding its pedagogical footprint. In 2018, Wilson et al. called for a systematic investigation of the hidden benefits of dissection [9]. When the second author first entered laboratory in the early 1970s, the manifest function of laboratory was teaching anatomical form and function. It did so utilizing two principal pedagogical vehicles: lecture and dissection. There was no formal recognition that anatomy might be involved in the emotional socialization of medical students—beyond some vague notions of the cadaver as a "first patient" or that laboratory sessions might help "needy" students to attain a level of detachment and objectivity they would subsequently need in dealing with clinical challenges such as disease, disfigurement, dying, and death. There also was little formal discussion about "respecting the body" and no consideration (per the quote that opens this section) that donors might have expectations of students. Today, anatomy has a new face—and a somewhat expanded mission. Today, anatomists teach using a dizzying array of tools including plastinated specimens, plastic models, interactive 3D atlases, virtual dissection, computerized models, body painting, poetry, songs, and spiritual readings. The goal of learning anatomical form and function, in turn, has been joined by the formal teaching and evaluation of issues such as professionalism, leadership, and teamwork [10]. Today, it is not unusual to read how laboratory sessions can manifestly function to teach compassion and respect and clinical ethics, promote social bonding among students, add to the humanistic goals of medical education, and directly contribute to laying a foundation for more clinical arenas such as end-of-life care [11–15]. In short, and with some degree of irony, the very topics that were off the formal curriculum radar in the 1970s (e.g., compassion, end-of-life care, professionalism) now are key arguments for why anatomy should continue to occupy a key and perhaps even expanded seat at the medical school educational table.

A recent study of body painting and anatomical teaching provides an interesting, and perhaps distorting, edge to HC thinking and anatomy education. In interviewing anatomy staff, Aka and colleagues [16] encountered interviewees who imagined body painting as a pedagogical vehicle by which they could intentionally "teach by stealth" and "use the hidden curriculum" to achieve certain curricular goals (e.g., developing professional scripts during a body painting session or learning about expected conduct without any explicit teaching or guidance). However, the HC, definitionally speaking, is both hidden and conveyed tacitly, and while a number of studies have remedially suggested first uncovering the HC and then embedding learned insights into the formal curriculum, no one to date has recast tacit learning as intentional. Such a recasting raises a number of questions. How does one evaluate a learning outcome that is formally unstated and intentionally invisible to students? In turn, how would students know if they are meeting such goals? Similarly, how would faculty know when these goals were being met? What happens to the pedagogical tool of reflection in the face of intentionally invisible teaching? Correspondingly, what happens to the integrity of learning environments under such subterfuge? Are such strategies even ethical? Finally, why would faculty even think that the HC they were referring in these interviews was the HC in either the medical education or anatomy literatures? Regardless of any such (good) intentions, the underlying theme that students learn best when they are less than fully aware of faculty intentions is an assumption that bears both analytic and pedagogical scrutiny regardless of the setting or stage of training. Whether curriculum deans and/or educational policy committees buy such a reformulated face is another matter, but what is clear is that the self-identity of anatomy has changed in the ensuing 40 years.

What has not changed, nor will it ever, is the enduring presence of the hidden curriculum within the landscape of medical education. For example, students have been dissecting in four-person teams since time immemorial—all the while tacitly internalizing lessons about teamwork in spite of the fact that "learning teamwork skills" was not a formal course objective until recently. Furthermore, the teamwork lessons being learned may not have been the teamwork faculty had in mind, had faculty been so inclined. Making teamwork a formal part of that course and then explicitly identifying the teamwork competencies students are expected to master thus become an important step in shifting one's learning about teamwork from an other-than-formal to the formal curriculum. Nonetheless, while undertaking such restructuring, whether about teamwork, professionalism, or some other learning objective, may make the anatomy experience more pedagogically coherent, it does not erase the HC from the learning milieu. The HC is still there. It may exist around other issues or take on other forms. It may surface as students attempt to "game" the specificities found in many competency-based curricula. Once again, we refer the reader to our first case example (above) and once again refer to the gap between formal faculty messages about the importance of dissection and the more tacit yet overwhelming powerful and negating messages students received via faculty examination policies. One does not "fix" the hidden curriculum. Instead, one works with it in conjunction with the formal curriculum to make the learning experience of students as positive and as reinforcing as possible.

For all of anatomy's successes, and there are many, challenges remain. Some are directly tied to those successes while others are tied to the ubiquitous presence of the HC. The fact that anatomy has deliberately broadened its pedagogical base to embrace other learning objectives in addition to anatomical form and function has placed anatomy at some odds with other basic science disciplines. As noted, in some detail by Jones [17], the internal success of building in "additional" topics such as teamwork, leadership, compassion, and professionalism has turned anatomy into a somewhat stand-alone and/or isolated player. Students may well (again referencing Jones) come out of anatomy fired up about the importance of compassion and/or professionalism as being critical to good doctoring, only to find that their other basic science courses never mention these topics. Subsequently, students may find themselves in clinical settings (e.g., preceptorships, etc.) where such topics also remain essentially unaddressed—or if addressed perhaps not assessed. In turn, students, ever searching to make sense of their learning environments as they develop peer-based rules about discriminating between the "unimportant," "somewhat important," and "really important" things they need to learn, may conclude that professionalism, compassion, etc., are "really not all that important after all"—because if they were (the reasoning continues), then they would be covered (via formal teaching and assessment) in these other courses. In sum, students can, and will, remain quite sensitive to what they see as the mixed messages they receive from faculty about their state and status within the educational environment and thus continuously vulnerable to feelings of cynicism, isolation, and burnout.

> And that's exactly how the hidden curriculum is taking us down—the hardest part about med school is being so isolated. *They* want us to believe that it's 'normal' to bottle up our emotions and power through every struggle. *They* teach that 'humble' is doing good deeds without talking about it. *They* want us to prioritize our own grades over relationships. And the more we stay quiet, the more *they* win. Well [expletives deleted] they. –A medical student blog [18]

The next challenge for anatomy, therefore, is to take its beachhead and seek to have other courses join in anatomy's effort to move beyond "the subject at hand" and begin to explore how the *entire* and *truly integrated* educational enterprise (which extends far beyond the formal curriculum) can be *devoted* to creating better doctors in service to the public. This means, among many other things, an end to subject matter balkanization—long a hallmark of medical education. Anatomy has shown how some of this broadening and integration can be done. Nonetheless, anatomy must find ways to share that vision and wisdom with the rest of its colleagues.

References

1. Becker HS, Geer B. Participant observation and interviewing: a comparison. Hum Organ. 1957;16:28–32.
2. Dueñas AN, Kirkness K, Finn GM. Uncovering hidden curricula: use of dark humor in anatomy labs and its implications for basic sciences education. Medical Science Educator. 2020;1–0.
3. Merton RK. Social theory and social structure. Glencoe: Free Press; 1957.
4. Lave J, Wenger E. Situated learning: legitimate peripheral participation. Cambridge: Cambridge University Press; 1991.
5. Halbesleben JR, Wakefield DS, Wakefield BJ. Workarounds in health care settings: literature review and research agenda. Health Care Manag Rev. 2008;33:2–12.
6. Flinders DJ, Noddings N, Thornton SJ. The null curriculum: its theoretical basis and practical implications. Curric Inq. 1986;16:33–42.
7. Hafferty FW. Cadaver stories and the emotional socialization of medical students. J Health Soc Behav. 1988;29:344–56.
8. A reader's comment to the online Time Magazine article "Doctor-in-training: anatomy lesson" by Laura Fitzpatrick. http://healthland.time.com/2013/01/29/doctor-in-training-anatomy-lesson/#ixzz2NLKDwOXM/. Accessed 5 May 2019.
9. Wilson AB, Miller CH, Klein BA, Taylor MA, Goodwin M, Boyle EK, et al. A meta-analysis of anatomy laboratory pedagogies. Clin Anat. 2018;31(1):122–33.
10. Krych AJ, March CN, Bryan RE, Peake BJ, Pawlina W, Carmichael SW. Reciprocal peer teaching: students teaching students in the gross anatomy laboratory. Clin Anat. 2005;18:296–301.

11. Charlton R. Education about death and dying at Otago University Medical School. N Z Med J. 1993;106:447–9.
12. Weeks SE, Harris EE, Kinzey WG. Human gross anatomy: a crucial time to encourage respect and compassion in students. Clin Anat. 1995;8: 69–79.
13. Rizzolo LJ. Human dissection: an approach to interweaving the traditional and humanistic goals of medical education. Anat Rec. 2002;269:242–8.
14. Jones DG. Reassessing the importance of dissection: a critique and elaboration. Clin Anat. 1997;10: 123–7.
15. Lempp HK. Perceptions of dissection by students in one medical school: beyond learning about anatomy: a qualitative study. Med Educ. 2005;39:318–25.
16. Aka JJ, Cookson NE, Hafferty F, Finn GM. Teaching by stealth: Utilising the hidden curriculum through body painting within anatomy education. Eur J Anat. 2018;22(2):173–82.
17. Jones TW. Creating a longitudinal environment of awareness: teaching professionalism outside the anatomy laboratory. Acad Med. 2013;88:304–8.
18. Medical student blog. http://www.sidenotelife.com/2013/03/ken-explains-hidden-curriculum-we-dont.html/. Accessed 1 Dec 2014.

Teaching Anatomical Sciences to Dental Students

48

Stephen McHanwell and Joanna Matthan

Introduction

Dentistry is a surgical science; consequently, there is a strong focus on structural biology throughout a dental degree. Students will typically be undertaking irreversible, and potentially painful, procedures on patients within 2–3 years of starting their course. There is a need to ensure they are equipped with sufficient anatomy to become safe and effective practitioners. The anatomical subjects studied in dentistry are broad. Dental students will be expected to study not only gross anatomy but also tooth morphology, embryology, postnatal development of the head and neck, and basic and specialized dental histology. They also require an understanding of relevant neuroanatomy, including pain pathways and coordination of jaw movements. These courses typically occupy a significant fraction of basic science teaching in dentistry alongside physiology, metabolism and nutrition, oral physiology, microbiology, and immunology. Usually, these subjects are taught in the early years of a dental program. However, because dental anatomy links vertically to many courses in the later years of dental studies including oral surgery, restorative dentistry, temporomandibular disorders, and human diseases, vertical integration across the 4 or 5 years of a dental program is also needed.

Many of the issues that confront teachers of anatomy to other groups, including medical students, present similar challenges to teachers of dental students [1]. Thus, provision of appropriately qualified teaching staff; the place of dissection or prosection versus other methods of teaching anatomy, including virtual resources and e- and blended learning; and choice of teaching strategies all raise issues for dental teachers [1]. Since these are issues discussed elsewhere in this book, we revisit them here only where the teaching of dental students raises special problems. Concerns have been frequently raised over the loss of teaching time in anatomy for medical students [1–3], but these pressures do not seem to have been felt to the same degree in dental courses. Given the pressure to provide students with clinical exposure at a much earlier stage than previously [4], this may change. Nevertheless, the fact that dentistry is a surgical science makes a strong case for adequate anatomy teaching throughout a dental course.

This chapter will discuss the place and content of dental gross anatomy courses at undergraduate level, the role of interprofessional education

S. McHanwell (✉)
School of Medical Education and School of Dental Sciences, Faculty of Medical Sciences, Newcastle University, Newcastle upon Tyne, UK
e-mail: stephen.mchanwell@ncl.ac.uk

J. Matthan
Faculty of Medical Sciences, School of Dental Sciences, Newcastle University, Newcastle upon Tyne, UK
e-mail: joanna.matthan2@ncl.ac.uk

L. K. Chan, W. Pawlina (eds.), *Teaching Anatomy*, https://doi.org/10.1007/978-3-030-43283-6_48

(IPE) in the dental curriculum, and how dental anatomy is taught and assessed. Brief mention will be made of anatomy in postgraduate dental teaching and in the teaching of professions complementary to dentistry. This chapter will be confined to an examination of the teaching of gross anatomy, embryology, and neuroanatomy. It will not cover the teaching of dental anatomy (tooth morphology) which is usually undertaken separately and is described elsewhere [5, 6].

Content of Dental Gross Anatomy Courses

Professional Body Guidelines

There appears to be little disagreement among course designers and teachers of dental courses that a knowledge of gross anatomy is crucial to a qualified dentist, although what and how much might be contested [7–12]. Clearly, knowledge of relevant gross anatomy is necessary for the safe administration of local anesthesia, understanding the spread of dental infection and its surgical treatment, undertaking simple surgical procedures such as submandibular sialolith removal or surgical extraction of teeth, or interpreting radiographs, just to give selected examples. Such clinical examples stress the relevance of dental anatomy to dental students and provide an important stimulus to learn [8]. The Education Committee of the British Association of Oral and Maxillofacial Surgeons has recognized the importance of anatomy to oral surgery in their surveys of undergraduate opinion of the oral surgery training they are receiving [13, 14]. However, in comparison to medicine, there have been relatively few attempts to define the detailed content of a dental gross anatomy course [4, 15, 16]. Guidelines in the USA were produced by the American Association of Dental Schools in 1981 and revised in 1993 [17, 18]. Until recently there have been no similar detailed guidelines in the UK or Europe, although increasingly core syllabuses for different disciplines are being generated to guide course designers with key anatomical content and recently a dental gross anatomy syllabus has been published [19, 20]. The General Dental Council (GDC) in its First 5 Years document, revised in 2012 as Preparing for Practice, sets out, only in the broadest terms, the content of the basic science components of dental courses in the UK [21, 22]. In the later Preparing for Practice document, the GDC states that relevant members of the dental team will, upon qualification, be able to "identify relevant and appropriate dental, oral, craniofacial and general anatomy and explain their application to patient management" [22]. This statement, although helpful in stressing the importance of relevance when considering the content of the basic science components of a dental curriculum, lacks precise detail to guide course designers and dental educators.

Defining Course Content

Given the absence of detail from professional bodies in the UK or up-to-date detail in case of the USA a rigorous survey of current professional opinion on the content of dental gross anatomy courses has been undertaken using a Delphi analysis for dental undergraduate anatomy in the UK by the present authors in collaboration with G. Finn, B. Moxham, and other colleagues [19, 20]. Indirect methods can also be used to determine the appropriate content of a dental gross anatomy course. A review of four of the standard dental anatomy textbooks shows a degree of consensus on what constitutes core content, while differing sometimes in precise detail. The anatomy of the head and neck is prominent in all four books, each advocating detailed coverage of the anatomy in these regions [23–26]. The requirement of dental students to study the head is self-evident, the anatomy of the neck in such detail perhaps slightly less so, but the ability to perform a clinical examination of the neck is a required skill for every dentist in the UK [27], and so, a knowledge of the anatomy of the region is not only necessary pedagogically but clinically required. There is less agreement on how much knowledge dental students must know of the anatomy of the rest of the body.

Three of the textbooks provide coverage of the anatomy of the thorax, and the anatomy of this region is covered in all UK dental courses with which the authors are familiar and in most US courses [7, 23–25]. Typically, some teaching is provided on the upper limb, although this may be as little as simply covering the anatomy relevant to intravenous therapy, and on the anatomy of the abdomen.

Many dental anatomy courses do not teach the pelvis, perineum, or lower limb. Clearly, dental students need to know less about these regions compared to the head and neck, but the majority, if not all, of dental courses provides teaching to their students on human diseases. This would suggest that some anatomy of the whole body is necessary, though the need to stress relevance through exemplification with relevant clinical examples would be important [8, 28]. This would also depend on the background of the students entering training. Not all dental schools require students to have basic science knowledge upon entry, and so, coverage of the anatomy and physiology of other regions of the body would be especially important for them.

The content of a dental anatomy syllabus needs to go beyond gross anatomy to include surface anatomy and some limited imaging anatomy. The neck, the face, the temporomandibular joint, and the oral cavity are four very obvious regions whose living anatomy needs to be mastered. While radiology is usually taught as a separate course in a dental degree, the use of appropriate images in the gross anatomy course may deepen the understanding of a region as well as provide preparation for more detailed courses to follow. Imaging technologies have been shown to be effective for aiding anatomy understanding and gaining student engagement with the content, and integration of anatomy and radiology can enhance image interpretation [29], granted that there is vociferous disagreement as to just how much of the latter is required of dental graduates.

One approach to defining the appropriate content aims in the gross anatomy component of a dental program is to ask about the overall intended program aims. If that overall aim is to train dental practitioners or technologists, then these content aims might be quite limited. If the overall aim is a broader one of producing dental practitioners educated to degree level who are also dental scientists, some of whom might want to carry out dental research, then these content aims will be necessarily broader. A similar question about the destination of dental graduates might be posed in relation to possible preparation for future study. The designing of syllabuses for postgraduate study has attracted attention [30–32]. Considerations of the design of postgraduate dental syllabuses raise the issue, as it has done in medicine, of the balance in anatomy teaching between what is needed at the undergraduate level and what should be reserved for postgraduate specialties.

Delivery of Gross Anatomy Teaching

Interprofessional Education in Dentistry

Achieving integration and collaboration between different healthcare providers may be one of the principal challenges facing healthcare educators today [33]. Anatomy education could offer the ideal setting for students across these professions to learn about each other and contribute to each other's learning [34]. Prior to the escalation of the role of IPE within healthcare professions' curricula, now championed by the Centre for Advancement of Interprofessional Education (CAIPE) to improve quality of care, health outcomes, and patient well-being [35], studies reported preconceived and stereotypical views of other professions, even at an undergraduate level [36]. A recent study looking at the approach to learning anatomy between speech science, medical, and dental students found that there are different characteristics of learning anatomy, for instance, the link between utilizing cadavers and a deep learning approach, which may be inherent to anatomy education rather than to a particular professional group [37], complicating the implementation of wide-scale IPE anatomy education. The implementation of IPE has been found to be least successful in anatomy and physiology teaching, mainly due to the aforementioned different teaching requirements [38, 39]. Nevertheless, despite numerous challenges [40] in implementing IPE into the anatomy education (not least for fear of "dumbing down" the anatomy content to suit an unequal/unmatched group), an evident political and educational [35, 39, 41, 42] push toward incorporation of IPE into several healthcare settings is palpable, necessitating reviewing and mapping dental curricula to find suitable areas of commonality [43–45].

Dental students are often taught as a distinct group. Where there is joint teaching with other groups, this is done interprofessionally with other members of the dental team, most often with dental therapists and hygienists [46]. Several elements of the dental anatomy course are also shared by other professions, as most students of healthcare professions require some knowledge of anatomical structures and three-dimensional relationships [46]. Although clinical applications and depth of material differ, the learning process to master this knowledge is essentially the same, and students may benefit from learning together in an interprofessional manner for selected areas of study [47]. Bakr et al. caution against isolating dental students from other students in different health-related disciplines, particularly with regard to the recent direction of IPE in anatomical sciences [48]. A study of an interprofessional intervention involving medical students, fourth-year dental students, and dental faculty revealed significant benefits for both medical and dental students [34]. Previous studies have, however, also shown some disadvantages in relation to the marginalization of dental students when taught in a shared program [49], and it may prove insurmountable to create equal-status contact between unequal-status groups where the location and curriculum may invariably be designed to meet the needs of predominantly one group of learners [33]. Carelessly implemented IPE interventions may serve to reinforce stereotypical views and prejudices toward other healthcare professions, rather than seek to eliminate them [36]; the anatomical knowledge and underpinning clinical relevance are also quite different [46]. Suitable IPE collaborations that map against the clinical anatomy of allied dental professionals, medical students, and other healthcare professionals are encouraged, with the aim of enhancing the learning experience and anatomical knowledge of *all* groups involved. For dental students, the authors believe, these may be best undertaken in the clinical years of the dental course.

Curriculum Structure and the Place of Anatomy

Though many dental courses continue to rely on didactic forms of teaching, an increasing number employ problem-based or scenario-based approaches. In virtually all dental schools where there is didactic teaching, anatomy courses are usually taught regionally. This may be as a part of a very traditional discipline-based strategy in which an anatomy course stands alongside other subject-based courses. Such courses are much less frequent and, instead, systems-based courses are more usual [7]. In such cases, head and neck anatomy may be taught as a separate stand-alone course or as a part of an oral biology or oral functions course. Histology may be taught as part of an anatomy course, within physiology, or as a component of the relevant systems-based course. Since engaging students to learn histology is often an issue in the UK, where fewer students study histology prior to starting their dental course, stressing functional or clinical relevance will be important. Furthermore, students often have no experience of using a microscope or interpreting the images it produces, and so the barriers to learning histology are greater.

There have been a number of studies that have attempted to assess the effectiveness of traditional, didactic teaching compared to problem-based learning. This is also the case for dentistry. A cross-sectional study by Last and colleagues, comparing the basic science knowledge of the final cohort of dental students enrolled on their conventional course with that of the cohort of students enrolled on the first year of their PBL course that replaced it, found no significant difference between the two cohorts [50]. A similar study looked at perceived preparedness of dental students for dental practice if enrolled on a PBL as compared to a traditional curriculum and again found no differences [51]. The limited conclusion one could draw from these two studies is that when looking at knowledge retention PBL courses perform no worse than traditional courses, but no better either. Hattie, in an extensive series of meta-analyses of problem-based learning more widely across the education sector, reached a more complex conclusion [52]. Hattie concluded that students on PBL courses performed less well or even negatively on measures of surface knowledge but better on deeper learning and understanding where surface learning had already occurred [52]. This is not a surprising conclusion, given the very different aims of the two approaches. It is of significance for anatomy, where significant surface learning has to occur in order that one can manipulate ideas and develop a constructed approach to learning. A study with dental students highlights the importance of context in helping students to understand clinical problems and move from the certainty of facts taught in the early part of an anatomy course to the uncertainty encountered in dental practice [53].

Teaching in the Dissecting Room

An important feature of any gross anatomy course, including dentistry, is the practical component and time spent interacting with materials in an active way. This continues to mean, in most cases, working in a dissecting room examining the human cadaver. In the past, this meant working as a group dissecting a formalin-fixed cadaver. This is not the place to enter the vigorous continuing debate over the value of dissection as a teaching tool in anatomy. As syllabuses have developed, the time available for a full course of dissection has diminished [1, 2, 15]. For dissection to be effective as a teaching tool, adequate time has to be allowed, and this is often simply not available. Added to that, many students arrive having no previous experience of dissection or the dexterity that it requires and, so, are far less well equipped to undertake dissection. Thus, professionally prosected specimens will be usually more capable of showing fuller range of detail than that produced by students. Dental students value the dissecting room experience as a positive part of their course but have also stated a preference for prosected specimens [54, 55]. The authors attributed this difference to the fact that dental students predominantly study the head and neck which are both more complex and difficult to dissect, and so there would be benefits to providing specimens prosected by experienced staff [55]. One often-cited benefit of dissection is the development of hand skills, but the type of skill developed during dissection seems unlikely to transfer to the skills needed in dental practice which are taught intensively elsewhere. Some schools are recognizing that a key disadvantage of formalin-fixed cadavers is their unrealistic nature and are moving to Thiel-fixed cadavers. The softer fixation can allow development of some skills relevant to dentistry, such as the administration of inferior alveolar nerve blocks, and this is increasingly being employed as a part of the pathway to learning this technique.

Benefits of Dissecting Room Teaching

Whether through dissection they undertake themselves or by examination of specimens prepared by experienced staff, exposure to human specimens, though still a simulation, does provide an authenticity and immediacy that no other teaching technique in anatomy can fully replicate. For most, this acts as a powerful stimulus to learning. Students gain other benefits working in a dissecting room setting. The same study referred to above also found that students rapidly acquired a high level of professional values in relation to working in the dissecting room [55]. The acquisition of professional values as another outcome of a gross anatomy course has been observed in other settings [56]. This positive and professional attitude of dental students to studying anatomy in the dissecting room is something seen by the present authors during many years of dental teaching. Studying anatomy partly through the medium of the cadaver has also been shown to have positive effects on the approaches students take to learning. One study found that students responding positively to a statement, "that cadaveric teaching was an important part of becoming doctor/dentist or healthcare professional," were more likely to be adopting deep approaches to the learning of anatomy [37]. Thus, working in the dissecting room would appear both to aid in developing professionalism and the ideal of deep learning approaches. As already noted above, deep approaches to learning, when balanced with surface learning, are necessary for developing abilities to predict and hypothesize which will, in turn, lead to more constructed understandings of the subject that will facilitate in students the ability to develop their own ideas [52]. This will be of benefit to students encountering real-life clinical problems later in their training.

Other Resources for Teaching

Cadaveric anatomy is not the only means of teaching anatomy practically; certainly, some current evidence would seem to suggest that cadaveric dissection may be no more or less effective than digital learning [57, 58]. Variety may be key to enhancing learning, especially when incorporating technology into anatomy teaching [59]. There is a range of software and other virtual resources available, although, given the relatively smaller size of the dental market, fewer of these are written with the needs of dentistry specifically in mind [60]. Different resources do, however, confer different advantages when teaching anatomy [1, 61, 62], and combining technology-enhanced learning (TEL) with cadaveric learning can be a very effective mode of teaching [63].

Radiology is usually studied as a separate dental discipline; consequently, there is usually little radiology taught within the gross anatomy component. However, students frequently experience conceptual difficulties in interpreting two-dimensional radiographic images in terms of three-dimensional anatomy, and so there are advantages to studying radiology alongside gross anatomy through the intermediary of three-dimensional reconstruction facilities offered by software such as the Visible Man package or newer technologies such as augmented reality [64]. Studies incorporating TEL in anatomy education fall short of the mark and are not comprehensive enough to draw any conclusions or recommendations from [65]. TEL resources must not be considered the panacea to plugging programmatic pedagogical flaws or student satisfaction concerns [59] and should be evidence-based where possible. Educators must be aware of the associated "magpie effect": technology innovations possess an allure that is hard to resist but lack the evidence to show they are any more effective than other, more traditional (and perhaps more cost-effective) methods. Other pedagogical techniques such as social media [66–68] and body painting have gained widespread publicity due to the "novelty factor" associated with them; these methods do offer opportunities for teaching anatomy in different ways [69–71], although their value in dental teaching is less explored. There are, however, some innovative examples at dental schools where face painting has been incorporated into clinics to revise and reinforce the underlying anatomy prior to undertaking procedures [69].

As in gross anatomy courses with other students [72, 73], experiments have been made with peer teaching with some success but also not without some difficulties [74]. Dental students, in common with many other students, find formative quizzes delivered during their gross anatomy course an effective way of stimulating learning [75]. 3D printing, Internet resources, virtual reality glasses, and other forms of augmented reality anatomy learning are fast making their way into anatomy education, providing educators with a veritable smorgasbord of educational tools from which to diversify their teaching [76]. Coupled with traditional anatomy textbooks, educators now have access to an extensive educational toolkit with which to enhance guided self-directed learning, but caution must be exercised when implementing new technologies which, despite numerous radical opportunities, may pose neglected and ill-considered challenges [59].

When Should Anatomy Be Taught?

The perception persists that healthcare students' anatomical knowledge is diminishing rapidly and that this may impact the delivery of effective and safe patient care [77]. The reasons are multifactorial and outside the remit of this paper but may have resulted from, among other things, almost complete neglect of authentic vertical integration of anatomy teaching [78]. Within problem-based medical curricula, true vertical integration is usually met due to delivery of authentic cases [10], but vertical integration in other curricular models requires collaboration across the preclinical and clinical years [79] and may prove more cumbersome to implement. Dentistry is a case in point.

In most dental courses, gross anatomy is taught within the first 2 years of the program and then not explicitly revisited again, perhaps until postgraduate-level examinations are undertaken. This is not the best approach to deliver a subject required throughout the program to analyze real-life clinical problems encountered in treating patients [80–83]. Unsurprisingly, perhaps, when retention of anatomical knowledge among dental students was tested, it was found that knowledge of anatomy declined over the duration of their course [80], hindering their ability to apply their anatomical knowledge to actual clinical problems [81].

Bergman suggests that incorporation of anatomy learning into curricula should follow a three-step approach [84]: (1) students should be taught the vocabulary of anatomy either via lectures or e-learning, (2) three-dimensional relationships of structures should be studied and undertaken in the dissecting room, and (3) discussions on clinical application of anatomical knowledge (to, for instance, explain signs and symptoms) can be done in tutorial group settings. This model suggests revisiting anatomy knowledge in a structured manner at varying stages of the curriculum. Finding time in the later stages of a dental course to deliver basic science teaching is not easy, given the importance of acquiring clinical competence to ensure patient safety; online courses have been utilized to reinforce the basic anatomy taught in the early years and to introduce students to its clinical application, quite effectively [46, 81, 82].

Relevance may be important in reinforcing the teaching of anatomy in the early years and as a means of explaining why it is important to learn it at a time when the clinical experience of students is limited [8]. However, in the early years, complex clinical examples can overwhelm students, and so there needs to be progressive elaboration of clinical input as the course progresses, developing learners' understandings of the importance of basic science as part of acquiring clinical competence [82]. In this "lined-up curriculum," basic sciences, pathology, and treatment for head and neck conditions and for the rest of the body are organized into a coherent story [82]. This underlines the importance of context in helping dental students to understand complex clinical problems and the need to move from the relative certainty of facts presented early on to the uncertainties when real-life patients are encountered [53, 82]. Joining up the curriculum in this way is also the means to ensure that anatomy teaching is integrated horizontally with relevant functional courses such as physiology and also with the relevant pathology such as they taught in a human diseases course [28].

It is worth bearing in mind that providing a contextual framework may not necessarily always equate with improved learning. A recent study by Bergman et al. [84] found that provision of a paper-patient case (context or relevance) when studying anatomical knowledge did not improve performance, despite findings from previous studies [85, 86]. Instead, the authors recommend systematically varying subject familiarity and prior knowledge, learning approaches and cognitive load, and searching for consistent interactions [84]. Nevertheless, vertical integration is a critical step to full engagement with basic sciences, but it must go beyond simple repetition of previously taught material. Rather, it means reviewing and summarizing and the elaboration through the use of progressively more complex clinical examples. Perhaps, though, the most effective route to vertical integration, where resources permit, is through the delivery of clinical reasoning sessions through team teaching, where basic and clinical teachers collaborate to deliver case presentations or scenario-based teaching, for example, in both the early and later stages of a course.

Tips for Vertical Integration of Dental Anatomical Content in the Clinical Years

- Identify anatomy learning outcomes from the basic science syllabus that have links to procedures.
- Link institutional learning outcomes to available national guidelines.
- Revisit national guidelines regularly to ensure most up-to-date requirements and competencies are delivered.
- "Shadow" key institutional collaborators and dental clinicians to identify key clinical anatomy required for teaching safe application of procedures.
- Make gross anatomy relevant to students by learning some of the common procedures associated with dentistry.
- Develop step-by-step instructions for anatomical structures encountered during the procedures (these may be different to those encountered in the early years).
- Develop bite-size digital learning packages revisiting specific anatomy content prior to teaching of procedural skills (e.g., dental blocks).
- Introduce short innovative sessions for revisiting key clinically relevant anatomical content (e.g., face painting of the relationships of the superficial and deep structures in the face and neck prior to undertaking a procedure).
- Feedforward from the basic anatomical sciences to elucidate how concepts will be revisited on a practical level later on in the course.
- Develop authentic clinical reasoning packages to pull together clinicoanatomical correlations required to safely and competently treat patients.
- Identify suitable, well-mapped, and well-matched interprofessional sessions to develop clinical anatomical understanding in a multidisciplinary manner.
- Offer "top-up" cadaveric anatomy sessions or refresher bite-size cadaveric sessions to keep anatomical knowledge at the forefront.
- Offer targeted dissection projects in the clinical years focusing on dental procedures to increase confidence levels and improve handling of surgical equipment, teamwork, and other professional skills.
- Have an institutional open door policy for dental students to allow access to dissection room facilities even past the basic science years.
- Offer refresher courses utilizing fresh frozen cadavers and surgical approaches.
- Offer regular formative assessment of anatomical knowledge linking to the clinical cases and clinic sessions encountered.
- Incorporate summative assessment of clinicoanatomical concepts throughout the duration of the degree course.
- Incorporate near-peer or peer teaching into anatomy sessions to benefit both learners and near-peer teachers.

Neuroanatomy and Embryology in the Dental Sciences

Neuroanatomy is an invariable part of any dental anatomy course, but students frequently struggle to see its role and purpose in a dental curriculum [87, 88]. Clearly, an understanding of pain requires knowledge of the relevant neuroanatomy, but dental students' knowledge of the subject needs to extend beyond that to include regulation of vital functions such as control of respiration and the cardiovascular system, mastication, swallowing, speech, and taste [89]. As part of any human diseases course, their knowledge needs to extend even further to understand the more common pathologies they are likely to encounter in their patients which would include, but not be limited to, stroke, trauma, and the more common neurodegenerative diseases, such as Alzheimer's and Parkinson's diseases. Curricular guidelines produced in 1992 by the Anatomical Sciences Section of the American Association of Dental Schools identified that a dental neuroscience course should familiarize students with the cellular structure of the nervous system, the structure and function of the peripheral and central nervous systems including major pathways, the innervation, somatic and autonomic, of the head and neck, and apply this knowledge to dental clinical practice [90]. As Kluber has identified, the overlap in content between neuroanatomy and gross anatomy courses means that coordination or integration of these two topics can bring significant educational benefits [91].

Embryology can be taught as part of a gross anatomy, histology, or oral biology course or independently [7]. Clinically and anatomically, there are important linkages to be made between the development of a region and its adult anatomy. In the case of a dental course, the anatomy of the head and neck is more readily comprehensible if the development of pharyngeal arches and clefts is understood, so there are advantages to be gained by teaching some early development of the embryo alongside that of the relevant gross anatomy. Later development may be best taught as part of a craniofacial development course in which normal and abnormal development can be studied together, allowing appropriate clinical linkages to be made. Students frequently find embryology challenging to learn, so stressing relevance to encourage engagement with the topic is particularly important [8, 75, 80]. Indeed, the need to stress relevance when teaching anatomy might be considered a theme running through this chapter and a point at which to conclude discussions about course content.

Assessment of Student Learning

Assessment of student learning is an extremely important part of what we do as teachers, so it will seem strange to have left this until last. The reason for this is simply that the modes of assessment used in dental courses are broadly similar to that of other anatomy courses. It will be found that a variety of objective tests are employed ranging from true/false answers at the very simplest through multiple-choice and extended matching item tests to more sophisticated approaches such as single-best-answer-type questions. These tests are increasingly being delivered to students in online formats. Objective tests will require some form of standard setting procedures to be employed when determining an appropriate pass mark. Tests requiring students to write free text responses vary. They can be short-answer-type questions where students' responses to questions can be single words, short phrases, or a few sentences. A more sophisticated approach involves some kind of semi-structured essay developing a theme or scenario. Essays, once the favored approach, are becoming less widely used. Practical examinations of the traditional steeplechase or spotter variety are frequently employed as tests of practical knowledge using either cadaveric specimens or images or a combination of the two. Resource-efficient and easily standardizable digitalized 3D spotters may well be a future trend in dental anatomy assessment. Thus, the range of assessment types in anatomical sciences in dentistry is no less wide than in courses for other groups.

Conclusions

1. Courses in anatomy for dental students include head and neck anatomy and anatomy of the thorax as well. Other regions of the body need to be taught for students to be able to understand the anatomical bases of human disease.

Practical Advice on Teaching a Dental Anatomy Course

- Avoid teaching dental students their anatomy in the early years as part of courses delivered to other professional groups (though there is great benefit to be derived from teaching members of the dental team together).
- Consider the content of the course and what is going to distinguish it from other courses that you teach. You are teaching future dental professionals; your course must be relevant yet coherent.
- Talk to members of your dental faculty, gain a sense of what is important, gather relevant clinical examples to illustrate your teaching, and understand how the whole dental course is structured.
- Ensure your course integrates with other relevant courses, especially embryology and neuroanatomy, if you are not teaching those topics as part of your own course.
- Choose your resources carefully, using materials written with the needs of dental students in mind wherever possible.
- Undertake discretionary team teaching with dentally qualified staff, utilizing the principles of effective clinical reasoning.
- Ensure anatomy is not just taught in year 1 of the program but spirals throughout the course and is assessed both formatively and summatively.
- Experiment with new technologies, but ensure they are grounded in solid educational evidence.
- Beware of the "magpie effect" and incorporating technology or newer pedagogies simply because of their novelty and before gaining student acceptance.

2. Courses in anatomy for dental students can be taught successfully in a variety of curricula structures, but, in a predominantly didactic course, one successful format can be that where regional anatomy is taught as part of the relevant integrated systems-based courses with head and neck anatomy delivered as either a stand-alone course or part of an oral structure and functions course.
3. Cadaveric teaching remains an important cornerstone of the teaching of anatomy to dental students not least because dentistry is a surgical science. Given limited time available, prosection-based teaching can be a very effective way for students to study the cadaver. Dissection can be utilized to support anatomical knowledge as well as other professional skills and may become more relevant in the later years of training.
4. It is important to ensure that the teaching of anatomy is vertically integrated with clinical teaching so that anatomical knowledge is reinforced later in the clinical course when students begin to treat patients and encounter complex clinical cases.
5. Horizontal integration of neuroanatomy teaching with that of the head and neck can enrich student learning and encourage engagement with neuroanatomy as a subject.
6. Utilizing a variety of teaching methods that are grounded in sound educational theory can enhance the learning experience for students. Technology-enhanced learning and interprofessional learning opportunities need to be carefully tailored to a suitable time and location within the course to avoid a teaching mismatch.

References

1. Sugand K, Abrahams P, Khurana A. The anatomy of anatomy: a review for its modernization. Anat Sci Educ. 2010;3:83–93.
2. Ahmed K, Rowland S, Patel V, Khan RS, Ashrafian H, Davies DC, et al. Is the structure of anatomy curriculum adequate for safe medical practice? Surgeon. 2010;8:318–24.
3. Collins J. Modern approaches to teaching and learning anatomy. Br Med J. 2008;337:665–7.
4. Elangovan S, Venugopalan SR, Srinivasan S, Karimbux NY, Weistroffer P, Allareddy V. Integration of basic-clinical sciences, PBL, CBL, and IPE in U.S. dental schools' curricula and a proposed inte-

grated curriculum model for the future. J Dent Educ. 2016;80(3):281–90.
5. Obrez A, Briggs C, Buckman J, Goldstein LM, Lamb C, Knight WG. Teaching clinically relevant dental anatomy in the dental curriculum: description and assessment of an innovative module. J Dent Educ. 2011;75:797–803.
6. Bogacki RE, Best A, Abbey LM. Equivalence study of a dental anatomy computer-assisted learning program. J Dent Educ. 2004;68:867–71.
7. Guttman GD. The current status of the anatomical sciences curriculum in US and Canadian dental schools. J Dent Educ. 2003;67:375–9.
8. Guttman GD, Ma TP, MacPherson BR. Making gross anatomy relevant to dental students. J Dent Educ. 2003;67:355–8.
9. Drake RL, McBride JM, Lachman N, Pawlina W. Medical education in the anatomical sciences: the winds of change continue to blow. Anat Sci Educ. 2009;2:253–9.
10. Bergman EM, Prince KJ, Drukker J, van der Vleuten CP, Scherpbier AJ. How much anatomy is enough? Anat Sci Educ. 2008;1:184–8.
11. Louw G, Eizenberg N, Carmichael SW. The place of anatomy in medical education: AMEE guide no 41. Med Teach. 2009;31:373–86.
12. Craig S, Tait N, Boers D, McAndrew D. Review of anatomy education in Australian and New Zealand medical schools. ANZ J Surg. 2010;80:212–6.
13. Macluskey M, Durham J, Bell A, Cowpe J, St Crean J, Dargue A, et al. A national survey of UK final year students' opinions of undergraduate oral surgery teaching. Eur J Dent Educ. 2012;16:e205–12.
14. Macluskey M, Shepherd S, Carter E, Bulsara Y, Durham JA, Bell A, et al. A national follow-up survey of UK graduates opinion of undergraduate oral surgery teaching. Eur J Dent Educ. 2016;20(3):174.
15. McHanwell S, Atkinson M, Davies DC, Dyball R, Morris J, Ockleford C, et al. A core syllabus in anatomy for medical students. Adding common sense to need to know. Eur J Anat. 2007;11:S151–66.
16. Educational Affairs Committee, American Association of Clinical Anatomists. A clinical anatomy curriculum for the medical student of the 21st century: gross anatomy. Clin Anat. 1996;9:71–99.
17. Horn SD, Susi FR, Sharawy M, LaFleche F, with American Association of Dental Schools. Curricular guidelines in gross anatomy. J Dent Educ. 1981;45:236–50.
18. Hiloowaka R, Hinton R, Innes P, Nava P, Williams V, with AADS Section on Anatomical Sciences. Curriculum guidelines for gross anatomy. J Dent Educ. 1993;57:461–8.
19. Moxham BJ, Plaisant O, Smith CF, Pawlina W, McHanwell S. An approach toward the development of core syllabuses for the anatomical sciences. Anat Sci Educ. 2014;7(4):302–11.
20. Matthan J, Cobb M, McHanwell S, Moxham BJ, Finn GM. The anatomical society's core anatomy syllabus for dental undergraduates. J Anat. 2020;236:737–51.
21. General Dental Council (UK). The first five years: the undergraduate dental curriculum. London: General Dental Council; 2008. https://www.gdc-uk.org/professionals/education. Accessed 9 Apr 2019.
22. General Dental Council (UK). Preparing for practice: dental team learning outcomes for registration. London: General Dental Council; 2012. http://www.gdc-uk.org/aboutus/education/pages/learning-outcomes.aspx. Accessed 8 Mar 2019.
23. Liebgott B. Anatomical basis of dentistry. 3rd ed. St. Louis: Mosby Elsevier; 2011.
24. Atkinson ME. Anatomy for dental students. 4th ed. Oxford: Oxford University Press; 2013.
25. Hiatt JL, Gartner LP. Textbook of head and neck anatomy. 4th ed. Baltimore: Lippincott Williams and Wilkins; 2010.
26. Berkovitz BKB, Holland GR, Moxham BJ. Oral anatomy and histology and embryology. 4th ed. Mosby International: Edinburgh; 2009.
27. National Health Service (Scotland). Oral health assessment and review. Dental clinical guidance. Scottish dental effectiveness programme. National Health Service. 2012. http://www.sdcep.org.uk/published-guidance/oral-health-assessment/. Accessed 8 Mar 2019.
28. Greenwood M, Seymour RA, Meechan JG. Textbook of human disease in dentistry. Oxford: Wiley-Blackwell; 2009.
29. Dettmer S, Tschernig T, Galanski M, Pabst R, Rierck B. Teaching surgery, radiology and anatomy together: the mix enhances motivation and comprehension. Surg Radiol Anat. 2010;32:791.
30. Rogers H, Sollecito T, Felix D, Yepes J, Williams M, D'Ambrosio J, et al. An international survey in postgraduate training in oral medicine. Oral Dis. 2011;17:95–8.
31. Napeñas JJ, Kujan O, Arduino PG, Sukumar S, Galvin S, Baričević M, et al. World workshop on oral medicine VI: controversies regarding dental management of medically complex patients: assessment of current recommendations. Oral Surg Oral Med Oral Pathol Oral Radiol. 2015;120(2):207–26.
32. Iacopino AM, Taft TB. Core curricula for postdoctoral dental students: recent problems, potential solutions, and a model for the future. J Dent Educ. 2007;71:1428–34.
33. Kreindler SA, Dowd DA, Dana Star N, Gottschalk T. Silos and social identity: the social identity approach as a framework for understanding and overcoming divisions in health care. Milbank Q. 2012;90(2):347–74.
34. Niekrash CE, Copes LE, Gonzalez RA. Frank Netter's legacy: interprofessional anatomy instruction. American Assoc Anat. 2015;8:348–59.
35. Barr H, Ford J, Gray R, Helme M, Hutchings M, Low H, et al. Interprofessional Education Guidelines. CAIPE. 2017. https://www.caipe.org/resources/publications/caipe-publications/caipe-2017-interprofessional-education-guidelines-barr-h-ford-j-gray-r-helme-m-hutchings-m-low-h-machin-reeves-s. Accessed 4 Mar 2019.

36. Tunstall-Pedoe S, Rink E, Hilton S. Student attitudes to undergraduate interprofessional education. J Interprof Care. 2003;17:161–72.
37. Smith CF, Martinez-Álvarez C, McHanwell S. The context of learning anatomy: does it make a difference? J Anat. 2014;224:270–8.
38. Barr H, Helme M, D'Avray L. Review of interprofessional education in the United Kingdom, 1997–2013. CAIPE. 2014. https://www.caipe.org/resources/publications/caipe-publications/caipe-2014-review-of-interprofessional-education-in-the-united-kingdom-1997-2013-in-brief-authored-by-barr-h-helme-m-davray-l. Accessed 4 Mar 2019.
39. Barr H, Gray R, Helme M, Low H, Reeves S. Steering the development of interprofessional education. J Interprof Care. 2016;30(5):549–52.
40. Smith CF, Hall S, Border S, Adds PJ, Finn GM. Interprofessional anatomy education in the United Kingdom and Ireland: perspectives from students and teachers. American Assoc Anat. 2015;8:360–70.
41. Thistlethwaite JE. Interprofessional education and the basic sciences: rationale and outcomes. Anat Sci Educ. 2015;8:299–304.
42. Guilding C, Hardisty J, Randles E, Statham L, Green A, Bhudia R, et al. Making it work: the feasibility and logistics of delivering large-scale interprofessional education to undergraduate healthcare students in a conference format. J Interprof Care. 2018;32(5):653–5.
43. Mitchell BS, McCrorie P, Sedgwick P. Student attitudes towards anatomy teaching and learning in a multiprofessional context. Med Educ. 2004;38:737–48.
44. Parsell G, Bligh J. The development of a questionnaire to assess the readiness of health care students for interprofessional learning (RIPLS). Med Educ. 1999;33:95–100.
45. Krause DA, Hollman JH, Pawlina W, Newcomer KL. Interprofessional education: collaboration or competition? A tale of two experiences. Curr Sports Med Rep. 2014;13:291–2.
46. McHanwell S. Teaching anatomical sciences to dental students. In: Chan LK, Pawlina W, editors. Teaching anatomy: a practical guide. Cham: Springer; 2015. p. 353–61.
47. Johnson EO, Charchanti AV, Troupis TG. Modernization of an anatomy class: from conceptualization to implementation. A case for integrated multimodal-multidisciplinary teaching. Anat Sci Educ. 2012;5:354–66.
48. Bakr MM, Thompson CM, Massadiq M. Anatomical sciences: a foundation for a solid learning experience in dental technology and dental prosthetics. Anat Sci Educ. 2016;10(4):395–404.
49. Ajjawi R, Hyde S, Roberts C, Nisbet G. Marginalisation of dental students in a shared medical programme. Med Educ. 2009;43:238–45.
50. Last KS, Appleton J, Stevenson H. Basic science knowledge of dental students on conventional and problem-based learning (PBL) courses at Liverpool. Eur J Dent Educ. 2001;5:148–54.
51. Yiu CKY, McGrath C, Bridges S, Corbet EF, Botelho M, Dyson J, et al. Graduates' perceived preparedness for dental practice from PBL and traditional curricula. J Dent Educ. 2011;75:1270–9.
52. Hattie J. Visible learning. Oxford: Routledge; 2011.
53. Kieser J, Livingstone V, Meldrum A. Professional storytelling in clinical dental anatomy teaching. Anat Sci Educ. 2008;1:84–9.
54. Redwood CJ, Townsend GC. The dead center of the dental curriculum: changing attitudes of dental students during dissection. J Dent Educ. 2011;75:1333–44.
55. Snelling J, Sahai A, Ellis H. Attitudes of medical and dental students to dissection. Clin Anat. 2003;16:165–72.
56. Pawlina W, Hromanik MJ, Milanese TR, Dierhising R, Viggiano TR, Carmichael S. Leadership and professionalism curriculum in the gross anatomy course. Ann Acad Med Singap. 2006;35:609–14.
57. McMenamin PG, McLachlan J, Wilson A, McBride JM, Pickering J, Evans DJR, et al. Do we really need cadavers anymore to learn anatomy in undergraduate medicine? Med Teach. 2018;40(10):1020–9.
58. Wilson AB, Miller CH, Klein BA, Taylor MA, Goodwin M, Boyle EK, et al. A meta-analysis of anatomy laboratory pedagogies. Clin Anat. 2018;31:122–33.
59. Finn GM, Matthan J. Pedagogical perspectives on the use of technology within medical curricula: moving away from norm driven implementation. In: Rae P, editor. Biomedical visualisation. Advances in experimental medicine and biology, vol. 1120. Cham: Springer; 2019. p. 55–64.
60. Berkovitz B, Kirsch C, Moxham BJ, Alusi G, Cheeseman T. 3D head and neck anatomy with special senses and basic neuroanatomy. London: Primal Pictures; 2007.
61. Ward PJ, Walker JJ. The influence of study methods and knowledge processing on academic success and long-term recall of anatomy learning by first-year veterinary students. Anat Sci Educ. 2008;1:68–74.
62. Hackett M, Proctor M. The effect of autostereoscopic holograms on anatomical knowledge: a randomised trial. Med Educ. 2018;52(11):1147–55.
63. Biasutto SN, Caussa LI, del Río LEC. Teaching anatomy: cadavers vs. computers? Ann Anat. 2006;188(2):187–90.
64. Kugelmann D, Stratmann L, Nühlen N, Bork F, Hoffmann S, Samarbarksh G, et al. An augmented reality magic mirror as additive teaching device for gross anatomy. Ann Anat. 2018;215:7177.
65. Clunie L, Morris NP, Joynes VC, Pickering JD. How comprehensive are research studies investigating the efficacy of technology-enhanced learning resources in anatomy education? A systematic review. Anat Sci Educ. 2018;11:303–19.
66. Keenan ID, Slater JD, Matthan J. Social media: insights for medical education from instructor perceptions and usage. MedEdPublish. 2018;7. https://www.mededpublish.org/manuscripts/1420. Accessed 16 Apr 2019.

67. Hennessy CM, Kirkpatrick E, Smith CF, Border S. Social media and anatomy education: using twitter to enhance the student learning experience in anatomy. Anat Sci Educ. 2016;9(6):505–15.
68. Pickering JD, Swinnerton BJ. Exploring the dimensions of medical student engagement with technology-enhanced learning resources and assessing the impact on assessment outcomes. Anat Sci Educ. 2019;12(2):117–28.
69. Nanjundaiah K, Chowdapurkar S. Body-painting: a tool which can be used to teach surface anatomy. J Clin Diagn Res. 2012;6(8):1405.
70. Barmaki R, Yu K, Pearlman R, Shingles R, Bork F, Osgood GM, et al. Enhancement of anatomical education using augmented reality: an empirical study of body painting. Anat Sci Educ. 2019;12:599–609.
71. Chapman JA, Williams A-MM, Allen MF. Anatomical body painting: an inexpensive, high-impact face-to-face practical for a generalist anatomy student cohort [conference extract]. Australian and New Zealand association of clinical anatomists (ANZACA) meeting. 9–11 December 2015; Adelaide, Australia. http://ecite.utas.edu.au/109511. Accessed 4 Mar 2019.
72. Ten Cate OT, Durning S. Dimensions and psychology of peer teaching in medical education. Med Teach. 2007;29(6):546–52.
73. Evans DJ, Cuffe T. Near-peer teaching in anatomy: an approach for deeper learning. Anat Sci Educ. 2009;2:227–33.
74. Bruecker JK, MacPherson BR. Benefits from peer teaching in the dental gross anatomy laboratory. Eur J Dent Educ. 2004;8:72–7.
75. Lee LMJ, Nagel RW, Gould DJ. The educational value of online mastery quizzes in a human anatomy course for first-year dental students. J Dent Educ. 2012;76:1195–9.
76. McMenamin PG, Eizenberg N, Buzzard A, Fogg Q, Lazarus M. A broad perspective on anatomy education: celebrating teaching diversity and innovations. Med J Aust. 2016;204(2):57.
77. Bergman EM. Discussing dissection in anatomy education. Perspect Med Educ. 2015;4(5):211–3.
78. Bergman EM, van der Vleuten CPM, Scherpbier AJJA. Why don't they know enough about anatomy? A narrative review. Med Teach. 2011;33(5):403–9.
79. Dahle LO, Brynhildsen J, Behrbohm Fallsberg M, Rundquist I, Hammar M. Pros and cons of vertical integration between clinical medicine and basic science within a problem-based undergraduate medical curriculum: examples and experiences from Linköping, Sweden. Med Teach. 2002;24:280–5.
80. Last KS, Appleton J, Ferguson DB, Stevenson H. The value of a questionnaire in assessing the acquisition and retention of basic science knowledge by dental students. Eur J Dent Educ. 2000;4:3–9.
81. Durham J, Brettell S, Summerside C, McHanwell S. "It's a good idea and just sort of needs a bit of fine tuning"—qualitative and quantitative evaluation of a vertically integrated virtual anatomy course for clinical undergraduates. Eur J Dent Educ. 2009;13:100–9.
82. Hendricson WD. Changes in educational methodologies in predoctoral dental education: funding the perfect intersection. J Dent Educ. 2012;76:118–41.
83. Rafai N, Lemos M, Kennes LN, Hawari A, Gerhardt-Szép S, Classen-Linke I. Anatomy meets dentistry! Linking anatomy and clinical practice in the preclinical dental curriculum. BMC Med Educ. 2016;16(1):305.
84. Bergman EM, de Bruin ABH, Vorstenbosch MA, Kooloos JG, Puts GC, Leppink J, et al. Effects of learning content in context on knowledge acquisition and recall: a pretest-posttest control group design. BMC Med Educ. 2015;15:133.
85. Koens F, Ten Cate OT, Custers EJ. Context-dependent memory in a meaningful environment for medical education: in the classroom and at the bedside. Adv Health Sci Educ Theory Pract. 2003;8(2):155–65.
86. Böckers A, Mayer C, Böckers TM. Does learning in clinical context in anatomical sciences improve examination results, learning motivation, or learning orientation? Anat Sci Educ. 2014;7(1):3–11.
87. Javaid MA, Chakraborty S, Cryan JF, Schellekens H, Toulouse A. Understanding neurophobia: reasons behind impaired understanding and learning of neuroanatomy in cross-disciplinary healthcare students. Anat Sci Educ. 2018;11:81–93.
88. Sotgiu MA, Mazzarello V, Bandiera P, Madeddu R, Montella A, Moxham B. Neuroanatomy, the Achilles' Heel of medical students. A systematic analysis of educational strategies for the teaching of neuroanatomy. Anat Sci Educ. 2020;13:107–16.
89. Sessle BJ, Bradley RM, Dubner R, Matthews B, Moller E. Dental neuroscience. New Dent. 1979;10:32–8.
90. Curriculum guidelines for neuroanatomy. Section on Anatomical Sciences of the American Association of Dental Schools. J Dent Educ. 1992;56:762–5.
91. Klueber KM. Neuroanatomy for the dentist in the twenty-first century. J Dent Educ. 2003;67:366–9.

Negotiation and Assessment as Tools for Tailoring Anatomy Courses to Allied Health Programs

49

Lawrence J. Rizzolo, William B. Stewart, Alexandria Garino, and Linda H. Pellico

Three questions perennially confront anatomists: What should be taught? Who should teach it? What pedagogy should be used? At medical schools, different disciplines compete for school resources and time in the curriculum. For allied health programs, there is an added dimension, because the tradition of a strong grounding in anatomy is not firmly established. For example, all accredited physician assistant (PA) programs must teach anatomy, even though 86% list anatomy as an entrance requirement [1]. However, the instructional hours devoted to anatomy (102.4 ± 56.1) vary greatly [2]. Accelerated graduate nursing programs enable students without a nursing degree to become a registered nurse after 1 year with an advance practice degree 2 years later [3]. Most of the 68 programs in the USA require anatomy as a prerequisite, but it appears only Yale University includes anatomy in its required curriculum. Yale's concern was that undergraduate courses fail to focus on the anatomy needed to be an advanced practice nurse. The common experience of the PA accreditation board and nursing programs [4] is that nonclinical university departments fail to ensure a consistent standard level of knowledge and are costly in terms of student dollars and time.

Each allied health field has its own needs, and there is little standardization within fields. Therefore, anatomists must become skilled negotiators by developing methods to answer the questions posed above [5–9]. Many individuals and groups have a stake in these questions, which means the success of your course depends upon your ability to identify and satisfy conflicting concerns. You will need to (1) reconcile the myriad demands placed on the limited time and resources possessed by you, your students, and your faculty collaborators and (2) negotiate for the time and resources required to build a course that integrates with the rest of the program's curriculum.

We found that negotiation with stakeholders led to course goals and two types of assessment that could shepherd the course toward achieving agreed upon outcomes [9–11]. Year-end summative assessments identify strengths and weaknesses but often fail to identify the cause of these attributes. Formative assessments provide

L. J. Rizzolo (✉) · W. B. Stewart
Department of Surgery, Yale University, New Haven, CT, USA
e-mail: lawrence.rizzolo@yale.edu

A. Garino
Physician Associate Program, Department of Medicine, Yale University, New Haven, CT, USA

L. H. Pellico
School of Nursing, Yale University, New Haven, CT, USA

L. K. Chan, W. Pawlina (eds.), *Teaching Anatomy*, https://doi.org/10.1007/978-3-030-43283-6_49

feedback during the course that identifies ways to develop the course, train faculty, and help students progress. We will discuss our medical school and PA programs first, because they are the most comprehensive and were the incubators where we developed and verified these principles [9, 10]. Our graduate nursing course had been a lightly attended elective until we applied the principles learned from redesigning the medical and PA programs. Our stakeholders were so pleased that anatomy has become a required course. A detailed outline of our methodology is available online [8]. Table 49.1 summarizes the various courses we teach, but this chapter will focus on the medical, PA, and nursing programs. We will discuss the needs analysis of our stakeholders, but you are encouraged to use this as a guide for your site-specific negotiations and analysis.

Negotiation Skills (See References for Complete Citations)

- General references [5–7]
- Anatomy-specific references [8, 11]

Identifying and Negotiating with Stakeholders: Agreeing on Goals, Cases, and Procedures

You have more stakeholders than you might realize, and their needs can be contradictory. The dean has concerns that include costs, resources, curricular hours, and overall curriculum. The dean's vision may include more than the educational mission, and those goals compete for the same resources. Department chairs may wish to support the dean's overarching goals for education, but they also have some goals for their department that conflict with educational goals.

The broader community includes accreditation bodies and licensing boards. Programs for advanced degrees might list your course as a prerequisite with the hope that you will accomplish certain goals. Often information about these goals is hard to obtain. We found it valuable to work backward by consulting the faculty who teach after our courses. What should a student need to know to graduate? To accomplish that goal, what should a student know before entering their last year and so forth? The approach reveals the anatomy these faculty stakeholders consider essential and how to prepare students. It partitions the subject matter into introductory and subsequent courses/clinical training and establishes responsibility for the deferred material.

Students and their anatomy instructors are essential stakeholders. Especially when engaging pedagogies that are foreign to them, students need to know how anatomy integrates with their career goals. Given the many demands placed on their time, are you asking too little or too much of them? How do they define being prepared for class and how does their answer compare to your definition? The faculty has analogous concerns if a new course places new burdens on them. Does your new course structure require training in new pedagogical skills, and if so, how will those new skills be acquired? This problem is especially acute if your faculty has other responsibilities such as maintaining a research laboratory or clinic income. No matter how well designed, the faculty's ability to implement your course will determine success or failure.

The negotiation process is slow. It begins with learning about the needs of the stakeholders and their responsibilities that may impede their desire to collaborate. After you complete this "needs analysis," you will be ready to devise solutions. In the second round of negotiations, your proposals will gain more traction the more stakeholders feel you understand their concerns. As long as you are flexible, they will develop a sense of ownership and provide you the help you need to succeed. Here are some examples of how we did our needs assessment for various programs and the points of negotiation that we uncovered.

Questions for stakeholders:

- *Interviews*: How is anatomy used in your discipline? How do you wish students were trained? What solutions would you recommend? What circumstances would allow you to help? Can we keep discussing this?
- *Focus groups*: What was your experience of your anatomy course? In what ways did it help your subsequent training? What would have been more helpful?
- *Surveys*: Short-answer questions based on results of interview and focus groups.
- *Follow-up questions for yourself to guide further negotiations*: What do I have that they need? What do they have that I need?
- The answers will help you choose cases and procedures, pedagogical methods, and assessment procedures.

Medicine and Physician Assistant Programs

Physician assistants are taught in the medical model. The basic approach was devised for the medical students and then adapted to the PA students. We first interviewed all of the section chiefs of each surgical department. An example of a conflict that emerged was the dean's desire to use more surgeons as teachers, which added to the burden of surgical departmental chairs, whose primary concern was patient care and clinical education. Surgeons wanted to help, because they felt that all students needed a better understanding of surgery and lacked sufficient anatomical training. At the same time, surgeons noted that the pressure to generate clinical income impeded their desire to be part of the solution. We also found that faculty from the medical departments were interested in advancing their own understanding of the anatomical basis of the diagnostic procedures, in order to better instruct trainees. Virtually all the clinicians interviewed favored presenting critical information in the context in which it would be used.

The interview results guided our negotiation strategy. Clinicians would welcome the opportunity to work with students early in their training, if the anatomy curriculum was more relevant to their practice. The opportunity to help design the course made them more willing to teach it. Interviews, focus groups, and surveys revealed that students favored a patient-centered curriculum and a student-centered course, which dovetailed with goals of the clinical faculty. This led to a course design based entirely on common clinical cases that students would certainly see during their clinical training. The new structure focused more on critical reasoning and skillful observation.

The course begins as the clinician begins with a patient by mimicking a physical exam. This is an exercise in surface anatomy. Later, dissection (or reopening the body wall in a prosection course) reveals whether the organs of that particular donor might have been enlarged or displaced due to disease. The problem presented to students is "what anatomy can be learned by physical examination." Think of this as just-in-time learning. More anatomy progressively unfolds as the student engages more complex problems. Subsequent laboratories and workshops are based on the anatomy that underlies the patient's history, physical exam, imaging studies, and medical or surgical resolution. The dissection procedures are modeled on classical, open surgical procedures and are freely accessible on the website "Anatomy Clinic" [12]. The result was a shorter course with more clinician involvement that focused on preparing students for the next step of training. Students were motivated by problem-solving exercises, practical application, and extensive formative assessments. They profited when problem-oriented modalities were integrated, including dissection, prosection, computer exercises, radiology, and small group discussion. Our work with stakeholders has had an added benefit. As each program revised its curriculum,

Table 49.1 Anatomy courses for clinical students at Yale University

Students	Curriculum developers	Delivery	Faculty	Laboratory experience	Summative examinations	Content emphasis
Medical	Surgical, medical, radiology, anatomy faculty	108 h (60 laboratory, 16 small group, 28 lecture, 4 test)	Anatomy, surgical, medical radiologic, medical students, residents	Full-body dissection	Anonymous, two written (pass/fail)	Clinical context, critical observation, problem-solving
Physician assistant	Surgical, medical, radiology, anatomy faculty	104 h (51 laboratory, 7 small group, 40 lecture, 3 laboratory test, 3 written test)	Anatomy	Full-body dissection	Three written/three laboratory (graded)	Clinical context, critical observation, problem-solving
Graduate nursing	Nursing/anatomy faculty	30 h (13 lecture, 8 anatomy laboratory, 6 clinical laboratory)	Anatomy, nursing (co-teaching)	*Anatomy laboratory*: prosection *Clinical laboratory*: skills	Written midterm and final (graded)	Clinical context, physical exam, nursing care/procedures
Surgical PA residents	Anatomy/ residency faculty	18 h (laboratory)	Anatomy	Partial-body dissection	No exam	Clinical procedures
Paramedics	Anatomy, paramedic faculty	15 h (laboratory)	Anatomy, paramedic (co-teaching)	Partial-body dissection	Anatomy tested in other course	Vital signs, emergency procedures, major organs
High school	Anatomy, clinical, high school faculty	15 h (laboratory/small group)	Anatomy, medical students	Prosection	Written tests at Yale and at high school—graded exams	Physical exam, disease, exercise

anatomy was viewed as a central feature that integrated well with the whole.

Ongoing negotiation with students revealed concerns about how to prepare and what was expected of them. We learned to develop (1) a close connection between stated objectives, workshop and laboratory activities, take-home messages for each activity, and exam questions; (2) interactive web activities that required students to make decisions, consult texts/web resources, and process feedback in order to proceed through the activity; and (3) a social structure that fostered teamwork and active participation by all students.

Weekly faculty meetings revealed that faculty shared student anxieties. They were unfamiliar with student-centered techniques. Interactions at the dissection table or small group workshop tended to follow teacher-centered mini-lecture rather than fostering student-centered exploration and group problem-solving. To help these essential stakeholders, we invited an educational expert to our faculty meetings to run development sessions that enabled faculty to meet the new demands placed upon them.

The PA course reflects the needs of its administrative stakeholders. PA students have less free time for out-of-class preparation than their medical school colleagues. The PA course is approximately the same length as the medical school course in terms of class hours, but the time is structured differently. PA students have fewer web-based assignments, and some time is shifted from laboratory and workshop to lecture format. Whereas clinical radiology is integrated into the medical school course, PAs have a stand-alone radiology course. The close interaction with the curriculum directors avoided duplication of effort and promoted integration with the rest of the curriculum.

Graduate Nursing Program

Our initial experience is a good example of what happens when you ignore stakeholders. Our course had been an elective, but because it failed to satisfy the needs of students, many dropped the course for lack of interest. We met with the nursing program director to develop a course directed specifically to graduate nursing students. The program director interviewed the directors of the nursing specialties to identify the critical concepts to be taught. Similar to the medical school and PA experience, we learned about the specific ways that anatomy would be used in this career path. Accordingly, lecture and laboratory exercises were based on the following organ systems: respiratory, cardiovascular, alimentary, urogenital, musculoskeletal, and nervous system. Lectures were typically followed by a prosection laboratory where the relevant structures could be examined in coordination with a nursing skills laboratory where the points of clinical care could be highlighted. Skills labs are based on the following diagnostic and invasive procedures: lung exam, airway management, chest tube placement and care, peripheral vascular exam, electrocardiogram, feeding tube placement and maintenance, urinary catheter placement and care, and the cranial nerve exam. Viewing prosections can be a passive, teacher-centered activity, but the integration with the lecture and skills laboratory promotes discussion as students come into the anatomy laboratory with questions in mind.

Team teaching during each class session was a critical component [4]. Advanced practice registered nurses (APRN) and anatomy faculty complemented each other's experience to integrate nursing skills with anatomy. Because clinical faculty and program administration were directly involved in the design and implementation, the course fully integrated with the nursing curriculum. The curriculum committee (another stakeholder) approved the new elective. Once it proved itself to all stakeholders, the course became required.

To do:

- Use the information gained from your stakeholders to develop a vision for your course.
- Design the elements of your course that will achieve your vision.
- Go back to your stakeholders to share your vision for their feedback: be flexible! The more stakeholders share in a common vision, the more helpful they will be as the project develops.

Selecting Pedagogical Modalities and Assessments

> To attain knowledge, add things every day.
> To attain wisdom, remove things every day.
> –Lao Tzu, mystic philosopher of ancient China and founder of Taoism

Many factors go into the selection of modalities and assessments. These will emerge from your negotiations with stakeholders. Students have different learning styles, but research indicates that students need to transition to the learning style favored by most professionals [13, 14]. Similarly, teachers have a preferred pedagogy, but they should be comfortable with all methods. Multiple pedagogies allow faculty to meet the students where they are and guide them toward the learning styles favored by clinicians. Is your course trying to transmit information or ask students to process information in a new way, e.g., three-dimensional reasoning? The first goal is suited to teacher-centered pedagogies, e.g., lectures and demonstrations. The second goal involves helping students gain comfort with unfamiliar learning styles and is more suited to student-centered pedagogies, e.g., workshops, dissection labs, and open discussions. Often a course is a balance between these goals and entails two challenges: (1) giving students clear expectations and the tools to achieve them and (2) helping faculty develop new skills.

Two types of assessment are required to evaluate your selections: (1) formative assessments to understand how your selections are working in real time and (2) summative assessments to determine whether you achieved the goals of your stakeholders. Both need to be designed at the time you are designing the course. Assessments are often designed as an afterthought, but such assessments are of minimal help to improve the course and tend to be unconvincing for your stakeholders.

Formative assessments address the problem that new courses are often flawed. Instead of asking whether a course element is working, a well-designed formative assessment identifies underlying problems and reveals feasible solutions while the course is running and changes can be made. Performance on class exams provides clues, but these need to be supplemented by surveys and focus groups that ask what students and faculty were experiencing and how they tried to resolve their anxieties and concerns. To facilitate group work, instructors used weekly faculty meetings to share experiences and brainstorm solutions. The meetings focused on techniques that foster and manage discussion in place of mini-lecture. A rubric was developed to help students and faculty identify the factors that promote group process. The approach provided both an assessment and fodder for the faculty development program. To facilitate group work among students in the medical school, students were grouped into "Learning Societies." A society consisted of five dissection teams of four students and a mentor. Societies worked together in laboratory and workshop. As one example, we made novel use of ungraded practical exams. Students were encouraged to take the exams in teams where they could discuss their logic and consult with an instructor when necessary. The approach was valued by faculty, because by listening to students discuss their problem-solving, faculty learned whether their focus on observation skills and reasoning with data was effective. It was valued by students, because peer teaching honed reasoning skills, and they learned how well they understood concepts relative to their peers.

To develop summative assessments that will convince your stakeholders, you need to know how they define a successful outcome and what assessment criteria would convince them of success. In our case, we needed written exams, based on clinical vignettes that tested the course goals of critical observation and interpretation. The exams used short essay questions that required students to interpret patient data and imaging studies, extended matching questions that required students to combine data from multiple images, and standard multiple-answer questions. You may need to collect new data on student behaviors, attitudes, and performance in your current course, as you are designing your new one. If you wait too long, you may find that you missed the opportunity to collect the necessary pre-intervention data. This approach allowed us to demonstrate that long-term recall of anatomy increased for both medical and PA students [9, 10]. Designing your summative assessment concurrent with the design of the course will help you define course objectives and inform your choice of pedagogy.

To do:

- Use stakeholder assessments to design formative and summative assessments that will convince stakeholders that you satisfied their needs. Do this before your course begins!
- Initially, avoid asking whether each course element works or not. Instead ask the following: What are students experiencing and how are they behaving? What are faculty experiencing and how are they behaving? For example, how do they prepare for sessions? Which resources do they use? What frustrations are participants experiencing and how do they overcome them?
- Use the answers to understand *why* successful elements worked and *why* unsuccessful elements failed. Revise your course accordingly until each element works properly. The cycle of revision and re-evaluation can take 3–5 years. Stick with it!

Medicine and Physician Assistant Programs

The Yale School of Medicine emphasizes independent learning, "honors system" exams, and professional responsibility. It puts a premium on small group, collaborative learning. The absence of grades puts a great burden on students and faculty to define and achieve success. The students are given substantial out-of-class time for independent study. By contrast, the PA program is run more along the lines of a college curriculum. For the anatomy course, both programs use "Anatomy Clinic" [12] to provide dissection instructions, but the pedagogy had to be modified to suit the environments of the medical and PA program.

The medical school environment required that we give students the tools and direction that foster productive use of their independent-study time. "Anatomy Clinic" is a research guide with highly interactive features [9, 12]. Students analyze physical exam or imaging data and receive instant feedback on short essay questions. Rather than present and quiz information, "Anatomy Clinic" is designed to develop habits of mind for problem-solving and processing spatial information. It provides students a sense of what it means to be prepared for a given workshop or laboratory session. Instructors follow a Socratic method and facilitate discussions rather than lecture. Students struggle with and master a related set of problems over a block of sessions. Lectures are more content and concept driven and are used to either introduce topics or consolidate a block of web, workshop, and dissection activities. As the medical school curriculum has evolved, our medical and PA courses have evolved with it. Using the same basic principals, "Anatomy Clinic" was replaced by a series of iBooks that are freely available on the on-line Apple Store using the search term "human anatomy video project series".

To adapt the course to the PA environment, the number of laboratory hours was decreased to allow more of that material to be presented in a lecture format. These laboratory-specific lectures were in addition to the standard 1-h lectures. More direction is given in the laboratory in the form of "must-see" lists. There is less reliance on web activities and workshop discussions. Although full-body dissection is still done, students work in groups of six instead of groups of four. Accordingly, there is a partial shift from student-centered toward teacher-centered pedagogy. This is mitigated by the small class size (36 students), which allows lectures to be more interactive.

PA students' mastery and long-term recall of the material were less substantive than the medical course but far superior to the previous PA course [10]. As with all choices, there are gains and losses. In this case, the changes from the medical school course resulted in less depth, but the gains were that students met programmatic goals with greater success and without compromising their attention to the other basic science courses.

Graduate Nursing Program

The hallmark of the program that emerged is coordinated nursing and anatomy labs. The shared interaction of an anatomist and an APRN in lecture and anatomy laboratory resulted in a unique opportunity to better understand how structure and structural relationships relate to clinical presentations and predictions of disease. In the anatomy labs, students view prosected donors to explore the anatomy that underlay the skills that had been presented in lecture. Because the laboratory groups were small and the corresponding nursing skills laboratory was taught the same day, the students were full of anatomical questions. This enabled the laboratory faculty to blend teacher-centered and student-centered pedagogies. The value of the APRN/anatomist collaboration cannot be overemphasized.

To do:
- Determine which pedagogies are best suited to your goals.
- Take an inventory of the strengths and weaknesses of you and your faculty with regard to these pedagogies.
- Find support through your school's department of education or faculty development for training that will address those weaknesses.

Conclusion

Implicit in this discussion is the need to approach the task of creating/revising a course with a bit of humility. Rather than presume you know what students need to know, find out. By learning about the common concerns and procedures that engage practitioners, you enhance your ability to recommend course content. After working on a gait activity with a rheumatologist, we learned about the issues that most concerned him, and he in turn was reminded of the anatomical basis of his trade. He concluded "Now I understand why it is I do what I do, and can now be a better instructor for my residents!" Negotiation with stakeholders is a two-way street where the sharing of information benefits everyone. You will earn the enthusiastic support of your administration and colleagues that will enable you to teach in the context that will benefit students' future needs. Armed with actual cases and procedures, you will be able to design hands-on activities that highlight critical anatomy. Team teaching with clinical partners will provide continuity of instruction and convince students that they are focused on essential content. Perhaps the most important outcome of this work is that you will develop a clear path from course objectives to class activities to assessment procedures. By examining how students and faculty engage these three elements, you will have a method for continually reassessing and improving your course(s).

References

1. ARC-PA. Accreditation Standards for Physician Assistant Education. 4th ed. Johns Creek, GA: Accreditation Review Commission on Education for the Physician Assistant, Inc.; 2012. http://www.arc-pa.org/acc_standards/.
2. PAEA. Twenty-fifth annual report on Physician Assistant Educational Programs in the United States, 2008–2009. Alexandria: Physician Assistant Education Association; 2010. http://www.paeaonline.org/index.php?ht=d/sp/i/243/pid/243.
3. AACN. Accelerated programs: the fast track to careers in nursing. Washington, DC: American Association of Colleges of Nursing; 2013. http://www.aacn.nche.edu/publications/issue-bulletin-accelerated-programs.
4. Larcombe J, Dick J. Who is best qualified to teach bioscience to nurses? Nurs Stand. 2003;17(51):38–44.
5. Fisher R, Ury W, Patton B. Getting to yes: negotiating agreement without giving in. 3rd ed. New York: Penguin; 2011. p. xxix, 204 p.
6. Heifetz RA, Linsky M. Leadership on the line: staying alive through the dangers of leading. Boston, MA: Harvard Business School Press; 2002. p. xi, 252 p.

7. Kegan R, Lahey LL. How the way we talk can change the way we work: seven languages for transformation. 1st ed. San Francisco: Jossey-Bass; 2001. p. ix, 241 p.
8. Rizzolo L, Rando W. Using assessments to energize your innovation. Educational roundtable breakfast. Bethesda: American Association of Anatomists Annual Meeting; 2010. http://www.anatomy.org/content/education-teaching-sessions.
9. Rizzolo LJ, Rando WC, O'Brien K, Haims A, Abrahams J, Stewart WB. Design, implementation, and evaluation of an innovative anatomy course. Anat Sci Educ. 2010;3:109–20.
10. Rizzolo LJ, Rando WC, O'Brien MK, Garino A, Stewart WB. Effectiveness of a shortened, clinically engaged anatomy course for physician assistant students. Anat Sci Educ. 2011;4(2):64–70.
11. Rizzolo LJ, Stewart WB, O'Brien M, Haims A, Rando W, Abrahams J, et al. Design principles for developing an efficient clinical anatomy course. Med Teach. 2006;28(2):142–51.
12. Rizzolo LJ. Anatomy clinic. New Haven: Yale University School of Medicine; 2013. http://www.anatomy.yalemedicine.org/.
13. Engels P, de Gara C. Learning styles of medical students, general surgery residents, and general surgeons: implications for surgical education. BMC Med Educ. 2010;10(1):51.
14. Felder RM, Brent R. Understanding student differences. J Eng Educ. 2005;94(1):57–72.

Teaching Anatomy to Students in a Physical Therapy Education Program

50

David A. Krause, Nathan J. Hellyer, and Beth A. Cloud-Biebl

Gross human anatomy is a foundational course within the academic curriculum of physical therapy (PT) programs with the preferred teaching method being human cadaveric dissection [1, 2]. The Commission on Accreditation in Physical Therapy Education (CAPTE) accredits physical therapy programs to assure quality [3]. CAPTE presumes a PT professional curriculum includes content and learning experiences in anatomy that prepare students for subsequent physical therapy courses and for clinical practice. Year 1 students at the Mayo Clinic Program in Physical Therapy are enrolled in a six-credit "Anatomy for Physical Therapists" course that meets daily during a 16-week fall semester. Lectures last for 1 hour and are held 3 days per week, whereas laboratory dissection sessions last for 3 hours twice a week with the major emphasis directed to anatomy of the limbs and back. The primary educational objective is to provide students with a broad understanding of functional anatomy so he or she can examine the musculoskeletal system and identify structures responsible for movement dysfunction.

Today's changing landscape in anatomy education has fostered a variety of active learning techniques designed to promote student engagement, while learning a large body of information [4]. Many current students prefer a variety of active learning schemes including fewer lectures, more multimedia presentations, and the opportunity to collaborate with classmates [5]. This chapter will describe how various dynamic learning activities can be used in the anatomy dissection laboratory to facilitate teaching and learning.

Reciprocal Peer Teaching

The term "reciprocal peer teaching" (RPT) was first used by Allen and Boraks [6] to describe an educational system where fellow classmates alternate roles between teacher and student. This can be an effective means for students learning gross anatomy while trying to master anatomical relationships [7, 8]. To supplement lecture and dissection experiences, groups of three or four students present weekly demonstrations to fellow classmates. The RPT assignment augments anatomical concepts presented in the classroom and subsequently observed at the dissection table. The demonstration topics are created by the anatomy instructor prior to the start of the course. For each RPT session, groups of students acting as teachers are given three or four learning objectives by the course faculty. Student teachers review the learning objectives with a faculty member prior to the scheduled demonstration.

D. A. Krause (✉) · N. J. Hellyer · B. A. Cloud-Biebl
Program in Physical Therapy, Mayo Clinic, Rochester, MN, USA
e-mail: Krause.david@mayo.edu; Hellyer.nathan@mayo.edu; Cloud.beth@mayo.edu

L. K. Chan, W. Pawlina (eds.), *Teaching Anatomy*, https://doi.org/10.1007/978-3-030-43283-6_50

The faculty member might offer suggestions about how the group might illustrate an anatomical concept on a prosected cadaver, skeleton, or plastic model. For example, one RPT exercise involves illustrating the six fibro-osseous compartments on the dorsal aspect of the wrist. The student group might begin their teaching presentation by reviewing the osteology of the dorsum of the distal radius and ulna using a skeleton. Student teachers point out the locations of the six extensor compartments and name the contents of each compartment. Next, the student teachers use the cadaver to illustrate the contents of each extensor compartment to their classmates, and lastly, students don gloves, palpate, and use markers to draw tendons coursing through the respective compartments on the gloved hand (Fig. 50.1). Similarly, gloved hands are used in other demonstrations to draw intrinsic muscles of the hand and nerve distributions. During the RPT experience, student teachers circulate throughout the laboratory and present their demonstration until all dissection groups are taught [9, 10]. Upon completion of the RPT experience, the students return to their dissection table and began that day's dissection assignment.

An advantage of peer teaching is that students are able to share their unique visions of anatomy. Creation of three-dimensional models as an example allows for novel perspectives and creative mastery of structural relationships. One particularly complex anatomical relationship is the inguinal canal which is formed by the conjoining of abdominal tendinous intersections. To illustrate, students fold paper into a rectangular box identifying the walls, roof, and floor of the inguinal canal and color code the blending of the transversus abdominis and oblique tendons. A light rope represents the spermatic cord (or round ligament of the uterus) passing through the canal, with a cutout for the superficial inguinal ring on the anterior wall and posterior wall cutout for the deep inguinal ring (Fig. 50.2). Student engagement with their model and sharing of their creation with peers increases ownership of the subject matter being learned and retained.

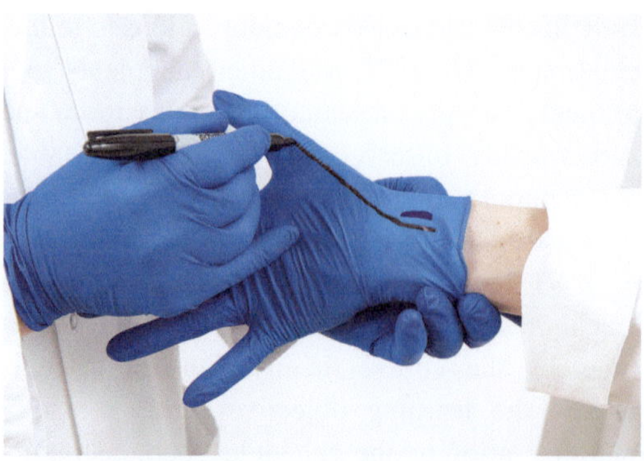

Fig. 50.1 Drawing the course of the extensor pollicis longus around the dorsal tubercle

Another example of RPT occurs when students teach peers about muscle actions. Students use a piece of string or a band to visualize the action line of a muscle when learning attachments and actions [11]. By holding the string at a muscle's attachment sites, the student teacher can help learners readily observe the relationship of the muscle to a joint's axis of motion. For example, students often have difficulty visualizing the obturator internus as a hip external rotator. Using a string, band, or rolled up towel and a human skeleton, a student acting as teacher holds one end of the string or towel on the obturator foramen within the pelvis, courses it around the lesser sciatic notch, and places the other end on the medial aspect of the greater trochanter (Fig. 50.3). The simulated muscle action line is observed to lie posterior to the hip joint's longitudinal axis and helps the learners visualize and understand how the obturator internus can externally rotate the femur on the pelvis.

With respect to teaching, students perceive RPT sessions as enjoyable, because group members assemble beforehand to pool ideas to form creative ways to teach their classmates. Groups of physical therapy students acting as teachers reported that they "dig a bit deeper" for the clinical relevance of anatomical relationships when preparing for their RPT demonstration. This supplementary preparation time promotes confidence when instructing their classmates. Verbal interactions among laboratory teammates while preparing for the RPT demonstrations promote cooperative learning among classmates. For example, a student as teacher might share an

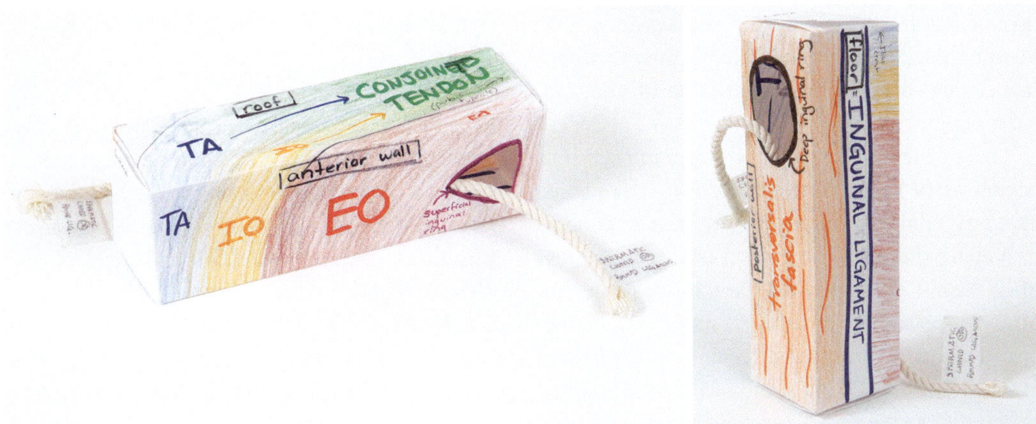

Fig. 50.2 Paper model of the inguinal canal

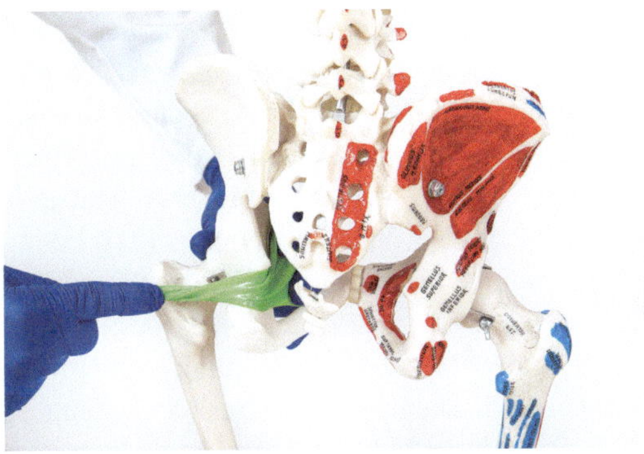

Fig. 50.3 Band used to illustrate the obturator internus

acrostic or acronym with his or her group members that he or she found useful when learning anatomical facts. Because of this sharing process, classmates experience a novel way to store and retrieve new information. After both teaching and learning from one another, students report that they are more likely to summon assistance from a peer regarding anatomical concepts.

Near-Peer Teaching

Educators contend that near-peer (NP) teaching promotes development of professional behaviors and is successful because peer tutors and tutees may communicate more effectively than do teachers and students [6, 12]. NP teaching is incorporated into the 16-week semester anatomy course as an educational method to enhance learning of anatomical concepts by Year 1 Doctor of Physical Therapy (DPT) students. Approximately six Year 2 DPT students are selected by the course coordinator to serve as NP teachers for Year 1 DPT students. These Year 2 students lead weekly review sessions which are each 60–90 minutes in length. The theme for each NP teaching session is developed based upon objectives contained in the course syllabus.

A common element of the weekly NP teaching review sessions is a learning packet of questions and quiz activities assembled by one of the NP teachers. The primary role of preparing the learning packet is rotated on a weekly basis among the NP teachers, yet it is collaborative. The individual responsible for the learning packet requests critical comments from his/her peers about the most appropriate way to assemble the learning materials. Prior to the weekly review session, the lead NP teacher or tutor distributes the packet without answers to the other tutors for their review and completion. Development of the learning packet typically requires 2–3 hours and involves selecting appropriate resources, reading and reviewing chapters from course textbooks, perusing previous anatomy notes, and preparing review questions and answers. NP teachers select "big concept" anatomical relationships they deemed foundational for understanding concepts to be covered in subsequent courses during the remainder of the DPT curriculum. NP teachers do not lecture during the review sessions. The learning packet provides talking points to stimulate critical thinking and discussion of anatomical topics.

One of the most popular NP teaching sessions occurs when Year 2 NP teachers design a practice anatomy laboratory practical for Year 1 students to prepare them for the graded practical. A class of 28 Year 1 students rotate among separate stations with structures identified on human cadaveric specimens or bones. Year 1 students are allowed 90 seconds per station to identify tagged structures. At the completion of the examination, an answer key is available, so Year 1 students can check their answers. By setting up a simulated laboratory practical, Year 2 NP teachers prepare their Year 1 counterparts for a challenging test experience and simultaneously reduce Year 1 students' test anxiety.

Year 1 students perceived the review sessions as well organized and informative. Year 2 NP tutors successfully reinforce anatomical concepts presented previously by the anatomy course faculty as they design the review packets according to weekly themes printed in the anatomy course syllabus. Year 1 students comment that Year 2 tutors introduced mnemonic devices that facilitate retrieval of functional anatomical relationships.

Communication and teaching are professional skills which develop during the anatomy NP teaching experience by Year 2 tutors. The Year 2 students recognize the weekly peer review sessions provide valuable opportunities to try various strategies to engage Year 1 students. Year 2 students use words such as "awesome," "empowering," "beneficial," and "two-way learning experience" to describe the benefits of NP teaching. All NP tutors acknowledge that the experience provides an opportunity to (1) accrue teaching experience, (2) deepen their understanding of gross anatomy, (3) assemble meaningful learning materials, and (4) strengthen time management skills. Furthermore, NP teaching is believed to ease the demands of modern curricula by using smaller groups despite the existence of low faculty-to-student ratios [13, 14].

The Use of an Audience Response System

An audience response system (ARS) has become popular among educators in medicine and the health professions as a means to engage listeners during a presentation [15–19]. The educator can periodically display a question and each member of the audience can respond anonymously using an ARS. Proponents of ARS contend that the technology (1) enhances student attention and learning during a lecture, (2) motivates reluctant students to interact since responses are anonymous, and (3) permits the presenter to rapidly assess the audience's knowledge, attitudes, or opinions [20]. According to the literature, the primary setting for the use of the interactive ARS system has been a classroom during a didactic presentation. Recently, we reported the usefulness of ARS during weekly planned non-lecture interactive near-peer (NP) teaching sessions during a semester course in gross anatomy for physical therapists [21].

The unique feature of each NP teaching/tutoring session is a nongraded 12–15 question quiz strategically placed at the beginning of the review session and taken by those Year 1 students in attendance. Year 2 physical therapy students construct the quiz using ARS technology and objectives located in the course syllabus. Questions focus on key anatomical relationships covered during the most recent lecture and laboratory sessions. During the ARS quiz, each NP teacher individually provides clarification and further instruction for his/her authored questions when indicated by the response results or prompted by questions from Year 1 DPT students.

Utilizations of an audience response system:

- Help students to review anatomical facts.
- Create interactive experience.
- Provide nongraded quizzes as formative feedback.
- Provide feedback to teachers on students' comprehension of important anatomical concepts.

Inserting the ARS quiz at the onset of a NP teaching session provides students with immediate formative feedback regarding his/her mastery of key anatomical relationships emphasized during the most recent classroom lectures and laboratory sessions. These self-assessment features permit each student to measure his/her preparedness for the upcoming examination and compare his/her performance to that of classmates. The majority of Year 1 DPT students perceives the anatomy sessions to be appealing because the ARS quiz makes effective use of their time and promotes verbal interaction between themselves and Year 2 NP teachers. NP teachers recognize that the ARS quiz provides an opportunity to (1) estimate Year 1 students' level of understanding of anatomical concepts, (2) gain confidence in personal preparation for teaching sessions, and (3) acquire indispensable teaching skills for subsequent use in the physical therapy clinical setting.

Clinical Skills in the Human Anatomy Laboratory

Performing specific clinical examination skills in gross anatomy is an activity that can be utilized to reinforce anatomical concepts. Learning of anatomy is a process of initial exposure to material followed by deeper understanding as a result of application [22]. Performing physical examination clinical tests learned in patient skills classes on human cadavers can challenge students to consider the association of anatomy with the specific tests. To illustrate the process, a clinical test is presented to the class. The mechanics of the test along with findings representing a positive or negative test are discussed. Next, laboratory activities are discussed, and then students break into small groups to conduct the specific activities. As an example, varus and valgus stress tests and the anterior drawer and Lachman's test are demonstrated and discussed. Students then break into smaller groups at their assigned cadaver. Given that the anterior cruciate ligament serves as a primary restraint to anterior translation of the tibia on the femur and as a secondary restraint to a valgus stress at the knee, structures can be sectioned in a sequential manner to demonstrate this function. After review of the regional anatomy, students identify ligaments of the knee. Valgus stress testing in 0° and 30° of knee flexion and the anterior drawer and Lachman's test are performed to demonstrate how the ligamentous structures are positioned to maintain stability of the knee with performance of the clinical tests. Next, the medial collateral ligament is sectioned. The same tests are performed, and the student is able to see that in extension, there is minimal increased medial opening with a valgus-producing force but at 30° there is marked opening. This illustrates the role of the anterior cruciate ligament as a secondary restraint to valgus stress. Lastly, the anterior cruciate ligament can be sectioned, and now medial opening with a valgus stress test occurs at both 0° and 30°, and both anterior drawer and Lachman's tests are positive. While the embalmed and dissected cadavers are relatively stiff, thus presenting a different feel than "normal" tissue, the concept of the various tests is well illustrated with this activity.

Performance of the clinical tests requires the student to review the pertinent anatomy, expose the anatomy through detailed dissection, and finally section ligamentous structures involved in the specific clinical test. In clinical skills classes, students typically perform tests on each other. This experience provides an excellent opportunity to experience a "normal" test; however, students have limited opportunities to experience pathology. While performing tests on cadavers is not the same as a patient in the clinical setting, students are able to get an appreciation of anatomic deficiencies which contribute to a positive test. In addition to the knee, this activity works well for ligament tests at the shoulder, elbow, thumb and fingers, and the ankle.

Painting and Drawing Anatomical Relationships

Hands-on approaches to learning anatomy, such as with body painting, drawing on white boards, and modeling with clay, have been shown to improve anatomical knowledge and increase student confidence [23]. Body painting is an activity used to facilitate understanding of anatomical concepts. Painting may be done on the skin or on clothing. Students tend to see body painting as an enjoyable and useful activity [24, 25]. Body painting for more sensitive regions of the body may be uncomfortable for some. We have used a combination of body painting and simple board drawings to reinforce anatomical relationships. The hand is an anatomic region that is challenging for some. We have utilized body painting on disposable gloves for learning purposes. Activities we have used include drawing the relationship of the extensor tendons coursing through the tunnels on the dorsum of the wrist and drawing the carpal bones, the vascular supply to the hand, nerve distribution in the hand, the intrinsic muscles of the hand, and the extensor mechanism. Each student can draw on their own "hand" or a partner's, while the instructor presents the anatomy of the hand and provides guidance in the drawing activity. One advantage of using gloves is the ability to "erase" marking by simply applying a new glove.

Areas such as the face, chest, and proximal thigh are more sensitive regions which may make students uncomfortable during body painting. Other regions may be difficult to draw on the skin. For these regions, we opt to use a whiteboard or drawing paper. Examples of these drawing activities include drawing the femoral triangle with the associated boundaries and contents, drawing the entire vascular supply to the lower extremity starting at the external iliac artery, drawing the anterior and medial thigh musculature, and drawing the compartments of the leg and their respective contents. These activities challenge the student to consider anatomy in terms of how structures relate to each other as they create their drawings.

Body painting may also be used to demonstrate the relationship of anatomy with orthopedic clinical tests and as a competency check for palpation skills as these skills are presented in the classroom setting. As an example, students can draw the lateral ligaments of the ankle on a partner's skin. Instructors can readily check if the placement of the markings is correct to confirm that a student's palpation is correct. With the accurate placement of the markings, the student is able to visualize the rationale behind specific technique tests such as the anterior drawer to test the anterior talofibular ligament and the inversion stress test to test the calcaneofibular ligament.

Interprofessional Education

Interprofessional education (IPE) in anatomy is joint learning between students in different health professions. IPE proposes to indoctrinate or engage students in collaborative learning in order to prepare them for collaborative practice [26]. Gross anatomy is particularly favorable for reciprocal learning between medical and physical therapy students as it is a foundational course in the curriculum of both programs. Medical and physical therapy students may apply anatomical knowledge differently in their practice settings with general medical practitioners more focused on vital organs and therapists focused on the various systems contributing to movement. As both practitioners must ultimately have knowledge of all systems, the education setting provides a prime opportunity to share expertise on complimentary areas and to learn from other professionals in a mutually respectful manner. For example, physical therapists are required to practice with an extensive knowledge of the musculoskeletal and peripheral nervous system anatomy, whereas physicians typically require more familiarity with the abdominal viscera. Therefore, curricula will have more emphasis placed on each of these respective areas. Peer learning allows each group of students to learn with and about another profession and to develop confidence in teaching in their area of expertise. This fosters mutual respect and knowledge of the other's repertoire of anatomical knowledge and passion. Most students acknowledge that IPE experience will likely help them to work interprofessionally in their careers. A benefit of IPE in anatomy is cost-effectiveness as space, faulty, and resources may be shared [27].

IPE begins with interdisciplinary faculty planning. Faculty teamwork is essential, not only in terms of modeling teamwork for students but also to foster successful experiences. Early tasks include identifying strengths and areas for opportunity in the individual programs, faculty availability, space, and laboratory resources. Clear identification of overlapping and unique knowledge within individual disciplines reveals opportunities for new objectives that may be achieved through joint interaction. Planning, assessment, and objectives together help to shape and organize the desired learning environment and experience. An introduction and orientation of interdisciplinary students through an informal social event can be used to encourage cross-interaction and collegiality between students [26]. Collection of student feedback through post-course surveys to assists faculty in planning future experiences.

There are several practical ways in which to employ IPE in a gross anatomy course. One way, if space and curricula allow, is to simply have both groups of students participate in the same lecture and laboratory course. This is often difficult to coordinate so other approaches may be more feasible. A flexible approach is to provide students with side-by-side peer learning opportunities in which students concurrently dissect selected body regions. One student group can dissect a body area and then provide an introductory tutorial to the other group before their dissection. For example, physical therapy students who have dissected the complex anatomy of the hand can demonstrate this to the medical students who may cover this area very quickly in didactic fashion. In return, the medical students can prosect the digestive organs and teach this area of anatomy to the physical therapy students.

Opportunity for IPE also exists through the use of professionals from differing disciplines. With physical therapy students, pathologists and surgeons as examples can bring not only expertise to the anatomy laboratory but a unique professional perspective. For example, pathologists are able to provide insight into a pathological kidney which may be present in a cadaver to which they are assigned. Through interactive discussion, students gain understanding of the clinical diagnosis and acquire insight into how pathologists assist in making the diagnosis. Surgeons may relate their approach to repairing an aneurysm of the abdominal aorta or performing a total knee arthroplasty. The student is energized by anticipating a condition that he or she may later treat or encounter in the clinic. Likewise, the surgeon can appreciate the enthusiasm of the future therapists who will be part of the care team.

Implementing interprofessional education:

- Identify strengths and weaknesses of individual programs.
- Identify opportunities for shared learning.
- Derive learning objectives.
- Organize experience to meet objectives.
- Orientate students to interdisciplinary teams or experience.
- Collect student feedback.

Conclusion

Teaching anatomy to physical therapy students involves a variety of teaching pedagogies whereby students teach anatomical concepts and relationships to classmates and students from other health professions using readily available resources. Practical teaching sessions incorporated information previously provided within a formal lecture format. During the semester course in gross anatomy, physical therapy students perceived the teaching opportunities valuable because they fostered communication skills with future patients and professional colleagues.

References

1. Mattingly GE, Barnes CE. Teaching human anatomy in physical therapy education in the United States: a survey. Phys Ther. 1994;74:720–7.
2. Thomas KJ, Denham BE, Dinolfo JD. Perceptions among occupational and physical therapy students of

a nontraditional methodology for teaching laboratory gross anatomy. Anat Sci Educ. 2011;4:71–7.
3. APTA. American Physical Therapy Association. Commission on accreditation in physical therapy education (CAPTE). Evaluative criteria for PT programs. 2013. http://www.capteonline.org/AccreditationHandbook.
4. Wright SJ. Student perceptions of an upper-level undergraduate human laboratory course without cadavers. Anat Sci Educ. 2012;5:146–57.
5. Roberts DH, Newman LR, Schwartzstein RM. Twelve tips for facilitating Millennials' teaching. Med Teach. 2012;34:274–8.
6. Allen AR, Boraks N. Peer tutoring: putting it to the test. Read Teach. 1978;32:274–8.
7. Kinirons AS, Reddin VM, Maguffin J. Effects of alternating dissection with peer teaching and faculty prosected cadaver demonstrations in a physical therapy and occupational therapy gross anatomy course. Anat Sci Educ. 2019;12:468–77
8. Youdas JW, Krause DA, Hellyer NJ, Hollman JH, Rindflesch AB. Perceived usefulness of reciprocal peer teaching among doctor of physical therapy students in the gross anatomy laboratory. J Phys Ther Educ. 2007;21:31–8.
9. Krych AJ, March CN, Bryan RE, Peake BJ, Pawlina W, Carmichael SW. Reciprocal peer teaching: students teaching students in the gross anatomy laboratory. Clin Anat. 2005;18:296–301.
10. Bentley BS, Hill RV. Objective and subjective assessments of reciprocal peer teaching in medical gross anatomy. Anat Sci Educ. 2009;2:143–9.
11. Neumann DA. Hip. In: Neumann DA, editor. Kinesiology of the musculoskeletal system. Foundations for rehabilitation. 2nd ed. St. Louis: Mosby; 2010. p. 465–519.
12. Alvarez S, Schultz JH. Professional and personal competency development in near-peer tutors of gross anatomy: a longitudinal mixed-methods study. Anat Sci Educ. 2019;12:129–37.
13. Durán CEP, Bahena EN, Rodríguez MG, Baca GJ, Uresti AS, Elizondo-Omaña E, et al. Neer-peer teaching in an anatomy course with a low faculty-to-student ratio. Anat Sci Educ. 2012;5:171–6.
14. Sytsma TT, Haller EP, Youdas JW, Krause DA, Hellyer NJ, Pawlina W, et al. Long-term effect of a short interprofessional education interaction between medical and physical therapy students. Anat Sci Educ. 2015;8(4):317–23.
15. Copeland HL, Longworth DL, Hewson MG, Stroller JK. Successful lecturing: a prospective study to validate attributes of the effective medical lecture. J Gen Intern Med. 2000;15:366–71.
16. Nasmith L, Steinert Y. The evaluation of a workshop to promote interactive learning. Teach Learn Med. 2001;13:43–8.
17. Schakow TE, Chavez M, Loya L, Friedman M. Audience response system: effect on learning in family medicine residents. Fam Med. 2004;36:494–504.
18. Streeter JL, Rybicki FJ. Education techniques for lifelong learning. A novel standard-compliant audience response system for medical education. Radiographics. 2006;26:1243–9.
19. Alexander CJ, Crescini WM, Juskewitch JE, Lachman N, Pawlina W. Assessing the integration of audience response system technology in teaching of anatomical sciences. Anat Sci Educ. 2009;2:160–6.
20. Collins J. Audience response systems: technology to engage learners. J Am Coll Radiol. 2008;5:993–1000.
21. Wait KR, Cloud BA, Forster LA, Jones TM, Nokleby JJ, Wolfe CR, et al. Use of an audience response system during peer teaching among physical therapy students in gross anatomy: perceptions of peer teachers and students. Anat Sci Educ. 2009;2:286–93.
22. Smith CF, Mathias HS. What impact does anatomy education have on clinical practice? Clin Anat. 2011;24:113–9.
23. Nicholson LL, Reed D, Chan C. An interactive, multimodal anatomy workshop improves academic performance in the health sciences: a cohort study. BMC Med Educ. 2016;16:7.
24. McMenamin PG. Body painting as a tool in clinical anatomy teaching. Anat Sci Educ. 2008;1:139–44.
25. Skinder-Meredith AE. Innovative activities for teaching anatomy of speech production. Anat Sci Educ. 2010;3:234–43.
26. Hamilton SS, Yuan BJ, Lachman N, Hellyer NJ, Krause DA, Hollman JH, et al. Interprofessional education in gross anatomy: experience with first-year medical and physical therapy students at Mayo Clinic. Anat Sci Educ. 2008;1:258–63.
27. Mitchell BS, McCrorie P, Sedgwick P. Student attitudes towards anatomy teaching and learning in a multiprofessional context. Med Educ. 2004;38:737–48.

Research Methods in Anatomy Education

51

Zubair Amin and Hamza Mohammad Abdulghani

Context of Research in Anatomy Education

The traditional view of scholarship with an inclination toward pure biomedical research and scientific publications in peer-reviewed journals has been challenged by many [1, 2]. A widely acclaimed expanded view of scholarship includes, in addition to scholarship of discovery or scientific research in common sense, scholarship of integration, scholarship of application, and scholarship of teaching [1]. Major universities now promote educational scholarship through a variety of means including teaching innovation funds, teaching awards, and creation of a separate career track for educators [2].

Anatomy educators form distinct groups with different yet somewhat overlapping needs and orientation to research. In a typical academic anatomy department affiliated with a university or a medical school, there are biomedical researchers with a secondary interest in educational research and anatomy educators with a primary focus in educational research. It is important to recognize that the needs, intensity, and expectations of educational research from these two groups are different and are higher for the anatomy educators.

An anatomy educator needs to prove his/her worth in scholarship through substantial educational research and might be expected to develop a planned, programmatic educational research on a long-term, sustainable basis. He/she might start with manageable, locally relevant studies and progress to more rigorous research methodologies that would allow generalization of study findings beyond his/her institute. In other words, the findings and conclusions from the research should generate interest from academics outside the researcher's own institute. The applicability of findings and conclusions therefore must be able to stand rigorous scrutiny.

We propose that action research and need-driven developmental research are attractive starting points for the vast majority of anatomy educators who are about to embark on educational scholarship. Educational action research is highly relevant to the local context and easily accessible to beginners as well as advanced researchers. We would argue that this should form an essential element of one's research scholarship repertoire. Therefore, the major focus in this chapter would be on educational action research with specific reference to the broad discipline of anatomy.

Z. Amin (✉)
Department of Pediatrics, Yong Loo Lin School of Medicine, National University of Singapore, National University Health System, Singapore, Singapore

Department of Neonatology, National University Hospital, National University Health System, Singapore, Singapore
e-mail: paeza@nus.edu.sg

H. M. Abdulghani
Department of Medical Education, Assessment and Evaluation Center, King Saud University, Riyadh, Saudi Arabia

L. K. Chan, W. Pawlina (eds.), *Teaching Anatomy*, https://doi.org/10.1007/978-3-030-43283-6_51

The primary purpose of educational action research is to "systematically investigate one's own teaching/learning facilitation practice, with the dual aim of improving that practice and contributing to theoretical knowledge in order to benefit student learning" [3]. Institutions and disciplines always value systematic analysis and review of teaching and learning to further improve educational practices and to inform others of such improvements. Such research provides academics an opportunity to engage in deeper pedagogical exploration to understand the process of learning and teaching.

Therefore, it is useful for researchers in this field to start with an introspection and critical reflection to identify our own and institutional needs, relationship with other peers, and possible utility and applicability of the attempted research. How we relate to a larger ecosystem, be it an institution or the subject domain, is important to determine how our research can contribute to the knowledge in anatomy education. A systematic investigation might provide results beneficial to an individual (e.g., why are my students not attending anatomy dissections?), to the department or discipline (e.g., does integration of anatomy with other subjects dilute students' core knowledge in clinical anatomy?), or to a larger organizational issue (e.g., can the impact from shortage of anatomy educators be minimized through employment of students trained as teachers?). It is imperative that we, as budding researchers, define the target audience or stakeholders of the action research. Early determination of the target would allow us to channel the message appropriately and to bring organizational changes necessary for long-term improvement.

Possible targets of educational action research:

- Student: an individual or a class
- Teacher: an individual or peers
- Curriculum: module or global, instructional and assessment methods, impact, and quality assurance issues
- Organization: departments, institutions, or professional bodies

Methodological Considerations in Educational Research

Educational research might include qualitative methods, semiqualitative methods, quantitative methods, or a combination of them. Qualitative methods include interviews, focus group discussion, narratives, observation, nominal group technique, and others. Although, in pure biomedical science literature, qualitative research is relatively uncommon, for many research questions in education, qualitative research methodologies are the only feasible way to conduct research. For logistic, pedagogical, and ethical reasons, it may not be possible to conduct quasi-experimental and experimental research, whereby students are exposed to a possibly inferior educational intervention.

Quantitative methods are more familiar to biomedical researchers working with natural sciences. Quantitative research methods typically include experimental or quasi-experimental design with narrower research questions, stricter control of variables, standardized interventions, and numeric data as outcomes. Methodological choice would depend upon the nature of the questions to be answered in the research, feasibility of the study, ethical and legal considerations, and familiarity of the researcher with a particular method. It is imperative to note that qualitative and quantitative research methods are not exclusionary; both research methods can be used together in parallel or in sequence to ask similar questions.

Sampling is an important issue in a qualitative as well as quantitative research. In a qualitative research, the emphasis is not on the number of the participants who will be included in the study but on the appropriateness of the sample to be included in the study. As the sample size in qualitative study is relatively small, the effect of having a biased group of participants in the sample is likely to be more magnified in qualitative research methods. Thus, exclusion and inclusion criteria of the participants need to be clearly defined. The validity and applicability of data generated from the qualitative research are very much dependent on the careful selection of the participants. See Table 51.1 for relative comparison between qualitative and quantitative methods.

Table 51.1 Differences between qualitative and quantitative methods

	Qualitative methods	Quantitative methods
Question	Exploratory	Confirmatory
Scope	Wide	Narrow
Hypothesis	Ill-defined	Well-defined
Setting	Naturalistic	Experimental
Variables	Uncontrolled	Manipulated
Sample size	Small	Large
Data	Texts	Numeric
Conclusion	Investigate causality	Suggest causality

Qualitative research methodologies:

- Action research
- Historical method/narrative research
- Ethnography
- Phenomenology
- Discourse analysis
- Grounded theory

Qualitative research data collection methods:

- Nominal group
- Interviews
- Focus groups
- Case studies
- Text analysis
- Observation

Examples of qualitative research questions:

- How can teachers emphasize the importance of anatomy to their students?
- How do new medical students perceive cadaveric dissection?
- What do students think about the benefits of learning anatomy for future practice?
- How do students learn anatomy in a small group with problem-based learning scenarios?
- How does linguistic ability of the students influence learning anatomy?
- How do students react to difficult anatomical terminology?

From Exploration to Confirmation

Educational research can be diverse; the classification and categorization depend on the nature of the research, context of research, and target participants. It is natural that in some situations more than one classification can be employed.

A useful approach to understand the nature of educational research might be based on the types of research questions that the researchers aim to address: whether the question is exploratory or confirmatory in nature. An exploratory research is often an appropriate starting point when the phenomenology under investigation is poorly understood or there is no strong theoretical underpinning. In the absence of such grounding, exploratory research deals with uncontrolled variables; there are no fixed educational interventions that can be employed in the research. As an extension, there could not be a hypothesis. Therefore, exploratory research is useful for emerging phenomena where a suitable educational intervention does not exist.

Let us consider an example from the literature. Pandey and Zimitat reported a research exploring medical students' approach to learning anatomy [4]. Students' own description of successful approaches to learning was correlated with their own approach to learning and quality of learning. This was a single institution-based study. They used multiple tools and data sources including study process questionnaire, structure of observed learning outcomes, and grades. They found that students' learning of anatomy involves various combinations of memorization, understanding, and visualization. This study could not have a strong theoretical model at the inception as there were little published data regarding students' approach to learning anatomy. Their research opens up the further possibility of confirmatory research from other institutes with similar situations.

Confirmatory research, on the other hand, aims to further validate an already known intervention. Researchers are already aware of potential educational interventions, the controlled and uncontrolled variables, and a possible hypothesis. The idea of confirmatory research often originates from exploratory studies or studies with limited number of participants or studies that originate in one or few institutes. Therefore, there is a need to field test the hypothesis with larger cohort of research participants with more diverse background.

Let us consider an example of confirmatory research from the anatomy education. Nnodim explored the superiority of prosection over more conventional cadaveric dissection with first-year preclinical students [5]. The control group dissected the cadaver and learned anatomy in a conventional manner, whereas the experimental group was excluded from dissection. The experimental group worked with prosected specimens using other innovative learning methods. The outcome measures include knowledge and practical tests and an opinion survey. The researcher also tracked the time spent on the learning. The researcher found that the experimental group, despite spending less time in learning anatomy, had better outcomes as measured by knowledge and practical tests. The hypothesis was expected from similar prior studies and confirmed again in this study.

To further illustrate the nature of exploratory and confirmatory research and how they can be applied in anatomy education, we have included two scenarios. In the first, SMA, a young educator, was confronted with the problem of students' unwillingness to access e-learning materials. In the second, BZA, an established educator, wanted to demonstrate superiority or at least equivalence of e-learning over conventional teaching and learning.

Example 1 SMA recently joined the anatomy department of a renowned university. Soon after joining the department, she was introduced to the vast body of e-learning materials in anatomy that are available freely in the web. Despite being in an institute with off and on campus Internet connectivity and a high laptop and smartphone usage, SMA was intrigued by the fact that students tend to favor printed books over e-learning materials. As she started to ponder the issue, she decided to engage in informal discussions with colleagues and students. Soon she realized that students have very different reasons not to prefer e-learning materials over printed books. She joined hands with an educator to explore the phenomenon in greater detail. Together, they decided to employ focus group discussion with an invited group of students and teachers to understand the barriers, enablers, and potential motivators for e-learning material usages in her institute.

In SMA's situation, the primary driver for research is exploratory in nature. The phenomenon in question is ill-defined; the variables and their relationship are poorly understood. There is no fixed hypothesis. Moreover, although there are similar studies in existing literature from other countries and institutes, she was reasonably hesitant to extrapolate the findings to her own situation as she recognizes from her discussions with students and teachers that there are certain unique cultural and sociological norms that cannot be ignored. For example, students' prior experience, validation from the teachers, relationship to the curriculum, and suitability of e-learning materials to the assessment are among some of the factors that determine students' usage of web-based e-learning.

Example 2 BZA has developed an e-learning module on locomotion with her colleagues. The newly developed module has self-study materials in the form of texts, audiovisuals, and self-assessment questions. The module adheres to the existing curriculum in terms of content coverage, sequence, and end-of-module assessment. She is interested to find out whether this e-learning module is better than the conventional classroom-based learning experience of the students. Her background literature search revealed many similar studies with variable results. She decides to conduct a confirmatory study with two experimental cohorts. There are several possible outcomes that can be measured to prove the worthiness of an e-learning module over the existing curriculum. Potential outcome measures might include the following:

(a) Usage rate of e-learning module over traditionally delivered module
(b) User satisfaction data comparing e-learning module and traditionally delivered module

(c) Result of end-of-module assessment between the two groups
(d) Extracted result on locomotion system from standardized national or international exam
(e) Performance of students on OSCE stations on locomotion during orthopedics posting

BZA decides to use a composite of several outcome measures as she is interested not only to know how students like her module but also how they perform in standardized examinations.

BZA's experimental design also calls for controlling of several variables such as comparability of prior academic performance between e-learning group and conventional module group. Can you identify other variables? How can you control these variables?

Ethical Considerations

Educational research, whether it is qualitative and quantitative, typically involves human subjects. As such, ethical consideration and approval from competent authority prior to the start of data collection are mandatory. An institutional review board may determine that a particular educational research proposal does not warrant a thorough review and may expedite the review process. However, that is a judgment of the institutional review board that the educational researcher must respect.

There are other ethical considerations to take note of. For example, student-related data such as test results, psychological profile, and socioeconomic data are highly sensitive in nature, and confidentiality must be maintained throughout. Students may feel compelled to participate in a study especially if the study is conducted by a faculty member of the same university. Direct observation and interview may open up many sensitive topics that should not be made public without prior consent. Finally, reputable medical educational journals now require ethical considerations and approval, if needed, prior to publication.

Descriptive, Correlational, and Causal Questions

Descriptive Questions

Descriptive questions are important in educational planning, resources identification, and allocation. For example, descriptive questions might compare the status of education in an individual institute or a selected group to a normative reference group or an international benchmark. Methods that are employed to answer descriptive questions might include semi-structured or structured interviews, internally or externally developed questionnaires, and attitudinal scales such as Likert's.

Examples of descriptive questions in anatomy education can include the following:

- How has the allocation of teaching hours in anatomy education evolved since the introduction of problem-based learning (PBL) curriculum in various medical schools? How does the allocation of teaching hours in anatomy vary between PBL and traditional medical schools? Do PBL-based medical schools employ more clinicians to teach anatomy as compared to non-PBL medical schools? Do the clinicians teaching in the anatomy course teach pure morphological anatomy or applied anatomy?
- How many medical schools in a given country routinely use cadavers and cadaveric dissection to teach anatomy? Is there a difference between private medical schools and public (governmental) in terms of their intensity of usages of cadavers in anatomy education? In case of a shortage of cadavers, how do the medical schools substitute cadaver-based dissections?
- What is the role of anatomy education in an integrated curriculum? How has anatomy education responded to curricular integration? What are some of the new roles adopted by the anatomy departments in anatomy education? How does the faculty perceive attempted curriculum integration? What motivates anatomy teachers toward greater curricular integration?

Correlational Questions

Correlational questions often try to establish correlation between two variables. The correlation could be either positive, negative, or zero. While conducting a correlational research, as a researcher, we might be interested to ask first:

1. Is this correlation expected?
2. Can this be explained by our contemporary understanding of education?
3. Are there any unexpected findings? If so, what are possible sources of biases?

Correlational questions can be answered by qualitative methods (e.g., naturalistic observation where researchers collect data unobtrusively without interfering or controlling variables), semiquantitative methods (e.g., survey instruments), or quantitative methods (e.g., examination performance on a standardized test).

We have to be careful in determining causation from correlation. A correlation or an association merely suggests that two variables are related; it does not prove that one variable resulted in change in the other. For example, an association can be found between students' performance in medical schools and the presence of a well-equipped anatomy museum. One can reasonably argue that although an association exists between the two variables, it might have resulted from other confounding factors. For example, medical schools with well-equipped anatomy museums are more likely to have a favorable student-teacher ratio, better dissection room, and students from higher socioeconomic background. Unless these, and other confounders, are addressed, the proposed causation could be merely speculative and often counterproductive.

Examples of correlational questions in anatomy education can include the following:

- Do students in PBL-based curriculum tend to have poorer knowledge in clinical anatomy compared to those in traditional curriculum?
- Does student satisfaction in medical education correlate with the availability and intensity of cadaveric dissection during anatomy teaching? Does student performance in relevant practical examinations improve with increased exposure to cadavers?
- Do the medical schools which employ clinicians to teach anatomy have students who are better in clinical anatomy? Is there a positive correlation between the number of integrated lectures and student performance in integrated examination?

Causal Questions

Causal research questions often try to establish cause-and-effect relationship in educational interventions. We can employ a variety of tools such as surveys, standardized assessments, and observations to evaluate outcomes of intervention. In causal design, researchers are interested to find out comparative efficiency of an intervention over another. Sometimes, the interest might be in finding out the efficiency of an intervention over no intervention.

It is important to recognize that educational outcomes rarely result from a single intervention. Educational outcomes are often dependent on multiple factors which might or might not originate in school or formal classroom settings. For example, we are interested to find out whether introduction of an innovative learning strategy, such as a web-based self-learning module, resulted in better knowledge acquisition as compared to naturalistic state. We can randomize the class into an intervention or a no-intervention group. However, it is very likely that any shortcoming of web-based learning would be compensated by the students through more rigorous learning utilizing other resources. There will be frequent and inevitable contamination between the two groups as well.

Examples of causal questions in anatomy education can include the following:

- Does student knowledge in clinical anatomy improve with the introduction of case-based discussion as compared to no case-based discussion? The hypothesis is that introduction of case-based discussion would result in better knowledge in clinical anatomy.

- Is student satisfaction in classroom teaching, as measured by a standardized questionnaire, similar between professional anatomy teachers and trained student-teachers? The hypothesis is that the trained student-teachers deliver equally satisfying teaching to the students as compared to professional teachers.
- Does an anatomy teaching module designed around prosected specimens result in equivalent test scores in standardized assessment as compared to the cadaveric dissection-based module? The hypothesis is that the two modules lead to similar results, and thereby the prosection-based module can be an alternative to the dissection-based module.

To do:

- Think of your local context; talk to the colleagues and students.
- Identify issues that need further exploration.
- Perform an extensive literature search and gather evidence related to the issues under investigation.
- Develop relevant questions, and classify them according to the abovementioned schema.
- Choose the most appropriate method for the questions.
- Consider the logistics and other requirements, and refine the questions and method, if needed.

Conclusion

The changing landscape of anatomy education provides us with an exciting opportunity to engage in meaningful and impactful scholarly research. For the vast majority of anatomy educators, need-driven action research is a pragmatic and logical stepping stone. Contextual variables and uniqueness are important considerations to explore before devising most pertinent research questions for the stakeholders. The nature of research questions is the most important factor to determine the choice of research methodology.

References

1. Boyer EL. Scholarship reconsidered: priorities for professoriate. Princeton: Carnegie Foundation for Advancement of Teaching; 1990.
2. Trigwell K, Marin E, Benjamin J, Prosser M. Scholarship of teaching: a model. High Educ Res Dev. 2000;19(2):155–68.
3. Norton LS. Action research in teaching and learning a practical guide to conducting pedagogical research in universities, vol. 59. Oxon: Routledge; 2009.
4. Pandey P, Zimitat C. Medical students' learning of anatomy: memorisation, understanding and visualisation. Med Educ. 2007;41:7–14.
5. Nnodim JO. Learning human anatomy: by dissection or from prosection? Med Educ. 1990;24:389–95.

Further Reading

Barbour RS. Making sense of focus group. Med Educ. 2005;39(7):742–50.. In depth discussion on focus group, a popular qualitative method in medical education.

Bordage G. Conceptual frameworks to illuminate and magnify. Med Educ. 2009;43:312–9.. A scholarly discussion highlighting the importance of thinking of conceptual framework in research.

Denzin NK, Lincoln YS. Handbook of qualitative research. 2nd ed. Thousand Oaks: Sage Publications; 2000.. A superb introduction to theory and practice of qualitative research.

Hanson JL, Balmer DF, Giardino AP. Qualitative research methods for medical educators. Ambul Pediatr. 2011;11:375–86.. A primer in qualitative research methodology. Available freely in the web at http://www.ambpeds.org/site/vlp/ESPMeetings/documents/HansonBalmerGiardinoQualMethods_Acad%20Peds_FINAL.pdf.

Scholarship of Teaching and Learning in Anatomy

52

Valerie Dean O'Loughlin

Ernest Boyer's 1990 publication, *Scholarship Reconsidered*: *Priorities of the Professoriate* [1], was among the first to state that academic scholarship should be defined more broadly than just the traditional research contributions. He listed four separate, yet overlapping, criteria or functions of the professoriate:

- The scholarship of discovery (i.e., what is traditionally defined as "research")
- The scholarship of integration (whereby connections are made across disciplines and how research and subject matter fit into a larger picture) (note, some consider this criterion to be an extension of the scholarship of discovery)
- The scholarship of application (which includes service to the discipline and applying the knowledge gained to real-life problems)
- The scholarship of teaching (whereby excellent teachers utilize effective pedagogical techniques to assist their students in becoming critical, lifelong learners)

Boyer emphasized that all four criteria (and not just the single "discovery" criterion) should be used when evaluating the academic.

V. D. O'Loughlin (✉)
Department of Anatomy and Cell Biology, Indiana University School of Medicine - Bloomington, Bloomington, IN, USA
e-mail: vdean@indiana.edu

Boyer noted that the scholarship of teaching also should refer to the review and assessment of teaching and the public dissemination of those findings. And in fact, in order for educational research to be viewed equal to more traditional scientific research, Boyer [1] stated that educational research must be presented publicly and/or published in peer-reviewed journals.

Since Boyer's original work, many academics have tried to further define the scholarship of teaching and develop models for its assessment. Glassick et al. [2] refined the description of exemplar scholarship of teaching works to include the following characteristics: clear goals, appropriate and rigorous methods, clear results, and reflective critique. Lee Shulman, former president of the Carnegie Foundation for the Advancement of Teaching, referred to "teaching as community property" [3] and has used this phrase to encourage educators to present their educational research findings both within and across disciplines. Trigwell et al. [4] surveyed 20 academics at a variety of Australian universities and through these responses developed a model to describe scholarship of teaching. This model states that individuals engaging in the scholarship of teaching are well informed by the general and discipline-specific literature on teaching, assess their own teaching, reflect on their teaching from the view of the student,

L. K. Chan, W. Pawlina (eds.), *Teaching Anatomy*, https://doi.org/10.1007/978-3-030-43283-6_52

and communicate their findings to their peers. The phrase "scholarship of teaching" has been expanded to "scholarship of teaching and learning (SoTL)" [5, 6] to include assessment and understanding of student learning.

Thus, an anatomy educator is not considered a scholar of teaching and learning unless she/he assesses the educational process and makes these findings public in the form of a presentation or a publication [5–9]. The previous chapter in this text and other publications [10, 11] provide extensive information about how an anatomy teaching question may be developed into an educational research project. In this chapter, I provide information about local, national, and international research opportunities in anatomy educational research. I describe and compare professional societies active in anatomy education. This chapter ends with a list of journals known to publish anatomy educational research manuscripts and guidelines for developing a suitable manuscript for review.

Local Educational Research Opportunities

Academics are familiar with the research methods for their specific discipline but may be less comfortable with developing an educational research project. If your institution has a faculty development center or a center for teaching and learning, contact them and ask them for advice as to how to proceed with your project. The faculty and staff affiliated with these centers are trained to help faculty assess their teaching and may be valuable collaborators. In addition, these centers may offer workshops on assessing teaching or even host scholarship of teaching talks. For example, Indiana University's Center for Innovative Teaching and Learning (CITL) has multiple instructional consultants on staff to assist with assessing student learning and developing an educational research project [12]. In addition, CITL oversees the university's scholarship of teaching and learning (SoTL) program [13], which brings in national SoTL speakers and offers faculty grants for educational research projects. The SoTL program also has invited local faculty members to present their educational research findings. A center for teaching and learning may present you with previously unknown resources on your campus and help introduce you to individuals from other disciplines that may help you with your educational research [11].

A well-developed educational research project often requires the collaboration with individuals from other disciplines [14, 15]. If your institution has a school of education, consider contacting some of the faculty there for advice regarding your project. The education faculty also may have graduate students who are looking for research projects on which to collaborate, and so you may find an enthusiastic coauthor. They may introduce you to qualitative research methods of which you may have been previously unaware. Your school's librarian may assist you with refining your background literature search [10] for your project. The math or statistics department may have faculty or graduate students willing to help you analyze your quantitative data.

As the previous chapter mentioned, it is essential you get human subjects approval from your IRB (Institutional Review Board) for your education research project. Education research projects typically involve the study of students (who are human subjects) or data from these students, hence the need for review. Your institution's IRB likely is located with other campus administrative offices, such as the Dean of Faculties or Sponsored Research Services. Sometimes, a group of institutions reports to a single IRB—again, contact your research offices to determine the specific guidelines and procedures you need to follow. Most education research journals will *not* publish your manuscript unless you can demonstrate that your study was reviewed by an IRB or similar administrative body—so make sure you do not skip this important step.

To do:

- Contact your institution's teaching and learning center (or faculty development center), and meet with a consultant to discuss your project.
- Determine if there are departments and faculty at your institution with whom you could collaborate, such as the math department or the School of Education.
- Work with a librarian at your institution to help run a thorough and efficient literature search related to your educational research project.
- Obtain human subjects approval from your Institutional Review Board (IRB) before you start your data collection.

Once you have completed your anatomy educational research project, consider presenting your findings to your department. Perhaps your department has a weekly colloquium; this would be an ideal first arena to present your findings. If your institution has teaching workshops and teaching presentations (ask your teaching and learning center—they will know!), consider presenting in this forum as well. Educators from other disciplines will be interested in your work and will want to apply your findings to their own classrooms. In so doing, you will be doing your part to promote SoTL work and help it become part of the campus culture at your institution, as Hutchings et al. [6] have promoted.

National and International Educational Research Opportunities

There are a variety of professional societies, and independent conferences support the dissemination of educational research. In this section, I subdivide these opportunities into societies active specifically in anatomy educational research and societies/conferences involved with the scholarship of teaching and learning in general.

Societies Active in Anatomy Educational Research

Within the past 15 years, a variety of new professional organizations have come into existence whose missions involve anatomy teaching and anatomy educational research. In addition, several anatomy societies with a much more extended history have recently expanded their foci to include anatomy education. Below is a list and descriptions of the main national and international societies that are heavily involved in anatomy educational research. (Please note that this list is not all inclusive and does not mention those anatomy societies that are active primarily in traditional scientific research rather than educational research.)

American Association for Anatomy (AAA) (https://www.anatomy.org)
AAA celebrated its 125th anniversary in 2013, and it continues to be a vibrant organization for both traditional scientific research and anatomy education. Its international membership consists of faculty from medical and professional schools as well as undergraduate institutions. AAA is one of the participating societies in the multidisciplinary Experimental Biology (EB) meeting, typically held every April. The AAA meetings have both scientific research and education tracks. Its anatomy education articles formerly were published in the *Anatomical Record, Part B: The New Anatomist.* Since 2008, AAA's education research journal is *Anatomical Sciences Education*, which is published in cooperation with the American Association of Clinical Anatomists and the Human Anatomy and Physiology Society. The journal's aim is to publish teaching innovations, review articles, viewpoints, and research in the anatomies [16]. Since its inception, *Anatomical Sciences Education* is considered one of the top tier journals to publish anatomy educational research.

American Association of Clinical Anatomists (AACA) (https://www.clinical-anatomy.org)
AACA encourages excellence in the teaching of anatomy and in developing research in clinical

anatomy. Its first annual meeting was in 1984, and the society typically meets yearly in June or July. Its official journal, *Clinical Anatomy*, publishes eight times yearly, and its goal is to provide a medium of exchange between anatomists and clinicians. Among the manuscripts the journal publishes are clinical anatomy or medical education articles. (This journal is also the official journal of the British Association of Clinical Anatomists, the Anatomical Society of Southern Africa, and the Australian and New Zealand Association of Clinical Anatomists.)

Anatomical Society (AS) (http://www.anatsoc.org.uk)

The AS was founded in 1887 but more recently promotes the study of both traditional (bench) research and anatomy educational research. Many of its members are from the United Kingdom and are involved in teaching and/or research in higher education. They host numerous symposia and an annual meeting in the United Kingdom, usually in July or August. Their official journal, *Journal of Anatomy*, will publish anatomy education articles only "if they are of a robust quantitative measure or will inform about international education policy" [17].

Australian and New Zealand Association of Clinical Anatomists (ANZACA) (http://anzaca.otago.ac.nz)

ANZACA promotes the science, teaching, and scholarship of clinical anatomy in Southeast Asia. Their annual meeting is held in late November or early December in Australia and New Zealand. Their 2013 conference theme explores technology-enhanced anatomy learning and research. Their journal is *Clinical Anatomy* (described previously with AACA).

British Association of Clinical Anatomists (BACA) (https://baca-anatomy.co.uk/)

BACA's mission is to advance the study and research of clinical anatomy. Founded in 1977, the organization hosts summer (July) and winter (December) meetings in the United Kingdom. Their journal is *Clinical Anatomy* (described previously with AACA).

Which professional societies are the best fit for you?

- Visit the organization's website and read its mission. Does it have educators that are similar to you in terms of student populations? Can you see yourself interacting with members of this organization?
- Examine the society's journal (if they have one). Does this journal publish manuscripts that align with your specific anatomy education interests? Can you see yourself publishing in this journal?
- Find out when and where a society holds its annual meeting. Would you be able to regularly attend their meetings and be an active participant?

Keep in mind many anatomists involved in educational research are members of multiple societies, because no one society addresses all of their interests. In this case, what may limit you is simply the cost of multiple memberships and attending multiple meetings.

Human Anatomy and Physiology Society (HAPS) (https://www.hapsweb.org)

HAPS' mission is to promote excellence in the teaching of human anatomy and physiology. Founded in 1989, its international membership consists of educators from 2- to 4-year colleges, universities, medical schools, and high schools. The majority of educators teach premedical and allied health (undergraduate) students. Their annual conference is held in late May in either the United States or Canada. HAPS also holds several regional conferences throughout the year in the United States. Their publication is *HAPS Educator*, and it publishes peer-reviewed articles three times a year on anatomy and physiology education, as well as educational research. Articles older than 1 year are open access through the organization's website.

International Association of Medical Sciences Educators (IAMSE) (http://www.iamse.org)
IAMSE originated in the late 1980s as part of a special interest group of the Association of American Medical Colleges. This special interest group developed into an independent society in 1997 and now boasts members from over 40 countries. IAMSE's vision is to ensure that medical education is grounded in science and to provide a multidisciplinary forum to discuss teaching innovations. Their meetings typically are held in mid-June to mid-July every year in either North American or international locales. The organization's journal is *Medical Science Educator* and encourages publication about basic science education, clinical teaching, and implementation of educational technology.

Societies Active in Scholarship of Teaching and Learning and Educational Research in General

You should not limit the dissemination of your anatomy educational research to just the anatomical societies. There are several international societies that encourage interdisciplinary communication of educational research. In so doing, these organizations promote the scholarship of teaching and learning (or SoTL) not only by making the educational research findings public but also by encouraging discussion of these topics among disciplines [5, 6].

Asian Medical Education Association (AMEA) (https://www.med.hku.hk/amea/)
AMEA, founded in 2001, is an institution-based association of mainly Asian medical schools. Its mission is to strengthen and promote good pedagogy and research on medical education, through the sharing of information and experience. It organizes an AMEA Symposium once every 2 years (conference dates vary) in different cities in the Asia-Pacific region.

Association for Medical Education in Europe (AMEE) (https://amee.org/home)
Founded in 1972, AMEE's mission is to promote communication and collaboration among medical educators in Europe. Its membership consists of faculty, physicians, administrators, and students in healthcare professions from over 90 countries. Their annual meeting is held in late August or early September in a European city. Their official journal, *Medical Teacher*, is published monthly. It disseminates information about curricular development and assessment in medical education, teaching methods, as well as medical education research.

International Society for the Scholarship of Teaching and Learning (IS-SOTL) (https://www.issotl.com/)
IS-SOTL was founded in 2004 in order to publicize and promote the discussion of scholarship of teaching and learning across national boundaries. The board of directors is composed of members from the United States, the United Kingdom, Europe, Canada, and Southeast Asia and Australia. Their journal, *Teaching and Learning Inquiry*, is published twice a year and disseminates educational research across an international forum. Their conference is held every October and typically alternates between a North American location and an Asian or European location.

Society for Teaching and Learning in Higher Education (STHLE) (https://www.stlhe.ca)
The mission of STHLE is to enhance teaching and learning in postsecondary education both within Canada and beyond. They support and promote educational research in a multitude of disciplines. STHLE has an annual conference every June in Canada.

In addition to these professional societies, there are several international educational conferences not associated with any one particular organization. For example, the Lilly International Conference in College Teaching [18] is held every November at Miami University in Oxford, Ohio. This interdisciplinary conference is attended by both faculty and faculty developers involved in the scholarship of teaching and learning. In addition, smaller versions of the Lilly Conference are held in different US locations throughout the year [19].

Other Conferences That Showcase Medical Education Research

In addition to the conferences associated with the societies mentioned in the previous sections, there are other conferences (not affiliated with a single society) that also showcase medical education research. These conferences offer additional opportunities to present your scholarship of teaching research.

American Association of Medical Colleges (AAMC); Medical Education Scholarship Research and Evaluation (MESRE) Section; Research in Medical Education (RIME) Conference (https://www.aamc.org/members/gea/gea_sections/mesre/)
The RIME conference is held within the larger AAMC conference every fall in the United States. It is organized by the AAMC, Medical Education Scholarship Research and Evaluation (MESRE) section, which is part of the AAMC Group on Educational Affairs (GEA). MESRE is involved with medical education and education research. (*Note, the GEA also offers several US regional conferences on medical education as well.*)

Asia Pacific Medical Education Conference (APMEC) (https://10times.com/apmec-singapore)
This international conference, founded in 2003, is hosted by the Yong Loo Lin School of Medicine at the National University of Singapore. The conference typically is held in January and is designed to provide an international perspective of medical education.

Ottawa Conferences on the Assessment of Competence in Medicine and the Healthcare Professions (http://www.ottawaconference.org)
The Ottawa conference was first established in 1985 and brought presenters from around the world to discuss the assessment of competence in both clinical and nonclinical domains throughout medical education. Contrary to its name, this conference has been held biennially at various international locations since its inception.

Publishing Opportunities in Anatomy Educational Research

In addition to making educational research public via posters and presentations, consider publishing scholarly work so it may be disseminated to an even larger audience [1–5]. The previous chapter of this text and other publications [8, 10, 11] provide guidelines for educational research methods and developing the final publication. Table 52.1 lists the main journals that have published anatomy educational research and scholarship of teaching manuscripts and do not charge a fee for authors to publish in their journal. Impact factors (IF) are listed for the journals that have them, but note that impact factors for education research journals typically are much lower than those for bench science research. The reason for this discrepancy is because there are many/fewer science education researchers than bench researchers, so the number of citations for these articles (which play a role in calculating an impact factor) is much lower. Also note that some of these journals are not indexed in the databases that calculate impact factors, which is why "N/A" is listed for some journal impact factors.

Beware of predatory publishers! The term *predatory publisher* was coined by librarian Jeff Beall in 2010 and refers to an open-access publisher that fraudulently appears to be scholarly and typically does not perform peer review and whose main intent is to accrue as much revenue as possible from authors [20]. Predatory publishers are not concerned with good science and stringent peer review of manuscripts, but with how much money they can make from publication fees.

Note that not all journals that have a publication fee are predatory publishers—several (e.g., *PLOS One, BMC Medical Education*) are reputable, undergo stringent peer review, and charge publishing fees only after an article has been accepted to help recoup the costs of being open access. Predatory publishers own journals whose titles are obscure to most and are more likely to randomly email you, asking you to publish in their journal. If you have never heard of the journal before and the journal requires money from you to publish, stay away! Additional informa-

Table 52.1 Journals that publish anatomy educational research manuscripts (open-access journals that charge the author a fee for publication are not listed in this table)

Journal title	Journal website and instructions for authors	2017 impact factor
Academic Medicine	https://journals.lww.com/academicmedicine/pages/default.aspx	4.801
Advances in Health Sciences Education	https://link.springer.com/journal/10459	2.552
Advances in Physiology Education	https://www.physiology.org/journal/advances	1.981
American Biology Teacher	https://nabt.org/Resources-American-Biology-Teacher	0.271
American Educational Research Journal	https://journals.sagepub.com/home/aer	2.462
Anatomical Sciences Education	https://onlinelibrary.wiley.com/journal/19359780	3.114
BioScience	https://www.aibs.org/bioscience/	5.876
Clinical Anatomy	https://onlinelibrary.wiley.com/journal/10982353	1.908
HAPS Educator	https://www.hapsweb.org/page/hapsed_home	N/A
Innovative Higher Education	https://link.springer.com/journal/10755	N/A
International Journal of Science Education	https://www.tandfonline.com/loi/tsed20	1.325
Journal of Biological Education	https://www.tandfonline.com/toc/rjbe20/current	0.663
Journal of College Science Teaching	https://www.nsta.org/college/	N/A
Journal of Nursing Education	https://www.healio.com/nursing/journals/jne	1.185
Journal of Research in Science Teaching	https://onlinelibrary.wiley.com/journal/10982736	3.210
Journal of Veterinary Medical Education	https://jvme.utpjournals.press/loi/jvme	0.850
Medical Education	https://onlinelibrary.wiley.com/journal/13652923	4.405
Medical Education Online	https://www.tandfonline.com/toc/zmeo20/current	1.440
Medical Science Educator	https://link.springer.com/journal/40670	N/A
Medical Teacher	http://www.medicalteacher.org/	N/A
Research in Science Education	https://link.springer.com/journal/11165	1.568
Science Education	https://onlinelibrary.wiley.com/journal/1098237x	3.035
Teaching and Learning in Medicine	https://www.tandfonline.com/toc/htlm20/current	1.292

tion about how to spot predatory publishers may be found in Beall, 2017 [20], and at Publons.com through their journal listings [21].

Advantages of publishing your anatomy educational research

- You disseminate your findings to a wider audience than an oral presentation or poster would.
- Your research becomes part of the permanent record for scholars.
- Your publication is considered "scholarly work" for promotion and tenure.
- Ensure you are publishing in a reputable journal—beware of predatory publishers!

Closely read the aims and scope of the journal (described on the journal's website), and scan through the past publications from several issues to determine if your manuscript is suitable for the journal in question. The decision of where to publish also will depend on the scope of your specific anatomy education project and the audience you wish to reach. Some journals are specific to medical education (e.g., *Medical Education*), while others such as *Anatomical Sciences Education* publish educational research articles from undergraduate (pre-medical), medical, and professional schools. Some journals may focus on anatomy projects from a particular geographic region of the world or be more suited to administrators rather than anatomy faculty. If your educational research project has important findings for other disciplines, consider publishing in an interdisciplinary journal such as *BioScience* or *Advances in Health Sciences Education*. In addition, do not assume a journal has to have "anatomy" in its title in order to publish anatomy educational research. For example, *Advances in Physiology Education* publishes research not only in physiology but also in anatomy, biochemistry, and other health professions fields.

Make sure you carefully read the "instructions to authors" section on the journal's website prior to submitting your manuscript. There the editors will provide you with specific information such as how references and tables should be formatted, any length restrictions for the manuscript, and any other guidelines about acceptable publications. Manuscripts that are not in proper format and do not follow the guidelines will be returned to the author without publishing. Most, if not all, of the journals listed above only allow for online submission of the manuscript. In addition, make sure you have at least one other colleague read over your manuscript prior to submission, to check for grammatical errors and to ensure the overall flow of the manuscript is acceptable. If your institution has a teaching and learning center or a school of education, consider asking individuals in these places to review your manuscript as well. These individuals likely have critiqued many similar manuscripts and may provide you with valuable suggestions you may not have considered.

The review process for your manuscript may take as little as a few weeks to as long as a few months or longer, depending on the journal and the manuscript in question. Journal editors typically have at least two or more individuals review a manuscript and give comments for the author. If your manuscript is rejected, do not despair. Carefully review the editor and reviewer comments to see how you may improve the publication. Have multiple local colleagues read your manuscript for further guidance. After you make the edits these individuals suggest, consider submitting the manuscript to a different journal that may be more appropriate. If the editor has not rejected the paper but does want edits, make sure you address all reviewer concerns and document how and where in the manuscript these concerns were addressed.

Checklist prior to submitting your manuscript to a journal:

- Determine the aims and scope of the journal are aligned with my research project.
- The journal in question publishes educational research projects similar to mine.
- Read the "instructions for authors" section of the journal website.
- Have at least one other colleague read your manuscript for general flow, grammatical issues, and whether the research questions were adequately addressed.
- Double-check the "instructions for authors" section to determine all journal guidelines were followed.

Conclusion

As you develop your anatomy educational research project, take advantage of the resources available to you. Contact faculty and staff from other departments, and consult with your center for teaching and learning to assist you in your project. Join one or more professional societies involved in educational research, and be active in the society's meetings. Consider presenting your research at a variety of local, national, and international venues. Finally, write up your research findings, and submit them for publication, so the field of anatomy education may learn from your research.

In order for your research to be considered scholarship of teaching and learning (SoTL), you must methodically assess your teaching and student learning and present your findings in the form of an oral presentation or written publication.

References

1. Boyer EL. Scholarship reconsidered: priorities of the professoriate. Princeton: The Carnegie Foundation for the Advancement of Teaching; 1990.
2. Glassick CE, Huber MT, Maeroff GI. Scholarship assessed: evaluation of the professoriate. San Francisco: Jossey-Bass; 1997.
3. Shulman LS. Teaching as community property: putting an end to pedagogical solitude. In: Shulman LS, editor. Teaching as community property: essays on higher education. San Francisco: Jossey-Bass; 2004. p. 139–44.
4. Trigwell K, Martin E, Benjamin J, Prosser M. Scholarship of teaching: a model. Higher Educ Res Dev. 2000;19(2):155–68.
5. Hutchings P, Shulman LS. The scholarship of teaching: new elaborations, new developments. Change. 1999;31:11–5.
6. Hutchings P, Huber MT, Ciccone A. The scholarship of teaching and learning reconsidered: institutional integration and impact. San Francisco: Jossey-Bass; 2011.
7. Kreber C. Conceptualizing the scholarship of teaching and identifying unresolved issues: the framework for this volume. In: Kreber C, editor. Scholarship revisited: perspectives on the scholarship of teaching. New directions for teaching and learning, vol. 86. San Francisco: Jossey-Bass; 2001. p. 1–18.
8. Richlin L. Scholarly teaching and the scholarship of teaching. In: Kreber C, editor. Scholarship revisited: perspectives on the scholarship of teaching. New directions for teaching and learning, vol. 86. San Francisco: Jossey-Bass; 2001. p. 57–68.
9. Bass RJ. The scholarship of teaching: what is the problem? Inventio. 1999;1(1):1–10.
10. O'Loughlin VD. A 'How-to' guide for developing a publishable scholarship of teaching project. Adv Physiol Educ. 2006;30:83–8.
11. Gurung RAR, Schwartz BM. Optimizing teaching and learning: practicing pedagogical research. Malden: Wiley; 2009.. http://onlinelibrary.wiley.com/book/10.1002/9781444305883.
12. Indiana University. Center for Innovative Teaching and Learning. 2019. https://citl.indiana.edu/.
13. Indiana University. Scholarship of teaching and learning. 2019. https://citl.indiana.edu/programs/scholarship-teaching-learning/.
14. Shulman LS. Professing educational scholarship. In: Lagemann E, Shulman LS, editors. Issues in education research: problems and possibilities. San Francisco: Jossey-Bass; 1999. p. 159–65.
15. Eisenhart M, DeHaan R. Doctoral preparation of scientifically based education researchers. Educ Res. 2005;34:3–13.
16. Drake RL, Pawlina W. Anatomical sciences education: a new arrival in the medical education field. Anat Sci Educ. 2008;1(1):1.
17. Anatomical Society. Journal of Anatomy. 2019. http://www.anatsoc.org.uk/journals/journal-of-anatomy.
18. Lilly International Conference on College Teaching. 2019. http://celt.miamioh.edu/lillycon/.
19. Lilly Conferences on College and University Teaching. 2019. https://www.lillyconferences.org/.
20. Beall J. Predatory journals, peer review, and educational research. New Horiz Adult Educ Hum Res Dev. 2017;29(1):54–8.
21. Publons. Beall's List. Gone but not lost. 2017. https://publons.com/blog/bealls-list-gone-but-not-lost/.

Index

L. K. Chan, W. Pawlina (eds.), *Teaching Anatomy*, https://doi.org/10.1007/978-3-030-43283-6

MIX
Papier aus verantwortungsvollen Quellen
Paper from responsible sources
FSC® C105338

If you have any concerns about our products,
you can contact us on
ProductSafety@springernature.com

In case Publisher is established outside the EU,
the EU authorized representative is:
Springer Nature Customer Service Center GmbH
Europaplatz 3, 69115 Heidelberg, Germany

Printed by Libri Plureos GmbH
in Hamburg, Germany